Lecture Notes in Computer Scie

T0238684

Commenced Publication in 1973
Founding and Former Series Editors:
Gerhard Goos, Juris Hartmanis, and Jan van Leeuwen

Chengan Guo   Zeng-Guang Hou
Zhigang Zeng (Eds.)

Advances in Neural Networks – ISNN 2013

10th International Symposium on Neural Networks
Dalian, China, July 4-6, 2013
Proceedings, Part II

 Springer

Volume Editors

Chengan Guo
Dalian University of Technology
School of Information and Communication Engineering
A 530, Chuangxinyuan Building
Dalian 116023, China
E-mail: cguo@dlut.edu.cn

Zeng-Guang Hou
Chinese Academy of Sciences
Institute of Automation
Beijing 100864, China
E-mail: zengguang.hou@mail.ia.ac.cn

Zhigang Zeng
Huazhong University of Science and Technology
School of Automation
Wuhan 430074, China
E-mail: zgzeng@mail.hust.edu.cn

ISSN 0302-9743   e-ISSN 1611-3349
ISBN 978-3-642-39067-8   e-ISBN 978-3-642-39068-5
DOI 10.1007/978-3-642-39068-5
Springer Heidelberg Dordrecht London New York

Library of Congress Control Number: 2013940936

CR Subject Classification (1998): F.1.1, I.5.1, I.2.6, I.2.8, I.2.10, I.2, I.4, I.5,
F.1, E.1, F.2

LNCS Sublibrary: SL 1 – Theoretical Computer Science and General Issues

Typesetting: Camera-ready by author, data conversion by Scientific Publishing Services, Chennai, India

Printed on acid-free paper

Springer is part of Springer Science+Business Media (www.springer.com)

Preface

This book and its sister volume collect the refereed papers presented at the 10th International Symposium on Neural Networks (ISNN 2013), held in Dalian, China, during July 4–6, 2013. Building on the success of the previous events, ISNN has become a well-established series of popular and high-quality conferences on neural network and its applications. The field of neural networks has evolved rapidly in recent years. It has become a fusion of a number of research areas in engineering, computer science, mathematics, artificial intelligence, operations research, systems theory, biology, and neuroscience. Neural networks have been widely applied for control, optimization, pattern recognition, signal/image processing, etc. ISNN aims at providing a high-level international forum for scientists, engineers, educators, as well as students to gather so as to present and discuss the latest progresses in neural network research and applications in diverse areas.

ISNN 2013 received a few hundred submissions from more than 22 countries and regions. Based on the rigorous peer reviews by the Program Committee members and the reviewers, 157 papers were selected for publications in the LNCS proceedings. These papers cover major topics of theoretical research, empirical study, and applications of neural networks.

In addition to the contributed papers, three distinguished scholars (Cesare Alippi, Polytechnic University of Milan, Italy; Derong Liu, Institute of Automation, Chinese Academy of Sciences, China; James Lo, University of Maryland - Baltimore County, USA) were invited to give plenary speeches, providing us with the recent hot topics, latest developments, and novel applications of neural networks. Furthermore, ISNN 2013 also featured two special sessions focusing on emerging topics in neural network research.

ISNN 2013 was sponsored by Dalian University of Technology and The Chinese University of Hong Kong, financially co-sponsored by the National Natural Science Foundation of China, and technically co-sponsored by the IEEE Computational Intelligence Society, IEEE Harbin Section, Asia Pacific Neural Network Assembly, European Neural Network Society, and International Neural Network Society.

We would like to express our sincere gratitude to all the Program Committee members and the reviewers of ISNN 2013 for their professional review of the papers and their expertise that guaranteed the high qualify of the technical program! We would also like to thank the publisher, Springer, for their cooperation in publishing the proceedings in the prestigious series of *Lecture Notes in Computer Science*. Moreover, we would like to express our heartfelt appreciation to the plenary and panel speakers for their vision and discussion of the latest research developments in the field as well as critical future research directions, opportunities, and challenges. Finally, we would like to thank all the speakers, authors, and participants for their great contribution and support that made ISNN 2013 a huge success.

July 2013

Chengan Guo
Zeng-Guang Hou
Zhigang Zeng

ISNN 2013 Organization

ISNN 2013 was organized and sponsored by Dalian University of Technology and The Chinese University of Hong Kong, financially co-sponsored by the National Natural Science Foundation of China, and technically co-sponsored by the IEEE Computational Intelligence Society, IEEE Harbin Section, Asia Pacific Neural Network Assembly, European Neural Network Society, and International Neural Network Society.

General Chair

Jun Wang   The Chinese University of Hong Kong, Hong Kong, China
Dalian University of Technology, Dalian, China

Advisory Chairs

Marios M. Polycarpou   University of Cyprus, Nicosia, Cyprus
Gary G. Yen   Oklahoma State University, Stillwater, USA

Steering Chairs

Derong Liu   Institute of Automation, Chinese Academy of Sciences, Beijing, China
Wei Wang   Dalian University of Technology, Dalian, China

Organizing Chairs

Min Han   Dalian University of Technology, Dalian, China
Dan Wang   Dalian Maritime University, Dalian, China

Program Chairs

Chengan Guo   Dalian University of Technology, Dalian, China
Zeng-Guang Hou   Institute of Automation, Chinese Academy of Sciences, Beijing, China
Zhigang Zeng   Huazhong University of Science and Technology, Wuhan, China

Special Sessions Chairs

Tieshan Li   Dalian Maritime University, Dalian, China
Zhanshan Wang   Northeast University, Shenyang, China

Publications Chairs

Jie Lian   Dalian University of Technology, Dalian, China
Danchi Jiang   University of Tasmania, Hobart, Australia

Publicity Chairs

Jinde Cao   Southeast University, Nanjing, China
Yi Shen   Huazhong University of Science and
                                      Technology, Wuhan, China

Registration Chairs

Jie Dong   Dalian University of Technology, Dalian, China
Shenshen Gu   Shanghai University, Shanghai, China
Qingshan Liu   Southeast University, Nanjing, China

Local Arrangements Chair

Jianchao Fan   National Ocean Environment Protection
                                      Research Institute, Dalian, China

Program Committe

Alma Y. Alanis	Jixiang Du	Junhao Hu
Tao Ban	Haibin Duan	He Huang
Gang Bao	Jianchao Fan	Tingwen Huang
Chee Seng Chan	Jian Feng	Amir Hussain
Jonathan Chan	Jian Fu	Danchi Jiang
Rosa Chan	Siyao Fu	Feng Jiang
Guici Chen	Wai-Keung Fung	Haijun Jiang
Mou Chen	Chengan Guo	Min Jiang
Shengyong Chen	Ping Guo	Shunshoku Kanae
Long Cheng	Shengbo Guo	Rhee Man Kil
Ruxandra Liana Costea	Qing-Long Han	Sungshin Kim
Chuangyin Dang	Hanlin He	Bo Li
Liang Deng	Zeng-Guang Hou	Chuandong Li
Mingcong Deng	Jinglu Hu	Kang Li

Tieshan Li
Yangmin Li
Yanling Li
Jinling Liang
Hualou Liang
Wudai Liao
Jing Liu
Ju Liu
Meiqin Liu
Shubao Liu
Xiaoming Liu
Wenlian Lu
Yanhong Luo
Jiancheng Lv
Jinwen Ma
Xiaobing Nie

Seiichi Ozawa
Qiankun Song
Norikazu Takahashi
Feng Wan
Cong Wang
Dianhui Wang
Jun Wang
Ning Wang
Xin Wang
Xiuqing Wang
Zhanshan Wang
Zhongsheng Wang
Tao Xiang
Bjingji Xu
Yingjie Yang
Wei Yao

Mao Ye
Yongqing Yang
Wen Yu
Wenwu Yu
Zhigang Zeng
Jie Zhang
Lei Zhang
Dongbin Zhao
Xingming Zhao
Zeng-Shun Zhao
Chunhou Zheng
Song Zhu
An-Min Zou

Additional Reviewers

Aarya, Isshaa
Abdurahman, Abdujelil
Bai, Yiming
Bao, Gang
Bao, Haibo
Bobrowski, Leon
Bonnin, Michele
Cao, Feilong
Chang, Xiaoheng
Chen, Fei
Chen, Juan
Chen, Yao
Chen, Yin
Cheng, Zunshui
Cui, Rongxin
Dai, Qun
Deng, Liang
Dogaru, Radu
Dong, Yongsheng
Duan, Haibin
Er, Meng Joo
Esmaiel, Hamada
Feng, Pengbo
Feng, Rongquan
Feng, Xiang

Fu, Siyao
Gan, Haitao
Garmsiri, Naghmeh
Gollas, Frank
Guo, Zhishan
He, Xing
Hu, Aihua
Hu, Bo
Hu, Jianqiang
Huang, He
Huang, Lei
Huang, Xia
Jeff, Fj
Jiang, Haijun
Jiang, Minghui
Jiang, Yunsheng
Jin, Junqi
Jinling, Wang
Lan, Jian
Leung, Carson K.
Li, Benchi
Li, Bing
Li, Fuhai
Li, Hongbo
Li, Hu

Li, Kelin
Li, Lulu
Li, Shihua
Li, Will
Li, Xiaolin
Li, Yang
Li, Yongming
Li, Zifu
Lian, Cheng
Liang, Hongjing
Liang, Jianyi
Liang, Jinling
Lin, Yanyan
Liu, Cheng
Liu, Huiyang
Liu, Miao
Liu, Qingshan
Liu, Xiao-Ming
Liu, Xiaoyang
Liu, Yan-Jun
Liu, Yang
Liu, Zhigang
Lu, Jianquan
Lu, Yan
Luo, Weilin

Ma, Hongbin
Maddahi, Yaser
Meng, Xiaoxuan
Miao, Baobin
Pan, Lijun
Pang, Shaoning
Peng, Zhouhua
Qi, Yongqiang
Qin, Chunbing
Qu, Kai
Sanguineti, Marcello
Shao, Zhifei
Shen, Jun
Song, Andy
Song, Qiang
Song, Qiankun
Sun, Liying
Sun, Ning
Sun, Yonghui
Sun, Yongzheng
Tang, Yang
Wang, Huanqing
Wang, Huiwei
Wang, Jinling
Wang, Junyi
Wang, Min
Wang, Ning
Wang, Shenghua

Wang, Shi-Ku
Wang, Xin
Wang, Xinzhe
Wang, Xiuqing
Wang, Yin-Xue
Wang, Yingchun
Wang, Yinxue
Wang, Zhanshan
Wang, Zhengxin
Wen, Guanghui
Wen, Shiping
Wong, Chi Man
Wong, Savio
Wu, Peng
Wu, Zhengtian
Xiao, Jian
Xiao, Tonglu
Xiong, Ping
Xu, Maozhi
Xu, Yong
Yang, Bo
Yang, Chenguang
Yang, Feisheng
Yang, Hua
Yang, Lu
Yang, Shaofu
Yang, Wengui
Yin, Jianchuan

Yu, Jian
Yu, Wenwu
Yuen, Kadoo
Zhang, Jianhai
Zhang, Jilie
Zhang, Long
Zhang, Ping
Zhang, Shuyi
Zhang, Weiwei
Zhang, Xianxia
Zhang, Xin
Zhang, Xuebo
Zhang, Yunong
Zhao, Hongyong
Zhao, Mingbo
Zhao, Ping
Zhao, Yu
Zhao, Yue
Zheng, Cheng-De
Zheng, Cong
Zheng, Qinling
Zhou, Bo
Zhou, Yingjiang
Zhu, Lei
Zhu, Quanxin
Zhu, Yanqiao
Zhu, Yuanheng

Table of Contents – Part II

Control, Robotics and Hardware

Bioinformatics and Biomedical Engineering, Brain-Like Systems and Brain-Computer Interfaces

Evolutionary Neural Networks, Hybrid Intelligent Systems

Data Mining and Knowledge Discovery

Other Applications of Neural Networks

Table of Contents – Part I

Computational Neuroscience and Cognitive Science

Neural Network Models, Learning Algorithms, Stability and Convergence Analysis

Kernel Methods, Large Margin Methods and SVM

Optimization Algorithms / Variational Methods

Feature Analysis, Clustering, Pattern Recognition and Classification

Vision Modeling and Image Processing

Optimal Tracking Control Scheme for Discrete-Time Nonlinear Systems with Approximation Errors

Qinglai Wei and Derong Liu*

The State Key Laboratory of Management and Control for Complex Systems
Institute of Automation, Chinese Academy of Sciences, Beijing 100190, China
{qinglai.wei,derong.liu}@ia.ac.cn

Abstract. In this paper, we aim to solve an infinite-time optimal tracking control problem for a class of discrete-time nonlinear systems using iterative adaptive dynamic programming (ADP) algorithm. When the iterative tracking control law and the iterative performance index function in each iteration cannot be accurately obtained, a new convergence analysis method is developed to obtain the convergence conditions of the iterative ADP algorithm according to the properties of the finite approximation errors. If the convergence conditions are satisfied, it is shown that the iterative performance index functions converge to a finite neighborhood of the greatest lower bound of all performance index functions under some mild assumptions. Neural networks are used to approximate the performance index function and compute the optimal tracking control policy, respectively, for facilitating the implementation of the iterative ADP algorithm. Finally, a simulation example is given to illustrate the performance of the present method.

Keywords: Adaptive dynamic programming, generalized value iteration, neural networks, optimal control, reinforcement learning.

1 Introduction

Optimal tracking problems of nonlinear systems have always been the key focus in the control field in the latest several decades. Adaptive dynamic programming (ADP), proposed by Werbos [13, 14], is a powerful tool to solve optimal control problems forward-in-time which is extensively applied to solve optimal tracking problems [7, 9, 15]. The idea of ADP is to approximate the optimal performance index function and the optimal control law by using function approximation structures. In recent years, ADP and related research have gained much attention from researchers [1, 2, 6]. In [4], ADP approaches were classified into several main schemes which are heuristic dynamic programming (HDP), action-dependent HDP (ADHDP), also known as Q-learning, dual heuristic dynamic programming (DHP) and action-dependent DHP (ADDHP), globalized DHP (GDHP), and ADGDHP. Iterative methods are also used in ADP to obtain the solution of Hamilton-Jacobi-Bellman (HJB) equation indirectly and have received

---

* This work was supported in part by the National Natural Science Foundation of China under Grants 61034002, 61233001, 61273140, in part by Beijing Natural Science Foundation under Grant 4132078, and in part by the Early Career Development Award of SKLMCCS.

C. Guo, Z.-G. Hou, and Z. Zeng (Eds.): ISNN 2013, Part II, LNCS 7952, pp. 1–10, 2013.

more and more attention [3, 5, 8, 10–12, 16, 17]. Value iteration algorithms are one class of the most primary and important iterative ADP algorithms. In [15], Zhang proposed a value iteration algorithm to solve optimal tracking problems which obtains good results. While in [15], the accurate performance index function and the accurate iterative control law are required in each iteration to make the algorithm convergent. For most real-world control systems, however, the accurate iterative control laws in the iterative ADP algorithms cannot be obtained. When the accurate iterative control laws cannot be obtained, the convergence properties in the accurate iterative ADP algorithms may be invalid. This means that the system states may not track the desired trajectory under the iterative control law. Till now, to the best of our knowledge, there are no discussions on the convergence properties of the iterative ADP algorithms for optimal tracking problems with approximation errors.

In this paper, we will develop a new ADP scheme for infinite-time optimal tracking control problems. The main contribution of this paper is that the optimal tracking control problems with finite approximation errors are solved effectively using the present iterative ADP algorithms. A new convergence analysis method is established and the least upper bound of the converged iterative performance index function is also presented. Furthermore, in order to facilitate the implementation of the iterative ADP algorithm, we use neural networks to obtain the iterative performance index function and the optimal tracking control law, respectively. Finally, a simulation example is given to show the effectiveness of the present iterative ADP algorithm.

2   Problem Statement

Consider the following affine nonlinear system

$$x(k+1) = f(x(k)) + g(x(k))u(x(k)) \tag{1}$$

where $x(k) \in \Re^n$, $f(x(k)) \in \Re^n$, $g(x(k)) \in \Re^{n \times m}$ and the input $u(k) \in \Re^m$. Here we assume that the system is controllable on $\Re^n$.

For infinite-time optimal tracking problem, the control objective is to design optimal control $u(x(k))$ for system (1) such that the state $x(k)$ tracks the specified desired trajectory $\eta(k) \in \Re^n$, $k = 0, 1, \ldots$. Define the tracking error as

$$z(k) = x(k) - \eta(k). \tag{2}$$

Define the following quadratic performance index function

$$J(z(0), \underline{u}_0) = \sum_{k=0}^{\infty} \left\{ z^T(k)Qz(k) + (u(k) - u_e(k))^T R(u(k) - u_e(k)) \right\}, \tag{3}$$

where $Q \in \Re^{n \times n}$, $R \in \Re^{m \times m}$ are positive definite matrices and $\underline{u}_0 = (u(0), u(1), \ldots)$. Let $U(z(k), v(k)) = z^T(k)Qz(k) + v^T(k)Rv(k)$ be the utility function, where $v(k) = u(k) - u_e(k)$. Let $u_e(k)$ denote the expected control, which can be given as

$$u_e(k) = g^{-1}(\eta(k))(\eta(k+1) - f(\eta(k))), \tag{4}$$

where $g^{-1}(\eta(k))g(\eta(k)) = I$, $I \in \Re^{m \times m}$ is the identity matrix. Combining (1) with (2), we can get

$$z(k+1) = f(z(k) + \eta(k)) - \eta(k+1) + g(z(k) + \eta(k))$$
$$\times \left(v(k) + g^{-1}(\eta(k))(f(\eta(k)) - \eta(k+1))\right). \quad (5)$$

Through the system transformation, we can see that the optimal tracking control problem for (1) has been changed into an optimal regulation problem for (5). The optimal performance index function is defined as

$$J^*(z(k)) = \inf_{\underline{v}_k} \{J(z(k), \underline{v}_k)\}, \quad (6)$$

where $\underline{v}_k = (v(k), v(k+1), \ldots)$. According to Bellman's principle of optimality, $J^*(z(k))$ satisfies the discrete-time HJB equation

$$J^*(z(k)) = \min_{v(k)} \{U(z(k), v(k)) + J^*(F(z(k), v(k)))\}. \quad (7)$$

Then the law of optimal single control vector can be expressed as

$$v^*(z(k)) = \arg\min_{v(k)} \{U(z(k), v(k)) + J^*(z(k+1))\}. \quad (8)$$

Hence, the HJB equation (7) can be written as

$$J^*(z(k)) = U(z(k), v^*(z(k))) + J^*(z(k+1)). \quad (9)$$

In [15], a greedy HDP iteration algorithm is proposed to solve the optimal control problem for (5), where the performance index and control law are updated by iterations, with the iteration number i increasing from 0 to ∞. First, the initial performance index function $V_0(z(k)) \equiv 0$. Then, for $i = 0, 1, \ldots$, the iterative control law $v_i(z(k))$ and iterative performance index function $V_{i+1}(z(k))$ are be computed the following two equations

$$v_i(z(k)) = \arg\min_{v(k)} \{z^T(k)Qz(k) + v^T(k)Rv(k) + V_i(z(k+1))\} \quad (10)$$

and

$$V_{i+1}(z(k)) = \min_{v(k)} \{z^T(k)Qz(k) + v^T(k)Rv(k) + V_i(z(k+1))\}$$
$$= z^T(k)Qz(k) + v_i^T(z(k))Rv_i(z(k)) + V_i(z(k+1)) \quad (11)$$

In [15], it was proved that the iterative performance index function $V_i(z(k))$ converges to $J^*(z(k))$, as $i \to \infty$. For the greedy HDP iteration algorithm, we can see that for $\forall i = 0, 1, \ldots$, the accurate iterative control law and accurate iterative performance index function must be obtained which guarantee the convergence of the iterative performance index function. In the real-world implementation, however, for $\forall i = 0, 1, \ldots$, the accurate iterative control law $v_i(z(k))$ and the iterative performance index function $V_{i+1}(z(k))$ are generally impossible to obtain. In this situation, the convergence of the iterative performance index function and iterative control law may be invalid and the iterative ADP algorithm may even be divergent. To overcome this difficulty, a new ADP analysis method must be developed based on the approximation errors.

3 Properties of the Iterative ADP Algorithm with Finite Approximation Errors

3.1 Derivation of the Greedy HDP Iteration Algorithm with Finite Approximation Errors

In the present iterative ADP algorithm, the performance index function and control law are updated by iterations, with the iteration index i increasing from 0 to ∞. For $i = 0$, let $V_0(z(k)) = 0$. The iterative control law $\hat{v}_0(x_k)$ can be computed as follows:

$$
\begin{aligned}
\hat{v}_0(z(k)) &= \arg\min_{v(k)} \left\{ z^T(k)Qz(k) + v^T(k)Rv(k) + \hat{V}_0(z(k+1)) \right\} + \rho_0(z(k)) \\
&= \arg\min_{v(k)} \left\{ z^T(k)Qz(k) + v^T(k)Rv(k) + \hat{V}_0(z(k+1)) \right\} + \rho_0(z(k))
\end{aligned}
\tag{12}
$$

where $\hat{V}_0(z(k+1)) = V_0(z(k+1))$. The iterative performance index function can be updated as

$$
\hat{V}_1(z(k)) = z^T(k)Qz(k) + \hat{v}_0^T(z(k))R\hat{v}_0(z(k)) + \hat{V}_0(z(k+1)) + \pi_0(z(k)). \tag{13}
$$

For $i = 1, 2, \ldots$, the iterative ADP algorithm will iterate between

$$
\begin{aligned}
\hat{v}_i(z(k)) &= \arg\min_{v(k)} \left\{ z^T(k)Qz(k) + v^T(k)Rv(k) + \hat{V}_i(z(k+1)) \right\} + \rho_i(z(k)) \\
&= \arg\min_{v(k)} \left\{ z^T(k)Qz(k) + v^T(k)Rv(k) + \hat{V}_i(z(k+1)) \right\} + \rho_i(z(k))
\end{aligned}
\tag{14}
$$

and

$$
\begin{aligned}
\hat{V}_{i+1}(z(k)) &= \min_{v(k)} \left\{ z^T(k)Qz(k) + v^T(k)Rv(k) + \hat{V}_i(z(k+1)) \right\} + \pi_i(z(k)) \\
&= z^T(k)Qz(k) + \hat{v}_i^T(z(k))R\hat{v}_i(z(k)) + \hat{V}_i(z(k+1)) + \pi_i(z(k)). \tag{15}
\end{aligned}
$$

3.2 Properties of the Iterative ADP Algorithm with Finite Approximation Errors

From the iterative ADP algorithm (12)–(15), we can see that for $\forall i = 1, 2, \ldots$, there exists an approximation error between the iterative performance index functions $\hat{V}_i(z(k))$ and $V_i(z(k))$. As the accurate iterative control law $v_i(z(k))$ cannot be obtained which means the iterative performance index functions $V_i(z(k))$ cannot be accurately obtained, then the detailed value of each approximation error is unknown and nearly impossible to obtain. It makes the property analysis of the iterative performance index function $\hat{V}_i(z(k))$ and iterative control law $\hat{v}_i(z(k))$ very difficult. So, in this subsection, a new "error bound" analysis method is established. The idea of the "error bound" analysis method is that for each iterative index $i = 0, 1 \ldots$, the least upper bound of the iterative performance index functions $\hat{V}_i(z(k))$ is analyzed, which avoids to analyze the value of $\hat{V}_i(z(k))$ directly. Using the "error bound" method, it can be proved that the iterative performance index functions $\hat{V}_i(z(k))$ can uniformly converge to a bounded neighborhood of optimal performance index function.

Define a new iterative performance index function as

$$\Gamma_i(z(k)) = \min_{v(k)}\{U(z(k), v(k)) + \hat{V}_{i-1}(z(k+1))\} \tag{16}$$

where $\hat{V}_i(z(k))$ is defined in (15) and $v(k)$ can accurately be obtained in $\Re^m$. Then, for $\forall i = 0, 1, \ldots$, there exists a finite constant $\sigma \geq 1$ that makes

$$\hat{V}_i(z(k)) \leq \sigma \Gamma_i(z(k)) \tag{17}$$

hold uniformly. Hence, we can give the following theorem.

Theorem 1. *For $\forall i = 0, 1, \ldots$, let $\Gamma_i(z(k))$ be expressed as (16) and $\hat{V}_i(z(k))$ be expressed as (15). Let $\gamma < \infty$ and $1 \leq \delta < \infty$ are both constance that make*

$$J^*(z(k+1)) \leq \gamma U(z(k), v(k)) \tag{18}$$

and

$$V_0(z(k)) \leq \delta J^*(z(k)) \tag{19}$$

hold uniformly. If there exists $1 \leq \sigma < \infty$ that makes (17) hold uniformly, then we have

$$\hat{V}_i(z(k)) \leq \sigma \left(1 + \sum_{j=1}^{i} \frac{\gamma^j \sigma^{j-1}(\sigma - 1)}{(\gamma + 1)^j} + \frac{\gamma^i \sigma^i (\delta - 1)}{(\gamma + 1)^i} \right) J^*(z(k)), \tag{20}$$

where we define $\sum\limits_{j}^{i}(\cdot) = 0$, for $\forall j > i$ and $i, j = 0, 1, \ldots$.

Proof. The theorem can be proved by mathematical induction. First, let $i = 0$. Then, (20) becomes

$$\hat{V}_0(z(k)) \leq \sigma \delta J^*(z(k)). \tag{21}$$

As $\hat{V}_0(z(k)) \leq \delta J^*(z(k))$, then we can obtain $\hat{V}_0(z(k)) \leq \delta J^*(z(k)) \leq \sigma \delta J^*(z(k))$, which obtains (21). So, the conclusion holds for $i = 0$.

Next, let $i = 1$. We have

$$\Gamma_1(z(k)) = \min_{v(k)} \left\{ U(z(k), v(k)) + \hat{V}_0(F(z(k), v(k))) \right\}$$

$$\leq \min_{v(k)} \left\{ U(z(k), v(k)) + \sigma\delta J^*(F(z(k), v(k))) \right\}$$

$$\leq \min_{v(k)} \left\{ \left(1 + \gamma\frac{\sigma\delta - 1}{\gamma + 1} \right) U(z(k), v(k)) \right.$$

$$\left. + \left(\sigma\delta - \frac{\sigma\delta - 1}{\gamma + 1} \right) J^*(F(z(k), v(k))) \right\}$$

$$= \left(1 + \gamma\frac{\sigma\delta - 1}{\gamma + 1} \right) \min_{v(k)} \left\{ U(z(k), v(k)) + J^*(F(z(k), v(k))) \right\}$$

$$= \left(1 + \frac{\gamma(\sigma - 1)}{\gamma + 1} + \frac{\gamma\sigma(\delta - 1)}{\gamma + 1} \right) J^*(z(k)). \tag{22}$$

According to (17), we can obtain

$$\hat{V}_1(z(k)) \leq \sigma \left(1 + \frac{\gamma(\sigma - 1)}{\gamma + 1} + \frac{\gamma\sigma(\delta - 1)}{\gamma + 1} \right) J^*(z(k)), \tag{23}$$

which shows that (20) holds for $i = 1$.

Assume that (20) holds for $i = l - 1$, where $l = 1, 2, \ldots$. Then, for $i = l$, we have

$$\Gamma_i(z(k))$$

$$= \min_{v(k)} \left\{ U(z(k), v(k)) + \hat{V}_{l-1}(F(z(k+1))) \right\}$$

$$\leq \min_{v(k)} \left\{ U(z(k), v(k)) + \sigma \left(1 + \sum_{j=1}^{l-1} \frac{\gamma^j \sigma^{j-1}(\sigma - 1)}{(\gamma + 1)^j} + \frac{\gamma^{l-1}\sigma^{l-1}(\delta - 1)}{(\gamma + 1)^{l-1}} \right) J^*(z(k)) \right\}$$

$$\leq \left(1 + \sum_{j=1}^{l} \frac{\gamma^j \sigma^{j-1}(\sigma - 1)}{(\gamma + 1)^j} + \frac{\gamma^l \sigma^l(\delta - 1)}{(\gamma + 1)^l} \right) \min_{v(k)} \left\{ U(z(k), v(k)) + J^*(z(k+1)) \right\}$$

$$= \left(1 + \sum_{j=1}^{l} \frac{\gamma^j \sigma^{j-1}(\sigma - 1)}{(\gamma + 1)^j} + \frac{\gamma^l \sigma^l(\delta - 1)}{(\gamma + 1)^l} \right) J^*(z(k)). \tag{24}$$

Then, according to (17), we can obtain (20) which proves the conclusion for $\forall i = 0, 1, \ldots$.

From (20), we can see that for arbitrary finite i, σ and δ, there exists a bounded error between the iterative performance index function $\hat{V}_i(z(k))$ and the optimal performance index function $J^*(z(k))$. While as $i \to \infty$, the bound of the approximation error may increase to infinity. Thus, in the following part, we will give the convergence properties of the iterative ADP algorithm (12)–(15) using error bound method.

Theorem 2. *Suppose Theorem 1 holds for* $\forall z(k) \in \Re^n$. *If for* $\gamma < \infty$ *and* $\sigma \geq 1$, *the inequality*

$$\sigma < \frac{\gamma + 1}{\gamma} \tag{25}$$

holds, then as $i \to \infty$, *the iterative performance index function* $\hat{V}_i(z(k))$ *in the iterative ADP algorithm (12)–(15) is uniformly convergent into a bounded neighborhood of the optimal performance index function* $J^*(z(k))$, *i.e.,*

$$\lim_{i \to \infty} \hat{V}_i(z(k)) = \hat{V}_\infty(z(k)) \leq \sigma \left(1 + \frac{\gamma(\sigma - 1)}{1 - \gamma(\sigma - 1)} \right) J^*(z(k)). \tag{26}$$

Proof. According to (24) in Theorem 1, we can see that for $j = 1, 2, \ldots$, the sequence $\left\{ \dfrac{\gamma^j \sigma^{j-1}(\sigma - 1)}{(\gamma + 1)^j} \right\}$ is a geometrical series. Then, (24) can be written as

$$\Gamma_i(z(k)) \leq \left(1 + \frac{\dfrac{\gamma(\sigma - 1)}{\gamma + 1} \left(1 - \left(\dfrac{\gamma\sigma}{\gamma + 1} \right)^i \right)}{1 - \dfrac{\gamma\sigma}{\gamma + 1}} + \frac{\gamma^i \sigma^i (\delta - 1)}{(\gamma + 1)^i} \right) J^*(z(k)). \tag{27}$$

As $i \to \infty$, if $1 \leq \sigma < \dfrac{\gamma + 1}{\gamma}$, then (27) becomes

$$\lim_{i \to \infty} \Gamma_i(z(k)) = \Gamma_\infty(z(k)) \leq \left(1 + \frac{\gamma(\sigma - 1)}{1 - \gamma(\sigma - 1)} \right) J^*(z(k)). \tag{28}$$

According to (17), let $i \to \infty$, we have

$$\hat{V}_\infty(z(k)) \leq \sigma \Gamma_\infty(z(k)). \tag{29}$$

Taking (29) and (28), we can obtain (26).

Corollary 1. *Suppose Theorem 1 holds for* $\forall z(k) \in \Re^n$. *If for* $\gamma < \infty$ *and* $\sigma \geq 1$, *the inequality (25) holds, then the iterative control law* $\hat{v}_i(z(k))$ *of the iterative ADP algorithm (12)–(15) is convergent, i.e.,*

$$\hat{v}_\infty(z(k)) = \lim_{i \to \infty} \hat{v}_i(z(k)) = \arg \min_{v(k)} \left\{ U(z(k), v(k)) + \hat{V}_\infty(z(k+1)) \right\}. \tag{30}$$

4 Simulation Study

Our example is chosen as the example in [15]. Consider the following affine nonlinear system

$$x(k + 1) = f(x(k)) + g(x(k))u(k) \tag{31}$$

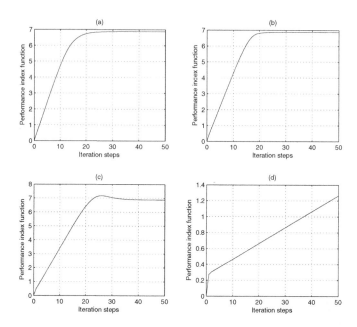

Fig. 1. Performance index functions. (a) $\sigma = 10^{-6}$. (b) $\sigma = 10^{-4}$. (c) $\sigma = 10^{-3}$. (d) $\sigma = 10^{-1}$.

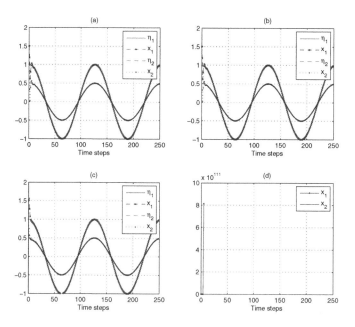

Fig. 2. State trajectories. (a) $\sigma = 10^{-6}$. (b) $\sigma = 10^{-4}$. (c) $\sigma = 10^{-3}$. (d) $\sigma = 10^{-1}$.

where $x(k) = [x_1(k)\ x_2(k)]^T$, $u(k) = [u_1(k)\ u_2(k)]^T$. Let the system function
$$f(x(k)) = \begin{bmatrix} 0.2x_1(k)\exp(x_2^2(k)) \\ 0.3x_2^3(k) \end{bmatrix}, \quad g(x(k)) = \begin{bmatrix} -x_1(k)x_2(k) & 0.1 \\ x_2(k) & -0.8x_1(k)x_2(k) \end{bmatrix}.$$
The desired trajectory is set to $\eta(k) = [\sin(k + \frac{\pi}{2})\ \ 0.5\cos(k)]^T$. The performance index function is defined as (3), where $Q = R = I$.

We use neural networks to implement the iterative ADP algorithm. The critic network and the action network are chosen as three-layer BP neural networks with the structures of 2–8–1 and 2–8–2, respectively. We choose four approximation errors $\sigma = 10^{-6}, \sigma = 10^{-4}, \sigma = 10^{-3}, \sigma = 10^{-1}$, the iterative performance index functions are shown in Fig. 1.

The state and corresponding control trajectories are shown in Figs. 2 and 3, respectively. From the approximation errors $\sigma = 10^{-6}, \sigma = 10^{-4}$, we can see that the performance index functions are monotonically increasing convergent. While for $\sigma = 10^{-3}$, the performance index function is not monotonically increasing but still is convergent. While for $\sigma = 10^{-1}$, we can see that the iterative performance index function is not convergent any more and the system is not stable.

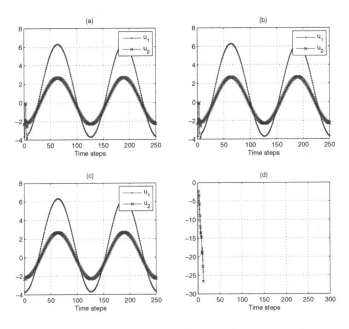

Fig. 3. Control trajectories. (a) $\sigma = 10^{-6}$. (b) $\sigma = 10^{-4}$. (c) $\sigma = 10^{-3}$. (d) $\sigma = 10^{-1}$.

5   Conclusions

In this paper, an effective ADP algorithm is developed to solve optimal tracking control problems for infinite horizon discrete-time nonlinear systems. The iterative performance index functions is proved to converge to the finite neighborhood of the optimal performance index function if the convergence conditions are satisfied. Finally, a simulation example is given to illustrate the performance of the present algorithm.

References

1. Abu-Khalaf, M., Lewis, F.L.: Nearly optimal control laws for nonlinear systems with saturating actuators using a neural network HJB approach. Automatica 41(5), 779–791 (2005)
2. Abu-Khalaf, M., Lewis, F.L., Huang, J.: Neurodynamic programming and zero-sum games for constrained control systems. IEEE Transactions on Neural Networks 19(7), 1243–1252 (2008)
3. Al-Tamimi, A., Abu-Khalaf, M., Lewis, F.L.: Adaptive critic designs for discrete-time zero-sum games with application to H_∞ control. IEEE Trans. Systems, Man, and Cybernetics-Part B: Cybernetics 37(7), 240–247 (2007)
4. Prokhorov, D.V., Wunsch, D.C.: Adaptive critic designs. IEEE Transactions on Neural Networks 8(5), 997–1007 (1997)
5. Hao, X., Jagannathan, S.: Model-free H_∞ stochastic optimal design for unknown linear networked control system zero-sum games via Q-learning. In: 2011 IEEE International Symposium on Intelligent Control (ISIC), Singapore, pp. 198–203 (2011)
6. Liu, D., Zhang, Y., Zhang, H.: A self-learning call admission control scheme for CDMA cellular networks. IEEE Transactions on Neural Networks 16(5), 1219–1228 (2005)
7. Tan, F., Liu, D., Guan, X., Xing, S.: Trajectory tracking control of nonholonomic mobile robot system based on unfalsified control theory. Control and Decision 25(6), 1693–1697 (2010)
8. Wang, F., Jin, N., Liu, D., Wei, Q.: Adaptive dynamic programming for finite-horizon optimal control of discrete-time nonlinear systems with ϵ-error bound. IEEE Transactions on Neural Networks 22(1), 24–36 (2011)
9. Wang, D., Liu, D., Wei, Q.: Finite-horizon neuro-optimal tracking control for a class of discrete-time nonlinear systems using adaptive dynamic programming approach. Neurocomputing 78(1), 14–22 (2012)
10. Wei, Q., Liu, D.: Nonlinear multi-person zero-sum differential games using iterative adaptive dynamic programming. In: 30th Chinese Control Conference (CCC), Yantai, China, pp. 2456–2461 (2011)
11. Wei, Q., Liu, D.: An iterative ϵ-optimal control scheme for a class of discrete-time nonlinear systems with unfixed initial state. Neural Networks 32, 236–244 (2012)
12. Wei, Q., Zhang, H., Dai, J.: Model-free multiobjective approximate dynamic programming for discrete-time nonlinear systems with general performance index functions. Neurocomputing 72(7-9), 1839–1848 (2009)
13. Werbos, P.J.: Advanced forecasting methods for global crisis warning and models of intelligence. General Systems Yearbook 22, 25–38 (1977)
14. Werbos, P.J.: A menu of designs for reinforcement learning over time. In: Miller, W.T., Sutton, R.S., Werbos, P.J. (eds.) Neural Networks for Control, pp. 67–95. MIT Press, Cambridge (1991)
15. Zhang, H., Wei, Q., Luo, Y.: A novel infinite-time optimal tracking control scheme for a class of discrete-time nonlinear systems via the greedy HDP iteration algorithm. IEEE Transactions on System, Man, and cybernetics-Part B: Cybernetics 38(4), 937–942 (2008)
16. Zhang, H., Song, R., Wei, Q.: Optimal tracking control for a class of nonlinear discrete-time systems with time delays based on heuristic dynamic programming. IEEE Transactions on Neural Networks 22(12), 1851–1862 (2011)
17. Zhang, H., Wei, Q., Liu, D.: An iterative adaptive dynamic programming method for solving a class of nonlinear zero-sum differential games. Automatica 47(1), 207–214 (2011)

Distributed Output Feedback Tracking Control of Uncertain Nonlinear Multi-Agent Systems with Unknown Input of Leader

Zhouhua Peng[1,*], Dan Wang[1,**], and Hongwei Zhang[2]

[1] School of Marine Engineering, Dalian Maritime University,
Dalian 116026, P.R. China
[2] School of Electrical Engineering, Southwest Jiaotong University,
Chendu 610031, P.R. China
{zhouhuapeng,dwangdl}@gmail.com, hwzhang@swjtu.edu.cn

Abstract. This paper considers the distributed tracking control of uncertain nonlinear multi-agent systems in the presence of unmeasured states and unknown input of the leader over an undirected network. By approximating the uncertain nonlinear dynamics via neural network and constructing a local observer to estimate the unmeasured states, distributed output feedback tracking controllers, with static and dynamic coupling gains, respectively, are proposed, based on the relative observed states of neighboring agents. It is proved that with the developed controllers, the state of each agent synchronizes to that of the leader for any undirected connected graphs even when only a fraction of the agents in the network have access to the state information of the leader, and the distributed tracking errors are uniformly ultimately bounded. A sufficient condition to the existence of the distributed controllers is that each agent is stabilizable and detectable. Future works include an extension to the directed network topologies.

Keywords: Distributed Control, Neural Networks, Nonlinear Multi-agent Systems, Output Feedback.

1 Introduction

Recent years have witnessed a great interest in consensus problem of multi-agent systems for its broad applications in engineering such as formation flight of satellites, formation control of unmanned vehicles, sensor networks, and so on. The key of consensus is to design a distributed control law based on local interactions such that the group agree on some value of interest [1–3]. In literature, a great deal of effort has been made on two consensus problems of multi-agent

---

* This work was supported in part by the National Nature Science Foundation of China under Grants 61273137, 51209026, 61074017, and in part by the Fundamental Research Funds for the Central Universities under Grants 2682013CX016, SWJTU11ZT06, 3132013037.
** Corresponding author.

C. Guo, Z.-G. Hou, and Z. Zeng (Eds.): ISNN 2013, Part II, LNCS 7952, pp. 11–18, 2013.

systems. One is *leaderless consensus*, and the other is *leader-follower consensus* where there exists an active leader who acts as a trajectory generator for the group to follow [4, 5]. In literature, the leader-follower consensus is also known as *distributed tracking* [11] or *cooperative tracking* [19].

During the past few years, distributed tracking control of multi-agent systems has been extensively studied by many researchers from different perspectives. In [6], a neighbor-based observer is proposed for tracking control of first-order linear multi-agent system under fixed and variable topologies. In [7], a consensus algorithm is developed for first-order linear systems with a time-varying leader dynamics. In [8, 9], distributed output regulation approaches are proposed for linear multi-agent systems. In [10], consensus of linear multi-agent systems and synchronization of complex networks are unified in a framework. In [11], cooperative tracking control for linear multi-agent systems with a leader of bounded unknown input is considered. In [12], leader-follower consensus algorithms are developed for both fixed and switching interaction topologies. In [13], a framework for cooperative tracking is proposed, including state feedback, observer and output feedback. Note that all aforementioned results are based on the exact linear system model, which may not be adequate to describe the practical agent dynamics in many real-world applications. Therefore, some researchers considered the distributed tracking control problem of nonlinear systems [15–20]. Distributed tracking of first-order nonlinear systems on strongly connected graphs is investigated in [15], where neural network is employed to account for the unknown nonlinear dynamics. This result is extended to second-order uncertain nonlinear systems in [16]. Neural network-based leader-following control of first-order multi-agent systems with uncertainties is presented in [17], and the result is extended to second-order case using backstepping technique. Robust consensus tracking of second-order nonlinear multi-agent systems is discussed in [18], where continuous distributed consensus protocols are developed to enable global asymptotic tracking performance. Distributed tracking control of high-order nonlinear systems is provided in [19], where the communication graph does not require to be connected. Synchronized tracking control of high-order nonlinear systems using state and output information is considered in [20]. Despite these results, there are still no results on distributed tracking control of nonlinear systems with non-Brunovsky canonical form, especially under the condition that only the partial state information can be measured.

In this paper, we focus on the distributed output feedback tracking control of uncertain nonlinear multi-agent systems with an unknown input of the leader over an undirected network. It is assumed that a fraction of agents have access to the state of the leader and the input of the leader is bounded. As for the distributed state feedback tracking control of uncertain nonlinear systems, the unknown input of the leader can be considered as a disturbance to be rejected by the neural network. However, this cannot be done in the context of the output feedback case due to the fact the unknown input cannot be merged into the uncertain part. By approximating the uncertain nonlinear dynamics via neural networks and constructing a local observer to estimate the unmeasured states, distributed output feedback tracking controllers, with static and dynamic coupling gains, respectively, are developed. It is proved that with the developed controllers, the state of each agent synchronize to that of the leader for any undirected connected networks, and distributed tracking errors are guaranteed to be uniformly ultimately bounded. A sufficient condition to the existence of the distributed controllers is that each agent is stabilizable and detectable. It is worth noting that the static tracking controller depends on the eigenvalues of the communication graph and the upper bound of the leader's control input. This restriction is removed by using adaptive coupling strategy which results in a fully distributed observer-based tracking controllers.

2   Preliminaries and Problem Formulation

2.1   Preliminaries

Notation. Throughout the paper, the Euclidean norm, Frobenius norm, minimum singular value, maximum singular value, and trace are denoted by $|| \cdot ||$, $|| \cdot ||_F$, $\underline{\sigma}(\cdot)$, $\overline{\sigma}(\cdot)$, and $tr\{\cdot\}$, respectively. A diagonal matrix is represented by $\text{diag}\{\lambda_1, ..., \lambda_N\}$ with λ_i being the ith diagonal element. An identity matrix of dimension N is denoted by I_N. The Kronecker product is denoted by $\otimes$ with the properties $(A \otimes B)^T = A^T \otimes B^T$, $\alpha(A \otimes B) = (\alpha A) \otimes B = A \otimes (\alpha B)$, $(A \otimes B)(C \otimes D) = (AC) \otimes (BD)$, where A, B, C, D are matrices and α is a scalar.

Graph Theory. Consider a network of multi-agent systems consisting of N agents and one leader. If each agent is considered as a node, the neighbor relation can be described by a *graph* $\mathcal{G} = \{\mathcal{V}, \mathcal{E}\}$, where $\mathcal{V} = \{n_1, ..., n_N\}$ is a node set and $\mathcal{E} = \{(n_i, n_j) \in \mathcal{V} \times \mathcal{V}\}$ is an edge set with the element (n_i, n_j) that describes the communication from node i to node j. The neighbor set of the node i is denoted by $\mathcal{N}_i = \{j | (n_j, n_i) \in \mathcal{E}\}$. Define an adjacency matrix $\mathcal{A} = [a_{ij}] \in \mathbb{R}^{N \times N}$ with $a_{ij} = 1$, if $(n_j, n_i) \in \mathcal{E}$, and $a_{ij} = 0$, otherwise. Define the in-degree matrix as $\mathcal{D} = \text{diag}\{d_1, ..., d_N\}$ with $d_i = \sum_{j \in \mathcal{N}_i} a_{ij}$. The Laplacian matrix $L = [l_{ij}] \in \mathbb{R}^{N \times N}$ associated with the graph $\mathcal{G}$ is defined as $L = \mathcal{D} - \mathcal{A}$. If $a_{ij} = a_{ji}$, for $i, j = 1, ..., N$, then the graph $\mathcal{G}$ is *undirected*. A *path* in a graph is an ordered sequence of nodes such that any two consecutive nodes in the sequence are an edge of the graph. An undirected graph is *connected* if there is a path between every pair of nodes. Finally, define a leader adjacency matrix as $\mathcal{A}_0 = \text{diag}\{a_{10}, ..., a_{N0}\}$, where $a_{i0} > 0$ if and only if the ith agent has access to the leader information; otherwise, $a_{i0} = 0$. For simplicity, denote $L + \mathcal{A}_0$ by H.

 Lemma 1 [6]. Suppose that the graph $\mathcal{G}$ is undirected and connected, and at least one agent has access to the leader. Then, H is positive definite.

2.2   Problem Formulation

Consider a network of uncertain nonlinear systems consisting of N agents and a leader. The dynamics of the ith agent is given by

$$\dot{x}_i = Ax_i + B[u_i + f_i(x_i)],$$
$$y_i = Cx_i, \tag{1}$$

where $x_i = [x_{i1}, ..., x_{in}]^T \in \mathbb{R}^n$ is the system state; $u_i \in \mathbb{R}^m$ is the control input; $f_i(x_i) \in \mathbb{R}^m$ is an unknown matched uncertainty; $A \in \mathbb{R}^{n \times n}, B \in \mathbb{R}^{n \times m}$, and $C \in \mathbb{R}^{n \times m}$ are known matrices, and the triple (A, B, C) is assumed to be stabilizable and detectable.

The dynamics of the leader is described by

$$\dot{x}_0 = Ax_0 + Br, \tag{2}$$

where $x_0 \in \mathbb{R}^n$ is the leader state, and $r \in \mathbb{R}^m$ is an unknown bounded input that bounded by $\|r\| \leq r_M$ with r_M being a constant. For any initial conditions, we assume that the solution x_0 exists for all $t \geq 0$.

The control objective of this paper is to design a distributed control law u_i for each agent (1) to track the leader (2) such that the state of each agent synchronizes to that of the leader, i.e., $x_i \to x_0$, with bounded synchronization errors.

To move on, we make use of the following assumption.

Assumption 1 [14, 19, 21, 22]. The matched uncertainty $f_i(x_i)$ can be linearly parameterized by a neural network (NN) as

$$f_i(x_i) = W_i^T \varphi_i(x_i) + \varepsilon_i, \quad \forall x_i \in D, \tag{3}$$

where $W_i \in \mathbb{R}^{s \times m}$ is an unknown constant ideal weight matrix and satisfies $\|W_i\| \leq W_{iM}$ with $W_{iM} \in \mathbb{R}$ a positive constant; $\varphi_i(\cdot) : \mathbb{R}^n \to \mathbb{R}^s$ is a known vector of the form $\varphi_i(x_i) = [\varphi_{i1}(x_i), \varphi_{i2}(x_i), ..., \varphi_{is}(x_i)]^T$ and satisfies $\|\varphi_i\| \leq \varphi_{iM}$ with $\varphi_{iM} \in \mathbb{R}$ a positive constant; ε_i is the approximation error satisfying $\|\varepsilon_i\| \leq \varepsilon_{iM}$ with $\varepsilon_{iM} \in \mathbb{R}$ a positive constant; D is a sufficiently large domain $D \subset \mathbb{R}^n$.

3 Distributed Output Feedback Tracking Control with Static Coupling

3.1 Controller Design

To begin with, the following distributed controller is proposed

$$u_i = u_{in} - u_{iad}. \tag{4}$$

The first term u_{in} is a nominal controller designed as

$$u_{in} = c_1 K \hat{e}_i + c_2 \text{sgn}(K \hat{e}_i), \tag{5}$$

where $c_1 \in \mathbb{R}, c_2 \in \mathbb{R}$ are positive coupling gains; $K \in \mathbb{R}^{m \times n}$ is a feedback matrix with

$$K = -B^T P, \tag{6}$$

where P is the unique positive definite solution to the following Riccati equation

$$A^T P + PA + Q - PBB^T P = 0, \tag{7}$$

where $Q \in \mathbb{R}^{n \times n}$ is positive definite; $\hat{e}_i$ is defined as

$$\hat{e}_i = \sum_{j=0}^{N} a_{ij}(\hat{x}_i - \hat{x}_j), \tag{8}$$

where $\hat{x}_i$ is an estimate of x_i obtained using a state observer described by

$$\begin{aligned} \dot{\hat{x}}_i &= A\hat{x}_i + Bu_{in} + F(y_i - \hat{y}_i), \\ \hat{y}_i &= C\hat{x}_i, \end{aligned} \tag{9}$$

where $F \in \mathbb{R}^{n \times p}$ is an observer gain matrix designed such that $A_e = A - FC$ is Hurwitz. Using (5), the observer dynamics of $\hat{x}_i$ can be expressed by

$$\dot{\hat{x}}_i = A\hat{x}_i + c_1 BK\hat{e}_i + c_2 B\mathrm{sgn}(K\hat{e}_i) + FC(x_i - \hat{x}_i). \tag{10}$$

The second term u_{iad} is an adaptive term that tries to compensate for the uncertainty $f_i(x_i)$ and

$$u_{iad} = \hat{W}_i^T \varphi_i(\hat{x}_i), \tag{11}$$

where $\hat{W}_i$ is an estimate of W_i that updated as

$$\dot{\hat{W}}_i = \Gamma_{iW}[\varphi_i(\hat{x}_i)\tilde{y}_i^T - \frac{1}{2\kappa}\varphi_i(\hat{x}_i)\varphi_i^T(\hat{x}_i)\hat{W}_i - k_W\hat{W}_i], \tag{12}$$

where $\tilde{y}_i = y_i - \hat{y}_i$; $\kappa \in \mathbb{R}$, $\Gamma_{iW} \in \mathbb{R}$, and $k_W \in \mathbb{R}$ are positive constants.

Denote an estimated error $\tilde{x}_i = x_i - \hat{x}_i$ whose dynamics with (1) and (9) can be written as

$$\dot{\tilde{x}}_i = A_e\tilde{x}_i + B[-\tilde{W}_i^T \varphi_i(\hat{x}_i) + W_i^T \tilde{\varphi}_i + \varepsilon_i], \tag{13}$$

where $\tilde{W}_i = \hat{W}_i - W_i$; $W_i^T \tilde{\varphi}_i = W_i^T[\varphi_i(x_i) - \varphi_i(\hat{x}_i)]$.

Let $\tilde{x} = [\tilde{x}_1^T, ..., \tilde{x}_N^T]^T$, $\tilde{W} = \mathrm{diag}\{\tilde{W}_1, ..., \tilde{W}_N\}$, $W = \mathrm{diag}\{W_1, ..., W_N\}$, $\varphi(\hat{x}) = [\varphi_1^T(\hat{x}_1), ..., \varphi_1^T(\hat{x}_N)]^T$, $\tilde{\varphi} = [\tilde{\varphi}_1^T, ..., \tilde{\varphi}_N^T]^T$, $\varepsilon = [\varepsilon_1^T, ..., \varepsilon_N^T]^T$. Then the N subsystem of (13) is written as

$$\dot{\tilde{x}} = (I_N \otimes A_e)\tilde{x} + (I_N \otimes B)[-\tilde{W}^T \varphi(\hat{x}) + W^T \tilde{\varphi} + \varepsilon], \tag{14}$$

Define an estimated state tracking error $\tilde{\delta}_i = \hat{x}_i - x_0$ whose time derivative along (2) and (9) is

$$\dot{\tilde{\delta}}_i = A\tilde{\delta}_i + c_1 BK\hat{e}_i + c_2 B\mathrm{sgn}(K\hat{e}_i) + FC\tilde{x}_i - Br. \tag{15}$$

To facilitate stability analysis, define a tracking error $\delta_i = x_i - x_0$ and let $\tilde{\delta} = [\tilde{\delta}_1^T, ..., \tilde{\delta}_N^T]^T$, $\tilde{x} = [\tilde{x}_1^T, ..., \tilde{x}_N^T]^T$, $\delta = [\delta_1^T, ..., \delta_N^T]^T$. Then the dynamics of $\tilde{\delta}$ can be written as

$$\dot{\tilde{\delta}} = (I_N \otimes A + c_1 H \otimes BK)\tilde{\delta} + c_2(I_N \otimes B)\text{sgn}((H \otimes K)\tilde{\delta})$$
$$+ (I_N \otimes FC)\tilde{x} - (\mathbf{1} \otimes B)r. \tag{16}$$

Finally, introduce a parameter dependent Riccati equation

$$A_e^T P_e + P_e A_e + Q_e + \kappa M M^T = 0$$
$$M = C^T - P_e B, \tag{17}$$

where $Q_e, P_e \in \mathbb{R}^{n \times n}$ are positive definite. $0 < \kappa \leq \kappa_{max}$ define the largest set within which there exists a positive definite solution for P_e.

3.2 Stability Analysis

Theorem 3.1. Consider the multi-agent systems (1) with the leader node (2) under Assumption 1. Suppose the network $\mathcal{G}$ is undirected and connected, and at least one agent has access to the leader node. Select the control law (4) with the coupling strength c_1, c_2 satisfying

$$c_1 \geq \frac{1}{2 \min_{i=1,...,N}(\lambda_i)}, \tag{18}$$

$$c_2 \geq r_M, \tag{19}$$

and the adaptive law (12), together with the state observer (10). Then, all signals in the closed-loop network are uniformly ultimately bounded, and the state estimate error $\tilde{x}$, the estimated state tracking error $\tilde{\delta}$ and the tracking error δ satisfy $\lim_{t \to \infty} \|\tilde{x}\| \leq \gamma_1, \lim_{t \to \infty} \|\tilde{\delta}\| \leq \gamma_2, \lim_{t \to \infty} \|\delta\| \leq \gamma_3$ for some constants $\gamma_1, \gamma_2, \gamma_3 \in \mathbb{R}$.

Proof. Omitted here due to the limited space.

4 Distributed Output Feedback Tracking Control with Dynamic Coupling

In the previous section, the distributed controller design depends on the minimal eigenvalue $\min_{i=1,...,N}(\lambda_i)$ of H and the upper bound r_M of the leader's input. However, the knowledge of minimal eigenvalue of the graph $\mathcal{G}$ belongs to the global information in the sense that each agent has to know the topology of the entire communication network to calculate it. On the other hand, the upper bound r_M depends on their own dynamics and may not be explicitly available to the followers. Hence, the controller given in the previous section cannot be implemented in a fully distributed manner. Accordingly, the objective of this section is to develop a fully distributed controller without requiring knowledge of $\min_{i=1,...,N}(\lambda_i)$ and r_M.

4.1 Controller Design

Similar to Section 3.1, we propose the following fully distributed control law

$$u_i = u_{if} - u_{iad}, \tag{20}$$

with

$$u_{if} = \varrho_i K \hat{e}_i + \varrho_i \mathrm{sgn}(K \hat{e}_i), \tag{21}$$

where $\varrho_i \in \mathbb{R}$ an adaptive coupling strength updated as

$$\dot{\varrho}_i = \Gamma_{i\varrho}[\hat{e}_i^T P B B^T P \hat{e}_i + \|K \hat{e}_i\|_1 - k_\varrho \varrho_i] \tag{22}$$

where $\Gamma_{i\varrho} \in \mathbb{R}, k_\varrho \in \mathbb{R}$ are positive constants. K, P and u_{iad} are defined in (6), (7) and (11), respectively.

4.2 Stability Analysis

Theorem 4.1. Consider the multi-agent systems (1) with the leader node (2) under Assumption 1. Suppose the network $\mathcal{G}$ is undirected and connected, and at least one agent has access to the leader node. Select the control law (20) with the adaptive law (12) and the adaptive coupling strength (22), together with the state observer (10). Then, all signals in the closed-loop network are uniformly ultimately bounded , and the state estimate error $\tilde{x}$, the estimated state tracking error $\tilde{\delta}$, and the tracking error δ satisfy $\lim_{t \to \infty} \|\tilde{x}\| \leq \gamma_4, \lim_{t \to \infty} \|\tilde{\delta}\| \leq \gamma_5, \lim_{t \to \infty} \|\delta\| \leq \gamma_6$, for some constants $\gamma_4, \gamma_5, \gamma_6 \in \mathbb{R}$.

Proof. Omitted here due to the limited space.

5 Conclusions

This paper considered the distributed tracking control of uncertain nonlinear multi-agent systems with unmeasured states and unknown input of leader. Distributed observer-based tracking controllers with static and dynamic coupling gains are developed, based on the observed states of neighboring agents. It is proved with the developed controllers, synchronization to the leader can be reached for any undirected connected graphs, and all signals in the closed-loop network are uniformly ultimately bounded. Further works includes an extension to the directed network topologies.

References

1. Fax, J.A., Murray, R.M.: Information flow and cooperative control of vehicle formations. IEEE Trans. Automa. Control 49(9), 1465–1476 (2004)
2. Jadbabaie, A., Lin, J., Morse, A.S.: Coordination of groups of mobile autonomous agents using nearest neighbor rules. IEEE Trans. Automa. Control 48(6), 988–1001 (2003)

3. Olfati-Saber, R., Fax, J.A., Murray, R.M.: Consensus and cooperation in networked multi-Agent systems. Proc. IEEE 95(1), 215–233 (2007)
4. Ren, W., Beard, R., Atkins, E.: Information consensus in multivehicle cooperative control. IEEE Control Syst. Mag. 27(2), 71–82 (2007)
5. Ren, W., Beard, R.: Consensus seeking in multiagent systems under dynamically changing interaction topologies. IEEE Trans. Automa. Control 50(5), 655–661 (2005)
6. Hong, Y.G., Hu, J.P., Gao, L.X.: Tracking control for multi-agent consensus with an active leader and variable topology. Automatica 42(7), 1177–1182 (2006)
7. Ren, W.: Multi-vehicle consensus with a time-varying reference state. Syst. & Control Lett. 56(7), 474–483 (2007)
8. Wang, X.L., Hong, Y.G., Huang, J., Jiang, Z.P.: A distributed control approach to a robust output regulation problem for multi-agent linear systems. IEEE Trans. Automa. Control 55(12), 2891–2895 (2010)
9. Hong, Y.G., Wang, X.L., Jiang, Z.P.: Distributed output regulation of leader-follower multi-agent systems. Int. J. Robust Nonlin. 23(1), 48–66 (2013)
10. Li, Z.K., Duan, Z.S., Chen, G.R., Huang, L.: Consensus of multiagent systems and synchronization of complex networks: A unified viewpoint. IEEE Trans. Circuits Syst. I, Reg. Papers 57(1), 213–224 (2010)
11. Li, Z.K., Liu, X.D., Ren, W., Xie, L.H.: Distributed tracking control for linear multi-Agent systems with a leader of bounded unknown input. IEEE Trans. Automa. Control 58(2), 518–523 (2013)
12. Ni, W., Cheng, D.Z.: Leader-following consensus of multi-agent systems under fixed and switching topologies. Syst. & Control Lett. 59(4), 209–217 (2010)
13. Zhang, H.W., Lewis, F.L., Das, A.: Optimal design for synchronization of cooperative systems: state feedback, observer and output feedback. IEEE Trans. Automa. Control 56(8), 1948–1952 (2011)
14. Hou, Z.G., Cheng, L., Tan, M.: Decentralized robust adaptive control for the multiagent system consensus problem using neural networks. IEEE Trans. Syst., Man, Cybern. Part B: Cybern. 39(3), 636–647 (2009)
15. Das, A., Lewis, F.L.: Distributed adaptive control for synchronization of unknown nonlinear networked systems. Automatica 26(12), 2014–2021 (2010)
16. Das, A., Lewis, F.L.: Cooperative adaptive control for synchronization of second-order systems with unknown nonlinearities. Int. J. Robust Nonlin. 21(13), 1509–1524 (2011)
17. Cheng, L., Hou, Z.G., Tan, M., Lin, Y.Z., Zhang, W.J.: Neural-network-based adaptive leader-following control for multiagent systems with uncertainties. IEEE Trans. Neural Netw. Learn. Syst. 21(8), 1351–1358 (2010)
18. Hu, G.Q.: Robust consensus tracking of a class of second-order multi-agent dynamic systems. Syst. & Control Lett. 61(1), 134–142 (2012)
19. Zhang, H.W., Lewis, F.L.: Adaptive cooperative tracking control of higher-order nonlinear systems with unknown dynamics. Automatica 48(7), 1432–1439 (2012)
20. Cui, R.X., Ren, B.B., Ge, S.S.: Synchronised tracking control of multi-agent system with high-order multi-agent systems. IET Control Theory Appl. 6(5), 603–614 (2012)
21. Zhang, H.W., Lewis, F.L., Qu, Z.H.: Lyapunov, adaptive, and optimal design techniques for cooperative systems on directed communication graphs. IEEE Trans. Ind. Electron. 59(7), 3026–3041 (2012)
22. Wang, D., Huang, J.: Neural network-based adaptive dynamic surface control for a class of uncertain nonlinear systems in strict-feedback form. IEEE Trans. Neural Netw. Learn. Syst. 6(1), 195–202 (2005)

Quadrotor Flight Control Parameters Optimization Based on Chaotic Estimation of Distribution Algorithm

Pei Chu[1] and Haibin Duan[1,2,*]

[1] Science and Technology on Aircraft Control Laboratory,
School of Automation Science and Electrical Engineering, Beihang University,
Beijing, 100191, P.R. China
[2] Provincial Key Laboratory for Information Processing Technology, Soochow University,
Suzhou 215006, P.R. China
chupei@asee.buaa.edu.cn, hbduan@buaa.edu.cn

Abstract. Quadrotor is a type of rotor craft that consists of four rotors and two pairs of counter-rotating, fixed-pitch blades located at the four corners of the body. The flight control parameters optimization is one of the key issues for quadrotor. Estimation of distribution algorithm is a new kind of evolutionary algorithm developed rapidly recently. However, low convergence speed and local optimum of the EDA are the main disadvantages that limit its further application. To overcome the disadvantages of EDA, a chaotic estimation of distribution algorithm is proposed in this paper. It is a combination of chaos theory and principles of estimation of distribution algorithm. Series of experimental comparison results are presented to show the feasibility, effectiveness and robustness of our proposed method. The results show that the proposed chaotic EDA can effectively improve both the global searching ability and the speed of convergence.

Keywords: quadrotor, estimation of distribution algorithm, chaos, flight control.

1   Introduction

Quadrotor is a type of rotorcraft that consists of four rotors and two pairs of counter-rotating, fixed-pitch blades located at the four corners of the body. The idea of using four rotors is realized as a full-scale helicopter as early as 1920s [1]. However, quadrotor is dynamically unstable and not widely developed in applications until the advance in computers and micro sensors. Flight control system of quadrotor is a complex MIMO nonlinear system with time-varying, strong-coupling characteristics[2].Though we can rely on small disturbance linearization equation to design the control system, the apparent coupling among the equations will make it difficult to set the parameters.

---

[*] Corresponding author.

C. Guo, Z.-G. Hou, and Z. Zeng (Eds.): ISNN 2013, Part II, LNCS 7952, pp. 19–26, 2013.

Estimation of distribution algorithm is novel kind of optimization algorithm. It is a combination of genetic algorithms and statistical learning [3]. Nowadays it has become a significant method dealing with programming problems such as the optimization of flight control system. Besides, estimation of distribution algorithms was put forward as a significant issue in almost every academic seminar such as ACMEVO, IEEE and CEC. Nevertheless, it can easily trap into the local optimum, hence would probably end up without finding a satisfying result. Considering the outstanding performance of chaos theory in jumping out of stagnation, we introduce it to improve the robustness of basic EDA algorithm, and the comparative experimental results testified that our proposed method manifests better performance than the basic EDA algorithm. We also applied the chaotic EDA to flight parameters optimization of quadrotor, whose type is AR.Drone.

The remainder of this paper is organized as follows. Section 2 introduces the modeling of the quadrotor. Subsequently, section 3 describes the chaotic estimation of distribution algorithm. Experimental results are given in section 4. Our concluding remarks are contained in section 5.Acknowledgements are contained at the end.

2   Modeling of the Quadrotor

AR.Drone is a Wi-Fi-controlled quadrotor with cameras attached to it which is developed by Parrot Inc [1]. It uses an ARM9 468MHz embedded microcontroller with 128M of RAM running the Linux operating system [4].The onboard downward Complementary Metal Oxide Semiconductor (CMOS) color camera provides RGB images in size of 320*240. An inertial system uses a 3-axisaccelerometer, 2-axis gyro and a single-axis yaw precision gyro. An ultrasonic altimeter with a range of 6 meters provides vertical stabilization. With a weight of 380g or 420git can maintain flight for about 12 minutes with a speed of 5m/s. Fig.1 shows the top view and side view of the quadrotor.

Fig. 1. Top view and side view of the quadrotor

1. Transform from angle to voltage[5]

$$u(t) = K_1\theta(t) \tag{1}$$

2. Transform from voltage to torque

The relationship between voltage and torque is:

$$\begin{cases} T_m \dfrac{dw(t)}{dt} + w(t) = K_m u(t) \\[2mm] M_m = K_2 \dfrac{u}{R} \\[2mm] M_f = fw(t) \\[2mm] M = M_m - M_f \end{cases} \tag{2}$$

3. Transform from torque to pneumatic tension

$$M = Fr \tag{3}$$

4. Transform from pneumatic tension to the tilt angle of the quadrotor

$$\begin{cases} M = J_m \dfrac{d^2\theta}{dt} \\[2mm] M = 2Xl \\[2mm] J_m = \dfrac{1}{3}m_1 l^2 + 2m_2 l^2 \end{cases} \tag{4}$$

From 1, 2, 3, 4 we can get the following result through Laplace transformation.

$$G(s) = G_1 G_2 G_3 G_4 = \frac{6226.s + 311330}{s^3 + 100s^2} \tag{5}$$

The PID controller architecture is shown in Fig. 2 [1].

The input of the controller is the errors, which can be obtained by our proposed chaotic EDA. The output of the controller is:

$$u = k_p e(t) + k_i \int_0^\infty e(t)dt + k_d \frac{de(t)}{dt} \tag{6}$$

Considering there is no apparent relationship between the inputs and the outputs of the flight control system and in order to avoid the overshoot, we choose the following objective function [12].

$$\begin{cases} J = \displaystyle\int_0^\infty (w_1 \,|\, e(t)\,| + w_2 u^2(t))dt + w_3 t_u, e(t) >= 0 \\[3mm] J = \displaystyle\int_0^\infty (w_1 \,|\, e(t)\,| + w_2 u^2(t) + w_4 \,|\, e(t)\,|)dt + w_3 t_u, e(t) < 0 \end{cases} \tag{7}$$

Where J is the objective function, $e(t)$ represents the error, $u(t)$ denotes the outputs of the controller, t_u means the rising time and w_1, w_2,w_3 and w_4 are the weights and $w_4 \gg w_1$.

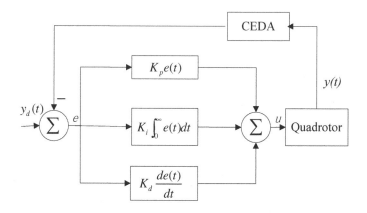

Fig. 2. The PID controller architecture

3   Chaotic Estimation of Distribution Algorithm(CEDA)

3.1   Principles of CEDA

The term of EDA alludes to a family of evolutionary algorithms which represents an alternative to the classical optimization methods in the area [6], [7], [8]. EDA gene-rates new population by establishing probability distribution model and generate new individuals based on the model. Indeed, this distribution is responsible for one of the main characteristics of these algorithms. The basic procedures can be shown as follows.

Given an n-dimensional probability vector model $P(x) = P(x_1, x_2,...,x_n) = (0.5, 0.5,...,0.5)$. Then generate initial population based on the model. We conduct the selection operation and select m<n individuals to update the model by the following formula:

$$P_{l+1}(x_i) = \frac{P_l(x_i)}{\sum\limits_{i=1}^{m} P_l(x_i)} \qquad (8)$$

Where P means the probability, l represents the evolutionary times and m denotes the better m individuals selected from the former population. Repeat the selection and updating operation until reaching the stopping criteria.

Chaos is the highly unstable motion of deterministic systems infinite phase space which often exists in nonlinear systems [9].Chaos theory is epitomized by the so-called 'butterfly effect' detailed by Lorenz [10]. Until now, chaotic behavior has already been observed in the laboratory in a variety of systems including electrical circuits, lasers, oscillating chemical reactions, fluid dynamics, as well as computer models of chaotic processes. Chaos theory has been applied to a number of fields, among which one of the most applications was in ecology, where dynamical systems have been used to show how population growth under density dependence can lead to chaotic dynamics. Sensitive dependence on initial conditions is not only observed in complex systems, but even in the simplest logistic equation. In the well-known logistic equation:

$$x_{n+1} = 4x_n(1 - x_n) \qquad (9)$$

Where $0<x_n<1$, a very small difference in the initial value of x would give rise to large difference in its long-time behavior, which is the basic characteristic of chaos. The track of chaotic variable can travel ergodically over the whole space of interest. The variation of the chaotic variable has a delicate inherent rule in spite of the fact that its variation looks like in disorder. Therefore, after each search round, we can conduct the chaotic search in the neighborhood of the current optimal parameters by listing a certain number of new generated parameters through chaotic process. In this way, we can make use of the ergodicity and irregularity of the chaotic variable to help the algorithm to jump out of the local optimum as well as finding the optimal parameters. The experimental results in section 4 show the efficiency of our algorithm.

3.2   CEDA Approach for Flight Control Parameters Optimization

Chaotic estimation of distribution algorithm (CEDA) is a combination of chaos theory and basic EDA. The CEDA is superior to the basic EDA mainly in the following aspects. The introduction of chaotic theory into basic EDA is an important improvement. EDA can converge fast, but sometimes the fast convergence happens in the first few iterations and relapses into a local optimum easily [11]. By introducing the chaos theory, we can avoid from the local optimum as well as to increase the speed of reaching the optimal solution. The detailed procedure of our proposed CEDA approach to the optimization of flight control parameters can be described as follows.

Step1: Initialize the detailed parameters of the estimation of distribution algorithms (EDA) such as the population size, coding length and so on.

Step2: Encode the variables in a proper way.

Step3: Initialize the probability distribution model $P_0(x)=P_0(x_1,x_2,...,x_n)$ =(0.5, 0.5,...,0.5). (l=0). Then generate the initial population including N individuals according to the $P_l(x)$.

Step4: Calculate the fitness of every individual according to formula (7) and select the best $M<=N$ individuals.

Step5: Conduct the chaotic search around the best solution based on formula (9). Among the engendered series of solutions, select the best one and use it to replace the former best solution.

Step6: Update the probability distribution model to $P_{l+1}(x)$ according to formula (8) and generate another population of new generation based on the model $P_{l+1}(x)$.
Step7: Echo the step4, step5 and step6 until reaching the stopping criteria.
Step8: Decode the variables and output the results.

4   Experimental Results

In order to investigate the feasibility and effectiveness of the chaotic estimation of distribution algorithm to the optimization of parameters, a series of comparative experiments have been conducted.

Our control object is the formula (5). The detailed parameters are set as follows. w_1=0.999, w_2=0.001, w_3=2.0, w_4=100. The population size is 30. The evolutionary times are 100. By means of Matlab, we can easily obtain the comparative results in Fig.3 and Fig.4.

It is noted that the 'EDA' in Fig.3 and Fig.4 represents the simulation results of basic EDA while 'CEDA' denotes the results of chaotic EDA. It turns out that our method performs better than the basic EDA. As is shown in Fig.3, the objective function can converge to a smaller range with a faster speed by CEDA compared with basic EDA. While in Fig.4, we can see clearly that using CEDA can make the quadrotor to track the given signal faster and more steadily.

From the experimental results, it is obvious that our improved EDA can jump outof the local optimum as well as speeding up the process of finding the optimal parameters. The experimental results proves that our proposed method is a more feasible and effective approach in solving the problem of optimization of flight control parameters.

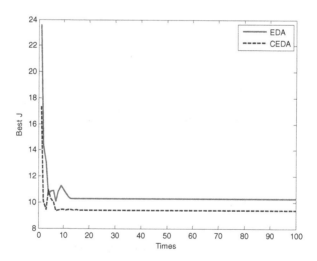

Fig. 3. Comparative objective function response curves by using EDA and CEDA

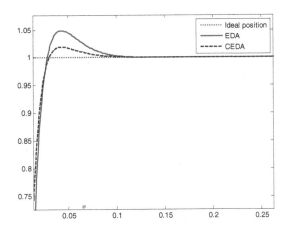

Fig. 4. Comparative results of step response curves by using EDA and CEDA

5   Conclusions

In this paper, an improved estimation of distribution algorithm-CEDA is proposed. The chaos theory is introduced into the basic EDA, and a better performance can be attained in this way. Comparative experimental results of the proposed CEDA and-basic EDA are also given to verify the feasibility and effectiveness of our proposed approach, which provide a more effective way for the optimization of flight control parameters.

Our future work will focus on applying our proposed CEDA to the actual flight control system.

Acknowledgements. This work was partially supported by Natural Science Foundation of China(NSFC) under grant #61273054 and #60975072, National Key Basic Research Program of China under grant #2013CB035503, Program for New Century Excellent Talents in University of China under grant #NCET-10-0021, Aeronautical Foundation of China under grant #20115151019, and YUYUAN Science and Technology Innovation Foundation for Undergraduate of Beihang University.

References

1. Bi, Y.C., Duan, H.B.: Implementation of Autonomous Visual Tracking and Landing for a Low-cost Quadrotor. Optik - International Journal for Light and Electron Optics (in press, 2013), http://dx.doi.org/10.1016/j.ijleo.2012.10.060
2. Hu, Q., Fei, Q., Wu, Q.H., Geng, Q.B.: Research and Application of Nonlinear Control Techniques for Quadrotor UAV. In: Chinese Control Conference, pp. 706–710. IEEE press, Hefei (2012)

3. Wright, A., Poli, R., Stephens, C., Langdon, W.B., Pulavarty, S.: An Estimation of Distribution Algorithm Based on Maximum Entropy. In: Deb, K., Tari, Z. (eds.) GECCO 2004. LNCS, vol. 3103, pp. 343–354. Springer, Heidelberg (2004)

4. Krajník, T., Vonásek, V., Fišer, D., Faigl, J.: AR-Drone as a Platform for Robotic Research and Education. In: Obdržálek, D., Gottscheber, A. (eds.) EUROBOT 2011. CCIS, vol. 161, pp. 172–186. Springer, Heidelberg (2011)

5. Baidu Library,
   `http://wenku.baidu.com/view/f34ec66b561252d380eb6ead.html`

6. Pedeo, L., Lozano, J.A.: Estimation of Distribution Algorithms. A New Tool for Evolutionary Computation. Kluwer Academic Publishers, Boston (2002)

7. Pelikan, M., Goldberg, D.E., Lobo, F.G.: A Survey of Optimization by Building and Using Probabilistic Models. Computational Optimization and Applications 21, 5–20 (2002)

8. Muhlenbein, H., Paap, G.: From recombination of genes to the estimation of distributions I. Binary parameters. In: Ebeling, W., Rechenberg, I., Voigt, H.-M., Schwefel, H.-P. (eds.) PPSN 1996. LNCS, vol. 1141, pp. 178–187. Springer, Heidelberg (1996)

9. Xu, C.F., Duan, H.B., Liu, F.: Chaotic Artificial Bee Colony Approach to Uninhabited Combat Air Vehicle (UCAV) Path Planning. Aerospace Science and Technology 14, 535–541 (2010)

10. Zi, F., Zhao, D.W., Zhang, K.: Image Pre-processing Algorithm Based on Lateral Inhibition. In: The 8th International Conference on Electronic Measurement and Instruments, pp. 701–705. IEEE press, Xi'an (2007)

11. Liu, F., Duan, H.B., Deng, Y.M.: A Chaotic Quantum-behaved Particle Swarm Optimization Based on Lateral Inhibition for Image Matching. Optik-International Journal for Light and Electron Optics 123, 1955–1960 (2012)

12. Sun, Y., Zhang, W.G., Zhang, M., Yin, W.: Optimization of Flight Controller Parameters Based on Chaotic PSO Algorithm of Adaptive Parameter Strategy. Journal of System Simulation 22, 1222–1225 (2012)

Stability Analysis
on Pattern-Based NN Control Systems

Feifei Yang and Cong Wang

College of Automation Science and Engineering,
South China University of Technology, Guangzhou 510641, P.R. China
yfflgw@163.com,
wangcong@scut.edu.cn

Abstract. This technical note introduces stability analysis on pattern-based neural network (NN) control systems. Firstly, different control situations are defined as dynamical patterns and are identified via deterministic learning (DL). When the dynamical pattern is correctly classified, the corresponding NN learning controller with knowledge or experience is selected. Secondly, by adopting a class of switching signals with average dwell time (ADT) property , it is shown that the NN learning controller can achieve small tracking errors and fast convergence rate with small control gains. These results will guarantee not only stability of the closed-loop systems, but also better performance in the aspects of time saving or energy saving. Finally, the theoretical analysis is supported by simulations.

Keywords: Convergence, Pattern-based, Average Dwell Time, Deterministic Learning, Uncertain Nonlinear System, RBF Neural Network.

1 Introduction

Pattern-based control is an interesting and challenging idea in control area. Pattern recognition has been studied in the control literature in the 1960s together with adaptive, learning and self-organizing systems, see for instance[1]. In that time, a pattern in control was defined as a control situation which was represented by a set of state variables. Information of a control situation learned during the process of closed-loop control was taken as a control experience. Pattern recognition techniques were proposed to classify different control situations. Based on the classification result, an experienced controller corresponding to the specific control situation was selected to control the system[2].

The idea of combining pattern recognition with control might be motivated naturally by human learning and control, in which pattern learning, recognition and control together play important roles. It has been observed that with sufficient practice a human can learn many highly complicated control tasks, and these tasks can be performed again and again by a proficient individual with little effort. The implementation of the idea in technology, however, is very difficult. One problem, which was indicated as early as in 1970 by Fu[2], is "learning

C. Guo, Z.-G. Hou, and Z. Zeng (Eds.): ISNN 2013, Part II, LNCS 7952, pp. 27–34, 2013.

in nonstationary or dynamic environments". This might be the most difficult problem in the area of adaptive and learning control systems. Other problems include representation, rapid recognition and classification of different patterns in control, i.e., control situations. It is obvious that conventional pattern recognition methods, e.g., representation of nonstationary state variables by using a finite number of different stationary patterns, and recognition techniques for identification and classification of stationary patterns, are not suitable to cope with these problems. A new framework is required to implement pattern-based learning, recognition and control in a unified way.

Recently, a deterministic learning (DL) theory was proposed for identification, recognition and control of nonlinear dynamical systems undergoing periodic or recurrent motions [5]. Elements of DL include (i) employment of the localized radial basis function (RBF) network, (ii) satisfaction of a partial persistent excitation(PE) condition along a periodic or recurrent orbit, and (iii) accurate RBF network identification or approximation of unknown nonlinear dynamics achieved in a local region along a recurrent orbit. This means that a partial true system model can be accurately identified. Further, rapid recognition of a test dynamical pattern from a set of training dynamical patterns is achieved by using the locally accurate NN approximation of system dynamics. The idea of pattern-based control is implemented as: First, for different training control tasks, the system dynamics corresponding to the training control tasks are identified via deterministic learning. Second, a set of training dynamical patterns is learned, and set of pattern-based NN controllers are constructed accordingly. Third, in a new control situation, a dynamical pattern is introduced that can be rapidly recognized by the set of training dynamical patterns. Based as the recognition, the corresponding pattern-based NN controller is selected to control the system. This NN controller can be able to achieve guaranteed stability and improved control performance. In fact, related researches mentioned above have already been fully detailed in [3,4] and in monograph [5], however, stability issues on pattern-based NN control systems have not yet been paid much more attention, i.e., although every NN controller can guarantee the subsystem stable, whether the overall closed-loop control systems are stable has remained unknown. This observation motivates the present study.

The main contributions of this paper are given as follows. First, By adopting a class of switching signals with ADT property, which means that the number of switches in a finite interval is bounded and the average time between consecutive switching is not less than a constant [6], it is shown that the NN learning controller can achieve better performance. Second, the synchronously switched stabilization problem of NN control systems with ADT is studied by designing a set of pattern-based NN controllers and by finding a set of switching signals with admissible ADT. The minimal ADT is obtained by multiple lyapunov function method.

The organization of this paper is as follows. In section 2, the problem formulation is described. Section 3 presents that the pattern-based closed-loop NN control systems constructed by NN controllers can be stabilized when satisfying the ADT property. In Section 4, simulation results are presented. Some conclusions are given in Section 5.

2   Problem Formulation

Consider the system

$$\dot{x}_1 = x_2,$$
$$\dot{x}_2 = f_\sigma(x) + u, \tag{1}$$

where $x = [x_1, x_2]^T \in R^2$ are the state variables. σ is a piecewise constant function of time, called a switching signal, which takes values in a finite set $K = \{1, \ldots, N\}$; $N > 1$ is the number of subsystems. At arbitrary time t, σ is dependent on t or $x(t)$, or both, or other logic rules. For a switching time sequence $0 < t_1 < t_2 < \cdots$, σ is continues from the right everywhere. When $t \in [t_k, t_{k+1})$, we say that $\sigma(t_k)$th subsystem is activated and the trajectory $x(t)$ of system (1) is the trajectory of the $\sigma(t_k)$th subsystem. $f_\sigma(x)$ are the unknown nonlinearities, corresponding to different operating environments such as a normal state and changes in system dynamics, faults in the system, sensor failures, and external disturbances. Firstly, we design the subsystem state $x(t)$ in all the environments to track a set of periodic or periodic-like reference orbits $x_d(t)$ generated from the following reference models

$$\dot{x}_{d_1} = x_{d_2},$$
$$\dot{x}_{d_2} = f_d(x_d), \tag{2}$$

or

$$\dot{x}_{d_1} = x_{d_2},$$
$$\dot{x}_{d_2} = f_d^m(x_d), \tag{3}$$

where $x_d = [x_{d_1}, x_{d_2}]^T$ is the system state, $f_d(\cdot)$ or $f_d^m(\cdot)(m = 1, \cdots, M)$ is the smooth nonlinear function and (2) is the case when the reference orbit remains unchanged but system dynamics $f_\sigma(x)$ changes due to some reasons, whereas (3) is another case when there are different reference tracking orbits x_d^m corresponding to changes in initial conditions or system parameters. Both of the two cases are taken into account in this paper. $u \in R$ is the system input generated by switching among a collection of NN controllers constructed as

$$u^k = -z_1 - c_2 z_2 - \overline{W}^{kT} S(Z) + \dot{\alpha}, \quad k \in K, \tag{4}$$

and satisfies $z_1 = x_1 - x_{d_1}$, $z_2 = x_2 - \alpha_1$, $\alpha = -c_1 z_1 + \dot{x}_{d_1}$, $\overline{W} = mean_{t \in [t_a, t_b]} \widehat{W}(t)$, here, $\widehat{W}$ is the estimate of NN optimal value W^* and meets $\dot{\widehat{W}} = \Gamma(S(Z)z_2 - \sigma \widehat{W})$, where $Z = [x_1, x_2]^T \in \Omega \subset R^2$ is the NN input, $\Gamma = \Gamma^T$ is a design matrix in diagonal form, $\sigma > 0$ is of small value and $c_1 > 0, c_2 > 0$ are control gains. Then we aims to find a set of admissible switching signals such that all signals in the closed-loop systems (1) remain bounded and the state tracking error $\widetilde{x} = x(t) - x_d(t)$ converges to exponentially to a small neighborhood around zero under the two cases described above.

3   Main Results

It has been well accepted that the multiple Lyapunov-like functions is an efficient stability analysis tool for switched systems [7], essentially for slowly switched systems with ADT. The key point on multiple Lyapunov-like functions is that each Lyapunov-like function constructed for every subsystem is generally considered to be decreasing. Our main results are based on this.

Assumption 1. *There exist continually differentiable functions* $V_p : R^n \to R, p \in K$, *positive constants* λ_0, μ, *and functions* $\alpha, \bar{\alpha}$ *of class* K_∞ *such that* $\frac{\partial V_p}{\partial x} F_p(x) \leq -2\lambda_0 V_p, \alpha(\|x\|) \leq V_p(x) \leq \bar{\alpha}(\|x\|), V_p(x) \leq \mu V_q(x)$, *for each* $x \in R^n$ *and* $p, q \in K$.

Assumption 2. *Reference orbits generated by (2) and (3) are considered to be covered by the RBF network we have constructed.*

Lemma 1. *([8]) If assumption 1 hold, system* $\dot{x} = f_\sigma(x)$ *is globally asymptotically stable for any switching signal that has the average dwell time property with* $\tau > log\mu/\lambda$, *where* τ *is the average dwell time.*

Lemma 2. *([9]) System* $\dot{x} = f_\sigma(x)$ *is globally asymptotically stable if and only if there exist a* C^∞ *function* $V : R^n \to [0, +\infty)$ *such that* $\alpha_1(|\xi|) \leq V(\xi) \leq \alpha_2(|\xi|)$, $\nabla V(\xi)f_\lambda(\xi) < -\alpha_3(|\xi|), \forall \xi \in R^n, \forall \lambda \in \Lambda$, *where* α_1 *and* α_2 *are* K_∞ *function,* α_3 *is a continues positive definite function.*

Theorem 1. *Consider the closed-loop system consisting of the plant (1), the reference model(2), and the neural learning controller(4) with neural weights* $\overline{W}$ *being given as* $\overline{W} = mean_{t \in [t_a, t_b]} \widehat{W}(t)$. *For initial condition* $x_d^p(0)(p \in K)$ *which generated the recurrent reference orbit, and with corresponding to initial condition* $x^p(0)$ *in a close vicinity of* φ_{d_ς}, *we have that (i) all signals in the closed-loop subsystem remain bounded and the state tracking error* $\widetilde{x}^p = x^p(t) - x_d^p(t)$ *exponentially converges to a small neighborhood around zero. (ii) the closed-loop systems (1) is locally asymptotically stable for any switching signal that has the average dwell time property with* $\tau > log\mu/\lambda$. *(iii) all signals in the closed-loop systems (1) remain bounded, and the state tracking error* $\widetilde{x} = x(t) - x_d(t)$ *exponentially converges to a small neighborhood around zero.*

Proof. (i) The derivatives of z_1^p and z_2^p of the pth subsystem are given as $\dot{z}_1^p = \dot{x}_1^p - \dot{x}_{d_1}^p = -c_1 z_1^p + z_2^p$ and $\dot{z}_2^p = f^p(x) + u^p - \dot{\alpha}_1^p = -z_1^p - c_2 z_2^p - \overline{W}^{pT} S(x) + f^p(x)$. Consider Lyapunov function candidate: $V_z^P = \frac{1}{2}z_1^{p2} + \frac{1}{2}z_2^{p2}, p \in K$. We will rewrite it as $V_z^P = \frac{1}{2}z_{p1}{}^2 + \frac{1}{2}z_{p2}{}^2, p \in K$ for concise. Then the derivative of V_z^P is $\dot{V}_z^p = -c_{p1}z_{p1}^2 - c_{p2}z_{p2}^2 - z_{p2}(\overline{W}^T S^p(x) - f^p(x))$. Because $-\frac{1}{2}c_{p2}z_{p2}^2 - z_{p2}(\overline{W}^T S^p(x) - f^p(x)) \leq \frac{\left|\overline{W}^T S^p(x) - f^p(x)\right|^2}{2c_{p2}}$, we have $\dot{V}_z^p \leq -c_{p1}z_{p1}^2 - \frac{1}{2}c_{p2}z_{p2}^2 + \left|\overline{W}^{pT} S(x) - f^p(x)\right|^2$, furthermore, $x_{p1} - x_{d_1} = z_{p1}, x_{p2} - x_{d_2} = z_{p2} - c_{p1}z_{p1}$, for all $\|x_p(t) - x_d^p(t)\| < d_p$, so there exists $d_{p1} > 0$ (with $\|d_p\| - \|d_{p1}\|$ small) such that $\|z_d\| < d_{p1}$, where $z_p = [z_{p1}, z_{p2}]^T$. Using the local knowledge stored in $\overline{W}$ corresponding to the training reference dynamical φ_d and the control system dynamics $f^p(x)$, that is, $dist(x, \varphi_d) < d \Rightarrow \left|\overline{W}^T S(x) - f^p(x)\right| < \xi_i^*$, we have $\dot{V}_z^p \leq -c_{p1}z_{p1}^2 - \frac{1}{2}c_{p2}z_{p2}^2 + \frac{\xi_i^{*2}}{2c_{p2}}$, which holds in a local region when $\|z_p\| < d_{p1}$. Choosing $c_{p1} \leq \frac{1}{2}c_{p2}$ and denoting $\delta_p := \frac{\xi_i^{*2}}{2c_{p2}}, \rho_p := \delta_p/2c_{p1} = \frac{\xi_i^{*2}}{4c_{p1}c_{p2}}$, then

$$0 \leq V_z^p < \delta_p + (V_z^p(0) - \delta_P)exp(-2c_{p1}t),$$ Hence, we have $\sum_{k=1}^{2} \frac{1}{2}z_k^{p2} < \delta_p + (V_z^p(0) - \delta_P)exp(-2c_{p1}t) < \delta_p + V_z^p(0)exp(-2c_{p1}t)$.

Since ξ_i^* is a small value thanks to the previous accurate learning as described in Section 2, ρ_p can be made very small without high control gains c_{p1} and c_{p2}. Thus, for initial condition $x_d^p(0)$ which generates the test reference pattern and with initial condition $x_p(0)$ satisfying $z_p(0) = [x_P(0) - x_d^p(0)] \in \Omega_{z_p 0} := \{z_p|V_z^p < \frac{1}{2}d_{p1}^2 - \rho_p\}$, which guarantees that $\|z_p(t)\| < d_{p1}$ and thus $\|x_p(t) - x_d^p(t)\| < d_P$. Consequently, the state x_p will remain bounded in the local region, in which the past experience is valid for use. Using (4), in which $\dot{\alpha}$ is bounded since every term is bounded, and $S(x_p)$ is bounded for all values of the NN input $z_P = x_p$, we conclude that control u is also bounded. Accordingly, all the signals in the closed-loop system remain bounded. We can also prove that every subsystem of the control systems (1) is locally exponentially stable in the same way.

(ii) In light of [9], i.e., Lemma 2, because all the systems in the family (1) are proved to be locally exponentially stable, then for each $p \in K$ there exists a Lyapunov function $V_p(x)$ that for some positive constants α_p, β_p and γ_p satisfies $\alpha_p(|x|) \leq V_p(x) \leq \beta_p(|x|)$, and $\nabla V_p(x)f_p(x) < -\gamma_p(|x|), \forall x \in R^n, p \in K$. Undoubtedly, we obtain $\nabla V_p(x)f_p(x) \leq -\lambda_p V_p(x), p \in K$, where $\lambda_p = \gamma_p/\beta_p$. This implies that $V_p(x(t_0 + \tau)) \leq exp(-\lambda_p\tau)V_p(x(t_0)), p \in K$. Provided that $\sigma(t) = p$ for almost all $[t_0, t_0 + \tau]$, σ taking on the value on $[t_0, t_1), [t_1, t_2), \cdots, [t_{k-1}, t_k)$, where $t_{i+1} - t_i \geq \tau, i = 0, 1, \cdots, k - 1$. From the above inequalities and $V_p(x(t_0 + \tau)) \leq exp(-\lambda_p\tau)V_p(x(t_0)), p \in K$ we have $V_1(t_k) \leq \frac{\beta_1\beta_2\cdots\beta_k}{\alpha_1,\alpha_2\cdots\alpha_k}exp(-(\lambda_1 + \lambda_2 + \cdots + \lambda_k)\tau)V_1(t_0)$. Then it is clear that if τ is large enough, we can make sure that $V_1(t_k) < V_1(t_0)$. We definite that $\mu := sup\left\{\frac{V_p(x)}{V_q(x)} : x \in R^n, p, q \in K\right\}$. It is obvious that $\mu < \infty$. Then the lower bound on τ is guaranteed by Lemma 2 with $\lambda_p = \lambda$ since K is a compact set.

(iii) We can immediately get the conclusion that the closed-loop system (1) is locally exponentially stable using the result of (i) and (ii), and thus all signals in the closed-loop system remain bounded, and the state tracking error $\tilde{x} = x(t) - x_d(t)$ converges to exponentially to a small neighborhood around zero. This concludes the proof. □

Remark 1. Indeed, accurate RBF network identification or approximation of unknown nonlinear dynamics is achieved in a local region along a recurrent orbit, thus only local exponentially stability can be established here.

Remark 2. Note that Theorem 1 implies, if a unknown pattern is recognized to be similar with one of the patterns of systems (1) and resemble another pattern as a result of environment change later, the closed-loop systems remain stable when satisfying the ADT property.

In practical systems, reference models sometimes may also happen to change for some reasons, which produce several other cases, such as same system tracks different recurrent reference models, which is similar to Theorem 1, hence, remark below describes another more complicated case.

Remark 3. Consider the closed-loop system consisting of the plant (1), the reference model(3), and the neural learning controller(4) with neural weights $\overline{W}$ being given as $\overline{W} = mean_{t\in[t_a,t_b]}\widehat{W}(t)$. For initial condition $x_d^p(0)(p \in K)$ which generated the recurrent reference orbit $\varphi_{d_{\zeta_p}}(p \in K)$, here, different reference models are designed not by initial conditions but by system parameters, and with corresponding to initial condition $x^p(0)$ in a close vicinity of $\varphi_{d_{\zeta_p}}$, we have that all signals in the closed-loop systems (1) remain bounded, and the state tracking error $\widetilde{x} = x(t) - x_d(t)$ exponentially converges to a small neighborhood around zero when satisfying the ADT property.

4   Numerical Example

In this section, a numerical example in a neural learning control system will be presented to demonstrate the potential and validity of our theoretical results. For the first case, we design the dynamical systems to track the same reference model. Another case will be designed to track different recurrent reference models.

In this simulation $K = \{1, 2, 3\}$ and learning control systems (1) with $f^1(x) = -x_1 + 0.7(1 - x_1^2)x_2, f^2(x) = (1 + x_2^2)x_1$, and $f^3(x) = x_1 - 1.5(1 + x_1^2)x_2$. The kth NN controller appropriately designed as (4), which is capable of learning autonomously every system during tracking control to a recurrent reference orbit.

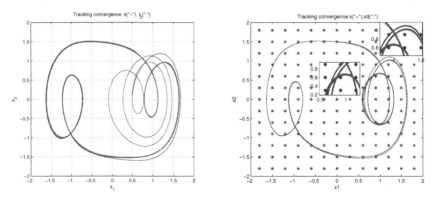

(a) Tracking convergence phase plane (b) Tracking convergence phase plane (dif-
(same reference model)   ferent reference models)

Fig. 1. Tracking convergence phase plane

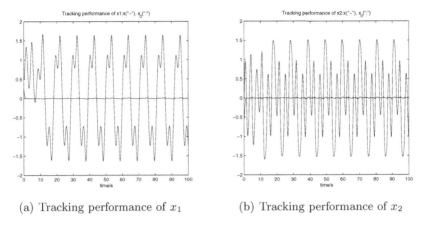

(a) Tracking performance of x_1 (b) Tracking performance of x_2

Fig. 2. Tracking performance of x (same reference model)

Remark 4. The learning here can be achieved from tracking control in a deterministic and autonomous way. The parameter convergence is trajectory-dependent, and the NN approximation of the closed-loop system dynamics is locally accurate along the tracking orbit. This kind of learning capability is very desirable in the advanced intelligent control system.

In the first case, the reference model (2) is applied with $f_d = 1.1x_{d1} - x_{d1}^3 - 0.55x_{d2} + 1.498cos(1.8t)$, which has phase plane trajectory of a period-2 limit cycle. Tracking convergence performance are showed in Fig. 1(a) and Fig. 2. In the second, the reference model (3) with $f_d^1(x_d) = 1.1x_{d1} - 1.0x_{d1}^3 - 0.4x_{d2} + 0.620cos(1.8t)$, $f_d^2(x_d) = 1.1x_{d1} - 1.0x_{d1}^3 - 0.55x_{d2} + 1.498cos(1.8t)$, and $f_d^3(x_d) = 1.1x_{d1} - 1.0x_{d1}^3 - 0.35x_{d2} + 1.498cos(1.8t)$. Tracking convergence performance are presented in Fig. 1(b) and Fig. 3 in this case.

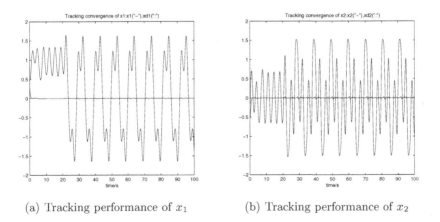

(a) Tracking performance of x_1 (b) Tracking performance of x_2

Fig. 3. Tracking performance of x (different reference models)

Both of the cases satisfy that the first system is activated for 20s, the second 30s, and the last 50s. We can choose $\mu = 1.5$ and $\lambda = 1$, so τ is calculated as 0.176, which is a small value. Then we can easily satisfy stability requirements.

5   Conclusion

Stability issues for pattern-based NN control systems via deterministic learning were investigated in this paper. The main results describe that closed-loop systems can be stabilized if it has the ADT property. Results were presented with three locally exponentially stable subsystems, which were designed to track the same reference model and different recurrent reference models, respectively. The simulation showed that the closed-loop systems were stable when the subsystem was activated long enough. Moreover, the results in this paper might be the first step towards pattern-based NN control via deterministic learning. In fact, there are still many problems to be solved on this topic, e.g., whether the results can be applied to more general nonlinear system, or how to deal with the asynchronous switching problems. These issues prove to motivate our further study.

Acknowledgments. This work was supported by the National Science Fund for Distinguished Young Scholars (Grant No. 61225014), and by the National Natural Science Foundation of China (Grant No. 60934001).

References

1. Tsypkin, K.Y.: Adaptation and Learning in Automatic Systems. Academic Press, New York (1971)
2. Fu, K.S.: Learning Control Systems-review and Outlook. IEEE Transactions on Automatic Control 15(2), 210–221 (1970)
3. Wang, C., Hill, D.J.: Learning from Neural Control. IEEE Transactions on Neural Networks 17(1), 30–46 (2006)
4. Wang, C., Liu, T., Wang, C.-H.: Deterministic Learning and Pattern-based NN Control. In: IEEE 22nd International Symposium on Intelligent Control, pp. 144–149. IEEE Press, Singapore (2007)
5. Wang, C., Hill, D.J.: Deterministic Learning Theory for Identification, Recognition and Control. CRC Press, New York (2009)
6. Hespanha, J.P., Morse, A.S.: Stability of Switched Systems with Average Dwell Time. In: 38th IEEE Conference on Decision and Control, pp. 2655–2660. IEEE Press, AZ (1999)
7. Branicky, M.S.: Multiple Lyapunov Functions and Other Analysis Tools for Switched and Hybrid Systems. IEEE Transactions on Automatic Control 43(4), 475–782 (1998)
8. Liberzon, D., Morse, A.S.: Basic Problems in Stability and Design of Switched Systems. IEEE Control Systems 19(5), 59–70 (1999)
9. Mancilla-Aguilar, J.L., Garcia, R.A.: A Converse Lyapunov Theorem for Nonlinear Switched Systems. Systems & Control Letters 41(1), 67–71 (2000)

Observer-Based H_∞ Fuzzy Control for T-S Fuzzy Neural Networks with Random Data Losses

Xiaoning Duan

Zhejiang Agriculture and Business College, Shaoxing, 312000, China
`xningduan@gmail.com`

Abstract. This paper investigates the observer-based H_∞ control problem for a class of discrete-time Takagi-Sugeno fuzzy neural networks with both random communication packet losses. The random data losses are described by a Bernoulli distributed white sequence that obeys a conditional probability distribution. In the presence of random packet losses, sufficient conditions for the existence of an observer-based feedback controller are derived, such that the closed-loop control system is asymptotically mean-square stable and preserves a guaranteed H_∞ performance. Finally, a numerical example is provided to illustrate the effectiveness of the developed theoretical results.

Keywords: Observer-based H_∞ control, Takagi-Sugeno fuzzy neural networks, Random packet losses.

1 Introduction

Since the theory of H_∞ control has proposed by Zames [1], much effort has been made in H_∞ controller design in order to guarantee desired stability [2]. However, this control is often based on the assumption that the entire state is available, which may not hold in many systems. Therefore, it is necessary to design observers that produce an estimate of the system state [3, 4].

Recently, networked control systems (NCSs) have been widely used in many areas such as industrial automation, unmanned vehicles, remote surgery, robots and so on, because of the lower cost of installation and implementation, simpler installation and maintenance, etc. [3, 5]. On the other hand, in NCSs, there are many new problems such as intermittent data packet losses, network-induced time delays, and communication constraints, etc. This paper is focused on observer-based feedback control of mixed delay systems subject to packet losses. As in practical applications, temporal failures may happen to actuators/sensors, therefore the delivered signals may be incomplete. Hence it is of engineering significance to design reliable controllers in the presence of possible actuator/sensor failures. The control problem of systems with packet losses has been studied in many recent papers [6–10]. It should be pointed out that, in the most of the existing literature, the random sensor-to-controller packet losses and the random controller-to-actuator packet losses have not been simultaneously considered. Because of the random packet losses, the system performance

C. Guo, Z.-G. Hou, and Z. Zeng (Eds.): ISNN 2013, Part II, LNCS 7952, pp. 35–44, 2013.

requirements such as the disturbance rejection attenuation have not obtained sufficiently study. Therefore, in this paper, we are motivated to develop a more reasonable model for a class of T-S fuzzy neural networks with packet losses in both channels from the sensor to the controller and from the controller to the actuator. And this will increase the difficulties in the controller design, because in most of the observer-based controller design problem, the control input of the observer is identical with the control input of the controlled objects, however, in the proposed design methods of the observer-based H_∞ control, these inputs are different as the existence of the random packet losses in the communication channel from the controller to the actuator.

Notation. $\mathbb{N}^+$ stands for the set of nonnegative integers; $\mathbb{R}^n$ and $\mathbb{R}^{n \times m}$ denote, respectively, the n dimensional Euclidean space and the set of all $n \times m$ real matrices. I_n is an n dimensional identity matrix. The notation $P > 0$ (≥ 0) means that P is positive definite(semi-definite). In symmetric block matrices or complex matrix expressions, we use an asterisk ($*$) to represent a term that is induced by symmetry and diag$\{\cdots\}$ stands for a block-diagonal matrix. Matrices, if their dimensions are not explicitly stated, are assumed to be compatible for algebraic operations. Moreover, we may fix a probability space $(\Omega, \mathcal{F}, \mathcal{P})$ where, $\mathcal{P}$, the probability measure, has total mass 1. Prob$\{\alpha\}$ means the occurrence probability of the event α, $\mathbb{E}\{x\}$ stands for the expectation of stochastic variable x. $L_2[0, +\infty)$ is the space of square integrable vectors. The notation $||.||$ stands for the usual $L_2[0, +\infty)$ norm while $|.|$ refers to the Euclidean vector norm. If A is a symmetric matrix, $\lambda_{\max}(A)$(respectively $\lambda_{\min}(A)$) denotes the largest (respectively, smallest) eigenvalue of A.

2 Problem Formulation

Consider the following discrete-time fuzzy neural network with mixed delays:
Plant Rule i: IF $\theta_1(k)$ is M_{i1} and $\theta_2(k)$ is M_{i2} and $\cdots$ $\theta_p(k)$ is M_{ip}, THEN

$$
\begin{cases}
x(k+1) = A_i x(k) + A_{di} x(k - d(k)) + A_{li} \sum_{m=1}^{+\infty} \mu_m x(k - m) \\
\qquad + B_i \omega(k) + D_i u(k), \\
y(k) = \alpha_k C_{1i} x(k) + C_{2i} \omega(k), \\
z(k) = E_{1i} x(k) + E_{2i} \omega(k), \\
x(k) = \phi(k), -\infty < k \leq 0,
\end{cases}
\tag{1}
$$

where $x(k) \in \mathbb{R}^n$ is the state vector; $y \in \mathbb{R}^p$ is the measured output vector with random communication packet loss, $u(k) \in \mathbb{R}^m$ is the control input; $z(k) \in \mathbb{R}^r$ is the output; $\omega(k) \in \mathbb{R}^q$ is the disturbance input, which belongs to $L_2[0, \infty)$; $d(k)$ denotes the time-varying delay with lower and upper bounds $\underline{d} \leq d(k) \leq \bar{d}, k \in \mathbb{N}^+$ where $\underline{d}, \bar{d}$ are known positive integers; $\phi(k)$ is the initial state of the system. The constants $\mu_m \geq 0 (m = 1, 2, \cdots)$ satisfy the following convergence conditions:

$$\bar{u} := \sum_{m=1}^{+\infty} \mu_m \leq \sum_{m=1}^{+\infty} m\mu_m < +\infty. \tag{2}$$

The stochastic variable $\alpha_k \in \mathbb{R}$ is a Bernoulli distributed white sequence with

$$\begin{cases} \text{Prob}\{\alpha_k = 1\} = \mathbb{E}\{\alpha_k\} := \alpha, \\ \text{Prob}\{\alpha_k = 0\} = 1 - \mathbb{E}\{\alpha_k\} := 1 - \alpha, \\ \text{var}\{\alpha_k\} = \mathbb{E}\{(\alpha_k - \bar{\alpha})^2\} = (1 - \alpha)\alpha = \alpha_1^2. \end{cases} \tag{3}$$

$A_i, A_{di}, A_{li}, B_i, C_{1i}, C_{2i}, D_i, E_{1i}, E_{2i}$ are known real constant matrices with appropriate dimensions, $i = 1, 2, \cdots, r$ and r is the number of IF-THEN rules, $\theta_k = [\theta_1(k), \cdots, \theta_p(k)]$ is known premise variable vector, M_{ij} is the fuzzy set, p is the premise variable number. Throughout the paper, it is assumed that the premise variable do not depend on the input variable $u(k)$ explicitly. Given a pair of $(x(k), u(k))$, the final output of the fuzzy neural network is inferred as

$$\begin{cases} x(k+1) = \sum_{i=1}^{r} h_i(\theta_k)\Big[A_i x(k) + A_{di} x(k - d(k)) + A_{li} \sum_{m=1}^{+\infty} \mu_m x(k - m) \\ \qquad\qquad + B_i \omega(k) + D_i u(k)\Big], \\ y(k) = \sum_{i=1}^{r} h_i(\theta_k)\Big[\alpha_k C_{1i} x(k) + C_{2i}\omega(k)\Big], \\ z(k) = \sum_{i=1}^{r} h_i(\theta_k)\Big[E_{1i} x(k) + E_{2i}\omega(k)\Big], \\ x(k) = \phi(k), -\infty < k \leq 0, \end{cases} \tag{4}$$

where $h_i(\theta_k) = \frac{w_i(\theta(k))}{\sum_{i=1}^{r} w_i(\theta(k))}$; $w_i(\theta(k)) = \prod_{j=1}^{p} M_{ij}(\theta_j(k))$, with $M_{ij}(\theta_j(k))$ representing the grade of membership of $\theta_j(k)$ in M_{ij}. Then, we will drop the argument of $h_i(\theta_k)$ for brevity. Therefore, for all k, $h_i \geq 0$, $\sum_{i=1}^{r} h_i = 1$.

The design of observer-based H_∞ controllers for system (4) is performed through the parallel distributed compensation, and the overall observer-based law is inferred as

$$\text{Observer}: \begin{cases} \hat{x}(k+1) = \sum_{i=1}^{r} h_i(\theta_k)\Big[A_i \hat{x}(k) + A_{di}\hat{x}(k - d(k)) + D_i \bar{u}(k) \\ \qquad\qquad + A_{li} \sum_{m=1}^{+\infty} \mu_m \hat{x}(k - m) + L_i(y(k) - \alpha C_{1i}\hat{x}(k))\Big], \\ \bar{u}(k) = \beta \hat{u}(k), \end{cases} \tag{5}$$

$$\text{Controller}: \begin{cases} \hat{u}(k) = -\sum_{i=1}^{r} h_i(\theta_k)K_i \hat{x}(k), \\ u(k) = \beta_k \hat{u}(k), \end{cases} \tag{6}$$

where $\hat{x}(k) \in \mathbb{R}^n$ is the state estimate of system (4), $\bar{u}(k) \in \mathbb{R}^m$ is the control input of the observer, $\hat{u}(k) \in \mathbb{R}^m$ is the control input without packet dropout,

$u(k) \in \mathbb{R}^m$ is the control input of the plant, $K_i \in \mathbb{R}^{m \times n}$ and $L_i \in \mathbb{R}^{n \times p}$ are the given controller and observer gains, respectively. K_i and L_i are two parameters to be determined. The stochastic variable $\beta_k \in \mathbb{R}$ is a Bernoulli distributed sequence with

$$\begin{cases} \text{Prob}\{\beta_k = 1\} = \mathbb{E}\{\beta_k\} := \beta, \\ \text{Prob}\{\beta_k = 0\} = 1 - \mathbb{E}\{\beta_k\} := 1 - \beta, \\ \text{var}\{\beta_k\} = \mathbb{E}\{(\beta_k - \bar{\beta})^2\} = (1 - \beta)\beta = \beta_1^2. \end{cases} \quad (7)$$

The stochastic variables α_k and β_k characterizes the possibilities of the links from sensor to controller and from controller to actuator. Therefore, the larger values of α_k and β_k, the higher the chances of successful transmission.

Remark 1. The packet losses from sensor to controller and from controller to actuator are simultaneously considered to be two Bernoulli distributed white sequences as described in [9, 11]. Recently, the control input of the observer is considered to be different from that of the controller in [4, 11]. Based on this new kind of observer, the observer-based H_∞ fuzzy control of discrete-time mixed delay systems with random packet losses will be investigated.

Now, define the state estimate error as

$$e(k) = x(k) - \hat{x}(k). \quad (8)$$

Substituting (5) and (6) into (4) and (8), then

$$\begin{cases} \eta(k+1) = \sum\limits_{i=1}^{r} \sum\limits_{j=1}^{r} h_i(\theta_k) h_j(\theta_k) \Big[\bar{A}_{ij} \eta(k) + (\beta_k - \beta) \bar{A}_{1ij} \eta(k) + \bar{A}_{di} \eta(k - d(k)) \\ \qquad\qquad + (\alpha_k - \alpha) \bar{A}_{2i} \eta(k) + \bar{A}_{li} \sum\limits_{m=1}^{+\infty} \mu_m \eta(k-m) + \bar{B} \omega(k) \Big], \\ z(k) = \sum\limits_{i=1}^{r} h_i(\theta_k) \Big[\bar{E}_{1i} \eta(t) + E_{2i} \omega(k) \Big], \end{cases} \quad (9)$$

where $\bar{E}_{1i} = [E_{1i} \ \ 0]$,

$$\eta(t) = \begin{bmatrix} x(t) \\ e(t) \end{bmatrix}, \ \bar{A}_{ij} = \begin{bmatrix} A_i - \beta D_i K_j & \beta D_i K_j \\ 0 & A_i - \alpha L_i C_{1i} \end{bmatrix}, \ \bar{A}_{1ij} = \begin{bmatrix} -D_i K_j & D_i K_j \\ -D_i K_j & D_i K_j \end{bmatrix},$$

$$\bar{A}_{2i} = \begin{bmatrix} 0 & 0 \\ -L_i C_{1i} & 0 \end{bmatrix}, \ \bar{A}_{di} = \begin{bmatrix} A_{di} & 0 \\ 0 & A_{di} \end{bmatrix}, \ \bar{A}_{li} = \begin{bmatrix} A_{li} & 0 \\ 0 & A_{li} \end{bmatrix}, \ \bar{B}_i = \begin{bmatrix} B_i \\ B_i - L_i C_{2i} \end{bmatrix}.$$

The purpose of this paper is to design the observer (6) and the observer-based controller (6) for system (4), such that, in the presence of the mixed delays, random packet losses and stochastic nonlinearities, the closed-loop system (9) is asymptotically mean-square stable and the H_∞ performance constraint is also satisfied. More specifically, we aim to establish some sufficient conditions under which the following two conditions are satisfied.

(Q1) System (9) is asymptotically mean-square stable.

(Q2) Under the zero-initial condition, for all nonzero $\omega(k)$, the controlled output $z(k)$ satisfies

$$\sum_{k=0}^{\infty} \mathbb{E}\{\|z(k)\|^2\} \leq \gamma^2 \sum_{k=0}^{\infty} \mathbb{E}\{\|\omega(k)\|^2\}. \tag{10}$$

3 Main Results

Theorem 1. Given a positive constant scalar $\gamma > 0$ and gain L_i, K_i, system (9) is asymptotically mean-square stable with its H_∞ norm being less than γ, if there exist matrices $P, Q, R, \hat{L}$ satisfying

$$\begin{bmatrix} \Gamma_{11} & 0 & 0 & 0 & \Gamma_{18} & \Gamma_{19} & \Gamma_{10} & \bar{E}_{1i}^T \\ * & -Q & 0 & 0 & \Gamma_{38} & 0 & 0 & 0 \\ * & * & -\frac{1}{\mu}R & 0 & \Gamma_{58} & 0 & 0 & 0 \\ * & * & * & -\gamma^2 I & \Gamma_{78} & 0 & 0 & E_{2i}^T \\ * & * & * & * & -P & 0 & 0 & 0 \\ * & * & * & * & * & -P & 0 & 0 \\ * & * & * & * & * & * & -P & 0 \\ * & * & * & * & * & * & * & -I \end{bmatrix} < 0, \tag{11}$$

where $P = \mathrm{diag}\{P_1, P_2\}$, $Q = \mathrm{diag}\{Q_1, Q_2\}$, $R = \mathrm{diag}\{R_1, R_2\}$ and $P_1 = P_1^T > 0, P_2 = P_2^T > 0, Q_1 = Q_1^T > 0, Q_2 = Q_2^T > 0, R_1 = R_1^T > 0, R_2 = R_2^T > 0$,

$$\Gamma_{11} = (\bar{d} - \underline{d} + 1)Q + \bar{\mu}R - P,$$

$$\Gamma_{18} = \begin{bmatrix} A_{ij}^T P_1 - \beta \hat{K}_j^T P_1 & 0 \\ \beta \hat{K}_j^T P_{1i} & A_{ij}^T P_2 - \alpha C_{1i}^T \hat{L}_i \end{bmatrix}, \quad \Gamma_{19} = \begin{bmatrix} -\beta_1 \hat{K}_j^T P_1 & -\beta_1 \hat{K}_j^T P_2 \\ \beta_1 \hat{K}_j^T P_1 & \beta_1 \hat{K}_j^T P_2 \end{bmatrix},$$

$$\Gamma_{10} = \begin{bmatrix} 0 & -\alpha_1 C_{1i}^T \hat{L}_i \\ 0 & 0 \end{bmatrix}, \quad \Gamma_{38} = \begin{bmatrix} A_{di}^T P_1 & 0 \\ 0 & A_{di}^T P_2 \end{bmatrix},$$

$$\Gamma_{58} = \begin{bmatrix} A_{li}^T P_1 & 0 \\ 0 & A_{li}^T P_2 \end{bmatrix}, \quad \Gamma_{78} = \begin{bmatrix} B_i^T P_1 & B_i^T P_2 - C_{2i}^T \hat{L}_i \end{bmatrix}.$$

Moreover, if the above conditions are feasible, then $K_j = D^{-1}\hat{K}_j$, $L_i = P_2^{-1}\hat{L}_i^T$.

Proof. Define $V(k) = \sum_{i=1}^{3} V_i(k)$, where $V_1(k) = \eta^T(k) P \eta(k)$,

$$V_2(k) = \sum_{i=k-d(k)}^{k-1} \eta^T(i) Q \eta(i) + \sum_{j=k-\bar{d}+1}^{k-\underline{d}} \sum_{i=j}^{k-1} \eta^T(i) Q \eta(i),$$

$V_3(k) = \sum_{m=1}^{+\infty} \mu_m \sum_{l=k-m}^{k-1} \eta^T(l) R \eta(l)$. Calculating the difference of $V(k)$ along system (9) and taking the mathematical expectation

$$\mathbb{E}\{\triangle V(k)\} = \sum_{i=1}^{3} \mathbb{E}\{\triangle V_i(k)\} = \sum_{i=1}^{3} \mathbb{E}\{V_i(k+1) - V_i(k)\},$$

where

$$
\begin{aligned}
\mathbb{E}\{\triangle V_1(k)\} = \mathbb{E}\Big\{ &\sum_{i=1}^{r}\sum_{j=1}^{r} h_i(\theta_k)h_j(\theta_k)\Big[\eta^T(k)\big(\bar{A}_{ij}^T P\bar{A}_{ij} + \beta_1^2\bar{A}_{1ij}^T P\bar{A}_{1ij} + \alpha_1^2\bar{A}_{2i}^T \\
&\times P\bar{A}_{2i} - P\big)\eta(k) + \eta^T(k-d(k))\bar{A}_{di}^T P\bar{A}_{di}\eta(k-d(k)) + \omega^T(k) \\
&\times \bar{B}_i^T P\bar{B}_i\omega(k) + (\sum_{m=1}^{+\infty}\mu_m\eta(k-m))^T\bar{A}_{lij}^T P\bar{A}_{li}(\sum_{m=1}^{+\infty}\mu_m\eta(k-m)) \\
&+ 2\eta^T(k)\bar{A}_{ij}^T P\bar{A}_{di}\eta(k-d(k)) + 2\eta^T(k)\bar{A}_{ij}^T P\bar{B}_i\omega(k) \\
&+ 2\eta^T(k)\bar{A}_{ij}^T P\bar{A}_{li}(\sum_{m=1}^{+\infty}\mu_m\eta(k-m)) + 2\eta^T(k-d(k))\bar{A}_{di}^T P\bar{B}_i\omega(k) \\
&+ 2\eta^T(k-d(k))\bar{A}_{di}^T P\bar{A}_{li}(\sum_{m=1}^{+\infty}\mu_m\eta(k-m)) \\
&+ 2(\sum_{m=1}^{+\infty}\mu_m\eta(k-m))^T\bar{A}_{lij}^T P\bar{B}_i\omega(k)\Big]\Big\},
\end{aligned}
$$

$$
\mathbb{E}\{\triangle V_2(k)\} \le \mathbb{E}\Big\{(\bar{d}-\underline{d}+1)\eta^T(k)Q\eta(k) - \eta^T(k-d(k))Q\eta(k-d(k))\Big]\Big\},
$$

$$
\mathbb{E}\{\triangle V_3(k)\} \le \mathbb{E}\{\bar{\mu}\eta^T(k)R\eta(k) - \frac{1}{\bar{\mu}}(\sum_{m=1}^{+\infty}\mu_m\eta(k-m))^T R(\sum_{m=1}^{+\infty}\mu_m\eta(k-m))\}.
$$

Then,

$$
\mathbb{E}\{\triangle V(k)\} \le \mathbb{E}\Big\{\sum_{i=1}^{r}\sum_{j=1}^{r} h_i(\theta_k)h_j(\theta_k)\Big[\zeta^T(k)\Pi\zeta(k)\Big]\Big\}, \tag{12}
$$

where

$$
\Pi = \begin{bmatrix} \Pi_{11} & \bar{A}_{ij}^T P\bar{A}_{di} & \bar{A}_{ij}^T P\bar{A}_{li} & \bar{A}_{ij}^T P\bar{B}_i \\ * & \Pi_{22} & \bar{A}_{di}^T P\bar{B}_i & \\ * & * & \Pi_{33} & \bar{A}_{li}^T P\bar{B}_i \\ * & * & * & \bar{B}_i^T P\bar{B}_i \end{bmatrix},
$$

$$
\Pi_{11} = \bar{A}_{ij}^T P\bar{A}_{ij} + \beta_1^2\bar{A}_{1ij}^T P\bar{A}_{1ij} + \alpha_1^2\bar{A}_{2i}^T P\bar{A}_{2i} + (\bar{d}-\underline{d}+1)Q + \bar{\mu}R - P
$$

$$
\Pi_{22} = \bar{A}_{di}^T P\bar{A}_{di} - Q, \qquad \Pi_{33} = \bar{A}_{l}^T P\bar{A}_{li} - \frac{1}{\bar{\mu}}R,
$$

$$
\zeta(k) = \Big[\eta^T(k)\ \eta^T(k-d(k))\ (\sum_{m=1}^{+\infty}\mu_m\eta(k-m))^T\ \omega^T(k)\Big]^T.
$$

From system (9),

$$
\begin{aligned}
\mathbb{E}\{z^T(k)z(k)\} = \sum_{i=1}^{r} h_i(\theta_k)\Big\{&\eta^T(k)\bar{E}_{1i}^T\bar{E}_{1i}\eta(k) + \eta^T(k)\bar{E}_{1i}^T E_{2i}\omega(k) \\
&+ \omega^T(k)E_{2i}^T\bar{E}_{1i}\eta(k) + \omega^T(k)E_{2i}^T E_{2i}\omega(k)\Big\}. \tag{13}
\end{aligned}
$$

From (12) and (13) and assuming zero initial condition,

$$J_N = \mathbb{E}\{\sum_{k=0}^{N}(z^T(k)z(k) - \gamma^2\omega^T(k)\omega(k))\}$$

$$\leq \mathbb{E}\left\{\sum_{i=1}^{r}\sum_{j=1}^{r}h_i(\theta_k)h_j(\theta_k)\left[\sum_{k=0}^{N}\zeta^T(k)\Phi\zeta(k)\right]\right\}, \tag{14}$$

where

$$\Phi = \begin{bmatrix} \Phi_{11} & \bar{A}_{ij}^T P\bar{A}_{di} & \bar{A}_{ij}^T P\bar{A}_{li} & \bar{A}^T P\bar{B}_i + \bar{E}_{1i}^T E_{2i} \\ * & \Phi_{22} & \bar{A}_{di}^T P\bar{A}_{li} & \bar{A}_{di}^T P\bar{B}_i \\ * & * & \Phi_{33} & \bar{A}_{li}^T P\bar{B}_i \\ * & * & * & \bar{B}_i^T P\bar{B}_i + E_{2i}^T E_{2i} - \gamma^2 I \end{bmatrix},$$

$$\Phi_{11} = \bar{A}_{ij}^T P\bar{A}_{ij} + \beta_1^2 \bar{A}_{1ij}^T P\bar{A}_{1ij} + \alpha_1^2 \bar{A}_{2i}^T P\bar{A}_{2i} + (\bar{d} - \underline{d} + 1)Q + \bar{\mu}R + \bar{E}_{1i}^T \bar{E}_{1i} - P,$$

$$\Phi_{22} = \bar{A}_{di}^T P\bar{A}_{di} - Q, \quad \Phi_{33} = \bar{A}_{li}^T P\bar{A}_{li} - \frac{1}{\bar{\mu}}R.$$

Define $\Psi = \text{diag}\{I\ I\ I\ I\ P^{-1}\ P^{-1}\ P^{-1}\ I\}$, Pre- and Post-multiply (11) by Ψ^T and Ψ, respectively. Then, from (9), $\Phi < 0$, subsequently $J_N < 0$. Therefore, system (9) is asymptotically stable in the mean square sense. This completes the proof.

Under the H_∞ performance constraint (10) with minimum γ, to design the observer-based H_∞ control law (6) for system (4), the optimization problem is $\min_{\Omega} \gamma$, s.t.(11), where $\Omega \in \{P_1 = P_1^T > 0, P_2 = P_2^T > 0, Q_2 = Q_2^T > 0, Q_1 = Q_1^T > 0, R_1 = R_1^T > 0, R_2 = R_2^T > 0, \text{and } \hat{L}\}$.

Remark 2. In Theorem 1, the observer-based H_∞ control problem has been solved for discrete-time mixed delay systems with random packet losses and stochastic nonlinearities. And we can easily obtain the observer-based feedback controller by solving the LMI equation given in (11) by the Matlab LMI Toolbox.

4   Numerical Example

In this section, a numerical example is used to demonstrate the effectiveness of the proposed observer-based H_∞ fuzzy control for a class of discrete-time mixed delay systems with random packet losses. Consider the fuzzy neural network:

Model Rule i: IF $x_1(k)$ is $h_i(x_1(k)), i = 1, 2$, THEN

$$\begin{cases} x(k+1) = A_i x(k) + A_{di}x(k - d(k)) + A_{li}\sum_{m=1}^{+\infty}\mu_m x(k - m) \\ \quad\quad\quad + B_i\omega(k) + D_i u(k), \\ y(k) = \alpha_k C_{1i}x(k) + C_{2i}\omega(k), \\ z(k) = E_{1i}x(k) + E_{2i}\omega(k), \end{cases} \tag{15}$$

where

$$A_1 = \begin{bmatrix} -0.042 & 0.806 \\ -1.040 & 0.100 \end{bmatrix}, \quad A_2 = \begin{bmatrix} -1.042 & -0.006 \\ 0.040 & 1.000 \end{bmatrix}, \quad A_{d1} = \begin{bmatrix} 0.200 & -0.101 \\ -0.101 & 0.100 \end{bmatrix},$$

$$A_{d2} = \begin{bmatrix} 0.100 & 0.001 \\ 0.001 & 0.200 \end{bmatrix}, \quad A_{l1} = \begin{bmatrix} -0.100 & -0.010 \\ -0.010 & 0.100 \end{bmatrix}, \quad A_{l2} = \begin{bmatrix} 0.100 & -0.000 \\ -0.000 & 0.300 \end{bmatrix},$$

$$B_1 = \begin{bmatrix} 0.100 & -0.100 \end{bmatrix}^T, \quad B_2 = \begin{bmatrix} 0.000 & 0.111 \end{bmatrix}^T, \quad D_1 = \begin{bmatrix} 0.005 & -0.250 \end{bmatrix}^T,$$

$$D_2 = \begin{bmatrix} 0.035 & -0.550 \end{bmatrix}^T, \quad C_{11} = \begin{bmatrix} -0.200 & 0.200 \end{bmatrix}, \quad C_{12} = \begin{bmatrix} -0.100 & 0.100 \end{bmatrix},$$

$$E_{11} = \begin{bmatrix} 0.100 & -0.200 \end{bmatrix}, \quad E_{12} = \begin{bmatrix} 0.100 & 0.200 \end{bmatrix}, \quad E_{21} = 0.100, \quad E_{22} = -0.100,$$

$$C_{21} = 0.200, \quad C_{22} = 0.500, \quad d(k) = 1 + \frac{1 + (-1)^k}{4}, \mu_m = 3^{-(3+m)}.$$

According to (2), $\bar{u} = \sum\limits_{m=1}^{+\infty} \mu_m = \frac{1}{54} < \sum\limits_{m=1}^{+\infty} m\mu_m = \frac{1}{36} < +\infty$. And it is easy to verify that $\underline{d} = 1$, $\bar{d} = 1.5$.

The design of observer-based H_∞ controllers is performed through the parallel distributed compensation, and the observer-based law is inferred as

Model Rule i: IF $x_1(k)$ is $h_i(x_1(k)), i = 1, 2$, THEN

$$\text{Observer}: \begin{cases} \hat{x}(k+1) = A_i\hat{x}(k) + A_{di}\hat{x}(k - d(k)) + A_{li}\sum\limits_{m=1}^{+\infty} \mu_m\hat{x}(k - m) \\ \qquad\qquad + D_i\bar{u}(k) + L_i(y(k) - \alpha C_{11}\hat{x}(k))], \\ \bar{u}(k) = \beta\hat{u}(k), \end{cases} \tag{16}$$

$$\text{Controller}: \begin{cases} \hat{u}(k) = -K_i\hat{x}(k), \\ u(k) = \beta_k\hat{u}(k), \end{cases} \tag{17}$$

In this example, two cases with different data loss probabilities have been considered and two corresponding controllers have been designed as shown in Table 1 by Theorem 1. Therefore the obtained results can be used to the system with random data losses in [6, 8]. However, the results on considering either the random sensor-to-controller packet losses or the random controller-to-actuator packet losses in [7, 8] are not suitable for this problem investigated.

The membership function is assumed to be

$$h_1(x_1(k)) = \begin{cases} 1, & x_1(k) \le -1, \\ 0.5 - 0.5x_1(k), & -1 \le x_1(k) \le 1, \\ 0, & x_1(k) \ge 1, \end{cases} \quad h_2(x_1(k)) = 1 - h_1(x_1(k)),$$

and the initial conditions of system (4) are supposed to be $x(0) = \begin{bmatrix} 0.5 & -0.5 \end{bmatrix}^T$, $\bar{x}(0) = \begin{bmatrix} 0 & 0 \end{bmatrix}^T$, the disturbance input is chosen as $\frac{1}{k^2}$. The state responses of the controlled system with different data loss probabilities are shown in Figs. 1 and 2, which demonstrate that the closed-loop system is asymptotically stable in the mean square sense and when the data losses become severer, the H_∞ performance γ becomes larger.

Table 1. Observer gain matrix L, controller gain matrix K and γ at the different data loss probabilities

α	β	K1, K2	L1, L2	γ
0.9	0.9	K1=[-7.5819 -4.2430] K2=[-0.0136 -0.2216]	L1=[-2.2604 2.8479] L2=[2.4555 8.1589]	0.7416
0.8	0.8	K1=[1.9818 -0.6012] K2=[-0.0178 -0.1555]	L1=[-1.5841 1.7970] L2=[3.3389 5.4351]	0.9220

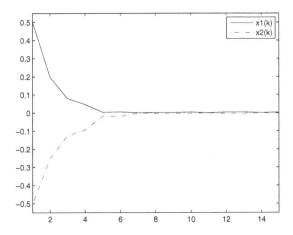

Fig. 1. The state responses of controlled system (4) with $\alpha = 0.9, \beta = 0.9, \gamma = 0.7416$

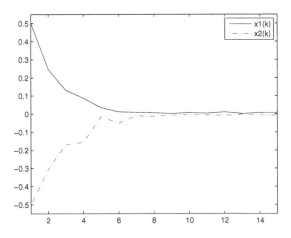

Fig. 2. The state responses of controlled system (4) with $\alpha = 0.8, \beta = 0.8, \gamma = 0.9220$

5   Conclusions

In this paper, the observer-based H_∞ fuzzy control has been studied for a class of discrete-time Takagi-Sugeno fuzzy neural networks with random packet dropouts. A new Lyapunov-Krasovskii functional, which is introduced to account for distributed and time-varying discrete-time delays, has been used to design the observer and controller, such that the closed-loop system is asymptotically mean-square stable. And the controller parameters can be obtained by solving certain LMIs. An illustrative example has been used to show the effectiveness of the proposed method.

References

1. Zames, G.: Feedback and Optimal Sensitivity: Model Reference Transformations, Multiplicative Seminorms and Approximate Inverses. IEEE Transactions on Automatic Control 26, 301–320 (1981)
2. Costa, O., Benites, G.: Linear Minimum Mean Square Filter for Discrete-time Linear Systems with Markov Jumps and Multiplicative Noises. Automatica 47, 466–476 (2011)
3. Chang, W., Wu, W., Ku, C.: H_∞ Constrained Fuzzy Control via State Observer Feedback for Discrete-time Takagi-Sugeno Fuzzy Systems with Multiplicative Noises. ISA Transactions 50, 207–212 (2011)
4. Li, J., Yuan, J., Lu, J.: Observer-based H_∞ Control for Networked Nonlinear Systems with Random Packet Losses. ISA Transactions 49, 39–46 (2010)
5. Antsaklis, P., Baillieul, J.: Special issue on Technology of networked control systems. IEEE Proceedings 95, 5–8 (2007)
6. de la peña, D., Christofides, P.: Output Feedback Control of Nonlinear Systems Subject to Sensor Data Losses. Systems and Control Letters 57, 631–642 (2008)
7. Wang, J., Yang, H.: H_∞ Control of a Class of Networked Control Systems with Time Delay and Packet Dropout. Applied Mathematics and Computation 217, 7469–7477 (2011)
8. Wang, W., Ho, D., Liu, X.: Variance-constrained Filtering for Uncertain Stochastic Systems with Missing Measurements. IEEE Transactions on Automatic control 48, 1254–1258 (2003)
9. Wang, Z., Liu, X., Yang, F., Ho, D., Liu, X.: Robust H_∞ Control for Networked Systems with Random Packet Losses. IEEE Transactions on Systems Man Cybernetics-B Cybernetics 37, 916–924 (2007)
10. Wu, J., Chen, T.: Design of Networked Control Systems with Packet Dropouts. IEEE Transactions on Automatic Control 52, 1314–1319 (2007)
11. Wan, X.B., Fang, H.J., Fu, S.: Observer-based Fault Detection for Networked Discrete-time Infinite-distributed Delay Systems with Packet Dropout. Applied Mathematical Modelling 36, 270–278 (2012)

Robust Adaptive Neural Network Control for Wheeled Inverted Pendulum with Input Saturation[*]

Enping Wei, Tieshan Li[**], and Yancai Hu

Navigational College, Dalian Maritime University, Dalian 116026, China
tieshanli@126.com

Abstract. In this paper, a novel control design is proposed for wheeled inverted pendulum with input saturation. Based on Lyapunov synthesis method, back-stepping design procedure and the Neural network (NN) approximation to the uncertainty of the system, the adaptive NN tracking controller is constructed by considering actuator saturation constraints. The stability analysis subject to the effect of input saturation constrains are conducted with the help of an auxiliary design system. The proposed controller guarantees uniformly ultimately bounded of all the signals in the closed-loop system, while the tracking error can be made arbitrarily small. Simulation studies are given to illustrate the effectiveness and the performance of the proposed scheme.

Keywords: wheeled inverted pendulum, backstepping design, neural network (NN), input saturation.

1 Introduction

Inverted pendulum system is widely concerned by scholars and experts. Because of its characteristics of absolute instability, multivariable, high-degree, strong coupling and nonlinearity, it is proved to be an ideal model to test the control theory and application. The research of control techniques for inverted pendulum has important practical meaning.

In the past several years, there were many literatures to study the inverted pendulum [1-3]. The main topic of this research was how to keep the pendulum balance at the upright position. The rail-cart structure was the most usual type in control experiments. Nowadays, many studies of extensions of the inverted pendulum control system have been proposed. The most popular problem is how to control a mobile wheeled inverted pendulum (WIP) system which the cart is no longer run on a guide

---

[*] This work was supported in part by the National Natural Science Foundation of China (Nos.51179019, 60874056), the Natural Science Foundation of Liaoning Province (No. 20102012) , the Program for Liaoning Excellent Talents in University(LNET) (Grant No.LR2012016), and the Applied Basic Research Program of Ministry of Transport of China.

[**] Corresponding author.

C. Guo, Z.-G. Hou, and Z. Zeng (Eds.): ISNN 2013, Part II, LNCS 7952, pp. 45–52, 2013.

rail. In [4], Newtonian approach and linearization method was proposed to design a controller for a mobile inverted pendulum. In [5], a dynamic model of the WIP was derived with respect to the wheel motor torques as input while taking the nonholonomic no-slip constrains into considerations. In [6], Jung S and Kim SS developed a mobile inverted pendulum using the neural network (NN) control combined with proportional integral derivative (PID) controller. In [7], a self-tuning PID control strategy, based on a deduced model, was proposed to stabilize a two-wheeled vehicle (TWV).

Neural networks were used to construct the control design of dynamic systems in many recent studies [8-14]. It is well known that NNs are powerful building blocks for a wide class of complex nonlinear system control strategies when model information is absent or when a controlled plant is a "black box" [8]. The ability of NNs to uniformly approximate arbitrary input–output linear or nonlinear mappings on closed subsets is extremely useful. Thus, NN-based controllers have been applied to compensate for effects of nonlinearities and system uncertainties in a control system to improve system stability, convergence and robustness. In control engineering, RBF neural networks are usually used as a tool for modelling nonlinear functions because of their good capabilities in function approximation.

Besides, in practice, actuator saturation is one of the most important non-smooth nonlinearities which usually appear in many industry control systems. This problem is of great importance because almost all practical control systems have limitations on the amplitudes of control inputs, and such limitations can cause serious deterioration of control performances and even destroy the stability of the control systems. If we ignore it, it can severely degrade the closed-loop system performance. In [15], Hyperbolic tangent function $g(v)=u_M \times \tanh(v/u_M)$ was used to handle the input saturation in wheeled inverted pendulum. But this method is based on the known control inputs. It will fail with the consideration of unobservable control inputs.

This work is motivated by the wheeled inverted pendulum control with the input saturation constraint. Based on RBF neural networks, a new algorithm is designed for the wheeled inverted pendulum control system with input saturation constraints. Unlike [15], this paper introduces the auxiliary design system to analyze and handle the effect of input saturation. The main objective in this paper is to design a novel controller considering the effect of input saturation.

2   Problem Formulation

Different from the mathematical model in [15], with its three degrees of freedom, the dynamic equation of WIP system is described as following [4]:

$$
\begin{cases}
\dot{x}_1(t) = x_2(t) \\
\dot{x}_2(t) = f(x,t) + g(x,t)u(t) + d(t) \\
y = x_1(t)
\end{cases}
\tag{1}
$$

where $x=\left[x_1(t),x_2(t)\right]^T=\left[x_1(t),\dot{x}_1(t)\right]^T$ is the state(angle) vector of the WIP system which is assumed to be available for measurement, $f(x,t)\in R$ is the nonlinear dynamic function, $g(x,t)\in R$ denotes the control gain of the system and $g(x,t)>0$ for all x and t; $u(t)\in R$ is the control input and $d(t)\in R$ denotes the unknown external disturbance.

Then let us recall the approximation property of the RBF neural networks. We can utilize $\theta_1^{*T}\varphi_1(x_1)$ to approximate the given function $f(x)$ with the approximation error bounded by $\mathcal{E}_m$, i.e.,

$$f(x)=\theta^{*T}\psi(x)+\varepsilon^* \tag{2}$$

With $|\varepsilon^*|\le\varepsilon_m$ where ε^* represents the network reconstruction error, i.e.,

$$\varepsilon^*=f(x)-\theta^{*T}\psi(x) \tag{3}$$

Since θ^* is unknown, we use the notation $\hat{\theta}$ to denote the estimation of θ^* and develop an adaptive law to update the parameter $\hat{\theta}$.

This paper considers the input saturation constraints on rudder u as follows:

$$-u_m\le u\le u_M \tag{4}$$

where u_m and u_M are the known lower limit and upper limit of the input saturation constrains of u, which satisfies,

$$u=sat(v)=\begin{cases}u_M, & if\ \ v>u_M\\ v, & if\ \ u_m\le v\le u_M\\ -u_m, & if\ \ v<-u_m\end{cases} \tag{5}$$

where v is the designed control input of the system.

The following assumptions are introduced:

Assumption 1: The reference $y_r(t)$ is a sufficiently smooth function of t, and y_r, $\dot{y}_r$, $\ddot{y}_r$ are bounded.

Assumption 2: The unknown external disturbance $d(t)\in R$ is bounded, i.e., there is a positive constant ϖ which satisfies $|d(t)|\le\varpi$.

In this paper, we also use the following lemma to solve the problem of unknown external interference $d(t)$.

Lemma 1. The following inequality holds for any $\delta>0$ and for any $\mu\in R$

$$0\le|\mu|-\mu\tanh\left(\frac{\mu}{\delta}\right)\le0.2785\delta \tag{6}$$

The proof of the above lemma follows after straightforward algebraic manipulation.

The control objective is to design a state-feedback controller with input saturation for the system (1), such that all the signals in the closed-loop system remain uniformly ultimately bounded (UUB), and the tracking error $z_1=x_1(t)-y_r(t)$ converges to a small neighborhood of the origin.

3   Controller Design and Stability Analysis

In this section, we develop a novel design procedure for the system (1) under the constraint of input saturation, by employing backstepping technique. It mainly includes 2 steps.

Step 1: Define the error variable $z_1 = x_1 - y_r$, and choose the intermediate stabilizing function α_2 as a virtual control law for the first subsystem. At the same time, define error variable $z_2 = x_2 - \alpha_2$, thus, considering(1), the time derivative of z_1 is

$$\dot{z}_1 = z_2 + \alpha_2 - \dot{y}_r \tag{7}$$

The virtual control law is chosen as

$$\alpha_2 = -c_1 z_1 + \dot{y}_r \tag{8}$$

where $c_1 > 0$, substituting(8) into(7), we obtain

$$\dot{z}_1 = -c_1 z_1 + z_2 \tag{9}$$

Consider the Lyapunov function candidate $V_1 = \dfrac{1}{2} z_1^2$, the time derivative of V_1 is

$$\dot{V}_1 = -c_1 z_1^2 + z_1 z_2 \tag{10}$$

The first term on the right-hand side is negative, and the second term will be considered in the next step.

Step2: Define $z_2 = x_2 - \alpha_2$, and consider (1) and differentiate z_2 with respect to time yields

$$\dot{z}_2 = \dot{x}_2 - \dot{\alpha}_2 = f(x) + g(x) \cdot u + d - \dot{\alpha}_2 \tag{11}$$

Where

$$\dot{\alpha}_2 = -c_1 \dot{z}_1 + \ddot{y}_r \tag{12}$$

For convenience of constraint effect analysis of the input saturation, the following auxiliary design system [16] is given as follows:

$$\dot{e} = \begin{cases} -c_{21} e - \dfrac{f(\cdot)}{\|e\|^2} \cdot e + g(u - v), & \|e\| \geq \varepsilon \\ 0, & \|e\| < \varepsilon \end{cases} \tag{13}$$

where $c_{21} > 0$, $f(\cdot) = f(z_2, \Delta u) = \left| z_2 \cdot g(\cdot) \cdot \Delta u \right| + \dfrac{1}{2} \Delta u^2$, $\Delta u = u - v$, ε is a small positive design parameter and e is a variable of the auxiliary design system introduced to ease the analysis of the effect of the input saturation. Control law v will be designed.

Consider the following Lyapunov function candidate

$$V_2 = V_1 + \frac{1}{2}z_2^2 + \frac{1}{2}\tilde{\theta}^T\Gamma^{-1}\tilde{\theta} + \frac{1}{2}e^2 \tag{14}$$

Invoking (10) and (11), the time derivative of V_2 is

$$
\begin{aligned}
\dot{V}_2 &= -c_1 z_1^2 + z_1 z_2 + z_2 \dot{z}_2 - \tilde{\theta}^T \Gamma^{-1}\dot{\hat{\theta}} + e\cdot\dot{e} \\
&= -c_1 z_1^2 + z_2\left[z_1 + \left(\tilde{\theta}^T + \hat{\theta}^T\right)\psi(x) + \varepsilon^* + g\cdot u + d - \dot{\alpha}_2 \right] \\
&\quad - \tilde{\theta}^T\Gamma^{-1}\dot{\hat{\theta}} + e\dot{e} \\
&= -c_1 z_1^2 + z_2\left[z_1 + \hat{\theta}^T\psi(x) + g\cdot u + \varepsilon^* + d - \dot{\alpha}_2 \right] \\
&\quad - \tilde{\theta}^T\Gamma^{-1}\left(\dot{\hat{\theta}} - \Gamma\psi(x)z_2\right) + e\dot{e}
\end{aligned}
\tag{15}
$$

It is clear that

$$e\cdot\dot{e} = -c_{21}e^2 - \frac{|z_2\cdot g\cdot\Delta u| + \frac{1}{2}\Delta u^2 g^2}{e^2}\cdot e^2 + g\Delta u\cdot e \tag{16}$$

$$\Delta u\cdot e \le \frac{1}{2}\Delta u^2 + \frac{1}{2}e^2 \tag{17}$$

Substituting (16) and (17) into (15), we have

$$
\begin{aligned}
\dot{V}_2 &\le -c_1 z_1^2 + z_2\left[z_1 + \hat{\theta}^T\psi(x) + gu + \varepsilon^* + d - \dot{\alpha}_2 \right] - (c_{21} - 0.5)e^2 \\
&\quad - \tilde{\theta}^T\Gamma^{-1}\left(\dot{\hat{\theta}} - \Gamma\psi(x)z_2\right) - |z_2\cdot g(u - v)| \\
&\le -c_1 z_1^2 + z_2\left[z_1 + \hat{\theta}^T\psi(x) + gv + \varepsilon^* + d - \dot{\alpha}_2 \right] \\
&\quad - \tilde{\theta}^T\Gamma^{-1}\left(\dot{\hat{\theta}} - \Gamma\psi(x)z_2\right) - (c_{21} - 0.5)e^2
\end{aligned}
\tag{18}
$$

Consider the input saturation effect, the control law is proposed as

$$v = \frac{1}{g}\left[-c_{20}(z_2 - e) - z_1 - \hat{\theta}^T\psi(x) + (-c_1\dot{z}_1 + \ddot{y}_d) - d^*\tanh\left|\frac{d^* z_2}{s}\right| \right] \tag{19}$$

where $c_{20} > 0$, $d^* = \varepsilon_m + d_m$, $|\varepsilon^*| \le \varepsilon_m$ and $|d(t)| \le d_m$. And it is clear that

$$z_2\cdot e \le \frac{1}{2}z_2^2 + \frac{1}{2}e^2 \tag{20}$$

Substituting (19) and (20) into (18), Using $2\tilde{\theta}_i^T\hat{\theta}_i \ge \|\tilde{\theta}_i\|^2 - \|\theta_i^*\|^2$ gives

$$
\begin{aligned}
\dot{V}_2 &\le -c_1 z_1^2 - 0.5c_{20}z_2^2 - (c_{21} - 0.5 - 0.5c_{20})e^2 + \sigma\tilde{\theta}^T\Gamma^{-1}\hat{\theta} + 0.2785s \\
&\le -c_1 z_1^2 - 0.5c_{20}z_2^2 - (c_{21} - 0.5 - 0.5c_{20})e^2 - \frac{1}{2}\sigma\tilde{\theta}^T\Gamma^{-1}\tilde{\theta} + \frac{1}{2}\sigma\theta^{*T}\Gamma^{-1}\theta^* + 0.2785s
\end{aligned}
\tag{21}
$$

Let $\quad C = 2\min\left\{c_1, 0.5c_{20}, c_{21} - 0.5 - 0.5c_{20}, \dfrac{1}{2}\sigma\right\}$, $\quad M = \dfrac{1}{2}\sigma\theta^{*T}\Gamma^{-1}\theta^* + 0.2785s$ then

$$\dot{V}_2 \le -CV_2 + M \qquad 0 \le V(t) \le \dfrac{M}{C} \qquad\qquad (22)$$

Actually, the Equation (22) means that $V(t)$ is bounded. Thus, all signals of the closed-loop system are proved uniformly ultimately bounded. Moreover, we can appropriately choose the design parameters $c_i, \Gamma_i^{-1}, \sigma$ etc. to make the tracking error arbitrarily small. This concludes the proof simply.

4   Numerical Simulation

In the simulation, the initial condition for $d = 5\sin(t)$. The initial conditions of x_1, x_2 are $1, 0$ respectively.

Choose the control parameters $c_1 = 77$, $c_{20} = 70$, $c_{21} = 15$, $\varepsilon = 0.01$, $d^* = 6$, $\Gamma_2 = \mathrm{diag}\{0.01\}$, $u_M = 60$. The initial value of e is 30, Neural networks $\hat{\theta}_2^T \xi_2(Z_2)$ contains 135 nodes (i.e., $l_2 = 135$), with centers $\mu_l (l = 1,...,l_l)$ evenly spaced in $[-4,4]\times[-4,4]\times[-6,6]$ and widths $\eta_l = 2(l = 1,...,l_2)$. The initial weight $\hat{\theta}_2(0) = 0.0$.

From Fig.1, we can see that after a short transit process, a good tracking and a small tracking error is obtained. Fig.2 is the trajectories of the control u with input saturation.

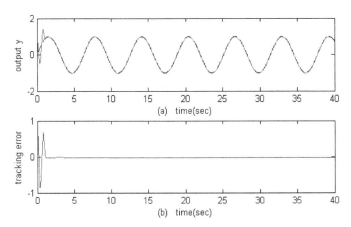

Fig. 1. Simulation results for wheeled inverted pendulum system: output y , reference and tracking error

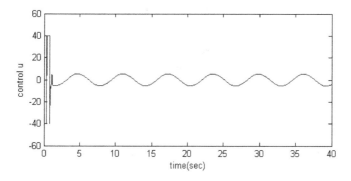

Fig. 2. Simulation results for wheeled inverted pendulum system: Control u

5   Conclusion

In this paper, the adaptive tracking control for the wheeled inverted pendulum has-been investigated. By employing the NN approximation to address the uncertainty of the system, an adaptive neural tracking controller has been explored. In addition, the effect of input saturation constrains is considered in this control design. All the sig-nals of the closed-loop system are guaranteed to be uniformly ultimately bounded. By adjusting the parameters, the tracking error may be made arbitrarily small. Simulation results for the driven inverted pendulum system are presented to demonstrate the good tracking performance of the proposed scheme.

References

1. El-Hawwary, M.I., Elshafei, A.L., Emara, H.M., Fattah, H.A.A.: Adaptive fuzzy control of the inverted pendulum problem. IEEE Control Syst. Technol. 14(6), 1135–1144 (2006)
2. Tao, C.W., Taur, J.S., Hsieh, T.W., Tsai, C.L.: Design of a fuzzy controller with fuzzy swing-up and parallel distributed pole assignment schemes for an inverted pendulum and cart system. IEEE Control Syst. Technol. 16(6), 1277–1288 (2008)
3. Takimoto, T., Yamamoto, S., Oku, H.: Washout control for manual operations. IEEE Control Syst. Technol. 16(6), 1169–1176 (2008)
4. Grasser, F., Arrigo, A., Colombi, S., Rufer, A.C.: JOE: a mobile, inverted pendulum. IEEE Trans. Ind. Electron 39(1), 107–114 (2002)
5. Pathak, K., Franch, J., Agrawal, S.K.: Velocity and position control of a wheeled inverted pendulum by partial feedback linearization. IEEE Trans. Robotics 21(3), 505–513 (2005)
6. Jung, S., Kim, S.S.: Control experiment of a wheel-driven mobile inverted pendulum using neural network. IEEE Trans. Control Syst. Technol. 16(2), 297–303 (2008)
7. Ren, T.J., Chen, T.C., Chen, C.J.: Motion control for a two-wheeled vehicle using a self-tuning PID controller. Control Eng. Practice 16(3), 365–375 (2008)
8. Agarwal, M.: A systematic classification of neural-networkbased control. IEEE Trans. Control Syst. Mag. 17(2), 75–93 (1997)

9. Khan, M., Rahman, M.A.: Development and implementation of a novel fault diagnostic and protection technique for IPM motor drives. IEEE Trans. Ind. Electron 56(1), 85–92 (2009)

10. Mazumdar, J., Harley, R.G.: Recurrent neural networks trained with backpropagation through time algorithm to estimate nonlinear load harmonic currents. IEEE Trans. Ind. Electron 55(9), 3484–3491 (2008)

11. Jemei, S., Hissel, D., Pera, M.C., Kauffmann, J.M.: A new modeling approach of embedded fuel-cell power generators based on artificial neural network. IEEE Trans. Ind. Electron 55(1), 437–447 (2008)

12. Hou, Z.G., Zou, A.: Adaptive control of an electrically driven nonholonomic mobile robot via backstepping and fuzzy approach. IEEE Transactions on Control Systems Technology 17(4), 803–815 (2009)

13. Albus, J.S.: A new approach to manipulator control the cerebellar model articulation controller (CMAC). Trans. ASME J. Dyn. Syst. Meas Control 97(3), 220–227 (1975)

14. Cheng, L., Hou, Z.G., Tan, M.: Adaptive neural network tracking control for manipulators with uncertain kinematics, dynamics and actuator model. Automatica 45(10), 2312–2318 (2009)

15. Wei, E.P., Li, T.S., Li, J.F., Hu, Y.C.: Neural Network-Based Adaptive Dynamic Surface Control for Inverted Pendulum System. In: The IEEE First International Conference on Cognitive Systems and Information Processing, CSIP 2012, pp. 56–63 (2012)

16. Chen, M., Ge, S.S., Choo, Y.: Neural network tracking control of ocean surface vessels with input saturation. In: Proc. of the 2009 IEEE International Conference on Automation and Logistics, ICAL 2009, pp. 85–89 (2009)

An Improved Learning Scheme
for Extracting T-S Fuzzy Rules from Data Samples[*]

Ning Wang[1,2], Xuming Wang[1], Yue Tan[1], Pingbo Shao[1], and Min Han[2]

[1] Marine Engineering College, Dalian Maritime University, Dalian 116026, China
n.wang.dmu.cn@gmail.com, wxuming_dmu@163.com
[2] Faculty of Electronic Information and Electrical Engineering,
Dalian University of Technology, Dalian 116024, China
minhan@dlut.edu.cn

Abstract. In this paper, we present an improved learning scheme for extracting T-S fuzzy rules from data samples, whereby a neuro-fuzzy architecture implements the T-S fuzzy system using ellipsoidal basis functions. The salient characteristics of this approach are as follows: 1) A novel structure learning algorithm incorporating a pruning strategy into new growth criteria is developed. 2) Compact fuzzy rules can be extracted from training data. 3) The linear least squares (LLS) method is employed to update consequent parameters, and thereby contributing to high approximation accuracy. Simulation studies and comprehensive comparisons with other well-known algorithms demonstrate the effective and superior performance of our proposed scheme in terms of compact structure and promising accuracy.

Keywords: extracting fuzzy rules, fuzzy neural networks, ellipsoidal basis function, structure learning, high approximation accuracy.

1 Introduction

Fuzzy logic can be used to express knowledge of domain experts, ill-defined and uncertain system and handle real-life situations that the classical control approach finds it difficult or impossible to tackle. Fuzzy system can approximate any continuous function on a compact set to any desired accuracy [1]. However, the conventional way of designing the fuzzy system has the following problems: 1) Parameter estimation, the determining of premise parameters and consequent parameters; 2) Structure identification, determining the membership function to partition input space and the numbers of the fuzzy rules. To circumvent these problems, many researchers merge neural networks into fuzzy systems, and thereby fuzzy neural networks (FNN) was proposed, which utilize the learning ability of

---

[*] This work is supported by the National Natural Science Foundation of China (51009017), Applied Basic Research Funds from Ministry of Transport of P. R. China (2012-329-225-060), China Postdoctoral Science Foundation (2012M520629), and Fundamental Research Funds for the Central Universities of China (2009QN025, 2011JC002 and 3132013025).

C. Guo, Z.-G. Hou, and Z. Zeng (Eds.): ISNN 2013, Part II, LNCS 7952, pp. 53–60, 2013.

neural networks to extract fuzzy rules and optimize parameters. The famous adaptive-network-based fuzzy inference system (ANFIS) [2] is known as a seminal work which takes advantages of both fuzzy logics and neural networks. However, the structure learning of ANFIS is mainly determined by expert knowledge. In [3]-[5], the author proposed the self-organizing learning scheme for structure learning. In the RAN [6], the method that adds new hidden neurons to the network based on the novelty of new data in the process of sequential learning. In addition, the RAN via the extended Kalman filter (RANEKF) [7] enhanced the performance of the RAN by adopting an extended Kalman filter instead of the LMS method. However, in the foregoing two approaches, inactive hidden nodes will be never removed once the new hidden units are sequentially generated. To overcome this drawback, the minimal resource allocation networks (MARN) [8] developed a pruning method to delete the inactive hidden neurons during the learning process. Chen *et al.* [9] proposed an orthogonal least squares (OLS) learning scheme to identify both structure and parameters. Recently, the dynamic fuzzy neural networks (DFNN) based on a RBF neural network [10] proposed a hierarchical on-line self-organizing learning algorithm. The parameters are adjusted by the LLS method and the structure can be self-adaptive by growing and pruning criteria. Furthermore, a generalized dynamic fuzzy neural networks (GDFNN) based on the ellipsoidal basis function (EBF) [11] was presented, which adopted a novel on-line parameter allocation mechanism based on ε-completeness to allocate the initial widths for each dimension of the input variables. However, the improved performance of the GDFNN is at the cost of slower learning speed. Corresponding improvements have been made by a fast and accurate online self-organizing scheme for parsimonious fuzzy neural networks (FAOS-DFNN) [12], whereby a novel structure learning algorithm incorporating a pruning strategy into new growth criteria was proposed, and in the parameter learning phase, all the parameter of hidden neurons are updated by the extended Kalman filter (EKF) method. However, the generalization capability is required to be enhanced. Lately, Wang *et al.* [13] further extended abovementioned fuzzy neural learning schemes to fast and parsimonious neuro-fuzzy systems employing generalized ellipsoidal basis functions (GEBF), whereby the promising performance of approximation and generalization is validated in system modeling and time-series prediction.

In this paper, we deal with the neuro-fuzzy learning scheme for extracting T-S fuzzy rules from data samples. It can increase the accuracy of function approximation and speed up the learning process with compact structures. To be specific, a generating criteria combined with the pruning strategy is proposed to learn fuzzy rules. The system starts with no hidden neurons corresponding to fuzzy rules, and the neurons can be generated restrictively based on the combined generation criteria, and thereby contributing to a parsimonious structure. After structure learning, the LLS method is employed to identify parameters in each learning epoch. Finally, simulation studies on function approximation demonstrate the effective performance of our learning scheme. Comprehensive comparisons with other well-known methods like ANFIS, RAN, OLS, DFNN, GDFNN, and FAOS-PFNN have been conducted to evaluate the superiority of our algorithm in terms of high accuracy and compact structure.

2   System Architecture

In this section, a four-layer fuzzy neural network, similar to [12], is employed to realize the T-S fuzzy system as follows:

Rule R_j: IF x_1 is A_{1j} and ...and x_i is A_{ij}...and x_r is A_{rj}; THEN y is w_j,

$$(j = 1,2,...,u;   i=1,2,...,r.)$$

Layer 1: Each node in layer 1 represents an input variable. Let r be the number of input variables.

Layer 2: Each node in layer 2 represents a membership function (MF). Each input variable has u membership functions which are in the form of Gaussian functions:

$$\mu_{ij}(x_i) = \exp\left[-\frac{(x_i - c_{ij})^2}{\sigma_{ij}^2}\right] \quad i = 1,2,...,r; \; j = 1,2,...,u. \tag{1}$$

where u_{ij} is jth membership function of x_i; c_{ij} is center of the jth Gaussian function of x_i; σ_{ij} is width of the jth Gaussian function of x_i.

Layer 3: Each node represents a possible IF-part (premise parameters) of fuzzy rules, if the T-norm selected to compute each rule's firing strength is multiplication, the output of the jth rule R_j is obtained,

$$\phi_j(x_1,...,x_i,...,x_r) = \exp\left[-\sum_{i=1}^{r}\frac{(x_i - c_{ij})^2}{\sigma_{ij}^2}\right] \quad j = 1,2,...,u. \tag{2}$$

Layer 4: Each node represents an output variable as the weighted summation of incoming signals and is given by:

$$y(x_1,...,x_i,...,x_r) = \sum_{j=1}^{u} w_j\phi_j \quad j = 1,2,...,u. \tag{3}$$

where y is the value of an output variable and w_j is the THEN-part (consequent parameters) or connection weight of the jth rule as follows:

$$w_j = k_{j0} + k_{j1}x_1 + k_{j2}x_2 + ... + k_{jr}x_r \quad j = 1,2,...,u. \tag{4}$$

3   Learning Scheme for Extracting Fuzzy Rules

In this section, the main idea of this system will be presented. Assume that there are n training data pairs, for each observation (X^k, t^k). X^k is the kth input vector and t^k is the desired output, the output y^k of this system can be obtained by (1)-(4). In the structure learning process, suppose that it has generated u hidden neurons in layer 3, to obtain a fast and accurate fuzzy neuron network, the online learning algorithm of the growth criteria incorporates into the pruning criterion is adopted. In parameter learning, the traditional LLS approach is used to adjust the consequent parameter.

3.1   Criteria of Rule Generation

3.1.1   System Error

System error is the deviation between output of this system and desired output, it's an important factor for determining whether to recruit a new rule or not. When the kth observation (X^k, t^k) incoming, the system error could be computed as follows:

$$\left\| e^k \right\| = \left\| t^k - y^k \right\|$$ (5)

If

$$\left\| e^k \right\| > k_e, \quad k_e = \max\left\{ e_{max} \beta^{k-1}, e_{min} \right\},$$ (6)

It means that the performance of the fuzzy neural network unsatisfied, a new fuzzy rule should be recruited.

Here, k_e is a predefined threshold during the learning process where e_{max} is the maximum error chosen, $\beta \in (0,1)$ is the convergence constant and e_{min} is the minimum error chosen which is the desired accuracy of output of this system.

3.1.2   Input Partition

The method of distance criteria will be adopted in this paper, distance criteria means the minimum distance between the new observation and the existing membership functions, the distance could be described as follows:

$$d_{kj} = \left\| X^k - C_j \right\|, \quad k = 1,2,...,n; \ j = 1,2,...,u.$$ (7)

where $X^k = [x_{1k}, x_{2k},...,x_{rk}]^T$ is the kth observation, $C_j = [c_{1j}, c_{2j},...,c_{rj}]^T$ is the center of the jth cluster. The minimum distance between kth observation and existing centers is obtained,

$$d_{k \min} = \min d_{kj}, \quad j = 1,2,...,u.$$ (8)

If

$$d_{\min} > k_d, \quad k_d = \max\left\{ d_{max} \gamma^{k-1}, d_{min} \right\}$$ (9)

It implies that the input space cannot be partitioned by the existing input membership functions well, so a new cluster should be considered or the premise parameters of the existing membership functions should be adjusted.

Where d_{max} is the maximum distance chosen; d_{min} is the minimum distance chosen, and γ is the decay constant and $\gamma \in (0,1)$.

3.1.3   Generation Criteria

Assume that there are n input-output data pairs, consider (3) as a special case of the linear regression model:

$$y(x_1,...,x_i,...,x_r) = \sum_{j=1}^{u} w_j \phi_j; \quad t^k = \sum_{j=1}^{u} w_j \phi_j + e^k \quad j = 1,2,...,u. \tag{10}$$

Rewriting in a compact form:

$$Y = \Psi W; \quad T = \Psi W + E \tag{11}$$

where $Y=[y^1,y^2,...,y^n]^T$ is the real output vector, $T=[t^1,t^2,...,t^n]^T$ is the desired output vector, $W=[w_1,w_2,...,w_n]$ is the weight vector, $E=[e^1,e^2,...,e^n]^T$ is the error vector, $\Psi=[\psi_1, \psi_2, ..., \psi_u]$ is the output matrix of layer 3 which is given by:

$$\Psi = \begin{bmatrix} \overset{\bullet}{\varphi_{11}} \cdots \varphi_{u1} \\ \vdots \ddots \vdots \\ \varphi_{1n} \cdots \varphi_{un} \end{bmatrix} \tag{12}$$

For any matrix Ψ if its row number is larger than the column number, we can transform Ψ into a set of orthogonal basis vector which make it possible to calculate an individual contribution to the desired output energy from each basis vector. Thus Ψ is decomposed into:

$$\Psi = PQ \tag{13}$$

where the $n{\times}u$ matrix $P=[p_1,p_2,...,p_u]$ has the same dimension as ψ with orthogonal columns and Q is an upper triangular $u{\times}u$ matrix.

Substituting (13) into (11) yields:

$$T = PQW + E = PG + E \tag{14}$$

where $G=[g_1,g_2,...,g_u]^T$, it could be calculated by the method of LLS. And each element is:

$$g_k = \frac{p_k^T T}{p_k^T p_k}, \quad k = 1,2,...,u. \tag{15}$$

And an error reduction rate(ERR) due to p_k can be defined as:

$$err_k = \frac{g_k^T p_k^T p_k}{T^T T} = \frac{(p_k^T T)^2}{p_k^T p_k T^T T}, \quad k = 1,2,...,u. \tag{16}$$

If err_k has the largest value, the corresponding p_k and T will be greatest, p_k is the most significant factor to the output.

Define the ERR an $(r+1){\times}u$ matrix $\Delta=(\delta_1, \delta_2, ...,\delta_u)$ whose elements can be obtained by (16) and the kth column of Δ is the total ERR corresponding to the ith RBF unit. Simultaneously, define

$$\eta_k = \sqrt{\frac{\delta_k^T \delta_k}{r+1}} \tag{17}$$

η_k represents the significance of the kth RBF unit. If

$$\eta_k < k_{err} \tag{18}$$

where k_{err} is the threshold chosen, it indicates that the RBF unit is less important and the unit will be pruned, otherwise the fuzzy neural network needs more hidden neurons and adjusts the free parameters to achieve high generation performance.

3.2   Parameter Adjustment

Note that after structure learning, the parameter learning is performed in the entire network regardless of whether the hidden neurons are newly generated or already existing.

When there are new hidden neurons generated, the allocation of new RBF unit parameters could be described as:

$$\begin{cases} C_k = X^k \\ \sigma_k = kd_{k\,\min} \end{cases} \tag{19}$$

where k is a predefined parameter which determines the overlapping degree.

After structure learning, premise parameter of all the hidden neurons are updated as follow:

$$\begin{cases} C_{u+1} = X^k \\ \sigma_{u+1} = k_w d_{k\,\min} \end{cases} \tag{20}$$

where k_w is an overlap factor determining the overlap of responses of the RBF unites.

Premise parameters are updated as above-mentioned, and the consequent parameter could be updated at the same time. For sake of LLS method can achieve fast computation and obtain optimal resolution. So it is adopted to adjust consequent parameters, it can be represented as follow:

$$Y = W\Psi$$
$$E = \|T - Y\| \tag{21}$$

In order to find an optimal coefficient vector W such that the error energy $E^T E$ is minimized as follows:

$$W^*\Psi = T. \tag{22}$$

The optimal W^* is in the following form:

$$W^* = T\Psi^+. \tag{23}$$

Where Ψ^+ is the pseudoinverse of Ψ, it could be calculated as follow:

$$\Psi^+ = \left(\Psi^T \Psi\right)^{-1} \Psi^T. \tag{24}$$

It shows that LLS method provides a computationally simple but efficient procedure for determining the weights.

4   Simulation Studies

In this section, the effectiveness of the proposed algorithm is demonstrated in function approximation. Some comparisons are made with other significant learning algorithm such as RBF-AFS, OLS, RAN, RANEKF, DFNN, GDFNN, FAOS-PFNN, *etc.*

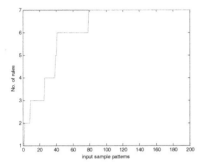
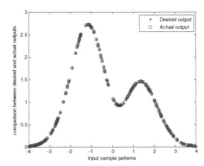

Fig. 1. Growth of neurons   **Fig. 2.** Hermite function and the system

Table 1. Comparison of Structure and approximationPerformance of Different Algorithms

Algorithms	Numbers of rules	RMSE
Our method	7	**0.0026**
D-FNN	6	0.0056
GDFNN	6	0.0097
OLS	7	0.0095
M-RAN	7	0.0090
RANEKF	13	0.0262

Consider the Hermite polynomial,

$$f(x) = 1.1(1 - x + 2x^2)\exp\left(-\frac{x^2}{2}\right) \tag{25}$$

Random sampling of interval [-4,4] is used to obtain 200 input-output data pairs for the training set. The parameters selected in this example are as follows: $d_{max} = 2$, $d_{min} = 0.1$, $\gamma = 0.977$, $\sigma_0 = 1.1$, $e_{max} = 2$, $e_{min} = 0.02$, $\beta = 0.95$, $k = 1.05$, $k_w = 1.01$, $k_{err} = 0.0035$. The growth of hidden neurons is showed in Fig. 1. Fig. 2 depicts the target function curve and the system approximation output, from which we can see that the resulting fuzzy neural network can approximate the original function well. It is evident that good approximation performance has been achieved. Comparisons of structure and performance of different algorithm are listed in Table 1. It can be seen that our method preserves remarkable accuracy with compact structure.

5   Conclusions

In this paper, a fast and accurate approach to extract fuzzy rules via neuro-fuzzy learning schemes is proposed. In structure learning, a novel growing criteria incorporating into the pruning strategy is presented, which decreases the computational burden and accelerate the learning speed. In addition, the LLS method has been applied to parameters identification in each learning epoch. Simulation studies show that a faster and more accurate fuzzy neural network could be self-constructed by the proposed algorithm. Comprehensive comparisons with other learning scheme indicate that the proposed approach is likely to offer tremendous advantages in approximation performance and compact structure.

References

1. Wang, L.X.: Adaptive Fuzzy System and Control: Design and Stability Analysis. Prentice-Hall, Englewood Cliffs (1994)
2. Jang, J.-S.R.: ANFIS: Adaptive-Network-Based Fuzzy Inference System. IEEE Trans. Syst., Man, Cybern. 22, 665–684 (1993)
3. Lin, C.T.: A Neural Fuzzy Control System with Structure and Parameter Learning. Fuzzy Sets Syst. 70, 183–212 (1995)
4. Cho, K.B., Wang, B.H.: Radial Basis Function Based Adaptive Fuzzy Systems and Their Applications to System Identification and Prediction. Fuzzy Sets Syst. 83, 325–339 (1996)
5. Juang, C.F., Lin, C.T.: An On-Line Self-Constructing Neural Fuzzy Inference Network and Its Applications. IEEE Trans. Fuzzy Syst. 6, 12–32 (1998)
6. Platt, J.: A Resource-Allocating Network for Function Interpolation. Neural Comput. 5, 213–225 (1991)
7. Kadirkamannathan, V., Niranjan, M.: A Function Estimation Approach to Sequential Learning with Neural Networks. Neural Comput. 5, 954–975 (1993)
8. Yingwei, L., Sundararajan, N., Saratchandran, P.: Performance Evaluation of a Sequential Minimal Radial Basis Function (RBF) Neural Networks Learning Algorithm. IEEE Trans. Neural Netw. 9, 308–318 (1998)
9. Chen, S., Cowan, C.F.N., Grant, P.M.: Orthogonal Least Squares Learning Algorithm for Radial Basis Function Network. IEEE Trans. Neural Netw. 2, 1411–1423 (1996)
10. Wu, S.-Q., Er, M.J.: Dynamic Fuzzy Neural Networks-A Novel Approach to Function Approximation. IEEE Trans. Syst., Man, Cybern., B, Cybern. 30, 358–564 (2000)
11. Wu, S.-Q., Er, M.J., Gao, Y.: A Fast Approach for Automatic Generation of Fuzzy Rules by Generalized Dynamic Fuzzy Neural Networks. IEEE Trans. Fuzzy Syst. 9, 578–594 (2001)
12. Wang, N., Er, M.J., Meng, X.Y.: A Fast and Accurate Online Self-Organizing Scheme for Parsimonious Fuzzy Neural Networks. Neurocomput. 72, 3818–3829 (2009)
13. Wang, N.: A Generalized Ellipsoidal Basis Function Based Online Self-Constructing Fuzzy Neural Network. Neural Process. Lett. 34, 13–37 (2011)

On the Equivalence between Generalized Ellipsoidal Basis Function Neural Networks and T-S Fuzzy Systems

Ning Wang[1,*], Min Han[2], Nuo Dong[1], Meng Joo Er[1,3], and Gangjian Liu[1]

[1] Marine Engineering College, Dalian Maritime University, Dalian 116026, China
n.wang.dmu.cn@gmail.com
[2] Faculty of EIEE, Dalian University of Technology, Dalian 116024, China
minhan@dlut.edu.cn
[3] School of EEE, Nanyang Technological University, Singapore 639798
emjer@ntu.edu.sg

Abstract. This paper deals with the functional equivalence between Generalized Ellipsoidal Basis Function based Neural Networks (GEBF-NN) and T-S fuzzy systems. Significant contributions are summarized as follows. 1) The GEBF-NN is equivalent to a T-S fuzzy system under the condition that the GEBF unit and the local model correspond to the premise and the consequence of the T-S fuzzy system. 2) The normalized (nonnormalized) GEBF-NN is equivalent to a normalized (nonnormalized) T-S fuzzy system using dissymmetrical Gaussian functions (DGF) as univariate membership functions and local models as consequent parts. 3) The equivalence between the normalized GEBF-NN and the nonnormalized T-S fuzzy system is established by employing GEBF units as multivariate membership functions of fuzzy rules. 4) These theoretical results would not only fertilize the learning schemes for fuzzy systems but also enhance the interpretability of neural networks, and thereby contributing to innovative neuro-fuzzy paradigms. Finally, numerical examples are conducted to illustrate the main results.

Keywords: generalized ellipsoidal basis function, neural network, T-S fuzzy system, functional equivalence.

1 Introduction

Many investigations have revealed that fuzzy inference systems (FIS) and neural networks (NN) can approximate any function to any desired accuracy provided that sufficient fuzzy rules or hidden neurons are available [1]. Innovative merger of the two paradigms has resulted in a powerful technology termed fuzzy neural

* This work is supported by the National Natural Science Foundation of China (under Grant 51009017), Applied Basic Research Funds from Ministry of Transport of P. R. China (under Grant 2012-329-225-060), China Postdoctoral Science Foundation (under Grant 2012M520629), and Fundamental Research Funds for the Central Universities of China (under Grant 2009QN025, 2011JC002 and 3132013025).

C. Guo, Z.-G. Hou, and Z. Zeng (Eds.): ISNN 2013, Part II, LNCS 7952, pp. 61–69, 2013.
© Springer-Verlag Berlin Heidelberg 2013

networks (FNN), which is designed to realize a FIS through the topology of NN, and thereby incorporating generic advantages of NN like massive parallelism, robustness, and learning ability into FIS [2]. Actually, the FNN mechanisms stem from the equivalence between FIS and NN. To our knowledge, Jang *et al.* [3] proposed the seminal work on the functional equivalence between fuzzy logic and radial basis function (RBF) neural networks under certain conditions, which ware relaxed by Andersen *et al.* [4]. The modified set of restrictions was applicable for a much wider range of FIS. Similarly, Hunt *et al.* [5] extended previous results to a class of generalized Gaussian RBF networks by removing the restrictions on standard RBF networks and fuzzy systems. Later on, Azeem *et al.* [6] developed the generalized ANFIS (GANFIS) by promising a generalized fuzzy model and considering a generalized RBF (GRBF) network, whereby the functional equivalent behavior was established. It is followed by the structure identification for the GANFIS based on the abovementioned equivalence, whereby he minimum number of fuzzy rules can be derived from corresponding GRBF units [7]. Furthermore, Li *et al.* [8] proved that any given FIS with the fuzzy partition satisfying Kronecker's property can be approximately represented by a feedforward neural network, and vice versa. In addition, Kolman *et al.* [9] introduced a novel Mamdani-type fuzzy model, referred to as the all-permutations fuzzy rule base, and showed that it was mathematically equivalent to a standard feedforward neural network. Aznarte *et al.* [10] disclosed functional equivalences between neural autoregressive paradigms and fuzzy rule-based systems in the framework of time series analysis. Besides, Barra *et al.* [11] demonstrated an exact mapping between the Hopfield network and the hybrid Boltzmann machine, where the hidden layer is analog and the visible layer is digital. It should be noted that the previous works mainly focus on various RBF-type neural networks and standard fuzzy inference systems, which would inevitably undertake restrictive conditions for equivalences. Lately, Wang [12] proposed a promising Generalized Ellipsoidal Basis Function based Fuzzy Neural Network (GEBF-FNN) which implements a T-S FIS. The GEBF-FNN algorithm could effectively partition the input space and optimize the corresponding weights within the topology of GEBF-FNN. Arguably, this method is an excellent candidate for nonlinear system modeling and control with unmodeled dynamics and/or uncertainties. However, the functional equivalence has not been explicitly revealed between GEBF-type networks and T-S fuzzy systems.

In this paper, we establish the functional equivalence between the GEBF-NN and T-S fuzzy systems, and thereby enhancing the their abilities of knowledge extractions and insertions. Firstly, the novel GEBF-NN using dissymmetrical Gaussian functions (DGF) with flexible widths and shapes, as well as employing local models as weights between GEBF units and output nodes is formulated in the MIMO form. Furthermore, normalized and nonnormalized GEBF-NNs are reasonably related to corresponding normalized and/or nonnormalized T-S fuzzy systems, whereby membership functions are relaxed to univariate or multivariate ones. Theoretical results demonstrate that the GEBF-NN is functionally equivalent to T-S fuzzy systems with high freedom and less restrictions.

2 Generalized Ellipsoidal Basis Function Neural Network

In this section, we briefly investigate the generalized ellipsoidal function neural network (GEBF-NN) so as to present the general network architecture in multi-input-multi-output (MIMO) form. The intuitive concept of the generalized ellipsoidal basis function (GEBF) is incorporated into the GEBF-NN topology which is expected to realize a T-S fuzzy system. The GEBF unit eliminates the symmetry restriction of previous RBF-based nodes projected into each dimension and thereby increasing the clustering flexibility in the input space.

The four-layer architecture of the GEBF-NN is described as follows.

Layer 1. This layer accepts inputs x_i, $i = 1, \cdots, r$ to the system via input nodes without any other computations.

Layer 2. Each node in this layer represents an univariate local receptive unit defined by the dissymmetrical Gaussian-like function (DGF) as follows:

$$DGF\left(x_i; c_{ij}, \sigma_{ij}(x_i), b_{ij}\right) = \exp\left(-\left|\frac{x_i - c_{ij}}{\sigma_{ij}(x_i)}\right|^{b_{ij}}\right) \tag{1}$$

$$\sigma_{ij}(x_i) = \begin{cases} \sigma_{ij}^R, & x_i \geq c_{ij} \\ \sigma_{ij}^L, & x_i < c_{ij} \end{cases} \tag{2}$$

where c_{ij}, σ_{ij}^L and σ_{ij}^R denote the center, left and right spreads of the jth DGF unit for the input variable x_i, respectively. And, b_{ij} controls the shape. Note that the DGF node permits nonidentical spreads with regard to the center in addition to shape freedom. As a consequence, the DGF unit can be preferably used as local receptive node with properties of asymmetry and flexibility.

Layer 3. Each node represents a multivariate local receptive unit taking the DGF units as inputs. The composed output for the GEBF unit is as follows:

$$GEBF\left(\mathbf{x}; \mathbf{c}_j, \sigma_j(\mathbf{x}), \mathbf{b}_j\right) = \prod_{i=1}^{r} DGF\left(x_i; c_{ij}, \sigma_{ij}(x_i), b_{ij}\right) \tag{3}$$

where, $\mathbf{x} = [x_1, x_2, \cdots, x_r]^T$, $\mathbf{c}_j = [c_{1j}, c_{2j}, \cdots, c_{rj}]^T$, $\sigma_j = [\sigma_{1j}(x_1), \sigma_{2j}(x_2), \cdots, \sigma_{rj}(x_r)]^T$ and $\mathbf{b}_j = [b_{1j}, b_{2j}, \cdots, b_{rj}]^T$ denote the input, center, dissymmetrical width and shape index vectors of the jth GEBF unit, respectively, and the dissymmetrical width $\sigma_{ij}(x_i)$ is defined by (2).

Layer 4. The output layer performs the summation of all the inputs weighted with local model $w_{kj}(\mathbf{x})$ given by,

$$w_{kj}(\mathbf{x}) = a_{kj}^{(0)} + a_{kj}^{(1)}x_1 + \cdots + a_{kj}^{(r)}x_r, \ j = 1, 2, \cdots, u, \ k = 1, 2, \cdots, s \tag{4}$$

where $a_{kj}^{(0)}, a_{kj}^{(1)}, \cdots, a_{kj}^{(r)}$ are corresponding weights for input variables. The uniform formulation for weighted output could be described as follows:

$$\mathbf{y} = \mathbf{f}(\mathbf{x}) \triangleq \mathbf{A}\mathbf{\Psi}(\mathbf{x}) \tag{5}$$

where,

$$\mathbf{A} = \begin{bmatrix} \mathbf{a}_{11}^{\mathrm{T}} & \mathbf{a}_{12}^{\mathrm{T}} & \cdots & \mathbf{a}_{1u}^{\mathrm{T}} \\ \mathbf{a}_{21}^{\mathrm{T}} & \mathbf{a}_{22}^{\mathrm{T}} & \cdots & \mathbf{a}_{2u}^{\mathrm{T}} \\ \vdots & & \ddots & \vdots \\ \mathbf{a}_{s1}^{\mathrm{T}} & \mathbf{a}_{s2}^{\mathrm{T}} & \cdots & \mathbf{a}_{su}^{\mathrm{T}} \end{bmatrix}_{s \times v} \tag{6}$$

$$\boldsymbol{\Psi}(\mathbf{x}) = \left[\boldsymbol{\psi}_1^{\mathrm{T}}(\mathbf{x}), \boldsymbol{\psi}_2^{\mathrm{T}}(\mathbf{x}), \cdots, \boldsymbol{\psi}_u^{\mathrm{T}}(\mathbf{x}) \right]^{\mathrm{T}} \tag{7}$$

$$\boldsymbol{\psi}_j(\mathbf{x}) = \left[\varphi_j(\mathbf{x}), \varphi_j(\mathbf{x}) x_1, \cdots, \varphi_j(\mathbf{x}) x_r \right]^{\mathrm{T}} \tag{8}$$

here, $\mathbf{y} = [y_1, y_2, \cdots, y_s]^{\mathrm{T}}$ is the output vector, $\mathbf{a}_{kj}^{\mathrm{T}} = [a_{kj}^{(0)}, a_{kj}^{(1)}, \cdots, a_{kj}^{(r)}]$ is the weight vector for the output variable y_k related to the jth GEBF unit, r, s and u are the numbers of input, output variables and GEBF units, respectively, and $v = (r+1)u$. $\varphi_j(\mathbf{x})$ corresponds to the activation level of the jth GEBF unit which is reasonably defined in two cases, i.e., nonnormalized and normalized,

$$\varphi_j(\mathbf{x}) = \begin{cases} GEBF\left(\mathbf{x}; \mathbf{c}_j, \sigma_j, \mathbf{b}_j\right) \Big/ \sum_{j=1}^{u} GEBF\left(\mathbf{x}; \mathbf{c}_j, \sigma_j, \mathbf{b}_j\right), & \text{normalized} \\ GEBF\left(\mathbf{x}; \mathbf{c}_j, \sigma_j, \mathbf{b}_j\right), & \text{nonnormalized} \end{cases} \tag{9}$$

3   T-S Fuzzy System

Consider the r-input and s-output system, the general fuzzy rule base for T-S fuzzy systems in the MIMO form is given by,

$$\text{Rule } j : \text{IF } x_1 \text{ is } A_{1j} \text{ and ... and } x_r \text{ is } A_{rj},$$
$$\text{THEN } y_k = \bar{\mathbf{a}}_{kj}^{\mathrm{T}} \bar{\mathbf{x}}, j = 1, \cdots, \bar{u}, \ k = 1, \cdots, s \tag{10}$$

where A_{ij} is the fuzzy set of the ith input variable x_i, $\bar{\mathbf{x}} = [1, x_1, \cdots, x_r]^{\mathrm{T}}$ and $\bar{\mathbf{a}}_{kj}^{\mathrm{T}} = [\bar{a}_{kj}^{(0)}, \bar{a}_{kj}^{(1)}, \cdots, \bar{a}_{kj}^{(r)}]$ are the augmented input and weight vectors for the output variable y_k in the jth fuzzy rule, r, s and $\bar{u}$ are the numbers of input, output variables and fuzzy rules, respectively.

Let $\mu_{ij}(x_i)$ be the membership function for fuzzy sets A_{ij}, the fire strength of the jth fuzzy rule in normalized and nonnormalized cases could be obtained,

$$\phi_j(\mathbf{x}) = \begin{cases} \bigwedge_{i=1}^{r} \mu_{ij}(x_i) \Big/ \sum_{j=1}^{\bar{u}} \bigwedge_{i=1}^{r} \mu_{ij}(x_i), & \text{normalized} \\ \bigwedge_{i=1}^{r} \mu_{ij}(x_i), & \text{nonnormalized} \end{cases} \tag{11}$$

where " $\wedge$ " denotes fuzzy conjunction operator, i.e., T-norm. Accordingly, the overall output for the foregoing T-S fuzzy system is defined as follows:

$$\mathbf{y} = \bar{\mathbf{f}}(\mathbf{x}) \triangleq \bar{\mathbf{A}} \bar{\boldsymbol{\Psi}}(\mathbf{x}) \tag{12}$$

where,

$$\bar{\mathbf{A}} = \begin{bmatrix} \bar{\mathbf{a}}_{11}^{\mathrm{T}} & \bar{\mathbf{a}}_{12}^{\mathrm{T}} & \cdots & \bar{\mathbf{a}}_{1\bar{u}}^{\mathrm{T}} \\ \bar{\mathbf{a}}_{21}^{\mathrm{T}} & \bar{\mathbf{a}}_{22}^{\mathrm{T}} & \cdots & \bar{\mathbf{a}}_{2\bar{u}}^{\mathrm{T}} \\ \vdots & \vdots & \ddots & \vdots \\ \bar{\mathbf{a}}_{s1}^{\mathrm{T}} & \bar{\mathbf{a}}_{s2}^{\mathrm{T}} & \cdots & \bar{\mathbf{a}}_{s\bar{u}}^{\mathrm{T}} \end{bmatrix}_{s \times \bar{v}} \tag{13}$$

$$\bar{\mathbf{\Psi}}(\mathbf{x}) = \left[\bar{\psi}_1^{\mathrm{T}}(\mathbf{x}), \bar{\psi}_2^{\mathrm{T}}(\mathbf{x}), \cdots, \bar{\psi}_{\bar{u}}^{\mathrm{T}}(\mathbf{x}) \right]^{\mathrm{T}} \tag{14}$$

$$\bar{\psi}_j(\mathbf{x}) = [\phi_j(\mathbf{x}), \phi_j(\mathbf{x})x_1, \cdots, \phi_j(\mathbf{x})x_r]^{\mathrm{T}} \tag{15}$$

4   Equivalence between GEBF-NN and T-S Fuzzy System

We are now in a position to establish the functional equivalence between the GEBF-NN and the T-S fuzzy system. Consider the overall outputs of the GEBF-NN and the general T-S fuzzy systems in MIMO form defined by (5)-(9) and (11)-(15), respectively, we present the preliminary result as follows.

Theorem 1. For the MIMO system with input $\mathbf{x} = [x_1, x_2, \cdots, x_r]^{\mathrm{T}}$ and output $\mathbf{y} = [y_1, y_2, \cdots, y_s]^{\mathrm{T}}$, the GEBF-NN defined by (5) is equivalent to T-S fuzzy systems (12) if the following conditions are satisfied:

$$\mathrm{C1}: \mathbf{A} = \bar{\mathbf{A}}, \text{ and } \mathrm{C2}: \varphi_j(\mathbf{x}) = \phi_j(\mathbf{x}), \; \forall \mathbf{x}, j \tag{16}$$

where matrices $\mathbf{A}$ and $\bar{\mathbf{A}}$ are defined by (6) and (13), activation (fire strenth) functions $\varphi_j(.)$ and $\phi_j(.)$ are defined by (9) and (11), respectively.   □

Proof. From the condition C1 in (16), we acquire $v = \bar{v}$ and $u = \bar{u}$ in addition to $\mathbf{a}_{kj} = \bar{\mathbf{a}}_{kj}$, $k = 1, 2, \cdots, s$, $j = 1, 2, \cdots, u$. It implies that the number of GEBF units in the GEBF-NN is equal to the number of fuzzy rules in the T-S fuzzy system.

Furthermore, under the condition C2 in (16), the vectors $\psi_j(\mathbf{x})$ and $\bar{\psi}_j(\mathbf{x})$ are functionally equivalent. With $u = \bar{u}$, we obtain $\mathbf{\Psi}(\mathbf{x}) \equiv \bar{\mathbf{\Psi}}(\mathbf{x})$.

As a consequence, one can get $\mathbf{A}\mathbf{\Psi}(\mathbf{x}) \equiv \bar{\mathbf{A}}\bar{\mathbf{\Psi}}(\mathbf{x})$, which implies $\mathbf{f}(\mathbf{x}) \equiv \bar{\mathbf{f}}(\mathbf{x})$. This concludes the proof.   □

Theorem 2. For the MIMO system with input $\mathbf{x} = [x_1, x_2, \cdots, x_r]^{\mathrm{T}}$ and output $\mathbf{y} = [y_1, y_2, \cdots, y_s]^{\mathrm{T}}$, the normalized (nonnormalized) GEBF-NN defined by (5) is equivalent to normalized (nonnormalized) T-S fuzzy systems (12) taking T-norm " $\wedge$ " as multiplication if both activation and firing strength, i.e., (9) and (11), use normalized (nonnormalized) method and the following conditions are satisfied:

$$\begin{cases} u = \bar{u} \\ \mathbf{a}_{kj} = \bar{\mathbf{a}}_{kj}, \; \forall k, \; j \\ \mu_{ij}(x_i) = DGF\left(x_i; c_{ij}, \sigma_{ij}(x_i), b_{ij}\right), \; \forall x_i, \; i, \; j \end{cases} \tag{17}$$

where u and $\bar{u}$ are numbers of GEBF units and fuzzy rules, $\mathbf{a}_{kj}$ and $\bar{\mathbf{a}}_{kj}$ are vector elements in matrices $\mathbf{A}$ and $\bar{\mathbf{A}}$ given by (6) and (13), respectively, and $\mu_{ij}(x_i)$ is membership function of fuzzy set A_{ij}.   □

Proof. From the first two equations of the condition (22), one can directly obtain $\mathbf{A} = \bar{\mathbf{A}}$ satisfying the condition C1 in Theorem 1.

If the T-norm "$\wedge$" of T-S fuzzy system is selected as multiplication, the firing strength for the jth fuzzy rule is defined as,

$$\phi_j(\mathbf{x}) = \begin{cases} \prod_{i=1}^{r} \mu_{ij}(x_i) \Big/ \sum_{j=1}^{\bar{u}} \prod_{i=1}^{r} \mu_{ij}(x_i), \text{ normalized} \\ \prod_{i=1}^{r} \mu_{ij}(x_i), \qquad\qquad\quad \text{nonnormalized} \end{cases} \tag{18}$$

Substituting the third equality of (22) into (18) and using (3), one can obtain

$$\phi_j(\mathbf{x}) = \begin{cases} GEBF(\mathbf{x}; \mathbf{c}_j, \sigma_j, \mathbf{b}_j) \Big/ \sum_{j=1}^{u} GEBF(\mathbf{x}; \mathbf{c}_j, \sigma_j, \mathbf{b}_j), \text{normalized} \\ GEBF(\mathbf{x}; \mathbf{c}_j, \sigma_j, \mathbf{b}_j), \qquad\qquad\qquad \text{nonnormalized} \end{cases} \tag{19}$$

Comparing with the activation levels of GEBF units $\varphi_j(\mathbf{x})$ defined by (9), if both activation and firing strength use normalized (nonnormalized) method, it holds that $\varphi_j(\mathbf{x}) = \phi_j(\mathbf{x})$, $\forall \mathbf{x}, j$, which coincides with the condition C2 in Theorem 1.

It follows that the conditions C1 and C2 are satisfied simultaneously if the equation (22) holds under minor constrains in the statement of Theorem 2. This concludes the proof.   □

Furthermore, if T-norm "$\wedge$" is used as multiplication, the T-S fuzzy system defined in (10) can be rewritten in the compact form:

$$\text{Rule } j: \text{ IF } \mathbf{x} \text{ is } A_j, \text{ THEN } y_k = \bar{\mathbf{a}}_{kj}^{\mathsf{T}} \bar{\mathbf{x}}, j = 1, \cdots, \bar{u} \tag{20}$$

where $A_j = A_{1j} \times A_{2j} \times \cdots \times A_{rj}$ is the jth multivariate fuzzy set. Accordingly, the firing strength for the jth fuzzy rule is defined as follows:

$$\phi_j(\mathbf{x}) = \begin{cases} \mu_j(\mathbf{x}) \Big/ \sum_{j=1}^{\bar{u}} \mu_j(\mathbf{x}), \quad \text{normalized} \\ \mu_j(\mathbf{x}), \qquad\qquad \text{nonnormalized} \end{cases} \tag{21}$$

where, $\mu_j(\mathbf{x})$ is the multivariate membership function of fuzzy set A_j.

In this case, we are in a position to obtain insightful equivalence between GEBF-NN and T-S fuzzy system defined by (20).

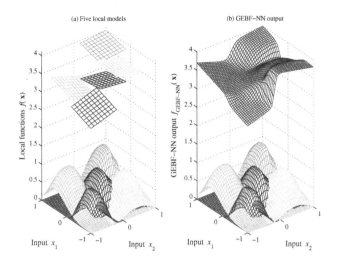

Fig. 1. (a) Five local models, and (b) GEBF-NN output

Theorem 3. For the MIMO system with input $\mathbf{x} = [x_1, x_2, \cdots, x_r]^{\mathrm{T}}$ and output $\mathbf{y} = [y_1, y_2, \cdots, y_s]^{\mathrm{T}}$, the *normalized* GEBF-NN defined by (5) is equivalent to *nonnormalized* T-S fuzzy systems (20) if the following conditions are satisfied:

$$
\begin{cases}
u = \bar{u} \\
\mathbf{a}_{kj} = \bar{\mathbf{a}}_{kj}, \ \forall k, \ j \\
\mu_j(\mathbf{x}) = GEBF\left(\mathbf{x}; \mathbf{c}_j, \sigma_j, \mathbf{b}_j\right) \Big/ \sum_{j=1}^{u} GEBF\left(\mathbf{x}; \mathbf{c}_j, \sigma_j, \mathbf{b}_j\right), \ \forall \mathbf{x}, \ j
\end{cases}
\tag{22}
$$

where u and $\bar{u}$ are numbers of GEBF units and fuzzy rules, $\mathbf{a}_{kj}$ and $\bar{\mathbf{a}}_{kj}$ are vector elements in matrices $\mathbf{A}$ and $\bar{\mathbf{A}}$ given by (6) and (13), respectively, and $\mu_j(\mathbf{x})$ is multivariate membership function of fuzzy set A_j. □

Proof. Comparing (19) with (21), the proof of this theorem is straightforward and omitted. □

5   Numerical Examples

As the illustrative example, a GEBF-NN with arbitrarily selected paramters (without optimization) is considered to demonstrate the equivalence between a specific GEBF-NN and the counterpart of T-S fuzzy systems. Without loss of generality, parameters of DGF units and weight vectors are chosen as follows: $u = 5$, $c_{11} = c_{21} = -0.8$, $c_{12} = c_{24} = -0.5$, $c_{13} = c_{23} = 0$, $c_{22} = c_{14} = 0.5$, $c_{15} = c_{25} = 0.8$, $\sigma_{ij}^{L}, \sigma_{ij}^{L} \in \{0.4, 0.5, 0.6\}$, $b_{ij} \in [1, 5]$, $i = 1, 2, j = 1, 2, \cdots, 5$; $\mathbf{a}_1^{\mathrm{T}} =$

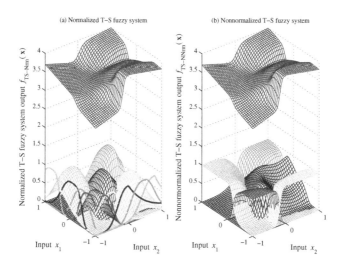

Fig. 2. (a) Normalized T-S fuzzy system output, and (b) Nonnormalized T-S fuzzy system output

$[3.5, 0.3, 0.3]$, $\mathbf{a}_2^{\mathrm{T}} = [4.1, 0.1, -0.2]$, $\mathbf{a}_3^{\mathrm{T}} = [3.5, -0.1, -0.1]$, $\mathbf{a}_4^{\mathrm{T}} = [3.1, 0.3, -0.3]$, $\mathbf{a}_5^{\mathrm{T}} = [3.7, 0.2, 0.1]$. With the weight vectors $\mathbf{a}_j$, five local models shown in Fig. 1(a) correspond to GEBF units. The overall output of the normalized GEBF-NN is steadily obtained by (5), which is shown in Fig. 1(b).

The overall output of equivalent normalized T-S fuzzy system with membership functions and input space partitioning is shown in Fig. 2(a), from which we can find that each fuzzy region partitioning the input space corresponds to a GEBF unit, and thereby contributing to the functionally equivalent output.

The resulting output of the nonnormalized T-S fuzzy system with membership functions directly partitioning the input space is shown in Fig. 2(b). Due to the nonnormalized calculation, each multivariate membership function derived from GEBF units is actually used as fuzzy partitions overlapped with each other. In this case, the nonnormalized T-S fuzzy system using equivalent multivariate fuzzy sets would reduce implication computations and preserve high flexibility. In addition, interpretable fuzzy rules could be directly extracted from the GEBF-NN.

6   Conclusions

In this paper, we establish the functional equivalence between T-S fuzzy systems and a specific type of four-layer neural networks, termed as Generalized Ellipsoidal Basis Function based Neural Network (GEBF-NN), under minor restrictions. To be specific, the GEBF-NN is equivalent to a T-S fuzzy system of which the premise firing strength and the consequence part are identical to the

GEBF unit and the local model simultaneously. Furthermore, we prove that the normalized (nonnormalized) GEBF-NN is equivalent to the normalized (nonnormalized) T-S fuzzy system with DGF units as univariate membership functions and local models as consequents. In addition, the equivalence between normalized GEBF-NN and nonnormalized T-S fuzzy system is obtained by employing GEBF units as multivariate membership functions. Finally, numerical examples illustrate proposed theoretical results.

References

1. Wang, N., Er, M.J., Meng, X.: A Fast and Accurate Online Self-Organizing Scheme for Parsimonious Fuzzy Neural Networks. Neurocomput. 72, 3818–3829 (2009)
2. Wang, N., Er, M.J., Meng, X.Y., Li, X.: An Online Self-Organizing Scheme for Parsimonious and Accurate Fuzzy Neural Networks. Int. J. Neural Syst. 20, 389–403 (2010)
3. Jang, J.R., Sun, C.T.: Functional Equivalence between Radial Basis Function Networks and Fuzzy Inference Systems. IEEE Trans. Neural Netw. 4, 156–159 (1993)
4. Anderson, H.C., Lotfi, A., Westphal, L.C., Jang, J.R.: Comments on "Functional Equivalence between Radial Basis Function Networks and Fuzzy Inference Systems". IEEE Trans. Neural Netw. 9, 1529–1531 (1998)
5. Hunt, K.J., Haas, R., Murray-Smith, R.: Extending the Functional Equivalence of Radial Basis Function Networks and Fuzzy Inference Systems. IEEE Trans. Neural Netw. 7, 776–781 (1996)
6. Azeem, M.F., Hanmandlu, M., Ahmad, N.: Generalization of Adaptive Neuro-Fuzzy Inference Systems. IEEE Trans. Neural Netw. 11, 1332–1346 (2000)
7. Azeem, M.F., Hanmandlu, M., Ahmad, N.: Structure Identification of Generalized Adaptive Neuro-Fuzzy Inference Systems. IEEE Trans. Fuzzy Syst. 11, 666–681 (2003)
8. Li, H.X., Chen, C.P.: The Equivalence between Fuzzy Logic Systems and Feedforward Neural Networks. IEEE Trans. Neural Netw. 11, 356–365 (2000)
9. Kolman, E., Margaliot, M.: Are Artificial Neural Networks White Boxes? IEEE Trans. Neural Netw. 16, 844–852 (2005)
10. Aznarte, J.L., Benitez, J.M.: Equivalences between Neural-Autoregressive Time Series Models and Fuzzy Systems. IEEE Trans. Neural Netw. 21, 1434–1444 (2010)
11. Barra, A., Bernacchia, A., Santucci, E., Contucci, P.: On the Equivalence of Hopfield Networks and Boltzmann Machines. Neural Netw. 34, 1–9 (2012)
12. Wang, N.: A Generalized Ellipsoidal Basis Function Based Online Self-Constructing Fuzzy Neural Network. Neural Process. Lett. 34, 13–37 (2011)

Adaptive NN Control for a Class of Strict-Feedback Discrete-Time Nonlinear Systems with Input Saturation

Xin Wang[1], Tieshan Li[1,*], Liyou Fang[1], and Bin Lin[2]

[1] Navigational College, Dalian Maritime University, Dalian 116026, China
[2] Department of Information Science and Technology,
Dalian Maritime University, Dalian 116026, China
luyunliunanting@yeah.net, tieshanli@126.com, binlin@dlmu.edu.cn

Abstract. In this paper, an adaptive neural network (NN) control scheme is proposed for a class of strict-feedback discrete-time nonlinear systems with input saturation. which is designed via backstepping technology and the approximation property of the HONNs, aimed to solve the the input saturation constraint and system uncertainty in many practical applications. The closed-loop system is proven to be uniformly ultimately bounded (UUB). At last, a simulation example is given to illustrate the effectiveness of the proposed algorithm.

Keywords: input saturation, discrete-time, adaptive control, backstepping, high-order neural networks (HONNs).

1   Introduction

In recently years, the research on the neural network control of various nonlinear uncertain systems has advanced significantly. In the literature of adaptive neural network control, neural networks (NNs) are primarily used as on-line approximators for the unknown nonlinearities due to their inherent approximation capabilities. By using the idea of backstepping design [1], several adaptive neural-networks control [2-6] have been presented for some classes of uncertain nonlinear strict-feedback systems.

However, the above mentioned methods are limited to the continuous-time domain, they are not directly applicabled to discrete-time systems due to the noncausal problem in the controller design procedures. Recently, the adaptive control via the universal approximators for uncertain discrete-time nonlinear systems has obtained many results. For example, the approach proposed in [7] was given to achieve the tracking control of a class of unknown nonlinear dynamic systems using a discrete-time NN controller. Subsequently, several elegant adaptive control schemes were studied in [8-13] for discrete-time nonlinear systems based on the approximation property of the

---

* This work was supported in part by the National Natural Science Foundation of China (Nos.5 1179019, 61001090), the Natural Science Foundation of Liaoning Province (No.20102012), the Program for Liaoning Excellent Talents in University of (LNET)(Grant No.LR2012016) and the Applied Basic Research Program of Ministry Transport of China.

C. Guo, Z.-G. Hou, and Z. Zeng (Eds.): ISNN 2013, Part II, LNCS 7952, pp. 70–78, 2013.

neural network. For instance, both the state and ouput feedback adaptive neural network controllers were presented for a class of discrete-time nonlinear systems in the strict-feedback form [8]. A novel approach in designing neural network based adaptive controllers for a class of nonlinear discrete-time systems was presented in [9].

For practical perspective, the input saturation may cause serious influence on system stability and performance[14]. Therefore, the effects of input saturation cannot be ignored in the controller design [15]. Recently, to solve the input saturation constraint, some adaptive continuous nonlinear systems with input saturtion has been addressed in [16-18]. But there are still little works for the discrete-time nonlinear system [19].

In this paper, adaptive neural network controller via backstepping is presented for a class of discrete-time nonlinear system with input saturation. During the controller design process, the HONNs are used to approximate the unknown nonlinear function in the system. With an aided design system of the input saturation, the input saturation constraint of the discrete-time nonlinear system is solved in the controller design. By using the lyapunov analysis method, the closed-loop systems are proven to be UUB, and the tracking error converges to a small neigborhood of the origin. At last, the simulation results show the effectiveness of the proposed method.

2   Problem Formulation and Preliminaries

2.1   Problem Formulation

Consider the following single-input single-output (SISO) discrete-time nonlinear system in strict-feedback form in [7]:

$$\begin{cases} x_i(k+1) = f_i(\overline{x}_i(k)) + g_i(\overline{x}_i(k))x_{i+1}(k) & i = 1,2...n-1 \\ x_n(k+1) = f_n(\overline{x}_n(k)) + g_n(\overline{x}_n(k))u(k) \\ y_k = x_1(k) \end{cases} \qquad (1)$$

Where $\overline{x}_i(k) = [x_1(k), x_2(k),...,x_i(k)]^T \in R^i, i = 1,2...n$, $u(k) \in R$ and $y_k \in R$ are the state variables, system input and output respectively; $f_i(\overline{x}_i(k))$ and $g_i(\overline{x}_i(k)), i = 1,2,...,n$ are unknow smooth functions.

The control objective is to design an adaptive NN controller for system (1) such that: (i) all the signals in the closed-loop system are UUB and (ii) the system output follows the desired reference signal $y_d(k)$. The desired reference signal $y_d(k) \in \Omega_y, \forall k > 0$ is smooth and known, where $\Omega_y := \{\chi | \chi = x_1\}$.

Assumption 2. The signal of $g_i(\overline{x}_i(k)), i = 1,2,...,n$ are known and there exist co-nstants $\underline{g}_i > 0$ and $\overline{g}_i > 0$ such that $\underline{g}_i \leq |g_i(\overline{x}_i(k))| \leq \overline{g}_i$, $\forall \overline{x}_n(k) \in \Omega$.

2.2   Input Saturation Constraint

Considering the input saturation constraints u satisfies $-u_{min} \leq u \leq u_{max}$, where u_{min} and u_{max} are the known lower limit and up limit of input constraints. Thus,

$$u = sat(v) = \begin{cases} u_{max}, & if \ v > u_{max} \\ v, & if \ -u_{min} \leq v \leq u_{max} \\ -u_{min}, if \ v < -u_{min} \end{cases} \tag{2}$$

where v is the designed controller input of the system .

Then , to be convenient to consider the affect of the input saturation constraint, the aided design system is considered as follows:

$$e(k+1) = \begin{cases} -c_1 e(k) - \dfrac{f(\cdot)}{e^2} \cdot e + (u-v), & |e| \geq \theta \\ 0, & |e| < \theta \end{cases} \tag{3}$$

where e is a state variable of the aided design system, θ is a small positive design parameters, $c_1 > 0$ also a design paramaters. and $f(\cdot) = f(\eta_n, \Delta u) = |\eta_n \cdot \Delta u| + \dfrac{1}{2} \Delta u^2 (\eta_n$ is a error variable in control design) is a aided design function.

3   Adaptive NN Control Design with Input Saturation Constraint

Consider the strict-feedback SISO nonlinear discrete-time system described in (1). Since assumption 1 is only valid on the compact set Ω , it is necessary to guarantee the system's states remaining in Ω for all time. We will design an adaptive control $u(k)$ for system (1) which makes system output y_k follow the desired reference signal $y_d(k)$, and simultaneously guarantees $\bar{x}_n(k) \in \Omega, \forall k > 0$ under the condition that $\bar{x}_n(0) \in \Omega$.

The strict-feedback form (1) is transformed to n-step head function as follows:

$$x_1(k+n) = F_1(\bar{x}_n(k)) + G_1(\bar{x}_n(k))x_2(k+n-1),$$
$$\vdots$$
$$x_{n-1}(k+2) = F_{n-1}(\bar{x}_n(k)) + G_{n-1}(\bar{x}_n(k))x_n(k+1)$$
$$x_n(k+1) = f_n(\bar{x}_n(k)) + g_n(\bar{x}_n(k))u(k)$$
$$y_k = x_1(k). \tag{4}$$

Form the definition of $G_i(\bar{x}_n(k))$ in each step, it is clear that the value of $G_i(\bar{x}_n(k))$ is the same as $G_i(\bar{x}_i(k))$, therefore $G_i(\bar{x}_n(k))$ satisfy:

$$\underline{g}_i \le G_i(\bar{x}_n(k)) \le \bar{g}_i, \forall \bar{x}_n(k) \in \Omega$$

Now we can construct the controller for (4) via backstepping method without the problem of causality contradiction. For convenience of analysis and discussion, for $i = 1, 2, \ldots, n-1$ let:

$$F_i(k) = F_i(\bar{x}_n(k)), G_i = G_i(\bar{x}_n(k)), f_n(k) = f_n(\bar{x}_n(k)), g_n(k) = g_n(\bar{x}_n(k))$$

Before further going, let $k_i = k - n + i, i = 1, 2 \ldots n - 1$

Step 1: For $\eta_1(k) = x_1(k) - y_d(k)$, its n th difference is given by

$$\eta_1(k+n) = F_1(k) + G_1(k)x_2(k+n-1) - y_d(k+n) \tag{5}$$

Considering $x_2(k+n-1)$ as a fictitious control, if we choose

$$x_2(k+n-1) = x_{2d}^*(k) = -\frac{1}{G_1(k)}[F_1(k) - y_d(k+n)] \tag{6}$$

It is obvious that $\eta_1(k+n) = 0$. Since $F_1(k)$ and $G_1(k)$ are unknown, they are not available for constructing a fictitious control $x_{2d}^*(k)$. However, $F_1(k)$ and $G_1(k)$ are function of system state $\bar{x}_n(k)$, therefore we can use HONN to approximate $x_{2d}^*(k)$ as follow:

$$x_{2d}^*(k) = W_1^{*T} S_1(z_1(k)) + \varepsilon_z(z_1(k)), \ z_1(k) = [\bar{x}_n^T(k), y_d(k+n)]^T \tag{7}$$

Letting $\hat{W}_1$ be the estimate of W_1^*, consider the direct adaptive fictitious control:

$$x_2(k+n-1) = x_{2f}(k) = \hat{W}_1^T(k)S_1(z_1(k)) \tag{8}$$

and the adaptive law as

$$\hat{W}_1(k+1) = (1 - \Gamma_1\sigma_1)\hat{W}_1(k_1) + \Gamma_1 S_1(z_1(k_1))\eta_1(k+1) \tag{9}$$

substituting (6), (7) and (8) into (5) yields

$$\eta_1(k+n) = G_1(k)[\tilde{W}_1(k)S_1(z_1(k)) - \varepsilon_{z1}] \tag{10}$$

Choose the lyapunov function candidate

$$V_1(k) = \frac{1}{\bar{g}_1}\eta_1^2(k) + \sum_{j=0}^{n-1} \tilde{W}_1^T(k_1+j)\Gamma_1^{-1}\tilde{W}_1(k_1+j), \ k_1 = k - n + 1 \tag{11}$$

Noting the fact that $\tilde{W}_1^T(k)S_1(z_1(k)) = \eta_1(k+1)/G_1(k_1) + \varepsilon_{z1}$, the first difference of (11) is

$$\Delta V_1 = \frac{1}{g_1}[\eta_1^2(k+1) - \eta_1^2(k)] + \tilde{W}_1^T(k+1)\Gamma_1^{-1}\tilde{W}_1(k+1) - \tilde{W}_1^T(k_1)\Gamma_1^{-1}\tilde{W}_1(k_1)$$

Step i : Simulated the procedure in step 1, for $\eta_i(k) = x_i(k) - x_{if}(k_{i-1})$, we can get the following direct adaptive fictitious controller and its adaptive law

$$x_{i+1}(k+n-i) = x_{(i+1)f}(k) = \hat{W}_i^T S_i(z_i(k))$$
$$\hat{W}_i(k+1) = (1 - \Gamma_i\sigma_i)\hat{W}_i(k_i) - \Gamma_i S_i(z_i(k_i))\eta_i(k+1)$$
$$z_i(k) = [\bar{x}_n^T(k), x_{if}(k)]^T \in \Omega_{zi} \subset R^{n+1} \qquad (12)$$

Then we can obtain

$$\eta_i(k+n-i+1) = G_i(k)[\tilde{W}_i^T(k)S_i(z_i(k)) - \varepsilon_{zi}] \qquad (13)$$

Choose the lyapunov function candidate as follows

$$V_i(k) = \sum_{j=1}^{i-1} V_j(k) + \frac{1}{g_i}\eta_i^2(k) + \sum_{j=0}^{n-i} \tilde{W}_i^T(k_i+j)\Gamma_i^{-1}\tilde{W}_i(k_i+j), \quad k_i = k-n+i.$$

Step n: For $\eta_n(k) = x_n(k) - x_f(k-1)$. its first difference is

$$\eta_n(k+1) = x_n(k+1) - x_{nf}(k) = f_n(k) + g_n(k)u(k) - x_{nf}(k) \qquad (14)$$

To deal with the affect of the input saturation constraint in this step, we employ the aided design system in (3). Now, consider the direct adaptive fictitious controller as

$$u(k) = \hat{W}_n^T S_n(z_n(k)) + e(k) \qquad (15)$$

and the adaptive law as follows:

$$\hat{W}_n(k+1) = (1 - \Gamma_n\sigma_n)\hat{W}_n(k) - \Gamma_n S_n(z_n(k))\eta_n(k+1) \qquad (16)$$

Substituting (15) and (16) into (14) , we can obtain

$$\eta_n(k+1) = g_n(k)[\tilde{W}_n^T(k)S_n(z_n(k)) + e(k) - \varepsilon_{zn}] \qquad (17)$$

Choose the lyapunov function candidate

$$V_n(k) = \sum_{j=1}^{n-1} V_j(k) + \frac{1}{g_n}\eta_n^2(k) + \tilde{W}_n^T(k)\Gamma_n^{-1}\tilde{W}_n(k). + \frac{1}{g_n}e^2(k) \qquad (18)$$

Noting the fact that: $\tilde{W}_n^T(k)S_n(z_n(k)) = \eta_n(k+1)/g_n(k) - e(k) + \varepsilon_{zn}$

The first difference of (18) with (16) and (17) is

$$\Delta V_n = \sum_{j=1}^{n-1} \Delta V_j + \frac{1}{\overline{g}_n}[\eta_n^2(k+1) - \eta_n^2(k)] + \tilde{W}_n^T(k+1)\Gamma_n^{-1}\tilde{W}_n(k+1)$$

$$- \tilde{W}_n^T(k)\Gamma_n^{-1}\tilde{W}_n(k) + \frac{1}{\overline{g}_n}[e^2(k+1) - e^2(k)]$$

$$\leq \sum_{j=1}^{n-1} \Delta V_j - \frac{1}{\overline{g}_n}[\eta_n^2(k+1) - \eta_n^2(k)] + 2\varepsilon_{zn}\eta_n(k+1) + 2e(k)\eta_n(k+1)$$

$$- 2\sigma_n\tilde{W}_n^T(k)\hat{W}_n(k) + S_n^T(z_n(k))\Gamma_n S_n(z_n(k))\eta_n^2(k+1)$$

$$+ 2\sigma_n\hat{W}_n^T(k)\Gamma_n S_n(z_n(k))\eta_n(k+1) + \sigma_n^2\hat{W}_n^T(k)\Gamma_n\hat{W}_n(k)$$

$$+ \frac{1}{\overline{g}_n}[e^2(k+1) - e^2(k)]$$

Noting the fact that:

$$S_j^T(z_j(k_j))S_j(z_j(k_j)) < l_j,$$

$$S_j^T(z_j(k_j))\Gamma_j S_j(z_j(k_j)) \leq \overline{\gamma}_j S_j^T(z_j(k_j))S_j(z_j(k_j)) \leq \overline{\gamma}_j l_j$$

$$2\varepsilon_{zj}\eta_j(k+1) \leq \frac{\overline{\gamma}_j\eta_j^2(k+1)}{\overline{g}_j} + \frac{\overline{g}_j\varepsilon_{zj}^2}{\overline{\gamma}_j}$$

$$2\sigma_j\hat{W}_j^T(k_j)\Gamma_j S_j(z_j(k_j))\eta_j(k+1) \leq \frac{\overline{\gamma}_j l_j \eta_j^2(k+1)}{\overline{g}_1} + \overline{g}_1\sigma_1^2\overline{\gamma}_1\left\|\hat{W}_1\right\|^2$$

$$2\tilde{W}_1^T(k_1)\hat{W}_1(k_1) = \left\|\tilde{W}_1(k_1)\right\|^2 + \left\|\hat{W}_1(k_1)\right\|^2 - \left\|W_1^*\right\|^2$$

$$e(k+1) = -[c_1 + f(\cdot)/e^2(k)]e(k) + \Delta u$$

And defining $c_2 = c_1 + f(\cdot)/e^2(k) > 0$, we can get

$$2e(k)\eta_n(k+1) \leq \frac{1}{c_2}[\frac{\overline{\gamma}_n\eta_n^2(k+1)}{\overline{g}_n} + \frac{\overline{g}_n}{\overline{\gamma}_n}\Delta u^2] + \frac{\overline{\gamma}_n}{c_2\overline{g}_n}\eta_n^2(k+1) + \frac{\overline{g}_n}{c_2\overline{\gamma}_n}e^2(k+1)$$

Then, we have

$$\Delta V_n \leq \sum_{j=1}^{n-1} \Delta V_j - \frac{\rho_n}{c_2\overline{g}_n}\eta_n^2(k+1) - \frac{1}{\overline{g}_n}\eta_n^2(k) + \beta_n - \frac{1}{\overline{g}_n}e^2(k)$$

$$\sigma_n(1 - \sigma_n\overline{\gamma}_n - \overline{g}_n\sigma_n\overline{\gamma}_n)\left\|\hat{W}_n(k)\right\|^2 + \frac{\overline{g}_n^2 + c_2\overline{\gamma}_n}{c_2\overline{\gamma}_n\overline{g}_n}e^2(k+1)$$

where $\rho_n = c_2 - c_2 \bar{\gamma}_n - \bar{\gamma}_n l_n - \bar{g}_n \bar{\gamma}_n l_n - 2\bar{\gamma}_n$

$$\beta_n = \beta_{n-1} + \frac{\bar{g}_n \varepsilon_{zn}^2}{\bar{\gamma}_n} + \frac{\bar{g}_n \Delta u^2}{c_2 \bar{g}_n} + \sigma_n \left\| W_n^* \right\|^2$$

If we choose the design parameters as follows

$$\bar{\gamma}_n < \frac{c_2}{c_2 + l_n + \bar{g}_n l_n + 2}, \quad -\frac{\bar{g}_n^2}{c_2} \leq \bar{\gamma}_n < 0, \quad \sigma_n < \frac{1}{(1 + \bar{g}_n)\bar{\gamma}_n} \tag{19}$$

then $\Delta V_n \leq 0$ once any one of the n errors satisfies $\left| \eta_j(k) \right| > \sqrt{\bar{g}_j \beta_n}$ and $j = 1, 2, \ldots, n$, This demonstrates that the tracking error $\eta_1(k), \eta_2(k), \ldots, \eta_n(k)$ are bounded for all $k \geq 0$.

Based on the procedure above, we can conclude that $\bar{x}_n(k+1) \in \Omega$ and $u(k)$ are bounded if $\bar{x}_n(k) \in \Omega$. Finally, if we initialize $\bar{x}_n(0) \in \Omega$, and choose the design parameters as (20), there exists a k^*, such that all errors asymptotically converge to zero, and NN weight errors are all bounded. This implies that the closed-loop system is UUB. and $\bar{x}_n(k) \in \Omega, \hat{W}_i, i = 1, 2 \ldots n$ will hold for all $k > 0$.

4   Simulation Studies

Now , we can apply adaptive NN control for a class of discrete-time nonlinear systems with input saturation to the ship heading control, the discerte-time ship heading control plant described by

$$\begin{cases} x_1(k+1) = x_2(k) \\ x_2(k+1) = -\frac{1}{T}(a_1 x_2(k) - a_2 x_2^3(k)) + \frac{K}{T}u(k) \\ y_k = x_1(k) \end{cases} \tag{20}$$

where a_1, a_2 denotes nonlinear coefficients of the ship, K, T are the parameter of rotary and trackability. It can be checked that Assumption 1 and 2 are satisfied. In ship heading control, the input saturation restrictions of the rudder angle is $35°$.

The initial condition for ship heading control states is $x_0 = [30, 0]^T$, and the H-ONN codes is $l_1 = 22, l_2 = 22$ .The simulation results are presented in Figs 1 and 2.

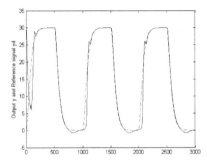 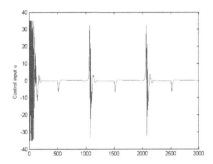

Fig. 1. The output y_k and the reference signal y_d **Fig 2.** The input of the controller u

5   Conclusion

By using the backstepping technique and the approximation property of the HONNs, as well as considering the input saturation, an adaptive control approach is proposed for a class of discrete-time nonlinear systems with input saturation. In this paper, the proposed controller solves the input saturation constraint and system uncertainty in practical applications, and all the signal of the resulting closed-loop system were guaranteed to be UUB, the tracking errors can be reduced to a small neighborhood of zero. The simulation example is proposed to show the performance of the presented scheme.

References

1. Krstic, M., Kanellakopoulos, I., Kokotovic, P.: Nonlinear and Adaptive Control Design. Wiley, New York (1995)
2. Polycarpou, M.M.: Stable adaptive neural control scheme for nonlinear systems. IEEE Trans. on Automatic Control 41, 447–451 (1996)
3. Polycarpou, M.M., Mears, M.J.: Stable adaptive tracking of uncertain systems using nonlin-early parametrized on-line approximators. International Journal of Control 70(3), 363–384 (1998)
4. Zhang, Y., Peng, P.Y., Jiang, Z.P.: Stable neural controller design for unknown nonlinear systems using backstepping. IEEE Trans. Neural Networks 11, 1347–1359 (2000)
5. Zhang, T., Ge, S.S., Hang, C.C.: Adaptive neural network control for strict-feedback nonlinear systems using backstepping design. Automatica 36, 1835–1846 (2000)
6. Ge, S.S., Wang, C.: Driect Adaptive NN Control of a class of nonlinear systems. IEEE Trans. on Neural Networks 13(1) (2002)
7. Ge, S.S., Li, G.Y., Lee, T.H.: Adaptive NN control for a class of strict-feedback discrete-time nonlinear systems. Automatica 39, 807–819 (2003)
8. Zhu, Q.M., Guo, L.Z.: Stable adaptive neuron control for nonlinear discrete-time systems. IEEE Trans. on Neural Networks 15(3), 653–662 (2004)
9. Vance, J., Jagannathan, S.: Discrete-time neural network output feedback control of nonlinear discrete-time systems in non-strict form. Automatica 44, 1020–1027 (2008)

10. Yang, C.G., Ge, S.S., Xiang, C., Chai, T.Y., Lee, T.H.: Output feedback NN control for two classes of discrete-time systems with unknown control directions in a unified approach. IEEE Trans. on Neural Networks 19(11), 1873–1886 (2008)
11. Ge, S.S., Yang, C.G., Lee, T.H.: Adaptive predictive control using neural network for a class of pure-feedback systems in discrete time. IEEE Trans. on Neural Networks 19(9), 1599–1614 (2008)
12. Chen, W.S.: Adaptive NN control for discrete-time pure-feedback systems with unknown control direction under amplitude and rate actuator constraints. ISA Transactions 48(3), 304–311 (2009)
13. Liu, Y.J.: Direst adaptive NN control for a class of discrete-time nonlinear strict-feedback systems. Neurocomputing 73, 2498–2505 (2010)
14. Chen, M., Ge, S.S., Choo, Y.: Neural network tracking control of ocean surface vessels with input saturation. In: Proc.of the 2009 IEEE International Conference on Automation and Logistics, pp. 85–89 (2009)
15. Li, J.F.: Design of Ship Course Autopilot in the Presence of input Saturation. The Master's Thesis of Dalian martime university (2012)
16. Polycarpou, M.M.: On-line approximation control of uncertain nonlinear systems: issues with control input saturation. In: The American Control Conference, June 4-6 (2003)
17. Yau, H.T., Chen, C.L.: Chaos control of Lorenz systems using adaptive controller with input saturation. Solitons and Fractals 34, 1567–1574 (2007)
18. Johnson, E.N., Calise, A.J.: Neural Network Adaptive Control of Systems with Input Saturation. In: American Control Conference, June 25-27 (2001)
19. He, P., Jagannathan, S.: Reinforcement learning neural-networkbased controller for nonlinear discrete-time systems with input constraints. IEEE Trans. on Systems, Man, and Cybernetics, Part B: Cybernetics 37(2), 425–436 (2007)

NN Based Adaptive Dynamic Surface Control for Fully Actuated AUV[*]

Baobin Miao[1], Tieshan Li[1,**], Weilin Luo[2], and Xiaori Gao[1]

[1] Navigational College, Dalian Maritime University, Dalian 116026, China
389147556@qq.com, tieshanli@126.com
[2] College of Mechanical Engineering and Automation, Fuzhou University, 350108, China
wlluofzu@gmail.com

Abstract. In this brief, we consider the problem of tracking a desired trajectory for fully actuated autonomous underwater vehicle (AUV), in the presence of external disturbance and model errors. Based on the backstepping method and Lypunov stability theorem, we introduce the dynamic surface control (DSC) technique to tackle the problem of "explosion of complexity" which existing in the traditional backstepping algorithm. Furthermore, the norm of the ideal weighting vector in neural network (NN) systems is considered as the estimation parameter, such that only one parameter is adjusted. The proposed controller guarantees uniform ultimate bounded (UUB) of all the signals in the closed-loop system, while the tracking error converges to a small neighborhood of the origin. Finally, simulation studies are given to illustrate the effectiveness of the proposed algorithm.

Keywords: underwater vehicle, function approximation, neural network, dynamic surface control, trajectory tracking.

1 Introduction

With the increased focus on subsea resources, activities are moving towards deeper and harsher ocean environments. The trajectory tracking of AUV has been required and developed for various applications such as scientific and environmental data gathering, inspection and construction of underwater structures etc[1]. An important issue of tracking control of AUV is the handling of unknown perturbations and model errors. In many practical applications on the control of uncertain nonlinear systems, neural network-based control methods are shown to be more efficient compared with

---

[*] This work was supported in part by the National Natural Science Foundation of China(Nos.51179019, 60874056, 51079031), the Natural Science Foundation of Liaoning Province (No. 20102012), the Program for Liaoning Excellent Talents in University(LNET) (Grant No.LR2012016), the Natural Science Foundation of Fujian Province (No. 2010J01004), the Applied Basic Research Program of Ministry of Transport of China, the Fundamental Research Funds for the Central Universities (No.3132013005).
[**] Corresponding author.

C. Guo, Z.-G. Hou, and Z. Zeng (Eds.): ISNN 2013, Part II, LNCS 7952, pp. 79–87, 2013.
© Springer-Verlag Berlin Heidelberg 2013

other modern control techniques due to their universal approximation capabilities [2]. However, a common weakness of these control schemes is that the number of updated parameters depends on the number of the neural network nodes. With an increase of the nodes, the number of parameters to be estimated will increase significantly, i.e. the problem of "dimensionality curse". This problem was first solved by Yang etc in there pioneering works [3-5], where some kinds of so-called "minimal learning parameter(MLP)" algorithms containing much less online adaptive parameters. As a result, only one parameter needs to be estimated online regardless of the number of the NN nodes. So it can guarantees that the computational burden of the algorithm drastically be reduced and the algorithm is convenient to implement in applications. On the other hand, the effect of the environmental disturbance is addressed by using the property of hyperbolic tangent function, so the AUV dynamic model system has a strong robustness.

Furthermore, with the development of a powerful recursive-design procedure, i.e., adaptive-backstepping technique, there are remarkable progresses for multi-input multi-output nonlinear systems [6], these schemes do not require matching conditions. But a major drawback with the backstepping technique is the problem of "explosion of complexity" [7], which is caused by the repeated differentiations of certain nonlinear functions. That is, the complexity of the controller grows drastically as the system order increases. To overcome this drawback, dynamic surface control technique was proposed by introducing a first-order filtering of the synthetic input at each step of the traditional backstepping approach [8-9]. There are also other means to eliminate the calculations of derivatives of virtual controls in typical backstepping design, such as a command filtered approach in [10].

Inspired by the previous work, a novel NN based adaptive DSC approach was proposed for tracking control of underwater vehicles with uncertainties. Especially, modeling errors are taken into account in the governing mathematical model, which were ignored in the past research. The operation of differentiation is replaced by introducing first-order filter, which would eliminate the explosion of complexity in the traditional backstepping design. Moreover, to reduce considerably the number of adjustable parameters in the controller design, the norm of the ideal weighting vector in NN systems is considered as the estimation parameter in the proposed method. The proposed controller guarantees that the computational burden of the algorithm can drastically be reduced and the algorithm is convenient to implement in applications.

2   Problem Formulation and Preliminaries

By use of Newtonian or Lagrangian mechanics analysis, the mathematical model for underwater vehicle motion of six degrees of freedom can be obtained in the general form [11]:

$$M(\eta)\ddot{\eta} + C(v,\eta)\dot{\eta} + G(\eta) + D(v,\eta)\dot{\eta} + \Delta(v,\eta) + \omega = \tau \qquad (1)$$

Where, $\eta = [x \ \ y \ \ z \ \ \phi \ \ \theta \ \ \psi]^T$ is the position and Euler angles vector with respect to earth-fixed coordinate system; $v = [u \ \ v \ \ w \ \ p \ \ q \ \ r]^T$ is the velocity vector with respect to body-fixed coordinate system; ω is an immeasurable environmental disturbance vector due to waves, sea current or cable traction etc. $\Delta(v, \eta)$ denotes the modeling errors vector or system perturbation; $D(v, \eta)$ denotes the general damping coefficients matrix derived from potential damping, skin friction, wave drift damping and damping due to vortex shedding etc. $M(\eta)$ is the inertia matrix, $C(v, \eta)$ is the so-called centripetal-Coriolis matrix. Both $M(\eta)$ and $C(v, \eta)$ are related to rigid-body dynamics and added mass forces and moments. $G(\eta)$ is the gravitational or restoring forces vector, τ denotes the external forces vector provided by rudders, thrusters or propellers etc.

3   Controller Design

The whole NN controller design procedure contains 2 steps, and the actual control law will be deduced at the second step. The norm of the ideal weighting vector in NN systems in considered as the estimation parameter instead of the elements of weight vector. Defining $\lambda_2^T = \|W_2^*\|^2$, since W_2^* is unknown, λ_2^T will be replaced by its estimation value in the following design procedure. Throughout this paper, let $\lambda_2 - \hat{\lambda}_2 = \tilde{\lambda}_2$. The detailed design procedure as follows.

Choosing $x_1 = \eta$; $x_2 = \dot{\eta}$. Eq.(1) can be rewritten as the following:

$$\dot{x}_1 = x_2$$
$$\dot{x}_2 = -M^{-1}(\eta)C(v, \eta)x_2 - M^{-1}(\eta)D(v, \eta)x_2 - M^{-1}(\eta)G(\eta) - M^{-1}(\eta)\omega - M^{-1}(\eta)\Delta(v, \eta) + M^{-1}(\eta)\tau$$

Step 1:
    Let

$$z_1 = x_1 - \eta_d \tag{2}$$

which is called the error surface with η_d as the desired trajectory. Then

$$\dot{z}_1 = \dot{x}_1 - \dot{\eta}_d = x_2 - \dot{\eta}_d \tag{3}$$

Let

$$z_2 = x_2 - a_2 \tag{4}$$

which is called the second error surface. Where a_2 is a new state variable and can be obtained by introducing a first-order filter with a time constant e_2 as follows.

$$e_2 \dot{a}_2 + a_2 = r_2 \quad a_2(0) = r_2(0) \tag{5}$$

Define

$$p_2 = a_2 - r_2 \tag{6}$$

Substituting Eq.(6) and Eq.(4) into Eq.(3) gives

$$\dot{z}_1 = z_2 + p_2 + r_2 - \dot{\eta}_d \tag{7}$$

Consider the following Lyapunov function candidate:

$$V_1 = \frac{1}{2} z_1^T z_1 + \frac{1}{2} p_2^T p_2 \tag{8}$$

The derivative of V_1 is

$$\dot{V}_1 = z_1^T \left(z_2 + p_2 + r_2 - \dot{\eta}_d \right) + p_2^T \dot{p}_2 \tag{9}$$

Choose the virtual control r_2 as follows:

$$r_2 = -k_1 z_1 + \dot{\eta}_d \tag{10}$$

Substituting Eq.(10) into Eq.(9) gives

$$\dot{V}_1 = z_1^T z_2 + z_1^T p_2 - k_1 \|z_1\|^2 + p_2^T \dot{p}_2$$
$$\leq \|z_1\|^2 + \frac{1}{4}\|z_2\|^2 + \|z_1\|^2 + \frac{1}{4}\|p_2\|^2 - k_1\|z_1\|^2 + p_2^T \dot{p}_2 \tag{11}$$

Choosing $k_1 \geq 2 + a_0$, where a_0 is a positive constant.

The Eq.(11) can be rewritten as follows:

$$\dot{V}_1 \leq \frac{1}{4}\|z_2\|^2 + \frac{1}{4}\|p_2\|^2 - a_0\|z_1\|^2 + p_2^T \dot{p}_2 \tag{12}$$

Substituting Eq.(10) into Eq.(6) gives

$$p_2 = a_2 + k_1 z_1 - \dot{\eta}_d \tag{13}$$

Noting that:

$$\dot{a}_2 = \frac{r_2 - a_2}{e_2} = -\frac{p_2}{e_2} \tag{14}$$

Gives

$$\dot{p}_2 = \dot{a}_2 + k_1 \dot{z}_1 - \ddot{\eta}_d$$
$$= -\frac{p_2}{e_2} + k_1 \dot{z}_1 - \ddot{\eta}_d \tag{15}$$

Let

$$B_2\left(z_1,z_2,p_2,\eta_d,\dot{\eta}_d,\ddot{\eta}_d\right)=k_1\dot{z}_1-\ddot{\eta}_d \tag{16}$$

Where $B_2(i)$ is a continuous function and has a maximum value $M_2(i)$ i.e. $\left|B_2(i)\right|\le M_2(i)$, please refer to [9] for details, $1\le i\le 6$.

Substituting Eq.(16) into Eq.(15) gives

$$\dot{p}_2=-\frac{p_2}{e_2}+B_2\left(z_1,z_2,p_2,\eta_d,\dot{\eta}_d,\ddot{\eta}_d\right) \tag{17}$$

Then Eq.(12) can be rewritten as the following:

$$\dot{V}_1\le\frac{1}{4}\|z_2\|^2+\frac{1}{4}\|p_2\|^2-\frac{\|p_2\|^2}{e_2}+p_2^TB_2-a_0\|z_1\|^2 \tag{18}$$

Let

$$\frac{1}{e_2}\ge\frac{1}{4}+\frac{\|M_2\|^2}{2b}+a_0 \tag{19}$$

Where $b>0$.

Using Young's inequality, it follows that

$$p_2^TB_2\le\frac{\|p_2\|^2\|B_2\|^2}{2b}+\frac{b}{2} \tag{20}$$

In view of Eq.(19) and Eq.(20), giving

$$\dot{V}_1\le-a_0\|z_1\|^2-a_0\|p_2\|^2+\frac{1}{4}\|z_2\|^2-\frac{\|M_2\|^2\|p_2\|^2}{2b}+\frac{\|p_2\|^2\|B_2\|^2}{2b}+\frac{b}{2} \tag{21}$$

Noting that $-\dfrac{\|M_2\|^2\|p_2\|^2}{2b}+\dfrac{\|p_2\|^2\|B_2\|^2}{2b}=-\dfrac{\|p_2\|^2}{2b}\left(\|M_2\|^2-\|B_2\|^2\right)\le0$,

then

$$\dot{V}_1\le-a_0\|z_1\|^2-a_0\|p_2\|^2+\frac{1}{4}\|z_2\|^2+\frac{b}{2} \tag{22}$$

Step 2: consider the following Lyapunov function candidate:

$$V_2=V_1+\frac{1}{2}z_2^Tz_2+\frac{1}{2}\tilde{\lambda}_2^T\Gamma_{22}^{-1}\tilde{\lambda}_2 \tag{23}$$

where $\Gamma_{22}=\Gamma_{22}^T>0$.

$$\begin{aligned}z_2^T\dot{z}_2=&-z_2^TM^{-1}(\eta)C(v,\eta)x_2-z_2^TM^{-1}(\eta)D(v,\eta)x_2-z_2^TM^{-1}(\eta)G(\eta)\\&-z_2^TM^{-1}(\eta)\omega-z_2^TM^{-1}(\eta)\Delta(v,\eta)+z_2^TM^{-1}(\eta)\tau-z_2^T\dot{a}_2\end{aligned} \tag{24}$$

If the RBFNN is used to approximate the $-M^{-1}(\eta)\Delta(v,\eta)$, we have

$$-M^{-1}(\eta)\Delta(v,\eta) = W_2^{*T}S_2(Z) + \varepsilon_2 \tag{25}$$

$$z_2^T W_2^{*T} S_2(Z) + z_2^T \varepsilon_2 \leq \frac{\lambda_2^T \|z_2\|^2 \|S_2(Z)\|^2}{2b_2^2} + \frac{b_2^2}{2} + \frac{\|z_2\|^2}{2} + \frac{\|\bar{\varepsilon}_2\|^2}{2} \tag{26}$$

The control law is proposed as

$$\tau = C(v,\eta)x_2 + D(v,\eta)x_2 + G(\eta) - \frac{M(\eta)z_2 \hat{\lambda}_2^T \|S_2(Z)\|^2}{2b_2^2} + M(\eta)\dot{a}_2$$

$$-k_2 M(\eta)z_2 - d\omega^* \tanh\left(\frac{z_2^T M^{-1}(\eta)d\omega^*}{s}\right) \tag{27}$$

where $s > 0$, then

$$z_2^T \dot{z}_2 \leq -z_2^T M^{-1}(\eta)\omega + \frac{\tilde{\lambda}_2^T \|z_2\|^2 \|S_2(Z)\|^2}{2b_2^2} + \frac{b_2^2}{2}$$

$$+ \frac{\|z_2\|^2}{2} + \frac{\|\bar{\varepsilon}_2\|^2}{2} - k_2 \|z_2\|^2 - z_2^T M^{-1}(\eta)d\omega^* \tanh\left(\frac{z_2^T M^{-1}(\eta)d\omega^*}{s}\right) \tag{28}$$

It is clear that

$$-z_2^T M^{-1}(\eta)\omega \leq \left|z_2^T M^{-1}(\eta)\omega\right| \leq \left|z_2^T M^{-1}(\eta)d\omega^*\right| \tag{29}$$

$$\left|z_2^T M^{-1}(\eta)d\omega^*\right| - z_2^T M^{-1}(\eta)d\omega^* \tanh\left(\frac{z_2^T M^{-1}(\eta)d\omega^*}{s}\right) \leq 0.2785s \tag{30}$$

The derivative of V_2 is

$$\dot{V}_2 \leq -a_0 \|z_1\|^2 - a_0 \|p_2\|^2 + \frac{1}{4}\|z_2\|^2 + \frac{b}{2} + \frac{\tilde{\lambda}_2^T \|z_2\|^2 \|S_2(Z)\|^2}{2b_2^2}$$

$$+ \frac{b_2^2}{2} + \frac{\|z_2\|^2}{2} + \frac{\|\bar{\varepsilon}_2\|^2}{2} - k_2 \|z_2\|^2 + 0.2785s - \tilde{\lambda}_2^T \Gamma_{22}^{-1} \dot{\hat{\lambda}}_2 \tag{31}$$

Let

$$\dot{\hat{\lambda}}_2 = \Gamma_{22}\left(\frac{\|z_2\|^2 \|S_2(Z)\|^2}{2b_2^2} - \sigma_2\left(\hat{\lambda}_2 - \lambda_2^0\right)\right) \tag{32}$$

Noting the following fact:

$$\sigma_2 \tilde{\lambda}_2^T \left(\hat{\lambda}_2 - \lambda_2^0 \right) \le -\frac{1}{2} \sigma_2 \tilde{\lambda}_2^T \tilde{\lambda}_2 + \frac{1}{2} \sigma_2 \left(\lambda_2 - \lambda_2^0 \right)^2 \tag{33}$$

$$-\frac{1}{2} \sigma_2 \tilde{\lambda}_2^T \tilde{\lambda}_2 \le -\frac{\sigma_2 \tilde{\lambda}_2^T \Gamma_{22}^{-1} \tilde{\lambda}_2}{2 \lambda_{max} \left(\Gamma_{22}^{-1} \right)} \tag{34}$$

Let $\dfrac{\sigma_2}{2\lambda_{max}\left(\Gamma_{22}^{-1}\right)} \ge a_0$, and choosing $k_2 \ge \dfrac{3}{4} + a_0$.

Invoking Eq.(32), Eq.(33) and Eq.(34), the Eq. (31) can be written as

$$\dot{V}_2 \le -2a_0 V_2 + C \tag{35}$$

where $C = \dfrac{b}{2} + \dfrac{b_2^2}{2} + \dfrac{\|\bar{\varepsilon}_2\|^2}{2} + 0.2785s + \dfrac{1}{2}\sigma_2\left(\lambda_2 - \lambda_2^0\right)^2$. Then

$$V_2(t) \le \frac{C}{2a_0} + \left(V_2(t_0) - \frac{C}{2a_0} \right) e^{-(t-t_0)} \tag{36}$$

It follows that, for any $\mu_1 > (C/a_0)^{1/2}$, there exists a constant $T > 0$ such that $\|z_1(t)\| \le \mu_1$ for all $t \ge t_0 + T$, and the tracking error can be made small, since $(C/a_0)^{1/2}$ can arbitrarily be made small if the design parameters are appropriately chosen.

4   Simulation Result

In this section, an example is given to show the efficiency of the proposed scheme. We will use the nonlinear model of the Naval Postgraduate School AUV II [11].

The reference trajectory is $\eta_d = [2\sin(t)\ 2\sin(2t)\ 20\sin(4t)\ 10\sin(0.25t)\ 10\sin(0.5t)\ 20\sin(2t)]^T$, and the initial trajectory is assumed 30% deviation from the desired. Modeling error is assumed $\Delta(v,\eta) = 6\sin\xi$ ($\xi = z_1 + \dot{z}_1$). The designed parameters of the above controller are given as $k_1 = 10$, $k_2 = 10$, $e_2 = 0.1$, $s = 0.1$, $\Gamma_{22} = diag\{0.5\}$, $\sigma_2 = 0.5$, the initial value of λ_2 is set zero.

It can be observed that the tracking performance is satisfactory under the time-varying disturbance and model errors using the proposed NN controller.

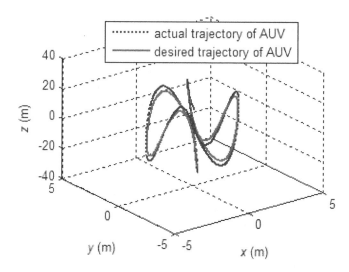

Fig. 1. Trajectory of AUV

5   Conclusions

In this brief, stable approximation-based tracking control has been designed for AUV in the presence of time-varying environmental disturbances, model errors. The controller is obtained by using Lyapunov function design in combination with DSC and RBFNN techniques. It has been shown that the closed-loop signals under the proposed control are uniformly ultimately bounded and the error signals can be made small through appropriate choice of control design parameters. On the other hand, the advantages of proposed algorithm are that the computational burden of the algorithm can drastically be reduced and the algorithm is convenient to implement in applications. Simulation results have demonstrated that the AUV is able to track a desired trajectory satisfactorily.

References

1. Luo, W.L., Zou, Z.J.: Neural network based robust control for trajectory tracking of underwater vehicle. China Ocean Engineering 21(2), 281–292 (2007)
2. Li, T.S., Li, R.H., Wang, D.: Adaptive neural control of nonlinear MIMO systems with unknown time delays. Neurocomputing 78(1), 83–88 (2012)
3. Yang, Y.S., Ren, J.S.: Adaptive fuzzy robust tracking controller design via small gain approach and its application. IEEE Trans. Fuzzy Syst. 11(6), 783–795 (2003)
4. Yang, Y.S., Feng, G., Ren, J.S.: A combined backstepping and small gain approach to robust adaptive fuzzy control for strict-feedback nonlinear systems. IEEE Trans. Syst., Man, Cybern. A, Syst., Humans 34(3), 406–420 (2004)

5. Yang, Y.S., Li, T.S., Wang, X.F.: Robust adaptive neural network control for strict-feedback nonlinear systems via small-gain approaches. In: Wang, J., Yi, Z., Żurada, J.M., Lu, B.-L., Yin, H. (eds.) ISNN 2006. LNCS, vol. 3972, pp. 888–897. Springer, Heidelberg (2006)
6. Gao, Y., Er, J.J.: Online adaptive fuzzy neural identification and control of a class of MIMO nonlinear systems. IEEE Trans. Fuzzy Syst. 11(4), 462–477 (2003)
7. Swaroop, D., Hedrick, J.K., Yip, P.P., Gerdes, J.C.: Dynamic surface control for a class of nonlinear systems. IEEE Trans. Automatic Control 45(10), 1893–1899 (2000)
8. Yip, P., Hedrick, J.K.: Adaptive dynamic surface control: a simplified algorithm for adaptive backstepping control of nonlinear systems. Int. J. Control 71(5), 959–979 (1998)
9. Wang, D., Huang, J.: Neural network-based adaptive dynamic surface control for a class of uncertain nonlinear systems in strict-feedback form. IEEE Transactions On Neural Networks 16(1), 195–202 (2005)
10. Zhang, H.M., Zhang, G.S.: Adaptive Backstepping Sliding Mode Control for Nonlinear Systems with Input Saturation. Trans. Tianjin Univ. 18, 46–51 (2012)
11. Fossen, T.I.: Guidance and control of ocean vehicles. John Wiley & Sons, New York (1998)

Bifurcation Control of a Fractional Order Hindmarsh-Rose Neuronal Model

Min Xiao[1,2]

[1] School of Mathematics and Information Technology,
Nanjing Xiaozhuang University, Nanjing 210017, China
[2] Department of Mathematics, Southeast University, Nanjing 210096, China
candymanxm2003@yahoo.com.cn

Abstract. This paper proposes to use a state feedback method to control the Hopf bifurcation for a fractional order Hindmarsh-Rose neuronal model. The order of the fractional order Hindmarsh-Rose neuronal model is chosen as the bifurcation parameter. The analysis shows that in the absences of the state feedback controller, the fractional order model loses stability via the Hopf bifurcation early, and can maintain the stability only in a certain domain of the gain parameter. When applying the state feedback controller to the model, the onset of the undesirable Hopf bifurcation is postponed. Thus, the stability domain is extended, and the model possesses the stability in a larger parameter range. Numerical simulations are given to justify the validity of the state feedback controller in bifurcation control.

Keywords: Fractional order Hindmarsh-Rose neuronal model, Hopf bifurcation, State feedback, Bifurcation control.

1 Introduction

Recently, the applications of fractional calculus [1] have grown rapidly, which has attracted fairly broad research activity, including viscoelastic systems [2], dielectric polarization [3], electrode-electrolyte polarization [4], electromagnetic waves [5], quantitative finance [6], and quantum evolution of complex systems [7].

In this paper, we consider the following fractional order system as the extended Hindmarsh-Rose neuronal model

$$D_t^\alpha x = c(x - \frac{x^3}{3} - y + I),$$
$$D_t^\alpha y = \frac{x^2 + dx - by + a}{c}, \tag{1}$$

where x and y denote the cell membrane potential and a recovery variable, respectively. I represents the external stimulus. a, b, c and d are parameters. D_t^α is the fractional derivative of x and y of order α ($0 < \alpha \le 1$) and in the Caputo

C. Guo, Z.-G. Hou, and Z. Zeng (Eds.): ISNN 2013, Part II, LNCS 7952, pp. 88–95, 2013.

sense is defined as

$$D_t^\alpha f(t) = \frac{1}{\Gamma(m-\alpha)} \int_0^t \frac{f^{(m)}(s)}{(t-s)^{\alpha-m+1}} ds, \tag{2}$$

where m is the first integer larger than α, i.e., $m-1 < \alpha \le m$ and $\Gamma(\cdot)$ is the gamma function. Using the order α as the bifurcation parameter, it has been found in [8] that a Hopf bifurcation may occur as the order α passes through a critical value in model (1), where a family of oscillations bifurcate from an equilibrium point.

Bifurcation control refers to the control of bifurcation properties of nonlinear dynamic systems, thereby resulting in some desired output behaviors of the systems. Typical objectives of bifurcation control include delaying the onset of an inherent bifurcation, stabilizing an unstable bifurcated solution or branch, and changing the critical values of an existing bifurcation [9]. Bifurcation control has attracted many researchers due to its promising potential applications in various areas: the prevention of voltage collapse in electric power systems, the stabilization of rotating stall and surge in axial flow compressors, the regulation of human heart rhythms and neuronal activity behavior, the elimination of seizing activities in human cerebral cortex, and so on [10]. Various bifurcation control approaches have been proposed in the literature [11,12,13]. Particularly, for the problem of relocating an inherent Hopf bifurcation, a dynamic state feedback control law incorporating a washout filter was proposed [11]. Later, a static state feedback controller with polynomial functions was developed to control Hopf bifurcations in the Lorenz and Rossler systems [13].

In this paper, we will apply an effective state feedback scheme to the fractional order Hindmarsh-Rose neuronal model (1) to control the Hopf bifurcation. We will show that the state feedback controller can increase the critical value of the Hopf bifurcation of the order, thereby guaranteeing the stability for large values of the order. It should be noted that the state feedback controller has been successfully used to autonomous systems on the Hopf bifurcation control [11,13], however, we first apply the state feedback controller to a fractional order system to realize the control of the Hopf bifurcation. To the best of the authors knowledge, such Hopf bifurcation control problem has not been investigated yet.

2 Bifurcation of Uncontrolled Model

In this section, the results of the stability and Hopf bifurcation for the fractional order Hindmarsh-Rose neuronal model (1) obtained in [8] are briefly reviewed here for convenience of comparison as well as for completeness.

Let $E_0 = (x_0, y_0)$ be the equilibrium of (1), i.e., it is the solution of equation

$$x - \frac{x^3}{3} - y + I = 0,$$
$$x^2 + dx - by + a = 0. \tag{3}$$

Theorem 1. *([8]) If $n > \frac{1}{4}m^2$ and $m < 0$, the equilibrium $E_0(x_0, y_0)$ of (1) is locally asymptotically stable when $\alpha \in (0, \alpha^*)$, and unstable when $\alpha > \alpha^*$, where*

$$m = \frac{b}{c} - c(1 - x_0^2), \qquad n = 2x_0 + d - b(1 - x_0^2), \qquad (4)$$

and

$$\alpha^* = \frac{2}{\pi} \arctan \frac{-\sqrt{4n^2 - m^2}}{m}. \qquad (5)$$

Theorem 2. *([8]) Suppose that $n > \frac{1}{4}m^2$ and $m < 0$. Then (1) undergoes a Hopf bifurcation at the equilibrium E_0 when the fractional order α passes through the critical value α^*, where α^* is defined by (4) and (5).*

Remark 1. For the fractional order Hindmarsh-Rose neuronal model (1), it can be seen from Theorems 1 and 2 that the Hopf bifurcation may occur as the order passes through the critical point α^*, where a family of oscillations bifurcate from an equilibrium point. Thus, in this case the stability is not guaranteed, which is in general not desirable. In this paper, we design a state feedback scheme to postpone the undesirable onset of α^*. Hence, the stability domain is extended, and model (1) possesses the stability in a larger parameter range.

3   Bifurcation Control by State Feedback

Following the general idea of the polynomial function controller [13], we propose a linear state feedback controller for the first equation of model (1) as follows:

$$u = -\gamma(x - x_0), \qquad (6)$$

where γ is a positive feedback gain parameter, which can be manipulated to control the Hopf bifurcation so as to achieve desirable behaviors,

Remark 2. The nonlinear state feedback controller (6) preserves the equilibrium point $E_0(x_0, y_0)$ of the fractional order Hindmarsh-Rose neuronal model (1). Thus, the bifurcation control can be realized without destroying the properties of the original model (1).

Remark 3. If the control objective is to relocate the onset of the Hopf bifurcation to a desired location, then the linear term is used only. The higher order terms can be added to influence the frequency and amplitude of the bifurcating oscillations.

Remark 4. Although the state feedback controller with polynomial functions has been successfully used to control the Hopf bifurcation in various autonomous systems [11,13,?], this is the first time that the state feedback controller is applied to a fractional order system to realize the control of the Hopf bifurcation.

With the nonlinear state feedback controller (6), the controlled fractional order Hindmarsh-Rose neuronal model (1) becomes

$$D_t^\alpha x = c(x - \frac{x^3}{3} - y + I) - \gamma(x - x_0),$$
$$D_t^\alpha y = \frac{x^2 + dx - by + a}{c}. \tag{7}$$

The Jacobian matrix of (7) at equilibrium E_0 is

$$J = \begin{bmatrix} c(1 - x_0^2) - \gamma & -c \\ \dfrac{2x_0 + d}{c} & -\dfrac{b}{c} \end{bmatrix}, \tag{8}$$

with the characteristic equation

$$\lambda^2 + m_c\lambda + n_c = 0, \tag{9}$$

where

$$m_c = \frac{b}{c} - c(1 - x_0^2) + \gamma, \qquad n_c = 2x_0 + d - b(1 - x_0^2) + \frac{b}{c}\gamma. \tag{10}$$

Theorem 3. *If $n_c > \frac{1}{4}m_c^2$ and $m_c < 0$, the equilibrium $E_0(x_0, y_0)$ of the controlled model (7) is locally asymptotically stable when $\alpha \in (0, \alpha_c^*)$, and unstable when $\alpha > \alpha_c^*$, where*

$$\alpha_c^* = \frac{2}{\pi} \arctan \frac{-\sqrt{4n_c^2 - m_c^2}}{m_c}. \tag{11}$$

The proof of Theorem 3 is similar to the proof of Theorem 1 in [8]. Hence, we omit the proof here.

Remark 5. If the controller u in (6) is removed from the controlled model (7), i.e., $\gamma = 0$, then (11) can be identical with the expression of α^* in Theorem 1. Therefore, α_c^* in (11) covers the value of α^* in Theorem 1. That is, α^* in Theorem 1 is a special case of α_c^* in (11) in the absence of the control.

Theorem 4. *Suppose that $n_c > \frac{1}{4}m_c^2$ and $m_c < 0$. Then the controlled model (7) exhibits a Hopf bifurcation at the equilibrium E_0 when the fractional order α passes through the critical value α_c^*, where α_c^* is defined by (11).*

The proof of Theorem 4 is similar to the proof of Theorem 2 in [8]. Hence, we omit the proof here.

Remark 6. Theorem 4 indicates that under the nonlinear state feedback control (6), one can delay the onset of α^* in Theorem 2 to α_c^* without changing the original equilibrium E_0 by choosing an appropriate feedback gain parameter value of γ. It should be noted that if one only needs to change the onset of the Hopf bifurcation, then a linear state feedback control with parameter γ is sufficient.

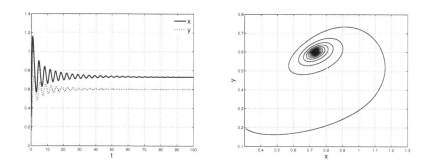

Fig. 1. Waveform plot and phase portrait of the uncontrolled model (1) with $a = -1$, $b = 1.4, c = 3, d = 1.8$ and $I = 0$. The equilibrium E_0 is asymptotically stable, where $\alpha = 0.8 < \alpha^* = 0.810427838252199$.

4   Numerical Simulations

In this section, we present numerical results to verify the analytical predictions obtained in the previous section, using the state feedback scheme to control the Hopf bifurcation of the fractional order Hindmarsh-Rose neuronal model (1).

For a consistent comparison, the same model (1) as used in [8] is discussed, with $a = -1, b = 1.4, c = 3, d = 1.8$ and $I = 0$. From (3), the uncontrolled model (1) has a unique nonzero equilibrium $E_0 = (0.7275, 0.5992)$. It follows from (5) that

$$\alpha^* = 0.810427838252199.$$

The dynamical behavior of this uncontrolled fractional order Hindmarsh-Rose neuronal model (1) is illustrated in Figures 1-3. From Theorems 1 and 2, it is shown that when $\alpha < \alpha^*$, trajectories converge to the equilibrium E_0 (see Figure 1), while as α is increased to pass through α^*, E_0 loses its stability and a Hopf bifurcation occurs (see Figures 2 and 3).

Now using our Theorems 3 and 4, we choose appropriate values of γ to control the Hopf bifurcation.

It is easy to see from Theorems 3 and 4 that for the linear state feedback control with an appropriate value of γ, we can delay the onset of the Hopf bifurcation. For example, by choosing

$$\gamma = 0.5,$$

we can apply (11) to obtain

$$\alpha_c^* = 0.915454296786782.$$

Note that the controlled fractional order Hindmarsh-Rose neuronal model (7) has the same equilibrium point as that of the original fractional order

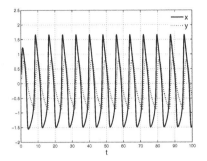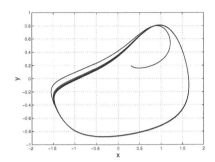

Fig. 2. Waveform plot and phase portrait of the uncontrolled model (1) with $a = -1$, $b = 1.4, c = 3, d = 1.8$ and $I = 0$. A oscillation bifurcates from the equilibrium E_0, where $\alpha = 0.88 > \alpha^* = 0.810427838252199$.

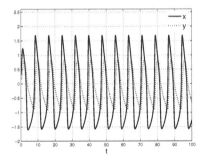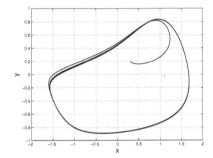

Fig. 3. Waveform plot and phase portrait of the uncontrolled model (1) with $a = -1$, $b = 1.4, c = 3, d = 1.8$ and $I = 0$. A oscillation bifurcates from the equilibrium E_0, where $\alpha = 0.9 > \alpha^* = 0.810427838252199$.

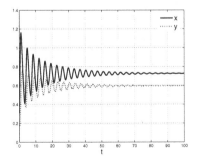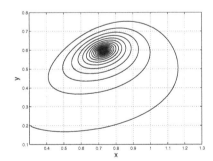

Fig. 4. Waveform plot and phase portrait of the controlled model (7) with $a = -1$, $b = 1.4, c = 3, d = 1.8, I = 0$ and $\gamma = 0.5$. The equilibrium E_0 is asymptotically stable, where $0.810427838252199 = \alpha^* < \alpha = 0.9 < \alpha_c^* = 0.915454296786782$.

Hindmarsh-Rose neuronal model (1), but the critical value α^* increases from 0.810427838252199 to 0.915454296786782, implying that the onset of the Hopf bifurcation is delayed.

Under the state feedback control with $\gamma = 0.5$, we choose $\alpha = 0.9 < \alpha_c^*$, which is the same value as that used in Figure 3. According to Theorem 4, we conclude that instead of having a Hopf bifurcation, the controlled fractional order Hindmarsh-Rose neuronal model (7) converges to the equilibrium point E_0, as shown in Figure 4.

Under the state feedback control with $\gamma = 0.5$, we choose $\alpha = 0.93 > \alpha_c^*$. From Theorem 4, the equilibrium point E_0 is unstable, as shown in Figure 5. It is seen that when α passes through the critical value $\alpha_c^* = 0.915454296786782$, a Hopf bifurcation occurs.

It can be shown that if we choose a larger value of γ, then the fractional order Hindmarsh-Rose neuronal model may not have a Hopf bifurcation even for larger values of α. This indicates that the state feedback controller can delay the onset

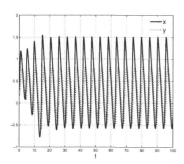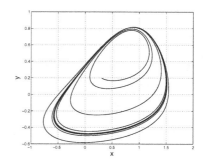

Fig. 5. Waveform plot and phase portrait of the controlled model (7) with $a = -1, b = 1.4, c = 3, d = 1.8, I = 0$ and $\gamma = 0.5$. A oscillation bifurcates from the equilibrium E_0, where $\alpha = 0.93 > \alpha_c^* = 0.915454296786782$.

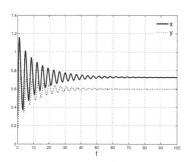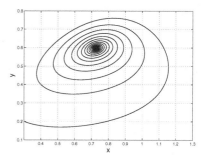

Fig. 6. Waveform plot and phase portrait of the controlled model (7) with $a = -1, b = 1.4, c = 3, d = 1.8, I = 0$ and $\gamma = 0.8$. The equilibrium E_0 is asymptotically stable, where $\alpha = 0.95 < \alpha_c^* = 0.973122179012338$.

of Hopf bifurcation, thus guaranteeing the stability for larger values of α. For example, when choosing $\gamma = 0.8$, the controlled fractional order Hindmarsh-Rose neuronal model (7) converges to the equilibrium solution E_0 if $\alpha < \alpha_c^* = 0.973122179012338$, as shown in Figure 6.

5   Concluding Remarks

In this paper, we have first applied the state feedback controller to a fractional order system to realize the control of the Hopf bifurcation. The linear state feedback control may be enough to use only for the fractional order Hindmarsh-Rose neuronal model in order to delay the onset of the Hopf bifurcation. Thus the stability is guaranteed for the large values of the order. This state feedback controller is valid for fractional order dynamical systems close to the bifurcation point.

Acknowledgement. This work was supported in part by the National Natural Science Foundation of China under Grant 61203232, and the Natural Science Foundation of Jiangsu Province of China under Grant BK2012072.

References

1. Podlubny, I.: Fractional Differential Equations. Academic Press, New York (1999)
2. Bagley, R.L., Calico, R.A.: Fractional Order State Equations for the Control of Viscoelastically Damped Structures. J. Guid. Control Dyn. 14, 304–311 (1991)
3. Sun, H.H., Abdelwahad, A.A., Onaral, B.: Linear Approximation of Transfer Function with a Pole of Fractional Order. IEEE Trans Autom. Control AC-29, 441–444 (1984)
4. Ichise, M., Nagayanagi, Y., Kojima, T.: An Analog Simulation of Noninteger Order Transfer Functions for Analysis of Electrode Process. J. Electroanal. Chem. 33, 253–265 (1971)
5. Heaviside, O.: Electromagnetic Theory. Chelsea, New York (1971)
6. Laskin, N.: Fractional Market Dynamics. Phys. A 287, 482–492 (2000)
7. Kusnezov, D., Bulgac, A., Dang, G.D.: Quantum Levy Processes and Fractional Kinetics. Phys. Rev. Lett. 82, 1136–1139 (1999)
8. Xiao, M.: Stability Analysis and Hopf-Type Bifurcation of A Fractional Order Hindmarsh-Rose Neuronal Model. In: Wang, J., Yen, G.G., Polycarpou, M.M. (eds.) ISNN 2012, Part I. LNCS, vol. 7367, pp. 217–224. Springer, Heidelberg (2012)
9. Abed, E.H., Fu, J.H.: Local Feedback Stabilization and Bifurcation Control: I. Hopf Bifurcation. Syst. Control Lett. 7, 11–17 (1986)
10. Chen, G.R., Moiola, J.L., Wang, H.O.: Bifurcation Control: Theories, Methods and Applications. Int. J. Bifurcation and Chaos 10, 511–548 (2000)
11. Wang, H., Abed, E.H.: Bifurcation Control of A Chaotic System. Automatica 31, 1213–1226 (1995)
12. Tesi, A., Abed, E.H., Genesio, R., Wang, H.O.: Harmonic Balance Analysis of Period-Doubling Bifurcations with Implications for Control of Nonlinear Dynamics. Automatica 32, 1255–1271 (1996)
13. Yu, P., Chen, G.R.: Hopf Bifurcation Control Using Nonlinear Feedback with Polynomial Functions. Int. J. Bifurcation and Chaos 14, 1683–1704 (2004)

Adaptive Neural Control for a Class of Large-Scale Pure-Feedback Nonlinear Systems

Huanqing Wang[1,2,*], Bing Chen[2], and Chong Lin[2]

[1] School of Mathematics and Physics, Bohai University,
Jinzhou 121000, Liaoning, P.R. China
[2] Institute of Complexity Science, Qingdao University,
Qingdao 266071, Shandong, P.R. China

Abstract. This paper considers the problem of adaptive neural decentralized control for pure-feedback nonlinear interconnected large-scale systems. Radical basis function (RBF) neural networks are used to model packaged unknown nonlinearities and backstepping is used to construct decentralized controller. The proposed control scheme can guarantee that all the signals in the closed-loop system are semi-globally uniformly ultimately bounded. A numerical example is provided to illustrate the effectiveness of the suggested approach.

Keywords: Pure-feedback nonlinear large-scale systems, Neural networks, Adaptive decentralized control, Backstepping approach.

1 Introduction

Large-scale systems being considered as a set of interconnected subsystems often exist in many practical systems, such as electric power systems, aerospace systems and multi-agent systems. Therefore, the study on stability analysis and control design of large-scale systems is an active topic in the control community in recent years. Because of the complexity of the control synthesis and physical restrictions on information exchange between subsystems, decentralized control strategy depending on local subsystem's state information is an efficient and effective way to achieve an objective for the whole large-scale systems. So far, many significant results on adaptive decentralized control for large-scale nonlinear systems have been reported in [1–3] and the references therein. In [1], an adaptive decentralized control scheme is proposed for a heating, ventilating, and air conditioning (HVAC) system. Afterwards, several decentralized control approaches are developed for interconnected nonlinear time-delay (or time-delay free) systems [2, 3]. By combining the adaptive backstepping control technique with fuzzy logic systems or neural networks, much research work has focused on the control design of large-scale systems with unknown continuous nonlinear functions [4–8]. In [9–11], some approximation-based adaptive decentralized control schemes are presented for non-affine nonlinear large-scale systems.

* This work is partially supported by the Natural Science Foundation of China (61074008, 61174033, 61104076, 11226139 and 11201037).

C. Guo, Z.-G. Hou, and Z. Zeng (Eds.): ISNN 2013, Part II, LNCS 7952, pp. 96–103, 2013.

Based on the above discussion, the problem of adaptive neural decentralized control is considered for a class of large-scale pure-feedback nonlinear systems. RBF neural networks are used to model packaged unknown nonlinearities, then an adaptive neural decentralized control scheme is constructed via backstepping. The proposed adaptive controller guarantees that all the signals in the closed-loop systems remain semi-globally uniformly ultimately bounded. In addition, the presented control scheme requires only one adaptive parameter that needs to be updated online for each subsystem. In this way, the computational burden is significantly alleviated. The simulation results are provided to further validate the effectiveness of the proposed control approach.

2   Problem Formulation and Preliminaries

In this paper, we consider a class of interconnected large-scale nonlinear pure-feedback systems with N subsystems, the ith $(i = 1, 2, \ldots, N)$ subsystem is described by:

$$\begin{cases} \dot{x}_{i,j} = f_{i,j}(\bar{x}_{i,j}, x_{i,j+1}) + h_{i,j}(\bar{y}), 1 \leq j \leq n_i - 1, \\ \dot{x}_{i,n_i} = f_{i,n_i}(\bar{x}_{i,n_i}, u_i) + h_{i,n_i}(\bar{y}), \\ y_i = x_{i,1}, \end{cases} \tag{1}$$

where $\bar{x}_{i,j} = [x_{i,1}, x_{i,2}, \ldots, x_{i,j}]^T$ and $\bar{y} = [y_1, y_2, \ldots, y_N]^T$. $x_i = [x_{i,1}, x_{i,2}, \ldots, x_{i,n_i}]^T \in R^{n_i}$, $u_i \in R$ and $y_i \in R$ are the state, the scalar control input and the scalar output of the ith nonlinear subsystem, respectively. $f_{i,j}(\cdot) : R^{j+1} \to R, (j = 1, 2, \ldots, n_i)$ are unknown nonlinear functions, $h_{i,j}(\cdot) : R^N \to R(j = 1, 2, \ldots, n_i)$ are unknown interconnections with $f_{i,j}(0) = h_{i,j}(0) = 0$.

Using mean value theorem [12], $f_{i,j}(\cdot)$ in (1) can be expressed as

$$\begin{aligned} f_{i,j}(\bar{x}_{i,j}, x_{i,j+1}) &= f_{i,j}(\bar{x}_{i,j}, x^0_{i,j+1}) + g_{\mu_{i,j}}(x_{i,j+1} - x^0_{i,j+1}), \\ f_{i,n_i}(\bar{x}_{i,n_i}, u_i) &= f_{i,n_i}(\bar{x}_{i,n_i}, u^0_i) + g_{\mu_{i,n_i}}(u_i - u^0_i), \end{aligned} \tag{2}$$

where $f_{i,j}(\cdot)$ is explicitly analyzed between $f_{i,j}(\bar{x}_{i,j}, x_{i,j+1})$ and $f_{i,j}(\bar{x}_{i,j}, x^0_{i,j+1})$, $g_{\mu_{i,j}} := g_{i,j}(\bar{x}_{i,j}, x_{\mu_{i,j}}) = \frac{\partial f_{i,j}(\bar{x}_{i,j}, x_{i,j+1})}{\partial x_{i,j+1}}|_{x_{i,j+1} = x_{\mu_{i,j}}}, x_{i,n_i+1} = u_i, x_{\mu_{i,j}} = \mu_{i,j} x_{i,j+1} + (1 - \mu_{i,j})x^0_{i,j+1}, 0 < \mu_{i,j} < 1, i = 1, 2, \ldots, N, j = 1, 2, \ldots, n_i$.

Furthermore, substituting (2) into (1) and choosing $x^0_{i,j+1} = 0, u^0_i = 0$ gives

$$\begin{cases} \dot{x}_{i,j} = g_{\mu_{i,j}} x_{i,j+1} + f_{i,j}(\bar{x}_{i,j}, 0) + h_{i,j}(\bar{y}), 1 \leq j \leq n_i - 1, \\ \dot{x}_{i,n_i} = g_{\mu_{i,n_i}} u_i + f_{i,n_i}(\bar{x}_{i,n_i}, 0) + h_{i,n_i}(\bar{y}), \\ y_i = x_{i,1}, \end{cases} \tag{3}$$

Assumption 1. The signs of $g_{\mu_{i,j}}, 1 \leq i \leq N, 1 \leq j \leq n_i$, do not change and without loss of generality, it is further assumed that there exist unknown constants b_m and b_M such that $0 < b_m \leq |g_{\mu_{i,j}}| \leq b_M < \infty$.

Assumption 2 [13]. For uncertain nonlinear functions $h_{i,j}(\bar{y})$ in (1), there exist unknown smooth functions $h_{i,j,l}(y_l)$ such that for $1 \leq i \leq N, 1 \leq j \leq n_i$, $|h_{i,j}(\bar{y})|^2 \leq \sum_{l=1}^N h^2_{i,j,l}(y_l)$, where $h_{i,j,l}(0) = 0, l = 1, 2, \ldots, N$.

Remark 1. Noting $h_{i,j,l}(y_l)$ in Assumption 2 are smooth functions with $h_{i,j,l}(0) = 0$, so there exist unknown smooth functions $\bar{h}_{i,j,l}(y_l)$ such that

$$|h_{i,j}(\bar{y})|^2 \leq \sum_{l=1}^{N} y_l^2 \bar{h}_{i,j,l}^2(y_l), \qquad (4)$$

In this note, the following RBF neural networks will be used to approximate any continuous function $f(Z) : R^n \to R$,

$$f_{nn}(Z) = W^T S(Z), \qquad (5)$$

where $Z \in \Omega_Z \subset R^q$ is the input vector with q being the neural networks input dimension, weight vector $W = [w_1, w_2, ..., w_l]^T \in R^l$, $l > 1$ is the neural networks node number, and $S(Z) = [s_1(Z), s_2(Z), ..., s_l(Z)]^T$ means the basis function vector with $s_i(Z)$ being chosen as the commonly used Gaussian function of the form $s_i(Z) = \exp[-\frac{(Z-\mu_i)^T(Z-\mu_i)}{\eta_i^2}], i = 1, 2, ..., l$, where $\mu_i = [\mu_{i1}, \mu_{i2}, ..., \mu_{iq}]^T$ is the center of the receptive field and η_i is the width of the Gaussian function. In [14], it has been indicated that with sufficiently large node number l, the RBF neural networks (5) can approximate any continuous function $f(Z)$ over a compact set $\Omega_Z \subset R^q$ to arbitrary any accuracy $\varepsilon > 0$ as $f(Z) = W^{*T} S(Z) + \delta(Z), \forall z \in \Omega_z \in R^q$, where W^* is the ideal constant weight vector and defined as $W^* := \arg\min_{W \in \bar{R}^l}\{\sup_{Z \in \Omega_Z} |f(Z) - W^T S(Z)|\}$, and $\delta(Z)$ denotes the approximation error and satisfies $| \delta(Z) | \leq \varepsilon$.

Lemma 1. [15]. Consider the Gaussian RBF networks (5). Let $\rho := \frac{1}{2} \min_{i \neq j} \|\mu_i - \mu_j\|$, then an upper bound of $\|S(Z)\|$ is taken as $\|S(Z)\| \leq \sum_{k=0}^{\infty} 3q$ $(k+2)^{q-1}e^{-2\rho^2 k^2/\eta^2} := s$.

3   Adaptive Neural Control Design

In this section, a backstepping-based design procedure is proposed. Both virtual control signals and adaption laws will be constructed in the following forms:

$$\alpha_{i,j}(Z_{i,j}) = -k_{i,j}z_{i,j} - \frac{1}{2a_{i,j}^2}z_{i,j}\hat{\theta}_i S_{i,j}^T(Z_{i,j})S_{i,j}(Z_{i,j}), \qquad (6)$$

$$\dot{\hat{\theta}}_i = \sum_{j=1}^{n_i} \frac{\lambda_i}{2a_{i,j}^2}z_{i,j}^2 S_{i,j}^T(Z_{i,j})S_{i,j}(Z_{i,j}) - \gamma_i\hat{\theta}_i, \qquad (7)$$

where $i = 1, 2, ..., N, j = 1, 2, ..., n_i$, $k_{i,j}$, $a_{i,j}$, λ_i and γ_i are positive design parameters, $Z_{i,1} = x_{i,1}, Z_{i,j} = [\bar{x}_{i,j}^T, \hat{\theta}_i]^T$ with $\bar{x}_{i,j} = [x_{i,1}, x_{i,2}, ..., x_{i,j}]^T$, and $z_{i,j}$ satisfy the following variable transformation:

$$z_{i,j} = x_{i,j} - \alpha_{i,j-1} \qquad (8)$$

with $\alpha_{i,0} = 0$. $\hat{\theta}_i$ is the estimation of an unknown constant θ_i which will be specified as

$$\theta_i = \max\{\frac{1}{b_m}\|W_{i,j}\|^2; j = 1, 2, \cdots, n_i\}, \qquad (9)$$

where $W_{i,j}$ will be specified later. Specially, α_{i,n_i} is the actual control input u_i.

Step 1. According to $z_{i,1} = x_{i,1}, z_{i,2} = x_{i,2} - \alpha_{i,1}$, it follows from (1) that

$$\dot{z}_{i,1} = g_{\mu_{i,1}} z_{i,2} + g_{\mu_{i,1}} \alpha_{i,1} + f_{i,j}(\bar{x}_{i,j}, 0) + h_{i,1}(\bar{y}). \tag{10}$$

Consider Lyapunov function candidate as $V_{i,1} = \frac{1}{2} z_{i,1}^2 = \frac{1}{2} y_i^2$. Then, the time derivative of V_1 along (10) is

$$\dot{V}_{i,1} = y_i (g_{\mu_{i,1}} z_{i,2} + g_{\mu_{i,1}} \alpha_{i,1} + f_{i,1}(\bar{x}_{i,1}, 0) + h_{i,1}(\bar{y})). \tag{11}$$

With the help of (4) and the completion of squares, one has

$$y_i h_{i,1}(\bar{y}) \le \frac{1}{2} y_i^2 + \frac{1}{2} \sum_{l=1}^{N} y_l^2 \bar{h}_{i,1,l}^2(y_l). \tag{12}$$

Substituting (12) into (11) yields

$$\dot{V}_{i,1} = y_i \left(g_{\mu_{i,1}} z_{i,2} + g_{\mu_{i,1}} \alpha_{i,1} + f_{i,1}(\bar{x}_{i,1}, 0) + \frac{1}{2} y_i \right) + \frac{1}{2} \sum_{l=1}^{N} y_l^2 \bar{h}_{i,1,l}^2(y_l). \tag{13}$$

Step $j(2 \le j \le n_i - 1)$. Using the coordinate transformation (8) gives

$$\dot{z}_{i,j} = g_{\mu_{i,j}} z_{i,j+1} + g_{\mu_{i,j}} \alpha_{i,j} + f_{i,j}(\bar{x}_{i,j}, 0) + h_{i,j}(\bar{y}) - \dot{\alpha}_{i,j-1}, \tag{14}$$

where

$$\dot{\alpha}_{i,j-1} = \sum_{k=1}^{j-1} \frac{\partial \alpha_{i,j-1}}{\partial x_{i,k}} (f_{i,k}(\bar{x}_{i,k+1}) + h_{i,k}(\bar{y})) + \frac{\partial \alpha_{i,j-1}}{\partial \hat{\theta}_i} \dot{\hat{\theta}}_i. \tag{15}$$

Take Lyapunov function $V_{i,j} = \frac{1}{2} z_{i,j}^2$, then the time derivative of $V_{i,j}$ is given by

$$\dot{V}_{i,j} = z_{i,j} \left(g_{\mu_{i,j}} z_{i,j+1} + g_{\mu_{i,j}} \alpha_{i,j} + f_{i,j}(\bar{x}_{i,j}, 0) + h_{i,j}(\bar{y}) - \dot{\alpha}_{i,j-1} \right). \tag{16}$$

Furthermore, similar to (12), we obtain

$$- z_{i,j} \sum_{k=1}^{j-1} \frac{\partial \alpha_{i,j-1}}{\partial x_{i,k}} h_{i,k}(\bar{y}) \le \frac{1}{2} z_{i,j}^2 \sum_{k=1}^{j-1} (\frac{\partial \alpha_{i,j-1}}{\partial x_{i,k}})^2 + \frac{1}{2} \sum_{k=1}^{j-1} \sum_{l=1}^{N} y_l^2 \bar{h}_{i,k,l}^2(y_l), \tag{17}$$

$$z_{i,j} h_{i,j}(\bar{y}) \le \frac{1}{2} z_{i,j}^2 + \frac{1}{2} \sum_{l=1}^{N} y_l^2 \bar{h}_{i,j,l}^2(y_l), \tag{18}$$

It immediately follows from (16) to (18) that

$$\dot{V}_{i,j} \le z_{i,j} \left(g_{\mu_{i,j}} z_{i,j+1} + g_{\mu_{i,j}} \alpha_{i,j} + f_{i,j}(\bar{x}_{i,j}, 0) - \sum_{k=1}^{j-1} \frac{\partial \alpha_{i,j-1}}{\partial x_{i,k}} f_{i,k}(\bar{x}_{i,k+1}) \right.$$
$$\left. + \frac{z_{i,j}}{2} + \frac{z_{i,j}}{2} \sum_{k=1}^{j-1} (\frac{\partial \alpha_{i,j-1}}{\partial x_{i,k}})^2 - \frac{\partial \alpha_{i,j-1}}{\partial \hat{\theta}_i} \dot{\hat{\theta}}_i \right) + \frac{1}{2} \sum_{k=1}^{j} \sum_{l=1}^{N} y_l^2 \bar{h}_{i,k,l}^2(y_l). \tag{19}$$

Step n_i. Similar to (14), the following result holds.

$$\dot{z}_{i,n_i} = g_{\mu_{i,n_i}} u_i + f_{i,n_i}(\bar{x}_{i,n_i}, 0) + h_{i,n_i}(\bar{y}) - \dot{\alpha}_{i,n_i-1}, \tag{20}$$

where $\dot{\alpha}_{i,n_i-1}$ is given in (15) with $j = n_i$. Consider Lyapunov function as $V_{i,n_i} = \frac{1}{2}z_{i,n_i}^2 + \frac{b_m}{2\lambda_i}\tilde{\theta}_i^2$, where $\tilde{\theta}_i = \theta_i - \hat{\theta}_i$ is the parameter error and λ_i is a positive design constant. Repeating the same derivations as (16)-(19) yields

$$\dot{V}_{i,n_i} \le z_{i,n_i}\left(g_{\mu_{i,n_i}}u_i + f_{i,n_i}(\bar{x}_{i,n_i}, 0) - \sum_{k=1}^{n_i-1}\frac{\partial\alpha_{i,n_i-1}}{\partial x_{i,k}}f_{i,k}(\bar{x}_{i,k+1}) - \frac{\partial\alpha_{i,n_i-1}}{\partial\hat{\theta}_i}\dot{\hat{\theta}}_i\right.$$
$$\left. + \frac{1}{2}z_{i,n_i}\sum_{k=1}^{n_i-1}\left(\frac{\partial\alpha_{i,n_i-1}}{\partial x_{i,k}}\right)^2 + \frac{1}{2}z_{i,n_i}\right) + \frac{1}{2}\sum_{k=1}^{n_i}\sum_{l=1}^{N}y_l^2\bar{h}_{i,k,l}^2(y_l) - \frac{b_m}{\lambda_i}\tilde{\theta}_i\dot{\hat{\theta}}_i. \tag{21}$$

Now, choose the Lyapunov function for the whole system as

$$V = \sum_{i=1}^{N}\sum_{j=1}^{n_i}V_{i,j} = \sum_{i=1}^{N}\left(\frac{1}{2}y_i^2 + \sum_{j=2}^{n_i}\frac{1}{2}z_{i,j}^2 + \frac{b_m}{2\lambda_i}\tilde{\theta}_i^2\right). \tag{22}$$

Subsequently, differentiating (22) and combining (13), (19) and (21), and using the following two inequalities

$$\frac{1}{2}\sum_{i=1}^{N}\sum_{s=1}^{n_i}\sum_{k=1}^{s}\sum_{l=1}^{N}y_l^2\bar{h}_{i,k,l}^2(y_l) = \frac{1}{2}\sum_{i=1}^{N}\sum_{l=1}^{N}\sum_{s=1}^{n_l}\sum_{k=1}^{s}y_i^2\bar{h}_{l,k,i}^2(y_i), \tag{23}$$

and

$$-\sum_{i=1}^{N}\sum_{j=2}^{n_i}z_{i,j}\frac{\partial\alpha_{i,j-1}}{\partial\hat{\theta}_i}\dot{\hat{\theta}}_i$$
$$\le \sum_{i=1}^{N}\sum_{j=2}^{n_i}z_{i,j}\frac{\partial\alpha_{i,j-1}}{\partial\hat{\theta}_i}\gamma_i\hat{\theta}_i - \sum_{i=1}^{N}\sum_{j=2}^{n_i}z_{i,j}\frac{\partial\alpha_{i,j-1}}{\partial\hat{\theta}_i}\sum_{k=1}^{j-1}\frac{\lambda_i}{2a_{i,k}^2}z_{i,k}^2S_{i,k}^TS_{i,k}$$
$$+ \sum_{i=1}^{N}\sum_{j=2}^{n_i}\frac{\lambda_i}{2a_{i,j}^2}z_{i,j}^2\left(\sum_{k=2}^{j}|z_{i,k}\frac{\partial\alpha_{i,k-1}}{\partial\hat{\theta}_i}|\right), \tag{24}$$

the result below holds.

$$\dot{V} \le \sum_{i=1}^{N}y_i\left(g_{\mu_{i,1}}\alpha_{i,1} + \bar{f}_{i,1}(Z_{i,1})\right) + \sum_{i=1}^{N}\sum_{j=2}^{n_i-1}z_{i,j}\left(g_{\mu_{i,j}}\alpha_{i,j} + \bar{f}_{i,j}(Z_{i,j})\right)$$
$$+ \sum_{i=1}^{N}z_{i,n_i}\left(g_{\mu_{i,n_i}}u_i + \bar{f}_{i,n_i}(Z_{i,n_i})\right) - \frac{1}{2}\sum_{i=1}^{N}\sum_{j=1}^{n_i}z_{i,j}^2 - \sum_{i=1}^{N}\frac{b_m}{\lambda_i}\tilde{\theta}_i\dot{\hat{\theta}}_i, \tag{25}$$

where the functions $\bar{f}_{i,j}(Z_{i,j}), i = 1, 2, \ldots, N$ are defined as

$$\bar{f}_{i,1}(Z_{i,1}) = f_{i,1}(\bar{x}_{i,1}, 0) + y_i + \frac{1}{2}y_i\sum_{l=1}^{N}\sum_{s=1}^{n_l}\sum_{k=1}^{s}\bar{h}_{l,k,i}^2(y_i), \tag{26}$$

$$\bar{f}_{i,j}(Z_{i,j}) = g_{\mu_{i,j}} z_{i,j-1} + f_{i,j}(\bar{x}_{i,j},0) - \sum_{k=1}^{j-1} \frac{\partial \alpha_{i,j-1}}{\partial x_{i,k}} f_{i,k}(\bar{x}_{i,k+1}) + z_{i,j}$$

$$+\frac{1}{2} z_{i,j} \sum_{k=1}^{j-1} (\frac{\partial \alpha_{i,j-1}}{\partial x_{i,k}})^2 + \frac{\partial \alpha_{i,j-1}}{\partial \hat{\theta}_i} \gamma_i \hat{\theta}_i - \frac{\partial \alpha_{i,j-1}}{\partial \hat{\theta}_i} \sum_{k=1}^{j-1} \frac{\lambda_i}{2a_{i,k}^2} z_{i,k}^2 S_{i,k}^T S_{i,k}$$

$$+\frac{\lambda_i}{2a_{i,j}^2} z_{i,j} (\sum_{k=2}^{j} |z_{i,k} \frac{\partial \alpha_{i,k-1}}{\partial \hat{\theta}_i}|), j = 2,\ldots, n_i, \qquad (27)$$

Then, $W_{i,j}^T S_{i,j}(Z_{i,j})$ is used to model $\bar{f}_{i,j}(Z_{i,j})$, such that, for any $\varepsilon_{i,j} > 0$,

$$\bar{f}_{i,j}(Z_{i,j}) = W_{i,j}^T S_{i,j}(Z_{i,j}) + \delta_{i,j}(Z_{i,j}), \qquad (28)$$

where $\delta_{i,j}(Z_{i,j})$ refers to the approximation error and satisfies $|\delta_{i,j}(Z_{i,j})| < \varepsilon_{i,j}$. Furthermore, by Young's inequality, one has

$$z_{i,j}\bar{f}_{i,j}(Z_{i,j}) \leq \frac{b_m}{2a_{i,j}^2} z_{i,j}^2 \theta_i S_{i,j}^T S_{i,j} + \frac{1}{2} a_{i,j}^2 + \frac{1}{2} z_{i,j}^2 + \frac{1}{2} \varepsilon_{i,j}^2, \qquad (29)$$

where $i = 1, 2, \ldots, N, j = 1, 2, \ldots, n_i$ and $\theta_i = \max\left\{\frac{1}{b_m}\|W_{i,j}\|^2; \ j = 1, 2, \cdots, n_i\right\}$. Now, choose the virtual control signals $\alpha_{i,j}$ in (6). Then, we have

$$z_{i,j}g_{\mu_{i,j}}\alpha_{i,j} \leq -k_{i,j}b_m z_{i,j}^2 - \frac{b_m}{2a_{i,j}^2} z_{i,j}^2 \hat{\theta}_i S_{i,j}^T S_{i,j}, 1 \leq i \leq N, 1 \leq j \leq n_i, \quad (30)$$

Further, combining (25) together with (29), (30)and (7) gives

$$\dot{V} \leq -\sum_{i=1}^{N} \left(\sum_{j=1}^{n_i} k_{i,j}b_m z_{i,j}^2 + \frac{\gamma_i b_m}{2\lambda_i} \tilde{\theta}_i^2\right) + \sum_{i=1}^{N} \sum_{j=1}^{n_i} \left(\frac{1}{2} a_{i,j}^2 + \frac{1}{2}\varepsilon_{i,j}^2 + \frac{\gamma_i b_m}{2\lambda_i}\theta_i^2\right), \quad (31)$$

where the inequality $\tilde{\theta}_i \hat{\theta}_i \leq -\frac{1}{2}\tilde{\theta}_i^2 + \frac{1}{2}\theta_i^2$ has been used in the above inequality. The main result of this paper will be summarized in the following theorem.

Theorem 1. Consider the large-scale nonlinear pure-feedback systems (1), the controller (6), and adaptive law (7) under Assumptions 1-2. Assume that for $1 \leq i \leq N, 1 \leq j \leq n_i$, the packaged unknown functions $\bar{f}_{i,j}(Z_{i,j})$ can be well approximated by the neural network $W_{i,j}^T S_{i,j}(Z_{i,j})$ in the sense that the approximation errors $\delta_{i,j}(Z_{i,j})$ are bounded, then for bounded initial conditions $[z_i^T(0), \hat{\theta}_i(0)]^T \in \Omega_0$, all the signals in the closed-loop system remain bounded and the error signals $z_{i,j}$ and $\tilde{\theta}_i$ eventually converge to the compact set Ω_s defined by

$$\Omega_s = \left\{ z_{i,j}, \tilde{\theta}_i \Big| |z_{i,j}| \leq \sqrt{2\frac{b_0}{a_0}}, |\tilde{\theta}_i| \leq \sqrt{\frac{2\lambda_i}{b_m}\frac{b_0}{a_0}}, 1 \leq i \leq N, 1 \leq j \leq n_i \right\}. \quad (32)$$

Proof: Let $a_0 = \min\{2k_{i,j}b_m, \gamma_i, i = 1, 2, \ldots, N, j = 1, 2, \ldots, n_i\}, b_0 = \sum_{i=1}^{N} \sum_{j=1}^{n_i}(\frac{1}{2}a_{i,j}^2 + \frac{1}{2}\varepsilon_{i,j}^2 + \frac{\gamma_i b_m}{2\lambda_i}\theta_i^2)$, then (31) can be rewritten as

$$\dot{V} \leq -a_0 V + b_0, t \geq 0. \qquad (33)$$

Next, from(33), the following inequality can be easily verified.

$$V(t) \leq \left(V(0) - \frac{b_0}{a_0} \right) e^{-a_0 t} + \frac{b_0}{a_0}, \forall t > 0, \tag{34}$$

which implies that $V(t) \leq V(0) + \frac{b_0}{a_0}, \forall t > 0$. Therefore, based on the definition of V in (22), we conclude that all the signals in the closed-loop system are bounded.

Furthermore, it is obtained that $V(t) \leq \frac{b_0}{a_0}, t \rightarrow +\infty$. Therefore, the error signals $z_{i,j}$ and $\tilde{\theta}_i$ eventually converge to the compact set Ω_s specified in (32).

4   Simulation Example

Consider the following interconnected pure-feedback nonlinear system.

$$\begin{cases} \dot{x}_{1,1} = (1 + \sin(x_{1,1}))x_{1,2} + x_{1,2}^3 + y_2^3 y_1, \\ \dot{x}_{1,2} = (2 + \cos(x_{1,1}x_{1,2}))u_1 + 0.2\cos(u_1) + y_1 y_2, \\ y_1 = x_{1,1}, \\ \dot{x}_{2,1} = (3 + \sin(x_{2,1}))x_{2,2} + 0.5x_{2,2}^5 + y_1\cos(y_2^2), \\ \dot{x}_{2,2} = (1 + x_{2,1}^2)u_2 + 0.1\sin(u_2) + y_2\ln(1 + y_1^2), \\ y_2 = x_{2,1}, \end{cases} \tag{35}$$

By using Theorem 1, choose the virtual control law $\alpha_{i,j}$ in (6) and the adaptive laws $\dot{\hat{\theta}}_i$ in (7). The simulation is run under the initial conditions $[x_{1,1}(0), x_{1,2}(0), x_{2,1}(0), x_{2,2}(0)]^T = [0.3, -0.2, 0.2, 0.4]^T$, and $[\hat{\theta}_1(0), \hat{\theta}_2(0)]^T = [0, 0]^T$. The design parameters are taken as $k_{1,1} = k_{1,2} = k_{2,1} = k_{2,2} = 6$, $a_{1,1} = a_{1,2} = a_{2,1} = a_{2,2} = 2, \gamma_1 = \gamma_2 = 1$, and $\lambda_1 = \lambda_2 = 2$. The details are shown in Figures 1-2.

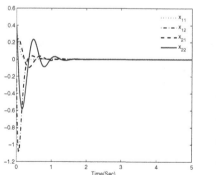
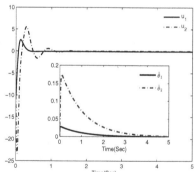

Fig. 1. State variables $x_{1,1}$, $x_{1,2}$, $x_{2,1}$ and $x_{2,2}$

Fig. 2. The variables u_1, u_2, $\hat{\theta}_1$ and $\hat{\theta}_2$

5   Conclusion

In this paper, an adaptive neural decentralized control scheme has been proposed for a class of large-scale pure-feedback nonlinear systems. The presented decentralized adaptive controller can ensure that all the signals in the closed-loop systems are bounded. Compared with the existing results, the main of advantage of the proposed controller is that it contains only one adaptive parameter needed to be estimated online for each subsystems. Simulation example is provided to further demonstrate the effectiveness of the proposed approach.

References

1. Zhang, H.G., Cai, L.: Decentralized nonlinear adaptive control of an HVAC system. IEEE Transactions on Systems, Man and Cybernetics Part C: Applications and Reviews 32, 493–498 (2002)
2. Zhou, J., Wen, C.Y.: Decentralized backstepping adaptive output tracking of interconnected nonlinear systems. IEEE Transactions on Automatic Control 53, 2378–2384 (2008)
3. Hua, C.C., Wang, Q.G., Guan, X.P.: Exponential stabilization controller design for interconnected time delay systems. Automatica 44, 2600–2606 (2008)
4. Tong, S.C., Liu, C.L., Li, Y.M.: Fuzzy adaptive decentralized control for large-scale nonlinear systems with dynamical uncertainties. IEEE Transactions on Fuzzy Systems 18, 845–861 (2010)
5. Tong, S.C., Li, Y.M., Zhang, H.G.: Adaptive neural network decentralized backstepping output-feedback control for nonlinear large-scale systems with time delays. IEEE Transactions on Neural Networks 22, 1073–1086 (2011)
6. Li, T.S., Li, R.H., Li, J.F.: Decentralized Adaptive Neural Control of Nonlinear Systems with Unknown Time Delays. Nonlinear Dynamics 67, 2017–2026 (2012)
7. Li, T.S., Li, R.H., Li, J.F.: Decentralized adaptive neural control of nonlinear interconnected large-scale systems with unknown time delays and input saturation. Neurocomputing 74, 2277–2283 (2011)
8. Chen, W.S., Li, J.M.: Decentralized output-feedback neural control for systems with unknown interconnections. IEEE Transactions on Systems, Man and Cybernetics Part B: Cybernetics 38, 258–266 (2008)
9. Karimi, B., Menhaj, M.B.: Non-affine nonlinear adaptive control of decentralized large-scale systems using neural networks. Information Sciences 180, 3335–3347 (2010)
10. Mao, Z.Z., Xiao, X.S.: Decentralized adaptive tracking control of nonaffine nonlinear large-scale systems with time delays. Information Sciences 181, 5291–5303 (2011)
11. Huang, Y.S., Wu, M.: Robust decentralized direct adaptive output feedback fuzzy control for a class of large-sale nonaffine nonlinear systems. Information Sciences 181, 2392–2404 (2011)
12. Apostol, T.M.: Mathematical analysis. Addison-Wesley, Reading (1963)
13. Li, J., Chen, W.S., Li, J.M.: Adaptive NN output-feedback decentralized stabilization for a class of large-scale stochastic nonlinear strict-feedback systems. International Journal of Robust and Nonlinear Control 21, 452–472 (2011)
14. Sanner, R.M., Slotine, J.E.: Gaussian networks for direct adaptive control. IEEE Transactions on Neural Network 3, 837–863 (1992)
15. Kurdila, A.J., Narcowich, F.J., Ward, J.D.: Persistency of excitation in identification using radial basis function approximants. SIAM Journal of Control and Optimization 33, 625–642 (1995)

Adaptive Synchronization
of Uncertain Chaotic Systems via Neural Network-Based
Dynamic Surface Control Design[*]

Liyou Fang, Tieshan Li[**], Xin Wang, and Xiaori Gao

Navigation College, Dalian Maritime University, Dalian 116026, China
tieshanli@126.com

Abstract. In this paper, the adaptive synchronization problem is investigated for a class of uncertain chaotic systems. By using the RBF networks to approximation unknown functions of the master system, an adaptive neural synchronization scheme is proposed with the combination of backstepping technique and dynamic surface control (DSC). This proposed method, similar to backstepping but with an important addition, can overcome the "explosion of complexity" of the traditional backstepping by introducing a first-order filtering. Thus, the closed-loop stability and asymptotic synchronization can be achieved. Finally, simulation results are presented to illustrate the effectiveness of the approach.

Keywords: synchronization, backstepping, dynamic surface, neural networks, chaotic system.

1   Introduction

Since the pioneering work of chaos control by Ott, Grebogi and York [1] in 1990 and presentation of synchronization of chaotic systems with each other by Pecora and Carrol [2] in the same year, Chaos control and synchronization has received increasing attention and has become a very active topic in nonlinear science. Various linear and nonlinear methods have emerged thereafter in search of more efficient algorithms for controlling and synchronizing identical and nonidentical chaotic systems [4-13].

In particular, backstepping design [3] have been recognized as one powerful design method to control and synchronize chaos, because it can guarantee global stabilities, tracking, and transient performance for a broad class of strict-feedback systems. In recent time, it has been employed for controlling, tracking and synchronization many chaotic systems [4-8] as well as hyperchaotic system [12]. However, a drawback with the backstepping technique is the problem of "explosion of complexity" [6]. This problem is caused by the repeated differentiations of certain nonlinear function.

[*] This work was supported in part by the National Natural Science Foundation of China (Nos.51179019, 60874056), the Natural Science Foundation of Liaoning Province (No. 20102012) , the Program for Liaoning Excellent Talents in University (LNET) (Grant No.LR2012016), the Fundamental Research Funds for the Central Universities (No.3132013005), and the Applied Basic Research Program of Ministry of Transport of China.

[**] Corresponding author.

C. Guo, Z.-G. Hou, and Z. Zeng (Eds.): ISNN 2013, Part II, LNCS 7952, pp. 104–111, 2013.

In this paper, we will further consider the problem of adaptive synchronization of uncertain nonlinear systems. The master system is any smooth, bounded, linear-parameter nonlinear system with key parameter unknown, while the slave system can be any smooth, nonlinear strict-feedback systems. Thus, we consider using radial basis function (RBF) [15][17] to approximate the unknown function of the master system. Furthermore, we introduce a new method called dynamic surface control (DSC) [14][15][16] to eliminate the problem of explosion of complexity. Global stability and asymptotic synchronization between the master and slave systems can be achieved. We are not only eliminated the problem of "explosion of complexity" by using this method, but also we have given a stability analysis which shows that control law can guarantee the tracking error arbitrarily small. Finally, the results are demonstrated in the simulation part.

The presentation is organized as follows. The problem formulation is presented in section 2. Adaptive dynamic surface control design procedure is presented in section 3. In section 4, we will give a stability analysis. In section 5, the simulation results are given to illustrate the effectiveness of the proposed approach. Section 6 contains the conclusion.

2   Problem Formulation

We consider the master system in the form of any smooth, bounded nonlinear chaotic system as

$$\dot{x}_{di} = f_{di}(x_d,t) + \theta^T F_{di}(x_d,t), \quad 1 \le i \le m \tag{1}$$

where $x_d = [x_{d1}, x_{d2}, \ldots, x_{dm}]^T \in R^m$ is the state vector; $f_{di}(\cdot)$ and $F_{di}(\cdot), i = 1, \ldots, m$ are unknown smooth nonlinear functions.

The slave system is in the form of strict-feedback nonlinear chaotic system as

$$\begin{aligned}
\dot{x}_1 &= g_1(x_1,t)x_2 + f_1(x_1,t), \\
&\vdots \\
\dot{x}_{n-1} &= g_{n-1}(x_1, \cdots, x_{n-1},t)x_{n-1} + f_{n-1}(x_1, \cdots, x_{n-1},t) \\
\dot{x}_n &= g_n(x,t)u + f_n(x,t), \quad n \le m
\end{aligned} \tag{2}$$

where $x = [x_1, x_2, \cdots, x_n]^T \in R^n$ and $u \in R$ are the state and control action, respectively; $g_i(\cdot) \ne 0, f_i(\cdot), i = 1, \ldots, n$ are known smooth nonlinear functions.

The control objective is to design an adaptive synchronization algorithm u, to guarantee global stability and force the state $x_1(t)$ of the slave system (2) to asymptotically synchronize with the state $x_{d1}(t)$ of the master system (1), to achieve $x_1(t) - x_{d1}(t) \to 0$, as $t \to \infty$.

In this paper, we employ RBF networks to approximate the unknown function in the master system. Let us first recall the approximation property of the RBF neural networks. The RBF neural networks take the form $\theta^T \xi(x)$ where $\theta \in R^N$ for some integer N is called weight vector, and $\xi(x) \in R^N$ is a vector valued function defined in R^N. Denote the components of $\xi(x)$ by $\rho_i(x), i = 1, \ldots, N$, then $\rho_i(x)$ is called a basis function. A commonly used basis function is the so-called Gaussian Function of the following form.

$$\rho_i(x) = \exp\left(\frac{-\|x - \xi_i\|}{\sigma^2}\right), \sigma \geq 0, i = 1, \ldots, N, \tag{3}$$

where $\zeta_i \in R^N, i = 1, \ldots, N$, are constant vectors called the center of the basis function. According to the approximation property of the RBF networks, given a continuous real valued function $f : \Omega \rightarrow R$ with $\Omega \in R^N$ a compact set, and any $\delta_m > 0$, by appropriately choosing $\sigma, \zeta_i \in R^N, i = 1, \ldots, N$, for some sufficiently large integer N, there exists an ideal weight vector $\theta^* \in R^N$ such that the RBF network $\theta^{*T}\xi(x)$ can approximate the given function f with the approximation error bounded by δ_m.

$$f(x) = \theta^{*T}\xi(x) + \delta^*, x \in \Omega \tag{4}$$

with $|\delta^*| \leq \delta_m$, where δ^* represents the network reconstruction error,

$$\delta^* = f(x) - \theta^{*T}\xi(x). \tag{5}$$

Since θ^* is unknown, we need to estimate θ^* online. We will use the notation $\hat{\theta}$ to denote the estimation of θ^* and develop an adaptive law to update the parameter $\hat{\theta}$.

3   Control Design

Step 1: we define the tracking error as the first error space as follows:

$$S_1 = x_1 - x_{d1} \tag{6}$$

Give a compact set $x_d \in R^m$, let θ_1^* and δ_1^* be such that for any $x_d \in R^m$

$$F_{d1}(x_d) = -\theta_1^{*T}\xi_1(x_d) - \delta_1^* \tag{7}$$

with $|\delta_1^*| \leq \delta_m$. The time derivative of (6) becomes

$$\dot{S}_1 = \dot{x}_1 - \dot{x}_{d1} = g_1 x_2 + f_1 - f_{d1} + \theta_1^{*T}\xi_1(x_d) + \delta_1^* \tag{8}$$

Choose a virtual control $\bar{x}_2$ as follows:

$$\bar{x}_2 = \frac{1}{g_1}\left[-g_1 x_2 - f_1 + f_{d1} - k_1 S_1 - \hat{\theta}_1^T \xi_1\right] \tag{9}$$

where $\hat{\theta}_1$ is the estimation of θ_1^* and is updated as follows:

$$\dot{\hat{\theta}}_1 = \Gamma_1 \xi_1(x_d) S_1 - \eta \Gamma_1 \hat{\theta}_1 \tag{10}$$

with any constant matrix $\Gamma_1 = \Gamma_1^T > 0$. Introduce a new state variable z_2 and let $\bar{x}_2$ pass through a first-order filter with time constant ϵ_2 to obtain z_2.

$$\in_2 \dot{z}_2 + z_2 = \overline{x}_2, \, z_2(0) = \overline{x}_2(0) \tag{11}$$

Step i: a similar design procedures is employed recursively at each step, $i = 2, \ldots, n-1$, Let

$$S_i = x_i - x_{di} - z_i \tag{12}$$

Give a compact set $x_d \in R^m$, for any $x_d \in R^m$ let θ_i^* and δ_i^* be such that

$$F_{di}(x_d) = -\theta_i^{*T} \xi_i(x_d) - \delta_i^* \tag{13}$$

with $\left| \delta_i^* \right| \le \delta_m$. The time derivative of (12) becomes

$$\dot{S}_i = \dot{x}_i - \dot{x}_{di} - \dot{z}_i = g_i x_{i+1} + f_i - f_{di} - F_{di} - \dot{z}_i \tag{14}$$

Choose a virtual control $\overline{x}_{i+1}$ as follows:

$$\overline{x}_{i+1} = \frac{1}{g_i} \left[-g_i x_{d(i+1)} - f_i + f_{di} + \dot{z}_i - k_i S_i - \hat{\theta}_i^T \xi_i \right] \tag{15}$$

where $\hat{\theta}_i$ is the estimation of θ_i^* and is updated as follows:

$$\dot{\hat{\theta}}_i = \Gamma_i \xi_i(x_d) S_i - \eta \Gamma_i \hat{\theta}_i \tag{16}$$

with any constant matrix $\Gamma_i = \Gamma_i^T > 0$. Introduce a new state variable z_{i+1} and let $\overline{x}_{i+1}$ pass through a first-order filter with time constant $\in_{i+1}$ to obtain z_{i+1}

$$\in_{i+1} \dot{z}_{i+1} + z_{i+1} = \overline{x}_{i+1}, \, z_{i+1}(0) = \overline{x}_{i+1}(0) \tag{17}$$

Step n: the final control law will be derived in this step. We consider the nth error surface as follows:

$$S_n = x_n - x_{dn} - z_n \tag{18}$$

Give a compact set $x_d \in R^m$ let θ_n^* and δ_n^* be such that for any $x_d \in R^m$

$$F_{dn}(x_d) = -\theta_n^{*T} \xi_n(x_d) - \delta_n^* \tag{19}$$

with $\left| \delta_i^* \right| \le \delta_m$. The time derivative of (18) becomes

$$\dot{S}_n = \dot{x}_n - \dot{x}_{dn} - \dot{z}_n = g_n u + f_n - f_{dn} + \theta_n^{*T} \xi_n(x_d) + \delta_n^* - \dot{z}_n \tag{20}$$

We specify the control law as follows:

$$u = \frac{1}{g_n} \left[-f_n + f_{dn} - \hat{\theta}_n^T \xi_n(x_d) + \dot{z}_n - k_n S_n \right] \tag{21}$$

where $\hat{\theta}_n$ is the estimation of θ_n^* and is updated as follows:

$$\dot{\hat{\theta}}_n = \Gamma_n \xi_n(x_d) S_n - \eta \Gamma_n \hat{\theta}_n \tag{22}$$

with any constant matrix $\Gamma_n = \Gamma_n^T > 0$.

4   Stability Analysis

Theorem1. We consider a close-loop uncertain chaotic system consisting of the plant (1) (2), the virtual control function (9) (15), the adaptive laws (10) (16) (22), and the control law (22). Give a compact set $x_d \in R^m$, let θ_i^* and δ_i^* be such that (13) for any $x_d \in R^m$ with $|\delta_i^*| \leq \delta_m$ assume there exists a known positive number θ_m such that, for all $i = 1, \ldots, n$, $\|\theta_i^*\| \leq \theta_m$.

In this section, we will establish that the closed-loop system possesses the uniformly ultimate boundedness property. To this end, we first define:

$$\tilde{\theta}_i = \hat{\theta}_i - \theta_i^*, i = 1, \ldots, n \tag{23}$$

$$y_j = z_j - \overline{x}_j, j = 2, \ldots, n \tag{24}$$

Then

$$\dot{y}_j = -\frac{y_j}{\epsilon_j} + B_j \tag{25}$$

where $B_j = -\dot{\overline{x}}_j$.

Proof: We define the following Lyapunov function candidate

$$V = \frac{1}{2}(\sum_{i=1}^{n} S_i^2 + \sum_{i=1}^{n} \tilde{\theta}_i^T \Gamma_i^{-1} \tilde{\theta}_i + \sum_{i=1}^{n} y_{i+1}^2) \tag{26}$$

$$\dot{V} = \sum_{i=1}^{n}(-k_i S_i^2) + \sum_{i=1}^{n-1}\left(g_i S_i S_{i+1} + g_i S_i y_{i+1} + \delta_i^* S_i\right) + \sum_{i=1}^{n} \tilde{\theta}_i \Gamma_i^{-1}\left(\dot{\hat{\theta}}_i - \Gamma_i \xi_i S_i\right) \\ + \sum_{i=1}^{n-1} y_{n+1} \dot{y}_{n=1} + \delta_n^* S_n \tag{27}$$

Using the following inequality $a^2 + b^2 \geq 2ab$, then

$$g_i S_i S_{i+1} \leq g_i^2 S_i^2 + \frac{S_{i+1}^2}{4}, \quad g_i S_i y_{i+1} \leq g_i^2 S_i^2 + \frac{y_{i+1}^2}{4},$$

$$\delta_i^* S_i \leq S_i^2 + \frac{\delta_i^{*2}}{4}, \delta_n^* S_n \leq S_n^2 + \frac{\delta_n^{*2}}{4} \tag{28}$$

$$\dot{V} \leq \sum_{i=1}^{n} -k_i S_i^2 + \sum_{i=1}^{n-1}\left(2g_i^2 S_i^2 + S_i^2 + \frac{S_{i+1}^2 + y_{i+1}^2 + \delta_i^{*2}}{4}\right) \\ + \sum_{i=1}^{n} \tilde{\theta}_i^T \Gamma_i^{-1}\left(\dot{\hat{\theta}}_i - \Gamma_i \xi_i S_i\right) + \sum_{i=1}^{n-1} y_{n+1} \dot{y}_{n+1} + S_n^2 + \frac{\delta_n^{*2}}{4} \\ = \left(2g_1^2 + 1 - k_1\right) S_1^2 + \sum_{i=2}^{n-1}\left(2g_i^2 + 1 - k_i + \frac{1}{4}\right) S_i^2 + \left(1 + \frac{1}{4} - k_n\right) S_n^2 \\ - \sum_{i=1}^{n} \eta \tilde{\theta}_i^T \hat{\theta}_i + \sum_{i=1}^{n-1} y_{i+1} \dot{y}_{i+1} + \sum_{i=1}^{n} \frac{\delta_i^2}{4} + \sum_{i=1}^{n-1} \frac{y_{i+1}^2}{4} \tag{29}$$

Define $2g_1^2 + 1 - k_1 \le -\alpha_0, 2g_i^2 + 1 - k_i + \dfrac{1}{4} \le -\alpha_0, 1 + \dfrac{1}{4} - k_n \le -\alpha_0$, where $\alpha_0 > 0$ and

$i = 2,\ldots,n-1$. Since $2\tilde{\theta}_i^T \hat{\theta}_i = 2\tilde{\theta}_i^T \left(\tilde{\theta}_i^T + \theta_i^* \right) \ge \left\| \tilde{\theta}_i \right\|^2 - \left\| \theta_i^* \right\|^2$, then

$$\dot{V} \le \sum_{i=1}^{n} \left(-\alpha_0 S_i^2 - \frac{\eta}{2\lambda_{max}\left(\Gamma_i^{-1} \right)} \tilde{\theta}_i^T \Gamma_i^{-1} \tilde{\theta}_i \right) + n \left(\frac{\delta_m^2}{4} + \frac{\eta}{2} \left\| \theta_m \right\|^2 \right)$$

$$+ \sum_{i=1}^{n-1} \left(\frac{1}{4} y_{i+1}^2 - \frac{y_{i+1}^2}{\in_{i+1}} + \left| B_{i+1} y_{i+1} \right| \right) \tag{30}$$

Let $\dfrac{\delta_m^2}{4} + \dfrac{\eta}{2} \left\| \theta_m \right\|^2 = e_M, \eta = 2\lambda_{max}\left(\Gamma_i^{-1} \right)\alpha_0, \dfrac{1}{4} - \dfrac{1}{\in_{i+1}} + \dfrac{M_{i+1}^2}{2T} = -\alpha_0$, and mentioning the

inequation $B_{i+1} y_{i+1} \le \dfrac{y_{i+1}^2 B_{i+1}^2}{2T} + \dfrac{T}{2} < \dfrac{y_{i+1}^2}{2T} M_{i+1}^2 + \dfrac{T}{2}$, with $T > 0$. Then

$$\dot{V} \le \sum_{i=1}^{n} \left(-\alpha_0 S_i^2 - \alpha_0 \tilde{\theta}_i^T \Gamma_i^{-1} \tilde{\theta}_i \right) + ne_M + \frac{(n-1)}{2} T + \sum_{i=1}^{n-1} -\alpha_0 y_{i+1}^2$$

$$= -2\alpha_0 V + ne_M + \frac{(n-1)}{2} T \tag{31}$$

Now, let $\alpha_0 > \left(ne_M + \dfrac{(n-1)}{2} T \right) \Big/ 2V$,then one can conclude $\dot{V} < 0$.

5   Simulation Studies

In the simulation studies, the master system is chosen as the chaotic Lorenz system, and the slave system is designed as the same Lorenz system as the master system except that systems parameters are different.

Consider the Lorenz system as the master system described as:

$$\begin{aligned} \dot{x}_{d1} &= a\left(x_{d2} - x_{d1} \right) \\ \dot{x}_{d2} &= bx_{d1} - x_{d1}x_{d3} - x_{d2} \\ \dot{x}_{d3} &= x_{d1}x_{d2} - cx_{d3} \end{aligned} \tag{32}$$

The slave system is chosen as:

$$\begin{aligned} \dot{x}_1 &= d\left(x_2 - x_1 \right) \\ \dot{x}_2 &= ex_1 - x_1x_3 - x_2 \\ \dot{x}_3 &= x_1x_2 - fx_3 \end{aligned} \tag{33}$$

where $a = 2, b = 1, c = 2, d = e = f = 1$.

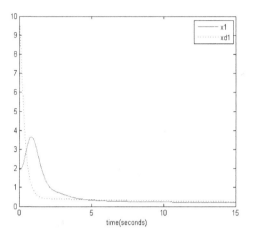

Fig. 1. $x_1(t)$ and $x_{d1}(t)$

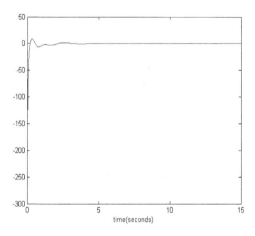

Fig. 2. boundedness of the control u

6   Conclusion

An approach for adaptive synchronization of uncertain chaotic systems via neural network has been presented in this paper. The adaptive control law has been designed by utilizing the dynamic surface control method, which can overcome the "explosion of complexity" of the traditional backstepping by introducing a first-order filtering. The simulation examples show that global stability and asymptotic synchronization can be achieved. This method is suitable for adaptive synchronization of uncertain chaotic systems in the chaos research literature.

References

1. Ott, E., Grebogi, C., Yorke, A.: Controlling chaos. Physical Review Letters 64, 1196–1199 (1990)
2. Pecora, L.M., Carroll, T.L.: Synchronization in chaotic systems. Physical Review Letters 64, 821–824 (1990)
3. Krstic, M., Kanellakopoulos, I., Kokotovic, P.V.: Nonlinear and Adaptive Control Design. Wiley, New York (1995)
4. Mascolo, S., Grassi, G.: Controlling chaotic dynamics using backstepping design with application to the Lorenz system and chua's circuit. Int. J. Bifurcation and Chaos 9, 1425–1434 (1999)
5. Idowu, B.A., Vincent, V.E., Njah, A.N.: Control and synchronization of chaos in nonlinear gyros via backstepping design. Int. J. of Nonlinear Science 5, 11–19 (2008)
6. Wang, C., Ge, S.S.: Adaptive synchronization of uncertain chaotic systems via backstepping design. Chaos Solitons and Fractals 12, 1199–1206 (2001)
7. Bowong, S., Kakmeni, F.M.: Synchronization of uncertain chaotic system via backstepping approach. Chaos Solitons and Fractals 21, 999–1011 (2004)
8. Wang, C., Ge, S.S.: Adaptive backstepping control of uncertain Lorenz system. Int. J. of Bifurcation and Chaos 11, 1115–1119 (2001)
9. Bai, E.W., Lonngren, K.E.: Synchronization of two lorenz systems using active control. Chaos Solitons and Fractals 8, 51–58 (1997)
10. Vincent, V.E.: Chaos synchronization using active control and backstepping control. Nonlinear Analysis: Modeling and Control 13, 253–261 (2008)
11. Yassen, M.T.: Controlling, synchronization and tracking chaotic liu system using active backstepping design. Physics Letters A 360, 582–587 (2007)
12. Zhang, H., Ma, X.K., Yang, Y., Xu, C.D.: Generalized synchronization of hyperchaos and chaos using active backstepping design. Chinese Physics 14, 1009–1963 (2005)
13. Chen, C.L., Yan, H.T., Peng, C.C.: Design of backstepping sliding mode controller for control of chaotic systems. Int. J. of Nonlinear Sciences and Numerical Simulation 8, 137–146 (2007)
14. Swaroop, D., Hedrick, J.K., Yip, P.P., Gerdes, J.C.: Dynamic surface control for a class of nonlinear systems. IEEE Trans. Autom. Control 45, 1893–1899 (2000)
15. Wang, D., Huang, J.: Neural network-based adaptive dynamic surface control for a class of uncertain nonlinear systems in strict-feedback form. IEEE Trans. Neural Networks 16, 195–202 (2005)
16. Han, S.I., Lee, J.M.: Adaptive fuzzy backstepping dynamic surface control for output-constrained non-smooth nonlinear dynamic system. International Journal of Control, Automation, and Systems 10, 684–696 (2012)
17. Sanner, R.M., Slotine, J.E.: Gaussian networks for direct adaptive control. IEEE Trans. Neural Netw. 3, 837–863 (1992)
18. Li, Y.H., Qiang, S., Zhuang, X.Y., Kaynak, O.: Robust and adaptive backstepping control for nonlinear systems using RBF neural networks. IEEE Trans. Neural Networks 15, 693–701 (2004)
19. Ge, S.S., Wang, C.: Direct adaptive nn control of a class of nonlinear systems. IEEE Trans. Neural Networks 13, 214–221 (2002)

Adaptive NN Dynamic Surface Control
for Stochastic Nonlinear Strict-Feedback Systems[*]

Zifu Li[1,2], Tieshan Li[1,*], and Xiaori Gao[1]

[1] Navigational College, Dalian Maritime University, Dalian 116026, China
[2] Navigation college, Jimei University, Xiamen, 361021, China
tieshanli@126.com

Abstract. Based on the dynamic surface control (DSC), an adaptive neural network control approach is proposed for a class of stochastic nonlinear strict-feedback systems in this paper. This approach simplifies the backstepping design and overcomes the problem of 'explosion of complexity' inherent in the backstepping method. The Lyapunov Stability analysis given in this paper shows that the control law can guarantee the solution of the closed-loop system uniformly ultimate boundedness (UUB) in probability. The simulation example is given to illustrate the effectiveness of the proposed control system.

Keywords: adaptive control, neural networks, dynamic surface control, stochastic nonlinear strict-feedback system.

1 Introduction

It is well known that stochastic disturbance often exist in many practical systems. Their existence is the source of instability for the control system. Efforts toward stabilization of stochastic nonlinear systems have been initiated in the work of Florchinger [1]-[6]. Pan and Basar [7] first derived a backstepping design for the stochastic nonlinear strict-feedback system. Since then, many interesting control schemes have been proposed by using the backstepping technical for different stochastic systems. With the Itô differentiation rule Deng and Krstić[8] gave a backstepping design for stochastic strict-feedback system with the form of quartic Lyapunov function.

Recently, the neural networks have been proved to be very useful tools for solving the control problem of uncertain systems. The main advantage is that the unknown nonlinear functions can be approximated by the neural networks. For this purpose, Chen [9] and Li [10] introduced the adaptive neural network control schemes to the output-feedback stochastic nonlinear strict-feedback systems, and designed a controller which was simpler than the existing results. Furthermore, in [11] a novel

---

[*] This work was supported in part by the National Natural Science Foundation of China (Nos.51179019, 60874056), the Natural Science Foundation of Liaoning Province (No. 20102012) , the Program for Liaoning Excellent Talents in University(LNET)(Grant No.LR2012016), and the Applied Basic Research Program of Ministry of Transport of China, the Fundamental Research Funds for the Central Universities (No. 3132013005).

[**] Corresponding author.

C. Guo, Z.-G. Hou, and Z. Zeng (Eds.): ISNN 2013, Part II, LNCS 7952, pp. 112–119, 2013.

direct adaptive neural network controller was proposed to control a class of stochastic system with completely unknown nonlinear functions. However, it still inherited the drawback with the backstepping technique, that is, the problem of 'explosion of complexity', which is caused by the repeated differentiations of certain nonlinear functions. For the purpose of overcoming the drawback of backstepping technique the dynamic surface control (DSC) technique was proposed by introducing a first-order filter of the synthetic input at each step of the traditional backstepping approach in [12]. In [13], by approximating the unknown nonlinear functions with radial basis function (RBF), the dynamic surface control technique was incorporated into the existing neural networks based on adaptive control design framework. In [14], the dynamic surface technique was incorporated into the decentralized control for a class of large-scale interconnected stochastic nonlinear system. Chen [15] combined the DSC technique with the adaptive neural network control technique in the output feedback stochastic nonlinear strict-feedback systems. However, the DSC technique was not incorporated into the state-feedback stochastic nonlinear strict-feedback system.

Motivated by the aforementioned discussion, in this paper, we will incorporate the DSC technique into the adaptive neural network control for the strict-feedback stochastic nonlinear systems. The main contributions lie in the following. It eliminates the problem of 'explosion of complexity'. The Lyapunov stability analysis is simpler than the former approaches.

2   Preliminaries and Problem Formulation

To establish stochastic stability as a preliminary, consider the following nonlinear stochastic system

$$dx = f(x)dt + \psi(x)dw \tag{1}$$

where $x \in R^n$ is the system state, w is an r-dimensional standard Brownian motion defined on the complete probability space (Ω, F, P) with Ω be a sample space, F being a σ-field. $f(x): R^n \rightarrow R^n$, $\Psi(x): R^n \rightarrow R^{n \times r}$ are locally Lipschitz continuous with $f(0) = 0$, $\Psi(0) = 0$.

Assumption 1. In this paper, the following RBF NN is used to approximate the continuous function, such as $h_{nn}(Z) = W^T S(Z)$, where $Z \in \Omega_Z \subset R^q$ is the input vector, weight vector $W = [w_1, w_2, \cdots, w_l]^T \in R^l$, $l > 1$ is the neural networks node number, and $S(Z) = [s_1(Z); \cdots, s_l(Z)]^T$ means the basis function vector, $s_i(Z)$ is the Gaussian function of the form $s_i(Z) = \exp\left[-(z - \mu_i)^T (z - \mu_i) / \varsigma^2\right]$, $i = 1, 2, \cdots, l$, where $\mu_i = [\mu_{i;}, \mu_{i2}, \cdots, \mu_{iq}]^T$ is the center of the receptive field and $\varsigma > 0$ are the width of the basis function.

It has been proven that neural network can approximate any continuous function over a compact set $\Omega_Z \subset R^q$ to arbitrary any accuracy as $h(Z) = W^{*T} S(Z) + \delta(Z)$, where W^* is the ideal constant weight vector and $\delta(Z)$ denotes the approximation error and satisfies $|\delta(Z)| \leq \varepsilon$.

Assumption 2 [17]. There exist constants b_m and b_M such that for $1 \leq i \leq n$, $\forall \bar{x}_i \in R^i$,
$0 < b_m \leq g_i(\bar{x}_i) \leq b_M < \infty$.

Assumption 3 [11]. For simplicity, an unknown constant θ is introduced which is specified as

$$\theta = \max\left\{\left\|W_i^*\right\|^2 / b_m \; ; \; i = 1, 2, \cdots, n\right\} \tag{2}$$

where $\left\|W_i^*\right\|$ denotes the norm of the ideal weight vector of the neural network.

Consider the following stochastic nonlinear strict-feedback system

$$\begin{cases} dx_i = (g(\bar{x}_i)x_{i+1} + f_i(\bar{x}_i))dt + \psi_i dw \;, \; 1 \leq i \leq n-1 \\ dx_n = (g(\bar{x}_i)u + f_i(\bar{x}_i))dt + \psi_i dw \\ y = x_1 \end{cases} \tag{3}$$

where $x = [x_1 ; \cdots , x_n]^T \in R^n$, $u \in R$, and $y \in R$ are the state variable, the control input, and the system output, respectively, $\bar{x}_i = [x_1 ; \cdots , x_i]^T \in R^i$, $f_i(\cdot)$, $g_i(\cdot): R^i \to R$, $\psi_i(\cdot): R^i \to R^r$, $(i = 1, \cdots, n)$ are unknown smooth nonlinear functions with $f_i(0) = 0$, $\psi_i(0) = 0 (1 \leq i \leq n)$ $(i = 1, 2, \cdots, n)$.

Lemma 1 [16]. For any given $V(x) \in C^2 (R^n \to R^+)$, associated with the stochastic differential equation (1) we define the differential operator L as follows:

$$LV(x) = \frac{\partial V}{\partial x} f(x) + \frac{1}{2} Tr\left\{\Psi^T \frac{\partial^2 V}{\partial x^2} \Psi\right\} \tag{4}$$

where $\frac{1}{2} Tr\left\{\Psi^T \frac{\partial^2 V}{\partial x^2} \Psi\right\}$ is called Itô correction term, in which the second-order differential $\partial^2 V / \partial x^2$ makes the controller design much more difficult than that of the deterministic case.

Lemma 2. Consider the stochastic system(1)and assume that $f(x)$, $\Psi(x)$ are C^1 in their arguments and $f(0)$, $g(0)$ are bounded uniformly in t. If their exist functions $V(x) \in C^2 (R^n \to R^+)$, $\mu_1(.)$, $\mu_2(.) \in K_\infty$, two constant $\gamma > 0$ and $\rho > 0$, such that $\mu_1(x) \leq V(x) \leq \mu_2(x)$ and $LV(x) \leq -\gamma V(x) + \rho$ for all $x \in R^n, t > 0$. Then the solution process of the system is bounded in probability.

Lemma 3. For any continuous function $f(x): R^n \to R$, $x = [x_1 ; \cdots , x_n]^T$, $f(0) = 0$, then there exist positive smooth functions $\eta_j(x_j): R \to R^+$, $j = 1, 2, \cdots, n$, such that $|f(x)| \leq \sum_{j=1}^n |x_j| \eta_j(x_j)$.

3   The DSC Controller Design

Step 1: Define the first error surface $z_1 = x_1 - y_r$, where y_r is the desired trajectory. Then we have

$$dz_1 = (g_1 x_2 + f_1 - \dot{y}_r)dt + \psi_1 dw \tag{5}$$

Define the virtual controller α_1 as follows

$$\alpha_1 = -k_1 z_1 - z_1^3 \hat{\theta} S_1^T(Z_1)S_1(Z_1)/2a_1^2 \tag{6}$$

where $k_1 > 0$ is a design constant, a_1 is also a design parameter, $\hat{\theta}$ is the estimated of θ which has been defined in (2). Introduce a new state variable $\bar{\alpha}_1$ and let α_1 pass through a first-order filter with time constant β_1 which will be chosen later to obtain $\bar{\alpha}_1$, which satisfies $\beta_1 \dot{\bar{\alpha}}_1 + \bar{\alpha}_1 = \alpha_1$, $\bar{\alpha}_1(0) = \alpha_1(0)$ .Then define the first filter error p, such as $p_1 = \bar{\alpha}_1 - \alpha_1$. The differential of p_1 is

$$dp_1 = (-(p_1/\beta_1) + B_1(z_1, S_1, p_1, \hat{\theta}, y_r, \dot{y}_r)dt + C_1(z_1, S_1, p_1, \hat{\theta}, y_r, \dot{y}_r)dw \tag{7}$$

where B_1 and $Tr\{C_1^T C_1\}$ are smooth function, which have their maximum denoted by M_1 and N_1 respectively. According Lemma 3 and $\psi_1(0) = 0$, thus there exist a function $\phi_{11}(.)$ such that $\psi_1(x_1) \leq z_1 \phi_{11}(x_1)$. Then, we have $(3/2)z_1^2 \psi_1^T \psi_1 \leq (3/2)z_1^4 \phi_{11}^2$, define a smooth function $\bar{f}_1$, let $\bar{f}_1 = f_1 + p_1 + 1.5 z_1 \phi_{11}^2 + 0.75 g_1^{(4/3)} z_1 - \dot{y}_r$. Then the RBF neural network can be used to approximate the unknown nonlinearities. This will be repeated in later design steps. Define $h_1(Z_1) = g_1^{-1} \bar{f}_1 = W_1^{*T} S_1(Z_1) + \delta_1(Z_1)$, $|\delta_1(Z_1)| \leq \varepsilon_1$, where $Z_1 \triangleq [x_1, \dot{y}_r]^T \subset R^2$, and $\delta_1(Z_1)$ is the approximation error. Then, using the 'Young's inequality', the following results can be gotten [11]

$$z_1^3 g_1 h_1(Z_1) \leq b_m z_1^6 \theta S_1^T(Z_1)S_1(Z_1)/2a_1 + 0.5a_1^2 b_M^2 + 0.75 g_1^{(4/3)} z_1^4 + 0.25 \varepsilon_1^4 \tag{8}$$

Step i $(2 \leq i \leq n-1)$: Define the i th error surface as $z_i = x_i - \bar{\alpha}_{i-1}$, the differential of it is $dz_i = (g_i x_{i+1} + f_i - \dot{\bar{\alpha}}_{i-1})dt + \psi_i(x_i)$, Define the virtual controller α_i as follows

$$\alpha_i = -k_i z_i - \frac{z_i^3 \hat{\theta} S_i^T(Z_i)S_i(Z_i)}{2a_i^2} \tag{9}$$

where $k_i > 0$ is a design constant, a_i is also a design parameter. Introduce a new state variable $\bar{\alpha}_i$ and let α_i pass through a first-order filter with time constant β_i which will be chosen later to obtain $\bar{\alpha}_i$, which satisfies $\beta_i \dot{\bar{\alpha}}_i + \bar{\alpha}_i = \alpha_i$, $\bar{\alpha}_i(0) = \alpha_i(0)$. Then define the first filter error p_i. Such as $p_i = \bar{\alpha}_i - \alpha_i$. According to the differentiation operator mentioned above, the differential of p_i is

$$dp_i = [(-\frac{p_i}{\beta_i}) + B_i(z_i, S_i, p_i, \hat{\theta}, \bar{\alpha}_i, \dot{\bar{\alpha}}_i)]dt + C_i(z_i, S_i, p_i, \hat{\theta}, \bar{\alpha}_i, \dot{\bar{\alpha}}_i)dw \tag{10}$$

Similar to the equation (7) mentioned above, B_i and $Tr\{C_i^T C_i\}$ have their maximum, denoted by M_i and N_i respectively. According to the lemma 3 it follows $\psi_i(x_i) \leq \sum_{l=1}^{i} |x_l| \varphi_{2l} \leq \sum_{l=1}^{i} z_l \varphi_{2l}$. Then we have

$$\frac{3}{2}z_i^3\psi_i^T(x_i)\psi_i(x_i) \leq \frac{3}{2}[iz_i^4\phi_{ii}^2 + r_i^2 + i^2 z_i^4 \frac{(\sum_{l=1}^{i-1} z_l^2\phi_{il}^2)^2}{r_i^2}] \tag{11}$$

Define a smooth function $\bar{f}_i$, let

$$\bar{f}_i = f_i + p_i + \frac{3}{2}[iz_i^4\phi_{ii}^2 + i^2 z_i^4 \frac{(\sum_{l=1}^{i-1} z_l^2\phi_{il}^2)^2}{r_i^2}] + \frac{1}{4}(3g^{(4/3)}z_i + g_{i-1}z_i) - \dot{\bar{\alpha}}_{i-1} \tag{12}$$

Since f_i, g_i, ϕ_{ii} and ϕ_{il} are unknown smooth function. $\dot{\bar{\alpha}}_{i-1}$ is in fact a scalar unknown nonlinear function. RBF neural network can be used to approximate the unknown nonlinearities. This will be repeated in later design steps. Define $h_i(Z_i) = g_i^{-1}\bar{f}_i = W_i^{*T}S_i(Z_i) + \delta_i(Z_i)$, $|\delta_i(Z_i)| \leq \varepsilon_i$, where $\delta_i(Z_i)$ is the approximation error and $Z_i \triangleq [x_1, x_2, \cdots, x_i, \dot{\bar{\alpha}}_{i-1}]^T \subset R^2$. At this step, by exploring the method utilized in (8), one has $z_i^3 g_i h_i(Z_i) \leq b_m z_i^6 S_i^T(Z_i)S_i(Z)/2a_i + b_M^2 g_i^2/2 + (3g_i^{(4/3)}z_i^4 + \varepsilon_i^4)/4$.

Step n: This is the final step. The final control law will be derived in this step. Define the n th error surface as $z_n = x_n - \bar{\alpha}_{n-1}$, the differential of z_n is

$$dz_n = (g_n u + f_n - \dot{\bar{\alpha}}_{n-1})dt + \psi_n(x_n)dw \tag{13}$$

Finally, let the final control u will be as follows

$$u = -k_n z_n - \frac{z_n^3 \hat{\theta}S_n^T(Z_n)S_n(Z_n)}{2a_n^2} \tag{14}$$

Similar to the equation (7), B_n and $Tr\{C_n^T C_n\}$ have their maximum, denoted by M_n and N_n respectively. According to equation $\psi_n(x_n) \leq \sum_{l=1}^n |x_l|\varphi_{2l} \leq \sum_{l=1}^n z_l\phi_{2l}$. Then

$$1.5z_n^3\psi_n^T(x_n)\psi_n(x_n) \leq 1.5[nz_n^4\phi_{nn}^2 + r_n^2 + n^2 z_n^4 (\sum_{l=1}^{n-1} z_l^2\phi_{nl}^2)^2/r_n^2] \tag{15}$$

define a smooth function $\bar{f}_n$, let

$$\bar{f}_n = f_n + p_n + \frac{3}{2}[nz_n^4\phi_{nn}^2 + \frac{1}{r_n^2}n^2 z_n^4(\sum_{l=1}^{n-1} z_l^2\phi_{nl}^2)^2] + \frac{1}{4}(3g_n^{(4/3)}z_n + g_{n-1}z_n) - \dot{\bar{\alpha}}_{n-1} \tag{16}$$

So define a smooth function $h_n(Z_n) = g_n^{-1}\bar{f}_n = W_n^{*T}S_n(Z_n) + \delta_n(Z_n)$, $|\delta_n(Z_n)| \leq \varepsilon_n$ where $Z_i \triangleq [x_1, x_2, \cdots, x_n, \dot{\bar{\alpha}}_{n-1}]^T \subset R^2$, and $\delta_n(Z_n)$ is the approximation error. Then

$$z_n^3 g_n h_n(Z_n) \leq \frac{b_m}{2a_n} z_n^6 S_n^T(Z_n)S_n(Z) + \frac{b_M^2}{2}g_n^2 + \frac{1}{4}(3g_n^{(4/3)}z_n^4 + \varepsilon_n^4) \tag{17}$$

4   Stability Analysis

Theorem 1[14]. Consider the system (5), and the above closed-loop systems, according to lemma 3, for any initial condition satisfying

$$\Pi = \left\{\sum_{i=1}^n z_i^4(0) + \sum_{i=1}^{n-1} p_i^4(0) + b_m\tilde{\theta}^2/2\lambda < M_0\right\} \tag{18}$$

where M_0 is any positive constant, then there exist the control parameters k_i, β_i such that all the signals in the closed-loop system are UUB in forth moment. Moreover, the ultimate boundedness of the above closed-loop signals can be tuned arbitrarily small by choosing suitable design parameters.

Proof Consider the following Lyapunov function candidate

$$V = (\sum_{i=1}^{n} z_i^4 + \sum_{i=1}^{n-1} p_i^4)/4 + b_m \tilde{\theta}^2/2\lambda \tag{19}$$

According to the Itô's differential rules, together with $x_{i+1} = z_{i+1} + p_i + \alpha_i$, and the adaptive law can designed as $\dot{\hat{\theta}} = \sum_{i=1}^{n} z_i^6 \tilde{\theta} S_i^T(Z_i) S_i(Z_i)/2a_i - k_0 \hat{\theta}$, the differential of the above function can be found as follows

$$LV \leq -\sum_{i=1}^{n-1}(k_i - 075.)b_m z_i^4 - k_n b_m z_n^4 - \sum_{i=1}^{n-1}[(1/\beta_i - (4/3)((\delta_i M_i)^{(4/3)} - (\zeta_i N_i)^2)]p_i^4$$
$$- k_0 b_m \tilde{\theta}\hat{\theta}/\lambda + \sum_{i=1}^{n}(a_i^2 b_M^2/2 + \varepsilon_i^4/4) + \sum_{i=1}^{n-1}(1/(2\zeta_i^2) + \delta_i^4/4) + \sum_{i=2}^{n} 3/2r_i^2 \tag{20}$$

where the design parameter $k_0 > 0$. As for the term $k_0 b_m \tilde{\theta}\hat{\theta}/\lambda$, which satisfies $k_0 b_m \tilde{\theta}\hat{\theta}/\lambda \leq -k_0 b_m(\tilde{\theta}^2 - \theta^2)/\lambda$. Then

$$LV \leq -\sum_{i=1}^{n-1}(k_i - 3/4)b_m z_i^4 - k_n b_m z_n^4 - \sum_{i=1}^{n-1}[1/\beta_i - 3(\delta_i M_i)^{(4/3)}/4 - 3(\zeta_i N_i)^2/4]p_i^4$$
$$- k_0 b_m(\tilde{\theta}^2 - \theta^2)/\lambda + \sum_{i=1}^{n}(a_i^2 b_M^2/2 + \varepsilon_i^4/4) + \sum_{i=1}^{n-1}(1/(2\zeta_i^2) + \delta_i^4/4) + \sum_{i=2}^{n} 3r_i^2/2 \tag{21}$$

Choose the design parameters $k_i, a_1, \beta_i, \delta_i, M_i, \zeta_i, N_i, \varepsilon_i, r_i, k_0$ $(i = 1, \cdots, n-1)$, such that $1/\beta_i - 3[(\delta_i M_i)^{(4/3)} - (\zeta_i N_i)^2]/4 = \beta_i^0 > 0$, $(k_i - 3/4)b_m = k_i^0 > 0$, $k_n = k_n^0 > 0$, $k_0 b_m \theta^2/\lambda + \sum_{i=1}^{n}(a_i^2 b_M^2/2 + \varepsilon_i^4/4) + \sum_{i=1}^{n-1}(1/(2\zeta_i^2) + \delta_i^4/4) + \sum_{i=2}^{n} 3r_i^2/2 = \rho$, substituting   the above equations into (21) yields

$$LV \leq -\sum_{i=1}^{n} k_i^0 z_i^4 - \sum_{i=1}^{n-1} \beta_i^0 p_i^4 - k_0 b_m \tilde{\theta}^2/\lambda + \rho \leq -\gamma V + \rho \tag{22}$$

where $\gamma = \min\{4k_i^0, 4\beta_i^0, k_0/2\} > 0$. According to the lemma 2 the above analysis on the closed-loop system means that all the signals in the system (1) are UUB in the sense of probability. This completes the proof.   □

5   Simulation Example

Consider the following third-order stochastic nonlinear system

$$\begin{cases} dx_1 = ((1+x_1^2)x_2 + x_1 \sin(x_1))dt + x_1^3 dw \\ dx_2 = ((1+x_2^2)x_3 - x_2 - 0.5x_2^3 - x_1^3 - \sqrt{x_1})dt + x_1 \cos(x_2)dw \\ dx_3 = ((1.5+\sin(x_1 x_2))u - 0.5x_3 - x_3^3/3 - x_2^2 x_3 - x_1/(1+x_1^2))dt + 3x_1 e^{-x_2^2} dw \\ y = x_1 \end{cases}$$

Based on the DSC controller design proposed in section 3, the virtual control laws and the true control law are designed as follows $\alpha_1 = -k_1 z_1 - z_1^3 \hat{\theta} S_1^T(Z_1)S_1(Z_1)/2a_1^2$ $\alpha_2 = -k_2 z_2 - z_2^3 \hat{\theta} S_2^T(Z_2)S_2(Z_2)2a_2^2$, $u = -k_3 z_3 - z_3^3 \hat{\theta} S_3^T(Z_3)S_3(Z_3)/2a_3^2$, where $Z_1 = z_1$, $Z_2 = [z_1, z_2, \hat{\theta}]^T$, $Z_3 = [z_1, z_2, z_3, \hat{\theta}]^T$, $z_1 = x_1 - y_r$, $z_2 = x_2 - \alpha_1$, the first-order filter $\beta_1 \dot{\bar{\alpha}}_1 + \bar{\alpha}_1 = \alpha_1$, $z_3 = x_3 - \alpha_2$, the second-order filter $\beta_2 \dot{\bar{\alpha}}_2 + \bar{\alpha}_2 = \alpha_2$, $\hat{\theta}$ is the adaptive law, which is designed as $\dot{\hat{\theta}} = \sum_{i=1}^3 z_i^6 \tilde{\theta} S_i^T(Z_i)S_i(Z_i) - k_0 \hat{\theta}/2a_i$.

If we chose the desired trajectory $y_r = \sin(t)$, the suitable parameters were chosen as $k_1 = 20$, $k_2 = 2.5$, $k_3 = 15$, $a_1 = 12$, $a_2 = 10$, $a_3 = 12$, $\lambda = 1.5$, $k_0 = 0.2$, $\beta_1 = 0.002$, $\beta_2 = 0.005$. The initial conditions are given by $[x_1(0), x_2(0), x_3(0)]^T = [0, 0.6, 0.4]^T$, $\hat{\theta}(0) = 0.1$, $\bar{\alpha}_1(0) = -0.3$, $\bar{\alpha}_2(0) = -1.2$. The simulation results are shown in Figs.1~4.

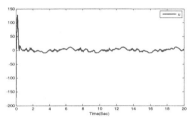

Fig. 1. The control input u

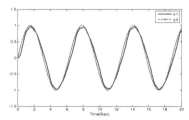

Fig. 2. The output of system

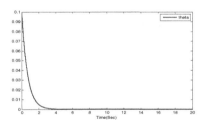

Fig. 3. The adaptive law $\hat{\theta}$

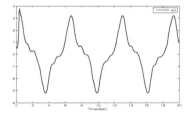

Fig. 4. The state of x_2

6   Conclusion

In this paper, the DSC technique has been proposed for a class of stochastic nonlinear strict-feedback systems. Using the proposed technique we can eliminate the problem of 'explosion of complexity', this stability analysis is simpler than that based on the backstepping method. The closed-loop system has been proved UUB. The effectiveness of the proposed approach has been verified by the simulation example.

References

1. Krstic, M., Kanellakopulos, I., Kocotovic, P.V., Nonlinear and Adaptive Control Design. Wiley, New York (1995)
2. Florchinger, P.: Lyapunov-like techniques for stochastic stability. SIAM J. Contr. Optim. 33, 1151–1169 (1995)
3. Ge, S.S., Wang, C., Direct Adaptive, N.N.: Control of a Class of Nonlinear Systems. IEEE Trans. Neural Networks 13(1), 214–221 (2002)
4. Yip, P.P., Hedrick, J.K.: Adaptive dynamic surface control: A simplified algorithm for adaptive backstepping control of nonlinear system. Int. J. Control. 71(5), 959–979 (1998)
5. Ji, H.B., Chen, Z.F., Xi, H.S.: Adaptive stabilization for stochastic parametric strict-feedback systems with Wiener noise of unknown covariance. International Journal of Systems Science 34(2), 123–127 (2003)
6. Mao, X.: LaSalle-type theorem for stochastic differential delay equations. J. Math. Anal. Appl. 236(2), 350–369 (1999)
7. Pan, Z.G., Basar, T.: Adaptive controller designs for tracking and disturbance attenuation in parametric strict-feedback nonlinear systems. IEEE Trans. Autom. Contro. 43(8), 1066–1083 (1998)
8. Deng, H., Kristic, M.: Stochastic nonlinear stabilization, Part I: a backstepping design. Syst. Control Lett. 32(3), 143–150 (1997)
9. Chen, W.S., Jiao, L.C., Li, J., Li, R.H., Adaptive, N.N.: Backstepping Output-Feedback Control for Stochastic Nonlinear Strict-Feedback Systems With Time-Varying Delays. IEEE Transactions on Systems, Man, and Cybernetics—Part B: Cybernetics 40(3), 939–950 (2010)
10. Li, J., Chen, W.S., Li, J.M., Fang, Y.Q.: Adaptive NN output-feedback stabilization for a class of stochastic nonlinear strict-feedback systems. ISA Transactions 48, 468–475 (2009)
11. Wang, H.Q., Chen, B., Lin, C.: Direct adaptive neural control for strict-feedback stochastic nonlinear systems. Nonlinear Dyn. 67, 2703–2718 (2012)
12. Swaroop, D., Gerdes, J.C., Yip, P.P., Hedrick, J.K.: Dynamic surface control of nonlinear systems. In: Proc. Amer. Control Conf., Albuquerque, NM, pp. 3028–3034 (June 1997)
13. Wang, D., Huang, J.: Neural network-based adaptive dynamic surface control for a class of uncertain nonlinear systems in strict-feedback form. IEEE Transactions on Neural Networks 16(1), 195–202 (2005)
14. Wang, R., Liu, Y.J., Tong, S.C.: Decentralized control of uncertain nonlinear stochastic systems based on DSC. Nonlinear Dyn. 64, 305–314 (2011)
15. Chen, W.S., Jiao, L.C., Du, Z.B.: Output-feedback adaptive dynamic surface control of stochastic non-linear systems using neural network. IET Control Theory Appl. 4(12), 3012–3021 (2010)
16. Krstic, M., Deng, H.: Stabilitization of nonlinear uncertain systems. Springer, London (1998)
17. Wang, Y.C., Zhang, H.G., Wang, Y.Z.: Fuzzy adaptive control of stochastic nonlinear systems with unknown virtual control gain function. Acta Autom. Sin. 32(2), 170–178 (2006)

Cooperative Tracking of Multiple Agents with Uncertain Nonlinear Dynamics and Fixed Time Delays

Rongxin Cui[1,*], Dong Cui[1], and Mou Chen[2]

[1] School of Marine Engineering
Northwestern Polytechnical University
Xi'an 710072, P.R. China
[2] College of Automation Engineering
Nanjing University of Aeronautics and Astronautics
Nanjing 210016, P.R. China
rongxin.cui@gmail.com

Abstract. In this paper, we focus on the cooperative tracking problem of multi-agent systems with nonlinear dynamics and communication time delays. Only a portion of agents can access the information of the desired trajectory and there are communication delays among the agents. Through designing an adaptive neural network based control law and constructing an appropriate Lyapunov-Krasovskii functional, it is proved that the tracking error of each agent converges to a neighborhood of zero. Simulation results are provided to show the effectiveness of the proposed algorithm.

Keywords: cooperative tracking, time delay, multi-agent system, Lyapunov-Krasovskii.

1 Introduction

Multi-agent systems (MAS), including multiple unmanned aerial vehicles (UAVs), autonomous underwater vehicles (AUVs) and unmanned ground vehicles (UGVs), etc., have been widely employed for implementing complex tasks due to their great advantages [1–5]. Various problems about MAS have been intensively investigated in the past few years. Most of these work are focused on the consensus problems, whose tasks are to find out an appropriate control law to make the system reach an agreement satisfying some certain properties.

In practical applications, time delays often occur in the transmission of information among the agents. There have been some works focusing on the consensus problem with time delay [6,7]. The effects of transmission delays in MAS with first-order dynamics have been discussed in [6], which shows the relationship between the eigenvalue of the Laplacian matrix and the stability of the first-order system with time delay. A second order consensus problem in a directed graph with non-uniform time delay is presented in [7]. The transmission delays in bilateral tele-operation systems have been discussed

---

[*] Corresponding author.

C. Guo, Z.-G. Hou, and Z. Zeng (Eds.): ISNN 2013, Part II, LNCS 7952, pp. 120–129, 2013.
© Springer-Verlag Berlin Heidelberg 2013

in [8, 9], which address specifically the stability analysis problem for a class of tele-operation systems. Formation control of a class of UAVs is presented in [10], which proposes different formation control schemes for various formation control situations with delayed communication. The methods in these works are restricted to the control of agents with known dynamics. For the agents with unknown dynamics, an adaptive fuzzy control is designed for the synchronization of nonlinear teleoperators in [11].

In this paper, we consider that the multi-agent system having following features: (i) the agent dynamics are high order with partially unknown dynamics; (ii) communication delays are exists among the agents; (iii) only portion of the agents can access the desired trajectory; and (iv) the leadership of the leader itself is unknown to all the others, and the leader can only affect the agents who can sense the leader. Under these conditions, we aim to illustrate that the system can obtain expected cooperative tracking when there exits time delays among the transmission. The previous work in [12] shows the adaptive neural network based control law is able to make the multiple agents track the desired trajectory in the case of ideal communication. In this work, we extend previous result to the case with constant time delay.

The remainder of the paper is organized as follows. The preliminaries and problem formulation are presented in Section 2 and Section 3, respectively. The control design is presented in Section 4, followed by simulation results in Section 5. Conclusions are drawn in Section 6.

2 Preliminaries

2.1 Agent Dynamics

The nonlinear dynamics of the agents studied in this work are described by

$$
\begin{aligned}
\dot{x}_i &= x_{i+1}, \ i = 1, \ldots, n-1 \\
\dot{x}_n &= f(x) + g(x)(u+d), \\
y &= x_1
\end{aligned}
\tag{1}
$$

where $x = [x_1, \ldots, x_n]^T \in \mathbb{R}^n$ are the generalized state variables of each agent; n shows the state dimension of the system; $y \in \mathbb{R}$ and $u \in \mathbb{R}$ are the system output and input respectively; the smooth function $f(\cdot)$ indicates the unknown dynamics with uncertainties; $g : \mathbb{R}^n \to \mathbb{R}$ is the open loop control gain of the system, which is an unknown function with certain properties, and d represents the external perturbation acting on the input channel.

Assumption 1. *The external disturbance d is uncertain bounded functions $d \in L_\infty$. That is, there exists unknown positive constant ϱ such that $|d(t)| \le \varrho < \infty$, where ϱ can be arbitrarily large.*

Assumption 2. *There exists a smooth function $\bar{g}(x)$ and a positive constant $\underline{g} > 0$, such that $\bar{g}(x) \ge g(x) > \underline{g} > 0$, $\forall(x) \in \mathbb{R}^n$. There exists a positive function $g_0(x)$ satisfying $|\dot{g}(x)/2g(x)| \le g_0(x)$, $\forall(x) \in \mathbb{R}^n$ as well. Without loss of generality, it is further assumed that the function $g(x)$ is positive $\forall(x) \in \mathbb{R}^n$.*

2.2   Neural Network Approximation

In control engineering, neural networks are widely employed to approximate the unknown nonlinear functions [13–16]. In this paper, linearly parameterized NN is used to approximate the continuous function $f_i(Z_i) : \mathbb{R}^q \to \mathbb{R}$ for i-th agent [17]

$$f_i(Z_i) = \theta_i^T \psi_i(Z_i) + \varepsilon_i(Z_i), \tag{2}$$

where the input vector $Z_i^T \in \mathbb{R}^q$, weight vector $\theta_i \in \mathbb{R}^l$, the NN node number $l > 1$, and $\psi_i(Z_i) \in \mathbb{R}^l$. Universal approximation results indicate that, if neural nodes are chosen sufficiently large, $\theta_i^T \psi_i(Z_i)$ can approximate any continuous function over a compact set $Z_i^T \in \Omega_{xi}$ to arbitrary degree of accuracy in the form of $f_i(Z_i) = \theta_i^{*T} \psi_i(Z_i) + \varepsilon_i(Z_i)$, $\forall Z_i \in \Omega_{xi} \subset \mathbb{R}^p$, where θ_i^* are the ideal constant weight vector, and $\varepsilon(Z_i)$ is the approximation error which is bounded over the compact set, i.e., $|\varepsilon_i(Z_i)| \leq \varepsilon_i^*$, $\forall Z_i \in \Omega_{xi}$ where $\varepsilon_i^* > 0$ is an unknown constant. The ideal weight vector θ_i^* is an artificial quantity required for analytical purposes. θ^* is defined as the value of θ_i that minimizes $|\varepsilon_i|$ for all $Z_i \in \Omega_{xi} \subset \mathbb{R}^p$, i.e., $\theta_i^* := \arg\min_{\theta_i \in \mathbb{R}^l} \left\{ \sup_{Z_i \in \Omega_{xi}} |f_i(Z_i) - \theta_i^T \psi_i(Z_i)| \right\}$.

3   Cooperative Tracking of Multiple Agents

In this paper, a weighted directed graph is used to model information exchange among the agents. In addition, we introduce a virtual agent, represented by v_0, and it can obtain the desired trajectory strictly. Then for the N agents network system, its graph $\mathcal{G}$ contains a node set $\mathcal{V} = \{v_0, v_1, \ldots, v_N\}$, and a weighted adjacent matrix $A^* = [a_{ij}^*] \in \mathbb{R}^{(N+1)\times(N+1)}$, where $a_{ij}^* > 0$ indicates that agent i can receive the information from agent j, otherwise $a_{ij} = 0$. Define a diagonal matrix $\Delta(\mathcal{G}) \in \mathbb{R}^{(N+1)\times(N+1)}$ with elements $\delta_{ii} = \sum_k a_{ik}$, and the normalized Laplacian of $\mathcal{G}$ as $L = I - A$, where the elements in the normalized adjacent matrix A are defined as $a_{ij} = a_{ij}^*/\delta_{ii}$, $\delta_{ii} \neq 0$, and $a_{ij} = a_{ij}^*$, $\delta_{ii} = 0$. With adding a virtual agent in the system, we call the graph $\mathcal{G}$ the *extended communication graph*. For each agent let $\mathcal{N}_i = \{v_j \in \mathcal{V} | a_{ij} > 0\}$ denotes the neighbor set of v_i.

Assumption 3. *The extended communication graph $\mathcal{G}$ has a spanning tree with the virtual agent as the root, and this virtual agent follows the desired trajectory strictly.*

The cooperative tracking control problem of multi-agent systems with time-delay studied in this work is formulated as follows. For a group of agents, the desired trajectory $y_d(t)$ and its derivations up to n-th order are bounded, and only a portion of the agents can sense the signals. Considering that constant time delay appears in the signal transmission among the agents, we design a control, using the full states of its neighbors and itself, such that the tracking error converges to a neighborhood of zero, i.e., $\lim_{t\to\infty} |y_i(t) - y_d(t)| = \bar{\varepsilon}$, where $\bar{\varepsilon} > 0$. At the same time, all closed-loop signals are to be kept bounded.

The desired trajectory $y_d(t)$ is generated by the following reference model: $\dot{x}_{dj} = x_{dj+1}$, $j = 1, \ldots, n-1$, $\dot{x}_{dn} = f_d(x_d, t)$, with $y_d = x_{d1}$, where $n \geq 2$ is a constant index, $x_d = [x_{d1}, \ldots, x_{dn}]^T \in \mathbb{R}^n$ are the states of reference system, $y_d \in \mathbb{R}$ is the system output.

Assumption 4. *The reference trajectory $y_d(t)$ and its n-th derivatives remain bounded, i.e., $x_d \in \Omega_d \subset \mathbb{R}^n$, $\forall t \geq 0$.*

4   Cooperative Tracking Control Design

In this section, we design the cooperative tracking control for each agent based on its neighbors' states. Since only a portion of the agents can access the information of the desired trajectory, the tracking control is designed based on the relative states with its neighbors. We assume that the information exchange between the i-th and the j-th agent is subject to a constant time delay $\tau \geq 0$.

Define the following error variables for each agent: $z_{i,1} = y_{i,1}(t) - y_{ir}(t)$, $z_{i,2} = \dot{z}_{i,1} = x_{i,2}(t) - \dot{y}_{ir}(t)$, ... $\dot{z}_{i,n} = z_{i,1}^{(n-1)} = x_{i,n-1}(t) - y_{ir}^{(n-1)}(t)$, where $y_{ir}(t) = \sum_{j \in \mathcal{N}_i} a_{ij} y_j(t - \tau)$, $y_{ir}^{(k)}(t) = \sum_{j \in \mathcal{N}_i} a_{ij} y_j^{(k)}(t - \tau)$, $k = 1, \cdots, n$, where a_{ij} is the element of the normalized adjacent matrix A of the extended communication graph $\mathcal{G}$.

For each agent, define a vector $\bar{z}_i = [z_{i,1}, \ldots, z_{i,n}]^T \in \mathbb{R}^n$, and a filtered tracking error $s_i(t) = [\Lambda^T \; 1]\bar{z}_i$, where $\Lambda = [\lambda_1, \ldots, \lambda_{n-1}]^T$ satisfies $p^{n-1} + \lambda_{n-1} p^{n-2} + \ldots + \lambda_1$ is Hurwitz. The first derivative of s_i is written as

$$\dot{s}_i(t) = f_i(x_i, \eta_i) + g_i(u_i + d_i) + [0 \; \Lambda^T]\bar{z}_i - y_{ir}^{(n)}(t). \tag{3}$$

Considering a scalar smooth function $V_{si} = \frac{1}{2g_i} s_i^2(t)$ and a Lyapunov-Krasovskii functional $V_{di} = \frac{\tau k_{iz}}{2} \int_{-\tau}^0 \int_{t+v}^t s_i^2(\sigma) d\sigma dv$, then we have

$$
\begin{aligned}
\dot{V}_{si} + \dot{V}_{di} = &-\left(g_0 + \frac{\dot{g}_i}{2g_i^2}\right) s_i^2(t) + s_i(t)(u_i + d_i) \\
&+ s_i(t) \frac{f_i(x_i, \eta_i) + [0 \; \Lambda^T]\bar{z}_i - y_{ir}^{(n)} + g_i g_0 s_i(t)}{g_i} \\
&+ \frac{\tau k_{iz}}{2}\left[\tau s_i^2 - \int_{t-\tau}^t s_i^2(\sigma) d\sigma\right].
\end{aligned}
$$

Due to the existence of the uncertain items, we use linearly parameterized NN to approximate the unknown nonlinear function $\bar{f}_i(x_i, \eta_i, \bar{z}_i) = \{f_i(x_i, \eta_i) + [0 \; \Lambda^T]\bar{z}_i - y_{ir}^{(\rho)}(t) + g_i g_0 s_i(t)\}/g_i$, which can be described as $\bar{f}_i(Z_i) = \theta_i^{*T} \varphi_i(Z_i) + \bar{\varepsilon}_i$, where θ^* is the ideal weighted vector, and $Z_i = [x_i, \eta_i, \bar{z}_i]^T$.

Considering the Lyapunov function candidate

$$V_i = V_{si} + V_{di} + \frac{1}{2\gamma_2}\tilde{\theta}_i^T \tilde{\theta}_i + \frac{1}{2\gamma_1}\tilde{\varphi}_i^2, \tag{4}$$

where γ_1 and γ_2 are the positive constants, $\tilde{\theta}_i = \hat{\theta}_i - \theta_i^*$, and $\tilde{\varphi}_i = \hat{\varphi}_i - \varphi_i^*$ are the estimated errors of parameters and the error bound, where $\hat{\theta}_i$ and $\hat{\varphi}_i$ are the estimation of θ_i^* and $\varphi_i^* = (\varrho_i + \bar{\varphi}_i)^2$ respectively. Then we have

$$
\begin{aligned}
\dot{V}_i &= -\frac{\dot{g}_i}{2g_i^2}s_i^2(t) + \frac{1}{g_i}s_i(t)\dot{s}_i(t) + \frac{1}{2\gamma_2}\tilde{\theta}_i^T\dot{\tilde{\theta}}_i + \frac{1}{2\gamma_1}\tilde{\varphi}_i\dot{\tilde{\varphi}}_i \\
&\quad + \frac{\tau k_{iz}}{2}\left[\tau s_i^2 - \int_{t-\tau}^t s_i^2(\sigma)d\sigma\right] \\
&= -\left(g_0 + \frac{\dot{g}_i}{2g_i^2}\right)s_i^2(t) + s_i(t)(u_i + d_i) + s_i(t)[\theta_i^{*T}\Psi_i(Z_i) + \bar{\varepsilon}_i] \\
&\quad + \frac{1}{2\gamma_2}\tilde{\theta}_i^T\tilde{\theta}_i + \frac{1}{2\gamma_1}\tilde{\varphi}_i^2 + \frac{\tau k_{iz}}{2}\left[\tau s_i^2 - \int_{t-\tau}^t s_i^2(\sigma)d\sigma\right].
\end{aligned}
\tag{5}
$$

The NN is constructed to approximate $\bar{f}_i(x_i, \eta_i, \bar{z}_i, y_{ir}^{(n)}) = \{f_i(x_i, \eta_i) + [0 \ \Lambda^T]\bar{z}_i - y_{ir}^{(n)}(t) + g_i g_0 s_i(t)\}/g_i$ on a whole, which avoids the possible singularity of the direct approximation of g_i. Select the following control u_i for each agent

$$
u_i = -\hat{\theta}_i^T\psi_i - k_{iz}\int_{t-\tau}^t s_i(\sigma)d\sigma - \frac{1}{2}\hat{\varphi}_i s_i - k_i s_i, \quad i = 1, \ldots, N.
\tag{6}
$$

The update laws for the parameters are designed as

$$
\begin{aligned}
\dot{\hat{\varphi}}_i &= -\gamma_1\left[-\frac{1}{2}(1 - \varpi_\varphi)s_i^2(t) + \sigma_1\hat{\varphi}_i\right] \\
\dot{\hat{\theta}}_i &= -\gamma_2\left[-\psi_i s_i(t) + \sigma_2\hat{\theta}_i\right],
\end{aligned}
\tag{7}
$$

where $\varpi_{\varphi_i} = 0$ if $|\hat{\varphi}_i| \le M_{\varphi_i}$ with M_{φ_i} is a designed positive constant, or 1 otherwise.

Remark 1. In this work, we assume that the time delay τ is constant and exactly known. Each agent has a register to store the history information received from its neighbors. The cases of unknown time delay or time-varying delay will be studied in future works.

By using Young's inequality, we have $-\sigma_2\tilde{\theta}_i^T\hat{\theta}_i \le -\frac{\sigma_2}{2}\|\tilde{\theta}_i\|^2 + \frac{\sigma_2}{2}\|\theta_i^*\|^2$, $-\sigma_1\tilde{\varphi}_i\hat{\varphi}_i \le -\frac{\sigma_1}{2}\tilde{\varphi}_i^2 + \frac{\sigma_1}{2}\varphi_i^{*2}$, and $(\varrho_i + \bar{\varepsilon}_i)s_i(t) \le \frac{1}{2} + \frac{1}{2}s_i^2\varphi_i^*$ Also Cauchy-Schwars's inequality lead us to write $(\int_{t-\tau}^t s_i(\sigma)d\sigma)^2 \le \tau\int_{t-\tau}^t s_i^2(\sigma)d\sigma$. Considering (6) and (7), the time derivative of V_i can be written as

$$
\begin{aligned}
\dot{V}_i &\le \frac{k_{iz}s_i^2}{2} + \frac{k_{iz}}{2}\left(\int_{t-\tau}^t s_i(\sigma)d\sigma\right)^2 - k_i s_i^2 - \frac{\sigma_1}{2}\tilde{\varphi}_i^2 - \frac{\sigma_2}{2}\|\tilde{\theta}_i\|^2 + c_{2i} \\
&\quad + \frac{\tau k_{iz}}{2}\left[\tau s_i^2 - \int_{t-\tau}^t s_i^2(\sigma)d\sigma\right] \\
&\le \frac{k_{iz}s_i^2}{2} + \frac{k_{iz}\tau}{2}\int_{t-\tau}^t s_i^2(\sigma)d\sigma - k_i s_i^2 - \frac{\sigma_1}{2}\tilde{\varphi}_i^2 - \frac{\sigma_2}{2}\|\tilde{\theta}_i\|^2 + c_{2i} \\
&\quad + \frac{\tau k_{iz}}{2}\left[\tau s_i^2 - \int_{t-\tau}^t s_i^2(\sigma)d\sigma\right] \\
&= -\frac{1}{2}[k_i - k_{iz}(\tau^2 + 1)]s_i^2 - \frac{\sigma_1}{2}\tilde{\varphi}_i^2 - \frac{\sigma_2}{2}\|\tilde{\theta}_i\|^2 + c_{2i},
\end{aligned}
\tag{8}
$$

where $c_{2i} = \frac{\sigma_2}{2}\|\theta_i^*\|^2 + \frac{\sigma_1}{2}\varphi_i^{*2} + \frac{1}{2}$.

Now choose constant positive parameters k_i, k_{iz}, and let $\lambda_i = \frac{1}{2}[k_i - k_{iz}(\tau^2 + 1)] > 0$. Define

$$\Omega_{si} = \left\{ s_i \,\middle|\, |s_i| \leq \sqrt{c_{2i}/\lambda_i} \right\},$$

$$\Omega_{\theta_i} = \left\{ (\tilde{\theta}_i, \tilde{\varphi}_i) \,\middle|\, \|\tilde{\theta}_i\| \leq \sqrt{2c_{2i}/\sigma_2}, |\tilde{\varphi}_i| \leq \sqrt{2c_{2i}/\sigma_1} \right\},$$

$$\Omega_{ei} = \left\{ (s_i, \tilde{\theta}_i, \tilde{\varphi}_i) \,\middle|\, \lambda_i s_i^2 + \frac{\sigma_2}{2}\tilde{\theta}_i^T \tilde{\theta}_i + \frac{\sigma_1}{2}\tilde{\varphi}_i^2 \leq c_{2i} \right\}.$$

Since c_{2i}, σ_1, σ_2, k_{id} and k_i are positive constants, we know that Ω_{si}, Ω_{θ_i} and Ω_{ei} are compact sets. Eq.(8) shows that $\dot{V}_i \leq 0$ once the errors are outside of the compact set Ω_{ei}. According to the standard Lyapunov theorem, we conclude that s_i, $\tilde{\theta}_i$, and $\tilde{\varphi}_i$ are bounded. From (8), it can be seen that V_i is strictly negative as long as s_i is outside the compact set Ω_{si}. Therefore, there exists a constant T_1 such that for $t > T_1$, the filtered tracking error s_i converges to Ω_{si}, that is to say, $s_i \leq \beta_{si}(k_i, k_{iz}, \gamma_1, \gamma_2, \sigma_1, \sigma_2, \theta_i^*, \varphi_i^*, \varepsilon_i^*) = \sqrt{c_{2i}/\lambda_i}$.

Define the error between i-th agent and the desired trajectory as $\tilde{y}_i(t) = y_i(t) - y_d(t) = y_i(t) - y_0(t)$, and the auxiliary states of each agent

$$\xi_i(t) = [\Lambda^T \ 1]Y_i, \tag{9}$$

$Y_i = [y_i, y_i^{(1)}, \ldots, y_i^{(n-1)}]^T$. The filtered error is denoted as $\tilde{\xi}_i(t) = \xi_i(t) - \xi_d(t) = \xi_i(t) - \xi_0(t)$. Also, define $\tilde{y} = [\tilde{y}_1, \ldots, \tilde{y}_N]^T$, $\tilde{\xi} = [\tilde{\xi}_1, \ldots, \tilde{\xi}_N]^T$.

Using the fact that $s_i(t) = \xi_i(t) - \sum_{j \in \mathcal{N}_i} a_{ij}\xi_j(t)$, we have $\tilde{\xi}_i = \xi_i - \xi_0 = \sum_{j \in \mathcal{N}_i} a_{ij}\xi_j + s_i - \xi_0$, $i = 1, \ldots, N$, and in the vector form $\tilde{\xi} = A\xi + s - \xi_0\mathbf{1}$, where $\mathbf{1} = [1, \ldots, 1]^T$, $s = [s_0, s_1, \ldots, s_N]^T$, and A is the normalized adjacency matrix of the extended communication graph. Through calculation, we can obtain $\tilde{\xi} = A\tilde{\xi} + s$. Under Assumption 3, we know that L is an invertible matrix according to Theorem 1 in [12], then we have $\tilde{\xi} = L^{-1}s$.

5 Simulation Studies

In this section, a simulation example is presented to demonstrate the effectiveness of the proposed synchronized tracking controller. Consider a network of four agents described in Fig. 1, and the agent dynamics is given by

$$\dot{x}_1 = x_2, \ \dot{x}_2 = x_3$$
$$\dot{x}_3 = 3\cos x_1 - x_3 \sin x_2 + x_2 x_3 + (2 - \sin x_3)(u + d) \tag{10}$$
$$y = x_1$$

The desired trajectory y_d is generated by $y_d = \frac{8}{s^3 + 6s^2 + 12s + 8} y_{\text{ref}}$, where

$$y_{\text{ref}}(t) = \begin{cases} 1, & \text{if } 0 \leq t \leq 10 \\ 0, & \text{if } 10 < t \leq 20 \\ 1, & \text{if } 20 < t \leq 30 \\ 0, & \text{if } t > 30 \end{cases} \tag{11}$$

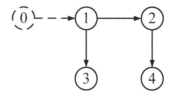

Fig. 1. Extended communication graph in the numerical example

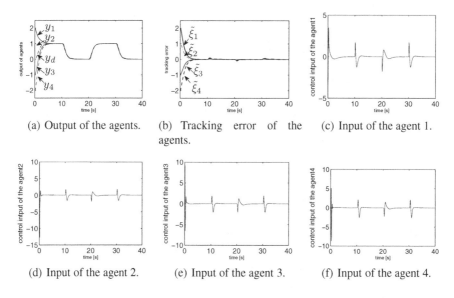

(a) Output of the agents.

(b) Tracking error of the agents.

(c) Input of the agent 1.

(d) Input of the agent 2.

(e) Input of the agent 3.

(f) Input of the agent 4.

Fig. 2. Outputs and tracking errors of each agent when time delay $\tau = 0.05$

In the simulation, both the nonlinear item, $f(x) = 3\cos x_1 - x_3 \sin x_2 + x_2 x_3$, and the open-loop control gain, $g(x) = (2 - \sin x_3)$, are considered as unknown. The external disturbance in the input channel of each agent is generated as $d_i(t) = 0.1 \sin(\frac{\pi t}{2i})$, $i = 1, \ldots, 4$. The control design parameters, and initial conditions are chosen as $\Lambda = [2, 3]^T$, $k_i = 3$, $k_{id} = 2$, $i = 1, \ldots, 4$, $\bar{\gamma}_1 = \bar{\gamma}_2 = 4$, $x_1(0) = [2, 0, 0, 0]^T$, $x_2(0) = [1, 0, 0, 0]^T$, $x_3(0) = [-1, 0, 0, 0]^T$, $x_4(0) = [-2, 0, 0, 0]^T$, $\hat{\theta}_i(0) = 0$, and $\hat{\psi}_i(0) = 0$. The saturation limits of the control are ± 20. In this example, there are 8 inputs of the NN for each agent, $x_{i,k}$, $z_{i,k}$, $k = 1, 2, 3$, and $y_{ir}^{(3)}$, and we employ two nodes for each input dimension of $\theta_i^T \psi(Z_i)$; thus we end up with 256 nodes (i.e., $l = 256$) with centers $\mu_k = 1.0$, $k = 1, \ldots, l$, evenly spaced in $[-3.0, 3.0] \times [-3.0, 3.0] \times [-3.0, 3.0] \times [-3.0, 3.0] \times [-3.0, 3.0] \times [-3.0, 3.0] \times [-3.0, 3.0] \times [-3.0, 3.0]$. The other NN control parameters are chosen as $\sigma_1 = 0.05$, $\gamma_1 = 1$, $\sigma_2 = 9 \times 10^{-4}$, and $\gamma_2 = 10^3$. In the simulation, three cases with different time delays are considered, including 0.05s, 0.2s and 0.8s.

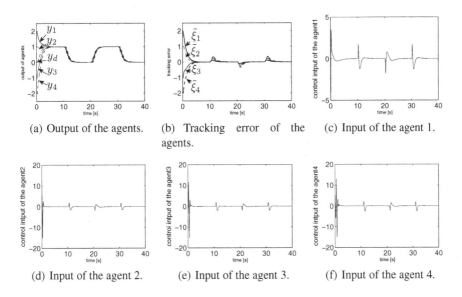

(a) Output of the agents.   (b) Tracking error of the agents.   (c) Input of the agent 1.

(d) Input of the agent 2.   (e) Input of the agent 3.   (f) Input of the agent 4.

Fig. 3. Outputs and tracking errors of each agent when time delay $\tau = 0.2$

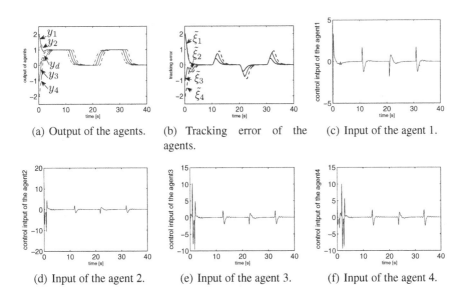

(a) Output of the agents.   (b) Tracking error of the agents.   (c) Input of the agent 1.

(d) Input of the agent 2.   (e) Input of the agent 3.   (f) Input of the agent 4.

Fig. 4. Outputs and tracking errors of each agent when time delay $\tau = 0.8$

Simulation results are shown in Fig. 2– Fig. 4. From these figures, we can find that the output of the agents are bounded. At the same time, the trajectory tracking error of each agent converges to a neighborhood of zero even though the time delay exists.

6   Conclusion

In this work, we have addressed the cooperative tracking problem of multiple agents with constant time delay. Based on the previous work, we design a new adaptive neural network based control and use a Lyapunov-Krasovskii functional to derive the delay-dependent stability criteria. Then we obtain the analytical results that the cooperative tracking of the controlled multi-agent systems with high order dynamics can be realized under the designed controllers. It has been shown that the tracking error of each agent asymptotically converges to an adjustable neighborhood of origin. Simulation results have shown the the the effectiveness of the proposed methods. Future works will focus on the cooperative tracking control problem with unknown and stochastic time delays.

Acknowledgments. This work was supported by the National Natural Science Foundation of China under Grant 51209174, the State Key Laboratory of Robotics and System (HIT) under grant SKLRS-2012-MS-04 and the Fundamental Research Program of Northwestern Polytechnical University (NPU) under Grant JCY20130113.

References

1. Li, Z., Tao, P., Ge, S.S., Adams, M.D., Wijesoma, W.S.: Robust adaptive control of cooperating mobile manipulators with relative motion. IEEE Transactions on System, Man, and Cybernetics, Part B (1), 103–116 (2009)
2. Li, Z., Cao, X., Ding, N.: Adaptive fuzzy control for synchronization of nonlinear teleoperators with stochastic time-varying communication delays. IEEE Transactions on Fuzzy System (4), 745–757 (2011)
3. Li, Z., Li, J., Kang, Y.: Adaptive robust coordinated control of multiple mobile manipulators interacting with rigid environments. Automatica 40(12), 2028–2034 (2010)
4. Yang, C., Ganesh, G., Haddadin, S., Parusel, S., Albu-Schaeffer, A., Burdet, E.: Human-like adaptation of force and impedance in stable and unstable interactions. IEEE Transactions on Robotics (5), 918–930 (2011)
5. Chen, M., Ge, S.S.: Robust attitude control of helicopters with actuator dynamics using nerual networks. In: IET Control Theory & Applications, pp. 2837–2854 (2010)
6. Olfati-Saber, R., Murray, R.M.: Consensus problems in networks of agents with switching topology and time-delays. IEEE Transactions on Automatic Control 49(9), 1520–1533 (2004)
7. Yang, W., Bertozzi, A.L., Wang, X.: Stability of a second order consensus algorithm with time delay. In: Proc. of 47th IEEE Conference on Decision and Control, pp. 2926–2931. IEEE (2008)
8. Hua, C.C., Liu, X.: Delay-dependent stability criteria of teleoperation systems with asymmetric time-varying delays. IEEE Transactions on Robotics 26(5), 925–932 (2010)
9. Chopra, N., Spong, M.W., Lozano, R.: Synchronization of bilateral teleoperators with time delay. Automatica 44(8), 2142–2148 (2008)

10. Abdessameud, A., Tayebi, A.: Formation control of vtol unmanned aerial vehicles with communication delays. Automatica 47(11), 2383–2394 (2011)
11. Li, Z., Cao, X.Q., Ding, N.: Adaptive fuzzy control for synchronization of nonlinear teleoperators with stochastic time-varying communication delays. IEEE Transactions on Fuzzy Systems 19(4), 745–757 (2011)
12. Cui, R., Ren, B., Ge, S.: Synchronised tracking control of multi-agent system with high order dynamics. IET Control Theory & Applications 6(5), 603–614 (2012)
13. Liu, Y.J., Chen, C.L.P., Wen, G.X., Tong, S.: Adaptive neural output feedback tracking control for a class of uncertain discrete-time nonlinear systems. IEEE Transactions on Neural Networks 22(7), 1162–1167 (2011)
14. Yang, C., Ge, S.S., Lee, T.H.: Output feedback adaptive control of a class of nonlinear discrete-time systems with unknown control directions. Automatica 45(1), 270–276 (2009)
15. Yang, C., Ge, S.S., Xiang, C., Chai, T., Lee, T.H.: Output feedback nn control for two classes of discrete-time systems with unknown control directions in a unified approach. IEEE Transactions on Neural Networks 19(11), 1873–1886 (2008)
16. Chen, M., Ge, S.S., Ren, B.: Adaptive tracking control of uncertain mimo nonlinear systems with input constraints. Automatica, 452–465 (2011)
17. Ge, S.S., Hang, C.C., Lee, T.T., Zhang, T.: Stable Adaptive Neural Network Control. Kluwer Academic, Boston (2002)

Adaptation Phase-Locked Loop Speed and Neuron PI Torque Control of Permanent Magnet Synchronous Motor*

Zhiqiang Wang[1], Jia Liu[1], and Dongsheng Yang[2]

[1] School of Instrument Science and Opto-electronics Engineering, Beihang University, Beijing 100191, China
[2] School of Information Science and Engineering, Northeastern University, Shenyang 110004, China

Abstract. This paper presents an excellent software phase-locked loop speed control system of permanent magnet synchronous motor (PMSM). A loop-gain adaptation scheme is developed using model reference adaptive system (MRAS) theory to suppress the torque disturbance which effect on motor speed. The following three points including accurate steady-state speed, fast transient response, and insensitivity to disturbance are especially important for speed control of permanent magnet synchronous motor. The software phase-locked loop (SPLL) technique has the significant ability to obtain precise speed regulation. When the feedback signal of the motor speed is synchronized with a reference signal, perfect speed regulation can be realized. The steady-state accuracy is about 0.02%~0.1% which is difficult to be obtained by conventional proportion integral differentiation (PID) speed control. But phase-locked loop system suffers from pool dynamics and limited lock range. The gain of SPLL has great effect on the performance of system. As the loop gain becomes larger, both the maximum speed error and load increase. If the loop gain varies according to the values of phase error and speed error, the SPLL system will be adaptive between the accuracy and sensitivity to load disturbance. A model reference adaptive system is designed to confine the transient phase error within the range of $[-2\pi\ 2\pi]$ at the present of torque disturbance. This means that the SPLL remains phase tracking. Also, in order to overcome the time varying and nonlinear of PMSM, and obtain the stable torque output, it is effective to utilize the neuron to seek the optimum controller parameter on line. Experiment results are presented to verify the validity of the proposed system.

Keywords: software phase-locked loop, model reference adaptive, single neuron control, permanent magnet synchronous motor.

1 Introduction

Phase-locked loop scheme has been widely used in control strategies of accurate motor speed control for its excellent characteristics, such as high steady state angular velocity, insensitivity to noise. A phase-locked regulator system, comprised of three

---

* Project Supported by National Natural Science Foundation of China.

C. Guo, Z.-G. Hou, and Z. Zeng (Eds.): ISNN 2013, Part II, LNCS 7952, pp. 130–138, 2013.

major functional subsystems: frequency/phase detector (PFD), loop filter (LF), PMSM and encoder.

The following three points are especially important speed control of PMSM: accurate steady-state speed; fast transient response; insensitivity to disturbance.

When the feedback signal of the motor speed is synchronized with a reference signal, perfect speed regulation can be realized, about 0.02%~0.1% of the steady-state accuracy which is difficult to be obtained by conventional PID speed control[1]. But phase-locked loop system suffers from pool dynamics and limited lock range. Furthermore, the motor speed is sensitive to torque disturbances[2]. In [3] and [4] presented dual-mode speed control, which includes PLL control and speed closed-loop control, is an efficient method to realize accurate speed control and perfect transient performance, but it is still a possibility that PLL may be out of lock at the presence of disturbance. In [5] and [6] a disturbance observer is designed to estimate disturbance torque. However, the mechanical parameters of motor must be known[7-9].

In the paper, adaptive SPLL speed control using MRAS theory is employed while the control of motor torque is designed by single neuron proportion integral PI control. The SPLL can be adapted to suppress the effect of torque disturbance on PMSM speed. The arithmetic does not need motor parameters; what's more, it is easy to be realized. Phase-locked steady speed control algorithm was realized with PMSM discrete control system based on DSP controller, which was applied to high-speed permanent magnet synchronous motor. Experiment results are presented to verify the validity of the proposed system.

2   Adaptive SPLL Control

The control system is shown in Fig.1. This system is roughly classified into two parts: the single neuron PI torque controller and the adaptive SPLL speed control. If the error between the speed command and the feedback motor speed is larger than a preset value, the single neuron PI torque controller operates in order to reduce the speed error quickly. When the error is smaller than preset value, system is switched to adaptive SPLL dual-closed loop control. An adaptation is used to suppress the effect of torque disturbance on motor speed.

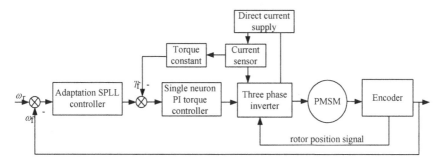

Fig. 1. Block diagram of the control system

2.1   SPLL Operation

When controller operates at SPLL mode, the block diagram of the drive system is shown in Fig.2. It comprises a phase-frequency detector (PFD), a loop filter (LF), loop-gain adaptation, single neuron PI torque controller, a high-speed PMSM and encoder. The combination of controller, PMSM and encoder woks as a voltage-controlled oscillator (VCO) in SPLL.

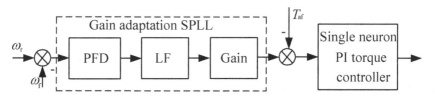

Fig. 2. Block diagram of SPLL motor speed regulations system

An adaptive SPLL controller in operation the transient phase error within the range of $[-2\pi\ 2\pi]$, in another case single neuron PI torque controller is operated.

2.2   SPLL Stability Analysis

Since the PFD always operates under the locked or small frequency error condition, its continuous model can be represented by $G_{PFD}(s)= k_d/s$, where k_d is the gain of PFD. In case a lead-lag filter is used, its transfer function is represented as $G_{LF}(s)=(\tau_d s+1)/(\tau_f s+1)$, where τ_d and τ_f is time constants. It has been indicated that the transient process of motor torque can be omitted due to the moment of inertia of the motor if the torque response is fast enough. The model of the PMSM can then be represented as $G_{PMSM}(s) = K_T/(Js+K_D)$, where J is the moment of inertia, K_T is a torque constant, K_D is the damp constant, the gain of loop K_A. The closed-loop transfer function of the system shown in Figure.2 is given by

$$W(s) = \frac{K_A k_d K_T s + K_A k_d K_T \tau_d}{\tau_f J s^3 + (K_D \tau_f + J)s^2 + (K_D + K_A k_d K_T)s + K_A k_d K_T \tau_d} \tag{1}$$

Then, the state equations of system is given by

$$\dot{X} = \begin{bmatrix} 0 & 1 & 1 \\ 0 & 0 & 1 \\ \dfrac{K_A k_d K_T \tau_d}{\tau_f J} & -\dfrac{K_D + K_A k_d K_T}{\tau_f J} & -\dfrac{K_D \tau_f + J}{\tau_f J} \end{bmatrix} X + \begin{bmatrix} 0 \\ 0 \\ 1 \end{bmatrix} u \tag{2}$$

$$Y = \begin{bmatrix} \dfrac{K_A k_d K_T \tau_d}{\tau_f J} & \dfrac{K_A k_d K_T}{\tau_f J} & 0 \end{bmatrix} X \tag{3}$$

By characteristic equation of matrix A

$$\det\left[\lambda I - A\right] = \begin{bmatrix} \lambda & -1 & -1 \\ 0 & \lambda & -1 \\ \dfrac{K_A k_d K_T \tau_d}{\tau_f J} & \dfrac{K_D + K_A k_d K_T}{\tau_f J} & \dfrac{K_D \tau_f + J}{\tau_f J} \end{bmatrix} = 0 \qquad (4)$$

It can be proved by Lyapunov method that the system is evolution stable for any K_A $k_d>0$ as long as $\tau_d>\tau_f$, in the range of $[-2\pi\ 2\pi]$.

2.3   Adaptive SPLL for PMSM Drive

The gain of SPLL has great effect on the performance of system. As the loop-gain becomes larger, both the maximum speed error and load ability increase, vice versa. If the loop-gain varies according to the values of phase error and speed error, the SPLL system will be adaptive between the accuracy and sensitivity to torque disturbance. The loop-gain will increase as the disturbance torque simultaneously. Thus, torque controller can provide enough torque command, and SPLL can be locked again quickly. Under even worse condition, the SPLL containing loop-gain adaptation can maintain locked at the torque which is too large for SPLL to be locked with original loop-gain.

A MRAS is designed to confine the transient phase error within the range of $[-2\pi\ 2\pi]$ at the present of torque disturbance. This means that the SPLL remains phase tracking and the linear model of PFD is still valid. The block diagram of the MRAS is shown in Figure 3.

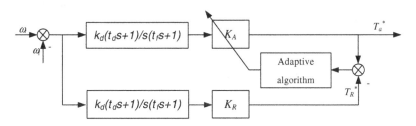

Fig. 3. Block diagram of the MRAS

The adjustable system is represented by

$$T_a^* + \tau_f T_a^{*'} = k_d K_A \theta_e + k_d \tau_d \omega_e \qquad (5)$$

where K_A is the gain to be adapted, θ_e is the phase error.

Reference model is designed as

$$T_R^* + \tau_f T_R^{*'} = k_d K_R \varphi_e + k_d \tau_d \omega_e \qquad (6)$$

where φ_e is the desired phase error which is a positive constant independent on the load torque. K_R is the gain of reference model. Reference model (6) expresses the desired performance of the system, i.e. the torque command must vary simultaneously according to the change of load torque. For the conventional PLL system, it changes phase error while the loop-gain maintains invariable. However, the reference model adapts its loop-gain K_R to provide the needed torque command while the phase error maintains unchanged. Although K_R is updated during practical control, it can be treated as a constant within every control period.

In order to derive the adaptive scheme of K_A, Lyapunov's theorem is utilized.

Theorem: Define the following Lyapunov function candidate

$$V = \frac{1}{2}[(T_a^* - T_R^*)^2 + \phi_e^2 (K_A - K_R)^2 / \xi]$$ (7)

where ξ is a positive constant. The time derivative of V becomes (7)

$$V' = -(T_a^* - T_R^*)^2 / \tau_f + K_A K_A' \phi_e (\phi_e - \theta_e) / \xi$$ (8)

During control process, $\theta_e > \varphi_e$ implies that the phase error of adjustable system is larger than desired, therefore K_A should be increased to reduce such phase error, i.e. $K_A' > 0$. On the contrary, $K_A' < 0$ for $\theta_e < \varphi_e$. This results in an negative value of $K_A'(\phi_e - \theta_e)$. Thus $V' < 0$ is satisfied. $T_a^* - T_R^*$ goes to zero according to Lyapunov's theorem.

3   Neuron PI Torque Control

The basic component of neural network is single neuron, which is a kind of practical self-learning function of PI controller. The structure of the controller is shown in Fig.4.

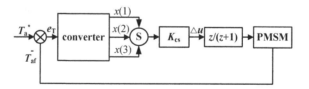

Fig. 4. The structure diagram of single neuron PI controller

In Fig.4, T_a^* and T_{af} is the controller input and output value respectively. And the grey part is the single neuron PI controller. The regulator input, $x_1(k)$, $x_2(k)$, $x_3(k)$, which is the necessary error function to study, is represented integral and proportional coefficient of the neurons. Due to the incremental structure, it can get functions as following:

$$\begin{cases} x_1(k) = T_a^*(k) - T_{af}(k) = e(k) \\ \quad x_2(k) = e(k) - e(k-1) \\ x_3(k) = e(k) - 2e(k-1) + e(k-2) \end{cases} \quad (9)$$

Through the adjustment of the weight value coefficient, the neurons can realize self learning function. And control algorithm is as follows,

$$u(k) = u(k-1) + K_{CS} \sum_{i=1}^{2} \overline{w_i(k)} x_i(k) \quad (10)$$

$$\overline{w_i(k)} = w_i(k) / \sum_{i=1}^{2} |w_j(k)| \quad 11)$$

where $w_i(k) = (i=1, 2)$ is the integral and proportional weight value, K_{CS} is the control parameters gain of the single neuron PI controller. Usually, self learning of single neuron PI uses the algorithm of supervised Hebb learning method [10-12]. The specific learning algorithm of single neuron PI weights adjustment is as following:

$$\begin{cases} w_1(k) = w_1(k-1) + \eta_1 z(k) u(k-1) x_1(k) \\ w_2(k) = w_2(k-1) + \eta_2 z(k) u(k-1) x_2(k) \end{cases} \quad (12)$$

where η_i ($i = 1, 2$) is the learning rate of integral and proportional, $z(k)$ is the signal of error. In formula (4), $z(k)$ is a teacher signal. The input of the motor and controller input $x_1(k)$, $x_2(k)$, $x_3(k)$ are integral and proportional items of the weights of the excitation input. According to the excitatory input, neurons can make the integral and proportional weights to reduce the error in the direction of the adjustment. Therefore, it can overcome the nonlinear and non-stationary in the course of adjustment of the motor speed control by using the self learning characteristics of neuron. But due to the structure characteristics of the learning algorithm, the integral term weights $w_1(k)$ of the change process is relative with the square of error. When the error appears too big for a long time, this parameter will become oversize, which can affect the learning effect, even emerging the overflow phenomenon.

4   Experiment Results

Control system has been verified by experiment shown in Fig.3. Implementation of the SPLL controller is based on TMS320F2812 digital signal controller. The moment of inertia of PMSM is $J=0.1 \mathrm{Kg \cdot m^2}$, damping constant $K_D=0.001 \mathrm{N \cdot m}/ 1000 \mathrm{r/min}$, torque coefficient $K_T=0.001178 \mathrm{N \cdot m/A}$. The high speed PMSM in 30000r/min steady speed curve in torque disturbance is revealed by Fig.5.

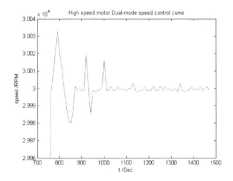

Fig. 5. PMSM 30000r/min steady speed curve in torque disturbance

The process of phase locking is shown in Fig.6 to Fig.8.

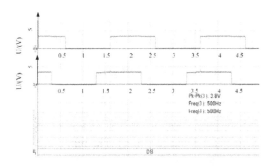

Fig. 6. Difference of output waveform when Feedback clock phase advance reference clock

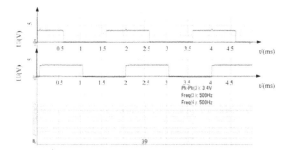

Fig. 7. Difference of output waveform when Feedback clock phase lags behind reference clock

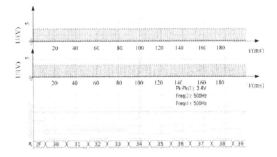

Fig. 8. Linear difference output when feedback clock phase lags behind reference clock

Fig.9 is the experiment curve of raising speed.

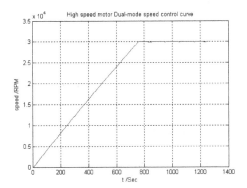

Fig. 9. Experiment curve of raising speed

5   Conclusion

An excellent speed control system for PMSM drives was proposed and verified by experiments in this paper. Firstly, a torque controller was designed by single neuron PI control method to obtain torque characteristic. For the lock range of SPLL motor control system was analyzed to determine the switching point. The stability of the SPLL systems was analyzed. A loop-gain adaptation scheme was developed using MRAS theory to suppress the effect of torque disturbance on PMSM speed. The adaptation scheme is easy to be realized. Secondly, the loop-gain adaptation control system possesses high performance including the precise steady-state speed regulation, quick dynamic response, robustness to the motor parameter variations and insensitivity to the torque disturbance. Finally, experiment results were accorded with theoretical analysis and verified the validity of the proposed system above.

References

[1] Jobe, K., Shoga, M., Koga, R.: A Systems-Oriented Single Event Effects Test Approach for High Speed Digital Phase-Locked Loops. IEEE Transactions on Nuclear Science 40(6), 2868–2873 (1996)

[2] Cheng, S., Tong, H., Silva-Martinez, J.: Steady-State Analysis of Phase-Locked Loops Using Binary Phase Detector. IEEE Transactions on Circuits and Systems—II: Express Briefs 54(6), 474–478 (2007)

[3] Hsu, J.-C., Su, C.: BIST for Measuring Clock Jitter of Charge-Pump Phase-Locked Loops. IEEE Transactions on Instrumentation and Measurement 57(2), 276–285 (2008)

[4] Wagdy, M.F., Cabrales, B.C.: A novel flash fast-locking digital phase-locked loop: design and simulations. IET Circuits Devices Syst. 3(5), 280–290 (2009)

[5] Teplinsky, A., Feely, O.: Phase-Jitter Dynamics of Digital Phase-Locked Loops: Part II. IEEE Transactions on Circuits and Systems—I: Fundamental theory and Applications 47(4), 458–473 (2000)

[6] Skiller, G., Huangp, D.: The Stationary Phase Error Distribution of a Digital Phase-Locked Loop. IEEE Transactions on Communications 48(6), 925–927 (2000)

[7] Loveless, T.D., Massengill, L.W., Bhuva, B.L., et al.: A Hardened-by-Design Technique for RF Digital Phase-Locked Loops. IEEE Transactions on Nuclear Science 53(6), 3432–3438 (2006)

[8] Lee, M., Heidari, M.E., Abidi, A.A.: A Low-Noise Wideband Digital Phase-Locked Loop Based on a Coarse–Fine Time-to-Digital Converter With Subpicose-cond Resolution. IEEE Journal of Solid-State Circuits 44(10), 2808–2816 (2009)

[9] Kim, J.: Adaptive-Bandwidth Phase-Locked Loop With Continuous Background Frequency Calibration. IEEE Transactions on Circuits and Systems-II: Express Briefs 56(3), 205–209 (2009)

[10] Zhao, J., Yang, Z., et al.: Indirect Vector Control Scheme for Linear Induction Motors Using Single Neuron PI Controllers with and without the End Effects, pp. 5263–5267. IEEE (2008)

[11] Wang, X., Hu, M., et al.: Single Neuron PID-PI Composite Speed Controller for Traveling wave Ultrasonic Motor. Micromotors, 19–22 (2008)

[12] Qi, C., Yang, Y.: Single Neuron Controller of VSCF Wind Power Generation System, pp. 493–497. IEEE (2009)

Global Tracking Control of a Wheeled Mobile Robot Using RBF Neural Networks*,**

Jian Wu, Dong Zhao, and Weisheng Chen***

Department of Mathematics, Xidian University, Xi'an 710071, China
wshchen@126.com

Abstract. In this paper, the global tracking control problem for a class of wheeled mobile robots is considered and a new adaptive position tracking control scheme is proposed where radial basis function (RBF) neural network (NN) is utilized to model the uncertainty. The feedback compensation scheme is obtained, where the information of reference position and real position of robot are both used as the NN input. Compered with the existing results, the main advantage is that the global stability of the closed-loop system can be ensured and the NN approximation domain can be determined based on the reference signal a prior. Finally, a simulation example is provided to demonstrate the effectiveness of the proposed control scheme.

Keywords: Wheeled Mobile Robot, Global Tracking, Adaptive Neural Network Control, Determination of Approximation Domain.

1 Introduction

Due to the well-known fact that neural networks (NNs), including radial basis function (RBF) NNs [1], high-order NNs [2], multilayer NNs [3], ect., have good approximation capabilities and adaptation abilities over a compact domain, they paly an important role in the control community, especially in uncertain nonlinear system control. During the past several decades, adaptive neural network control (ANNC) has evolved as a powerful methodology for uncertain nonlinear systems. A large amount of progress in ANNC has been obtained in theory and practical applications, see. e.g., [1,4,5] and the reference therein.

On the other hand, there exist many difficult problems in controlling mobile robots because of the inevitable uncertainties in practical situations. Therefore, ANNC schemes have widely been employed to solving the tracking control problem of mobile robots in the past several years and lots of interesting results have been obtained (e.g., see [6-13]). In [11], the authors investigated asymptotic adaptive NN tracking control of nonholonomic mobile robot formations

---

* This work is supported by National Natural Science Foundation of China (61174213,61203074), the Program for New Century Excellent Talents in University (NCET-10-0665).
** Fundamental Research Funds for the Central Universities (K5051370014).
*** Corresponding author.

C. Guo, Z.-G. Hou, and Z. Zeng (Eds.): ISNN 2013, Part II, LNCS 7952, pp. 139–146, 2013.
© Springer-Verlag Berlin Heidelberg 2013

and proved the global asymptotic stability of the followers. Chaitanya [12] resolved the speed jump problem in the tracking control of a mobile robot using a single layer neural networks (NNs) structure. Moreover, there are other approaches to study the control problem of robotic systems with uncertainties in the existing literature (e.g., see [14],[15]).

However, the above adaptive control has inherent drawbacks, we know NN universal approximation property holds only over a compact set, so recently there are a few results solving the global stability of closed-loop systems (e.g., see [16]). Meanwhile, we can not find appropriate methods to determine the RBF NN approximation domain since it relies on the uncertain position information. Motivated by the aforementioned discusses, we will attempt to solve this problem. First, according to the bound of reference signals, we determine the RBF NN approximation domain, then designing a feedforward compensation scheme and a feedback compensation scheme such that closed-loop systems reach to be globally stable, respectively.

2   Preliminaries and Problem Description

2.1   RBF NN Approximation

NNs have been found to be very useful for controlling nonlinear systems with uncertainties. In particular, RBF NN is usually used as a tool for modeling nonlinear functions due to its simple structure and good approximation ability[14]. Therefore, in this paper, we employ the RBF NN to approximate an unknown continuous function $f(\chi), \chi \in \Omega \subseteq R^l$, where Ω is a compact set, i.e.,

$$f(\chi) = \Phi^{\mathrm{T}}(\chi)W + \epsilon(\chi) \tag{1}$$

where $\Phi(\chi) : \Omega \to R^m$ is a vector-valued function and the neural node number $m > 1$. The components of $\Phi(\chi)$, denoted by $s_k(\chi), 1 \leq k \leq m$, are called the basis functions that are commonly chosen as Gaussian functions with the form

$$s_k(\chi) = \exp\left[-\frac{(\chi - \mu_k)^{\mathrm{T}}(\chi - \mu_k)}{\eta_k^2}\right], \ k = 1, 2, ...m \tag{2}$$

where $\mu_k \in \Omega$ is a constant vector which is called the center of $s_k(\chi)$, and $\eta_k > 0$ is a real number which is called the width of $s_k(\chi)$. The optimal weight vector $W = [w_1 \ ... \ w_m]^{\mathrm{T}}$ is defined as

$$W := \arg\min_{\hat{W} \in R^l} \left\{ \sup_{\chi \in \Omega} |f(\chi) - \Phi^{\mathrm{T}}(\chi)\hat{W}| \right\} \tag{3}$$

and $\epsilon(\chi)$ is the inherent neural networks approximation error, which can be decreased by increasing the neural network number m.

2.2   Main Lemmas

The following four lemmas are critical and usually used to design ANNC scheme and analyze the closed-loop system stability.

Lemma 1 [17]. Young inequality states that if a and b are nonnegative real numbers and p and q are positive real numbers such that $\frac{1}{p} + \frac{1}{q} = 1$, then

$ab \leq \frac{a^p}{p} + \frac{b^q}{q}$, which also gives rise to the so-called Young's inequality with ε (valid for every $\varepsilon > 0$), sometimes called the Peter-Paul inequality: $ab \leq \frac{a^2}{2\varepsilon} + \frac{\varepsilon b^2}{2}$.

Lemma 2 [18]. Let function $V(t) \geq 0$ be a continuous function defined $\forall t \geq 0$ and bounded, and $\dot{V}(t) \leq -\bar{\gamma} V(t) + \bar{\kappa}$, where γ and κ are positive constants, then $V(t) \leq V(0)e^{-\bar{\gamma}t} + \frac{\bar{\kappa}}{\bar{\gamma}}(1 - e^{-\bar{\gamma}t})$.

Lemma 3 [19]. The following inequality holds for any $\varepsilon > 0$ and for any $\eta \in R$ $0 \leq |\eta| - \eta \tanh\left(\frac{\eta}{\varepsilon}\right) \leq \kappa\varepsilon$ where κ is a constant that satisfies $\kappa = e^{-(\kappa+1)}$,i.e.,$\kappa = 0.2785$.

Lemma 4. The function $m(t)$ is a continuous switching function given by

$$m(t) = \begin{cases} 0, & ||t|| < r_a \\ 1, & ||t|| > r_b \\ \frac{||t||-r_a}{r_b-r_a}, & r_a \leq ||t|| \leq r_b \end{cases}$$

with $\Omega_p = \{0 \leq ||t|| \leq r_b\}$ being the neural active region.

2.3   System Description and Problem Formation

In this paper, we consider a car-like wheeled mobile robot shown in Fig.1[20], where P_o denotes the midpoint between the two driving wheels, P_c denotes the

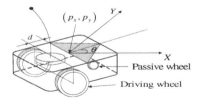

Fig. 1. The two-wheel-driven mobile robot

center of mass of the mobile robot, b is the half of the width of the mobile robot, r is the radius of the wheel, d is the distance from P_o to P_c, m is the mass of the mobile robot, I_0 is the moment of inertia of the body, (x,y) is the coordinates of P_c, and θ is the heading angle of the mobile robot. If we take the center of mass as the robot's position, the dynamic of robot is expressed as

$$M(q)\ddot{q} + V(q,\dot{q})\dot{q} + F(\dot{q}) + \tau_d = J^{\mathrm{T}}(q)\lambda + B(q)\tau \tag{4}$$

where $q = [x\ y\ \theta]^{\mathrm{T}}$ is the general coordinates of a robot moving on a plane, $M(q) \in R^{3\times3}$ is a symmetric positive definite inertial matrix,$V(q,\dot{q}) \in R^{3\times3}$ is the centripetal and coriolis matrix, $F(\dot{q}) \in R^3$ denotes the the surface friction, $J(q) \in R^{3\times1}$ is the matrix associated with the nonholonomic constraints. $B(q) \in R^{3\times2}$ is the input transformation matrix, τ_d is a vector of disturbances including unmodeled dynamics, $\tau \in R^2$ is the torque vector, and λ is the scalar of constraint forces. All these matrices are given by

$$M(q) = \begin{bmatrix} m & 0 & md\sin\theta \\ 0 & m & -md\cos\theta \\ md\sin\theta & -md\cos\theta & I_0 + md^2 \end{bmatrix}, V(q,\dot q) = \begin{bmatrix} 0 & 0 & 0 \\ 0 & 0 & 0 \\ md\dot\theta\cos\theta & md\dot\theta\sin\theta & 0 \end{bmatrix},$$

$$\tau = [\tau_l \ \ \tau_r]^{\mathrm T}, B = \begin{bmatrix} \cos\theta & \cos\theta \\ \sin\theta & \sin\theta \\ b & -b \end{bmatrix}, J(q) = [\sin\theta \ \ -\cos\theta \ \ d].$$

Normally, the nonholonomic constraints can be expressed as $J(q)\dot q = 0$, or,

$$\dot x \sin\theta - \dot y \cos\theta + d\dot\theta = 0. \tag{5}$$

From constraint equations, we know that the mass center is regarded as position. Therefore, from equation (5) we can express the second order derive of θ as:

$$\ddot\theta = \frac{1}{d}(\ddot y\cos\theta - \ddot x\sin\theta) + \frac{1}{2d^2}(\dot x^2 - \dot y^2)\sin2\theta - \frac{1}{d^2}\dot x\dot y(\cos^2\theta - \sin^2\theta). \tag{6}$$

We can easily find a full rank matrix $S(q)$ which is formed by the vectors spanning the null space of constraint matrix $J(q)$ below

$$S(q)^{\mathrm T}J(q)^{\mathrm T} = 0, \tag{7}$$

where $S(q) = \begin{bmatrix} \cos\theta & -d\sin\theta \\ \sin\theta & d\cos\theta \\ 0 & 1 \end{bmatrix}$. Multiplying both sides of equation (4) with $S^{\mathrm T}$ to eliminate nonholomic constraint forces λ, we can rewrite the dynamic equation

$$S^{\mathrm T}M(q)\ddot q + S^{\mathrm T}V(q,\dot q)\dot q + \bar F + \bar\tau_d = S^{\mathrm T}B(q)\tau, \tag{8}$$

where $\bar F(\dot q, \theta) = S^{\mathrm T}F(\dot q)$, $\bar\tau_d = S^{\mathrm T}\tau_d$.

On the left side of (8), the first two terms are rewritten as

$$S^{\mathrm T}M\ddot q + S^{\mathrm T}V\dot q = \begin{bmatrix} m\cos\theta & m\sin\theta \\ -\frac{I}{d}\sin\theta & \frac{I}{d}\cos\theta \end{bmatrix}\begin{bmatrix} \ddot x \\ \ddot y \end{bmatrix}$$
$$+ (\dot x\sin\theta - \dot y\cos\theta)\begin{bmatrix} \frac{m}{d}\sin\theta & -\frac{m}{d}\cos\theta \\ -\frac{I}{d^2}\cos\theta & \frac{I}{d^2}\sin\theta \end{bmatrix}\begin{bmatrix} \dot x \\ \dot y \end{bmatrix} \tag{9}$$

where $I = I_0 + md^2$. Then we define a new coordinate $p = [x\ y]^{\mathrm T}$ which only denotes the position of robot and further define $T = \begin{bmatrix} \cos\theta & \sin\theta \\ -\sin\theta & \cos\theta \end{bmatrix}, M_0 = \begin{bmatrix} m & 0 \\ 0 & \frac{I}{d} \end{bmatrix}$. Therefore, equation (8) can be rewritten as

$$M_0 T\ddot p + M_0\dot T\dot p = S^{\mathrm T}B\tau - \bar F - \bar\tau_d. \tag{10}$$

In the equation (10), the description of robot's dynamics seems to be ignoring the heading angle. In fact, all information of the robot is included in the equation (10). The control objective in this paper is formulated as follows. For a given reference signal p_r whose first and second derivatives are continuous and bounded, design a neural-network-based torque τ such that the tracking error $e(t) = p(t) - p_r(t)$ converges to a neighborhood around the origin whose size can be made arbitrarily small by adjusting the design parameters.

3   Globally Stable ANNC Scheme by Using RBF NN as Feedback Compensator

In this section, we develop a new ANNC scheme. Compered with the conventional ANNC schemes where only the semi-globally stable results can be obtained and the NN approximation domain cannot be determined, the main advantage of this new control strategy is that the global stability of the closed-loop systems can be ensured and the NN approximation domain can be determined based on the reference signal a prior. Define $\tilde{z} = Tz$ where a filtered error z as

$$z = \dot{e} + \lambda e \tag{11}$$

with λ being a positive design parameter. Then, the dynamic (10) can be further transformed into

$$M_0 \dot{\tilde{z}} = S^T B \tau + f(x) - \overline{\tau}_d \tag{12}$$

where $f(x) = [f_1(x), f_2(x)]^T$ is an unknown vector-valued function given by

$$f_1(x) = -m[\cos\theta \ \sin\theta]^T \ddot{p}_r - \frac{m}{d}[\sin\theta \ -\cos\theta]^T \dot{p}[\sin\theta \ -\cos\theta]^T \dot{p}_r, \quad (13)$$

$$f_2(x) = -\frac{I}{d}[-\sin\theta \ \cos\theta]^T \ddot{p}_r - \frac{I}{d^2}[\sin\theta \ -\cos\theta]^T \dot{p}[\cos\theta \ \sin\theta]^T \dot{p}_r, \quad (14)$$

with $x = [\dot{p} \ \theta \ \ddot{p}_r \ \dot{p}_r]^T$.

Next, we employ an RBF NN to approximate the unknown function $f(x)$ directly. To overcome the aforementioned drawbacks of the conventional ANNC schemes, we adopt the switching approach. Firstly, considering p will ultimately track p_r, we use the bound of p_r to determine the domain of p denoted by Ω_p. Furthermore, when p is inside Ω_p, the ANNC scheme works, otherwise, the control scheme will be switched to the robust adaptive scheme. Specifically, the control law is given by

$$\tau = (S^T B)^{-1} \left[-k\tilde{z} - (1 - m(\dot{p}))\Phi^T(x)\hat{W} - m(\dot{p})\left(\hat{\varrho}\tilde{z} + \hat{\nu}\tanh\frac{\tilde{z}}{\varsigma} \right) \right] \tag{15}$$

where $k > 0$ is a design parameter, $\varsigma > 0$ is a constant, $\hat{\varrho}$ and $\hat{\nu}$ denote the estimates of ϱ and ν defined later; $\Phi(x) = diag\{\Phi_1(x), \Phi_2(x)\}$ and $\hat{W}$ denotes the estimate of $W = [W_1^T, W_2^T]^T$. The adaptive laws are chosen as follows

$$\dot{\hat{W}} = \Gamma[(1 - m(\dot{p}))\Phi(x)\tilde{z} - \sigma\hat{W}] \tag{16}$$

$$\dot{\varrho} = \gamma[m(\dot{p})\tilde{z}^2 - \sigma\hat{\varrho}] \tag{17}$$

$$\dot{\nu} = \gamma[m(\dot{p})\tilde{z}^\mathrm{T} \tanh\frac{\tilde{z}}{\varsigma} - \sigma\hat{\nu}] \tag{18}$$

where σ and γ are positive design parameters, $\Gamma > 0$ is an adaptation matrix. Defining $\tilde{W} = W - \hat{W}$, $\tilde{\varrho} = \varrho - \hat{\varrho}$ and $\tilde{\nu} = \nu - \hat{\nu}$, we have

$$\dot{e} = T^{-1}\tilde{z} - \lambda e \tag{19}$$

$$M_0\dot{\tilde{z}} = -k\tilde{z} - m(\dot{p})\left[\hat{\varrho}\tilde{z} + \hat{\nu}\tanh\frac{\tilde{z}}{\varsigma} - f(x_r) - \Delta(x)\right]$$
$$+ (1 - m(\dot{p}))[\Phi^\mathrm{T}(x)\tilde{W} + \epsilon(x)] - \bar{\tau}_d \tag{20}$$

$$\dot{\tilde{W}} = -\Gamma[(1 - m(\dot{p}))\Phi(x)\tilde{z} - \sigma\hat{W}] \tag{21}$$

$$\dot{\tilde{\varrho}} = -\gamma[m(\dot{p})\tilde{z}^2 - \sigma\hat{\varrho}] \tag{22}$$

$$\dot{\tilde{\nu}} = -\gamma\left[m(\dot{p})\tilde{z}^\mathrm{T}\tanh\frac{\tilde{z}}{\varsigma} - \sigma\hat{\nu}\right]. \tag{23}$$

Theorem 1. Consider the closed-loop system consisting of the plant (10), the practical law (15) and the adaptive laws (16)-(18). For bounded initial conditions, all the closed-loop signals are globally uniformly ultimately bounded and the position tracking error $e(t)$ converges to a small neighborhood around the origin by choosing appropriate design parameter.

Proof. Consider the following Lyapunov function

$$V = \frac{1}{2}e^\mathrm{T}e + \frac{1}{2}\tilde{z}^\mathrm{T}M_0\tilde{z} + \frac{1}{2}\tilde{W}^\mathrm{T}\Gamma^{-1}\tilde{W} + \frac{1}{2\gamma}\tilde{\varrho}^2 + \frac{1}{2\gamma}\tilde{\nu}^2. \tag{24}$$

The time derivative of (24) can be calculated as follows

$$\dot{V} = -\lambda e^\mathrm{T}e + e^\mathrm{T}T^{-1}\tilde{z} - k\tilde{z}^\mathrm{T}\tilde{z} - m(\dot{p})\varrho\tilde{z}^\mathrm{T}\tilde{z} + m(\dot{p})\left[\tilde{z}^\mathrm{T}f(x_r) - \tilde{z}^\mathrm{T}\nu\tanh\frac{\tilde{z}}{\varsigma}\right]$$
$$+ m(\dot{p})\tilde{z}^\mathrm{T}\Delta(x) + (1 - m(\dot{p}))\tilde{z}^\mathrm{T}\epsilon(x) - \tilde{z}^\mathrm{T}\bar{\tau}_d + \sigma\tilde{W}^\mathrm{T}\hat{W} + \sigma\tilde{\varrho}\hat{\varrho} + \sigma\tilde{\nu}\hat{\nu}. \tag{25}$$

Using Young's inequality, we easily have

$$e^\mathrm{T}T^{-1}\tilde{z} \le \frac{\lambda}{4}e^\mathrm{T}e + \frac{1}{\lambda}\tilde{z}^\mathrm{T}\tilde{z} \tag{26}$$

$$|\tilde{z}^\mathrm{T}\Delta(x)| \le (\rho|\dot{p}_r| + \rho^2|\dot{p}_r|^2)\lambda\tilde{z}^\mathrm{T}\tilde{z} + \frac{\lambda}{4}e^\mathrm{T}e \tag{27}$$

$$\tilde{z}^\mathrm{T}[\epsilon(x) - \bar{\tau}_d] \le \frac{1}{4}\tilde{z}^\mathrm{T}\tilde{z} + [\epsilon(x) - \bar{\tau}_d]^2. \tag{28}$$

Substituting (26)-(28) into (25) yields

$$\dot{V} \le -\frac{\lambda}{2}e^\mathrm{T}e - \left(k - \frac{1}{2} - \frac{1}{\lambda}\right)\tilde{z}^\mathrm{T}\tilde{z} - \frac{\sigma}{2}\tilde{W}^\mathrm{T}\tilde{W} - \frac{\sigma}{2}\tilde{\varrho}^2 - \frac{\sigma}{2}\tilde{\nu}^2$$
$$+ \left[2\nu c_z\varsigma + \epsilon^2(x) + \bar{\tau}_d^2 + \frac{\sigma}{2}W^\mathrm{T}W + \frac{\sigma}{2}\varrho^2 + \frac{\sigma}{2}\nu^2\right] \le -\bar{\varsigma}V + \bar{\delta} \tag{29}$$

where $c_z = 0.2785, \|f(x_r)\| \leq \nu, \bar{\varsigma} := \min\{\lambda, \sigma/\lambda_{\max}\{\Gamma\}, (2k-1-\frac{2}{\lambda})/\lambda_{\max}\{M_0\}, \sigma\gamma\}, \bar{\delta} = \max\{2\nu c_z\varsigma + \epsilon^2(x) + \bar{\tau}_d^2 + \frac{\sigma}{2}W^TW + \frac{\sigma}{2}\varrho^2 + \frac{\sigma}{2}\nu^2\}, \varrho = \max\{\rho|\dot{p}_r| + \rho^2|\dot{p}_r|^2\}\lambda$. Based on Lemma 2 and inequality (29), it can be easily seen that all the closed-loop signals are bounded and the position tracking error $e(t)$ converges to a small neighborhood around the origin.

4   Simulation Study

In this section, choose the reference signal $x_r(t) = \cos t, y_r(t) = \sin t$, and the design parameters used in this simulation are given as follows: $m = 3.45$ kg, $d = 0.5$ m, $I = 2.59$ kg$\cdot$m$^2, r_a = 1.3, r_b = 1.5, \Gamma = 2, \gamma = 1, \sigma = 0.0001, k = 2, \lambda = 1, \varsigma = 0.1$. NN node numbers $N = 2187$, the initial values of robot's position $x(0) = 5, y(0) = -6$ outside the neural active region. NN input $\chi = [\dot{x}, \dot{y}, \theta, \dot{x}_r, \dot{y}_r, \ddot{x}_r, \ddot{y}_r]^T$. The simulation results are shown in Fig. 2. from which it can be seen that indeed the position tracking error $e(t)$ converges to a small neighborhood around the origin.

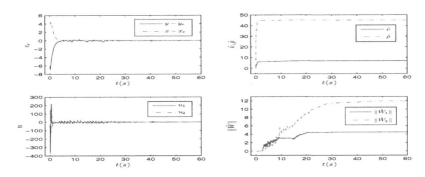

Fig. 2. Simulation curves of the control scheme with NN as feedback compensator

5   Conclusions

In this paper, a new adaptive position tracking control scheme is developed for a class of wheeled mobile robots. The so-called feedback compensation scheme is proposed. The main advantage of the obtained ANNC scheme is that the global stability of the closed-loop systems can be guaranteed and the NN approximation domain also can be determined based on the reference signal a prior.

References

1. Ge, S.S., Hang, C.C., Lee, T.H., Zhang, T.: Stable Adaptive Neural Network Control. Kluwer, Norwell (2001)
2. Kosmatopoulos, E.B., Polycarpou, M.M., Christodoulou, M.A., Ioannou, P.A.: High-order neural network structures for identification of dynamical systems. IEEE Trans. Neural Netw. 6, 422–431 (1995)

3. Lewis, F.L., Jagannathan, S., Yesildirek, A.: Neural Network Control of Robot Manipulators and Nonlinear Systems. Taylor and Francis, New York (1999)
4. White, D.A., Sofge, D.A.: Handbook of Intelligent Control: Neural, Fuzzy, and Adaptive Applications. Van Nostrand and Reinhold, New York (1993)
5. Ge, S.S., Lee, T.H., Harris, C.J.: Adaptive Neural Network Control of Robotic Manipulators. World Scientific, London (1998)
6. Das, T., et al.: Simple neuron-based adaptive controller for a nonholonomic mobile robot including actuator dynamics. Neurocomput. 69, 2140–2151 (2006)
7. Yoo, S.J., et al.: Indirect adaptive control of nonlinear dynamic systems using self recurrent wavelet neural network via adaptive learning rates. Inf. Sci. 177, 3074–3098 (2007)
8. Bugeja, M.K., Fabri, S.G., Camilleri, L.: Dual adaptive dynamic control of mobile robots using neural networks. IEEE Trans. Syst., Man, Cybern. B, Cybern. 39, 129–141 (2009)
9. Park, B.S., Yoo, S.J., Park, J.B., Choi, Y.H.: Adaptive neural sliding mode control of nonholonomic wheeled mobile robots with model uncertainty. IEEE Trans. Control. Syst. Technol. 17, 207–214 (2009)
10. Liu, Y., Li, Y.: Sliding mode adaptive neural-network control for nonholonomic mobile modular manipulators. J. Intel. Robot. Syst. 44, 203–224 (2005)
11. Dierks, T., Jagannathan, S.: Asymptotic Adaptive Neural Network Tracking Control of Nonholonomic Mobile Robot Formations. J. Intel. Robot. Syst. 56, 153–176 (2009)
12. Chaitanya, V.S.: Full-state tracking control of a mobile robot using neural networks. Int. J. Neur. Syst. 15, 403–414 (2005)
13. Fierro, R., Lewis, F.L.: Control of a nonholonomic mobile robot using neural networks. IEEE Trans. Neural Netw. 9, 589–600 (1998)
14. Sousa, C.D., et al.: Adaptive control for mobile robot using wavelet networks. IEEE Trans. Syst. Man, Cybern. B, Cybern. 32, 493–504 (2002)
15. Yoo, S.J., et al.: Adaptive dynamic surface control of flexible-joint robots using self-recurrent wavelet neural networks. IEEE Trans. Syst., Man, Cybern. B, Cybern. 36, 1342–1355 (2006)
16. Chen, W.S., Jiao, L.C., Wu, J.S.: Globally stable adaptive robust tracking control using RBF neural networks as feedforward compensators. Neural Comput. Appl. 21, 351–363 (2012)
17. Krstić, M., Deng, H.: Stabilization of nonlinear uncertain systems. Springer, London (1998)
18. Ioannou, P.A., Sun, J.: Robust adaptive control. Prentice-hall, Englewood Cliffs (1995)
19. Polycarpou, M.M.: Stable adaptive neural control schemes for nonlinear systems. IEEE Trans. Automat. Control. 41, 447–451 (1996)
20. Chen, X., Li, Y.: Smooth formation navigation of multiple mobile robots for avoiding moving obstacles. Int. J. of Contr., Automat. Syst. 4, 466–479 (2006)

Adaptive NN Tracking Control
of Double Inverted Pendulums with Input Saturation

Wenlian Yang[*], Junfeng Wu, Song Yang, and Ye Tao

Educational Technology & Computing Center, Dalian Ocean University, Dalian, P.R. China
lotusyangwl@163.com

Abstract. In this paper, the adaptive control problem with input saturation is investigated for double inverted pendulums. Based on Lyapunov stability theory and backstepping technique, incorporating dynamic surface control (DSC) technique into neural network based adaptive control, an adaptive neural controller is developed by explicitly considering uncertainties, unknown disturbances and input saturation. An auxiliary system is presented to tackle input saturation, and the states of auxiliary design system are utilized to develop the tracking control. It is proved that all the signals in the closed-loop system are uniformly ultimately bounded (UUB) via Lyapunov analysis. Finally, simulation studies are given to demonstrate the effectiveness of the proposed method.

Keywords: double inverted pendulums (DIPs), nerural network, saturation.

1   Introduction

The double inverted pendulums system is the most typical representation of nonlinear control applications which appear in most existing physical systems. Considering its unique characteristics, it is used to verify new control schemes in many control laboratories [1]. The control objective consists of swinging up and balancing it (them) about the vertical [2-3]. However, they are applicable when the system model behaves with complicated certainty or completely known structures only. The research of inverted pendulum (s) has been studied greatly for investigating effectiveness of various kinds of control schemes and demonstrating ideas emerging in the area of nonlinear control [4-6]. In addition, actuator saturation is one of the most important non-smooth nonlinearities which usually appear in the applications. The limitations on the amplitudes of control inputs can cause serious deterioration of control performances and even destroy the stability of the control systems. More recently, the analysis and design of control systems with input saturation nonlinearities have been studied in [7-8]. In [7], an auxiliary design system was introduced to ease the effect of input saturation constraints for vessels.

Motivated by the above observations, based on DSC technique [9] and NN adaptive control, a neural adaptive controller is developed for DIPs with input saturation. The "explosion of complexity" and controller singularity problems are avoided by using of

---

[*] Corresponding author.

C. Guo, Z.-G. Hou, and Z. Zeng (Eds.): ISNN 2013, Part II, LNCS 7952, pp. 147–154, 2013.
© Springer-Verlag Berlin Heidelberg 2013

DSC technique and utilizing a special property of the affine term, respectively. For handling the input saturation, the auxiliary design system is introduced to analyze the effect of input saturation, and states of auxiliary design system are used to design tracking control for DIPs.

2   Problem Formulation and Preliminaries

The DIPs system composed of two interconnected subsystems can be described by the following TITO systems [10]:

$$
\begin{aligned}
&\dot{x}_{1,1}=x_{1,2}, \\
&\dot{x}_{1,2}=\frac{m_1 gr}{J_1}\sin(x_{1,1})-\frac{k}{J_1}x_{1,1}+\frac{u_1}{J_1}+\frac{k}{J_1}x_{2,1}+\frac{v_1}{J_1} \\
&\dot{x}_{2,1}=x_{2,2}, \\
&\dot{x}_{2,2}=\frac{m_2 gr}{J_2}\sin(x_{2,1})-\frac{k}{J_2}x_{2,1}+\frac{u_2}{J_2}+\frac{k}{J_2}x_{1,1}+\frac{v_2}{J_2}
\end{aligned}
\tag{1}
$$

where, each pendulum may be positioned by a torque input u_1, $i=1,2$, applied by a servomotor at its base. v_i, $i=1,2$, are the torque disturbances. It is assumed that both θ_i and $\dot{\theta}_i$ are available to the i th controller for $i=1,2$. $x_{i,1}=\theta_i$ being the angular displacement of the i th pendulums from the vertical reference[10].

The torsional spring is relaxed when the pendulums are all in the upright position. And the origin $x_{1,1}=x_{1,2}=x_{2,1}=x_{2,2}=0$ is the equilibrium point of this system. The objective is to drive the angular position of each pendulum to track a reference signal $y_{i,d}$, $i=1,2$ under the torque disturbances.

Before the beginning of controller design, we convert the system (1) into a more general TITO form as follows

$$
\begin{aligned}
&\dot{x}_{i,1}=f_{i,1}(x_{i,1})+g_{i,1}(x_{i,1})x_{i,2}+\Delta_{i,1}(t,x) \\
&\dot{x}_{i,2}=f_{i,2}(\overline{x}_{i,2})+g_{i,2}(\overline{x}_{i,2})u_i+\Delta_{i,2}(t,x) \\
&y_{i,1}=x_{i,1}, i=1,2
\end{aligned}
\tag{2}
$$

where $\overline{x}_{i,2}=[x_{1,1},x_{1,2},x_{2,1}]^T$, $x=[x_{1,1},x_{1,2},x_{2,1},x_{2,2}]^T$ are vectors of the system states. $y_{i,1}$ is the output of the system. This general system (2) will be used as a design model later. And we assumed that $f_{i,j}$ and $g_{i,j}$ represent unknown nonlinear smooth functions, respectively, where $g_{i,j}$ is referred to as virtual control gain function.

In this paper, considering the presence of input saturation constraints on u_i as follows

$$
u_i=sat(v_i)=\begin{cases}u_{iM}, & if\ v_i>u_{iM} \\ v_i, & if\ -u_{im}\le v_i\le u_{im} \\ -u_{im}, & if\ v_i\le -u_{im}\end{cases}
\tag{3}
$$

where v_i is the designed control input of ith subsystem. u_{im} and u_{iM} are the known lower limit and upper limit of the input saturation constraints of u_i, respectively.

The control objective is to develop an adaptive NN controller such that all solutions of the resulting closed-loop system are SGUUB, and the tracking error $z_{i,1} = y_{i,1}(t) - y_{i,d}(t)$, $i = 1, 2$ can be rendered arbitrary small.

In this paper, the following assumptions are made on the system (2).

Assumption 1. There exist constants $g_{i,j1} \geq g_{i,j0} > 0$ and $g_{i,jd} > 0$, such that $g_{i,j1} \geq g_{i,j} \geq g_{i,j0}$ and $\left| \dot{g}_{i,j} \left(\bar{x}_{i,j} \right) \right| \leq g_{i,jd}$, $\forall \bar{x}_{i,j} \in \Omega \subset R^n$.

Assumption 2. $\left| \Delta_{i,j} \right| \leq d_{i,j}$ with $d_{i,j}$ being some constants.

Assumption 3. The given reference signal $y_{i,d}(t)$ is a sufficiently smooth function of t, $y_{i,d}(t)$, $\dot{y}_{i,d}(t)$ and $\ddot{y}_{i,d}(t)$ are bounded, that is, there exists a known positive constant $B_{i,0}$, such that $\Pi_{i,0} := \{ (y_{i,d}, \dot{y}_{i,d}, \ddot{y}_{i,d}) : y_{i,d}^2 + \dot{y}_{i,d}^2 + \ddot{y}_{i,d}^2 \leq B_{i,0} \}$ with $i = 1, 2$.

In this paper, the following RBF NN is used to approximate the continuous function. such as $h_{nn}(Z) = W^T S(Z)$, where $Z \in \Omega_Z \subset R^q$ is the input vector, weight vector $W = [w_1, w_2 ; \cdots, w_l]^T \in R^l$, $l > 1$ is the neural networks node number, and $S(Z) = [s_1(Z), s_2(Z) ; \cdots, s_l(Z)]^T$ means the basis function vector, $s_i(Z)$ is the Gaussian function of the form as follows

$$s_i(Z) = \exp\left[-(z - \mu_i)^T (z - \mu_i) / \varsigma^2 \right], \quad i = 1, 2, \cdots, l \tag{4}$$

where $\mu_i = [\mu_{i,}, \mu_{i2}, \cdots, \mu_{iq}]^T$ is the center of the receptive field, and $\varsigma > 0$ is the width of the basis function.

It has been proven that neural network can approximate any continuous function over a compact set $\Omega_Z \subset R^q$ to arbitrary any accuracy as

$$h(Z) = \theta^{*T} S(Z) + \delta^* \tag{5}$$

where θ^* is the ideal constant weight vector and δ^* denotes the approximation error.

3   Controller Design and Stability Analysis

3.1   Controller Design

Step 1: Define the error variable $z_{1,1} = x_{1,1} - y_{1,d}$, considering (2), the time derivative of $z_{1,1}$ is

$$\dot{z}_{1,1} = f_{1,1}\left(x_{1,1} \right) + g_{1,1}\left(x_{1,1} \right) x_{1,2} + \Delta_{1,1} - \dot{y}_{1,d} . \tag{6}$$

According to the neural network approximation in (5), we can define

$$h_{1,1}(Z_{1,1}) = \left(f_{1,1}(x_{1,1}) - \dot{y}_{1,d} \right) \big/ g_{1,1} = \theta_{1,1}^{*T} \xi_{1,1}(Z_{1,1}) + \delta_{1,1}^{*} \tag{7}$$

where $Z_1 = \left[x_{1,1}, \dot{y}_{1,d} \right]^{T} \subset R^2$. Now, choose the virtual control as follows

$$\alpha_{1,2} = -c_{1,1} z_{1,1} - \hat{\theta}_{1,1}^{T} \xi_{1,1}(Z_{1,1}) - D_{1,1} \tanh(D_{1,1} z_{1,1} / s_{1,1}) \tag{8}$$

where design parameters $c_{1,1}$, $s_{1,1} > 0$, $D_{1,1} = d_{1,1} / g_{1,10}$, $\hat{\theta}_{1,1}$ is the estimation of $\theta_{1,1}^{*}$ and is updated as follows

$$\dot{\hat{\theta}}_{1,1} = \Gamma_{1,1} \left[\xi_{1,1}(Z_{1,1}) z_{1,1} - \sigma_{1,1} \hat{\theta}_{1,1} \right] \tag{9}$$

where $\Gamma_{1,1} = \Gamma_{1,1}^{T} > 0$, $\sigma_{1,1} > 0$.

Introduce a first-order filter $\beta_{1,2}$ with a time constant $\tau_{1,2}$ as follows

$$\tau_{1,2} \dot{\beta}_{1,2} + \beta_{1,2} = \alpha_{1,2} \quad \beta_{1,2}(0) = \alpha_{1,2}(0) \tag{10}$$

Define $z_{1,2} = x_{1,2} - \beta_{1,2}$, then

$$\dot{z}_{1,1} = f_{1,1}(x_{1,1}) + g_{1,1}(x_{1,1})(z_{1,2} + \beta_{1,2}) + \Delta_{1,1} - \dot{y}_{1,d} \tag{11}$$

And define

$$\eta_{1,2} = \beta_{1,2} - \alpha_{1,2} = \hat{\theta}_{1,1}^{T} \xi_{1,1}(Z_{1,1}) + c_{1,1} z_{1,1} + D_{1,1} \tanh(D_{1,1} z_{1,1} / s_{1,1}) + \beta_{1,2} \tag{12}$$

Substituting (12) and (16) into (15), we have

$$\dot{z}_{1,1} = g_{1,1}(x_{1,1})(z_{1,2} - \tilde{\theta}_{1,1}^{T} \xi_{1,1}(Z_{1,1}) - c_{1,1} z_{1,1} - D_{1,1} \tanh(D_{1,1} z_{1,1} / s_{1,1}) + \delta_{1,1}^{*} + \eta_{1,2}) + \Delta_{1,1} \tag{13}$$

By defining the filter output error $\eta_{1,2} = \beta_{1,2} - \alpha_{1,2}$, it yields $\dot{\beta}_{1,2} = -\eta_{1,2} / \tau_{1,2}$ and

$$\begin{aligned} \dot{\eta}_{1,2} &= -\frac{\eta_{1,2}}{\tau_{1,2}} + \left(-\frac{\partial \alpha_{1,2}}{\partial x_{1,1}} \dot{x}_{1,1} - \frac{\partial \alpha_{1,2}}{\partial z_{1,1}} \dot{z}_{1,1} - \frac{\partial \alpha_{1,2}}{\partial \hat{\theta}_{1,1}} \dot{\hat{\theta}}_{1,1} - \frac{\partial \alpha_{1,2}}{\partial y_{1,d}} \dot{y}_{1,d} - \frac{\partial \alpha_{1,2}}{\partial \dot{y}_{1,d}} \ddot{y}_{1,d} \right) \\ &= -\frac{\eta_{1,2}}{\tau_{1,2}} + B_{1,2}(z_{1,1}, z_{1,2}, \eta_{1,2}, \hat{\theta}_{1,1}, y_{1,d}, \dot{y}_{1,d}, \ddot{y}_{1,d}) \end{aligned} \tag{14}$$

where $B_{1,2}(\cdot)$ is a continuous function and has a maximum value $M_{1,2}$.

Step 2: In this step, the final control law will be derived. Consider

$$\dot{x}_{1,2} = f_{1,2}(\bar{x}_{1,2}) + g_{1,2}(\bar{x}_{1,2}) u_1 + \Delta_{1,2} \tag{15}$$

Similarly, define

$$h_{1,2}(Z_{1,2}) = \left(f_{1,2}(x_{1,2}) - \dot{\beta}_{1,2} \right) \big/ g_{1,2} = \theta_{1,2}^{*T} \xi_{1,2}(Z_{1,2}) + \delta_{1,2}^{*} \tag{16}$$

where $Z_{1,2} \triangleq \left[x_{1,1}, x_{1,2}, x_{2,1}, \dot{\beta}_{1,2} \right]^{T} \subset R^4$.

For convenience of making constraint effect analysis of the input saturation, the following auxiliary design system is given as follows.

$$\dot{e}_1 = \begin{cases} -k_1 e_1 - (f(\cdot)/e_1^2)\cdot e_1 + (u_1 - v_1), & |e_1| \geq \varepsilon_1 \\ 0, & |e_1| < \varepsilon_1 \end{cases} \tag{17}$$

where $f(\cdot) = f(z_{1,2}, \Delta u) = |z_{1,2} \cdot \Delta u_1| + \dfrac{1}{2}\Delta u_1^2$, $k_1 > 0$, $\Delta u_1 = u_1 - v_1$, ε_1 is a small positive design parameter and e_1 is a variable of the auxiliary design system introduced to ease the analysis of the effect of the input saturation.

Considering the input saturation effect, the control law is proposed as

$$v_1 = -c_{1,2}z_{1,2} + e_1 - \hat{\theta}_{1,2}^T \xi_{1,2}(Z_{1,2}) - D_{1,2}\tanh(D_{1,2}z_{1,2}/s_{1,2}) \tag{18}$$

where $c_{1,2}$, $s_{1,2} > 0$, $D_{1,2} = d_{1,2} / g_{1,20}$, $\hat{\theta}_{1,2}$ is the estimation of $\theta_{1,2}^*$ and updated as

$$\dot{\hat{\theta}}_{1,2} = \Gamma_{1,2}\left[\xi_{1,2}(Z_{1,2})z_{1,2} - \sigma_{1,2}\hat{\theta}_{1,2}\right] \tag{19}$$

where $\Gamma_{1,2} = \Gamma_{1,2}^T > 0, \sigma_{1,2} > 0$.

From $z_{1,2} = x_{1,2} - \beta_{1,2}$,

$$\dot{z}_{1,2} = f_{1,2}(\overline{x}_{1,2}) + g_{1,2}(\overline{x}_{1,2})(v_1 + \Delta u_1) + \Delta_{1,2} - \dot{\beta}_{1,2} \tag{20}$$

Substituting (20), (22) into (24), we have

$$\dot{z}_{1,2} = g_{1,2}(\overline{x}_{1,2})(-\tilde{\theta}_{1,2}^T \xi_{1,2}(Z_{1,2}) - c_{1,2}z_{1,2} + e_1 + \Delta u_1 - D_{1,2}\tanh(\frac{D_{1,2}z_{1,2}}{s_{1,2}}) + \delta_{1,2}^*) + \Delta_{1,2} \tag{21}$$

3.2   Stability Analysis

Theorem1. Consider the closed-loop system, the final controller (18), auxiliary design system (17) and the updated laws (9), (19), under the Assumptions 1-3, there exist $c_{i,j}, \tau_{i,2}, \lambda_{max}\{\Gamma_{i,j}\}, s_{i,j}, k_i, \varepsilon_i$ and $\sigma_{i,j}$ such that all the signals in the closed-loop system are uniformly ultimately bounded. Furthermore, given any $\mu_{i,1} > 0$, we can tune our controller parameters such that the output error $z_{i,1} = y_{i,1}(t) - y_{i,d}(t)$ satisfies $\lim_{t\to\infty}|z_{i,1}(t)| = \mu_{i,1}$.

Proof. Choose the Lyapunov function candidate as

$$V = \left(\sum\nolimits_{j=1}^2 z_{i,j}^2 / g_{i,j}(\overline{x}_{i,j}) + \sum\nolimits_{j=1}^2 \tilde{\theta}_{i,j}^T \Gamma_{i,j}^{-1}\tilde{\theta}_{i,j} + \eta_{i,2}^2 + e_i^2\right)/2 \tag{22}$$

By mentioning $x_{i,2} = z_{i,2} + \beta_{i,2}$, $\beta_{i,2} = \eta_{i,2} + \alpha_{i,2}$, $\Delta u_i \cdot e_i \leq (\Delta u_i^2 + e_i^2)/2$ and $e_i \cdot \dot{e}_i = -k_i e_i^2 - (|z_{i,2} \cdot \Delta u_i| + \Delta u_i^2/2)/e_i^2 \cdot e_i^2 + \Delta u_i \cdot e_i$, the time derivative of V along the system trajectories is

$$\dot{V} \le -c_{i,1}z_{i,1}^2 + z_{i,1}z_{i,2} + z_{i,1}\eta_{i,2} + z_{i,1}\delta_{i,1}^* - \dot{g}_{i,1}(\overline{x}_{i,1})z_{i,1}^2 / 2g_{i,1}^2(\overline{x}_{i,1})$$
$$+ z_{i,2}\delta_{i,2}^* - (c_{i,2} - 0.5)z_{i,2}^2 - \dot{g}_{i,2}(\overline{x}_{i,2})z_{i,2}^2 / 2g_{i,2}^2(\overline{x}_{i,2}) - (k_i - 1)e_i^2$$
$$+ \sum_{j=1}^2 (0.2785s_{i,j}) + (-\eta_{i,2}^2 / \tau_{i,2} + |\eta_{i,2}B_{i,2}|) + \sum_{j=1}^2 (-\sigma_{i,j}\tilde{\theta}_{i,j}^T\hat{\theta}_{i,j}) \qquad (23)$$

Let $c_{i,j} = c_{i,j0} + c_{i,j1}$, with $c_{i,j0}$ and $c_{i,j1} > 0$. By choosing $c_{i,j0}$ such that $c_{i,j0}^* = (c_{i,j0} - g_{i,jd}(\overline{x}_{i,j})/2g_{i,j0}^2(\overline{x}_{i,j})) > 0$. After some manipulation, one has

$$\dot{V} \le -c_{i,11}z_{i,1}^2 + 3z_{i,1}^2 + z_{i,2}^2/4 + \eta_{i,2}^2/4 + \delta_{i,1}^{*2}/4 - c_{i,10}^*z_{i,1}^2$$
$$+ \delta_{i,2}^{*2}/4 + z_{i,2}^2 - c_{i,21}z_{i,2}^2 - c_{i,20}^*z_{i,2}^2 + 0.5z_{i,2}^2 - (k_i - 1)e_i^2$$
$$+ \sum_{j=1}^2 (0.2785s_{i,j}) + (-\eta_{i,2}^2/\tau_{i,2} + |\eta_{i,2}B_{i,2}|) + \sum_{j=1}^2 (-\sigma_{i,j}\tilde{\theta}_{i,j}^T\hat{\theta}_{i,j}) \qquad (24)$$

Choose $c_{i,11} = 3 + \alpha_0 - c_{i,10}^*$, $c_{i,21} = 1.25 + \alpha_0 - c_{i,20}^* + 0.5$, where α_0 is a positive constant, and noting that $2\tilde{\theta}_{i,j}^T\hat{\theta}_{i,j} \ge \left\|\tilde{\theta}_{i,j}\right\|^2 - \left\|\hat{\theta}_{i,j}^*\right\|^2$, then one has

$$\dot{V} \le \sum_{j=1}^2 \left(-\alpha_0 z_{i,j}^2 - (\sigma_{i,j}/2\lambda_{max}(\Gamma_{i,j}^{-1}))\tilde{\theta}_{i,j}^T\Gamma_{i,j}^{-1}\tilde{\theta}_{i,j} + \delta_{i,j}^{*2}/4 + \sigma_{i,j}\left\|\theta_{i,j}^*\right\|^2/2\right)$$
$$+ \sum_{j=1}^2 (0.2785s_{i,j}) + (\eta_{i,2}^2/4 - \eta_{i,2}^2/\tau_{i,2} + |\eta_{i,2}B_{i,2}|) - (k_i - 1)e_i^2 \qquad (25)$$

Let $(1/4)\delta_{i,j}^{*2} + (\sigma_{i,j}/2)\left\|\theta_{i,j}^*\right\|^2 = e_{i,j}$, $1/\tau_{i,2} = (1/4) + (M_{i,2}^2/2\kappa) + \alpha_0 g_{i,j0}$. Noting the fact $|\delta_{i,j}^*| \le \delta_m$ and $\left\|\theta_{i,j}^*\right\| \le \theta_M$ gives $e_{i,j} \le (1/4)\delta_m^2 + (\sigma_{i,j}/2)\theta_M^2 = e_M$. Also, for any positive number κ, $(\eta_{i,2}^2 B_{i,2}^2/2\kappa) + (\kappa/2) \ge |\eta_{i,2}B_{i,2}|$. Since on the level set defined by $|B_{i,2}| < M_{i,2}$, $V(z_{i,1}, z_{i,2}, \eta_{i,2}, \tilde{\theta}_{i,1}, \tilde{\theta}_{i,2}) = p_i$. Then

$$\dot{V} \le \sum_{j=1}^2 (-\alpha_0 z_{i,j}^2) + \sum_{j=1}^2 \left[-(\sigma_{i,j}/2\lambda_{max}(\Gamma_{i,j}^{-1}))\tilde{\theta}_{i,j}^T\Gamma_{i,j}^{-1}\tilde{\theta}_{i,j}\right] + 2e_M^*$$
$$+ \sum_{j=1}^2 (0.2785s_{i,j}) + \kappa/2 - \alpha_0 g_{i,j0}\eta_{i,2}^2 - (k_i - 1)e_i^2 \qquad (26)$$

If we choose $\alpha_0 \ge C/(2g_{i,j0})$, where C is a positive constant, and choose $\sigma_{i,j}, k_i$ and $\lambda_{max}(\Gamma_{i,j})$ such that $\sigma_{i,j} \ge C\lambda_{max}(\Gamma_{i,j})$, $(k_i - 1) \ge C/2$, $i = 1, 2$. Let $D = 2e_M + \sum_{j=1}^2 (0.2785s_{i,j}) + \kappa/2$. Then from (26) we have the following inequality:

$$\dot{V} \le -\left(\sum_{j=1}^2 Cz_{i,j}^2/(g_{i,j}) + \sum_{j=1}^2 C\tilde{\theta}_{i,j}^T\Gamma_{i,j}^{-1}\tilde{\theta}_{i,j} + C\eta_{i,2}^2 + Ce_i^2\right)/2 + D \le -CV + D \qquad (27)$$

Actually, the Equation (27) means that $V(t)$ is bounded. Thus, all signals of the closed-loop system are bounded. This concludes the proof simply.

4   Simulation Results

In the simulation, the system (1) is uses for simulation model, i.e., $f_{i,1} = 0$, $f_{1,2} = (m_1 gr\sin(x_{1,1}) - kx_{1,1} + kx_{2,1})/J_1$, $f_{2,2} = m_2 gr\sin(x_{2,1}) - kx_{2,1}/J_2 + kx_{1,1}/J_1$, $g_{i,1} = 1$,

$g_{i,2} = 1/J_i$, $\Delta_{i,1} = 0$, $\Delta_{i,2} = v_i/J_i$, $v_1(t) = v_2(t) = 5\sin(2\pi t)$ with $i = 1, 2$ in (2), and the reference signals $y_{1,d} = y_{2,d} = \sin(t)$. The parameters of the pendulums are $m_1 = 2$ kg, $m_2 = 2.5$ kg, $J_1 = 2$ kg, $J_2 = 2.5$ kg $k = 2$ Nm/rad, and $r = 1$ m.

We choose neural networks $\hat{\theta}_{i,1}^T \xi_{i,1}(Z_{i,1})$ contains 25 nodes (i.e., $l_{i,1} = 25$), with centers $\mu_l(l = 1, \ldots, l_{i,1})$ evenly spaced in $[-4, 4] \times [-4, 4]$, and widths $\eta_l = 2(l = 1, \ldots, l_{i,1})$. Neural networks $\hat{\theta}_{i,2}^T \xi_{i,2}(Z_{i,2})$ contains 135 nodes (i.e., $l_{i,2} = 135$), with centers $\mu_l(l = 1, \ldots, l_{i,2})$ evenly spaced in $[-4, 4] \times [-4, 4] \times [-4, 4] \times [-6, 6]$ and widths $\eta_l = 2(l = 1, \ldots, l_{i,2})$. The initial weights $\hat{\theta}_{i,1}(0) = \hat{\theta}_{i,2}(0) = 0$ ($i = 1, 2$). The initial conditions $\left[x_{1,1}(0), x_{2,1}(0)\right]^T = \left[30°, -30°\right]^T$, $\left[x_{1,2}(0), x_{2,2}(0)\right]^T = \left[0, 0\right]^T$, $e_1(0) = e_2(0) = 7$, $\mathrm{sat}(u_1) = [-26, 26]$, $\mathrm{sat}(u_2) = [-30, 30]$.

The controllers parameters are chosen as $\Gamma_{i,j} = \{0.01,\ 0.01,\ 0.01,\ 0.01\}$, $c_{i,j} = \{10,\ 80,\ 10,\ 80\}$, $\sigma_{i,j} = \{15,\ 15,\ 15,\ 15\}$, $s_{i,j} = \{0.5,\ 0.5,\ 0.5,\ 0.5\}$ with $i = 1, 2$, $j = 1, 2$, $k_1 = k_2 = 5$, $\varepsilon_1 = \varepsilon_2 = 0.001$ and $\tau_{1,2} = \tau_{2,2} = 0.5$. The effectiveness and good performance of he proposed algorithm are illustrated in Figure 1-2.

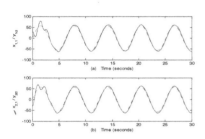

Fig. 1. DIPs angular positions   Fig. 2. DIPs controllers

5   Conclusions

The trajectory tracking control problem has been studied for DIP under constraints of input saturation. A direct adaptive neural network control scheme has been proposed by combining DSC technique and the approximation of RBF NNs. With the help of an auxiliary design system, the effect of input saturation has been considered adequately in the control design procedure. In addition, both the "explosion of complexity" problem and the controller singularity problem are removed in the proposed algorithm. And it is shown that the stability of the closed-loop system is guaranteed. Finally, simulation results are given to illustrate the effectiveness of the proposed tracking control scheme.

References

1. Benjanarasuth, T., Nundrakwang, S.: Hybrid controller for rotational inverted pendulu systems. In: SICE Annual Conference, pp. 1889–1894. The University Electro-Communications, Japan (2008)
2. Angeli, D.: Almost global stabilization of the inverted pendulum via continuous state feedback. Automatica 37(7), 1103–1108 (2001)
3. Xin, X., Kaneda, M.: Analysis of the energy-based control for swinging up two pendu-lums. IEEE Trans. On Automatic and Control 50(5), 679–684 (2005)
4. Matsuda, N., Izutsu, M., Furuta, K.: Simultaneous swinging-up and stabilization of double Furuta pendulums. In: Proc. of the SICE Annual Conference, pp. 110–115 (2007)
5. Mayhew, C.G., Teel, A.R.: Global asymptotic stabilization of the inverted equilibrium manifold of the 3-D pendulum by hybrid feedback. In: 49th IEEE Conference on Decision and Control, Atlanta, GA, USA, pp. 679–684 (2010)
6. Li, T.S., Li, W., Bu, R.X.: A novel decentralized adaptive NN tracking control for double inverted pendulums. Int. Journal of Modelling, Identification and Control 13(4), 269–277 (2011)
7. Zhou, J., Er, M.J., Zhou, Y.: Adaptive neural network control of uncertain nonlinear systems in the presence of input saturation. In: 9th International Conference on Control, Automation, Robotics and Vision, pp. 1–5 (2006)
8. Li, J.F., Li, T.S.: Design of ship's course autopilot with input saturation. ICIC Express Letters 5(10), 3779–3784 (2011)
9. Swaroop, D., Hedrick, J.K., Yip, P.P., Gerdes, J.C.: Dynamic surface control for a class of nonlinear systems. IEEE Trans. Autom. Control 45(10), 1893–1899 (2000)
10. Zhang, H.B., Li, C.G., Liao, X.F.: Stability analysis and H∞ controller design of fuzzy large-scale systems based on piecewise Lyapunov functions. IEEE Trans. Syst., Man, Cybern. B, Cybern. 36(3), 685–698 (2006)

Identification and Control of PMSM
Using Adaptive BP-PID Neural Network

Chao Cai[1], Fufei Chu[1], Zhanshan Wang[1], and Kaili Jia[2]

[1] School of Information Science and Engineering, Northeastern University,
Liaoning, 110004, People's Republic of China
[2] Shandong Liaocheng Power Supply Company
zhanshan_wang@163.com

Abstract. The control system of the permanent magnet synchronous motor (PMSM) has the characteristics of nonlinear and strong coupling. Therefore, In order to improve the control precision, the paper presents a novel approach of speed control for PMSM using adaptive BP (back-propagations)-PID neural network. The approach consists of two parts: on-line identification based on BP neural network and the adaptive PID controller. Lyapunov theory is used to prove the stability of the control scheme. Simulation results show that this control method can improve the dynamical performance and enhance the static precision of the speed system.

Keywords: PMSM, adaptive, BP-PID, Lyapunov.

1   Introduction

Comparing with other motors, PMSM has its own excellent control properties. It has been applied to social activities popularly [1]. Hence the d,q model of the PMSM can be derived from the well-known model of the synchronous machine [2].Though PMSM have been widely used as precision control motors. There are certain limitations of PMSM drives such as lack of robustness, overload capability, narrow speed range.

Neural Network is used effectively for the identification and control of nonlinear dynamical system [3-7]. The neural network plays a key role on system identification and speed control, as it has many advantageous features including parallel and distributed processing, efficient mapping between inputs and outputs without an exact system model [8]. An approach of speed control for PMSM using on-line self tuning artificial neural network is presented in [4]. An artificial neural network speed controller is developed and integrated with the vector control scheme of the PMSM drive. Different Topologies are used to identify the PMSM motor in [6, 7].Nonlinear dynamics of the motor and the load are captured by the neural network perfectly. In [9], a RBF neural network and model reference self-adaptive control method for PMSM is proposed. Compared with RBF neural network, the BP neural network can contain different hidden layers and it has the character of global convergence. In [10], based on the mathematical model of PMSM, a combination of an BP neural network and a

C. Guo, Z.-G. Hou, and Z. Zeng (Eds.): ISNN 2013, Part II, LNCS 7952, pp. 155–162, 2013.

general PID controller is used in its speed control system, but in that paper the author only gives the calculation process. The convergence result of the control scheme is not proved. In this paper, we give the Lyapunov Stability Analyses of the control scheme.

The paper is organized as follows. Section 2 explains the mathematical models of a PMSM and the volt-amper stator d, q equations in the rotor reference frame of the PMSM. Section 3 presents the approach of speed control for PMSM using adaptive BP-PID neural network. It consists of two parts: on-line identification based on BP neural network and the adaptive PID controller. In section 4, we construct a Lyapunov function to prove the stability of the control scheme. The simulation results which show the effectiveness of the proposed control scheme are given in Section 5.

2   PMSM Model

The use of the neural network in any system does not require the model of the system under study. In this work, unlike the conventional approach of controlling the speed, the control technique is incorporated with the neural network to obtain highest torque sensitivity of the PMSM drive. The control strategy is formulated in the synchronously rotating reference frame.

The mathematical models of a PMSM and the volt-amper stator d, q equations in the rotor reference frame of the PMSM are [5, 6]

$$u_d = Ri_d + p\psi_d - \omega_r\psi_q,\tag{1}$$

$$u_q = Ri_q + p\psi_q - \omega_r\psi_d,\tag{2}$$

where

$$\psi_q = L_q i_q,\tag{3}$$

and

$$\psi_d = L_d i_d + \psi_{df}.\tag{4}$$

u_q and u_d are the d,q-axis voltages, i_d and i_q are the d,q-axis stator currents, L_d and L_q are the d,q-axis inductance, ψ_d and ψ_q are the-axis flux linkages, R is the stator resistance, ψ_{df} is the constant magnet flux linkage produced by permanent magnet rotor, ω_r is the motor speed and P is the number of pole pairs.

The developed electric torque can be expressed as

$$T_e = \frac{3}{2}P[\psi_{df}i_q + (L_d - L_q)i_d i_q],\tag{5}$$

$$T_e = \frac{3}{2}P[\psi_{df}i_q] = K_T i_q,\tag{6}$$

K_T, K_b are motor torque constants.

In order to derive the training data for the neural network and apply the control algorithms, a discrete-time PMSM model is developed by combining the equations (1) to (6). Then we derive the discrete-time model of PMSM.The resulting space equation is

$$\omega_r(k+1) = \alpha\omega_r(k) + \beta\omega_r(k-1) + \chi\omega_r^2(k) + \delta\omega_r^2(k-1) + \varepsilon u(k) + \xi \quad (7)$$

Where $\alpha, \beta, \chi, \delta, \varepsilon$ are constants based on motor parameter as well as sampling period T is given in the appendix.

3   Adaptive BPNN-PID Control of PMSM

The ability of BP neural network to approximate large classes of nonlinear functions accurately makes it prime candidate for use in dynamic models for the representation of nonlinear plants. The model of discrete-time PMSM introduced above can be described by the following nonlinear difference equations;

$$\omega_r(k+1) = N[\omega_r(k), \omega_r(k-1)] + \varepsilon u(k), \quad (8)$$

where

$$N[\omega_r(k), \omega_r(k-1)] = \alpha\omega_r(k) + \beta\omega_r(k-1) + \chi\omega_r^2(k) + \delta\omega_r^2(k-1). \quad (9)$$

A BP neural network is a layered network consisting of an input layer, an output layer and at least one layer of nonlinear processing elements. The BPNN predictor estimates the motor speed as

$$\hat{\omega}_r(k+1) = N[\omega_r(k), \omega_r(k-1)] + \varepsilon u(k). \quad (10)$$

The trained BP neural network is applied as series-parallel type identifier to estimate the value of the function $N[.]$. Where $N[.]$ is a Neural Network with $N_{1,10,1}$.

Once a design of the BP neural network structure is done, the next step is to determine the weights and biases of the BPNN through training to achieve the specific target with the given inputs. The back-propagation training algorithm is used for this purpose which is based on the principle of minimization of a cost function of the error between the outputs and the target. Consider the discrete-time system given by (8)

The error index $E_1(k)$ should be defined as

$$E_1(k) = \frac{1}{2}[\omega_r(k) - \hat{\omega}_r(k)] = \frac{1}{2}e_1^2(k), \quad (11)$$

where $e_1(k) = \omega_r(k) - \hat{\omega}_r(k)$ is a learning error between the target and network output at time k .The partial derivatives of the error index $E(k)$,with respect to the weight of the network, is obtained using the dynamic neural model as following,

$$\frac{\partial E(k)}{\partial \omega_i(k)} = -e_1(k)o_i(k).$$ (12)

The PID controller should be constituted by a dynamic neural network structure. It consists of numerical integrator, time delay link Z^{-1}, and adaptive linear neurons. The output $u(k)$ can be represented as

$$u(k)$$ (13)

$$= k_p e(k) + k_i \sum_{j=0}^{k} e(j) + k_d[e(k) - e(k-1)]'$$

$$e(k) = r(k) - \omega_r(k),$$ (14)

Where $r(k)$ is the command input. The input of the adaptive linear neuron is given as

$$c_1(k) = e(k)$$ (15)

$$c_2(k) = \sum_{j=0}^{k} e(k)$$

$$c_3(k) = \Delta e(k) = e(k) - e(k-1)$$

The output can be also represented as

$$u(k) = v_1 c_1(k) + v_2 c_2(k) + v_3 c_3(k).$$ (16)

$v_i (i = 1, 2, 3)$ is the weight of the PID controller. The error index $E_2(k)$ is defined as

$$E_2(k) = \frac{1}{2}[r(k) - \hat{\omega}_r(k)]^2 = \frac{1}{2}e_2^2(k).$$ (17)

4   Lyapunov Stability Analyses

In this section, Lyapunov theory is used to prove the stability of the control scheme [10,11].While the step size η of neural network satisfy following requirement,

$$2(AA^T)^{-1} > \eta > 0.$$ (18)

the tracking error between the system output and the reference command will converge to a specified constant ε. The property of the control scheme can be proved.

Proof. Let $J(x)$ be a Lyapunov function

$$J(k) = \frac{1}{2}\sum_{i=1}^{k} e^2(i). \tag{19}$$

As the learning process lead to $V(x)$ change as follows

$$\Delta J(k) = \frac{1}{2}[\sum_{i=1}^{k+1} e^2(i) - \sum_{i=1}^{k} e^2(i)] = \frac{1}{2}\sum_{i=0}^{k}[e^2(i+1) - \sum_{i=1}^{k} e^2(i)]. \tag{20}$$

We define $e(0) = 0$.Then

$$e(k+1) = e(k) + \Delta e(k) = e(k) + [\frac{\partial e(k)}{\partial \omega(k)}]^T \Delta v(k). \tag{21}$$

As

$$\Delta v(k) = -\eta \frac{\partial E(k)}{\partial v(k)} = -\eta e(k) \frac{\partial e(k)}{\partial \Delta u(k)} \cdot \frac{\partial \Delta u(k)}{\partial v(k)}. \tag{22}$$

$$A = [\frac{\partial e(k)}{\partial v(k)}]^T = [\frac{\partial e(k)}{\partial \Delta u(k)} \cdot \frac{\partial \Delta u(k)}{\partial v(k)}]^T. \tag{23}$$

Then

$$\Delta e(k) = -\eta A A^T e(k). \tag{24}$$

It can be concluded from equations (22)

$$\Delta v(k) = \frac{1}{2}\sum_{i=0}^{k}(-2e(i)\eta A A^T e(i) + \eta^2 A A^T (A^T e(i))^T (A^T e(i))) \tag{25}$$

$$= -\frac{1}{2}\sum_{i=0}^{k}(A^T e(i))^T (2\eta - \eta^2 A A^T)(A^T e(i))$$

By the Lyapunov stability theory: if $\Delta J(k) < 0$ the whole system is stable, the system that satisfies the following properties:

$$2\eta - \eta^2 A A^T > 0, \tag{26}$$

η should satisfies if the system is stable

$$2(A A^T)^{-1} > \eta > 0. \tag{27}$$

As $\Delta J(k) < 0$, we can conclude

$$\frac{1}{2}e^2(k+1) < \frac{1}{2}e^2(k),\qquad(28)$$

then

$$\lim_{k\to\infty} e(k) \le \varepsilon .\qquad(29)$$

for some specified constants $\varepsilon \ge 0$.

5   Simulation Results of the PMSM System

In this section, the simulation results of the PMSM drive system are presented to verify the feasibility of the proposed control scheme under the operating conditions. First, under no-load, the dynamic performance of the drive system is given the speed command of 1800(Rad/sec).

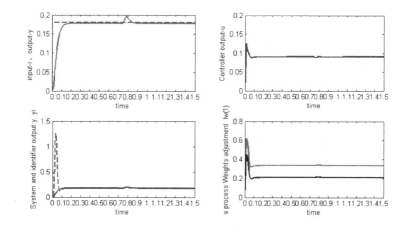

Fig. 1. Rotor speed response for a reference speed of 1800 rad/sec when a load of 2N·m

The disturbance rejection capabilities have been checked when a load of 2N·m is applied to the PMSM at t=0.75s-0.80s. The BP neural network adjusts its weights and biases to this changing circumstance of sudden load so that the system responses according to the reference speed.

Unlike the conventional controller, the on-line adaptive property of BPNN-PID controller reduces the possibility of large speed oscillation due to the sudden application of load. Figure 2 show the contrast of the BPNN controller and the conventional PID controller.

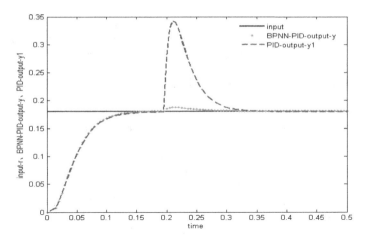

Fig. 2. The contrast of the BPNN-PID controller and the traditional PID controller

These figures clearly illustrate good dynamic performances in command tracking and load regulation performance which is realized by the controller.

6   Conclusion

In this paper, a novel approach of speed control for PMSM using adaptive BP-PID neural network is presented. To achieve accurate trajectory control of the speed of PMSM, the approach combine an on-line discrimination based on BP neural network with an adaptive PID controller. Lyapunov theory is used to prove the stability of the control scheme. Finally, Simulation results are used to validate the proposed approach.

Acknowledgements. This work was supported by the National Natural Science Foundation of China under Grants 61074073 and 61034005, Program for New Century Excellent Talents in University of China (NCET-10-0306), and the Fundamental Research Funds for the Central Universities under grants N110504001 and N100104102.

References

1. Jiang, J.: Control of permanent magnet synchronous motor. Science & Technology Vision 6, 182–183 (2012)
2. Lin, F.J., Shyu, K.K., Lin, Y.S.: Variable structure adaptive control for PM synchronous servo motor drive. IEE Pro. Ecelectric Power Application 146, 173–185 (1999)
3. Chen, F.C., Khalil, H.K.: Adaptive control of a class of nonlinear discrete-time systems using neural networks. IEEE Trans. Automat. Contr. 72, 791–807 (1995)
4. Rahman, M.A., Hoque, M.A.: On-line adaptive artificial neural network based vector control of permanent magnet synchronous motor. IEEE Trans. On Energy Conversion 13, 311–318 (1998)

5. Huang, Y., Xu, Y., Wu, L.: Research on DC motor PID control system based on BP neural network. J. of China Three Gorges Univ. 32, 29–32 (2010)
6. Kumar, R., Gupta, R.A., Bansal, A.K.: Identification and control of PMSM using artificial neural network. In: IEEE International Symposium on Industrial Electronics, pp. 30–35. Curran Associates Inc., Chennai (2007)
7. Kumar, R., Gupta, R.A., Bansal, A.K.: Novel topologies identification and control of pmsm using artificial neural network. In: IET-UK International Conference on Information and Communication Technology in Electrical Sciences, pp. 75–78 (2007)
8. Narenda, K.S., Parthasaratny, K.: Identification and control of dynamical system using neural networks. IEEE Trans. Neural Networks 1, 4–27 (1990)
9. Cai, Z., Tang, Z., Ma, S.: RBF neural network based on-line discrimination and model reference self-adaptive control for permanent magnet synchronous motors. East China Electric Power 36, 108–112 (2008)
10. Wang, Z., Qu, B.: Application of BP neural network in PMSM. Computer Simulation 26, 155–157 (2009)
11. Huo, L.: The stability of artificial neural network intelligent control system. Journal of Taiyuan University of Technology 31, 683–685 (2000)
12. Ge, B., Jiang, J.: Model algorithmic control for permanent magnet synchronous motor drive. Proceedings of the CSEE 19, 27–32 (1999)

Adaptive NN Control
for a Class of Chemical Reactor Systems

Dong-Juan Li[1] and Li Tang[2]

[1] School of Chemical and Environmental Engineering, Liaoning University of Technology,
Jinzhou, Liaoning, 121001, China
[2] College of Science, Liaoning University of Technology,
Jinzhou, Liaoning, 121001, China
ldjuan@126.com

Abstract. An adaptive control algorithm is applied to controlling a class of SISO continuous stirred tank reactor (CSTR) system in discrete-time. The considered systems belong to pure-feedback form where the unknown dead-zone and it is first to control this class of systems. Radial basis function neural networks (RBFNN) are used to approximate the unknown functions and the mean value theorem is exploited in the design. Based on the Lyapunov analysis method, it is proven that all the signals of the resulting closed-loop system are guaranteed to be semi-global uniformly ultimately bounded (SGUUB) and the tracking error can be reduced to a small compact set. A simulation example is studied to verify the effectiveness of the approach.

Keywords: Discrete-time system, CSTR control, adaptive predictive control, the neural networks, nonlinear systems.

1 Introduction

In the chemical engineering, the chemical reactor systems can be often found [1]. Recently, the control problem of the systems with unknown functions has been studied and many papers have been published one after the other [2-5] by using the fuzzy systems and the neural networks. Specifically, the stability of the CSTR systems has attracted much attention and obtained many achievements [6-8].

However, the results proposed in [6-8] are focused on nonlinear continuous systems. They can not be used directly to stabilize the discrete-time systems. Some results have been obtained in [9] for nonlinear discrete-time systems by using the neural networks. The results in [9] ignored the effect of the input nonlinearity. The input nonlinearities are an important in the practice and the nonlinear continuous systems with dead zone were stabilized in [10,11]. For nonlinear discrete-time systems, an adaptive control algorithm has been obtained in [12]. However, these results are required to satisfy the matching condition or strict feedback form.

This paper will study control problem for CSTR discrete-time systems in pure-feedback. The systems contain the unknown functions, the external disturbance and unknown dead zone. The systems are transformed into a prediction form. An adaptive

C. Guo, Z.-G. Hou, and Z. Zeng (Eds.): ISNN 2013, Part II, LNCS 7952, pp. 163–169, 2013.

NN controller and adaptation laws are constructed for transformed systems. The NNs are used to approximate the unknown functions and a compensator is used to compensate for the dead zone input. By constructing Lyapunov function, the stability of the closed-loop system is proven. The simulation example is given to verify that the approach is feasible for chemical process systems.

2   Problem Statement

Consider the SISO discrete-time CSTR system with unknown dead-zone

$$
\begin{cases}
z_1(k+1) = z_1(k) + \big[-z_1(k) + C_a(1 - z_1(k)) \\
\qquad \times e^{\gamma z_2(k)/(\gamma + z_2(k))} \big] T \\
z_2(k+1) = z_2(k) + \big[-z_2(k) \\
\qquad + BC_a(1 - z_1(k)) e^{\frac{\gamma z_2(k)}{\gamma + z_2(k)}} \\
\qquad - C_r(z_2(k) - D(u(k))) \big] T + d(k) \\
y = z_1(k)
\end{cases}
\tag{1}
$$

where $z_1(k)$ is the dimensionless concentration at time instant k and the dimensionless temperature at time instant k is denoted by $z_2(k)$, C_a and B are Damkohler Number and dimensionless heat of reaction, respectively; C_r denotes dimensionless cooling rate, and sampling period is T, the term $d(k)$ is added to represent an external disturbance, which is bounded by its constant $\bar{d}$, that is, $|d(k)| < \bar{d}$. $D(u(k))$ denotes the dimensionless coolant temperature, which is related to the control input $u(k)$ through the non-symmetric dead-zone. The non-symmetric dead-zone $D(u(k))$ is described as

$$
D(u) = \begin{cases}
m_r(u(k) - b_r), & if\ u(k) \geq b_r \\
0, & if\ -b_l < u(k) < b_r \\
m_l(u(k) + b_l), & if\ u(k) \leq -b_l
\end{cases}
\tag{2}
$$

where m_r, m_l, b_r and b_l are positive.

The parameters m_r and m_l stand for the right and left slope of the dead-zone characteristic and are known. The parameters b_r and b_l represent the breakpoints of the input nonlinearity.

The dead-zone input can be written as:

$$
D(u) = m(k)u(k) + b(k)
\tag{3}
$$

where

$$m(k) = \begin{cases} m_r(k), & \text{if } u(k) > 0 \\ m_l(k), & \text{if } u(k) \le 0 \end{cases} \tag{4}$$

and

$$b(k) = \begin{cases} -m_r(k)b_r(k), & \text{if } u(k) \ge b_r \\ -m(k)u(k), & \text{if } -b_l < u(k) < b_r \\ m_l(k)b_l(k), & \text{if } u(k) \le -b_l \end{cases} \tag{5}$$

It is obvious that $m(k)$ is known, and let $\max(m_r b_r, m_l b_l) = \overline{b}$ and $\min(m_r b_r, m_l b_l) = \underline{b}$. In this paper, we will use RBFNN to approximate the unknown functions of the system and their approximation property is described in the following subsection.

The control objective is to make an adaptive RBFNN control algorithm so that the tracking error converges to a small compact set and all the signals in the closed-loop are SGUUB.

3   System Transformations

In order to conveniently express, we denote

$$F_1(z_1(k), z_2(k)) = z_1(k) + \left[-z_1(k) + C_a \times (1 - z_1(k)) e^{\frac{\gamma z_2(k)}{\gamma + z_2(k)}} \right] T \tag{6}$$

$$F_2(z_1(k), z_2(k), D(u(k)))$$
$$= z_2(k) + \left[-z_2(k) + BC_a(1 - z_1(k)) \times e^{\frac{\gamma z_2(k)}{\gamma + z_2(k)}} - \alpha(z_2(k) - D(u(k))) \right] T \tag{7}$$

Then, (1) can be expressed as:

$$\begin{cases} z_1(k+1) = F_1(z_1(k), z_2(k)) \\ z_2(k+1) = d(k) + F_2(z_1(k), z_2(k), D(u(k))) \\ y = z_1(k) \end{cases} \tag{8}$$

It is observed that the system as shown in the above becomes a pure-feedback discrete-time system. For the system (1) and the dead-zone (3), it is necessary for the reference trajectory to satisfy the following assumption.

Assumption 1. There exist constants $\overline{F_i} > \underline{F_i} > 0$ such that $\underline{F_i} \le F_i \le \overline{F_i}, i = 1, 2$, where

$$F_1 = \frac{\partial F_1\left(z_1(k), z_2(k)\right)}{\partial z_2(k)} \qquad F_2 = \frac{\partial F_2\left(z_1(k), z_2(k), D\left(u(k)\right)\right)}{\partial u(k)}$$

Similar to the transformed procedure, we obtain

$$y(k+2) = \Psi_l\left(\overline{z}_2(k), D\left(u(k)\right)\right) + d_l(k) \tag{9}$$

where

$$\begin{aligned}
\Psi_l\left(\overline{z}_2(k), D\left(u(k)\right)\right) &= \Psi\left(\overline{z}_2(k), D\left(u(k)\right), 0\right) \\
d_l(k) &= \Psi\left(\overline{z}_2(k), D\left(u(k)\right), d(k)\right) - \Psi\left(\overline{z}_2(k), D\left(u(k)\right), 0\right)
\end{aligned} \tag{10}$$

There exists a certain finite constant L_d satisfies

$$\left|d_l(k)\right| \le L_d \left|d(k)\right| \le \overline{d}_l \tag{11}$$

where $\overline{d}_l = L_d \overline{d}$.

According to Assumption 1, one has

$$0 < \underline{F} \le \frac{\partial \Psi_l\left(\overline{z}_2(k), D\left(u(k)\right)\right)}{\partial D\left(u(k)\right)} = F \le \overline{F} \tag{12}$$

Define $\eta(k) = y(k) - y_d(k)$ and we obtain

$$\eta(k+2) = \Psi_l\left(\overline{z}_2(k), D\left(u(k)\right)\right) + d_l(k) - y_d(k+2) \tag{13}$$

Due to (12), we have

$$\frac{\partial \left(\Psi_l\left(\overline{z}_2(k), D\left(u(k)\right)\right) - y_d(k+2)\right)}{\partial D\left(u(k)\right)} = F > 0 \tag{14}$$

An ideal control is selected as

$$\Psi_l\left(\overline{z}_2(k), D_l^*\left(\xi(k)\right)\right) - y_d(k+2) = 0 \tag{15}$$

where $\xi(k) = \left[\overline{z}_2(k), y_d(k+2)\right]^T \in \Omega_\xi \subset R^3$.

Substituting the ideal control as shown in (15) into (13) leads to

$$\eta(k+2) = d_l(k) \tag{16}$$

This implies that $\eta(k+2) = 0$ if $d_l(k) = 0$. Furthermore, it can be obtained that $d_l(k)$ is bounded. Consider (15), subtracting and adding $\Psi_l\left(\overline{z}_2(k), D_l^*\left(\xi(k)\right)\right)$ on the right side of (13) results in

$$\eta(k+2) = \Psi_I\left(\overline{z}_2(k), D(u(k))\right) - \Psi_I\left(\overline{z}_2(k), D_I^*\left(\xi(k)\right)\right) + d_I(k) \qquad (17)$$

By using the mean-value theorem, we obtain

$$\eta(k+2) = \frac{\partial\Psi\left(\overline{z}_2(k), D(u(k))\right)}{\partial D(u(k))}\Bigg|_{D(u(k))=D_0(u(k))} \times \left(D(u(k)) - D_I^*\left(\xi(k)\right)\right) + d_I(k) \quad (18)$$

where $\Psi\left(\overline{z}_2(k), D(u(k))\right)$ is differentiable and

$D_0\left(u(k)\right) = \theta D\left(u(k)\right) + (1-\theta) D_I^*\left(u(k)\right)$ with $0 \le \theta \le 1$. Throughout this paper, for convenience, it has the following abbreviation

$$G(k) = \frac{\partial\Psi\left(\overline{z}_2(k), D(u(k))\right)}{\partial D(u(k))}\Bigg|_{D(u(k))=D_0(u(k))}.$$

Then, it has

$$\eta(k+2) = G(k) \times \left(D(u(k)) - D_I^*\left(\xi(k)\right)\right) + d_I(k) \qquad (19)$$

Similarly, it can be shown that

$$\eta(k+1) = \left(D(u(k_1)) - D_I^*\left(\xi(k_1)\right)\right) \times G(k_1) + d_I(k_1), \ k_1 = k-1 \qquad (20)$$

We can use RBFNN to approximate the ideal control with dead-zone input as

$$D_I^*\left(\xi(k)\right) = \omega_I^{*T}\tau\left(\xi(k)\right) + \varepsilon\left(\xi(k)\right) \qquad (21)$$

where $\tau\left(\xi(k)\right) \in R^{s_I}$ and $\varepsilon\left(\xi(k)\right)$ is the RBFNN approximation error. As we all known, the error can be surely arbitrary small by increasing RBFNN neurons number s_I. Because ω_I^* is unknown, then, let $\hat{\omega}_I(k)$ estimate it and define the estimation error as $\tilde{\omega}_I(k) = \hat{\omega}_I(k) - \omega_I^*$.

The controller is designed as

$$u = \frac{1}{m_{\min}\overline{F}}\eta(k) + \frac{1}{m_{\min}}\hat{\omega}_I^T\tau\left(\xi(k)\right) \qquad (22)$$

Chose the adaptive law as

$$\hat{\omega}_I(k+1) = \hat{\omega}_I(k_1) - \gamma_I\tau\left(\xi(k_1)\right) \times \eta(k+1) - \sigma_I\hat{\omega}_I(k_1) \qquad (23)$$

We have

$$\eta(k+1) = G(k_1)\frac{1+q}{\overline{F}}\eta(k) + G(k_1)\tilde{\omega}_I(k_1) \times \tau\left(\xi(k_1)\right)$$

$$+ qG(k_1)\hat{\omega}_I(k_1)\tau\left(\xi(k_1)\right) - G(k_1)\varepsilon\left(\xi(k_1)\right) + b(k)G(k_1) + d_I(k)$$

Choose a positive-definite function $V(k)$ as $V(k) = V_1(k) + V_2(k)$ where $V_1(k) = \eta^2(k)$ and $V_2(k) = \dfrac{\overline{F}}{F}\tilde{\omega}_l^T(k)\tilde{\omega}_l(k)$.

By a series of operations, the difference of $V(k)$ is

$$V(k+1) \le \frac{\overline{F}}{\gamma_l}(1-\sigma_l)\tilde{\omega}_l^T(k_1)\tilde{\omega}_l(k_1)$$

$$+\left(\overline{F} + \overline{F}\gamma_l\lambda_\tau + \frac{\overline{F}(\sigma_l+q)^2}{\sigma_l} + \sigma_l - \frac{1}{2}\right)\times\eta^2(k+1)$$

$$+\gamma_l\frac{(1+q)^2}{\overline{F}\sigma_l}\eta^2(k_1) + \overline{Fb}^2 - \frac{\overline{F}}{\gamma_l}\sigma_l(1-\sigma_l-\gamma_l\lambda_\tau)\hat{\omega}^2(k_1) + 2\overline{F}^2\mu_0^2 + \frac{\overline{F}}{\gamma_l}\sigma_l\omega^{*2}$$

Theorem 1. Consider the CSTR system (1). Under Assumption 1, consisting of the control law (22), and RBFNN adaptive law (23), we can conclude that all the signals in the closed-loop are SGUUB, and the tracking error and RBFNN weight estimation error are bounded.

Proof. Let $\beta_1 = \overline{F} + \overline{F}\gamma_l\lambda_\tau + \dfrac{\overline{F}(\sigma_l+q)^2}{\sigma_l} + \sigma_l - \dfrac{1}{2}$ and $\beta_2 = \sigma_l(1-\sigma_l-\gamma_l\lambda_\tau)$. If we choose $\beta_1 < 0$, $\beta_2 > 0$, $0 < \overline{c} < 1$, we have

$$V(k+1) \le \overline{c}V_1(k+1) + (1-\sigma_l)V_2(k+1) + b_M$$

By using Lemma 1 in [9], the tracking error and the estimation error are bounded, i.e., $\limsup\limits_{k\to\infty}\{V(k)\} \le \dfrac{b_M}{1-\overline{c}}$.

4   Conclusion

An adaptive predictive control algorithm is proposed for CSTR systems with dead-zone input. The controller was developed to solve the problem of dimensionless coolant temperature for a class of uncertain nonlinear discrete time CSTR system in a pure-feedback form. RBFNN is used to approximate the unknown functions and the mean value theorem is exploited to transform the pure-feedback form into a predictor form. It is proven that all the signals of the resulting closed-loop system are guaranteed to be SGUUB and the tracking error can be reduced to a small compact set. A`simulation example is studied to verify the effectiveness of the approach.

5   Funding

This work was supported by the Natural Science Foundation of China [grant no. 61104017]; the Program for Liaoning Excellent Talents in University [grant no. LJQ2011064]; The Foundation of Educational Department of Liaoning Province.

References

1. Aguilar, R.: Sliding-mode observer for uncertainty estimation in a class of chemical reactor: A differential-algebraic approach. Chemical Product and Process Modeling 2, 1934–2689 (2007)
2. Ge, S.S., Wang, C.: Adaptive NN of uncertain nonlinear pure-feedback systems. Automatica 38, 671–682 (2002)
3. Liu, Y.J., Wang, W.: Adaptive Fuzzy Control for a Class of Uncertain Nonaffine Nonlinear Systems. Information Sciences 177, 3901–3917 (2007)
4. Zhang, H.G., Luo, Y.H., Liu, D.R.: Neural network-based near-optimal control for a class of discrete-time affine nonlinear systems with control constraint. IEEE Transactions on Neural Networks 20, 1490–1503 (2009)
5. Liu, Y.J., Wang, W., Tong, S.C., Liu, Y.S.: Robust Adaptive Tracking Control for Nonlinear Systems Based on Bounds of Fuzzy Approximation Parameters. IEEE Transactions on Systems, Man, and Cybernetics, Part A: Systems and Humans 40, 170–184 (2010)
6. Ge, S.S., Hang, C.C., Zhang, T.: Nonlinear adaptive control using neural networks and its application to CSTR systems. Journal of Process Control 9, 313–323 (1999)
7. Zhang, H.G., Cai, L.L.: Nonlinear adaptive control using the Fourier integral and its application to CSTR systems. IEEE Transactions on Systems, Man, and Cybernetics, Part B: Cybernetics 32, 367–372 (2002)
8. Salehi, S., Shahrokhi, M.: Adaptive fuzzy backstepping approach for temperature control of continuous stirred tank reactors. Fuzzy Sets and Systems 160, 1804–1818 (2009)
9. Ge, S.S., Yang, C.G., Lee, T.H.: Adaptive predictive control using neural network for a class of pure-feedback systems in discrete time. IEEE Trans. Neural Netw. 19, 1599–1614 (2008)
10. Chen, M., Ge, S.S., Ren, B.B.: Robust Adaptive Neural Network Control for a Class of Uncertain MIMO Nonlinear Systems with Input Nonlinearities. IEEE Transactions on Neural Networks 21, 796–812 (2010)
11. Tong, S.C., Li, Y.M.: Adaptive Fuzzy Output Feedback Tracking Backstepping Control of Strict-Feedback Nonlinear Systems With Unknown Dead Zones. IEEE Transactions on Fuzzy Systems 20, 168–180 (2012)
12. Deolia, V.K., Purwar, S., Sharma, T.N.: Backstepping Control of Discrete-Time Nonlinear System Under Unknown Dead-zone Constraint. In: International Conference on Communication Systems and Network Technologies, pp. 344–349 (2011)

Application-Oriented Adaptive Neural Networks Design for Ship's Linear-Tracking Control

Wei Li, Jun Ning, and Zhengjiang Liu

Navigation College, Dalian Maritime University
{8235673,775729160}@qq.com, 15840682086@139.com

Abstract. By employing Radial Basis Function (RBF) Neural Networks (NN) to approximate uncertain functions, an application-oriented adaptive neural networks design for ship linear-tracking control was brought in based on dynamic surface control (DSC) and minimal-learning-parameter (MLP) algorithm. With less learning parameters and reduced computation load, the proposed algorithm can avoid the possible controller singularity problem and the trouble caused by "explosion of complexity" in traditional backstepping methods is removed, so it is convenient to be implemented in applications. In addition, the boundedness stability of the closed-loop system is guaranteed and the tracking error can be made arbitrarily small. Simulation results on ocean-going training ship 'YULONG' are shown to validate the effectiveness and the performance of the proposed algorithm.

Keywords: RBF Neural Networks, DSC, MLP, Linear-Tracking Control, Backstepping.

1 Introduction

Out while a vessel is traveling via way-points at constant cruise speed, what we want to see the most is that the ship can sail along the predetermined tracking lines. So ship linear-tracking control has become the most urgent and direct desire for a long time [1]-[2]. Traditional heading autopilot can't fully meet people's expectations for the reason that it can't directly control ship tracking error during the voyage [3]. Automatic heading helm can be regarded as the middle excessive substitute of the automatic tracking helm.

As we know, ship motion shows large inertia, long time delay, nonlinear characteristics, and vulnerable to model parameter perturbation, load condition and outside interference effect such as the winds, waves and currents, so it makes the controller design much difficult. In order to solve the problem, intelligent control which has the ability of adaptive, self-learning, self-optimization, self-adjusting have received considerable attention [4]- [6].

In this paper, based on DSC and MLP techniques proposed in [6]-[7], we developed an oriented adaptive neural networks design for ship linear-tracking control, which can circumvent the 'explosion of complexity' inherently in the conventional backstepping technique.

C. Guo, Z.-G. Hou, and Z. Zeng (Eds.): ISNN 2013, Part II, LNCS 7952, pp. 170–177, 2013.
© Springer-Verlag Berlin Heidelberg 2013

2   Problem Formulation

According to the references [6], we introduce the nonlinear ship straight-line motion equation plus the rudder actuator dynamics in the following form:

$$
\begin{cases}
\dot{y} = U\sin(\psi) \\
\dot{\psi} = r \\
\dot{r} = \eta_2(r) + g_2\delta + d \\
\dot{\delta} = -\dfrac{1}{T_E}\delta + \left(\dfrac{k_E}{T_E}\right)\delta_E
\end{cases}
\tag{1}
$$

where y denotes the sway displacement (cross-track error), ψ denotes the heading angle, r denotes the yaw rate, U denotes the cruise speed respectively. δ denotes the control rudder angle, d denotes uncertain external perturbations which is bounded, $\eta_2(r)$ is an unknown nonlinear function for r. g_2 is the control gain. One the other hand, we know that the ship's rudder actuator dynamics controller are designed as the fourth equation shown, where T_E and k_E are the time delay constant and the control gain of the rudder actuator, respectively. δ_E is the order angle of rudder. In order to make it   much easier, the coordinate transformation are defined as follows:

$$
x_1 = \psi + \arcsin\left(\frac{ky}{\sqrt{1+(ky)^2}}\right)
\tag{2}
$$

According to the formula above, where the parameters k is positive, we can see that if the system state x_1 and the yaw angle ψ can be stabilized, then cross-track error y will be stabilized. Due to the equation transformation, we can obtain a class of nonlinear uncertain system as follows:

$$
\begin{cases}
\dot{x}_1 = \eta_1 + g_1 x_2 + w_1 \\
\dot{x}_2 = \eta_2 + g_2 x_3 + w_2 \\
\dot{x}_3 = \eta_3 + g_3 u + w_3 \\
y = x_1
\end{cases}
\tag{3}
$$

$$
x_1 = \psi + \arcsin\left(\frac{ky}{\sqrt{1+(ky)^2}}\right) \;,\; x_2 = r \;,\; x_3 = \delta \;,\; g_1 = 1 \;,\; g_3 = \frac{K_E}{T_E} \;,
$$

$$
\eta_1 = \frac{k}{1+(ky)^2}U\sin\psi \;,\; \eta_3 = -\frac{1}{T_E}x_3 \;,\; d_2 = w \;,\; d_1 = d_3 = 0 \;,\; u = \delta_E \text{ and}
$$

$x = [x_1, x_2, x_3]^T \in R^q$ is the system state vector. $u, y \in R$ are the system's input and the output, respectively. Now we bring in the following assumptions.

Assumption 1. 1) The absolute value of the unknown virtual control-gain function g_i is positive. 2) For the purpose of analysis easier, without loss of generality, we make further assumption that:

$$0 < g_{min} \le |g_i| \le g_{max}$$

(4)

While g_{min} and g_{max} are the lower and upper bound of $|g_i|$, respectively.

Assumption 2. The reference signal $y_r(t)$ is a sufficiently smooth function of t, and y_r, $\dot{y}_r$, and $\ddot{y}_r$ are bounded, that means, there exists a positive constant Y_0 such that: $\Pi_0 := \left\{ (y_r, \dot{y}_r, \ddot{y}_r) : (y_r)^2 + (\dot{y}_r)^2 + (\ddot{y}_r)^2 < Y_0 \right\}$.

Assumption 3. $|w|$ is bounded which means, there exists a positive unknown constant γ_i, $|w_i| < \gamma_i, i = 1, \cdots, n$.

3   RBF Neural Network

It has been proved that we can use the following RBF neural networks to approximate an arbitrary smooth function F(x): $R^q \to R$:

$$F(x) = \vartheta^{\mathrm{T}} S(x)$$

(5)

Where $x \in \Omega_x \subset R^q$, weight vector $\vartheta = [\vartheta_1 ... \vartheta_l]^{\mathrm{T}} \in R^l$, $S(x) = [S_1(x), ... S_l(x)]^{\mathrm{T}}$, and the NNs node number $l > 1$. We usually choose $S_i(x)$ as the Gaussian function.

$$S_i(x) = \frac{1}{\sqrt{2\pi}\sigma_i} \exp\left[\frac{-(x - u_i)^T (x - u_i)}{\sigma_i^2} \right]$$

(6)

where $i = 1, ...l$, and $u_i = \left[u_{i_1} .... u_{i_l} \right]^{\mathrm{T}}$ is the center of the receptive field, σ_i is the width of the Gaussian function . Any continuous function with $\forall x \in \Omega_x \subset R^q$ can be approximated as

$$F(x) = \vartheta^{*\mathrm{T}} S(x) + \varepsilon, \forall x \in \Omega_x$$

(7)

where ϑ^* is the ideal weight vector, ε is the approximation error with an assumption of $|\varepsilon| \le \varepsilon^*$, where the unknown constant $\varepsilon^* > 0$ for all $Z \in \Omega_Z$. With the estimate value $\hat{\vartheta}$ minimizing $|\varepsilon|$ for all $x, x \in \Omega_x$, the ideal weight vector ϑ^* can be typically defined as

$$\vartheta^* = \arg \min_{\vartheta \in R^1} \left\{ \sup_{Z \in \Omega_Z} \left| F(x) - \vartheta^T S(x) \right| \right\}$$

(8)

Now, we bring in an assumption and a very important lemma.

Assumption 4. For the sake of simplicity, let ε_i^* be an unknown upper bound of the approximation errors $\varepsilon_i, i = 1, 2$.

Lemma 1[7]. For any given real continuous function $f(x)$ with $f(0) = 0$, if the continuous function separation technique [8] and the RBF NN approximation technique are used, then $f(x)$ can be denote as

$$f(x) = \overline{S}(x) A x$$

(9)

where $\overline{S}(x) = \left[1, S(x) \right] = \left[1, s_1(x), s_2(x), \cdots s_l(x) \right]$, $A^T = \left[\varepsilon, \vartheta^T \right]$, $\varepsilon^T = \left[\varepsilon_1, \varepsilon_2, \cdots \varepsilon_n \right]$ is a vector of the approximation error, and ϑ is a weight matrix.

$$\vartheta = \begin{bmatrix} \vartheta_{11}^* & \vartheta_{12}^* & \cdots & \vartheta_{1n}^* \\ \vartheta_{21}^* & \vartheta_{22}^* & \cdots & \vartheta_{2n}^* \\ \vdots & \vdots & \cdots & \vdots \\ \vartheta_{l1}^* & \vartheta_{l2}^* & \cdots & \vartheta_{ln}^* \end{bmatrix}$$

4   Control Design

In this part, we will incorporate the DSC-MLP technique and the RBF neural networks to develop an adaptive tracking control design scheme for the system (1). Define $\lambda_i = g_{\min}^{-1} \max \{ b_i^2, \xi_i^2 \}$, $\hat{\lambda}_i$ are estimates of λ_i, and choose the adaptive law for $\hat{\lambda}_i$ as follows:

$$\dot{\hat{\lambda}}_i = \Gamma_i \left[\frac{1}{4\gamma_i^2} S_i(\overline{x}_i) S_i^T(\overline{x}_i) s_i^2 + \frac{\varphi^2(\overline{x}_i)}{4\beta_i^2} s_i^2 - \sigma_i (\hat{\lambda}_i - \lambda_i^0) \right]$$

(10)

where γ_i, β_i, Γ_i, σ_i and λ_i^0 are design constants.

Step 1: According to Lemma 1, define the tracking error $s_1 = x_1 - y_r$, we have

$$\dot{s}_1 = g_1 x_2 + \eta_1(x_1) + w_1 - \dot{y}_r \tag{11}$$

$$\eta_1(x_1) = S_1(x_1)A_1 x_1 + \varepsilon_1 = S_1(x_1)A_1 s_1 + S_1(x_1)A_1 y_r + \varepsilon_1 \tag{12}$$

Where ε_1 denote the approximation error. Let $b_1 = \|A_1\|$, the normalized term $A_1^m = \dfrac{A_1}{\|A_1\|} = \dfrac{A_1}{b_1}$, and $\theta_1 = A_1^m s_1$, then we have

$$\eta_1(x_1) = b_1 S_1(x_1)\theta_1 + S_1(x_1)A_1 y_r + \varepsilon_1 \tag{13}$$

$$\dot{s}_1 = g_1 x_2 + b_1 S_1 \theta_1 + \Delta_1 - \dot{y}_r \tag{14}$$

where $\Delta_1 = S_1(x_1)A_1 y_r + \varepsilon_1 + w_1$, then we have

$$\|\Delta_1\| \le \|S_1(x_1)A_1 y_r + \varepsilon_1^* + w_1\| \le g_{\min}\xi_1\varphi_1(x_1) \tag{15}$$

where, $\xi_1 = g_{\min}^{-1} \max(\|A_1 y_r\|, \|\varepsilon_1^* + w_1\|)$ and $\phi_1(x_1) = 1 + \|S_1\|$. It is obvious that $\|\Delta_1\|$ is bounded, because the bound of y_r, ε_1^* and w_1. Now choose the virtual controller α_2 for x_2 as

$$\alpha_2 = -k_1 s_1 + \dot{y}_r - \frac{\varphi^2(x_1)}{4\beta_1^2} s_1 - \frac{\hat{\lambda}_1}{4\gamma_1^2} S_1(x_1)S_1^T(x_1)s_1 \tag{16}$$

For the sake of avoiding calculation explosion caused by repeated derivations, the DSC technique is developed and virtual control law is replaced by its estimation using the following first-order filter with the time constant of τ_2.

$$\tau_2 \dot{z}_2 + z_2 = \alpha_2, z_2(0) = \alpha_2(0) \tag{17}$$

By defining the output error of this filter as $y_2 = z_2 - \alpha_2$, it yields

$$\dot{y}_2 = \dot{z}_2 - \dot{\alpha}_2 = -\frac{y_2}{\tau_2} + (-\frac{\partial \alpha_2}{\partial s_1}\dot{s}_1 - \frac{\partial \alpha_2}{\partial x_1}\dot{x}_1 - \frac{\partial \alpha_2}{\partial \hat{\lambda}_1}\dot{\hat{\lambda}}_1 - \frac{\partial \alpha_2}{\partial \hat{\vartheta}_1}\dot{\hat{\vartheta}}_1)$$

$$= -\frac{y_2}{\tau_2} + B_2(s_1, s_2, y_2, \hat{\lambda}_1, y_r, \dot{y}_r, \ddot{y}_r) \tag{18}$$

Step 2: Define the error variable $s_2 = x_2 - z_2$, Similar to step 1, we have

$$\dot{s}_2 = g_2 x_3 + \eta_2(x_2) + w_2 - \dot{z}_2 \tag{19}$$

$$\eta_2(x_2) = S_2 A_2 \bar{x}_2^T + \mathcal{E}_2 = S_2 A_2 \begin{bmatrix} s_1 + y_r \\ s_2 + z_2 \end{bmatrix} + \mathcal{E}_2 = b_2 S_2 \theta_2 + \Delta_2' \tag{20}$$

where $b_2 = \|A_2\|$, the normalized term $A_2^m = \frac{A_2}{\|A_2\|}$, $\theta_2 = A_2^m s_2$ with

$\bar{s}_i = [s_1, s_2]^T$, and $\Delta_2' = S_2(\bar{x}_2)A_2 y_r + S_2(\bar{x}_2)A_2 z_2 + \mathcal{E}_2$, then we have

$$\dot{s}_2 = g_2(\bar{x}_2)x_3 + b_2 S_2(\bar{x}_2)\theta_2 + \Delta_2 - \dot{z}_2 \tag{21}$$

$$\|\Delta_2\| \le \|S_2 A_2 y_r + S_2 A_2 z_2 + \mathcal{E}_2 + w_2\| \le g_{min}\xi_2\varphi_2 \tag{22}$$

$\xi_2 = g_{min}^{-1} \max(\|A_2 y_r\|, \|A_2 z_2\|, \|\mathcal{E}_2^* + w_2\|)$ and $\phi(\bar{x}_2) = 1 + \|S_2(\bar{x}_2)\|$, also $\|\Delta_2\|$ is bounded. Then we choose virtual controller α_3 for x_3 as

$$\alpha_3 = -k_2 s_2 + \dot{z}_2 - \frac{\varphi^2(\bar{x}_2)}{4\beta_2^2}s_2 - \frac{\hat{\lambda}_2}{4\gamma_2^2}S_2(\bar{x}_2)S_2^T(\bar{x}_2)s_2 \tag{23}$$

Similar to step 1, we have

$$\tau_3 \dot{z}_3 + z_3 = \alpha_3, z_3(0) = \alpha_3(0) \tag{24}$$

$$\dot{y}_3 = \dot{z}_3 - \dot{\alpha}_3 = -\frac{y_3}{\tau_3} + (-\frac{\partial \alpha_3}{\partial s_2}\dot{s}_2 - \frac{\partial \alpha_3}{\partial x_2}\dot{x}_2 - \frac{\partial \alpha_3}{\partial \hat{\lambda}_2}\dot{\hat{\lambda}}_2 - \frac{\partial \alpha_3}{\partial \hat{\vartheta}_2}\dot{\hat{\vartheta}}_2 + \ddot{y}_r)$$

$$= -\frac{y_3}{\tau_3} + B_3(s_1, s_2, s_3, y_2, y_3, \hat{\lambda}_1, \hat{\lambda}_2, y_r, \dot{y}_r, \ddot{y}_r) \tag{25}$$

Step 3: Define the error variable $s_3 = x_3 - z_3$, similarly to above, we have

$$\dot{s}_3 = g_3 u + \eta_3(x_3) + w_3 - \dot{z}_3 \qquad (26)$$

$$\eta_3(x_3) = S_3 A_3 \bar{x}_3^T + \varepsilon_3 = S_3 A_3 \begin{bmatrix} s_1 + y_r \\ s_2 + z_2 \\ s_3 + z_3 \end{bmatrix} + \varepsilon_3 = b_3 S_3 \theta_3 + \Delta_3' \qquad (27)$$

where $b_3 = \|A_3\|$, the normalized term $A_3^m = \dfrac{A_3}{\|A_3\|}$, $\theta_3 = A_3^m s_3$ with

$\bar{s}_i = [s_1, s_2, s_3]^T$, and $\Delta_3' = S_3(\bar{x}_3) A_3 y_r + S_3(\bar{x}_3) A_3 z_2 + S_3(\bar{x}_3) A_3 z_3 + \varepsilon_3$, then

$$\dot{s}_3 = g_3(\bar{x}_3)\mu + b_3 S_3(\bar{x}_3)\theta_3 + \Delta_3 - \dot{z}_3 \qquad (28)$$

$$\|\Delta_3\| \le \|S_3 A_3 y_r + S_3 A_3 z_2 + S_3 A_3 z_3 + \varepsilon_3 + w_3\| \le g_{\min} \xi_3 \varphi_3 \qquad (29)$$

$\xi_3 = g_{\min}^{-1} \max(\|A_3 y_r\|, \|A_3 z_2\|, \|A_3 z_3\|, \|\varepsilon_3^* + w_3\|)$ and $\phi(\bar{x}_3) = 1 + \|S_3(\bar{x}_3)\|$,

$\|\Delta_3\|$ is also bounded. Then we choose virtual control law μ as

$$u = -k_3 s_3 + \dot{z}_3 - \frac{\varphi^2(\bar{x}_3)}{4\beta_3^2} s_3 - \frac{\hat{\lambda}_3}{4\gamma_3^2} S_3(\bar{x}_3) S_3^T(\bar{x}_3) s_3 \qquad (30)$$

Remark 1: It can be observed that, for the closed-loop control system, there are only one learning parameter, which is independent of order of the subsystem, to be updated online in the controller u and the virtual controller α_i, no matter how many input variables there are in the subsystem, thus the well-known curse of dimensionality is circumvented by the proposed algorithm in this paper.

Proof. The proof is similar to that in [9]

5   Application Examples

In this section, simulation results are based on an ocean going training vessel YULONG. The initial conditions for x_0, y_0, ψ_0 and r_0 are $[0m, 0m, -0.2, 0]$. The desired reference signal $y_r = 0$, and design parameters are chosen as $k = 0.002$, $k_1 = 0.5$, $k_2 = 3.5$, $k_3 = 10$. The external disturbance signal is chosen as $w = 0.001(1 + \sin(0.1t))$. $\Gamma_1 = \Gamma_2 = \Gamma_3 = 0.2$, $\sigma_1 = \sigma_2 = \sigma_3 = 0.005$, $\tau_2 = \tau_3 = 0.5$. The initial values of the weights vectors $\hat{\lambda}_i^0$, $i = 1, 2, 3$ are zero. The simulation results are shown in Figs. 1

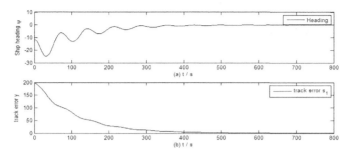

Fig. 1. (a)   heading ψ , (b) cross-track error y

6   Conclusion

In this paper, ship linear-tracking control problem has been considered. By combining the DSC technique and MLP algorithm, a scheme of adaptive neural networks control on ship linear-path following is developed based on Lyapunov stability theory. The proposed scheme can reduce the the computation load dramatically and force the ship to follow a desired path, so it is much easier to implement in applications.

References

1. Pettersen, K.Y., Lefeber, E.: Way-point tracking control of ships. In: Proc. of 40th IEEE CDC, Orlando, USA, pp. 940–945 (December 2001)
2. Ghommam, J., Mnif, F., Benali, A., Derbel, N.: Nonsingular Serret-Frenet based path following control for an underactuated surface vessel. Journal of Dynamic Systems, Measurement and Control 131(2), 1–8 (2009)
3. Lin, Y., Chen, X., Zhou, T.: Nonlinear PID Design for Ship Course Autopilot Control. Marine Electric & Electronic Technology 29(6), 46–48 (2009)
4. Xu, J.H., Liu, Y.J.: The application of Hybrid intelligence system for ship-tracking control, pp. 53–61. Master dissertation of Shanghai Maritime University (2006)
5. Li, T.S., Yu, B., Hong, B.G.: A Novel Adaptive Fuzzy Design for Path Following for Underactuated Ships with Actuator Dynamics. In: ICIEA 2009, pp. 2796–2800 (2009)
6. Yang, Y.S., Ren, J.S.: Adaptive fuzzy robust tracking controller design via small gain approach and its application. IEEE Trans. Fuzzy Syst. 11(6), 783–795 (2003)
7. Yang, Y.S., Feng, G., Ren, J.S.: A combined backstepping and small-gain approach to robust adaptive fuzzy control for strict-feedback nonlinear systems. IEEE Trans. Syst. Mon. Cyhern. A. Syst, Humans 34(3), 406–420 (2004)
8. Wang, D., Hang, J.: Neural network-based adaptive dynamic surface control for a class of uncertain nonlinear systems in strict-feedback form. IEEE Trans. Neural Netw. 16(1), 195–202 (2005)
9. Li, W., Ning, J., Liu, Z., Li, T.: Adaptive Neural Networks Control on Ship's Linear-Path Following. In: Huang, T., Zeng, Z., Li, C., Leung, C.S. (eds.) ICONIP 2012, Part V. LNCS, vol. 7667, pp. 418–427. Springer, Heidelberg (2012)

Adaptive Synchronization for Stochastic Markovian Jump Neural Networks with Mode-Dependent Delays

Cheng-De Zheng[*], Xixi Lv, and Zhanshan Wang

School of Science, Dalian Jiaotong University, Dalian, 116028, P.R. China
School of Information Science and Engineering, Northeastern University,
Shenyang, 110004, P.R. China

Abstract. This paper studies the adaptive synchronization problem for a kind of stochastic Markovian jump neural networks with mode-dependent and unbounded distributed delays. By virtue of the Lyapunov stability theory and the stochastic analysis technique, a generalized LaSalle-type invariance principle for stochastic Markovian differential delay equations is utilized to investigate the globally almost surely asymptotical stability of the error dynamical system in the mean-square sense.

Keywords: Adaptive synchronization, stochastic perturbation, mode-dependent delay, Markovian jump.

1 Introduction

Since the pioneering work [5] of Pecora and Carroll in 1990, control and synchronization of chaotic systems have become an important topic during the past decades. There exist many benefits of having synchronization or chaos synchronization in some engineering applications, such as secure communication, chaos generators design, chemical reactions, biological systems, information science, and so on. Many excellent papers and monographs on synchronization of chaotic systems with or without time delays have been published [2,6]. Variety of alternative schemes for ensuring the synchronization have been proposed, such as adaptive design control, feedback control, complete synchronization control, impulsive control, anti-synchronization control, and projective synchronization control. Because of the finite switching speed of amplifiers and the inherent communication time of neurons, time delays are frequently encountered in various engineering, biological, and economic systems. It has been revealed that time delay may cause periodic oscillations, bifurcation and chaotic attractors and

---

[*] This work was supported by the National Natural Science Foundation of China (61074073, 61034005, 61273022), Program for New Century Excellent Talents in University of China (NCET-10-0306), and the Fundamental Research Funds for the Central Universities under Grants N110504001.

C. Guo, Z.-G. Hou, and Z. Zeng (Eds.): ISNN 2013, Part II, LNCS 7952, pp. 178–185, 2013.

so on. Thus synchronization of delayed chaotic neural networks has become an important research area.

Motivated by the preceding discussions, our objective in this paper is to study the adaptive synchronization problem for a kind of stochastic Markovian jump neural networks with mode-dependent and unbounded distributed delays. By employing the Lyapunov stability theory, by virtue of stochastic analysis, a generalized LaSalle-type invariance principle for stochastic Markovian differential delay equations is utilized to investigate the globally almost surely asymptotical stability of the error dynamical system in the mean-square sense.

Notation. Throughout this paper, let W^T, W^{-1} denote the transpose and the inverse of a square matrix W, respectively. Let $W > 0(< 0)$ denote a positive (negative) definite symmetric matrix, I denotes the identity matrix of appropriate dimension, the symbol "*" denotes a block that is readily inferred by symmetry. The shorthand $\text{col}\{M_1, M_2, ..., M_k\}$ denotes a column matrix with the matrices $M_1, M_2, ..., M_k$. $\text{diag}\{\cdot\}$ stands for a diagonal or block-diagonal matrix. For $\tau > 0, \mathcal{C}([-\tau, 0]; \mathbb{R}^n)$ denotes the family of continuous functions ϕ from $[-\tau, 0]$ to $\mathbb{R}^n$ with the norm $||\phi|| = \sup_{-\tau \leq s \leq 0} |\phi(s)|$. Moreover, let $(\Omega, \mathbb{F}, \mathbb{P})$ be a complete probability space with a filtration $\{\mathbb{F}_t\}_{t \geq 0}$ satisfying the usual conditions and $\mathbb{E}\{\cdot\}$ representing the mathematical expectation. Denote by $\mathcal{C}^p_{\mathbb{F}_0}([-\tau, 0]; \mathbb{R}^n)$ the family of all bounded, $\mathbb{F}_0$-measurable, $\mathcal{C}([-\tau, 0]; \mathbb{R}^n)$-valued random variables $\xi = \{\xi(s) : -\tau \leq s \leq 0\}$ such that $\sup_{-\tau \leq s \leq 0} \mathbb{E}|\xi(s)|^p < \infty$. $||\cdot||$ stands for the Euclidean norm; Matrices, if not explicitly stated, are assumed to have compatible dimensions.

2   Problem Description and Preliminaries

We consider the following neural networks with mixed time delays

$$
\mathrm{d}x(t) = \Big[- C(\eta(t))x(t) + A(\eta(t))\hat{f}(x(t))
$$

$$
+ B(\eta(t))\hat{f}(x(t - \tau(t, \eta(t)))) + D(\eta(t)) \int_{-\infty}^t K(t - s)\hat{f}(x(s))\mathrm{d}s + J \Big] \mathrm{d}t,
\tag{1}
$$

$$
x(t) = \varphi_1(t), \qquad\qquad t \in (-\infty, 0],
$$

where $x(t) = [x_1(t), x_2(t), ..., x_n(t)]^T \in \mathbb{R}^n$ denotes the state of the i-th neuron at time t, the positive diagonal matrix $C(\eta(t))$ is the self-feedback term, $A(\eta(t)), B(\eta(t)), D(\eta(t)) \in \mathbb{R}^{n \times n}$ are the interconnection matrices representing the weight coefficients of the neurons. $\hat{f}(x(t)) = [\hat{f}_1(x_1(t)), \hat{f}_2(x_2(t)), ..., \hat{f}_n(x_n(t))]^T \in \mathbb{R}^n$ denotes the neural activation function, the bounded function $\tau(t, \eta(t))$ represents unknown time-varying delay with $0 \leq \tau(t, \eta(t)) \leq \bar{\tau}(\eta(t)) \leq \bar{\tau}, \dot{\tau}(t, \eta(t)) \leq \tau_d(\eta(t)) \leq \tau_d$, where $\bar{\tau}(\eta(t)), \bar{\tau}$ are positive scalars. $J = [J_1, J_2, ..., J_n]^T$ is an external input, $\varphi_1(t)$ is a real-valued initial vector function that is continuous on the interval $(-\infty, 0]$. $K(t - s) = \text{diag}\{k_1(t - s), k_2(t - s), ..., k_n(t - s)\}$ denotes the delay kernel. It is assumed that $k_i(\cdot)$ is a real value non-negative continuous function defined in $[0, \infty)$ satisfying

$$\int_0^\infty k_i(s)\mathrm{d}s = 1, \ i = 1, 2, ..., n.$$

$\{\eta(t), t \geq 0\}$ is a homogeneous, finite-state Markovian process with right continuous trajectories and taking values in finite set $\mathcal{N} = \{1, 2, ..., N\}$ with given probability space $(\Omega, \mathbb{F}, \mathbb{P})$ and the initial model η_0. Let $\Pi = [\pi_{ij}]_{N \times N}$ denote the transition rate matrix with transition probability:

$$\mathbb{P}(\eta(t + \delta) = j | \eta(t) = i) = \begin{cases} \pi_{ij}\delta + o(\delta), & i \neq j, \\ 1 + \pi_{ii}\delta + o(\delta), & i = j, \end{cases}$$

where $\delta > 0$, $\lim_{\delta \to 0^+} \frac{o(\delta)}{\delta} = 0$ and π_{ij} is the transition rate from mode i to mode j satisfying $\pi_{ij} \geq 0$ for $i \neq j$ with

$$\pi_{ii} = -\sum_{j=1, j \neq i}^{N} \pi_{ij}, \quad i, j \in \mathcal{N}.$$

For convenience, each possible value of $\eta(t)$ is denoted by $i(i \in \mathcal{N})$ in the sequel. Then we have

$$C_i = C(\eta(t)), \quad A_i = A(\eta(t)), \quad B_i = B(\eta(t)), \quad D_i = D(\eta(t)), \quad \tau_i(t) = \tau(t, \eta(t)).$$

Throughout this paper, the following assumption is made on the neuron activation functions:

Assumption 1. Each neural activation function $\hat{f}_j(\cdot)$ is bounded and there exist real constants σ_j^-, σ_j^+ such that

$$\sigma_j^- \leq \frac{\hat{f}_j(\xi) - \hat{f}_j(\zeta)}{\xi - \zeta} \leq \sigma_j^+, \quad \forall \xi, \zeta \in \mathbb{R}, \ \xi \neq \zeta, \ j = 1, 2, ..., n.$$

For notational simplicity, we denote
$\Sigma_1 = \text{diag}\{\sigma_1^-, \sigma_2^-, ..., \sigma_n^-\}, \Sigma_2 = \text{diag}\{\sigma_1^+, \sigma_2^+, ..., \sigma_n^+\}$.
In order to observe the synchronization behavior of system (1), we construct the response system as follows

$$\mathrm{d}y(t) = \left[-C_i y(t) + A_i \hat{f}(y(t)) + B_i \hat{f}(y(t - \tau_i(t))) + D_i \int_{-\infty}^t K(t - s)\hat{f}(y(s))\mathrm{d}s \right.$$
$$\left. + J + u(t) \right]\mathrm{d}t + v_i(t, y(t) - x(t), y(t - \tau_i(t)) - x(t - \tau_i(t)))\mathrm{d}\omega(t),$$

$$\tag{2}$$

$$y(t) = \varphi_2(t), \quad\quad\quad t \in (-\infty, 0],$$

where $u(t)$ is an appropriate control input that will be designed in order to obtain a certain control objective. $w(t)$ is a one-dimensional Brown motion defined on a complete probability space $(\Omega, \mathbb{F}, \mathbb{P})$ with a natural filtration $\{\mathbb{F}_t\}_{t \geq 0}$, and $v_i : \mathbb{R}^+ \times \mathbb{R}^n \times \mathbb{R}^n \to \mathbb{R}^n$ is the noise intensity vector. This type of stochastic perturbation can be regarded as a result from the occurrence of external random fluctuation and other probabilistic causes.

Let $e(t) = y(t) - x(t)$ be the error state, it yields the synchronization error dynamical systems as follows:

$$
\begin{aligned}
de(t) &= \Big[-C_i e(t) + A_i f(e(t)) + B_i f(e(t - \tau_i(t))) \\
&\quad + D_i \int_{-\infty}^t K(t-s) f(e(s)) ds + Z(t) e(t) \Big] dt + v_i(t, e(t), e(t - \tau_i(t))) dw(t), \\
&\doteq \rho_i(t) dt + v_i(t) dw(t), \\
e(t) &= \varphi(t), \qquad\qquad t \in (-\infty, 0],
\end{aligned}
\tag{3}
$$

where $f(e(t)) = \hat{f}(y(t)) - \hat{f}(x(t))$, $\varphi(t) = \varphi_2(t) - \varphi_1(t)$. From Assumption 1, it is easy to derive that

$$
f_j(0) = 0, \quad \sigma_j^- \leq \frac{f_j(s)}{s} \leq \sigma_j^+, \quad \forall\, s \in \mathbb{R},\ s \neq 0, \quad j = 1, 2, ..., n.
\tag{4}
$$

Furthermore, we make the following assumption:

Assumption 2. The noise intensity vector is assumed to be of the form:

$$
v_i(t) = E_i e(t) + F_i e(t - \tau_i(t)),
\tag{5}
$$

where E_i, F_i are known real matrices.

Instead of the usual linear feedback, in this paper, we consider the following feedback controller:

$$
u(t) = Z(t) e(t),
\tag{6}
$$

where the feedback strength $Z(t) = \text{diag}\{z_1(t), z_2(t), ..., z_n(t)\}$ is updated by the following law:

$$
\dot{z}_j(t) = -\gamma_j e_j^2(t),
\tag{7}
$$

where $\gamma_j > 0$ is an arbitrary constant, $j = 1, 2, ..., n$.

3   Main Result

As well known, Itô's formula plays important role in the stability analysis of stochastic Markovian systems and we cite some related results here [1]. Consider a general stochastic Markovian delay system

$$
dz(t) = f(t, z(t), z(t - \kappa), \eta(t)) dt + g(t, z(t), z(t - \kappa), \eta(t)) dw(t),
\tag{8}
$$

on $t \geq t_0$ with initial value $z(t_0) = z_0 \in \mathbb{R}^n$, where $\kappa > 0$ is time delay, $f : \mathbb{R}^+ \times \mathbb{R}^n \times \mathbb{R}^n \times \mathcal{N} \to \mathbb{R}^n$ and $g : \mathbb{R}^+ \times \mathbb{R}^n \times \mathbb{R}^n \times \mathcal{N} \to \mathbb{R}^{n+m}$. Let $\mathcal{C}^{2,1}(\mathbb{R}^+ \times \mathbb{R}^n \times \mathbb{R}^n \times \mathcal{N}, \mathbb{R}^+)$ denote the family of all nonnegative functions $V(t, z, v, \eta(t))$ on $\mathbb{R}^+ \times \mathbb{R}^n \times \mathbb{R}^n \times \mathcal{N}$ which are continuously twice differentiable in z, v and once differentiable in t. Let $\mathcal{L}$ be the weak infinitesimal generator of the random process $\{z(t), \eta(t)\}_{t \geq t_0}$ along the system (8) (see [3]), i.e.

$$\mathcal{L}V(t, z_t, v_t, i) := \lim_{\delta \to 0^+} \left[\mathbb{E}\{V(t+\delta, z_{t+\delta}, v_{t+\delta}, \eta(t+\delta)) | z_t, v_t, \eta(t) = i\} \right.$$
$$\left. - V(t, z_t, v_t, \eta(t) = i) \right], \quad (9)$$

then, by the Dynkin's formula, one can get

$$\mathbb{E}V(t, z(t), v(t), i) = \mathbb{E}V(t_0, z(t_0), v(t_0), i) + \mathbb{E}\int_{t_0}^t \mathcal{L}V(s, z(s), v(s), i)\mathrm{d}s.$$

Similar to Lemma 1 of [4], we can obtain a generalized LaSalle-type invariance principle for stochastic Markovian differential delay equations (8) stated as follows.

Lemma 1. Assume that system (8) exists a unique solution $z(t, \xi)$ on $t > 0$ for any given initial data $\{z(\theta) : -\kappa \leq \theta \leq 0\} = \xi \in \mathcal{C}^p_{\mathbb{F}_0}([-\tau, 0]; \mathbb{R}^n)$, moreover, both $f(t, z, v, \eta(t))$ and $g(t, z, v, \eta(t))$ are locally bounded in (z, v) and uniformly bounded in t. If there are a function $V \in \mathcal{C}^{2,1}(\mathbb{R}^+ \times \mathbb{R}^n \times \mathbb{R}^n \times \mathcal{N}, \mathbb{R}^+), \chi \in L^1(\mathbb{R}^+, \mathbb{R}^+)$ and $\psi_1, \psi_2 \in \mathcal{C}(\mathbb{R}^n, \mathbb{R}^+)$ such that

$$\mathbb{E}\mathcal{L}V(t, z, v, \eta(t)) \leq \chi(t) - \psi_1(z) + \psi_2(v), \quad (t, z, v, \eta(t)) \in \mathbb{R}^+ \times \mathbb{R}^n \times \mathbb{R}^n \times \mathcal{N},$$
$$\psi_1(z) \geq \psi_2(z), \quad \forall z \neq 0,$$
$$\lim_{||z|| \to \infty} \inf_{0 \leq t < \infty} V(t, z, v, \eta(t)) = \infty.$$

Then

$$\lim_{t \to \infty} z(t, \xi) = 0 \quad a.s.$$

for every $\xi \in \mathcal{C}^p_{\mathbb{F}_0}([-\tau, 0]; \mathbb{R}^n)$.

In order to get the main result, we propose the following lemma:

Lemma 2. For each $i \in \mathcal{N}$, we have the following equalities

$$\mathcal{L}\left\{ \int_{t-\tau(t,\eta(t))}^t x(s)^T Q(\eta(t)) x(s)\mathrm{d}s \right\}$$
$$= x(t)^T Q_i x(t) - (1 - \dot{\tau}_i(t)) x(t - \tau_i(t))^T Q_i x(t - \tau_i(t))$$
$$+ \sum_{j=1}^N \pi_{ij} \left\{ \int_{t-\tau_i(t)}^t x(s)^T Q_j x(s)\mathrm{d}s + \tau_j(t) x(t - \tau_i(t))^T Q_i x(t - \tau_i(t)) \right\}, \quad (10)$$

$$\pounds\left\{\int_{t-\tau_i(t)}^{t}\int_{\theta}^{t}x(s)^T Rx(s)\mathrm{d}s\mathrm{d}\theta\right\}$$

$$=\tau_i(t)x(t)^T Rx(t) - (1-\dot{\tau}_i(t))\int_{t-\tau_i(t)}^{t}x(s)^T Rx(s)\mathrm{d}s$$

$$+\sum_{j=1}^{N}\pi_{ij}\tau_j(t)\int_{t-\tau_i(t)}^{t}x(s)^T Rx(s)\mathrm{d}s. \tag{11}$$

Now, we begin to state our main result.

Theorem 1. *Consider the system (3) satisfying Assumption 1, the drive system (1) and the response system (2) can be synchronized for any $0 \le \tau_i(t) \le \bar{\tau}_i \le \bar{\tau}$, $\dot{\tau}_i(t) \le \tau_{di} < 1$, if there exist symmetric definite positive matrices $Q_{1i}, Q_{3i}, Q_{4i}, R_1, R_3, R_4, X, H$, diagonal positive matrices P_i, S_i, U_i, W and positive number α and any real matrices Q_{2i}, R_2 satisfying the following inequalities*

$$\mathcal{Q}_i \equiv \begin{bmatrix} Q_{1i} & Q_{2i} \\ * & Q_{3i} \end{bmatrix} > 0, \tag{12}$$

$$\mathcal{R} \equiv \begin{bmatrix} R_1 & R_2 \\ * & R_3 \end{bmatrix} > 0, \tag{13}$$

$$\sum_{j=1}^{N}(\pi_{ij}\mathcal{Q}_j + \pi'_{ij}\bar{\tau}_j\mathcal{R}) \le (1-\tau_{di})\mathcal{R}, \tag{14}$$

$$\sum_{j=1}^{N}\pi_{ij}Q_{4j} \le R_4, \tag{15}$$

$$\Omega_i = \begin{bmatrix} \Omega_{11i} & E_i^T(P_i+\bar{\tau}H)G_i & H & \Omega_{14i} & P_iB_i & P_iD_i & H \\ * & \Omega_{22i} & 0 & 0 & \Omega_{25i} & 0 & 0 \\ * & * & -Q_{4i}-H & 0 & 0 & 0 & H \\ * & * & * & \Omega_{44i} & 0 & 0 & 0 \\ * & * & * & * & \Omega_{55i} & 0 & 0 \\ * & * & * & * & * & -W & 0 \\ * & * & * & * & * & * & -H \end{bmatrix} < 0, \tag{16}$$

where

$$\Omega_{11i} = -P_iC_i - C_iP_i - 2\alpha P_i + E_i^T(P_i+\bar{\tau}H)E_i - H + X$$

$$+ \sum_{j=1}^{N}\pi_{ij}P_j - \Sigma_1\Sigma_2 S_i + Q_{1i} + \bar{\tau}_iR_1 + Q_{4i} + \bar{\tau}R_4,$$

$$\Omega_{14i} = P_iA_i + \frac{1}{2}(\Sigma_1+\Sigma_2)S_i + Q_{2i} + \bar{\tau}_iR_2,$$

$$\Omega_{22i} = G_i^T(P_i+\bar{\tau}H)G_i - \Sigma_1\Sigma_2 U_i - X - (1-\tau_{di})Q_{1i} + \sum_{j=1}^{N}\pi'_{ij}\bar{\tau}_jQ_{1i},$$

$$\Omega_{25i} = \frac{1}{2}(\Sigma_1 + \Sigma_2)U_i - (1 - \tau_{di})Q_{2i} + \sum_{j=1}^{N} \pi'_{ij}\bar{\tau}_j Q_{2i},$$

$$\Omega_{44i} = W - S_i + Q_{3i} + \bar{\tau}_i R_3, \quad \Omega_{55i} = -U_i - (1 - \tau_{di})Q_{3i} + \sum_{j=1}^{N} \pi'_{ij}\bar{\tau}_j Q_{3i},$$

with $\pi'_{ij} = \max\{\pi_{ij}, 0\}$.

Proof. Define the following Lyapunov-Krasovskii functional:

$$V(t, e_t, i) = e(t)^T P_i e(t) + \sum_{j=1}^{n} w_j \int_0^\infty k_j(s) \int_{t-s}^t f_j^2(e_j(\theta)) d\theta ds$$

$$+ \int_{t-\tau_i(t)}^t \xi(s)^T \mathcal{Q}_i \xi(s) ds + \int_{t-\tau_i(t)}^t \int_\theta^t \xi(s)^T \mathcal{R}\xi(s) ds d\theta$$

$$+ \int_{t-\bar{\tau}}^t e(s)^T Q_{4i} e(s) ds + \int_{t-\bar{\tau}}^t \int_\theta^t e(s)^T R_4 e(s) ds d\theta$$

$$+ \int_{t-\bar{\tau}}^t \int_\theta^t v_i(s)^T H v_i(s) ds d\theta + \sum_{j=1}^{n} \frac{p_{ji}}{\gamma_j}(z_j(t) + \alpha)^2,$$

with $\xi(s) = \mathrm{col}\{x(s), f(x(s))\}$, $W = \mathrm{diag}\{w_1, w_2, ..., w_n\}$, and α being a large positive constant which can be determined arbitrary.

Based on Lemma 2, by using the well-known Itô's differential formula, calculating the weak infinitesimal generator along the trajectory of (3) results in

$$\pounds V_i = 2e(t)^T P_i \rho_i(t) + \mathrm{trace}\left[v_i(t)^T P_i v_i(t)\right] + \sum_{j=1}^{N} \pi_{ij} e(t)^T P_j e(t)$$

$$+ 2\sum_{j=1}^{n} \frac{p_{ji}}{\gamma_j}(z_j(t) + \alpha)\dot{z}_j(t) + \sum_{j=1}^{n} w_j \int_0^\infty k_j(s)\left[f_j^2(e_j(t)) - f_j^2(e_j(t-s))\right] ds$$

$$+ \xi(t)^T \mathcal{Q}_i \xi(t) - (1 - \dot{\tau}_i(t))\xi(t - \tau_i(t))^T \mathcal{Q}_i \xi(t - \tau_i(t))$$

$$+ \sum_{j=1}^{N} \pi_{ij}\left\{\int_{t-\tau_i(t)}^t \xi(s)^T \mathcal{Q}_j \xi(s) ds + \tau_j(t)\xi(t - \tau_i(t))^T \mathcal{Q}_i \xi(t - \tau_i(t))\right\}$$

$$+ \tau_i(t)\xi(t)^T \mathcal{R}\xi(t) - (1 - \dot{\tau}_i(t))\int_{t-\tau_i(t)}^t \xi(s)^T \mathcal{R}\xi(s) ds$$

$$+ \sum_{j=1}^{N} \pi_{ij}\tau_j(t)\int_{t-\tau_i(t)}^t \xi(s)^T \mathcal{R}\xi(s) ds + e(t)^T Q_{4i} e(t)$$

$$- e(t - \bar{\tau})^T Q_{4i} e(t - \bar{\tau}) + \sum_{j=1}^{N} \pi_{ij}\int_{t-\bar{\tau}}^t e(s)^T Q_{4j} e(s) ds + \bar{\tau} e(t)^T R_4 e(t)$$

$$- \int_{t-\bar{\tau}}^t e(s)^T R_4 e(s) ds + \bar{\tau} v_i(t)^T H v_i(t) - \int_{t-\bar{\tau}}^t v_i(s)^T H v_i(s) ds.$$

According to Assumption 1, by the Leibniz-Newton formula, we get

$$\mathbb{E}\pounds V_i \leq \zeta_i(t)^T \Omega_i \zeta_i(t) - e(t)^T X e(t) + e(t - \tau_i(t))^T X e(t - \tau_i(t)),$$

where

$$\zeta_i(t) = \mathrm{col}\Big\{ e(t), e(t - \tau_i(t)), e(t - \bar{\tau}), f(e(t)),$$

$$f(e(t - \tau_i(t))), \int_{-\infty}^{t} K(t - s) f(e(s)) \mathrm{d}s, \int_{t-\bar{\tau}}^{t} \rho_i(s) \mathrm{d}s \Big\}.$$

The constant α plays an important role in making the matrix Ω_i negative definite. In fact, it can be chosen so big that the matrix Ω_i is negative definite.

From Eq. (16), we have

$$\mathbb{E}\pounds V_i \leq -e(t)^T (X + \lambda_i I) e(t) + e(t - \tau_i(t))^T (X - \lambda_i I) e(t - \tau_i(t))$$

$$\doteq -\psi_1(e(t)) + \psi_2(e(t - \tau_i(t))),$$

where λ_i denote the largest eigenvalue of the matrix Ω_i. Obviously, $\psi_1(e(t)) > \psi_2(e(t))$ for any $e(t) \neq 0$. Therefore, applying Lemma 1, we can conclude that the two coupled delayed neural networks (1) and (2) can be synchronized for almost every initial data.

4   Conclusion

In this paper, an adaptive feedback controller is proposed for the complete synchronization of stochastic Markovian jump neural networks with mode-dependent and unbounded distributed delays. A generalized LaSalle-type invariance principle for stochastic Markovian differential delay equations is employed to investigate the globally almost surely asymptotical stability of the error dynamical system, that is to say, the complete synchronization can be almost surely achieved.

References

1. Arnold, L.: Stochastic Differential Equations: Theory and Applications. Wiley, New York (1972)
2. Liu, Z., Lv, S., Zhong, S., Ye, M.: pth moment exponential synchronization analysis for a class of stochastic neural networks with mixed delays. Commun. Nonlinear Sci. Numer. Simulat. 15, 1899–1909 (2010)
3. Mao, X.: Exponential stability of stochastic delay interval systems with Markovian switching. IEEE Trans. Autom. Contr. 47(10), 1604–1612 (2002)
4. Mao, X.: A note on the LaSalle-type theorems for stochastic differential delay equations. J. Math. Anal. Appl. 268, 125–142 (2002)
5. Pecora, L., Carroll, T.: Synchronization in chaotic systems. Phys. Rev. Lett. 64, 821–824 (1990)
6. Zhang, Q., Lu, J.: Chaos synchronization of a new chaotic system via nonlinear control. Chaos, Solitons, Fractals 37, 175–179 (2008)

Neural Network H_∞ Tracking Control of Nonlinear Systems Using GHJI Method

Derong Liu, Yuzhu Huang, and Qinglai Wei

State Key Laboratory of Management and Control for Complex Systems
Institute of Automation, Chinese Academy of Sciences, Beijing 100190, China
{derong.liu,yuzhu.huang,qinglai.wei}@ia.ac.cn

Abstract. In this paper, an H_∞ optimal tracking control scheme based on generalized Hamilton-Jacobi-Isaacs (GHJI) equation is developed for discrete-time (DT) affine nonlinear systems. First, via system transformation, the optimal tracking problem is transformed into an optimal regulation problem with respect to the state tracking error. Second, with regard to the converted regulation problem, in order to obtain the H_∞ tracking control, the corresponding GHJI equation is formulated, and then the L_2-gain analysis of the closed-loop nonlinear system are employed. Third, an iterative algorithm based on the GHJI equation by using neural networks (NNs) is introduced to solve the optimal control. Finally, simulation results are presented to demonstrate the effectiveness of the proposed scheme.

Keywords: H_∞ tracking control, neural networks, generalized Hamilton-Jacobi-Isaacs, reinforcement learning, infinite horizon.

1 Introduction

As is well known, H_∞ optimal control has played an important role in modern robust optimal control [1–3]. For solving H_∞ nonlinear control problems, the difficulty is the requirement to solve the nonlinear Hamilton-Jacobi-Isaacs (HJI) equation which is usually too difficult to solve analytically. In recent years, in order to solve the optimal tracking control and H_∞ control problems, several methods have been proposed to obtain approximate solutions of the HJI equation [4–10]. In [4], Zhang et al. gave a novel infinite-horizon optimal tracking control scheme based on greedy heuristic dynamic programming (HDP) algorithm for discrete-time (DT) nonlinear systems with the requirement of the whole known system dynamics. In [5], Huang et al. developed an approximation approach to solve the HJI equation in terms of the Taylor series, and computed the corresponding coefficients by using a sequence of linear algebraic equations. In [6], Mehraeen et al. proposed an iterative approach to obtain the optimal solutions based on the generalized HJI (GHJI) equation for discrete-time (DT) nonlinear system by using neural networks (NNs).

However, to the best of our knowledge, there is still no result for solving the optimal tracking control problems for affine nonlinear DT systems via GHJI

C. Guo, Z.-G. Hou, and Z. Zeng (Eds.): ISNN 2013, Part II, LNCS 7952, pp. 186–195, 2013.

method. Upon completion of the transformation, the tracking problem is converted to a regulation problem, then, the corresponding HJI and GHJI equation are formulated. In order to obtain the H_∞ optimal control, an iterative algorithm is used to solve the GHJI equation by using NNs. Moreover, additional considerations are required when solving the GHJI equation to ensure the existence of a saddle-point in the zero-sum two-player game [3]. Finally, simulation results are presented to confirm the validity of the proposed optimal tracking control scheme.

2   Problem Formulation

Considering the affine DT nonlinear system given by

$$x_{k+1} = f(x_k) + g(x_k)u_k + h(x_k)w_k, \tag{1}$$

where $x_k \in \mathbb{R}^n$ is the state vector, $u_k \in \mathbb{R}^m$ is the control input, and $w_k \in \mathbb{R}^M$ is the disturbance, $f(\cdot) \in \mathbb{R}^n$, $g(\cdot) \in \mathbb{R}^{n \times m}$ and $h(\cdot) \in \mathbb{R}^{n \times M}$ are smooth functions defined in a neighborhood of the origin. Assume that $f(\cdot)$ and $g(\cdot)$ are Lipschitz continuous on a set Ω in $\mathbb{R}^n$ with $f(0) = 0$, and there exists a matrix function $g^{-1}(x_k) \in \mathbb{R}^{m \times n}$ such that $g^{-1}(x_k)g(x_k) = I \in \mathbb{R}^{m \times m}$ where I is the identity matrix.

For the optimal tracking control problem, the control objective is to find the optimal control u_k^*, so as to make the nonlinear system (1) to track a reference (desired) trajectory r_k in an optimal manner. The reference trajectory r_k is generated by the following autonomous system as

$$r_{k+1} = \phi(r_k), \tag{2}$$

where $r_k \in \mathbb{R}^n$ and $\phi(r_k) \in \mathbb{R}^n$, and it is assumed that the mapping between the state x_k and the desired trajectory r_k is one-to-one. Then, we define the state tracking error as $e_k = x_k - r_k$. Based on [4], we define the steady control corresponding to the desired trajectory r_k as

$$u_{dk} = g^{-1}(r_k)(\phi(r_k) - f(r_k) - h(r_k)w_k), \tag{3}$$

where $g^{-1}(r_k)g(r_k) = I$ and $I \in \mathbb{R}^{m \times m}$ is the identity matrix. Then, by subtracting the steady control u_{dk} from the actual control u_k, we define $u(e_k) = u_{ek}$ as

$$u_{ek} = u_k - u_{dk}. \tag{4}$$

Considering (1)–(4), the tracking error e_{k+1} is expressed as

$$
\begin{aligned}
e_{k+1} &= x_{k+1} - r_{k+1} \\
&= f(e_k + r_k) + g(e_k + r_k)u_{ek} + h(e_k + r_k)w_k + g(e_k + r_k)g^{-1}(r_k)(\phi(r_k) \\
&\quad - f(r_k) - h(r_k)w_k) - \phi(r_k).
\end{aligned} \tag{5}
$$

For the convenience of analysis, (5) is rewritten as

$$e_{k+1} = F(e_k, u_{ek}) = f_{ek} + g_{ek}u_{ek} + h_{ek}w_k, \tag{6}$$

where $f_{ek} = f(e_k + r_k) + g(e_k + r_k)g^{-1}(r_k)(\phi(r_k) - f(r_k) - h(r_k)w_k) - \phi(r_k)$, $g_{ek} = g(e_k + r_k)$ and $h_{ek} = h(e_k + r_k)$. For H_∞ optimal tracking control problem, the control objective is to find the control which can minimize the infinite horizon value or cost function

$$J(e_k, u_{ek}, w_k) = \sum_{j=k}^{\infty} (e_j^T Q e_j + u_{ej}^T R u_{ej} - \gamma^2 w_j^T P w_j) \tag{7}$$

in the presence of worst case disturbance w_k, where $r(x_j, u_{ej}, w_j) = x_j^T Q x_j + u_{ej}^T R u_{ej} - \gamma^2 w_j^T P w_j$ with Q, R and P are positive definite matrices, and γ is a constant. The quadratic value or cost function can not only force the system state to follow the reference, but also force the system input to be close to the steady control in maintaining the state to its reference value. Therefore, it is reasonable to consider the problem of solving the optimal tracking control u_k^* of the system (1) can be converted into solving the optimal feedback control u_{ek}^* for the new system (6) with respect to (7). For the optimal control problem, it should be noted that u_{ek} must not only stabilize the system (6) on Ω but also guarantee that (7) is finite, i.e., the control must be admissible [7].

In the following, based on [2], the problem of disturbance attenuation can be addressed by using the L_2-gain of the nonlinear system. Meanwhile, from [3], the two-player zero-sum differential game has a unique solution if a game theoretic saddle point exists, i.e., if the Nash condition holds

$$\min_{u_{ek}} \max_{w_k} J(e_0, u_{ek}, w_k) = \max_{w_k} \min_{u_{ek}} J(e_0, u_{ek}, w_k). \tag{8}$$

The H_∞ control problem can be referred to as a two-player zero-sum differential game where one player u_{ek} tries to minimize the cost function while the other w_k tries to maximize it. It is reasonable that the two policies u_{ek}^* and w_k^* are the optimal control and worst case disturbance of (7), respectively, such that $J(u_{ek}^*, w_k) \leq J(u_{ek}^*, w_k^*) \leq J(u_{ek}, w_k^*)$ for all u_{ek} and w_k. The pair (u_{ek}^*, w_k^*) then becomes the saddle-point solution of the optimization problem. A solution for HJI equation exists (saddle-point existence) if and only if there exists a smooth function $V^*(e_k)$ such that the HJI equation

$$V^*(e_k) = \min_{u_{ek}} \max_{w_k} \sum_{i=k}^{\infty} r(x_i, u_{ei}, w_i) = \max_{w_k} \min_{u_{ek}} \sum_{i=k}^{\infty} r(x_i, u_{ei}, w_i)$$

$$= e_k^T Q e_k + u_{ek}^{*T} R u_{ek}^* - \gamma^2 w_k^{*T} P w_k^* + V(f_{ek} + g_{ek}u_{ek}^* + h_{ek}w_k^*). \tag{9}$$

3   Solving H_∞ Optimal Control Based on GHJI

Considering the optimization problem (9), the DT HJI equation becomes

$$V^*(f_{ek} + g_{ek}u_{ek}^* + h_{ek}w_k^*) - V^*(e_k) + e_k^T Q e_k + u_{ek}^{*T} R u_{ek}^* - \gamma^2 w_k^{*T} P w_k^* = 0, \tag{10}$$

where u_{ek}^* and w_k^* are the optimal control and worst disturbance, respectively. Thus, the Hamiltonian function can be defined as

$$
\begin{aligned}
H(e_k, & u_{ek}, w_k) \\
&= V(f_{ek} + g_{ek}u_{ek} + h_{ek}w_k) - V(e_k) + e_k^T Q e_k + u_{ek}^T R u_{ek} - \gamma^2 w_k^T P w_k.
\end{aligned}
\tag{11}
$$

According to (9), u_{ek}^* and w_k^* can be obtained by applying the first-order necessary conditions, i.e,

$$
u_{ek}^* = -\frac{1}{2} R^{-1} g_{ek}^T \frac{\partial V^*(e_{k+1})}{\partial e_{k+1}}, \qquad w_k^* = \frac{1}{2\gamma^2} P^{-1} h_{ek}^T \frac{\partial V^*(e_{k+1})}{\partial e_{k+1}}.
\tag{12}
$$

Then, substituting (12) into (11), the HJI equation becomes

$$
\begin{aligned}
0 = & V^*(e_{k+1}) - V^*(e_k) + \frac{1}{4} \frac{\partial V^{*T}(e_{k+1})}{\partial e_{k+1}} (g_{ek} R^{-1} g_{ek}^T - h_{ek} P^{-1} h_{ek}^T) \\
& \cdot \frac{\partial V^*(e_{k+1})}{\partial e_{k+1}} + e_k^T Q e_k.
\end{aligned}
\tag{13}
$$

From (12), it is clear that u_k^* and w_k^* are solved if the optimal value function $V^*(e_{k+1})$ can be calculated from (13). However, it is generally difficult to solve the optimal value function. In [7], a generalized Hamilton-Jacobi-Bellman (GHJB) formulation-based NN approach is proposed to solve the single-player HJB optimization problem [8–10]. In this paper, we extend the effort in [7] to deal with the HJI equation.

In the following, based on [7], the DT GHJI equation is derived as

$$
\frac{1}{2} \triangle e_k^T \nabla^2 V(e_k) \triangle e_k + \nabla V(e_k)^T \triangle e_k + e_k^T Q e_k + u_{ek}^T R u_{ek} - \gamma^2 w_k^T P w_k = 0
$$

$$
V(e_k)|_{e_k=0} = 0,
\tag{14}
$$

where $\triangle e_k = f_{ek} + g_{ek}u_{ek} + h_{ek}w_k - e_k$, $\nabla V(e_k)$ and $\nabla^2 V(e_k)$ are the gradient vector and Hessian matrix of $V(e_k)$, respectively. With regard to (14), in order to obtain the optimal control u_{ek}^* and worst disturbance w_k^*, the pre-Hamiltonian function for the system (6) is defined as

$$
\begin{aligned}
H(e_k, u_{ek}, w_k) = & \frac{1}{2} \triangle e_k^T \nabla^2 V(e_k) \triangle e_k + \nabla V(e_k)^T \triangle e_k + e_k^T Q e_k + u_{ek}^T R u_{ek} \\
& - \gamma^2 w_k^T P w_k.
\end{aligned}
\tag{15}
$$

Note that when $H(e_k, u_{ek}^*.w_k^*) = 0$, the GHJI equation results where u_{ek}^* and w_k^* are the optimal control and worse case disturbance to be obtained. The optimal control input u_{ek}^* and worst case disturbance w_k^* can be found by differentiating the pre-Hamiltonian function (15) with respect to u_{ek} and w_k, respectively, which yields

$$
u_k^* = -\left(g_{ek}^T \nabla^2 V^*(e_k) g_{ek} + 2R\right)^{-1} g_{ek}^T \left(\nabla V^*(e_k) + \nabla^2 V^*(e_k)(f_{ek} + h_{ek}w_k^* - e_k)\right)
\tag{16}
$$

and

$$w_k^* = \left(2\gamma^2 P - h_{ek}^T \nabla^2 V^*(e_k) h_{ek}\right)^{-1} h_{ek}^T \left(\nabla V^*(e_k) + \nabla^2 V^*(e_k)(f_{ek} + g_{ek} u_{ek}^* - e_k)\right).$$
(17)

Before proceeding, the following theorem is required to demonstrate that the optimal control and worst disturbance (16) can ensure the existence of a saddle-point in the zero-sum two-player game.

Theorem 1. *Let the pair (u_{ek}, w_k^*) be an arbitrary admissible control and the worst disturbance provided by (16) for system (6). In addition, let the pair (u_{ek}^*, w_k) be the optimal control provided by (16) and an arbitrary disturbance for system (6). Then, the Hamiltonian function (15) satisfies $H(e_k, u_{ek}^*, w_k) \leq H(e_k, u_{ek}^*, w_k^*) \leq H(e_k, u_{ek}, w_k^*)$.*

Proof. The proof is shown in two steps. First, we show that $H(e_k, u_{ek}, w_k^*) - H(e_k, u_{ek}^*, w_k^*) \geq 0$. Note that $H(e_k, u_{ek}, w_k^*)$ and $H(e_k, u_{ek}^*, w_k^*)$ are nothing but the pre-Hamiltonian function (15) rewritten in terms of u_{ek}^* and w_k^*, we obtain

$$H(e_k, u_{ek}, w_k^*) - H(e_k, u_{ek}^*, w_k^*)$$
$$= (f_{ek} + h_{ek} w_k^* - e_k)^T \nabla^2 V(e_k) g_{ek}(u_{ek} - u_{ek}^*) + \frac{1}{2}(g_{ek} u_{ek})^T \nabla^2 V(e_k) g_{ek} u_{ek}$$
$$- \frac{1}{2}(g_{ek} u_{ek}^*)^T \nabla^2 V(e_k) g_{ek} u_{ek}^* + \nabla V(e_k) \cdot g_{ek}(u_{ek} - u_{ek}^*) + u_{ek}^T R u_{ek} - u_{ek}^{*T} R u_{ek}^*$$
(18)

Combining these terms, (18) can be rewritten as

$$H(e_k, u_{ek}, w_k^*) - H(e_k, u_{ek}^*, w_k^*)$$
$$= \left(\nabla V(e_k) + (f_{ek} + h_{ek} w_k^* - e_k)\nabla^2 V(e_k)\right) g_{ek}(u_{ek} - u_{ek}^*) + \frac{1}{2}u_{ek}^T \Gamma_u u_{ek}$$
$$- \frac{1}{2}u_{ek}^{*T} \Gamma_u u_{ek}^*$$
(19)

where $\Gamma_u = g_{ek}^T \nabla^2 V(e_k) g_{ek} + 2R > 0$. Considering (16), we have the relation

$$\left(g_{ek}^T(\nabla^2 V(e_k)) g_{ek} + 2R\right) u_{ek}^* = -g_{ek}^T \left(\nabla V(e_k) + \nabla^2 V(e_k)(f_{ek} + h_{ek} w_k^* - e_k)\right).$$
(20)

Substituting (20) into (19), we have

$$H(e_k, u_{ek}, w_k^*) - H(e_k, u_{ek}^*, w_k^*) = \frac{1}{2}(u_{ek} - u_{ek}^*)^T \Gamma_u (u_{ek} - u_{ek}^*).$$
(21)

So, with Γ_u being positive definite, we have $H(e_k, u_{ek}, w_k^*) - H(e_k, u_{ek}^*, w_k^*) \geq 0$. Second, we show that $H(e_k, u_{ek}^*, w_k) - H(e_k, u_{ek}^*, w_k^*) \leq 0$. Similar to (18), we have

$$H(e_k, u_{ek}^*, w_k) - H(e_k, u_{ek}^*, w_k^*)$$
$$= \left(\nabla V(e_k) + (f_{ek} + g_{ek} u_{ek}^* - e_k)^T \nabla^2 V(e_k)\right) h_{ek}(w_k - w_k^*)$$
$$+ \frac{1}{2}w_k^T \Gamma_w w_k - \frac{1}{2}w_k^{*T} \Gamma_w w_k^*$$
(22)

where $\Gamma_w = h_{ek}^T \nabla^2 V(e_k) h_{ek} - 2\gamma^2 P$. Considering (17), we have

$$H(e_k, u_{ek}^*, w_k) - H(e_k, u_{ek}^*, w_k^*) = \frac{1}{2}(w_k - w_k^*)^T \Gamma_w (w_k - w_k^*). \qquad (23)$$

Since $2\gamma^2 P - h_{ek}^T \nabla^2 V(e_k) h_{ek} > 0$, we obtain $H(e_k, u_{ek}^*, w_k) - H(e_k, u_{ek}^*, w_k^*) \leq 0$. Thus, based on the above analysis, we can get

$$H(e_k, u_{ek}^*, w_k) \leq H(e_k, u_{ek}^*, w_k^*) \leq H(e_k, u_{ek}, w_k^*). \qquad (24)$$

Next, an iterative algorithm is developed to update the control input and disturbance which consists of a sequential set of updates for the disturbance $w_k^{(i,j)}$ in an inner loop with index j accompanied by a sequential set of updates for the control input u_{ek}^i in an outer loop with index i. Let u_{ek}^0 be an initial admissible control. The algorithm starts with setting $u_{ek}^i = u_{ek}^0$ and $w_k^{(i,0)} = 0$ for $i = 0$. For convenience, in the sequel, denote $V_k = V(e_k)$. Then, the pre-Hamiltonian equation (15) is solved for $V_k^{(i,j)}$ as

$$\nabla V_k^{(i,j)}(f_{ek} + g_{ek}u_{ek}^i + h_{ek}w_k^{(i,j)} - e_k) + \frac{1}{2}(f_{ek} + g_{ek}u_{ek}^i + h_{ek}w_k^{(i,j)} - e_k)^T \nabla^2 V_k^{(i,j)}$$
$$\cdot (f_{ek} + g_{ek}u_k^i + h_{ek}w_k^{(i,j)} - e_k) + e_k^T Q e_k + u_{ek}^{i^T} R u_{ek}^i - \gamma^2 w_k^{(i,j)^T} P w_k^{(i,j)} = 0. \qquad (25)$$

The disturbance $w_k^{(i,j)}$ is updated by using (17) and written as

$$w_k^{(i,j+1)} = \left(2\gamma^2 P - h_{ek}^T \nabla^2 V_k^{(i,j)} h_{ek}\right)^{-1} h_{ek}^T \left(\nabla V_k^{(i,j)} + \nabla^2 V_k^{(i,j)}(f_{ek} + g_{ek}u_{ek}^i - e_k)\right). \qquad (26)$$

The inner loop j proceeds until it converges such that $V_k^{(i,j)} = V_k^{(i,j+1)} = V_k^{(i,\infty)}$. Next, u_k^i is updated according to (16) and written as

$$u_{ek}^{(i+1)} = -\left(g_{ek}^T \nabla^2 V_k^{(i,\infty)} g_{ek} + 2R\right)^{-1} g_{ek}^T \left(\nabla V_k^{(i,\infty)} + \nabla^2 V_k^{(i,\infty)}(f_{ek} + h_{ek}w_k^i - e_k)\right). \qquad (27)$$

Then, the value function is found by solving (25) for $V_k^{(i,j)}$. Similar to the inner loop, the outer loop i proceeds until it converges such that $V_k^{(i,\infty)} = V_k^{(i+1,\infty)} = V_k^{(\infty,\infty)}$.

Theorem 2. *Let u_{ek}^i be an initial admissible control input for pair (i,j) for system (6) on the set Ω. Then, iterating between (25) and (26) ensures $V_k^{(i,j)}$ is monotonically increasing until the worst disturbance for the control input u_{ek}^i is found, i.e., $V_k^{(i,j)} \leq V_k^{(i,j+1)} \leq V_k^{(i,\infty)}$ and $\lim_{j \to \infty} V_k^{(i,j)} = V_k^{(i,\infty)}$.*

Next, the convergence of the outer loop i is discussed.

Theorem 3. *Let u_{ek}^i be an initial admissible control input for pair (i,j) for the system (6) on the set Ω. If the controller is updated by (27), then, the system $e_{k+1} = f_{ek} + g_{ek}u_{ek}^{i+1} + h_{ek}w_k^{(i,\infty)}$ has the L_2-gain less than or equal to γ. Furthermore, $V_k^{(i,\infty)} \geq V_k^{(i+1,\infty)} \geq V_k^*$ and $\lim_{i \to \infty} V_k^{(i,\infty)} = V_k^*$ where V_k^* solves the GHJI equation (14).*

The satability of the system $e_{k+1} = f_{ek} + g_{ek}u_{ek}^{i+1} + h_{ek}w^j$ is shown in the following theorem.

Theorem 4. *Let u_{ek}^i be an initial admissible control for pair (i, j) for system (6) on the set Ω. Let the proposed successive approximation procedure of updates for the disturbance $w_k^{(i,j)}$ in the inner loop j and updates for the control u_{ek}^i in the outer loop i be performed. Then, for all j and i, $e_{k+1} = f_{ek} + g_{ek}u_{ek}^{i+1} + h_{ek}w^j$ is asymptotically stable on Ω.*

For space reasons, we will present the details of the proof of Theorems 2–4 in a future paper.

4 H_∞ Optimal Controller Design Using NNs

In this section, a multilayer feedforward NN is used to approximate the solution $V_k^{(i,j)}$ of the GHJI equation. Using the NN approximation property [8] in compact set Ω, we can approximate $V^{(i,j)}$ with an NN as

$$V^{(i,j)} \approx V_L(e) = \sum_{j=1}^{L} \omega_j \sigma_j(e) = W_L^T \bar{\sigma}_L(e), \tag{28}$$

where $\sigma_j(e)$ is activation function, and $\sigma_j(0) = 0$. $W_L(e) = [\omega_1, \omega_2, \cdots, \omega_L]$ is the vector of NN weights, $\bar{\sigma}_L(e) = [\sigma_1(e), \sigma_2(e), \cdots, \sigma_L(e)]$ is the vector of activation functions and L is the number of hidden layer neurons.

The NN weights are tuned to minimize the residual error in least square method. With $GHJI(V_k^{(i,j)}, u_{ek}^i, w_k^{(i,j)}) = 0$, $V_k^{(i,j)}$ is replaced by V_L to obtain the residual error, i.e, $GHJI(V_L^{(i,j)} = \sum_{j=1}^{L} \omega_j \sigma_j(e), u_{ek}^i, w_k^{(i,j)}) = \xi_L$.

To find the least squares solution, the method of weighted residuals is used [8]. We have

$$W_L = -\langle \Theta, \Theta \rangle^{-1} \langle e^T Q e + u_e^{iT} R u_e^i - \gamma^2 w^{(i,j)^T} P w^{(i,j)}, \Theta \rangle, \tag{29}$$

where $\Theta = \nabla \bar{\sigma}_L \triangle e + \frac{1}{2} \triangle e^T \nabla^2 \bar{\sigma}_L \triangle e$,

$$\nabla \bar{\sigma}_L = [\partial \sigma_1(e)/\partial e, \partial \sigma_2(e)/\partial e, \cdots, \partial \sigma_L(e)/\partial e]^T,$$

$$\nabla^2 \bar{\sigma}_L = [\partial^2 \sigma_1(e)/\partial e^2, \partial^2 \sigma_2(e)/\partial e^2, \cdots, \partial^2 \sigma_L(e)/\partial e^2]^T,$$

$$\triangle e = f_e(e) + g_e(e)u_e(e)^i + h_e(e)w(e)^{(i,j)} - e.$$

Lemma 1. *If the set $\{\sigma_j(e)\}_1^L$ is linearly independent and $u \in \Omega_u$, then the set $\{\nabla \bar{\sigma}_L \triangle e + \frac{1}{2} \triangle e^T \nabla^2 \bar{\sigma}_L \triangle e\}_1^L$ is also linearly independent.*

From Lemma 1, $\langle \Theta, \Theta \rangle$ is invertible. Thus, a unique solution for W_L exists. In addition, the inner products in (29) can be approximated as $\langle a(e), b(e) \rangle = \int_\Omega a(e)b(e)de = \sum_{i=1}^N a(e)b(e)\delta(e)$ where $\delta(e) = e_i - e_{i-1}$ is chosen small in Ω and N is large. By employing a mesh in the set Ω where the mesh size is $\delta(e)$, the NN weights can be found as

$$W_L = -(X^T X)^{-1} X^T Y, \tag{30}$$

where X and Y are defined as

$$X = [\Theta|_{e=e_1}, \cdots, \Theta|_{e=e_p}]^T, \tag{31}$$

$$Y = \begin{bmatrix} e^T Qe + u_e^{i^T} Ru_e^i - \gamma^2 w^{(i,j)^T} Pw^{(i,j)}|_{e=e_1} \\ \vdots \\ e^T Qe + u_e^{i^T} Ru_e^i - \gamma^2 w^{(i,j)^T} Pw^{(i,j)}|_{e=e_p} \end{bmatrix}, \tag{32}$$

and p is the number of points of the mesh.

5 Simulation Study

In this section, the example is derived from [9] with some modifications. Considering the following nonlinear system:

$$x_{k+1} = f(x_k) + g(x_k)u_k + h_k(x_k)w_k, \tag{33}$$

where $x_k = [x_{1k}\ x_{2k}]^T$ and $u_k = [u_{1k}\ u_{2k}]^T$ are the system state and the control input, respectively, w_k is the disturbance. The corresponding $f(x_k)$, $g(x_k)$ and $h(x_k)$ are given as:

$$f(x_k) = \begin{bmatrix} -\sin(0.5x_{2k})x_{1k}^2 \\ -\cos(1.4x_{2k})\sin(0.9x_{1k}) \end{bmatrix},$$

$$g(x_k) = \begin{bmatrix} 1 & 0 \\ 0 & 1 \end{bmatrix}, \qquad h(x_k) = \begin{bmatrix} x_{1k} \\ x_{1k}x_{2k} \end{bmatrix}.$$

The reference trajectory for the above system is selected as

$$r_k = \begin{bmatrix} \sin(0.25k) \\ \cos(0.25k) \end{bmatrix}.$$

The parameters of the cost function are chosen as $Q = 0.8I$, $R = P = I$, where I denotes the identity matrix with suitable dimensions, $\gamma = 20$. The state of the tracking system is initialized to be $x_0 = [0.8\ 0.5]^T$. In the simulation performed, the NN is going to be trained on the region $(-1.5 \le x_{1k} \le 1.5, -1.5 \le x_{2k} \le 1.5)$ with the mesh size being chosen to be 0.05. Based on [7], the activation functions of the NN are chosen as $\bar{\sigma}_L = [e_{1k}^2, e_{1k}e_{2k}, e_{2k}^2, e_{1k}^4, e_{1k}^3 e_{2k}, e_{1k}^2 e_{2k}^2, e_{1k}e_{2k}^3, e_{2k}^4, e_{1k}^6, e_{1k}^5 e_{2k}, e_{1k}^4 e_{2k}^2, e_{1k}^3 e_{2k}^3, e_{1k}^2 e_{2k}^4, e_{1k}e_{2k}^5, e_{2k}^6]$. Select the initial admissible

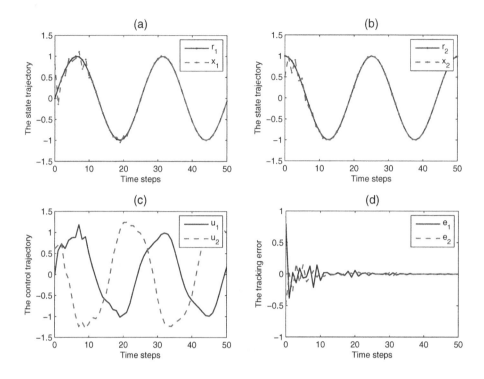

Fig. 1. Simulation results of the H_∞ tracking control scheme

control $u_{ek}^0 = [0.1e_{1k}-0.6e_{2k}, -0.2e_{1k}+0.35e_{2k}]^T$ and the disturbance $w_k^{(i,0)} = 0$, then update the control input and the disturbance by (27) and (26), respectively, while the value function $V_k^{(i,j)}$ is approximated by the NN whose weights are obtained from (30).

In addition, we set $\varepsilon = 10^{-6}$ to terminate the iterations in i and j. Let $i = 20$ and $j = 20$. Then, after completing the offline training, the final NN weights are $W_L = [1.330, 0.268, -0.295, -0.026, 1.569, 2.453, -1.487, 0.628, -0.058, -0.634, -0.767, -0.068, -1.442, 1.233, -0.382]^T$. Then, we apply the optimal tracking control policy to the system with a disturbance $w_k = sin(10k)e^{-0.1k}$ for 50 time steps and obtain the relevant simulation results. The obtained state curves are shown in Fig. 1(a) and Fig. 1(b), where the corresponding reference trajectories are also plotted for assessing the tracking performance. The tracking control curves and the tracking errors are shown in Fig. 1(c) and Fig. 1(d), respectively. These simulation results verify the excellent performance of the H_∞ tracking controller developed based on GHJI method.

6   Conclusion

In this paper, an H_∞ optimal tracking control scheme based on GHJI is developed for affine nonlinear systems. Via system transformation, the optimal tracking problem is transformed into an optimal regulation problem with respect to the state tracking error. For the converted regulation problem, the corresponding GHJI equation is formulated, and then the L_2-gain analysis of the closed-loop nonlinear system are employed. In order to obtain the H_∞ tracking control, an iterative algorithm based on the GHJI by using NNs is introduced. Simulation results demonstrate the effectiveness of the proposed scheme.

Acknowledgment. This work was supported in part by the National Natural Science Foundation of China under Grants 61034002, 61233001, and 61273140.

References

1. Zhou, K., Doyle, J.C., Glover, K.: Robust and Optimal Control. Prentice Hall, New Jersey (1996)
2. Van Der Schaft, A.J.: L_2-gain analysis of nonlinear systems and nonlinear state feedback H_∞ control. IEEE Transactions on Automatic Control 37(6), 770–784 (1992)
3. Basar, T., Bernard, P.: H_∞ Optimal Control and Related Minimax Design Problems, MA. Birkhauser, Boston (1995)
4. Zhang, H.G., Wei, Q.L., Luo, Y.H.: A novel infinite-time optimal tracking control scheme for a class of discrete-time nonlinear systems via the greedy HDP iteration algorithm. IEEE Transactions on Systems, Man, and Cybernatics–Part B: Cybernatics 38(4), 937–942 (2008)
5. Huang, J., Lin, C.F.: Numerical approach to computing nonlinear H_∞ control laws. Journal of Guidance, Control, and Dynamics 18(5), 989–994 (1995)
6. Mehraeen, S., Dierks, T., Jagannathan, S., Crowl, M.L.: Zero-sum two-player game theoretic formulation of affine nonlinear discrete-time systems using neural networks. In: Proceedings of the International Joint Conference on Neural Networks, pp. 761–769 (2010)
7. Chen, Z., Jagannathan, S.: Generalized Hamilton-Jacobi-Bellman formulation-based neural network control of affine nonlinear discrete-time systems. IEEE Transactions on Neural Networks 19(1), 90–106 (2008)
8. Jagannathan, S.: Neural Network Control of Nonlinear Discrete-time Systems, FL. CRC Press, Boca Raton (2006)
9. Dierks, T., Thumati, B.T., Jagannathan, S.: Optimal control of unknown affine nonlinear discrete-time systems using offline-trained neural networks with proof of convergence. Neural Networks 22(5-6), 851–860 (2009)
10. Wang, D., Liu, D.R., Wei, Q.L.: Finite-horizon neuro-optimal tracking cont rol for a class of discrete-time nonlinear systems using adaptive dynamic programming approach. Neurocomputing 78(15), 14–22 (2012)

Adaptive Neural Control
for Uncertain Attitude Dynamics
of Near-Space Vehicles with Oblique Wing

Mou Chen and Qing-xian Wu

College of Automation Engineering
Nanjing University of Aeronautics and Astronautics
Nanjing, China, 210016
{chenmou,wuqingxian}@nuaa.edu.cn

Abstract. In this paper, the adaptive neural attitude control is developed for near-space vehicles with the oblique wing (NSVOW) via using the sliding mode disturbance observer technique. The radial basis function neural network (RBFNN) is employed to approximate the unknown system uncertainty. Then, the sliding mode disturbance observer is designed to estimate the unknown external disturbance and the unknown neural network approximation error. Using outputs of the sliding mode disturbance observer and the RBFNN, the adaptive neural attitude control is proposed for NSVOWs. The stability of the closed-loop system is proved using the Lyapunov analysis. Finally, simulation results are presented to illustrate the effectiveness of the proposed adaptive neural attitude control scheme.

Keywords: NSVOW, Attitude control, Sliding mode disturbance observer, Adaptive neural control.

1   Introduction

The near-space vehicle (NSV) is one kind of new aerospace vehicles which attracted much concern around the world in recent years. Since the NSV has special characteristics compared with traditional flight vehicles, it has the potential military and civilian dual-use values [1-3]. To implement various tasks, the efficient flight control schemes need to be developed for the NSV. However, the NSV has the highly coupled control channels, unknown time-varying disturbances and strong nonlinearity which will further increase the design difficulty of the robust flight control scheme [3]. In the recent decade, many flight control schemes have been developed for the NSV.

According to the flight speed of the NSV, the flight envelop can be divided into the low-speed flight region and the high-speed flight region. To meet the different requirements for aerodynamic characteristic at low speed and high speed, the concept of oblique wing has been proposed by R.T. Jones in 1958 [4-6]. The research results suggest that the aircraft with oblique wings can offer many advantages at high transonic and low supersonic speeds. For example, the NSV

C. Guo, Z.-G. Hou, and Z. Zeng (Eds.): ISNN 2013, Part II, LNCS 7952, pp. 196–203, 2013.
© Springer-Verlag Berlin Heidelberg 2013

with an oblique wing (NSVOW) can change wing skew angle in accordance with flight status and can achieve optimal aerodynamic efficiency in subsonic, transonic and supersonic speeds in order to reduce the fuel consumption [7]. There are some literatures to address the concept of the oblique wing. However, a few work has been reported for the NSVOW.

To tackle the unknown system uncertainty, neural networks (NNs) as the universal function approximator have been widely used in the control system design of uncertain nonlinear systems [8-10]. In this paper, the RBFNN is introduced to approximate the unknown system uncertainty of the NSVOW. However, the neural network approximation error and the time-varying external disturbance of the NSV need to be efficiently handled to achieve satisfactory control performance. To improve the anti-disturbance ability of the NSVOW, the disturbance-observer-based adaptive neural attitude control strategy needs to be further investigated [11, 12]. Sliding mode disturbance observer (SMDO) has been extensively studied due to without depending on complete knowledge of the bounded disturbance mathematical model [13, 14]. In this paper, the sliding mode disturbance observer is employed to estimate the compounded disturbance and the adaptive neural control attitude control is designed using outputs of disturbance observer and the RBFNN for the NSVOW.

This work is motivated by the adaptive neural attitude control for the NSVOW with the unknown system uncertainty and the unknown time-varying external disturbance. The organization of this paper is as follows. The problem statement is given in Section 2. Section 3 describes the design of adaptive neural attitude control for the NSVOW. Simulation studies are provided in Section 4 to demonstrate the effectiveness of the adaptive neural attitude control approach of the NSVOW, followed by some concluding remarks in Section 5.

2   Problem Statement

To design the adaptive neural attitude controller for the NSVOW, we firstly consider the following general multi-input and multi-output nonlinear system:

$$\dot{x} = F(x) + \Delta F(x) + G(x)u + d(t)$$
$$y = x \tag{1}$$

where $x \in R^{n \times 1}$ and $y \in R^{n \times 1}$ are the system state and the system output, respectively. $F(x) \in R^{n \times 1}$ and $G(x) \in R^{n \times m}$ are continuous function vector and matrix, respectively. $u \in R^{m \times 1}$ is the system control input. $\Delta F(x) \in R^{n \times 1}$ is the system uncertainty and $d(t) \in R^{n \times 1}$ is the unknown external disturbance.

The system uncertainty $\Delta F(x)$ is approximated using the RBFNN which can be expressed as

$$\Delta F(x) = W^{*T} S(x) + \varepsilon^*(x) \tag{2}$$

where W^* are the optimal weights and $\varepsilon^*(x)$ is the smallest approximation error. $S(Z) = [s_1(Z), s_2(Z), \ldots, s_p(Z)]^T \in R^p$ and the Gaussian radial basis function is

$$s_i(x) = \exp[-(x - c_i)^T(x - c_i)/b_i^2], \ i = 1, 2, \ldots, p \qquad (3)$$

where c_i and b_i are the center and width of the neural cell of the i-th hidden layer.

Using the RBFNN to approximate the unknown system uncertainty $\Delta F(x)$ and invoking (2), we have

$$\begin{aligned}\dot{x} &= F(x) + W^{*T}S(Z) + \varepsilon^*(Z) + G(x)u + d(t) \\ &= F(x) + W^{*T}S(Z) + G(x)u + D(t)\end{aligned} \qquad (4)$$

where $D(t) = \varepsilon^*(Z) + d(t)$ denotes the system compounded disturbance.

For the continuous desired attitude y_d, the control objective is that the adaptive neural attitude control u is proposed to ensure the tracking error asymptotically convergent in the presence of system uncertainties and the time-varying unknown external disturbances. To proceed the design of sliding mode disturbance-observer-based adaptive neural attitude control for the uncertain attitude dynamic of the NSVOW (4), the following assumption is required:

Assumption 1: There exists positive constants δ_i such that the system compounded disturbance is bounded, i.e., $|D_i| \leq \delta_i, i = 1, 2, \ldots, n$ where $\delta_i > 0$.

3 Adaptive Neural Attitude Control Based on Sliding Mode Observer

To develop the sliding mode disturbance observer based adaptive neural attitude control, the tracking error is defined as

$$e = y - y_d = x - y_d \qquad (5)$$

Differentiating (5) and considering (4), we have

$$\dot{e} = F(x) + W^{*T}S(Z) + G(x)u + D(t) - \dot{y}_d \qquad (6)$$

For the unknown compounded disturbance $D(t)$ shown in (6), the sliding mode disturbance observer is employed to estimate it. To design the sliding mode disturbance observer, an auxiliary variable is proposed as

$$\sigma = z - e \qquad (7)$$

where $\sigma = [\sigma_1, \ldots, \sigma_n]^T \in R^{n \times 1}$ and $z = [z_1, \ldots, z_n]^T \in R^{n \times 1}$.

The variable z is designed as

$$\dot{z} = -\Lambda\sigma - \Gamma\text{sign}(\sigma) + \hat{W}^T S(Z) + G(x)u(t) - \dot{y}_d \qquad (8)$$

where $\Lambda = \text{diag}\{\Lambda_i\}_{n\times n} > 0$ is a designed diagonal matrix, $\Gamma = \text{diag}\{\delta_i + |F_i|\}_{n\times n}$, $\text{sign}(\sigma) = [\text{sign}(\sigma_1), \ldots, \text{sign}(\sigma_n)]^T$ and F_i is the ith element of $F(x)$ and $i = 1, 2, \ldots, n$.

Utilizing the auxiliary variable σ, the disturbance estimate $\hat{D}$ of the the sliding mode disturbance observer is designed as

$$\hat{D} = -\Lambda\sigma - \Gamma\text{sign}(\sigma) - F(x) \tag{9}$$

Differentiating (7), and considering (6) and (8) yields

$$\begin{aligned}
\dot{\sigma} = \dot{z} - \dot{e} &= -\Lambda\sigma - \Gamma\text{sign}(\sigma) + \hat{W}^T S(Z) + G(x)u(t) - \dot{y}_d \\
&\quad - (F(x) + W^{*T}S(Z) + G(x)u + D(t) - \dot{y}_d) \\
&= -\Lambda\sigma - \Gamma\text{sign}(\sigma) - (F(x) + D(t)) + \tilde{W}^T S(Z)
\end{aligned} \tag{10}$$

where $\tilde{W} = \hat{W} - W^*$.

Invoking (10) and the definition of the Γ, we obtain

$$\begin{aligned}
\sigma^T\dot{\sigma} &= -\sigma^T\Lambda\sigma - \sigma^T\Gamma\text{sign}(\sigma) - \sigma^T(F(x) + D(t)) + \sigma^T\tilde{W}^T S(Z) \\
&\leq -\sigma^T\Lambda\sigma - \sum_{i=1}^{n}\Gamma_i|\sigma_i| + \sum_{i=1}^{n}\Gamma_i|\sigma_i| + \sigma^T\tilde{W}^T S(Z) \\
&= -\sigma^T\Lambda\sigma + \sigma^T\tilde{W}^T S(Z)
\end{aligned} \tag{11}$$

Defining $\tilde{D} = D - \hat{D}$ and considering (6), (8) and (9), we have

$$\begin{aligned}
\tilde{D} = D - \hat{D} \\
&= \dot{e} - F(x) - G(x)u(t) - W^{*T}S(Z) + \dot{y}_d - (-\Lambda\sigma - \Gamma\text{sign}(\sigma) - F(x)) \\
&= \dot{e} - F(x) - G(x)u(t) - W^{*T}S(Z) + \dot{y}_d \\
&\quad - (\dot{z} - F(x) - G(x)u(t) - \hat{W}^T S(Z) + \dot{y}_d) \\
&= \dot{e} - \dot{z} + \tilde{W}^T S(Z) = -\dot{\sigma} + \tilde{W}^T S(Z)
\end{aligned} \tag{12}$$

Using outputs of the RBFNN and the sliding mode disturbance observer, the adaptive neural attitude control law is proposed as

$$u = -G^\dagger(Ke + F(x) + \hat{W}^T S(Z) + \hat{D} - \dot{y}_d + \tau\text{sign}(e)) \tag{13}$$

where $G^\dagger = MG(x)^T(G(x)MG(x)^T)^{-1}$, $K = K^T > 0$, $\tau = \text{diag}\{|\hat{D}_i| + \delta_i\}$, $\text{sign}(e) = [\text{sign}(e_1), \ldots, \text{sign}(e_n)]^T$ and M is a design matrix.

Substituting (13) into (6) yields

$$\begin{aligned}
\dot{e} &= -Ke - \hat{W}^T S(Z) + W^{*T}S(Z) - \hat{D}(t) + D(t) - \tau\text{sign}(e) \\
&= -Ke - \tilde{W}^T S(Z) + \tilde{D} - \tau\text{sign}(e)
\end{aligned} \tag{14}$$

where $\tilde{D} = D - \hat{D}$.

Considering (10) and (12), (14) can be written as

$$\dot{e} = -Ke - \tilde{W}^T S(Z) - \hat{D} + D(t) - \tau\text{sign}(e) \tag{15}$$

According to (15) and the definition of the τ, we have

$$
\begin{aligned}
e^T \dot{e} &= -e^T K e - e^T \tilde{W}^T S(Z) - e^T \hat{D} + e^T D(t) - e^T \tau \text{sign}(e) \\
&\leq -e^T K e - e^T \tilde{W}^T S(Z)
\end{aligned}
\tag{16}
$$

The parameter updated law of $\hat{\theta}_i$ is designed as

$$
\dot{\hat{W}}_i = \gamma_i S_i^T(x)(e_i - \sigma_i)
\tag{17}
$$

where $\gamma_i > 0$ is a design parameter.

Above analysis and design of the adaptive neural attitude control based on the disturbance observer can be summarized in the following theorem.

Theorem 1. *Considering the uncertain attitude dynamics (4) of the near-space vehicle with oblique wing, the sliding mode disturbance observer is designed in accordance with (7), (8) and (9). The updated law of the RBFNN is chosen as (17). Based on outputs of the RBFNN and the sliding mode disturbance observer, the adaptive neural attitude control law is proposed as (13). Then, the asymptotical convergence of all closed-loop signals can be guaranteed under the designed adaptive neural attitude control law.*

Proof. Let the Lyapunov function candidate be

$$
V = \frac{1}{2}\sigma^T \sigma + \frac{1}{2}e^T e + \sum_{i=1}^{n} \frac{1}{2\gamma_i} \tilde{W}_i^2
\tag{18}
$$

where $\tilde{W}_i = \hat{W}_i - W_i^*$.

Invoking (11) and (16), the time derivative of V is given by

$$
\begin{aligned}
\dot{V} &= \sigma^T \dot{\sigma} + e^T \dot{e} + \sum_{i=1}^{n} \frac{1}{\gamma_i} \tilde{W}_i \dot{\tilde{W}}_i \\
&\leq -\sigma^T \Lambda \sigma + \sigma^T \tilde{W}^T S(Z) - e^T K e - e^T \tilde{W}^T S(Z) + \sum_{i=1}^{n} \frac{1}{\gamma_i} \tilde{W}_i \dot{\tilde{W}}_i
\end{aligned}
\tag{19}
$$

Invoking the parameter updated law (17), we have

$$
\dot{V} \leq -\sigma^T \Lambda \sigma - e^T K e
\tag{20}
$$

Form (20), we obtain that all signals of the closed-loop system are asymptotically convergent with $t \to \infty$.

Remark 1. For developing the sliding mode disturbance observer to estimate the system compounded disturbance, the bounded assumption of the system compounded disturbance is needed. For the NSVOW, the neural network approximation error and the external disturbance are always bounded. Thus, the Assumption 1 is reasonable.

4   Simulation Example

In this section, we use the developed adaptive neural control scheme based on sliding mode disturbance observer to design the attitude controller for the NSVOW. Consider the attitude dynamics of the NSVOW with unknown system uncertainty and the unknown external disturbance in the form of

$$\dot{x}_f = F_f\left(\overline{x}_f\right) + \Delta F_f\left(\overline{x}_f\right) + G_f\left(\overline{x}_f\right)\overline{u}_f + D_f$$
$$y_1 = x_f \tag{21}$$
$$\dot{x}_s = F_s\left(\overline{x}_{s1}\right) + \Delta F_s\left(\overline{x}_{s1}\right) + G_{s1}\left(\overline{x}_{s1}\right)\overline{u}_s + D_s$$
$$y_2 = x_s \tag{22}$$

where $x_f = [p, q, r]^T$, $x_s = [\alpha, \beta, \mu]^T$, $\overline{x}_{s1} = [V, \gamma, \alpha, \beta, \mu]$, $\overline{u}_f = [\delta_a, \delta_e, \delta_r, \delta_y, \delta_z]$, $\overline{x}_f = [T, V, \gamma, \alpha, \beta, \mu, p, q, r]$, $\overline{x}_{s2} = [p, q, r]$, $\overline{u}_s = [p_c, q_c, r_c]$. ΔF_f and ΔF_s are system uncertainties. D_f and D_s are external disturbances. All definitions of variables and the detailed expressions of matrices $F_f(\overline{x}_f)$, $G_f(\overline{x}_f)$, $F_s(\overline{x}_{s1})$, $G_{s1}(\overline{x}_{s1})$ and $G_{s2}(\overline{x}_{s1})$ can be founded in [1].

To illustrate the effectiveness of the developed adaptive neural attitude control for the uncertain NSVOW (21) and (22), the sliding mode disturbance observer is designed according to (7), (8) and (9). The updated law of the RBFNN is chosen as (17). Using outputs of the RBFNN and the sliding mode disturbance observer, the adaptive neural attitude law is proposed as (13). The initial conditions of the attitude and the attitude angular velocity are arbitrarily chosen as $\alpha_0 = 2°$, $\beta_0 = 0°$, $\mu_0 = 0°$ and $p_0 = q_0 = r_0 = 0$ degree/s. We assume $D_s = 0.2D_f$ and the unknown time-varying disturbance moments D_f imposed on the NSV are given by [1]

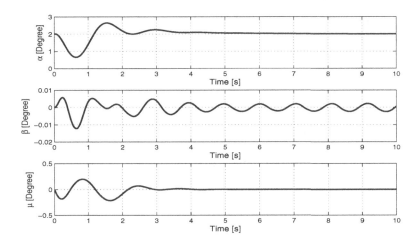

Fig. 1. Attitude Angle Response of the NSVOW

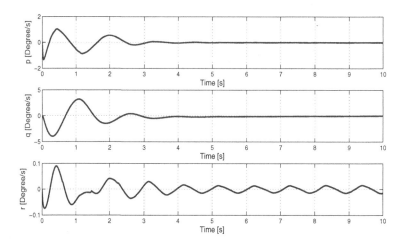

Fig. 2. Attitude Angle Velocity Response of the NSVOW

$$D_{f1}(t) = 180000(\sin(6t) + 0.2))Nm$$
$$D_{f2}(t) = 200000(\sin(5t) + 0.25))Nm$$
$$D_{f3}(t) = 200000(\sin(6t) + 0.15))Nm$$

The attitude angle response and the attitude angle velocity response are given in Fig.1 and Fig.2. In accordance with Fig.1 and Fig.2, we note that the tracking performance is satisfactory and the tracking error converges to zero under the developed neural attitude control for the uncertain NSVOW in the presence of the time-varying external disturbance. Based on these simulation results, we can obtain that the proposed disturbance-observer-based adaptive neural attitude control is valid for the uncertain dynamics of the NSVOW with the time-varying external disturbance and the unknown system uncertainty.

5   Conclusion

In this paper, the adaptive neural attitude control has been proposed for the uncertain attitude dynamics of the NSVOW. To improve the disturbance attenuation ability and attitude control robustness, the sliding mode disturbance observer has been designed to estimate the compounded disturbance which combines the external disturbance with the NN approximation error. Using outputs of the RBFNN and the sliding mode disturbance observer, the adaptive attitude control has been presented for the uncertain attitude dynamics of the NSVOW. Simulation results have been presented to illustrate the effectiveness of the proposed adaptive control scheme.

Acknowledgments. This work is partially supported by National Natural Science Foundation of China (Granted Number:61174102), Program for New Century Excellent Talents in University of China (Granted Number: NCET-11-0830), Jiangsu Natural Science Foundation of China (Granted Number: SBK2011069), the Project-sponsored by SRF for ROCS, SEM., and A Project Funded by the Priority Academic Program Development of Jiangsu Higher Education Institutions.

References

1. Du, Y.L., Wu, Q.X., Jiang, C.S., et al.: Adaptive functional link network control of near-space vehicles with dynamical uncertainties. Journal of System Engineering and Electronics 21(5), 868–876 (2010)
2. Jiang, B., Gao, Z.F., Shi, P., et al.: Adaptive fault-tolerant tracking control of near space vehicle using Takagi-Sugeno fuzzy models. IEEE Transactions on Fuzzy System 18(5), 1000–1007 (2010)
3. Xu, Y.F., Jiang, B., Tao, G., et al.: Fault tolerant control for a class of nonlinear systems with application to near space vehicle. Circuits System Signal Process 30(3), 655–672 (2011)
4. Desktop Aeronautics, Inc., Oblique Flying Wing: An Introduction and Whiter Paper
5. Enns, D.F., Bugajski, D.J., Klepl, M.J.: Flight control for the F-8 oblique wing research aircraft. In: American Control Conferrence, Minneapolis, USA, pp. 81–86 (June 1987)
6. Clark, R.N., LeTron, X.J.Y.: Oblique wing aircraft flight control system. Journal of Aircraft 12(2), 201–208 (1989)
7. Pang, J., Mei, R., Chen, M.: Modeling and control for near-Space vehicles with oblique wing. In: WCICA 2012, pp. 1773–1778 (2012)
8. Chen, M., Ge, S.S., How, B.: Robust adaptive neural network control for a class of uncertain MIMO nonlinear systems with input nonlinearities. IEEE Transactions on Neural Networks 21(5), 796–812 (2010)
9. Chen, M., Ge, S.S., Ren, B.B.: Robust attitude control of helicopters with actuator dynamics using neural networks. IET Control Theory & Application 4(12), 2837–2854 (2010)
10. Chen, M., Jiang, C.S., Jiang, B., Wu, Q.X.: Robust control for a class of uncertain time delay nonlinear system based on sliding mode observer. Neural Computing and Applications 19, 945–951 (2010)
11. Chen, M., Wu, Q.X., Cui, R.X.: Terminal sliding mode tracking control for a class of SISO uncertain nonlinear systems. ISA Transaction 52, 198–206 (2013)
12. Chen, M., Jiang, B.: Robust attitude control of near space vehicles with time-varying disturbances. International Journal of Control, Automation, and Systems 11, 82–187 (2013)
13. Hall, C.E., Shtessel, Y.B.: Sliding mode disturbance observer-based control for a reusable launch vehicle. In: AIAA Guidance, Navigation, and Control Conference and Exhibit, pp. 1–26 (August 2005)
14. Lu, Y.S.: Sliding-mode disturbance observer with switching-gain adaptation and its application to optical disk drives. IEEE Transactions on Industrial Electronics 56(9), 3743–3750 (2009)

EMG-Based Neural Network Control of an Upper-Limb Power-Assist Exoskeleton Robot⋆

Hang Su[1], Zhijun Li[1,⋆⋆], Guanglin Li[2], and Chenguang Yang[3]

[1] Key Lab of Autonomous System and Network Control
College of Automation Science and Engineering
South China University of Technology, Guangzhou, China
zjli@ieee.org
[2] Shenzhen Institutes of Advanced Technology
Chinese Academy of Sciences Shenzhen, Guangdong, China
gl.li@siat.ac.cn
[3] School of Computing and Mathematics
Plymouth University, U.K., PL4 8AA
chenguang.yang@plymouth.ac.uk

Abstract. The paper presents the electromyogram (EMG)-based neural network control of an upper-limb power-assist exoskeleton robot, which is proposed to control the robot in accordance with the user's motion intention. The upper limb rehabilitation exoskeleton is with high precision for co-manipulation tasks of human and robot because of its backdrivability, precise positioning capabilities, and zero backlash due to its harmonic drive transmission (HDT). The novelty of this work is the development of an adaptive neural network modeling and control approach to handle the unknown parameters of the harmonic drive transmission in the robot to facilitate motion control. We have conducted the experiments on human subject to identify the various parameters of the harmonic drive system combining sEMG information signals.

Keywords: Neural network control, sEMG, Harmonic Drive Transmission.

1 Introduction

In several countries, the increasing aging population and the decreasing working proportion has attracted much attention. In order to solve the problems associated with elders, disabled and weak people, many kinds of robot-assisted exoskeleton have been developed [1], [2]. The upper-limb robot-assisted exoskeleton

⋆ This work is supported in part by the Natural Science Foundation of China under Grants 61174045 and 61111130208, the International Science and Technology Cooperation Program of China under Grant 2011DFA10950, and the Fundamental Research Funds for the Central Universities under Grant 2011ZZ0104, and the Program for New Century Excellent Talents in University No. NCET-12-0195.
⋆⋆ Corresponding author.

C. Guo, Z.-G. Hou, and Z. Zeng (Eds.): ISNN 2013, Part II, LNCS 7952, pp. 204–211, 2013.

has mainly focused on restoring arm functions to achieve many activities of daily living (ADL) such as writing/typing, personal hygiene, with the use of devices such as the MIT-MANUS [5], [3], the ARMguide [4], NeReBot[7] or the MIME [6]. Numerous promising results has been yielded that illustrate the potential of robots to complement traditional assist in physical rehabilitation. Moreover, it is important for the robotic exoskeleton to be controlled in accordance with the human bio-signal feedback. In order to activate the robot according to the user's motion intention in real time, the robot must to understand the motion intention of the user [8]. In [9], the inverse model of the exoskeleton robot can be used to derive the joint torques. For rehabilitation, designated motion is basically generated with motion controller for the user before their motion.

The developed upper-limb robot-assisted rehabilitation is with high precision for co-manipulation tasks between human and robot. However, the development of an accurate dynamic model of the robot is extremely challenging because of the compliance and oscillations inherent in harmonic drive systems. Modeling of robot dynamics for the purpose of trajectory tracking using low-feedback gains has been studied previously for industrial manipulators. The novelty of this work is the development of rehabilitation robot and adaptive neural network modeling the parameters of a harmonic drive transmission in the robot to facilitate motion control. We have conducted the experiments on human body to identify the various parameters of the harmonic drive system combining sEMG information signals.

Fig. 1. Final version of exoskeleton

2   The Development of Upper Limb Exoskeleton

The developed robot follows the kinematic structure of the human upper limb and spans the elbow and wrist joints (see Fig. 1). It exhibits three degrees- of-freedom corresponding to elbow flexion-extension, forearm pronation-supination, and wrist flexion-extension, which is based upon the behavior of those physiological joints as hinges. Articulation of the developed exoskeleton is achieved about five single-axis revolute joints. The exoskeletal joints are labeled 1 through 5 from proximal to distal in the order.

3   Boosting-Based EMG Patterns Classification

MCLPBoost is inherently of multi-class for the random decision tree. Combining classifiers is a classic trick in pattern recognition field to make robust classifiers. Data comes in records of form $(X, Y) = (x_1, \cdots, x_n, Y)$, where the dependent variable, Y, is the target variable that we are trying to classify and the vector X is composed of the input variables, x_1, x_2, x_3 etc... A tree can be "learned" by splitting the source set into subsets based on an attribute value test. The details of train and prediction process are illustrated in Figs. 2 and 3, respectively. Before training, we first get the feature vector of a sample, which is labeled manually.

Fig. 2. The train process of MCLPBoost   **Fig. 3.** The prediction process of MCLP-Boost

Then, each tree is updated by each sample k times, where k is generated by Poisson distribution. Each tree is initialized as only one root node. A node contains the label density of each class, denoted by $p = \{p_1, \cdots, p_K\}$, where K is the number of classes. As a result, the density set p will be separated into two groups: $p^l = \{p_1^l, \cdots, p_K^l\}$ and $p^r = \{p_1^r, \cdots, p_K^r\}$. In order to select the best test, a measurement is given as $Score = m_l \times \sum_{i=1}^{K} p_i^l(1 - p_i^l) + m_r \times \sum_{i=1}^{K} p_i^r(1 - p_i^r)$, where $Score$ represents the importance of the test inversely; m_l and m_r denote the number of samples falling left and right respectively.

The linear programming problem in Fig. 2 is different from the original LP-Boost algorithm for its ability of training online $\min_{w_t, \xi} C \sum_{k \neq y} \xi_k + \|w_t\|_1$, s.t. $\forall n$: $w_{t,n} \geq 0, \forall k \neq y$: $\xi_k \geq 0, \forall k \neq y$: $(G_t(y, \cdot) - G_t(k, \cdot))w_t + \xi_k \geq 1$, where C is a designed parameter to limit the overfitting problem. In order to solve the optimization problem, its augmented Lagrangian dual formulation can be described in Eq. (3).

$$\max_{w_t, d_{y'}^n} d_{y'}^n + \sum_{n=1}^{N} w_{t,n}(1 - d_{y'}^n \Delta G_{y'}(n) - \zeta_n) - \frac{1}{2\theta} \sum_{n=1}^{N} (1 - d_{y'}^n \Delta G_{y'}(n) - \zeta_n)^2 \quad (1)$$

$$s.t. \ \forall n : \ \zeta_n \geq 0, w_{t,n} \geq 0 \quad (2)$$

$$0 \leq d_{y'} n \leq C \quad (3)$$

where $\Delta G_{y'}(n) = G(y, n) - G(y', n)$, $d_{y'}^n$ is the sample weight corresponding to the n-th weak learner which outputs the smallest margin on the non-target class y', ζ is a new set of slack variables, and $\theta > 0$ is a designed constant. In an iterative way, when a new sample arrives, for the first weak learner, we just assign a constant as its sample weight, but for training the n-th ($n > 1$) weak learner, we compute $d_{y'}^n$ by

$$q_j = 1 - d_{y'}^n \Delta G_{y'}(j) - \theta w_{t,j} \tag{4}$$

$$f = d_{y'}^n + \nu_d (1 + \frac{1}{\theta} \sum_{j=1, q_j < 0}^{n-1} q_j \Delta G_{y'}(j)) \tag{5}$$

$$d_{y'}^n = \max(0, \min(C, f)) \tag{6}$$

where q_j and f are temporary variables for calculating $d_{y'}^n$, and ν_d is the dual learning rate. Once the sample weight is obtained, we can calculate each weight of the weak learners.

$$\forall n : z_n = w_{t,n} - \nu_p(1 - d_{y'}^n \Delta G_{y'}(j)) \tag{7}$$

$$w_{t,n} = \max(0, z_n) \tag{8}$$

where z_n is a temporary variable to calculate $w_{t,n}$, ν_p is the learning rate for the primal. In Eq. (8), it's obvious that the weight of the weak learner, whose margin $\Delta G_{y'}(j))$ is lager, will get a bigger weight.

In contrast to the classifier training process, the prediction (Fig. 3) is relatively simpler. It can be summarized in the following equations.

$$p_n(k|x) = \frac{1}{T} \sum_{t=1}^{T} p_t(k|x) p(k|x) = \frac{1}{N} \sum_{n=1}^{N} w_i p_n(k|x) O(x) = arg \max_k p(k|x) \tag{9}$$

where T denotes the number of trees in a forest; p_t and p_n denote the confidence of the output of a tree and a forest (i.e., one weak learner), respectively, p is the confidence of final output, and O is the prediction.

4   Control Development

The MIMO nonlinear system dynamics can be described as

$$Y^{(r)} = F(x) + G(x)U \tag{10}$$

where $F(x)$ and $G(x)$ is unknown nonlinear function, U and Y is input and output vectors. Assume that $G(x)$ is a positive definite matrix, the desired position is $y_{di}(t)$ and its derivative of η_i order exists, thus define

$$e_i(t) = y_{di}(t) - y_i(t); \tag{11}$$

$$S_i(t) = (\frac{d}{d_t} + \lambda_i)^{\eta_i - 1} e_i(t), \lambda_i > 0; \tag{12}$$

and if $S_i(t) \to 0$, then we also have $e_i(t) \to 0$. According to Newton's binomial theorem, we can decompose $\dot{S}_i$ and get

$$\dot{S} = \zeta - F(x) - G(x)u; \tag{13}$$

$$u = G^{-1}(x)(-F(x) + \zeta + K_0 S); \tag{14}$$

where $\zeta = y_{d_i}^{r_i} + \sum_{j=1}^{r_i-1} \frac{(r_i-1)!}{(r_i-j)!(j-1)!} e_i^{(j)}(t) \lambda_i^{r_i-j}$ and $K_0 = diag[k_{01}, k_{02}, \dots, k_{0p}]$, $k_{0i} > 0$. Substitute Eq.(13) with Eq.(14) we get $\dot{S}(t) = -K_0 S(t)$. The solution of the differential equation is $s_i(t) = s_i(0)e^{-k_{0i}t}$, with $t \to \infty, s_i(t) \to 0$. If we known the nonlinear function $F(x)$ and $G(x)$, we can get the control output u easily; If not, we can design a fuzzy function to approach the two functions as $y(x) = \xi^T(x)\Theta$; where $\Theta = [\theta^1, \dots, \theta^M]^T$ is parameters vector, $\xi(x) = [\xi_1(x), \dots, \xi_M(x)]$, and

$$\xi_l(x) = \frac{\prod_{i=1}^{n} \mu_{F_i^l}(x_i)}{\sum_{l=1}^{M}(\prod_{i=1}^{n} \mu_{F_i^l}(x_i))} \tag{15}$$

The output of the system $y(x)$ converges gradually to the unknown nonlinear function $f_i(x)$ and $g_{ij}(x)$. θ_{f_i} and $\theta_{g_{ij}}$ are the corresponding adaptive regulator parameters. Define the optimal approximation parameters as $\theta_{f_i}^*$ and $\theta_{g_{ij}}^*$. Then we design the $\hat{F}(x, \theta_f)$ and $\hat{G}(x, \theta_g)$ to substitute the $F(x)$ and $G(x)$.

The adaptive control law Θ of the corresponding fuzzy system can be defined as follows

$$\dot{\Theta}_{f_i} = -\eta_{f_i} \xi_{f_i}(x) s_i; \tag{16}$$

$$\dot{\theta}_{g_{ij}} = -\eta_{g_{ij}} \xi_{g_{ij}}(x) s_i u_{cj}; \tag{17}$$

where $\eta_{f_i} > 0, \eta_{g_{ij}} > 0$. Then we substitute $\hat{F}(x, \theta_f)$ and $\hat{G}(x, \theta_g)$ into Eq.(14); we have the control equation as

$$u_c = \hat{G}^{-1}(x)(-\hat{F}(x) + \zeta + K_0 S); \tag{18}$$

To ensure the $\hat{G}(x, \theta_g)$ is nonsingular, we substitute it with a generalized inverse $\chi = \hat{G}^T(x, \theta_g)[\tau_0 I_p + \hat{G}(x, \theta_g)\hat{G}^T(x, \theta_g)]^{-1}$; where τ_0 is a small positive real number randomly, I_p is a unit matrix; Thus, the control input can be expressed as Eq.(19) and the control development is done.

$$u_c = \chi(-\hat{F}(x) + \zeta + K_0 S); \tag{19}$$

To reduce the modeling errors, we take robust control u_r. Thus $u = u_c + u_r$. Where $u_r = \frac{s|s^T|(\bar{\varepsilon}_f + \bar{\varepsilon}_g|u_c| + |u_0|)}{\sigma_0\|s\|^2 + \delta}$; $u_0 = \varepsilon_0[\varepsilon_0 I_p + \hat{G}(x, \theta_g)\hat{G}^T(x, \theta_g)]^{-1}(-\hat{F}(x, \theta_f) + \zeta + K_0 s)$; and the δ is a time-variable parameter.

$$\dot{\delta} = -\eta_0 \frac{|s^T|(\bar{\varepsilon}_f + \bar{\varepsilon}_g|u_c| + |u_0|)}{\sigma_0\|s\|^2 + \delta}; \tag{20}$$

where $\eta_0 > 0, \delta(0) > 0$.

5   Experiments

In this section, the human limb itself, can be used as control interface for the exoskeletons. EMG signals correspond to muscle activity when the muscle contracts. We can obtain sEMG signal from the sEMG sensor system which is fixed with signal amplifier. The amplifier could receive 8 channel of sEMG signal at the same time. The recorded sEMG signal of the wrist and elbow joint has been pre-processsde (e.g. amplification, filtering) and post-processed (e.g. smoothing).

5.1   Data Processing

The process of data processing in our experiment can be divided into three phases: feature extraction, feature reduction and classification. In our experiment, the data collected stored in the file folder of Record are used as training data and the data collected online are used as test data. And the classification result consists of three states which was labeled as $0, 1, 2$ in the experiment. And it's can be concluded that all emg signals recorded in the procedure have been classified correctly (see Fig. 4 and 5). Then we use the generated classification to determine the motion of the corresponding motor. In the paper, the desired trajectory is set $q_d = A * (1 - \cos t)$ when the muscle of corresponding joint flexion flex and $q_d = -A * (1 - \cos t)$ when it extent, where q_d is the desired trajectory, A is the positive amplitude of the trajectory. The experiment results using the neural network control are shown in Figs. 6–9. The trajectories of the weights Θ are listed in Figs.11–13. For comparison, the experiments using PD control are shown in Figs. 14–17. From these comparison, we can see the neural network is with good performance.

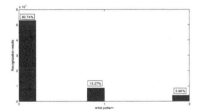
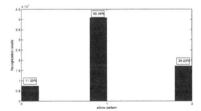

Fig. 4. The result of classification of wrist   **Fig. 5.** The result of classification of elbow

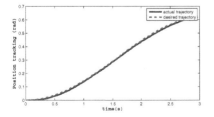
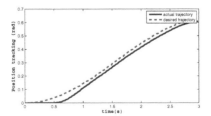

Fig. 6. The tracking of upward wrist   **Fig. 7.** The tracking of upward elbow

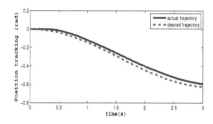

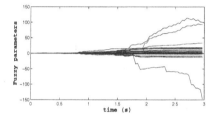

Fig. 8. The tracking of downward wrist   **Fig. 9.** The tracking of downward elbow

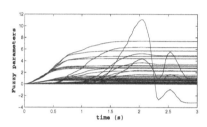

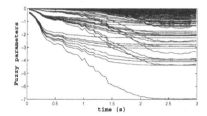

Fig. 10. The weights Θ of upward wrist   **Fig. 11.** The weights Θ of upward elbow

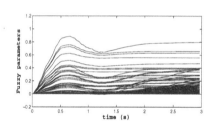

Fig. 12. The weights Θ of downward wrist   **Fig. 13.** The weights Θ of downward elbow

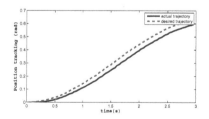

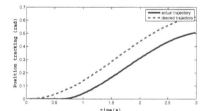

Fig. 14. The PD tracking of upward wrist   **Fig. 15.** The PD tracking of upward elbow

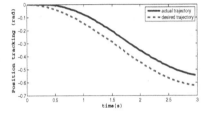 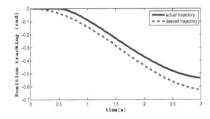

Fig. 16. The PD tracking of downward wrist   **Fig. 17.** The PD tracking of downward elbow

6   Conclusions

In this paper, the electromyogram (EMG)-based neural network control of an upper-limb power-assist exoskeleton robot has been developed. The development of an adaptive neural network modeling and control approach to handle the unknown parameters of the harmonic drive transmission in the robot to facilitate motion control. We have conducted the experiments on human subject to identify the various parameters of the harmonic drive system combining sEMG information signals.

References

1. Hallett, M.: Recent advances in stroke rehabilitation. Neurorehabil. Neural Repair 16, 211–217 (2002)
2. Lang, C., Schieber, M.: Differential impairment of individuated finger movements in humans after damage to the motor cortex or the corticospinal tract. J. Neurophysiol. 90, 1160–1170 (2003)
3. Masia, L., Krebs, H., Cappa, P., Hogan, N.: Design, characterization, and impedance limits of a hand robot. In: Proc. IEEE Int. Conf. Robotic Rehabil. (ICORR), pp. 1085–1089 (June 2007)
4. Kahn, L.E., Rymer, W.Z., Reinkensmeyer, D.J.: Adaptive assistance for guided force training in chronic stroke. In: Proceedings of the 26th Annual International Conference of the IEEE Engineering in Medicine and Biology Society, San Francisco, CA, USA, pp. 2722–2725 (2004)
5. Krebs, H.I., Hogan, N., Aisen, M.L., Volpe, B.T.: Robot-aided neurorahabilitation. IEEE Transactions on Rehabilitation Engineering 6(1), 75–87 (1998)
6. Lum, P.S., Burga, C.G., Loos, M.V., Sho, P.C., Majmundar, M., Yap, R.: MIME robotic device for upper-limb neurorehabilitation in subacute stroke subjects: A follow-up study. J. Rehabil. Res. Dev. 43(5), 631–642 (2006)
7. Masiero, S., Celia, A., Rosati, G., Armani, M.: Robotic-assisted rehabilitation of the upper limb after acute stroke. Arch. Phys. Med. Rehabil. 88(2), 142–149 (2007)
8. Kong, K., Jeon, D.: Design and control of an exoskeleton for the elderly and patients. IEEE/ASME Trans. Mechatronics 11(4), 428–432 (2006)
9. Kazerooni, H., Steger, R.: The Berkeley lower extremity exoskeleton. Trans. ASME, J. Dyn. Syst., Meas. Control 128(1), 14–25 (2006)

Neural Network Based Direct Adaptive Backstepping Method for Fin Stabilizer System[*]

Weiwei Bai, Tieshan Li[**], Xiaori Gao, and Khin Thuzar Myint

Navigation College, Dalian Maritime University, Dalian 116026, China
tieshanli@126.com

Abstract. Based on the backstepping method and the neural networks (NNs) technique, a direct adaptive controller is proposed for a class of nonlinear fin stabilizer system in this paper. This approach overcomes the uncertainty in the nonlinear fin stabilizer system and solves the problems of mismatch and controller singularity. The stability analysis shows that all the signals of the closed-loop system are uniformly ultimate boundedness (UUB). A simulation example is given to illustrate the effectiveness of the proposed method.

Keywords: fin stabilizer, direct adaptive control, neural networks, backstepping.

1 Introduction

It is one of the dangerous phenomenon that ship rolls in moderate and rough beam seas with large amplitude rolling motion. This will reduce the comfort and safety of the voyage. So, how to reduce the rolling motion of ship has been a hotspot in the ship control area. In the past few decades, various devices were applied to reduce the large amplitude rolling motion of ship such as bilge keels, antiroll tanks, etc. Due to the limited effectiveness of these devices, the fin stabilizer became the popular device in use [1]. Numerous studies on ship stabilization by using fin controllers have been conducted since 1940s. For example, Allan [2] investigated the required moment against the upsetting moment of regular sea with model fin tests. The effectiveness of active stabilizers in the two trial ships was developed theoretically with reasonable accuracy by Conolly [3]. For the purpose of further improving the effectiveness of fin stabilizer which mainly depends on the control strategy, many researches were carried out. In [4], the conventional PID controller was extended to the nonlinear constrained optimization. By using this technique, the optimal PID could match the classical PID

---

[*]This work was supported in part by the National Natural Science Foundation of China (Nos.51179019, 60874056), the Natural Science Foundation of Liaoning Province (No. 20102012) and the Program for Liaoning Excellent Talents in University (LNET) (Grant No.LR2012016), the Fundamental Research Funds for the Central Universities (No. 3132013005), and the Applied Basic Research Program of Ministry of Transport of China.
[**] Corresponding author.

C. Guo, Z.-G. Hou, and Z. Zeng (Eds.): ISNN 2013, Part II, LNCS 7952, pp. 212–219, 2013.

in the low frequency (LF) ranges, and in the high frequency (HF) ranges it was able to substantially reduce the HF peak. However, the PID controller based on the linear model was difficult to control the rolling motion caused by the nonlinear dynamic in rough sea. In [5], based on the backstepping and the closed-loop gain shaping algorithms a nonlinear robust controller was proposed, which solved the problems of mismatch of the system and nonlinearity of external disturbances. Nevertheless, the methods mentioned above could not avoid the model uncertainties of the system.

Recently, neural networks(NNs) have received considerable attention for its approximation ability in ship control area. NNs were used to track the parametric perturbation in the functions of fin stabilizer system and showed good results in [6-8]. Recently, Hassan [9] presented a controller combined neural network and PID control for roll control of ship with small draught. These methods mentioned above overcame the effects of parameter perturbations on the fin stabilizer system. In addition, an indirect adaptive backstepping based radial basis function (RBF) neural network method was proposed without building accurate mathematical model in [10], where the system can obtain better anti-rolling effectiveness and robustness by using the control strategy, but it suffered from a potential risk of controller singularity.

Motivated by the above observations, a direct adaptive NN controller for uncertain nonlinear fin stabilizer system is proposed in this paper by combining adaptive backstepping design method with NN control design framework. This method overcomes the problems of mismatch and controller singularity, and at the same time, avoids the model uncertainties in the fin stabilizer system.

The rest of this paper is organized as follows. In section 2, the nonlinear model of the fin stabilizer and radial basic function neural network are introduced. The controller design and stability analysis are presented in section 3. In section 4, simulation results are used to illustrate our approach. The conclusions are included in section 5.

2   Problem Formulation

2.1   Fin Stabilize Model

The nonlinear model of the fin stabilizer

$$(I_{xx} + J_{xx})\ddot{\phi} + \delta_N \dot{\phi} + \delta_W \dot{\phi}|\dot{\phi}| + Wh\phi\left[1 - (\phi/\phi_v)^2\right] = M_c + M_W \tag{1}$$

where ϕ is rolling angle of ship, I_{xx} and J_{xx} are the inertia moments and the added inertia moments of the own ship, δ_N, δ_W are the damping factors, W is the tonnage of ship, h is initial metacentric height, φ_v is flooding angle, M_c is control moment of the fin stabilizer, M_W is the moment of sea wave act on ship. I_{xx} , J_{xx} , δ_N, δ_W and M_c can be denoted

$$I_{xx} + J_{xx} = \frac{WB^2}{g}(0.3085 + \frac{0.0227B}{d} - \frac{0.0043L}{100})^2 , \delta_N = \frac{2c_1\sqrt{Wh(I_{xx} + J_{xx})}}{\pi} ,$$

$$\delta_W = \frac{3c_2(I_{xx}+J_{xx})}{4} \quad , M_c = -\rho v^2 A_f l_f C_L^\alpha (\alpha_f + \frac{\dot{\phi} l_f}{v}),$$

where g is gravity acceleration, B is the width of ship, L is length between tow-column of ship, d is draught, c_1, c_2 are test coefficient, ρ is the fluid density, v is the ship speed, A_f is the area of fin, l_f is the acting force arm of fin stabilizer, C_L^α is the lift coefficient of fin stabilizer, α_f is the rotation angle of fin stabilizer. Then

$$\ddot{\varphi} = a_1\varphi + a_2\varphi^3 + a_3\dot{\varphi} + a_4\dot{\varphi}|\dot{\varphi}| + b\alpha_f + f_W \tag{2}$$

where a_1, a_2, a_3, a_4, b are coefficients, f_W is the disturbance of sea wave.

2.2   The State Space Model of Fin Stabilizer

Choosing $x = [x_1, x_2]^T = [\varphi, \dot{\varphi}]^T$ as the state variable, $y = \varphi = x_1$ as the output variable, $u = \alpha_f$ as the input. We consider the uncertainties of the fin system, formula (2) can be transformed into the state space model as follows:

$$\begin{cases} \dot{x}_1 = x_2 \\ \dot{x}_2 = g_2(\bar{x}_2)u + f_2(\bar{x}_2) + d \\ y = x_1 \end{cases} \tag{3}$$

where $f_2(\bar{x}_2)$, $g_2(\bar{x}_2)$ are the unknown functions, d denotes the external disturbances.

For the development of the control laws, the following assumptions are made.

Assumption 1. $g_2(\cdot)$ is unknown nonlinear smooth function. The signs of $g_2(\cdot)$ is known, and there exist constants $g_{21} > g_{20} > 0$ such that $g_{21} > |g_2(\cdot)| > g_{20}$, $\forall \bar{x}_i \in \Omega \subset R^n$. And there exist constants $g_{2d} > 0$ such that $|\dot{g}_i(\cdot)| \le g_{id}$, $\forall \bar{x}_i \in \Omega \subset R^n$. So $g_i(\cdot)$ are strictly either positive or negative. Without losing generality, we assume $g_{21} > g_2(\bar{x}_2) > g_{20} > 0, \forall \bar{x}_i \in \Omega \subset R^n$.

Assumption 2. The disturbance d has an upper bound, that is, there exists an unknown constant $\bar{\omega} > 0,$ which satisfies $|d| \le \bar{\omega}$.

The following lemma is helpful to prove stability, we will use it later.

Lemma 1. If $V(t,x)$ is positive definite, and $\dot{V} \le -k_1 V + k_2$, where $k_1 \ge 0$, $k_2 \ge 0$ are bounded constant, then

$$V(t,x) \le \frac{k_2}{k_1} + \left(V(0) - \frac{k_2}{k_1}\right)e^{-k_1 t}$$

2.3   RBF Neural Network

RBF neural networks belong to a class of linearly parameterized networks. For comprehensive treatment of neural networks approximation, see [11]. RBF neural networks can be described as $w^T S(z)$ with input vector $z \in R^n$, weight vector $w \in R^l$, node number l, and basis function vector $S(z) \in R^l$. Universal approximation results indicate that, if l is chosen sufficiently large, then $w^T S(z)$ can approximate any continuous function to any desired accuracy over a compact set. In this paper, we use the following RBF neural networks to approximate a smooth function $h(z): R^q \to R$,

$$h_{nn}(z) = w^T S(z)$$

where the input vector $z \in \Omega \in R^n$, weight vector $w = [w_1, w_2, ..., w_l]^T \in R^l$, the neural network node number $l > 1$, and $S(z) = [s_1(z), s_2(z), ..., s_l(z)]^T$, with $s_i(z)$ being chosen as the commonly used Gaussian functions, which have the form $s_i(z) = \exp[\dfrac{-(z - \mu_i)^T (z - \mu_i)}{\eta_i^2}], i = 1, 2, ..., l$; where $\mu_i = [\mu_{i1}, \mu_{i2}, ..., \mu_{in}]^T$ is the center of the receptive field and η_i is the width of the Gaussian functions.

It is well known that for an unknown continuous nonlinear function $f(x)$, by using RBF NNs approximation over the compact sets Ω, one can obtain

$$f(x) = w^{*T} S(x) + \varepsilon. \ \forall x \in \Omega \subseteq R^n$$

Where $S(x)$ is the basis function vector, ε is the approximation error, which has an unknown upper bound ε^*, and w^* is an unknown ideal constant weight vector.

The ideal weight vector w^* is an "artificial" quantity required only for analytical purposes. Typically, w^* is chosen as the value of w that minimizes $|\varepsilon|$ for all $x \in \Omega$, where $\Omega \subseteq R^n$ is a compact set, i.e.,

$$w^* := \arg \min_{w \in R^n} \left\{ \sup_{x \in \Omega} \left| f(x) - w^T S(x) \right| \right\}$$

3   Controller Design and Stability Analysis

Step1.Consider the first equation of the system (3), define the error variable $z_1 = x_1$, and choose the intermediate stabilizing function α_2 as a virtual control law for the first subsystem. At the same time, define error variable $z_2 = x_2 - \alpha_2$, the time derivative of z_1 is

$$\dot{z}_1 = z_2 + \alpha_2 \tag{4}$$

The virtual control law is chosen as

$$\alpha_2 = -c_1 z_1 \tag{5}$$

where $c_1 > 0$, substituting(5) into (4), we obtain

$$\dot{z}_1 = -c_1 z_1 + z_2 \tag{6}$$

Consider the Lyapunov function candidate $V_1 = \dfrac{1}{2} z_1^2$ and the derivative of V_1 is

$$\dot{V}_1 = -c_1 z_1^2 + z_1 z_2 \tag{7}$$

Step2. Define $z_2 = x_2 - \alpha_2$, and differentiate z_2 with respect to time yields

$$\dot{z}_2 = g_2(\bar{x}_2)u + f_2(\bar{x}_2) + d - \dot{\alpha}_2 \tag{8}$$

Since $f_2(\bar{x}_2)$ and $g_2(\bar{x}_2)$ are unknown, $\dot{\alpha}_2$ is in fact an unknown scalar nonlinear function. Now, define $h_2(\bar{x}_2)$ and constructing an RBF neural network $\theta_2^{*T} \xi_2(\bar{x}_2)$ to approximate $h_2(\bar{x}_2)$ as follows

$$h_2(\bar{x}_2) \overset{\Delta}{=} \frac{1}{g_2(\bar{x}_2)} (f_2(\bar{x}_2) + d - \dot{\alpha}_2) = \theta_2^{*T} \xi_2(\bar{x}_2) + \delta_2^* \tag{9}$$

where θ_2^* denotes the ideal constant weights, and δ_2^* is the approximation error with constant $\delta_2^* > 0$. Then, (8) becomes

$$\dot{z}_2 = g_2(\bar{x}_2)\left(u + \theta_2^{*T} \xi_2(\bar{x}_2) + \delta_2^*\right) \tag{10}$$

Choosing the control law as

$$u = -c_2 z_2 - \hat{\theta}_2^T \xi_2(\bar{x}_2) \tag{11}$$

where $\hat{\theta}_2$ is the parameter estimate of θ_2^*, Let $\tilde{\theta}_2 = \hat{\theta}_2 - \theta_2^*$ be the parameter error. Then, we have

$$\dot{z}_2 = g_2(\bar{x}_2)(-c_2 z_2 - \tilde{\theta}_2^T \xi_2(\bar{x}_2) + \delta_2^*) \tag{12}$$

Consider the following Lyapunov function candidate

$$V_2 = V_1 + \frac{1}{2} \frac{z_2^2}{g_2(\bar{x}_2)} + \frac{1}{2} \tilde{\theta}_2^T \Gamma_2^{-1} \tilde{\theta}_2 \tag{13}$$

where $\Gamma_2 = \Gamma_2^T > 0$ is an adaptation gain matrix. The derivative of V_2 is

$$\dot{V}_2 = \dot{V}_1 - c_2 z_2^2 + z_2 \delta_2^* + \tilde{\theta}_2 \Gamma_2^{-1}(\dot{\hat{\theta}}_2 - \Gamma_2 \xi_2 z_2) + \frac{z_2^2 \dot{g}_2(\bar{x}_2)}{2(g_2(\bar{x}_2))^2} \tag{14}$$

Consider the following adaptation law:

$$\dot{\hat{\theta}}_2 = \Gamma_2(\xi_2 z_2 - \sigma_2 \hat{\theta}_2) \tag{15}$$

where $\sigma_2 > 0$ is a small constant. By using (7), (12) and (15), the derivative of V_2 becomes

$$\dot{V}_2 = -c_1 z_1^2 - c_2 z_2^2 + z_1 z_2 + z_2 \delta_2^* - \sigma_2 \tilde{\theta}_2^T \hat{\theta}_2 + \frac{z_2^2 \dot{g}_2(\bar{x}_2)}{2(g_2(\bar{x}_2))^2} \tag{16}$$

Mentioning the facts

$$z_1 z_2 \le z_1^2 + \frac{1}{4} z_2^2; \quad z_2 \delta_2^* \le z_2^2 + \frac{1}{4} \delta_2^{*2}; \quad 2\tilde{\theta}_2^T \hat{\theta}_2 \ge \left\| \tilde{\theta}_2 \right\|^2 - \left\| \theta_2^* \right\|^2;$$

Then, we have

$$\dot{V}_2 \le -(c_1 - 1)z_1^2 - (c_2 - \frac{5}{4})z_2^2 + \frac{1}{4}\delta_2^{*2} - \frac{\sigma_2}{2}(\left\| \theta_i \right\|^2 - \left\| \theta_i^* \right\|^2) + \frac{z_2^2 \dot{g}_2(\bar{x}_2)}{2 g_2^2(\bar{x}_2)} \tag{17}$$

Note that $-(c_2 z_2^2 + \dot{g}_2(\bar{x}_2)z_2^2 / 2g_2^2(\bar{x}_2)) \le -(c_2 z_2^2 + \dot{g}_{2d}(\bar{x}_2)z_2^2 / 2g_2^2(\bar{x}_2))$, we can choose c_2 such that $c_2^* \overset{\Delta}{=} (c_2 - g_2(\bar{x}_2)) / (2g_{20}^2(\bar{x}_2)) > 0$.

Let $c_1 + 3 = c_{11}$, $c_2^* - 5/4 = c_{21}$, $\frac{1}{4}\delta_2^{*2} + \frac{\sigma_2}{2}\left\| \theta_2^* \right\|^2 = e_2$, and pay attention to $\left| \delta_2^* \right| \le \delta_M$ and $\left\| \theta_2^* \right\| \le \theta_M$, we can get $e_i \le (1/4)\delta_m^2 + (\sigma_i/2)\theta_M^2 = e_M = D$. Then

$$\dot{V}_2 \le -c_{11} z_1^2 - c_{21} z_2^2 - \frac{\sigma_2}{2\lambda_{max}(\Gamma_2^{-1})}\tilde{\theta}_2^T \Gamma_2^{-1}\tilde{\theta}_2 + D \tag{18}$$

let $c_{11} = \frac{C}{2}$, $c_{21} = \frac{C}{2g_2(\bar{x}_2)}$, $\frac{\sigma_2}{2\lambda_{max}(\Gamma_2^{-1})} = \frac{C}{2}$, where $C > 0$ is a bound constant which satisfies $\sigma_2 \ge C\lambda_{max}(\Gamma_2^{-1})$. Then, it follows

$$\dot{V}_2 \le -CV_2 + D \tag{19}$$

Obviously, the Equation (19) means that $V_2(t)$ is bounded (please refer to [12] for details) according to the assumptions and lemma 1. So, it is clear that the bounded stability of the closed-loop system(3) is guaranteed, and the tracking error can be made arbitrarily small by appropriate selections of the design parameters[12].

4   Simulation Research

The simulation is based on a container ship, the parameters of the ship, such as, the length is 175m, the width of ship is 25.4m, draught is 8.5m, tonnage 21120t, the area of fin is $20.2\,m^2$, the acting force arm of fin stabilizer is 14.88m, flooding angle is $43°$, initial metacentric height is 1m, designed speed is 7.71m/s. the lift coefficient of fin stabilizer is 3.39.

In simulation, the RBF NNs contains 135nodes, with centers evenly spaced in $[-4\times4]\times[-4\times4]$ and widths $\eta_i = 2\ (i = 1\cdots25)$ and $\eta_j = 2\ (j = 1\cdots135)$. We choose the control system parameters: $c_1 = 3$, $c_2 = 5$, $\Gamma_2 = dig\{0.2\}$, $\sigma_2 = 0.2$, the disturbance is $d = 0.35*\sin(0.5t)$. Simulation results in Figure.1-2 illustrate the control performance of the proposed scheme.

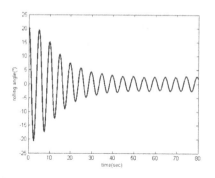   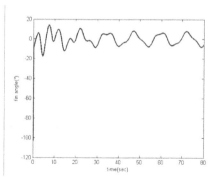

Fig. 1. Ship rolling angle   **Fig. 2.** The fin angle

5   Conclusions

In this paper, a neural network-based direct adaptive control scheme is proposed for uncertain nonlinear fin stabilizer system. The proposed scheme can solve the problems of mismatch and controller singularity without building accurate mathematical model, and can obtain better anti-rolling effectiveness and robustness. All the signals in the closed loop of the nonlinear fin stabilizer system are guaranteed to be UUB in the sense of Lyapunov stability, and the tracking error can be made arbitrarily small by adjusting the parameters in the control law. The simulation shows that good tracking performance can be obtained by the proposed approach.

References

1. Karakas, S.C., Sc, M.: Control design of fin roll stabilization in beam seas based on Lyapunov's direct method. Maritime Research 73(2), 25–30 (2012)
2. Allan, J.F.: Stabilisation of ships by activated fins. Transactions of the Royal Institution of Naval Architects RINA 87, 123–159 (1945)
3. Conolly, J.E.: Rolling and its stabilization by active fins. Transactions of the Royal Institution of Naval Architects RINA 111, 21–48 (1968)
4. Hickey, N.A., Johnson, M.A., Katebi, M.R., Grimble, M.J.: PID Controller Optimisation for Fin Roll Stabilisation. In: International Conference on Control Applications Kohala Coast-bland of Hawai'i, Hawai'i, USA, vol. 2, pp. 1785–1790 (1999)
5. Wang, X., Zhang, X.: Fin stabilizer control based on backstepping and closed-loop gain shaping algorithms. Journal of Dalian Maritime University 3(34), 89–92 (2008)
6. Fuat, A.: Internal model control using neural network for ship roll stabilization. Journal of Marine Science and Technology 15(2), 141–147 (2007)
7. Kang, T.Z., Luo, D., Yan, X., Niu, Y.: The Study of Neural Network Control System of the Fin Stabilizer. Ship Egineering 34(2), 278–301 (2012)
8. Fang, M., Zhuo, Y.: The application of the self-tuning neural network PID controller on the ship roll reduction in random waves. Ocean Engineering 37(7), 529–538 (2010)
9. Hassan, G., Fatemeh, D., Parviz, G., Babak, O.: Neural network-PID controller for roll fin stabilizer. Polish Maritime Research 17(2), 23–28 (2010)
10. Zhang, Y., Shi, W., Yin, L.: Adaptive backstepping and sliding model control of fin stabilizer based on the RBF neural network. Proceedings-Electric Power Applications 149(3), 184–194 (2002)
11. Ge, S.S., Hang, C.C., Zhang, T.: Stable adaptive neural network control. Kluwer Academic Publishers (2002)
12. Qu, Z.: Robust Control of Nonlinear Uncertain Systems. Wiley, New York (1998)

Output Feedback Adaptive Robust NN Control for a Class of Nonlinear Discrete-Time Systems

Xin Wang[1], Tieshan Li[1,*], Liyou Fang[1], and Bin Lin[2]

[1] Navigational College, Dalian Maritime University, Dalian 116026, China
[2] Department of Information Science and Technology, Dalian Maritime University, Dalian 116026, China
luyunliunanting@yeah.net, tieshanli@126.com, binlin@dlmu.edu.cn

Abstract. In this paper, output feedback direct adaptive robust NN control is investigated for a class of nonlinear discrete-time systems in strict-feedback form. To construct output feedback control, the original strict-feedback system is transformed into a cascade form, which the output feedback of the nonlinear discrete-time system can be carried out. Then with employment of the inputs and outputs, the output feedback direct adaptive robust NN control is developed. The HONNs is exploited to approximate unknown function, and a stable adaptive NN controller is synthesized. The proposed algorithm improves the rubostness of the discrete-time nonlinear systems. It is proven that all the signals in closed-loop system are uniformly ultimately bounded (UUB). A simulation example is presented to illustrate the effectiveness of the proposed algorithm.

Keywords: output feedback, rubostness, direct adaptive NN control, discrete-time nonlinear system.

1 Introduction

The last decade has witnessed an ever increasing research in adaptive neural network (NN) control, since the NNs have been proven to be particularly useful for controlling nonlinear systems with nonlinearly parameterized uncertainties owing to their universal approximation property [1]. In general, NNs are used as function approximators to approximate some suitable uncertainties appearing in controllers or systems. Through years of progress, much significant development has been achieved in continuous-time nonlinear systems in strict-feedback form [2-9] via backstepping technology. For example, adaptive NN control via backstepping design was presented for a class of minimun phase nonlinear systems with known relative degree in [2]. and the dynamic surface control (DSC) technique was employed to simplify the the backstepping

---

* This work was supported in part by the National Natural Science Foundation of China (Nos.5 1179019, 61001090), the Natural Science Foundation of Liaoning Province (No.20102012), the Program for Liaoning Excellent Talents in University of (LNET)(Grant No.LR2012016) and the Applied Basic Research Program of Ministry Transport of China.

C. Guo, Z.-G. Hou, and Z. Zeng (Eds.): ISNN 2013, Part II, LNCS 7952, pp. 220–227, 2013.
© Springer-Verlag Berlin Heidelberg 2013

design for nonlinear systems in strict-feedback form by overcoming the problem of "explos-ion of complexity" in [7].

Comparing to the nonlinear continuous-time systems indicated in the above mentioned papers, adaptive control is less developed for nonlinear discrete-time systems. The reason lies in that the linearity property of the derivative of a Lyapunov function in continuous-time is not persent in the difference of Lyapunov function in the discrete-time [10]. As a consequence, many elegant control schemes for continuous-time systems may be not suitable for discrete-time systems. However, there are still considerable advances in adaptive NN control for discrete-time systems, for example, see [11-14] and the references therein. To solve the noncausal problem, the approach that "looks ahead" was proposed in [11] for parameter strict-feedback discrete-time systems. In [12], adaptive NN backstepping design has been applied to the transformed strict-feedback discrete-time systems without noncausal problem. The result has also been extended to multi-input and multi-output (MIMO) systems in [14].

In the adaptive control domain, robustness has been an active topic of research in nonlinear systems. To enhance the robustness of the adaptive control system, the bound on approximation error will be estimated and the estimation can be adjusted in [15-17]. However, it is difficult to apply these results to control the discrete-time systems due to the noncausal problem. Very recently, in [18], a novel states-feedback robust NN algorithem was proposed to solve the obstacle of the application for robustness in discrete-time nonlinear systems. However, the control scheme becomes infeasible when some system states are unmeasurable. For such problems, several output feedback control schemes in discrete-time were developed by using the backstepping design in [19-23]. For example, the output feedback control for multiple-input-multiple-output (MIMO) was successfully proposed in [19], and then the ideas was applied to MIMO systems with unknown control directions [20]. For pure-feedback systems, the discrete-time system was transformed into an input-output predictor model by investigating the relationship between outputs and states, then the output feedback control was studied in [21-22], Similarly to above schemes, the output feedback controller for NARMAX systems was also investigated in [23]. Then a novel observer output feedback for non-strict feedback systems was developed in [24], where the observer can estimate the states with a small bounded error by using the measured output.

In this paper, we will study the output feedback direct adaptive robust NN control for a class of discrete-time nonlinear systems. Based on the previous research and study, the adaptive robust NN controller is proposed for output feedback discrete-time nonlinear systems, which marge the adaptive robust algorithm and adaptive neural network algorithm directly in the control law. By only utilizing the measurement of inputs and outputs, the output feedback adaptive control is relatively easy to implement and the controller structure is simple. During the controller design process, the HONNs are used to approximate the unknown nonlinear functions. All the signals in the closed-loop system are guaranteed to be UUB and the system output can track the reference signal within a bounded compact set. A simulation example is utilized to show the effectiveness of the proposed approach.

2   Problem Formulation and Preliminaries

2.1   System Description

Consider the following single-input-single-output (SISO) discrete-time nonlinear systems in the strict-feedback form:

$$\begin{cases} \xi_i(k+1) = f_i(\bar{\xi}_i(k)) + g_i(\bar{\xi}_i(k))\xi_{i+1}(k), & i = 1,2,\ldots,n-1, \\ \xi_n(k+1) = f_n(\bar{\xi}_n(k)) + g_n(\bar{\xi}_n(k))u(k) \\ y_k = x_1(k) \end{cases} \tag{1}$$

where $\bar{\xi}_i(k) = [\xi_1(k), \xi_2(k), \ldots, \xi_i(k)]^T \in R^i$, $i = 1,2,\ldots,n$, $u(k) \in R$ and $y_k \in R$ are the state variables, system input and output respectively; $f_i(\bar{\xi}_i(k))$ and $g_i(\bar{\xi}_i(k))$, $i = 1,2,\ldots,n$ are unknow smooth functions.

The control objective is to design an adaptive robust NN controller for the system and the output y_k follows the desired reference signal $y_d(k)$ within a small compact set, where $y_d(k) \in \Omega_y, \forall k > 0$ is smooth and known bounded function with $\Omega_y := \{\chi | \chi = \xi_1\}$.

Assumption 1. the signal of $g_i(\bar{\xi}_i(k))$, $i = 1,2,\ldots,n$ are known and there exist constants $\underline{g_i} > 0$ and $\bar{g}_i > 0$ such that $\underline{g_i} \leq |g_i(\bar{\xi}_i(k))| \leq \bar{g}_i$, $\forall \bar{\xi}_n(k) \in \Omega \subset R^n$.

3   Output Feedback Adaptive Robust NN Controller Design

If we consider the original system description as an one-step ahead predictor, we can transform the one-step ahead predictor into an equivalent maximum n-step ahead predictor, which can predict the future states $\xi_1(k+n)$, $\xi_2(k+n-1)$, $\ldots$, $\xi_n(k+1)$; then, the causality contradiction is avoided when the controller is constructed based on the maximum n-step ahead prediction [12]. The original system is transformed into the following form:

$$y_{k+n} = x_n(k+1) = f_0(\underline{z}(k)) + g_0(\underline{z}(k))u_k \tag{2}$$

where $f_0(\underline{z}(k)) := F_0(\underline{z}(k)) = f([x_1(k), \psi_2(\underline{z}(k)), \ldots, \psi_n(\underline{z}(k))]^T)$,

$g_0(\underline{z}(k)) := G_0(\underline{z}(k)) = g([x_1(k), \psi_2(\underline{z}(k)), \ldots, \psi_n(\underline{z}(k))]^T)$.

Defining $e_y(k) = y_k - y_d(k)$, then the tracking error dynamics can given by:

$$e_y(k+n) = -y_d(k+n) + f_0(\underline{z}(k)) + g_0(\underline{z}(k))u_k \tag{3}$$

This paper uses HONNs to approximate the known functions in (1). Supposing that the nonlinear function $f_0(\underline{z}(k))$ and $g_0(\underline{z}(k))$ are known exactly, we present a desired control $\bar{u}_k^*$, such that the output y_k follows the desired trajectory $y_d(k)$ in deadbeat step:

$$\bar{u}_k^* = -\frac{1}{g_0(\underline{z}(k))}(f_0(\underline{z}(k)) - y_d(k+n)) \tag{4}$$

Substituting the desired control $\bar{u}_k^*$ into error dynamics equation (3), we obtain $e_y(k+n) = 0$. This means that after n steps, we have $e_y(k) = 0$. Therefore, $\bar{u}_k^*$ is a n-step deadbeat control. Accordingly, the desired control $\bar{u}_k^*$ can be expressed as $\bar{u}_k^* = \bar{u}^*(\bar{z}(k))$, $\bar{z}(k) = [\underline{z}^T(k), y_d(k+n)]^T \in \Omega_{\bar{z}} \subset R^{2n}$. where $\Omega_{\bar{z}} = \{(\underline{y}(k), \underline{u}_{k-1}, y_d)|\underline{u}_{k-1}(k) \in \Omega_u, \underline{y}(k) \in \Omega_y, y_d \in \Omega_y\}$ Since $f_0(\underline{z}(k))$ and $g_0(\underline{z}(k))$ are unknown, they are not available for constructing control $\bar{u}_k^*$. However, $f_0(\underline{z}(k))$ and $g_0(\underline{z}(k))$ are function of $\underline{z}(k)$, therefore, we can use HONN to approximate $\bar{u}_k^*$ as follows:

$$\bar{u}^*(\bar{z}) = W^{*T}(k)S(\bar{z}(k)) + \varepsilon_{\bar{z}}, \quad \forall \bar{z} \in \Omega_{\bar{z}} \tag{5}$$

where $\varepsilon_{\bar{z}}$ is the NN estimation error satisfying $|\varepsilon_{\bar{z}}| < \varepsilon_0$. $\hat{W}(k), \hat{\delta}(k)$ denote the estimation of W^*, δ. Let $\tilde{W}(k) = \hat{W}(k) - W^*(k)$, $\tilde{\delta}(k) = \hat{\delta}(k) - \delta$.

If we choose the direct adaptive control law as follows

$$u = \hat{W}(k)S(\bar{z}(k)) + \hat{\delta}(k) \tag{6}$$

and the updating law as follows

$$\hat{W}(k+1) = \hat{W}(k) - \Gamma[S(\bar{z}(k))e_y(k+1) + \sigma\hat{W}(k)]$$
$$\hat{\delta}(k+1) = \hat{\delta}(k) - B[e_y(k+1) + \beta\hat{\delta}(k)] \tag{7}$$

Substituting (4), (5) and (6) into (3), we have

$$e_y(k+n) = g_0(\underline{z}(k))[\tilde{W}^T(k)S(\bar{z}(k)) + \hat{\delta}(k) - \varepsilon_{\bar{z}}] \tag{8}$$

Choose the following Lyapunov function candidate:

$$V(k) = \frac{1}{g}e_y^2(k) + \sum_{j=0}^{n-1}\tilde{W}^T(k+j)\Gamma^{-1}\tilde{W}(k+j) + \sum_{j=0}^{n-1}B^{-1}\tilde{\delta}^2(k+j) \tag{9}$$

Based on the fact that $\widetilde{W}^T(k)S(\overline{z}(k)) = \dfrac{e_y(k+1)}{g_0(\underline{z}(k))} - \hat{\delta}(k) + \varepsilon_{\overline{z}}$.

The difference of (9) along (7) and (8) is given:

$$\Delta V = \frac{1}{\overline{g}}[e_y^2(k+1) - e_y^2(k)] + \widetilde{W}^T(k+1)\Gamma^{-1}\widetilde{W}(k+1) - \widetilde{W}^T(k)\Gamma^{-1}\widetilde{W}(k)$$

$$+ B^{-1}\widetilde{\delta}^2(k+1) - B^{-1}\widetilde{\delta}^2(k)$$

$$\leq -\frac{1}{\overline{g}}e_y^2(k+1) - \frac{1}{\overline{g}}e_y^2(k) - 2\varepsilon_{\overline{z}}e_y(k+1) - 2\sigma\widetilde{W}^T(k)\hat{W}(k) +$$

$$(S^T(\overline{z}(k)))\Gamma S(\overline{z}(k)) \times e_y^2(k+1) + 2\sigma(\hat{W}^T(k))\Gamma S(\overline{z}(k))e_y(k+1)$$

$$+ \sigma^2(\hat{W}^T(k))\Gamma\hat{W}(k) + 2\delta e_y(k+1) - 2\beta\tilde{\delta}(k)\hat{\delta}(k) + Be_y^2(k+1)$$

$$+ 2B\beta\hat{\delta}(k)e_y(k+1) + B\beta^2\hat{\delta}^2(k)$$

Using the fact that:

$$S^T(\overline{z}(k))S(\overline{z}(k)) < l; \ (S(\overline{z}(k)))^T\Gamma S(\overline{z}(k)) \leq \gamma l;$$

$$-2\varepsilon_{\overline{z}}e_y(k+1) \leq \frac{\gamma}{\overline{g}}e_y^2(k+1) + \frac{\overline{g}\varepsilon_{\overline{z}}^2}{\gamma};$$

$$2\sigma(\hat{W}^T(k))\Gamma S(\overline{z}(k))e_y(k+1) \leq \frac{\gamma l}{\overline{g}}e_y^2(k+1) + \overline{g}\sigma^2\gamma\left\|\hat{W}(k)\right\|^2$$

$$2\widetilde{W}^T(k)\hat{W}(k) = \left\|\widetilde{W}(k)\right\|^2 + \left\|\hat{W}(k)\right\|^2 - \left\|W^*(k)\right\|^2$$

$$\sigma^2(\hat{W}^T(k))\Gamma\hat{W}(k) \leq \sigma^2\gamma\left\|\hat{W}(k)\right\|^2;$$

$$2B\beta\hat{\delta}(k)e_y(k+1) \leq Be_y^2(k+1) + B\beta^2\hat{\delta}^2(k);$$

$$2\tilde{\delta}(k)\hat{\delta}(k) = \tilde{\delta}^2(k) + \hat{\delta}^2(k) - \delta^2; \quad 2\delta e_y(k+1) \leq \frac{B}{\overline{g}}e_y^2(k+1) + \frac{\overline{g}}{B}\delta^2$$

we obtain

$$\Delta V \leq -\frac{\rho}{2\overline{g}}e_y^2(k+1) - \sigma(1 - \sigma\gamma - \overline{g}\sigma\gamma)\left\|\hat{W}(k)\right\|^2 - \frac{\omega}{2\overline{g}}e_y^2(k+1)$$

$$-\beta(1 - 2B\beta)\hat{\delta}(k) - \frac{1}{\overline{g}_1}e_y^2(k) + \theta$$

(10)

where $\rho = 1 - 2\gamma - 2\gamma l - 2\overline{g}\gamma l$, $\omega = 1 - 2B - 4B\overline{g}$, and

$$\theta = \frac{\overline{g}}{B}\delta^2 + \beta\delta^2 + \sigma\left\|W^*(k)\right\|^2 + \frac{\overline{g}\varepsilon_{\overline{z}}}{\gamma}$$

In (10), if we choose the design parameters to satisfy

$$\gamma < \frac{1}{2 + 2l + 2\overline{g}l}, B < \frac{1}{2 + 4\overline{g}}, \sigma < \frac{1}{\gamma + \overline{g}\gamma}, \beta < \frac{1}{2B} \tag{11}$$

then it is obvious that $\Delta V < 0$ once $\left| e_y(k) \right| > \sqrt{\overline{g}\theta}$. This implies the boundedness of $V(k)$ for all $k \geq 0$, which leads to the boundedness of $e_y(k)$. Furthermore, the tracking error $e_y(k)$ will asymptotically converge to the compact set denoted by $\varepsilon \leq \sqrt{\overline{g}\beta}$. Due to negativity of ΔV, we can conclude that $y_{k+1} \in \Omega_y$ if all past outputs $y_{k-j} \in \Omega_y, j = 0, \ldots, n-1$ and compact set ε is small enough.

We can use the same techniques as in [17] to show that the NN weight error stays in a small compact set Ω_{we}, and $u_k \in L_\infty$. Finally, if we initialize state $y_0 \in \Omega_{y0}$, $\widetilde{W}(0) \in \Omega_{w0}$, and we choose suitable parameters γ, σ according to (11) to make ε small enough, there exists a contant k^* such that all tracking errors asymptotically converges to Ω_{we} for all $k > k^*$. This implies that the closed-loop system is UUB [12]. Then $y_k \in \Omega_y$ and $\hat{W}(k) \in L_\infty$ will hold for all $k > 0$.

4   Simulation

To demonstrate the effectiveness of the proposed schemes, consider the following nonlinear discrete-time SISO plant described by [18]:

$$\begin{cases} \xi_1(k+1) = f_1(\xi_1(k)) + 0.3\xi_2(k) \\ \xi_2(k+1) = f_2(\overline{\xi}_2(k)) + u(k) \\ y_k = \xi_1(k) \end{cases} \tag{12}$$

where $f_1(\xi_1(k)) = \dfrac{1.4\xi_1^2(k)}{1 + \xi_1^2(k)}$ and $f_2(\overline{\xi}(k)) = \dfrac{\xi_1(k)}{1 + \xi_1^2(k) + \xi_2^2(k)}$. It can be checked that Assumption 1 is satisfied. The tracking objective is to make the output y_k following a desired reference signal $y_d(k) = \dfrac{1}{2}\sin(k\pi/20) + \dfrac{1}{2}\sin(k\pi/10)$.

The initial condition for system states is $x(0) = [0.7, 0]^T$, and the initial conditions of the adaptive laws are $\hat{W}(0) = 0, \hat{\delta}(0) = 0$. Other controller parameters are chosen as $l = 29$, $\Gamma = 0.08I$, $B = 0.4$. The simulation results are presented in Fig.1, 2, 3 and 4.

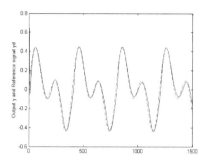

Fig. 1. The output y and the reference signal y_d

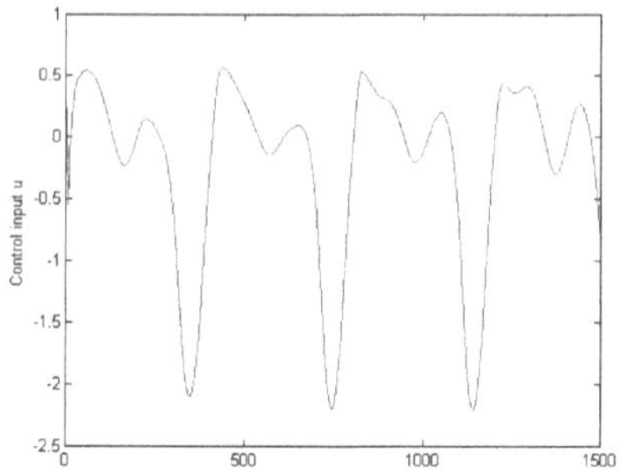

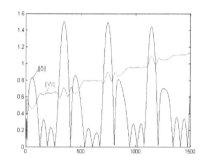

Fig. 2. The input of the controller u

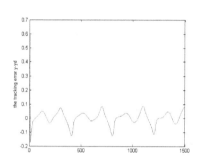

Fig. 3. The tracking error $y_k - y_d$

Fig. 4. The trajectories of $\left\|\hat{W}(k)\right\|$ and $\left\|\hat{\delta}(k)\right\|$

5   Conclusion

By using the approximation property of the neurel network, we have proposed a output feedback robust adaptive control scheme for a class of nonlinear discrete-time systems. The proposed direct adaptive robust NN control guarantees the boundedness of all the closed-loop signals and achieves perfect asymptotocal tracking performance. A simulation example was used to demonstrate the feasibility of the proposed controller.

References

1. Narendra, K.S., Parthasarathy, K.: Identification and control of dynamic systems using neural networks. IEEE Transactions on Neural Networks 1(1), 4–27 (1990)
2. Zhang, Y., Peng, P.Y., Jiang, Z.P.: Stable neural controller design for unknown nonlinear systems using backstepping. IEEE Transactions on Neural Networks 11(6), 1347–1360 (2000)
3. Ge, S.S., Wang, C.: Direct Adaptive NN Control of a Class of Nonlinear Systems. IEEE Transactions on Neural Networks 13(1), 214–221 (2002)
4. Ge, S.S., Wang, C.: Adaptive neural control of uncertain MIMO nonlinear systems. IEEE Transactions on Neural Networks 15(3), 674–692 (2004)

5. Polycarpou, M.M.: Stable adaptive neural control scheme for nonlinear systems. IEEE Transactions on Automatic Control 41(3), 447–451 (1996)
6. Ge, S.S., Hong, F.: Adaptive neural control of nonlinear time-delay systems with unknown virtual control coefficients. IEEE Transactions on Systems, Man, and Cybernetics-Part B: Cybernetics 34(1), 499–516 (2004)
7. Wang, D., Huang, J.: Neural network-based adaptive dynamic surface control for a class of uncertain nonlinear systems in strict-feedback form. IEEE Transactions on Neural Networks 16(1), 195–202 (2005)
8. Choi, J.Y., Farrell, J.A.: Adaptive observer backstepping control using neural networks. IEEE Transactions on Neural Networks 12(5), 1103–1112 (2001)
9. Wang, M., Zhang, S.Y., Chen, B.: Direct adaptive neural control for stabilization of nonlinear time-delay systems. Science China Information Sciences 53, 800–812 (2010)
10. Song, Y., Grizzle, J.W.: Adaptive output-feedback control of a class of discrete-time nonlinear systems. In: Proceedings of American Control Conference, pp. 1359–1364 (1993)
11. Yeh, P.C., Kokotovic, P.V.: Adaptive control of a class of nonlinear discrete-time systems. International Journal of Control 2(62), 303–324 (1995)
12. Ge, S.S., Li, G.Y., Lee, T.H.: Adaptive NN control for a class of strick-feedback discrete-time nonlinear systems. Automatica 39(5), 807–819 (2003)
13. Ge, S.S., Lee, T.H., Li, G.Y., Zhang, J.: Adaptive NN control for a class of discrete-time nonlinear systems. International Journal of Control 76(4), 334–354 (2003)
14. Ge, S.S., Zhang, J., Lee, T.H.: Adaptive neural networks control for a class of MIMO nonlinear systems with disturbances in discrete-time. IEEE Transactions on Systems, Man, and Cybernetics-Part B: Cybernetics 34(4), 1630–1645 (2004)
15. Ge, S.S., Wang, J.: Robust adaptive neural control for a class of perturbed strict feedback nonlinear systems. IEEE Transaction on Neural Networks 13(6), 1409–1419 (2002)
16. Diaz, D.V., Tang, Y.: Adaptive robust fuzzy control of nonlinear systems. IEEE Transactions on Systems, Man, and Cybernetics-Part B: Cybernetics 34(3), 1596–1601 (2004)
17. Liu, Y.J., Wang, W., Tong, S.C., Liu, Y.S.: Robust adaptive tracking control for nonlinear systems based on bounds of fuzzy approximation parameters. IEEE Transactions on Systems, Man and Cybernetic-part A: Systems and Humans 40(1), 170–184 (2010)
18. Wen, G.X., Liu, Y.J.: Direct adaptive robust NN control for a class of discrete-time nonlinear strict-feedback SISO systems. Neural Comput. & Applic. 21, 1423–1431 (2012)
19. Zhang, J., Ge, S.S., Lee, T.L.: Output feedback control of a class of discrete MIMO nonlinear systems with triangular form inputs. IEEE Transactions on Neural Networks 16(6), 1491–1503 (2005)
20. Li, Y.N., Yang, C.G., Ge, S.S.: Adaptive Output Feedback NN Control of a Class of Discrete-Time MIMO Nonlinear Systems With Unknown Control Directions. IEEE Transactions on Systems, Man, and Cybernetics-Part B: Cybernetics 41(2), 507–517 (2011)
21. Ge, S.S., Yang, C.G.: Adaptive predictive control using neural network for a class of pure-feedback systems in discrete time. IEEE Transactions on Neural Networks 19(9), 1599–1614 (2008)
22. Liu, Y.J., Philip Chen, C.L., Wen, G.X.: Adaptive Neural Output Feedback Tracking Control for a Class of Uncertain Discrete-Time Nonlinear Systems. IEEE Transactions on Neural networks 22(7), 1162–1167 (2011)
23. Yang, C.G., Ge, S.S., Xiang, C., Chai, T.Y., Lee, T.H.: Output feedback NN control for two classes of discrete-time systems with unknown control directions in a unified approach. IEEE Transactions on Neural Networks 19(11), 1873–1886 (2008)
24. Wen, G.X., Liu, Y.J., Tong, S.C., Li, X.L.: Adaptive neural output feedback control of nonli-near discrete-time systems. Nonlinear Dynamics 65(1-2), 65–75 (2011)

Robust Fin Control for Ship Roll Stabilization by Using Functional-Link Neural Networks

Weilin Luo[1], Wenjing Lv[1], and Zaojian Zou[2]

[1] College of Mechanical Engineering and Automation, Fuzhou University,
Fujian 350108, China
[2] School of Naval Architecture, Ocean and Civil Engineering, Shanghai Jiao Tong
University, Shanghai 200240, China

Abstract. To reduce the roll of a surface ship, a robust fin controller based on functional-link neural networks is proposed. The plant consists of the ship roll dynamics and that of the fin actuators. Modeling errors and the environmental disturbance induced by waves are considered in the cascaded roll system, which are identified by the neural networks. Lyapunov function is employed in the controller design, which guarantees the stability of the fin stabilizer. Numerical simulation demonstrates the good performance of the roll reduction based on the controller proposed.

Keywords: Surface ship, roll reduction, fin stabilizer, functional-link neural networks, uncertainties.

1 Introduction

The ship motion is usually characterized by nonlinearity, large inertia, and time-delay. In addition, some unpredictable environmental disturbance might exert an influence on a moving ship, which includes wind, waves, current and obstacles such as the other ships, bridge piers, banks and floating or fixed offshore platforms. The nonlinearities and uncertainties in the system of ship dynamics make it challenging to achieve a robust and accurate control system, especially for the uncertainties. Generally, it is difficult to describe these uncertainties in accurate mathematical forms, which usually refer to parameter errors (e.g. the hydrodynamic derivative errors), ignored high-order modes (e.g. hydrodynamic forces or moments more than third-order), unmodelled dynamics (e.g. thrust and torque losses), and external disturbances (e.g. environmental forces and moments). Simple control strategies such as PID control and feedback linearization control can no longer satisfy the requirements of robustness and accuracy. During last decades, many advanced robust control schemes have been developed for ship control. For example, the sliding mode variable structure method was proposed to the nonlinear ship steering [1] and course-keeping control [2]. The parameter adaptive control was adopted in the trajectory tracking [3] and dynamic positioning of surface ships [4]. The H-infinity robust control strategy was proposed for the rudder roll controller [5] and dynamic ship positioning [6]. The fuzzy control

C. Guo, Z.-G. Hou, and Z. Zeng (Eds.): ISNN 2013, Part II, LNCS 7952, pp. 228–237, 2013.

was adopted in the ship steering [7] and track-keeping [8]. The Line-of-sight based approach was proposed to the path-following control for surface ships [9],[10]. The generalized model predictive control was proposed for the path following of underactuated ships [11]. Artificial neural networks (ANN) were proposed to the roll stabilization [12] and autopilot of an autonomous surface vessel [13], etc. Among these control strategies, ANN technique presents a distinctive way to deal with the uncertainties in a ship motion model. First of all, it has the minimum requirement of mathematical model. The dynamics of ship motion can be described by a simple model. Secondly, owing to the excellent learning and nonlinear mapping abilities of ANN, uncertainties can be identified and compensated. The precision and robustness of controller can be guaranteed.

Backpropagation neural networks (BPNN) and gradient descended algorithm are in common use when ANN is applied to the ship control. Such an ANN structure is time-consuming and the solution is apt to be a local optimal one. This paper proposes an on-line feedforward functional-link neural network (FLNN) to obtain a robust fin stabilizer for surface ship roll motion. Uncertainties in the ship motion dynamics are identified by FLNN. The characteristics of fin actuator are also considered for the sake of reasonable application of control force. To guarantee the stability, Lyapunov function is introduced in controller design.

2   Problem Formulations

During last decades, ship roll stabilization has been increasingly paid attention to in consideration of cargo safety, the effectiveness of the crew, and the passenger comfort. Usually, four measures for roll reduction are available, including bilge keels, anti-rolling tanks, passive and active fin stabilizer, and rudder-roll stabilization [3]. For high speed vessels, the fin stabilizer has been proved as effective way to reduce the roll motion.

2.1   Ship Roll Dynamics

Without loss of generality, a second-order roll equation with an active fin moment and wave induced moment is considered

$$(I_x - K_p)\ddot{\phi} + f_1(\phi,\dot{\phi}) + f_2(\phi) = M_w + M_f, \tag{1}$$

where I_x is the inertia moment, $-K_p$ the added inertia moment, ϕ the roll angle, $f_1(\phi,\dot{\phi})$ the damping moment, $f_2(\phi)$ the restoring moment [14], M_w the wave induced moment. M_f the active fin moment, which can be expressed as

$$M_f = -\rho U^2 A_f l_f C_L \alpha, \tag{2}$$

where ρ is the water density, U the ship speed, A_f the fin area, l_f the length of chord, C_L the lift coefficient, α the fin angle [15].

Taking uncertainties into account, the equation (1) can be written as

$$(I_x - K_p)\ddot{\phi} + f_1(\phi,\dot{\phi}) + f_2(\phi) + \varDelta(\phi,\dot{\phi},\alpha) - M_w = K_\alpha \alpha, \tag{3}$$

where $K_\alpha = -\rho U^2 A_f l_f C_L$, the uncertain term $\varDelta(\phi,\dot{\phi},\alpha)$ denotes the unknown modeling error, which might result from parameter perturbation, ignored higher modes, and some unmodelled dynamics. Usually, taking such an uncertainty into account is helpful for accurate modeling and control of ship motion, especially when the ship is sailing in a complicated sea condition.

2.2   Fin Actuator Dynamics

In the research on fin stabilizer, various advanced control schemes have been proposed, however many of them ignored the effect of fin actuator dynamics. It is reasonable for the case of a rapid response in fin actuator, which also means the time constant in fin actuator is much smaller than that in ship response. Nevertheless, suppose both time constants are approximated, ignoring the dynamics of fin actuator might not only degrade the performance of close-loop system, but also exert a bad influence on ship stability.

Usually, the dynamics of a fin actuator can be described as a first-order linear system [15]

$$T_e \dot{\alpha} = K_e \alpha_c - \alpha, \tag{4}$$

where T_e is the actuator's time constant, K_e the control input gain, α_c the demanded fin angle. Similar to the system (3), an uncertain term can be added to equation (4),

$$T_e \dot{\alpha} + \alpha + \varDelta_2(\alpha,\dot{\alpha}) = K_e \alpha_c, \tag{5}$$

where $\varDelta_2(\alpha,\dot{\alpha})$ denotes the unknown modeling error.

For a cascaded system composed of the equations (3) and (5), it is obvious that the terminal or actual control input is α_c while the actual fin angle α is an interim state variable.

3   Controller Design

The control design adopts the backstepping method and Lyapunov stability theory, which can guarantee the robustness of close-loop system. For certainties in the plant, linearization feedback design is employed in consideration of the control accuracy; with regards to the uncertainties including the modeling errors and the environmental disturbance, i.e. the wave induced moment, an on-line functional-link neural network (FLNN) is adopted.

3.1   Feedback Linearization Design

The feedback linearization control is applied to obtain a linear error system with respect to tracking errors. First of all, a virtual or desired control input to system (3) can be defined as

$$\alpha_d = \frac{1}{K_\alpha}\left(f_1(\phi,\dot\phi)+f_2(\phi)-u_1\right),\tag{6}$$

where u_1 is introduced as an auxiliary controller to deal with the uncertainties including $\Delta(\phi,\dot\phi,\alpha)$ and the wave induced moment M_w. By defining a tracking error signal and a filtered error signal as

$$\xi = \alpha_d - \alpha, \eta = \dot\phi + \lambda\phi(\forall\lambda > 0),$$

an error system w.r.t. the system (3) can be obtained as

$$(I_x - K_p)\dot\eta = \lambda(I_x - K_p)\dot\phi - K_\alpha\xi - \Delta_1(\phi,\dot\phi,\alpha)+M_w - u_1.\tag{7}$$

For the system (5), the control input is designed as

$$\alpha_c = \frac{1}{K_e}(\alpha_d + u_2),\tag{8}$$

where u_2 is also introduced as an auxiliary controller to deal with the uncertainties. Another error system w.r.t the system (5) can be obtained

$$T_e\dot\xi = T_e\dot\alpha_d - \xi + \Delta_2(\alpha,\dot\alpha) - u_2.\tag{9}$$

Thus, a cascaded error system composed of (7) and (9) is obtained. Obviously, it is a linear system with reference to the error signals aforementioned. Appropriate designs of the two auxiliary controllers, i.e. u_1, u_2, will answer for the performance of the tracking system. They can be obtained by using the method of Lyapunov recursive function method.

3.2   Lyapunov Function Method

Lyapunov stability theory is used to the controller design. By using this method, the controller design observes a systematic and simple procedure. Meanwhile, the robustness and stability of the tracking system can be guaranteed [16].

In consideration of the systems (7) and (9), a positively definite Lyapunov function candidate is given as

$$V_0 = \frac{1}{2}\phi^2 + \frac{1}{2}(I_x - K_p)\eta^2 + \frac{1}{2}T_e\xi^2.\tag{10}$$

The derivative of V_0 is

$$\dot{V}_0 = -\lambda\phi^2 + \eta\left(\lambda(I_x - K_p)\dot{\phi} + \phi - K_\alpha\xi - \Delta_1 + M_w - u_1\right) + \xi\left(T_e\dot{\alpha}_d - \xi + \Delta_2 - u_2\right). \quad (11)$$

For certainties in the right hand side of the above equation, the inverse dynamical compensation is preferable for the sake of simplicity and control accuracy. For uncertainties, including Δ_1, Δ_2 and M_w, they will be compensated by NN. Furthermore, because it is tedious and difficult to obtain the explicit expression of $\dot{\alpha}_d$, it is also identified by NN.

3.3   Functional-Link Neural Network Identification

Functional-link neural network is a universal approximation of nonlinear functions with any accuracy provided the activation function is selected as basic or squashing one and appropriate number of the hidden layer nodes exist [17]. Compared to the popular backpropagation NN, FLNN provides a simpler structure and the on-line tuning algorithms of weight are available.

Assume that the uncertainties in (11) are approximated by FLNN

$$-K_\alpha\xi - \Delta_1 + M_w = W_1^\mathrm{T} g(h_1) + \varepsilon_1, \quad (12)$$

$$T_e\dot{\alpha}_d + \Delta_2 = W_2^\mathrm{T} g(h_2) + \varepsilon_2, \quad (13)$$

where $W_i\ (i = 1, 2)$ is the "ideal" weight vector and satisfies $\|W_i\|_F \le W_{iM}\ (W_{iM} > 0)$, $\|\cdot\|_F$ denotes the Frobenius norm with the form $\|A\|_F = \sqrt{\sum_{i=1}^{m}\sum_{j=1}^{n} a_{ij}^2}, \forall A = (a_{ij})_{m\times n}$, $g(\cdot)$ is the activation function of hidden layer usually taken as squeeze functions, e.g. a sigmoid or radius basis function, etc. h_i is the preprocessed input vector, ε_i is the reconstruction error and will satisfy $\|\eta_i\| \le \eta_{iN}\ (\eta_{iN} > 0)$, $\|\cdot\|$ denotes the Euclidean norm. Note that since the ideal weight W_i is unknown beforehand, its approximation is used in the real-time control. The convergence of weight error can be guaranteed by the upcoming stability proof.

The two auxiliary controllers can be designed as

$$u_1 = \lambda(I_x - K_p)\dot{\phi} + \phi + W_{1e}^\mathrm{T} g(h_1) + \lambda_1(I_x - K_p)\eta, \quad (14)$$

$$u_2 = -\xi + W_{2e}^\mathrm{T} g(h_2) + \lambda_2 T_e\xi, \quad (15)$$

where λ_i is a positive control gain, W_{ie} the updated weight vector. Substituting the above two equations into (11) and incorporating the definitions (12) and (13), the derivative (11) becomes

$$\dot{V}_0 = -\lambda\phi^2 - \lambda_1(I_x - K_p)\eta^2 - \lambda_2 T_e\xi^2 + \eta\tilde{W}_1^\mathrm{T} g(h_1) + \xi\tilde{W}_2^\mathrm{T} g(h_2) + \eta\varepsilon_1 + \xi\varepsilon_2. \quad (16)$$

where $\tilde{W}_i = W_i - W_{ie}$ is the weight error vector. To guarantee its convergence, a stepping Lyapunov function candidate is defined as

$$V_1 = V_0 + \frac{1}{2k_1}\left(tr\{\tilde{W}_1^T\tilde{W}_1\} + tr\{\tilde{W}_2^T\tilde{W}_2\}\right). \tag{17}$$

where $tr\{\bullet\}$ is the trace of a matrix and satisfies $tr\{A^TA\} = \|A\|_F^2$. And the tuning algorithm of the updated weight vector W_{ie} is designed as

$$\dot{W}_{1e} = k_1 g(h_1)\eta - k_2\|\eta\|W_{1e}, \tag{18}$$

$$\dot{W}_{2e} = k_1 g(h_2)\xi - k_2\|\sigma\|W_{2e}, \tag{19}$$

where $k_{1,2}$ are positive constants and an augmented error vector is introduced as

$$\sigma = \begin{bmatrix} \eta & \xi \end{bmatrix}^T. \tag{20}$$

3.4   Stability Analysis

The derivatives of the last two terms in the right-hand side of the equality (17) are

$$\frac{d}{dt}\left(\frac{1}{2k_1}(tr\{\tilde{W}_1^T\tilde{W}_1\} + tr\{\tilde{W}_2^T\tilde{W}_2\})\right) = -\eta\tilde{W}_1^Tg(h_1) - \xi\tilde{W}_2^Tg(h_2)$$
$$+ \frac{k_2}{k_1}\left(\|\eta\|tr\{\tilde{W}_1^T(W_1 - \tilde{W}_1)\} + \|\sigma\|tr\{\tilde{W}_2^T(W_2 - \tilde{W}_2)\}\right). \tag{21}$$

Introducing a constant $0 \leq \mu \leq 1$, the last two terms in the right-hand side of the above equality result in the identity

$$\frac{k_2}{k_1}\left(\|\eta\|tr\{\tilde{W}_1^T(W_1 - \tilde{W}_1)\} + \|\sigma\|tr\{\tilde{W}_2^T(W_2 - \tilde{W}_2)\}\right)$$
$$\triangleq -(1-\mu)\frac{k_2}{k_1}(\|\eta\|tr\{\tilde{W}_1^T\tilde{W}_1\} + \|\sigma\|tr\{\tilde{W}_2^T\tilde{W}_2\}) \tag{22}$$
$$+ \frac{k_2}{k_1}\left(\|\eta\|tr\{\tilde{W}_1^T(W_1 - \mu\tilde{W}_1)\} + \|\sigma\|tr\{\tilde{W}_2^T(W_2 - \mu\tilde{W}_2)\}\right).$$

The derivative of (17) becomes

$$\dot{V}_1 = -(1-\mu)\lambda\phi^2 - (1-\mu)\lambda_1(I_x - K_p)\eta^2 - (1-\mu)\lambda_2 T_e\xi^2$$
$$-(1-\mu)\frac{k_2}{k_1}\sum_{i=1}^{2}\|\beta_i\|tr\{\tilde{W}_i^T\tilde{W}_i\} - \mu\left(\lambda\phi^2 + \lambda_1(I_x - K_p)\eta^2 + \lambda_2 T_e\xi^2\right) \tag{23}$$
$$+ \frac{k_2}{k_1}\sum_{i=1}^{2}\|\beta_i\|tr\{\tilde{W}_i^T(W_i - \mu\tilde{W}_i)\} + \eta\varepsilon_1 + \xi\varepsilon_2.$$

where $\beta_1 = \eta, \beta_2 = \sigma$. Suppose $\lambda_0 = \min\{(1-\mu)\lambda,(1-\mu)\lambda_1,(1-\mu)\lambda_2,(1-\mu)k_2\|\beta_i\|\}$, and in consideration of the definition of V_1 in (17) and the above equality, the following inequalities hold true

$$-(1-\mu)\left(\lambda\phi^2 + (I_x - K_p)\eta^2 + \lambda_2 T_e\xi^2 + \frac{k_2}{k_1}\sum_{i=1}^{2}\|\beta_i\|tr\{\tilde{\boldsymbol{W}}_i^{\mathrm{T}}\tilde{\boldsymbol{W}}_i\}\right) \leq -2\lambda_0 V_1. \qquad (24)$$

$$b_1\|\boldsymbol{\sigma}\|^2 \leq \mu\left(\lambda\phi^2 + \lambda_1(I_x - K_p)\eta^2 + \lambda_2 T_e\xi^2\right) \leq b_2\|\boldsymbol{\sigma}\|^2, \forall 0 < b_1 < b_2, \qquad (25)$$

$$\frac{k_2}{k_1}\sum_{i=1}^{2}\|\beta_i\|tr\{\tilde{\boldsymbol{W}}_i^{\mathrm{T}}(\boldsymbol{W}_i - \mu\tilde{\boldsymbol{W}}_i)\} \leq \frac{k_2\|\boldsymbol{\sigma}\|}{4k_1\mu}\sum_{i=1}^{2}W_{iM}^2, \qquad (26)$$

$$\eta\varepsilon_1 + \xi\varepsilon_2 \leq b_3\|\boldsymbol{\sigma}\|, \forall b_3 > 0. \qquad (27)$$

where $(I_x - K_p) > 0, T_e > 0$. Thus, it holds

$$\begin{aligned}
\dot{V}_1 &\leq -2\lambda_0 V_1 - b_1\|\boldsymbol{\sigma}\|^2 + \frac{k_2\|\boldsymbol{\sigma}\|}{4k_1\mu}\sum_{i=1}^{2}W_{iM}^2 + b_3\|\boldsymbol{\sigma}\| \\
&= -2\lambda_0 V_1 + \|\boldsymbol{\sigma}\|\left(\frac{k_2}{4k_1\mu}\sum_{i=1}^{2}W_{iM}^2 + b_3 - b_1\|\boldsymbol{\sigma}\|\right).
\end{aligned} \qquad (28)$$

Given a compact set $U_\varsigma = \{\boldsymbol{\sigma}\,|\,\|\boldsymbol{\sigma}\| \leq b_\varsigma, b_\varsigma \in \mathbb{R}^+\}$, if $\boldsymbol{\sigma}$ is in U_ς at any time, the above inequality can be reduced to

$$\dot{V}_1 \leq -2\lambda_0 V_1 + b_\varsigma\left(\frac{k_2}{4k_1\mu}\sum_{i=1}^{2}W_{iM}^2 + b_3\right). \qquad (29)$$

It can be proven that the system is uniformly ultimately bounded (UUB) stable. If $\boldsymbol{\sigma}$ is out of U_ς at a time, by appropriate selection of parameters, one has

$$\|\boldsymbol{\sigma}\| > b_\varsigma > \frac{k_2}{4k_1\mu b_1}\sum_{i=1}^{2}W_{iM}^2 + \frac{b_3}{b_1}, \qquad (30)$$

The inequality (28) becomes

$$\dot{V}_1 \leq -2\lambda_0 V_1. \qquad (31)$$

Obviously, the system is stable.

4   Simulation

Numerical simulation is conducted to verify the validity of the controller designed. The model is adopted as given in [15], in which the environmental disturbance induced by waves was described by

$$M_w = F_w \sin(\omega_e t), \qquad (32)$$

where F_w is the wave amplitude, ω_e is the encounter frequency, in this paper, it is assumed stochastically distributing in 0.3~1.3rad/sec, the other model parameters in the (3) and (5) are $T_e = 2s, K_e = 1, K_\alpha / (I_x - K_p) = -0.043$. The sigmoid function is selected as the activation function $g(\cdot)$ in NN. The input variable vector to the FLNN is selected as $h_1 = h_2 = \begin{bmatrix} \phi & \dot{\phi} & 1 \end{bmatrix}^{\mathrm{T}}$, the initial weights are set zero. Suppose an initial roll deviation exists, i.e. $\phi(0) = 10^\circ$, and the modeling errors $\Delta_1 = 0.1\eta, \Delta_2 = 0.1\xi$. By designing the controller gains $\lambda_1 = 10, \lambda_2 = 1$, the time histories of the roll angle, roll rate, fin angle and the disturbance induced by waves can be obtained, as shown in Figure 1.

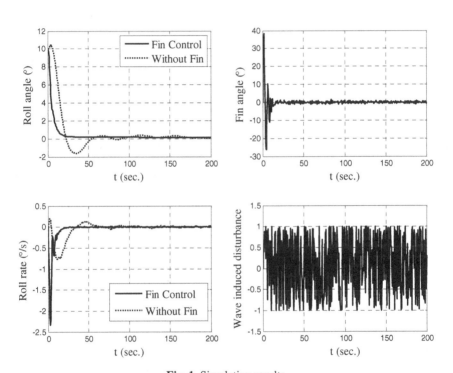

Fig. 1. Simulation results

As it can be seen from the simulation results, good performance of roll reduction is achieved by using the fin controller proposed. Comparatively, without fin, the roll response will oscillate and no convergence can be guaranteed.

5   Conclusions

Study on the fin control for ship roll reduction is presented. Uncertainties including the environmental disturbance induced by waves and modeling errors are considered in a cascaded system consisting of the roll dynamics and the fin actuator dynamics. On-line functional-link neural networks are used to identify the uncertainties. Based on the NN compensation, a robust roll controller is obtained by using the Lyapunov function method. Numerical simulation results demonstrate the validity of the fin controller proposed.

Acknowledgments. This work was supported by the National Natural Science Foundation of China (Grant No. 51279106) and the Program for New Century Excellent Talents in University of Fujian Province (Grant No. JA12015).

References

1. Perera, L.P., Guedes Soares, C.: Pre-filtered Sliding Mode Control for Nonlinear Ship Steering Associated with Disturbances. Ocean Engineering 51, 49–62 (2012)
2. Carletti, C., Gasparri, A., Longhi, S., UliviVapnik, G.: Simultaneous Roll Damping and Course Keeping via Sliding Mode Control for a Marine Vessel in Seaway. In: Proceedings of the 18th IFAC World Congress, Milano, Italy (August-September 2011)
3. Fossen, T.I.: Guidance and Control of Ocean Vehicles. Wiley, New York (1994)
4. Do, K.D.: Global Robust and Adaptive Output Feedback Dynamic Positioning of Surface Ships. Journal of Marine Science and Application 10(3), 325–332 (2011)
5. Tanguy, H., Lebret, G., Doucy, O.: Multi-objective Optimisation of PID and H1 Fin/Rudder Roll Controllers. In: Proceedings of the 5th IFAC Conference on Manoeuvring and Control of Marine Craft (MCMC 2003), Girona, Spain, pp. 179–184 (September 2003)
6. Katebi, M.R., Grimble, M.J., Zhang, Y.: H∞ Robust Control Design for Dynamic Ship Positioning. IEE Proceedings of Control Theory and Applications 114(2), 110–120 (2011)
7. Rigatosa, G., Tzafestas, S.: Adaptive Fuzzy Control for the Ship Steering Problem. Mechatronics 16(8), 479–489 (2006)
8. Velagica, J., Vukicb, Z., Omerdic, E.: Adaptive Fuzzy Ship Autopilot for Track-keeping. Control Engineering Practice 34, 2074–2085 (2007)
9. Moreiraa, L., Fossen, T.I., Guedes Soares, C.: Path Following Control System for a Tanker Ship Model. Ocean Engineering 16(8), 479–489 (2007)
10. Oh, S.-R., Sun, J.: Path Following of Underactuated Marine Surface Vessels Using Line-of-sight Based Model Predictive Control. Ocean Engineering 37(2-3), 289–295 (2010)
11. Wang, X.F., Zou, Z.J., Li, T.S., Luo, W.L.: Path Following Control of Underactuated Ships Based on Nonswitch Analytic Model Predictive Control. Journal of Control Theory and Applications 8(4), 429–434 (2010)
12. Alarçïn, F.: Internal Model Control Using Neural Network for Ship Roll Stabilization. Journal of Marine Science and Technology 15(2), 141–147 (2007)
13. Sharma, S.K., Naeem, W., Sutton, R.: An Autopilot Based on a local Control Network Design for an Unmanned Surface Vehicle. The Journal of Navigation 65, 281–301 (2012)
14. Taylan, M.: The Effect of Nonlinear Damping and Restoring in Ship Roll. Ocean Engineering 27, 921–932 (2000)

15. Yang, Y., Jiang, B.: Variable Structure Robust Fin Control for Ship Roll Stabilization with Actuator System. In: Proceeding of the 2004 American Control Conference, Boston, Massachusetts, USA, pp. 5212–5217 (2004)
16. Ishii, C., Shen, T.L., Qu, Z.H.: Lyapunov Recursive Design of Robust Adaptive Tracking Control with L2-gain Performance for Electrically-driven Robot Manipulators. Int. J. Con. 74(8), 811–828 (2001)
17. Kwan, C., Lewis, F.L., Dawson, D.M.: Robust Neural-network Control of Rigid-link Electrically Driven Robots. IEEE Trans. Neu. Net. 9(4), 581–588 (1998)

DSC Approach to Robust Adaptive NN Tracking Control for a Class of SISO Systems[*]

Wei Li, Jun Ning, and Renhai Yu

Navigation College, Dalian Maritime University
8235673@qq.com, 15840682086@139.com, yurenhai1982@163.com

Abstract. In this paper, by employing Radial Basis Function (RBF) Neural Networks (NN) to approximate uncertain functions, the robust adaptive neural networks design for a class of SISO systems was brought in based on dynamic surface control (DSC) and minimal-learning-parameter (MLP) algorithm. With less learning parameters and reduced computation load, the proposed algorithm can avoid the possible controller singularity problem and the trouble caused by "explosion of complexity" in traditional backstepping methods is removed, so it is convenient to be implemented in applications. In addition, it is proved that all the signals of the closed-loop system are uniformly ultimately bounded(UUB), and simulation results on ocean-going training ship 'YULONG' are shown to validate the effectiveness and the performance of the proposed algorithm.

Keywords: RBF Neural Networks, DSC, MLP, Adaptive Control, Backstepping.

1 Introduction

In the past decades, there has been a rapid growth of research efforts aimed at the development of systematic design methods for the adaptive control of SISO nonlinear systems with parametric uncertainty. Many remarkable results have been obtained. It is shown that the backstepping approach has become a powerful tool in the adaptive control area for nonlinear systems,[1]-[3].

However, there exist two drawbacks in aforementioned works, which restrict their further uses in real applications. The first limitation is the "explosion of learning parameters"[4]-[6]. The other limitation is the "explosion of complexity" [7]. In this paper, based on DSC and MLP techniques proposed in [8]-[9], a robust adaptive neural networks design for a class of SISO systems is developed, which can circumvent the 'explosion of complexity' inherently in the conventional backstepping technique. Also the proposed algorithm is easily implemented and will guarantee the stability of the closed-loop control system.

---

[*] This work was supported in part by the National Natural Science Foundation of China (No.51179019), the Natural Science Foundation of Liaoning Province (No. 20102012) and the Program for Liaoning Excellent Talents in University (LNET).

C. Guo, Z.-G. Hou, and Z. Zeng (Eds.): ISNN 2013, Part II, LNCS 7952, pp. 238–245, 2013.

2   Problem Formulation

Consider an uncertain nonlinear dynamic system in the following form

$$\begin{cases} \dot{x}_i = x_{i+1} + \Delta_i(t,x) \\ \dot{x}_n = f_n(\overline{x}_n) + g_n(\overline{x}_n)u + \Delta_n(t,x) \\ y = x_1 \end{cases} \qquad (1)$$

where $x = [x_1, x_2, ..., x_n]^T \in R^n$ is the system state vector. $u, y \in R$ are the system's input and the output, respectively. $\Delta_i(t,x)$ indicates non-linear uncertain disturbance, and $f_n(\overline{x}_n), g_n(\overline{x}_n)$ represent unstructured non-linear smooth functions, respectively, where $g_n(\overline{x}_n)$ is referred to as virtual control gain function. Now we bring in the following assumptions.

Assumption 1. 1) The absolute value of the unknown virtual control-gain function g_n is positive. 2) For the purpose of analysis easier, without loss of generality, we make further assumption that:

$$0 < g_{min} \leq |g_n| \leq g_{max} \qquad (2)$$

Assumption 2. The reference signal $y_r(t)$ is a sufficiently smooth function of t, and y_r, $\dot{y}_r$, and $\ddot{y}_r$ are bounded, that means, there exists a positive constant Y_0 such that: $\Pi_0 := \left\{ (y_r, \dot{y}_r, \ddot{y}_r) : (y_r)^2 + (\dot{y}_r)^2 + (\ddot{y}_r)^2 < Y_0 \right\}$.

Assumption 3. $|\Delta_i|$ is bounded which means, there exists a positive unknown constant γ_i, $|\Delta_i| < \gamma_i, i = 1, \cdots, n$.

The control objective is to find an adaptive NN tracking controller for (4) such that all the solutions of the resulting closed-loop system are SGUUB and the tracking error $s_1 = y_1 - y_r$ can be rendered small.

3   RBF Neural Network

It has been proved that we can use the following RBF neural networks to approximate an arbitrary smooth function F(x): $R^q \rightarrow R$:

$$F(x) = \vartheta^T S(x) \qquad (3)$$

Where $x \in \Omega_x \subset R^q$, $\vartheta = [\vartheta_1 \dots \vartheta_l]^{\mathrm{T}} \in R^l$, $S(x) = [S_1(x), \dots S_l(x)]^{\mathrm{T}}$, and the NNs node number $l > 1$. We usually choose $S_i(x)$ as the Gaussian function.

$$S_i(x) = \frac{1}{\sqrt{2\pi}\sigma_i} \exp\left[\frac{-(x - u_i)^T (x - u_i)}{\sigma_i^2}\right] \tag{4}$$

where $i = 1, \dots l$, and $u_i = \left[u_{i_1} \dots u_{i_l}\right]^{\mathrm{T}}$ is the center of the receptive field, σ_i is the width of the Gaussian function. Any continuous function with $\forall x \in \Omega_x \subset R^q$ can be approximated as

$$F(x) = \vartheta^{*\mathrm{T}} S(x) + \varepsilon, \forall x \in \Omega_x \tag{5}$$

where ϑ^* is the ideal weight vector, ε is the approximation error with an assumption of $|\varepsilon| \le \varepsilon^*$, where the unknown constant $\varepsilon^* > 0$ for all $Z \in \Omega_Z$. With the estimate value $\hat{\vartheta}$ minimizing $|\varepsilon|$ for all $x, x \in \Omega_x$, ϑ^* can be defined as

$$\vartheta^* = \arg\min_{\vartheta \in R}\left\{\sup_{Z \in \Omega_Z} \left|F(x) - \vartheta^{\mathrm{T}}S(x)\right|\right\} \tag{6}$$

Assumption 4: For the sake of simplicity, let ε_i^* be an unknown upper bound of the approximation errors $\varepsilon_i, i = 1, 2$.

Lemma 1[6]. For any given real continuous function $f(x)$ with $f(0) = 0$, if the continuous function separation technique [9]and the RBF NN approximation technique are used, then $f(x)$ can be denote as

$$f(x) = \overline{S}(x) Ax \tag{7}$$

where $\overline{S}(x) = [1, S(x)] = [1, s_1(x), s_2(x), \dots s_l(x)]$, $\varepsilon^{\mathrm{T}} = [\varepsilon_1, \varepsilon_2, \dots \varepsilon_n]$, $A^{\mathrm{T}} = [\varepsilon, \vartheta^{\mathrm{T}}]$, is a vector of the approximation error, and ϑ is a weight matrix.

$$\vartheta = \begin{bmatrix} \vartheta_{11}^* & \vartheta_{12}^* & \cdots & \vartheta_{1n}^* \\ \vartheta_{21}^* & \vartheta_{22}^* & \cdots & \vartheta_{2n}^* \\ \vdots & \vdots & \cdots & \vdots \\ \vartheta_{l1}^* & \vartheta_{l2}^* & \cdots & \vartheta_{ln}^* \end{bmatrix}$$

4   Control Design

In this part, we will incorporate the DSC-MLP technique and the RBF neural networks to develop an adaptive tracking control design scheme for the system (1). Define $\lambda_i = g_{min}^{-1} \max\{b_i^2, \xi_i^2\}$, $\hat{\lambda}_i$ are estimates of λ_i, and choose the adaptive law for $\hat{\lambda}_i$ as follows

$$\dot{\hat{\lambda}}_i = \Gamma_i \left[\frac{1}{4\gamma_i^2} S_i(\overline{x}_i) S_i^T(\overline{x}_i) s_i^2 + \frac{\varphi^2(\overline{x}_i)}{4\beta_i^2} s_i^2 - \sigma_i(\hat{\lambda}_i - \lambda_i^0) \right] \tag{8}$$

Step 1:   Define the tracking error $s_1 = x_1 - y_r$, its derivative is

$$\dot{s}_1 = x_2 + \Delta_1(t, x) - \dot{y}_r \tag{9}$$

According to Lemma 1 given above, we have

$$\Delta_1(t, x) = S_1(x_1) A_1 x_1 + \varepsilon_1 = S_1(x_1) A_1 s_1 + S_1(x_1) A_1 y_r + \varepsilon_1 \tag{10}$$

where ε_1 denote the approximation error. Let $b_1 = \|A_1\|$, the normalized term $A_1^m = \dfrac{A_1}{\|A_1\|} = \dfrac{A_1}{b_1}$, and $\theta_1 = A_1^m s_1$, then we have

$$\Delta_1(t, x) = b_1 S_1(x_1) \theta_1 + S_1(x_1) A_1 y_r + \varepsilon_1 \tag{11}$$

$$\dot{s}_1 = x_2 + b_1 S_1 \theta_1 + w_1 - \dot{y}_r \tag{12}$$

$$\|w_1\| \le \|S_1(x_1) A_1 y_r + \varepsilon_1^*\| \le g_{min} \xi_1 \varphi(x_1) \tag{13}$$

where, $\xi_1 = g_{min}^{-1} \max(\|A_1 y_r\|, \|\varepsilon_1^*\|)$ and $\varphi(x_1) = 1 + \|S_1\|$. It is obvious that $\|w_1\|$ is bounded, because the bound of y_r, ε_1^*. Now choose the virtual controller α_2 for x_2 as

$$\alpha_2 = -k_1 s_1 + \dot{y}_r - \frac{\varphi^2(x_1)}{4\beta_1^2} s_1 - \frac{\hat{\lambda}_1}{4\gamma_1^2} S_1(\overline{x}_1) S_1^T(x_1) s_1 \tag{14}$$

For the sake of avoiding calculation explosion caused by repeated derivations, the DSC technique is developed and virtual control law is replaced by its estimation using the following first-order filter with the time constant of τ_2.

$$\tau_2 \dot{z}_2 + z_2 = \alpha_2, z_2(0) = \alpha_2(0) \tag{15}$$

By defining the output error of this filter as $y_2 = z_2 - \alpha_2$, it yields

$$\dot{y}_2 = \dot{z}_2 - \dot{\alpha}_2 = -\frac{y_2}{\tau_2} + (-\frac{\partial \alpha_2}{\partial s_1}\dot{s}_1 - \frac{\partial \alpha_2}{\partial x_1}\dot{x}_1 - \frac{\partial \alpha_2}{\partial \hat{\lambda}_1}\dot{\hat{\lambda}}_1 - \frac{\partial \alpha_2}{\partial \hat{\vartheta}_1}\dot{\hat{\vartheta}}_1 + \ddot{y}_d)$$

$$= -\frac{y_2}{\tau_2} + B_2(s_1, s_2, y_2, \hat{\lambda}_1, y_r, \dot{y}_r, \ddot{y}_r)$$

(16)

where $B_2(\cdot)$ is a smooth bounded function and has maximum value M_2 [7].

Step i: $(2 \leq i \leq n-1)$ A similar procedure is recursively employed for each step i, $i = 2,\ldots,n-1$. Similar to step 1, define the error variable $s_i = x_i - z_i$,

$$\dot{s}_i = x_{i+1} + \Delta_i(t,x) - z_i$$

(17)

$$\Delta_i(t,x) = S_i(\overline{x}_i)A_i\begin{bmatrix} s_1 + y_r \\ s_2 + z_2 \\ \vdots \\ s_i + z_i \end{bmatrix} + \varepsilon_1 = b_i S_i(\overline{x}_i)\theta_i + w_i$$

(18)

where $b_i = \|A_i\|$, the normalized term $A_i^m = \dfrac{A_i}{\|A_i\|}$, $\theta_i = A_i^m s_i$ with $\overline{s}_i = [s_1, s_2,\ldots, s_i]^T$, and $w_i = S_i(\overline{x}_i)A_i y_r + S_i(\overline{x}_i)\sum_{j=2}^{i} A_i z_j + \varepsilon_i$, then we have

$$\dot{s}_i = x_{i+1} + b_i S_i(\overline{x}_i)\theta_i + w_i - \dot{z}_i$$

(19)

$$\|w_i\| \leq \|S_1(x_1)A_1 y_r + S_i(\overline{x}_i)\sum_{j=2}^{i} A_i z_j + \varepsilon_i^*\| \leq g_{\min}\xi_i\varphi(\overline{x}_i)$$

(20)

With $\xi_i = g_{\min}^{-1}\max(\|A_i y_r\|, \|\sum_{j=2}^{i} A_i z_j\|, \|\varepsilon_i^*\|)$ and $\varphi(\overline{x}_i) = 1 + \|S_i(\overline{x}_i)\|$.
Now, choose the virtual controller α_i for x_i as

$$\alpha_i = -k_i s_i + \dot{z}_i - \frac{\varphi^2(\overline{x}_i)}{4\beta_i^2}s_i - \frac{\hat{\lambda}_2}{4\gamma_i^2}S_i(\overline{x}_i)S_i^T(\overline{x}_i)s_i$$

(21)

Then, we introduce a first-order filter virtual z_i and let α_i pass through it with time constant τ_i, by defining the output error $y_i = z_i - \alpha_i$, it yields

$$\tau_i \dot{z}_i + z_i = \alpha_i, z_i(0) = \alpha_i(0)$$

(22)

$$\dot{y}_i = \dot{z}_i - \dot{\alpha}_i = -\frac{y_i}{\tau_i} + B_i(\overline{s}_i, y_2,\dots, y_i, \overline{\lambda}_i, y_r, \dot{y}_r, \ddot{y}_r) \tag{23}$$

where $B_i(\cdot)$ is a smooth bounded function, and has a maximum value M_i.

Step n: Define the error variable $s_n = x_n - z_n$, its derivative is

$$\dot{s}_n = g_n(\overline{x}_n)u + f_n(\overline{x}_n) + \Delta_n(t,x) - \dot{z}_n \tag{24}$$

Similarly to above, we approximate h_n, $h_n = f_n(\overline{x}_n) + \Delta_n(t,x)$,

$$h_n = S_n(\overline{x}_n)A_n \begin{bmatrix} s_1 + y_r \\ s_2 + z_2 \\ \vdots \\ s_n + z_n \end{bmatrix} + \varepsilon_n = b_n S_n(\overline{x}_n)\theta_n + w_n \tag{25}$$

where $b_n = \|A_n\|$, the normalized term $A_n^m = \dfrac{A_n}{\|A_n\|}$, $\theta_n = A_n^m s_n$ with

$\overline{s}_n = [s_1, s_2, \dots s_n]^T$, and $w_n = S_n(\overline{x}_n)A_n y_r + S_n(\overline{x}_n)\sum_{j=2}^i A_j z_j + \varepsilon_n^*$, then we have

$$\dot{s}_n = g_n(\overline{x}_n)u + b_n S_n(\overline{x}_n)\theta_n + w_n - \dot{z}_n \tag{26}$$

$$\|w_n\| \le \|S_1(x_1)A_1 y_r + S_n(\overline{x}_n)\sum_{j=2}^i A_j z_j + \varepsilon_n^*\| \le g_{\min}\xi_n\varphi(\overline{x}_n) \tag{27}$$

$\xi_n = g_{\min}^{-1}\max(\|A_n y_r\|, \|\sum_{j=2}^i A_j z_j\|, \|\varepsilon_n^*\|)$ and $\varphi(\overline{x}_n) = 1 + \|S_n(\overline{x}_n)\|$, Then we choose virtual control law μ as

$$u = -k_n s_n + \dot{z}_n - \frac{\varphi^2(\overline{x}_n)}{4\beta_n^2}s_n - \frac{\hat{\lambda}_n}{4\gamma_n^2}S_n(\overline{x}_n)S_n^T(\overline{x}_n)s_n \tag{28}$$

Remark 1. It can be observed that, for the closed-loop control system, there are only one learning parameter, which is independent of order of the subsystem, to be updated online in the controller u and the virtual controllers α_i, no matter how many input variables there are in the subsystem and how many rules are used in the constructed RBF neural networks system. Thus, the well-known curse of dimensionality is circumvented by the proposed algorithm in this paper.

Proof. The proof is similar to that in [10]

5   Application Examples

In this section, we will validate the control performance of the proposed controller via an application example, i.e., the control design of a ship autopilot. Before designing the autopilots, we will describe the dynamics of the ship as follows [5]:

$$\ddot{\psi} + \frac{K}{T} H(\dot{\psi}) = \frac{K}{T} \delta' \tag{29}$$

Where K is the gain in per second and T denote time constant in seconds. $H(\dot{\psi})$ is a nonlinear function of ψ. The function $H(\dot{\psi})$ can be found from the relationship between δ' and ψ in steady state such that $\ddot{\psi} = \dot{\psi} = \dot{\delta}' = 0$. It is known that the "spiral test" has shown that $H(\dot{\psi})$ can be approximated by

$$H(\dot{\psi}) = a_1 \dot{\psi} + a_2 \dot{\psi}^3 + a_3 \dot{\psi}^5 + \dots \tag{30}$$

Where $a_i, i = 1,2,3,\dots$, are real-valued constants. In this paper, let $a_i, i = 1,2,3,\dots$ be not equal to 0, and the nonlinear model (30) is used as the design model for designing the proposed controller in the sequel. Also we have

$$\dot{\delta}' = -\frac{1}{T_E} \delta' + \frac{K_E}{T_E} \delta'_E \tag{31}$$

Where T_E and K_E are the time delay constant and the control gain of the rudder actuator. δ'_E is the order angle of the rudder. Also we have

$$\ddot{\psi}_m(t) + 0.1\dot{\psi}_m(t) + 0.0025\psi_m(t) = 0.0025\psi_r(t) \tag{32}$$

Where ψ_m specifies the desired system performance for the ship heading $\psi(t)$. Letting $x_1 = \psi, x_2 = \dot{\psi}, x_3 = \delta'$ and $u = \delta'_E$ one can obtain

$$\begin{cases} \dot{x}_1 = x_2 \\ \dot{x}_2 = f_2(x_2) + g_2 x_3 \\ \dot{x}_3 = f_3(x_3) + g_3 u \end{cases} \tag{33}$$

Where $f_2 = -(K/T)H(x_2)$, $g_2 = -(K/T)$, $f_3 = -(1/T_E)x_3$, $g_3 = -(K_E/T_E)$, $g_1 = 1, f_1(x_1) = 0$. We assume that f_2, f_3 are unknown. The parameters are chosen as $k_1 = 0.1$, $k_2 = 40, k_3 = 5$.. $\Gamma_1 = \Gamma_2 = \Gamma_3 = 2$, $\sigma_1 = \sigma_2 = \sigma_3 = 0.001$, $\gamma = 0.5$ $\tau_2 = \tau_3 = 100$. The initial values of the weights vectors $\hat{\lambda}_i^0, i = 1,2,3$ are zero. Simulation results are shown in Figs.1

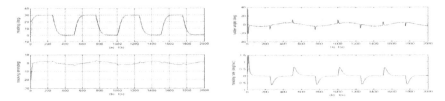

Fig. 1. (a) ship's heading $\psi(t)$ and its desired trajectory $\psi_m(t)$, (b) tracking error. (a') the control rudder angle δ'_E, (b') ship's heading rate $\dot{\psi}_m(t)$

6   Conclusion

In this paper, a robust adaptive neural networks design for a class of SISO systems was brought in based on dynamic surface control and minimal-learning-parameter algorithm. Simulation shows that the proposed scheme can reduce the the computation load dramatically and force the ship to follow a desired course, so it is much easier to implement in applications.

References

1. Krstic, M., Kanellakopoulos, I., Kokotovic, P.V.: Nonlinear and Adaptive Control Design. Wiley, New York (1995)
2. Krstic, M., Kanellakopoulos, I., Kokotovic, P.V.: Adaptive nonlinear control without over parameterization. Syst. Control Lett. 19(3), 177–185 (1992)
3. Kanellakopoulos, I.: Passive adaptive control of nonlinear systems. Int. J. Adapt. Control Signal Process. 7(5), 339–352 (1993)
4. Fischle, K., Schroder, D.: An improved stable adaptive fuzzy control method. IEEE Trans. Fuzzy Syst. 7(1), 27–40 (1999)
5. Yang, Y.S., Ren, J.S.: Adaptive fuzzy robust tracking controller design via small gain approach and its application. IEEE Trans. Fuzzy Syst. 11(6), 783–795 (2003)
6. Yang, Y., Li, T., Wang, X.: Robust adaptive neural network control for strict-feedback nonlinear systems via small-gain approaches. In: Wang, J., Yi, Z., Żurada, J.M., Lu, B.-L., Yin, H. (eds.) ISNN 2006. LNCS, vol. 3972, pp. 888–897. Springer, Heidelberg (2006)
7. Wang, D., Huang, J.: Neural Network-Based Adaptive Dynamic Surface Control for a Class of Uncertain Nonlinear Systems in Strict-Feedback Form. IEEE Trans. on Neural Networks 16(1), 195–202 (2005)
8. Li, T.S., Yu, B., Hong, B.G.: A Novel Adaptive Fuzzy Design for Path Following for Underactuated Ships with Actuator Dynamics. In: ICIEA 2009, pp. 2796–2800 (2009)
9. Liu, C., Li, T.S., Chen, N.X.: Dynamic surface control and minimal learning parameter(DSC-MLP)design of a ship's autopilot with rudder dynamics. Journal of Harbin Engineering University 33(1), 10–14 (2012)
10. Li, T.S., Li, W., Luo, W.L.: DSC approach to robust adaptive NN tracking control for a class of MIMO systems. Int. J. Modeling, Identification and Control 11 (2010)

Integrated Intelligent Control Method of Coke Oven Collector Pressure

Jiesheng Wang[1,2], Xianwen Gao[1], Lin Liu[2], and Guannan Liu[3]

[1] College of Information Science and Engineering, Northeastern University,
Shenyang 110014, China
[2] School of Electronic and Information Engineering,
University of Science & Technology Liaoning, Anshan 114044, China
[3] Liaoning Anshan Power Supply Company, Anshan 114044, China
wang_jiesheng@126.com, gaoxianwen@mail.neu.edu.cn,
623032121@qq.com, lgn0209@163.com

Abstract. Based on the data-driven modeling theory, the integrated modeling and intelligent control method of the coke oven collector pressure is carried out in the paper. The system includes the regression predictive model of coke oven global collector pressure based on support vector machine (SVM), the subtractive clustering algorithm based operation pattern extraction and migration reconfiguration strategy and the self-tuning PID decoupling controller based on the improved glowworm swarm optimization (GSO) algorithm of the coke oven collector pressure. Simulation results and industrial application experiments clearly show the feasibility and effectiveness of control methods and satisfy the real-time control requirements of the coke oven collector pressure system.

Keywords: Coke Oven Collector Pressure, Support Vector Machine, Operation Pattern, Glowworm Swarm Optimization Algorithm, Subtractive Clustering, PID Controller.

1   Introduction

China is a big coke production country, in which the coke production accounts is about 36% of the total production of coke in the world, and the coke export accounts for more than 50% of total coke export trade in the world [1]. Coke oven collector pressures are important process parameters in the coke production process, which will be affected by many aspects, such as coal feed, reversing, gas amount, blower suction, valve opening degree, and so on. All factors will make the collector pressure fluctuate continuously. Meanwhile, this influence is dynamic and uncertain to make the collector pressure have the complex nature of the time-varying, nonlinear and coupling. When the collector pressure is negative, the outside air will enter into the furnace and directly affects the gas quality, the coke oven life and the environment. It also relates to the safe operation of the condensation blower equipment. When the pressure is too big, there will be the phenomenon of "run smoke sparkle" that reduces the shortage of gas recovery.

C. Guo, Z.-G. Hou, and Z. Zeng (Eds.): ISNN 2013, Part II, LNCS 7952, pp. 246–252, 2013.

Domestic and foreign scholars have made many achievements in the field of the coke oven collector pressure control [2-6]. Based on the data-driven modeling theory, the integrated modeling and intelligent control method of the coke oven collector pressure is carried out in the paper. Simulation results and industrial application experiments clearly show the feasibility and effectiveness of control methods and satisfy the real-time control requirements of the coke oven collector pressure system.

2   Intelligent Control Strategy of Coke Oven Collector Pressure

2.1   Technique Flowchart

The technique flowchart of coke oven production process is shown in Figure 1.

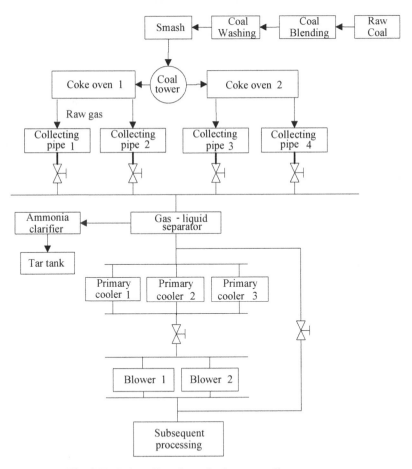

Fig. 1. Technique flowchart of coke oven collector pressure

The metallurgical coke utilizes the large coke oven, whose furnace is divided into three parts of the combustion chamber, the carbonization chamber and the regenerator. Large coke ovens generally have 40-100 carbonization chambers. Both sides of each carbonization chamber are the heating flues, and the regenerative chamber is under them. After air and gas entering into the regenerative chamber to be preheated, they enter the combustion chamber for combustion. The coal temperature is elevated and the chemical reaction is made to produce coal gas.

In the coking process, a large amount of by-product coke oven gas is produced. The gas shortage of about 850℃ overflowing from the carbonization chamber enters into the rising tube, which is cooled to about 80-100℃ by spraying the circulating ammonia at the bridge tube. The gas is fed into the gas-liquid separator through the suction catheter and the collector pipe. Then the coal gas was exported from the top of the gas-liquid separator to enter into the primary cooler. After pressurizing by the gas blower from the primary colder, the gas is conveyed to the purification and recovery processes.

2.2   Intelligent Control Strategy

The block diagram of the data-driven integrated modeling and intelligent control strategy of the coke oven collector pressure system is shown in Figure 2. The system includes the regression predictive model of coke oven global collector pressure based on support vector machine, the subtractive clustering algorithm based operation pattern extraction and migration reconfiguration strategy and the self-tuning PID decoupling controller based on the improved glowworm swarm optimization algorithm of the coke oven collector pressure.

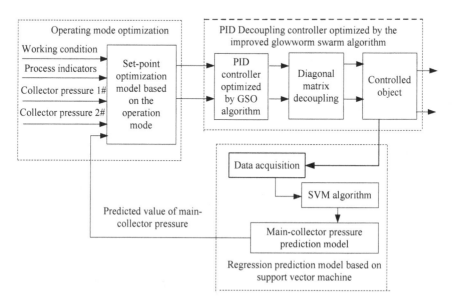

Fig. 2. Strategy diagram of pressure control system

The integrated modeling and intelligent control system of grinding process includes the adaptive wavelet neural network soft-sensor model of economic and technique indexes, the optimized set-point model utilizing case-based reasoning technology and the self-tuning PID decoupling controller based on the ISFLA. Firstly, the milling granularity and the discharge ratio predicted by the soft-sensor model are named as the input parameters of the set-point model. Then, through the case-based reasoning, the milling ore feed ratio and the water feed velocity of the pump pool are optimized. Finally, the self-tuning PID decoupling controller is adopted to achieve the optimized control on the milling discharge ratio and milling granularity ultimately.

2.3 Regression Predictive Method of Coke Oven Global Collector Pressure Based on Support Vector Machine

Because the coke oven gas pipeline brings the big lag, the change of the coke oven gas gathering in main-collector pressure has lagged far behind than the sub-collector pressure. So the butterfly valves can not be adjusted in time. According to the principle of empirical risk minimization (ERM) and structural risk minimization (SRM) for machine learning, Vapnik proposed a support vector machine (SVM) method shown in Figure 3 which can be used for pattern classification and nonlinear regression.

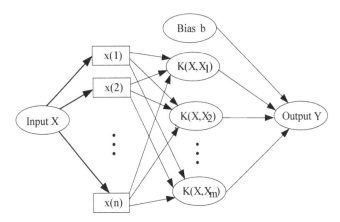

Fig. 3. Skeleton of support vector machine

SVM achieves the input space mapping to a high dimensional feature space based on the kernel functions. This paper puts forward a regression predictive method of the coke oven main-collector pressure based on SVM regression method. Complexity and lengthy of coke oven gas collector pipelines decided the obsoleteness of the main-collector pipe pressure. Therefore, the selected main-collector pressure is predicted based on the pressures of 12 sub-collectors. The optimum parameters obtained by using K-CV method are used to train SVM model and the regression prediction was done for the original data.

2.4   Set-Point Optimization of Coke Oven Main-Collector Pressure Based on Subtractive Clustering

On the basis of the collector pressure control characteristics, a large number of historical datum and the operating experiences in the production process, this paper proposed an operation mode extraction and migration reconstruction method of coke oven collector pressure based on the subtractive clustering method. Mode discovery and mode rule acquisition were realized based on the subtraction clustering method to form the optimized operation mode database finally, which is used to optimize the pressure set-point. The operation mode reconstruction strategy based on the model migration thoughts is utilized to realize the correction of operation mode. The framework of the operation mode optimization based on the subtractive clustering method is shown in Figure 4.

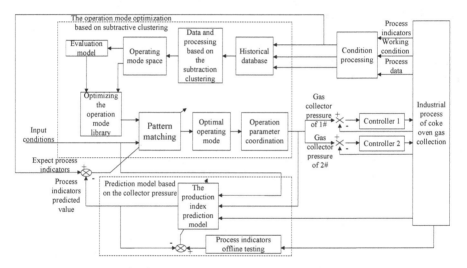

Fig. 4. Skeleton of optimized operation pattern

2.5   Self-tuning PID Controller Based on the Improved GSO Algorithm

According to the technique of coke oven collector pressure control system, the structure of the identification model is selected as the first order inertia delay model. The structure of self-tuning multivariable PID decoupling controller based on the improved GSO algorithm is shown in Figure 5. It is consisted of the system controller, the decoupling compensator based on the diagonal matrix method and the coupling controlled object. The parameters of the PID controller are optimized by the improved GSO algorithm [7]. The Z-N PID tuning method and the proposed improved GSO algorithm under the fitness functions of ISE, IAE, ITAE and ITSE are adopted to control the decoupled systems for comparison [8]. Through the comparison we can get the control performances of the proposed PID decoupling controllers under four different fitness functions.

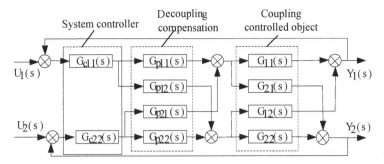

Fig. 5. Schematic flowchart of multivariable PID decoupling controller

3   Industrial Application Experiments

The Figure 6 shows the real-time control curves of coke oven collector pressure under the control strategy proposed in this paper.

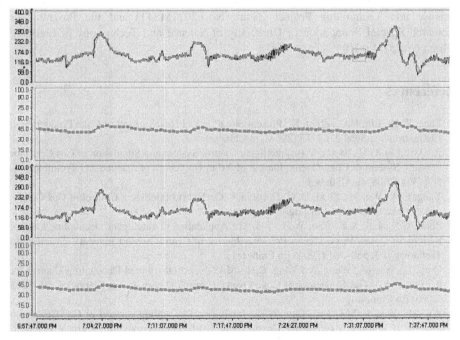

Fig. 6. Real-time control curves of coke oven collector pressure

It can be seen form the Figure 6 that the fluctuation of the coke oven collector pressure significantly reduced under the control of the proposed control strategy. The control accuracy has greatly improved. The pressure is controlled in the range of the technological requirements. The simulation and industrial application experiment results indicated that the proposed control strategy has many characteristics of good dynamic and steady performance, strong robustness and the adaptability of the various working conditions.

4   Conclusions

Based on the technological characteristics, operation experience and the accumulated large number of historical datum of the coke oven collector pressure, a complex industrial controlled object, an integrated automation and control system is proposed. Simulation and industrial experimental results show that the proposed data-driven integrated modeling and intelligent control methods have better feasibility and effectiveness for the coke oven collector pressure system. In future, this method could be extended to deal with the other industrial process. In addition, the design of a better optimization method is an important issue that merits for future study.

Acknowledgments. This work is partially supported by the Program for China Postdoctoral Science Foundation (Grant No. 20110491510), the Program for Liaoning Excellent Talents in University (Grant No. LJQ2011027), the Program for Anshan Science and Technology Project (Grant No. 2011MS11) and the Program for Research Special Foundation of University of Science and Technology of Liaoning (Grant No. 2011zx10).

References

1. Zhen, W.H., Liu, H.C., Zhou, K.: Present Situation and Development of Coke Production in China. Iron and Steel 3, 67–73 (2004) (in Chinese)
2. Ji, Q.C., Liu, G.H., Meng, Y.B.: Intelligent Control System and Simulation of Gas Collector Pressure System of Coke Ovens. Journal of Xi'an University of Science and Technology 3, 362–364 (2005) (in Chinese)
3. Yang, C.H., Wu, M., Shen, D.Y., Deconinck, G.: Hybrid Intelligent Control of Gas Collectors of Coke Ovens. Control Engineering Practice 7, 725–733 (2001)
4. Zhou, G.X., Lai, X.Z., Cao, W.H., Wu, M.: Application of Intelligent Decoupling Control System for Coke-oven Collector Pressure. Journal of Central South University (Science and Technology) 3, 558–561 (2006) (in Chinese)
5. Qing, B., Wu, M., Wang, X., Yang, C.H.: MAS-based Distributed Decoupling Control for the Pressure of Gas Collectors of Coke Ovens. Control Theory & Applications 6, 961–965 (2006) (in Chinese)
6. Wu, M., Yan, J., She, J.H., Cao, W.H.: Intelligent Decoupling Control of Gas Collection Process of Multiple Asymmetric Coke Ovens. IEEE Transactions on Industrial Electronics 7, 2782–2790 (2009)
7. Krishnanand, K.N., Ghose, D.: Glowworm Swarm Optimization for Simultaneous Capture of Multiple Local Optima of Multimodal Functions. Swarm Intelligence 2, 87–124 (2009)
8. Chanchal, D., Rajani, K.M.: An Improved Auto-tuning Scheme for PID Controllers. ISA Transactions 4, 396–409 (2009)

DSC Approach to Adaptive NN Control
on a Ship Linear-Path Following

Wei Li, Zhihui Li, and Jun Ning[*]

Navigation College, Dalian Maritime University
8235673@qq.com, _lizhihui1650@163.com, 15840682086@139.com

Abstract. The problem of ship linear path-keeping control is discussed. By employing radial based function neural network (RBF NN) to approximate uncertain nonlinear system functions, and by combining dynamic surface control (DSC) with backstepping technique and Nussbaum gain approach, the algorithm can not only overcome both the "explosion of complexity" problem inherent in the backstepping method and the possible "controller singularity" problem, but also reduce dramatically the number of on-line learning parameters, thus the algorithm can reduce the computation load of the algorithm correspondingly and make it easy in actual implementation. The stability analysis shows that all closed-loop signals will be semi-global uniformly ultimately bounded (SGUUB), when the tracking error converge to a small neighborhood around the origin through appropriately choosing design constants. Finally, simulation results are presented to show the effectiveness of the proposed algorithm.

Keywords: ship, track-keeping control, dynamic surface control (DSC), neural network (NN), Nussbaum gain.

1 Introduction

On the marine transport, it is common that merchant ships keep a long straight navigation to save time, distance and cost. In the early twenties of last century, classical control theory, namely, PID technique [1], was applied to course control of ship. Later, the autopilot based on model reference adaptive control and maximal variance self-tuning [3] was proposed to keep navigation straight. However, the movement of ship shows the characteristics of nonlinearity, large delay, uncertainties and so on, and can be effected by outside interference, such as wind, waves and flow. Considering the characteristics of ship, the above control schemes are not satisfactory.

---

[*] This work was supported in part by the National Natural Science Foundation of China (No.51179019), the Natural Science Foundation of Liaoning Province (No. 20102012) and the Program for Liaoning Excellent Talents in University (LNET).

C. Guo, Z.-G. Hou, and Z. Zeng (Eds.): ISNN 2013, Part II, LNCS 7952, pp. 253–259, 2013.

In recent two decades, in [4], the presented backstepping method was used in the system that steers the ship on its course to secure course stabilization. Later, by combining neural networks (NN) [5] or fuzzy [6] systems as approximators with the backstepping method, an adaptive fuzzy or NN control algorithm was applied on autopilot about the problem of uncertain nonlinear system functions [7]. However, there exists a drawback of 'explosion of complexity' within the conventional backstepping technique in all aforementioned works, which is caused by the repeated differentiations of certain non-linear functions as the order of the system increases.

In this paper, by combining the DSC technique [8] with the Nussbaum function [9], an adaptive NN control algorithm is proposed to steer the ship. This proposed algorithm not only guarantees the stability of the closed-loop control system, but also is easily implemented. The simulation results on the ocean-going training ship "YULONG" are given to demonstrate the performance of the proposed scheme.

2   Problem Formulation

In this passage, we introduce the nonlinear ship straight-line motion mathematical model equation in the following form:

$$\dot{y} = U \sin \psi$$
$$\dot{\psi} = r \tag{1}$$
$$\dot{r} = -\frac{1}{T}r - \frac{\alpha}{T}r + \frac{K}{T}u + \omega .$$

Where y denotes the sway displacement (cross-track error), ψ denotes the heading angle, r denotes the yaw rate, U denotes the cruise speed respectively, $u = \delta$ denotes the control rudder angle, T, K and α denote the parameter of ship model which is unknown, ω denotes the outside interference signal caused by wind, waves, and flow. In order to make the controller design work much easier, the coordinate transformation is defined as follows:

$$x_1 = \psi + \arcsin\left(\frac{ky}{\sqrt{1+(ky)^2}}\right), x_2 = r ,$$

where $k > 0$ is the parameter designed by the author. If x_1 and y could be calm, then ψ is also calm. In control designing for the system (1), we can ignore the subsystem $\dot{y}$ – of the stable zero dynamic [10] and obtain the second order standard form as follows:

$$\dot{x}_1 = f_1 + x_2$$
$$\dot{x}_2 = f_2 + g_2(\overline{x}_2)u + \omega(t, x) \quad . \tag{2}$$
$$y = x_1$$

x_1, x_2 denote the state variables, u and y denote the input and the output, $f_i(.)$, $i = 1,2$ is an unknown continuous function, $g_2(\overline{x}_2)$ is referred to as the virtual control gain function, $\omega(t, x)$ indicates non-linear uncertain disturbance.

Assumption 1. The uncertain virtual control gain functions $g_2(\overline{x}_2)$ are confined within a certain range such that $|g_2| \leq G_2$. G_2 is only for the need of analysis later.

Assumption 2. $\omega(t, x)$ is bounded which means there exists a positive unknown constant d, such that $|\omega(t, x)| \leq d$.

Assumption 3. The reference signal $y_r(t)$ is a sufficiently smooth function of t. $y_r, \dot{y}_r, \ddot{y}_r$ are bounded, that is, there exists a positive constant B_0, such that $\Pi_0 = \left\{ (y_r, \dot{y}_r, \ddot{y}_r) : y_r^2 + \dot{y}_r^2 + \ddot{y}_r^2 \leq B_0 \right\}$.

3   Controller Design

Definition 1. Continue function $N(\kappa): R \to R$ is defined as Nussbaum-type function, if there are some properties as follows $\lim_{s \to +\infty} \inf \frac{1}{s} \int_0^s N(\kappa) d\kappa = -\infty$,

$$\lim_{s \to +\infty} \inf \frac{1}{s} \int_0^s N(\kappa) d\kappa = -\infty.$$

Lemma 1. Suppose that functions $V(.)$, $\kappa(.)$ are smooth in region $[0, t_f)$ and satisfy $V(t) \geq 0$, the function $N(.)$ is assumed as a smooth Nussbaum-type function, If

$$V(t) \leq c_0 + e^{-c_1} \int_0^t g(x(\tau)) N(\kappa) \dot{\kappa} e^{-c_1} d\tau + e^{-c_1} \int_0^t \kappa e^{-c_1} d\tau,$$

where c_0 is a constant, $c_1 > 0$, $g(x(\tau))$ is a time-varying function, $V(.)$, $\kappa(.)$ and $\int_0^t g(x(\tau)) N(\kappa) \dot{\kappa} d\tau$ are bounded in region $[0, t_f)$ respectively.

Note 1. Basing on the proposition 2 in [11], if the closed-loop system is bounded, then, $t_f = \infty$.

Considering the nonlinearity of the system, we use the following RBF NN to approximate a smooth function.

Step 1. Consider the subsystem of $\dot{x}_1 - $, and define the tracking error variable:

$$s_1 = x_1 - y_r . \tag{3}$$

y_r is the tracking signal for x_1.

$$x_{2R} = -\hat{\theta}_1^T \xi_1(x_1) - k_1 s_1 + \dot{y}_r, \tag{4}$$

$$\dot{\hat{\theta}}_1 = \Gamma_1 \xi_1(x_1) s_1 - \sigma_1 \Gamma_1 \hat{\theta}_1. \tag{5}$$

where Γ_1 is positive definite and symmetrical matrix, k_1, σ_1 are positive design constants.

Now, introduce a variable z_2 and let x_{2R} pass through a first-order filter :

$$\eta_2 \dot{z}_2 + z_2 = x_{2R}, z_2(0) = x_{2R}(0). \tag{6}$$

Step 2: Define the tracking error variable:

$$s_2 = x_2 - z_2. \tag{7}$$

Choose a controller law u, and the updated laws are as follows:

$$u = N(\kappa_2) \left(k_2 s_2 + \hat{\theta}_2^T \xi_2(x_2) + s_2 \frac{\rho_2^2(x)}{c_2} - \dot{z}_2 \right), \tag{8}$$

$$N(\kappa_2) = (1/9)\kappa_2^2 \cos\left((\pi/2)\kappa_2\right), \tag{9}$$

$$\dot{\kappa}_2 = \left(k_2 s_2 + \hat{\theta}_2^T \xi_2(x_2) + s_2 \frac{\rho_2^2(x)}{c_2} - \dot{z}_2 \right) s_2, \tag{10}$$

$$\dot{\hat{\theta}}_2 = \Gamma_2 \left[\xi_2(x_2) s_2 - \sigma_2 \left(\hat{\theta}_2 - \theta_2^0 \right) \right] \tag{11}$$

$N(\kappa_2)$ is Nussbaum-type function in this paper and κ_2 is the variable for $N(\kappa_2)$.

4   Stability Analysis

The output error is defined as follows:

$$y_2 = z_2 - x_{2R}. \tag{12}$$

Consider the Lyapunov function candidate:

$$V_1 = \frac{1}{2}s_1^2 + \frac{1}{2}\tilde{\theta}_1^T\Gamma_1^{-1}\tilde{\theta}_1 + \frac{1}{2}y_2^2. \tag{13}$$

The derivate of the Lyapunov function can be found as follows:

$$\dot{V}_1 \leq -\alpha_{01}\left(s_1^2 + \tilde{\theta}_1^T\Gamma_1^{-1}\tilde{\theta}_1\right) + e_{M1} + \frac{\tau}{2} - \alpha_0 y_2^2. \tag{14}$$
$$= -\alpha_{01}V_1 + e_{M1}$$

where $\alpha_{01} = k_1 - 3$. Let $\alpha_{01} > \dfrac{e_{M1} + \dfrac{\tau}{2}}{2P}$, then $\dot{V}_1 < 0$.

Consider the situation of step 2, and choose the Lyapunov function candidate:

$$V_2 = \frac{1}{2}s_2^2 + \frac{1}{2}\tilde{\theta}_2^T\Gamma_2^{-1}\tilde{\theta}_2. \tag{15}$$

The derivate of the Lyapunov function can be found as follows:

$$\dot{V}_2 \leq -2\alpha_{02}V_2 + (g_2N(\kappa_2)+1)\dot{\kappa}_2 + \delta_2. \tag{16}$$

where $\dfrac{C_2}{2} + \dfrac{\sigma_2}{2}\left|\theta_n^* - \theta_n^0\right|^2 = e_2 = e_{M2}, \dfrac{\sigma_2}{2\lambda_{max}(\Gamma_2^{-1})} = \alpha_{02}, k_2 = \alpha_{02}, \delta_2 = e_{M2}$.

Based on the lemma 1 and note 1, there exist $k_i, c_i, \eta_i, \sigma_i, \Gamma_i$, $i = 1, 2$, so that the solution of the subsystem is uniformly ultimately bounded and all closed-loop signals are semi-global uniformly ultimately bounded (SGUUB).

5   Application Examples

In this section, the oceangoing training vessel YULONG is used as the simulation example to prove the proposed scheme above. The initial conditions are for $y_0 = 500m, \psi_0 = -10°$, the outside disturbed signal is $\omega = 0.01$, the Nussbaum function is $N(\kappa_2) = (1/9)\kappa_2^2\cos((\pi/2)\kappa_2)$, $\kappa(0) = 0.4 * \pi$. The simulation results on Matlab are shown on figure 1-4.

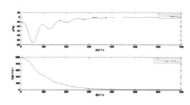

Fig. 1. (a) heading angle ψ , (b) cross-track error y **Fig. 2.** Control rudder angle u

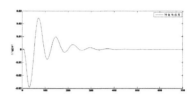

Fig. 3. (a) The variable of Nussbaum function,   **Fig. 4.** The yaw rate r
(b) Nussbaum function

6   Conclusion

In this paper, by employing radial based function neural network (RBF NN) to approximate the ship's nonlinear system functions, and by employing Nussbaum gain approach to overcome the possible "controller singularity" problem for the uncertain virtual control gain function of the ship, and by combining dynamic surface control (DSC) with backstepping technique to overcome the "explosion of complexity" problem inherent in the backstepping method, this method reduces dramatically the number of on-line learning parameters .This simulation results on ocean-going vessel not only prove the effectiveness of this method, but also show that it's easy to be applied in actual implementation.

References

1. Moradi, M.H., Katebi, M.R.: Predictive PID control for ship autopilot design. In: Proc. Control Applications in Marine Systems CAMS, Glasgow, UK, pp. 375–380 (2001)
2. Kallstrom, C.G., et al.: Adaptive autopilots for tankers. Automatica. J 15(2), 241–254 (1979)
3. Do, K.D., Jiang, Z.P., Pan, J.: Robust adaptive path following of underactuated ships. Automatica 40(6), 929–944 (2004)
4. Ge, S.S., Wang, C.: Adaptive neural control of uncertain MIMO nonlinear systems. IEEE Trans. on Neural Networks 15(3), 674–692 (2004)
5. Yesidirek, A., Lewis, F.L.: Feedback linearization using neural networks. Automatica 31(11), 1659–1664 (1995)

6. Yang, Y.S., Zhou, C., Ren, J.S.: Model reference adaptive robust fuzzy control for ship steering autopilot with uncertain nonlinear systems. Applied Soft Computing 3, 305–316 (2003)
7. Swaroop, D., Gerdes, J.C., Yip, P.P., Hedrick, J.K.: Dynamic surface control of nonlinear systems. In: IEEE Proceedings of the American Control Conference, Albuquerque, New Mexico, pp. 3028–3034 (1997)
8. Li, T.S., Zou, Z.J., Luo, W.L.: DSC approach to Robust adaptive NN control for a class of nonlinear system. Acta Automatica Sinica 34(11), 1424–1430 (2008) (in Chinese)
9. Li, T.S., Yu, B., Hong, B.G.: A Novel Adaptive Fuzzy Design for Path Following for Underactuated Ships with Actuator Dynamics. In: ICIEA, pp. 2796–2800 (2009)
10. Zhang, T.P., Ge, S.S.: Adaptive neural control of MIMO nonlinear state time-varying delay systems with unknown dead-zones and gain signs. Automatica 43, 1021–1033 (2007)

Observer-Based Adaptive Neural Networks Control of Nonlinear Pure Feedback Systems with Hysteresis

Yongming Li[1,2,*], Shaocheng Tong[1], and Tieshan Li[2]

[1] College of Science, Liaoning University of Technology,
Jinzhou, Liaoning, 121001, P.R. China
[2] Navigation College, Dalian Maritime University,
Dalian, Liaoning, 116026, P.R. China
l_y_m_2004@163.com, jztsc@sohu.com, tieshanli@126.com

Abstract. In this paper, the problem of adaptive neural output feedback control is investigated for a class of uncertain nonlinear pure feedback systems with unknown backlash-like hysteresis. In the design, RBF neural networks are used to approximate the nonlinear functions of systems, and a neural state observer is designed to estimate the unmeasured states. By utilizing the neural state observer, and combining the backstepping technique with adaptive control design, an observer-based adaptive neural output feedback control approach is developed. It is proved that the proposed control approach can guarantee that all the signals in the closed-loop system are semi-globally uniformly ultimately bounded (SUUB), and both observer error and tracking error can converge to a small neighborhood of the origin.

Keywords: Output feedback, adaptive neural control, backlash-like hysteresis, backstepping design, stability analysis.

1 Introduction

In the past decades, the control of nonlinear systems preceded by hysteresis has been a challenging and yet rewarding problem. The main reason is that hysteresis can be encountered in a wide range of physical systems and devices [1]. On the other hand, since the hysteresis nonlinearity is non-differentiable, the system performance is often severely deteriorated and usually exhibits undesirable inaccuracies or oscillations and even instability [2].

Recently, in order to control uncertain nonlinear systems with unknown backlash-like hysteresis, many adaptive controllers have been developed by backstepping technique. For example, [3-4] proposed adaptive state feedback control designs for a class of uncertain nonlinear systems with unknown backlash-like hysteresis, while, [5] proposed an adaptive fuzzy output feedback controller for

---

* Corresponding author.

C. Guo, Z.-G. Hou, and Z. Zeng (Eds.): ISNN 2013, Part II, LNCS 7952, pp. 260–267, 2013.
© Springer-Verlag Berlin Heidelberg 2013

a class of uncertain nonlinear systems preceded by unknown backlash-like hysteresis. However, the backstepping control methods in [3]-[5] all focus on the uncertain nonlinear systems in strict-feedback form, there are few results available in the literature on the nonlinear systems in pure-feedback form. As stated in [6], a nonlinear pure-feedback system has no affine appearance of the state variables to be used as virtual controls and the actual control just like a strict-feedback nonlinear systems, which makes the backstepping control design and the stability of the closed-loop system are more difficult and challenging. Motivated by the above observations, an adaptive neural output-feedback control approach is presented for a class of uncertain nonlinear pure feedback systems, preceded by unknown backlash-like hysteresis and without the measurements of the states.

2   Problem Formulations and Preliminaries

Consider a class of SISO n-th order nonlinear systems in the following form:

$$
\begin{cases}
\dot{x}_1 = F_1(\underline{x}_2) \\
\dot{x}_2 = F_2(\underline{x}_3) \\
\quad \vdots \\
\dot{x}_{n-1} = F_{n-1}(\underline{x}_n) \\
\dot{x}_n = F_n(\underline{x}_n) + \phi(v) \\
y = x_1
\end{cases}
\tag{1}
$$

where $\underline{x}_i = [x_1, \cdots, x_i]^T \in R^i, (i = 1, \cdots, n)$ is the state vector of the system, and $y \in R$ is the output, respectively. $F_i(\cdot)$ is an unknown smooth nonlinear function; $v \in R$ is the control input and $\phi(v)$ denotes hysteresis type of nonlinearity. This paper assumes that the states of the system (1) are unknown and only the output y is available for measurement.

According to [2], the control input v and the hysteresis type of nonlinearity $\phi(v)$ in system (1) can be described by

$$
\frac{d\phi}{dt} = \alpha \left| \frac{dv}{dt} \right| (cv - \phi) + B_1 \frac{dv}{dt}
\tag{2}
$$

where a, c and B_1 are constants, satisfying $c > B_1$. Let

$$
\begin{cases}
F_i(\underline{x}_{i+1}) = f_i(\underline{x}_i, x_{i+1}) + x_{i+1}, i = 1, 2, \ldots, n-2 \\
F_{n-1}(\underline{x}_n) = f_{n-1}(\underline{x}_{n-1}, \frac{1}{c}x_n) + \frac{1}{c}x_n \\
F_n(\underline{x}_n) = f_n(\underline{x}_{n-1}, \frac{1}{c}x_n)
\end{cases}
$$

The nonlinear pure-feedback system (1) is equivalent to the following system

$$
\begin{cases}
\dot{x}_i = f_i(\underline{x}_i, x_{i+1}) + x_{i+1}, i = 1, 2, \ldots, n-2 \\
\dot{x}_{n-1} = f_{n-1}(\underline{x}_{n-1}, \frac{1}{c}x_n) + \frac{1}{c}x_n \\
\dot{x}_n = f_n(\underline{x}_{n-1}, \frac{1}{c}x_n) + \varphi(v) \\
y = x_1
\end{cases}
\tag{3}
$$

Based on the analysis in [2], (2) can be solved explicitly as

$$
\phi(v) = cv(t) + d_1(v),
$$
$$
d_1(v) = [\phi_0 - cv_0]e^{-\alpha(v-v_0)\mathrm{sgn}\dot{v}} + e^{-\alpha v \mathrm{sgn}\dot{v}} \int_{v_0}^v [B_1 - c]e^{\alpha\eta\mathrm{sgn}\dot{v}}d\eta
\tag{4}
$$

where $v(0) = v_0$ and $\phi(v_0) = \phi_0$.

Based on above solution it is shown in [2] that $d_1(v)$ is bounded. Thus using (4), (3) can be reformulated as

$$
\begin{cases}
\dot{x}_i = f_i(\underline{x}_i, x_{i+1}) + x_{i+1}, i = 1, 2, \ldots, n-2 \\
\dot{x}_{n-1} = f_{n-1}(\underline{x}_{n-1}, \frac{1}{c}x_n) + \frac{1}{c}x_n \\
\dot{x}_n = f_n(\underline{x}_{n-1}, \frac{1}{c}x_n) + cv(t) + d_1(v) \\
y = x_1
\end{cases}
\tag{5}
$$

Define

$$
\begin{cases}
\chi_i = x_i, i = 1, 2, \ldots, n-1 \\
\chi_n = \frac{1}{c}x_n
\end{cases}
\tag{6}
$$

From (5) and (6), we have

$$
\begin{cases}
\dot{\chi}_i = f_i(\underline{\chi}_i, \chi_{i+1}) + \chi_{i+1}, i = 1, 2, \ldots, n-1 \\
\dot{\chi}_n = \frac{1}{c}f_n(\underline{\chi}_n) + v(t) + \frac{1}{c}d_1(v) \\
y = \chi_1
\end{cases}
\tag{7}
$$

Assumption 1. There exists a known constant L_i such that

$$
\left| f_i(\underline{\chi}_i) - f_i(\underline{\hat{\chi}}_i) \right| \le L_i \left\| \underline{\chi}_i - \underline{\hat{\chi}}_i \right\|, i = 1, 2, \ldots, n
$$

where $\underline{\hat{\chi}}_i = [\hat{\chi}_1, \hat{\chi}_2, \ldots, \hat{\chi}_i]^{\mathrm{T}}$ is the estimate of $\underline{\chi}_i = [\chi_1, \chi_2, \ldots, \chi_i]^{\mathrm{T}}$.

Control Objective. Our control objective is to design an adaptive neural networks output control controller such that all the signals involved in the closed-loop system are bounded, the observer error is as small as the desired, and the output y tracks the reference signal $y_r(t)$ within a neighborhood of zero.

Rewrite (7) as

$$
\begin{cases}
\dot{\chi}_1 = f_1(\hat{\chi}_1, \hat{\chi}_{2,f}) + \chi_2 + \Delta f_1 \\[4pt]
\dot{\chi}_2 = f_2(\hat{\underline{\chi}}_2, \hat{\chi}_{3,f}) + \chi_3 + \Delta f_2 \\[4pt]
\quad \vdots \\[4pt]
\dot{\chi}_{n-1} = f_{n-1}(\hat{\underline{\chi}}_{n-1}, \hat{\chi}_{n,f}) + \chi_n + \Delta f_{n-1} \\[4pt]
\dot{\chi}_n = \frac{1}{c} f_n(\hat{\underline{\chi}}_n) + v(t) + \frac{1}{c} d_1(v) + \frac{1}{c}\Delta f_n \\[4pt]
y = \chi_1
\end{cases}
\tag{8}
$$

where $\Delta f_i = f_i(\underline{\chi}_i, \chi_{i+1}) - f_i(\hat{\underline{\chi}}_i, \hat{\chi}_{i+1,f})$, $i = 1, 2, \ldots, n-1$, $\Delta f_n = f_n(\underline{\chi}_n) - f_n(\hat{\underline{\chi}}_n)$; $\hat{\underline{\chi}}_i$ is the estimates of $\underline{\chi}_i$, $\hat{\chi}_{i,f}$ is the filtered signal defined by [6]

$$
\hat{\chi}_{i,f} = H_L(s)\hat{\chi}_i
\tag{9}
$$

where $H_L(s)$ is a Butterworth low-pass filter (LPF), the corresponding filter parameters of Butterworth filters with the cutoff frequency $\omega_C = 1\mathrm{rad/s}$ for different values of n.

RBF neural networks are universal approximators, i.e., they can approximate any smooth function on a compact space, thus we can assume that the nonlinear terms in (8) can be approximated as

$$
\hat{f}_i(\hat{\underline{\chi}}_i, \hat{\chi}_{i+1,f} \,|\, \theta_i) = \theta_i^{\mathrm{T}} \varphi_i(\hat{\underline{\chi}}_i, \hat{\chi}_{i+1,f}), \; 1 \le i \le n
\tag{10}
$$

where $\hat{\underline{\chi}}_{n+1,f} = 0$. The optimal parameter vector θ_i^* is defined as

$$
\theta_i^* = \arg \min_{\theta_i \in \Omega_i} \left[\sup_{(\hat{\underline{\chi}}_i, \hat{\chi}_{i+1,f}) \in U_i} \left| \hat{f}_i(\hat{\underline{\chi}}_i, \hat{\chi}_{i+1,f} \,|\, \theta_i) - f_i(\hat{\underline{\chi}}_i, \hat{\chi}_{i+1,f}) \right| \right]
$$

where Ω_i and U_i are compact regions for θ_i and $(\hat{\underline{\chi}}_i, \hat{\chi}_{i+1,f})$, respectively. Also the minimum approximation error ε_i is defined as

$$
\varepsilon_i = f_i(\hat{\underline{\chi}}_i, \hat{\chi}_{i+1,f}) - \hat{f}_i(\hat{\underline{\chi}}_i, \hat{\chi}_{i+1,f} \,|\, \theta_i^*), \; \varepsilon_n = \frac{1}{c} f_n(\hat{\underline{\chi}}_n) - \hat{f}_n(\hat{\underline{\chi}}_n \,|\, \theta_n^*)
\tag{11}
$$

Assumption 2. There exist unknown constants ε_i^* and τ_i ($\tau_n = 0$), such that $|\varepsilon_i| \le \varepsilon_i^*$ and $|\hat{\chi}_{i+1} - \hat{\chi}_{i+1,f}| \le \tau_i$, $i = 1, 2, \ldots, n$.

By (10) and (11), System (8) can be rewritten as

$$
\begin{cases}
\dot{\chi}_1 = \theta_1^{*\mathrm{T}} \varphi_1(\hat{\chi}_1, \hat{\chi}_{2,f}) + \varepsilon_1(\hat{\chi}_1, \hat{\chi}_{2,f}) + \chi_2 + \Delta f_1 \\[4pt]
\dot{\chi}_2 = \theta_2^{*\mathrm{T}} \varphi_2(\hat{\underline{\chi}}_2, \hat{\chi}_{3,f}) + \varepsilon_2(\hat{\underline{\chi}}_2, \hat{\chi}_{3,f}) + \chi_3 + \Delta f_2 \\[4pt]
\quad \vdots \\[4pt]
\dot{\chi}_{n-1} = \theta_{n-1}^{*\mathrm{T}} \varphi_{n-1}(\hat{\underline{\chi}}_{n-1}, \hat{\chi}_{n,f}) + \varepsilon_{n-1}(\hat{\underline{\chi}}_{n-1}, \hat{\chi}_{n,f}) + \chi_n + \Delta f_{n-1} \\[4pt]
\dot{\chi}_n = \theta_n^{*\mathrm{T}} \varphi_n(\hat{\underline{\chi}}_n) + \varepsilon_n(\hat{\underline{\chi}}_n) + v(t) + \frac{1}{c} d_1(v) + \frac{1}{c}\Delta f_n \\[4pt]
y = \chi_1
\end{cases}
\tag{12}
$$

3 Neural State Observer Design

Note that the states $x_2, x_3, \ldots, x_{n-1}$ and x_n in system (1) are not available for feedback, therefore, a state observer should be established to estimate the states, and then neural networks adaptive output feedback control scheme is investigated.

In this paper, a neural state observer is designed for (12) as follows

$$
\begin{cases}
\dot{\hat{\chi}}_1 = \hat{\chi}_2 + \theta_1^{\mathrm{T}} \varphi_1(\underline{\hat{\chi}}_1, \hat{\chi}_{2,f}) + k_1(y - \hat{\chi}_1) \\
\dot{\hat{\chi}}_2 = \hat{\chi}_3 + \theta_2^{\mathrm{T}} \varphi_2(\underline{\hat{\chi}}_2, \hat{\chi}_{3,f}) + k_2(y - \hat{\chi}_1) \\
\vdots \\
\dot{\hat{\chi}}_{n-1} = \hat{\chi}_n + \theta_{n-1}^{\mathrm{T}} \varphi_{n-1}(\underline{\hat{\chi}}_{n-1}, \hat{\chi}_{n,f}) + k_{n-1}(y - \hat{\chi}_1) \\
\dot{\hat{\chi}}_n = v(t) + \theta_n^{\mathrm{T}} \varphi_n(\underline{\hat{\chi}}_n) + k_n(y - \hat{\chi}_1) \\
\hat{y} = \hat{\chi}_1
\end{cases}
\tag{13}
$$

Rewriting (13) in the following form:

$$
\begin{cases}
\dot{\hat{\chi}} = A\hat{\chi} + Ky + \bar{F} + E_n v(t) \\
\hat{y} = E_1^{\mathrm{T}} \hat{\chi}
\end{cases}
\tag{14}
$$

where $\hat{\chi} = [\hat{\chi}_1, \cdots, \hat{\chi}_n]^{\mathrm{T}}$, $A = \begin{bmatrix} -k_1 & & \\ \vdots & & I_{n-1} \\ -k_n & \cdots & 0 \end{bmatrix}$, $\bar{F} = [\theta_1^{\mathrm{T}} \varphi_1(\hat{\chi}_1, \hat{\chi}_{2,f}), \cdots, \theta_n^{\mathrm{T}} \varphi_n$ $(\underline{\hat{\chi}}_n)]^{\mathrm{T}}$, $K = [k_1, \cdots, k_n]^{\mathrm{T}}$, $E_1^{\mathrm{T}} = [1, 0, \cdots, 0]$ and $E_n^{\mathrm{T}} = [0, \cdots, 0, 1]$.

The coefficient k_i is chosen such that the polynomial $p(s) = s^n + k_1 s^{n-1} + \cdots + k_{n-1}s + k_n$ is a Hurwitz. Thus, given a $Q^{\mathrm{T}} = Q > 0$, there exists a positive definite matrix $P^{\mathrm{T}} = P > 0$ such that

$$
A^{\mathrm{T}} P + PA = -Q
\tag{15}
$$

Let $e = \chi - \hat{\chi} = [e_1, \cdots, e_n]^{\mathrm{T}}$ be observer error, then from (11)-(12), we have the observer errors equation

$$
\dot{e} = Ae + \varepsilon + d + \Delta f + \tilde{\Theta}
\tag{16}
$$

where $\Delta f = [\Delta f_1, \cdots, 1/c\Delta f_n]^{\mathrm{T}}$, $\varepsilon = [\varepsilon_1, \cdots, \varepsilon_n]^{\mathrm{T}}$ and $d = [0, \cdots, 0, 1/cd_1(v)]^{\mathrm{T}}$; $\tilde{\Theta} = [\tilde{\theta}_1^{\mathrm{T}} \varphi_1(\underline{\hat{\chi}}_1, \hat{\chi}_{2,f}), \cdots, \tilde{\theta}_n^{\mathrm{T}} \varphi_n(\underline{\hat{\chi}}_n)]^{\mathrm{T}}$ and $\tilde{\theta}_i = \theta_i^* - \theta_i$ $(i = 1, \ldots, n)$.

Consider the following Lyapunov candidate V_0 for (16) as

$$
V_0 = e^{\mathrm{T}} P e
\tag{17}
$$

The time derivative of V_0 along the solutions of (16) is

$$
\dot{V}_0 = -e^{\mathrm{T}} Q e + 2e^{\mathrm{T}} P(\varepsilon + d + \Delta f + \tilde{\Theta})
\tag{18}
$$

Using Young's inequality, Assumptions 1 and 1, we have

$$2e^{\mathrm{T}} P\varepsilon + 2e^{\mathrm{T}} Pd \le 2\|e\|^2 + \|P\|^2 \|\varepsilon^*\|^2 + \|P\|^2 \|d_1^*\|^2 \tag{19}$$

$$2e^{\mathrm{T}} P\tilde{\Theta} \le \|e\|^2 + \|P\|^2 \sum_{l=1}^{n} \tilde{\theta}_l^T \tilde{\theta}_l \tag{20}$$

$$2e^{\mathrm{T}} P\Delta f \le \|e\|^2 + \|P\|^2 (\sum_{i=1}^{n} L_i^2 \|e\|^2 + \sum_{i=1}^{n-1} L_i^2 \tau_i^2) \le r_0 \|e\|^2 + M_0 \tag{21}$$

where $r_0 = 1 + \|P\|^2 \sum_{i=1}^{n} L_i^2$, $M_0 = \|P\|^2 \sum_{i=1}^{n-1} L_i^2 \tau_i^2$, $\varepsilon^* = [\varepsilon_1^*, \cdots, \varepsilon_n^*]^{\mathrm{T}}$ and $|d_1| \le d_1^*$. with d_1^* is a positive unknown constant. Substituting the equations (19)-(21) into (18), we obtain

$$\dot{V}_0 \le -e^{\mathrm{T}} Qe + (r_0+3)\|e\|^2 + \|P\|^2 \|\varepsilon^*\|^2 + \|P\|^2 \|d_1^*\|^2 + M_0 + \|P\|^2 \sum_{l=1}^{n} \tilde{\theta}_l^T \tilde{\theta}_l \tag{22}$$

4   Adaptive Neural Control Design and Stability Analysis

The n-step adaptive neural networks backstepping output feedback control design is based on the following change of coordinates:

$$z_1 = y - y_r$$
$$z_i = \hat{\chi}_i - \alpha_{i-1}, \quad i = 2, \cdots, n \tag{23}$$

where α_{i-1} is called the intermediate control function, which will be given later.

Step 1: Consider the following Lyapunov function candidate:

$$V_1 = V_0 + \frac{1}{2} z_1^2 + \frac{1}{2\gamma_1} \tilde{\theta}_1^T \tilde{\theta}_1 + \frac{1}{2\bar{\gamma}_1} \tilde{\varepsilon}_1^2 \tag{24}$$

where $\gamma_1 > 0$ and $\bar{\gamma}_1 > 0$ are design constants. $\hat{\varepsilon}_1$ is the estimate of ε_1^* and $\tilde{\varepsilon}_1 = \varepsilon_1^* - \hat{\varepsilon}_1$.

Choose intermediate control function α_1, adaptation functions θ_1 and $\hat{\varepsilon}_1$ as

$$\alpha_1 = -c_1 z_1 - z_1 - \theta_1^{\mathrm{T}} \varphi_1(\hat{\chi}_1, \hat{\chi}_{2,f}) + \dot{y}_r - \hat{\varepsilon}_1 \tanh(z_1/\kappa) \tag{25}$$

$$\dot{\theta}_1 = \gamma_1 \varphi_1(\hat{\chi}_1, \hat{\chi}_{2,f}) z_1 - \sigma \theta_1 \tag{26}$$

$$\dot{\hat{\varepsilon}}_1 = \bar{\gamma}_1 z_1 \tanh(z_1/\kappa) - \bar{\sigma} \hat{\varepsilon}_1 \tag{27}$$

where $c_1 > 0$, $\kappa > 0$, $\sigma > 0$ and $\bar{\sigma} > 0$ are design parameters. We have

$$\dot{V}_1 \le -r_1 \|e\|^2 + \|P\|^2 \sum_{l=1}^{n} \tilde{\theta}_l^T \tilde{\theta}_l - c_1 z_1^2 + z_1 z_2 + \frac{\sigma}{\gamma_1} \tilde{\theta}_1^T \theta_1 + \frac{\bar{\sigma}}{\bar{\gamma}_1} \tilde{\varepsilon}_1 \hat{\varepsilon}_1 + M_1 \tag{28}$$

where $r_1 = \lambda_{\min}(Q) - r_0 - \frac{7}{2} - L_1^2$ and $M_1 = \|P\|^2\|\varepsilon^*\|^2 + \|P\|^2\|d_1^*\|^2 + M_0 + \varepsilon_1^*\kappa' + (L_1\tau_1)^2$.

Step i $(i = 2, \cdots, n)$: The time derivative of χ_i is

$$\dot{z}_i = \hat{\chi}_{i+1} + H_i - \frac{\partial\alpha_{i-1}}{\partial y}e_2 - \frac{\partial\alpha_{i-1}}{\partial y}(\tilde{\theta}_1^T\varphi_1(\hat{\chi}_1, \hat{\chi}_{2,f}) + \varepsilon_1 + \Delta f_1)$$
$$+ \tilde{\theta}_i^T\varphi_i(\hat{\underline{\chi}}_i, \hat{\chi}_{i+1,f}) - \tilde{\theta}_i^T\varphi_i(\underline{\chi}_i, \hat{\chi}_{i+1,f}) \tag{29}$$

where $\hat{\chi}_{n+1,f} = 0$ and $H_i = k_ie_1 + \theta_i^T\varphi_i(\hat{\underline{\chi}}_i, \hat{\chi}_{i+1,f}) - \sum\limits_{k=1}^{i-1}\frac{\partial\alpha_{i-1}}{\partial\hat{\chi}_k}\dot{\hat{\chi}}_k - \frac{\partial\alpha_{i-1}}{\partial\hat{\varepsilon}_1}\dot{\hat{\varepsilon}}_1 - \sum\limits_{k=1}^{i-1}\frac{\partial\alpha_{i-1}}{\partial\theta_k}\dot{\theta}_k - \sum\limits_{k=1}^{i}\frac{\partial\alpha_{i-1}}{\partial y_r^{(k-1)}}y_r^{(k)} - \frac{\partial\alpha_{i-1}}{\partial y}[\hat{\chi}_2 + \theta_1^T\varphi_1(\hat{\chi}_1, \hat{\chi}_{2,f})]$.

Consider the following Lyapunov function candidate

$$V_i = V_{i-1} + \frac{1}{2}z_i^2 + \frac{1}{2\gamma_i}\tilde{\theta}_i^T\tilde{\theta}_i \tag{30}$$

where $\gamma_i > 0$ is a design constant. We can obtain

$$\dot{V}_i \leq -r_i\|e\|^2 + \|P\|^2\sum_{l=1}^{n}\tilde{\theta}_l^T\tilde{\theta}_l + \frac{i-1}{2}\tilde{\theta}_1^T\tilde{\theta}_1 + \frac{1}{2}\sum_{l=1}^{i}\tilde{\theta}_l^T\tilde{\theta}_l - c_1z_1^2 - \sum_{l=2}^{i-1}c_lz_l^2$$
$$+ \sum_{l=1}^{i-1}\frac{\sigma}{\gamma_l}\tilde{\theta}_l^T\theta_l + \frac{\bar{\sigma}}{\bar{\gamma}_1}\tilde{\varepsilon}_1\hat{\varepsilon}_1 + M_i + z_i[z_{i-1} + z_{i+1} + \alpha_i + H_i$$
$$+ 2(\frac{\partial\alpha_{i-1}}{\partial y})^2z_i)] + \frac{1}{\gamma_i}\tilde{\theta}_i^T(\gamma_iz_i\varphi_i(\hat{\underline{\chi}}_i, \hat{\chi}_{i+1,f}) - \dot{\theta}_i) \tag{31}$$

where $r_i = r_{i-1} - L_1^2 - \frac{1}{2}$ and $M_i = M_{i-1} + \frac{1}{2}\varepsilon_1^{*2} + L_1^2\tau_1^2$.

Choose α_i $(\alpha_n = v)$, adaptation function θ_i as

$$\alpha_i = -z_{i-1} - c_iz_i - 2(\frac{\partial\alpha_{i-1}}{\partial y})^2z_i - H_i \tag{32}$$

$$\dot{\theta}_i = \gamma_iz_i\varphi_i(\hat{\underline{\chi}}_i, \hat{\chi}_{i+1,f}) - \sigma\theta_i \tag{33}$$

Substituting (32)-(33) into (31), we have

$$\dot{V}_i \leq -r_i\|e\|^2 + \|P\|^2\sum_{l=1}^{n}\tilde{\theta}_l^T\tilde{\theta}_l + \frac{i-1}{2}\tilde{\theta}_1^T\tilde{\theta}_1 + \frac{1}{2}\sum_{l=1}^{i}\tilde{\theta}_l^T\tilde{\theta}_l - c_1z_1^2$$
$$- \sum_{l=2}^{i}c_lz_l^2 + \sum_{l=1}^{i}\frac{\sigma}{\gamma_l}\tilde{\theta}_l^T\theta_l + \frac{\bar{\sigma}}{\bar{\gamma}_1}\tilde{\varepsilon}_1\hat{\varepsilon}_1 + M_i + z_iz_{i+1} \tag{34}$$

where $z_{n+1} = 0$. For $i = n$, we have

$$\dot{V}_n \leq -r_n\|e\|^2 - c_1\frac{z_1^2}{k_{b_1}^2 - z_1^2} - \sum_{l=2}^{n}c_lz_l^2 - (\frac{\sigma}{2\gamma_1} - \frac{1}{2} - \|P\|^2 - \frac{n-1}{2})\tilde{\theta}_1^T\tilde{\theta}_1$$
$$- \sum_{l=2}^{n}(\frac{\sigma}{2\gamma_l} - \frac{1}{2} - \|P\|^2)\tilde{\theta}_l^T\tilde{\theta}_l - \frac{\bar{\sigma}}{2\bar{\gamma}_1}\tilde{\varepsilon}_1^2 + \lambda \tag{35}$$

where $\lambda = \sum\limits_{l=1}^{n}\frac{\sigma}{2\gamma_l}\theta_l^{*T}\theta_l^* + \frac{\bar{\sigma}}{2\bar{\gamma}_1}\varepsilon_1^{*2} + M_n$.

Let $c = \min\{2r_n/\lambda_{\min}(P), 2c_k, 2\gamma_1(\frac{\sigma}{2\gamma_1} - \frac{1}{2} - \|P\|^2 - \frac{n-1}{2}), 2\gamma_l(\frac{\sigma}{2\gamma_l} - \frac{1}{2} - \|P\|^2), \bar{\sigma}\}$, $k = 1, \ldots, n$, $l = 2, \ldots, n$. Then (35) becomes

$$\dot{V}_n \leq -cV_n + \lambda \tag{36}$$

By (36), it can be proved that all the signals in the closed-loop system are SUUB.

5   Conclusions

For a class of uncertain nonlinear pure feedback systems without the measurements of the states and with unknown backlash-like hysteresis, an adaptive neural output feedback control approach has been developed. The proposed control scheme mainly solved three problems. First, the proposed controlled system is feedback nonlinear system. Second, the proposed control scheme does not require that all the states of the system are measured directly. Third, the problem of unknown backlash-like hysteresis can be overcome. It is proved that the proposed control approach can guarantee that all the signals of the closed-loop system are SUUB, and both the observer and the tracking errors can be made as small as desired by appropriate choice of design parameters.

Acknowledgments. This work was supported in part by the National Natural Science Foundation of China (Nos.61203008, 61074014, 51179019), the Program for Liaoning Innovative Research Team in University (No. LT2012013), the Program for Liaoning Excellent Talents in University (No.LR2012016), the Natural Science Foundation of Liaoning Province (No. 20102012), Liaoning Bai Qian Wan Talent Program (No. 2012921055).

References

1. Su, C.Y., Stepanenko, Y., Svoboda, J., Leung, T.P.: Robust adaptive control of a class of nonlinear systems with unknown backlash-like hysteresis. IEEE Trans. Autom. Contr. 45, 2427–2432 (2000)
2. Shahnazi, R., Pariz, N., Kamyad, A.V.: Adaptive fuzzy output feedback control for a class of uncertain nonlinear systems with unknown backlash-like hysteresis. Commun. Nonlinear Sci. Numer. Simulat. 15, 2206–2221 (2010)
3. Zhou, J., Wen, C.Y., Zhang, Y.: Adaptive backstepping control of a class of uncertain nonlinear systems with unknown backlash-like hysteresis. IEEE Trans. Autom. Contr. 49, 1751–1758 (2004)
4. Shen, Q.K., Zhang, T.P., Zhou, C.Y.: Decentralized adaptive fuzzy control of time-delayed interconnected systems with unknown backlash-like hysteresis. J. Syst. Engin. and Elec. 19, 1235–1242 (2008)
5. Li, Y.M., Tong, S.C., Li, T.S.: Adaptive fuzzy output feedback control of uncertain nonlinear systems with unknown backlash-like hysteresis. Inf. Sci. 198, 130–146 (2012)
6. Zou, A.M., Hou, Z.G.: Adaptive control of a class of nonlinear pure-feedback systems using fuzzy backstepping approach. IEEE Trans. Fuzzy Syst. 16, 886–897 (2008)

Active Disturbance Rejection Control
on Path Following for Underactuated Ships

Ronghui Li[1,*], Tieshan Li[1], Qingling Zheng[2], and Xiaori Gao[1]

[1] Navigation College of Dalian Maritime University
No.1 Linghai Road, 116026, Dalian, China
lironghui@163.com, tieshanli@126.com
[2] Center of Advanced Control Technologies Department of Electrical Computer Engineering,
Cleveland State University Cleveland, Ohio, USA
qinlingzheng@gmail.com

Abstract. To solve the path following problem of underactuated surface ships with internal dynamic uncertainties and external disturbances, an Active-Disturbance-Rejection Control (ADRC) controller is introduced to steer the ship to follow the desired path. Drift angle compensation is added to the controller by designing a coordinate transformation equation. The cross track static error caused by wind and current is overcome. Simulations were carried out on a fully nonlinear hydrodynamic model of a training ship to validate the stability and excellent robustness of the proposed controller.

Keywords: underactuated surface ships, ADRC, path following.

1   Introduction

The surface ship usually has large inertia, large time lag, nonlinearity, and under-actuated characteristics with a tracking motion that is strongly influenced by the model parameter perturbations as well as the effects of wind, waves, and current flow disturbances. So the design of a ship-tracking controller with high performance has been very challenging work.

In recent years, significant research has paid much attention to surface ships since they have fewer actuators than degrees of freedom to be controlled, the constraint on the acceleration is non-integrable, and the system is not transformable into an equivalent system without drifts[1]. In some literature, several methods have been proposed to deal with the uncertainties of the system and the external perturbations. An adaptive robust controller combining Nussbaum gain technique with a backstepping approach was developed in [2] to cope with ship straight-line tracking control system

---

* This work is supported by National Natural Science Foundation of China (Grant No.51179019), Natural Science Foundation of Liaoning Province (Grant No.20102012), the Program for Liaoning Excellent Talents in University(LNET)(Grant No.LR2012016). the Fundamental Research Funds for the Central Universities (Grant No.2012QN013, 3132013005).

C. Guo, Z.-G. Hou, and Z. Zeng (Eds.): ISNN 2013, Part II, LNCS 7952, pp. 268–275, 2013.

with parametric uncertainties and unknown control gain coefficients without a priori knowledge of its sign. Recently, output-feedback trajectory tracking control and stabilization of an underactuated omni-directional intelligent navigator were addressed in [3]. A full state-feedback solution was obtained in [4] which removed the assumption that the mass and damping matrices of the ships are diagonal, although the nonlinear damping terms cannot be included. In [5], a global controller was presented without velocity measurements for feedback. To deal with nonlinear damping coefficients, an adaptive observer was used to estimate the inaccuracies. Integral actions are added to the controller to compensate for a constant bias of environmental disturbances. In order to avoid the need of explicit knowledge of the detailed ship dynamics, application to the marine field of techniques of neural network, fuzzy logic control and other Artificial Intelligence (AI) were also investigated in recent years.

However, in most of these works, the uncertainty of external perturbation of the non-negligible ocean current was seldom explicitly involved. When a surface ship is proceeding under perturbation of a cross ocean constant direction current or wind, it is necessary to maintain a deliberate deviation angle known as "drift angle and leeway".

Based upon the above observations, this paper aims at developing a ship-tracking controller with improved performance in adaptation and robustness by employing the ADRC technique due to its independence on accurate mathematical model of the plant and its ability to compensate for the internal and external disturbances dynamically [6,7,8]. In [9], a ship nonlinear ADRC tracking controller was designed by employing two nonlinear cascaded ADRC controllers]. In [10], a straight line tracking control scheme was accomplished by adding a tracking differentiator while the feedback control law comprised of the errors of the two controlled variables instead of only one in the usual ADRC scheme. However, in the work of [9] and[10], only the Norbbin model was used in the course of design and simulation while ignoring ship drift angle leaving the controller unable to deal with the cross track static error caused by the constant direction wind and current.

In this paper, a new design method of ADRC controller based ship tracking control is proposed by using dynamic, nonlinear MMG model while considering the drift angle of ship. The drift angle compensation is added to the controller by means of designing a coordinate transformation equation. The controller can automatically seek "drift angle and leeway" and the input control quantity of rudder angle. The cross track static error resulting from constant direction wind and current was removed.

2   Ship Motion Control Model

In this paper, in order to design ADRC ship tracking controller, the following design model of ship tracking control is used,

$$
\begin{cases}
\dot{y} = u \sin\psi + v \cos\psi \\
\dot{\psi} = r \\
\dot{r} = f(\psi, r, w) + b\delta
\end{cases}
\tag{1}
$$

where, x, y and ψ are the longitudinal displacement, lateral displacement and heading angle, respectively, in the earth- fixed frame, u ,v and r are longitudinal, lateral velocities over ground, and yaw angular rate in the ship-fixed frame, respectively $f(\psi, r, w)$ is a multivariable function of both the states and external disturbances as well as time. w is the combined external disturbance from wind and current. $b>0$ is the design parameter.

   In practice, when an underactuated surface ship travels at sea, its rudder angle is the only control input used to follow a desired path and to steer a comparatively steady course. However, cross track cannot be regulated to zero by coordinate transformation for the sake of rudder angle under drift caused by wind and current, and it must be compensated by a loxodrome (or sideslip compensation) since no sway control means are available. Because of this, the equilibrium point of the system is not at the origin of transformed coordinates but at a drifting point since the wind and current is time and regional variant. Moreover, the only measurable state variables are the ship's position and heading in Earth fixed coordinates. The path following problem is rephrased as the stabilization to zero of a suitable scalar path error function based on basic knowledge of the steering feature.

3   ADRC Controller Design

3.1   ADRC Structure and Its Algorithm

For the sake of simplicity, consider a second order plant

$$\begin{cases} \dot{x}_1 = x_2 \\ \dot{x}_2 = f(x_1, x_2, w, t) + bu \\ y = x_1 \end{cases} \tag{2}$$

where, y and u are output and input, respectively, w is the external disturbances, b is control gain, $f(x_1, x_2, w, t)$ is the total external and internal disturbances function. The plant in (2) is written in state equation form

$$\begin{cases} \dot{x}_1 = x_2 \\ \dot{x}_2 = x_3 + b_0 u \\ \dot{x}_3 = h \\ y = x_1 \end{cases} \tag{3}$$

where $x_3 = f = f(x_1, x_2, w, t)$ is added as an augmented state, and $\dot{x}_3 = h = \dot{f}$ as unknown disturbance. Now f can be estimated using a state observer based on the state space model

$$\begin{cases} \dot{x} = Ax + Bu + Eh \\ y = Cx \end{cases} \tag{4}$$

where,

$$A = \begin{pmatrix} 0 & 1 & 0 \\ 0 & 0 & 1 \\ 0 & 0 & 0 \end{pmatrix}, \ B = \begin{pmatrix} 0 \\ b_0 \\ 0 \end{pmatrix}, \ C = \begin{bmatrix} 1 & 0 & 0 \end{bmatrix}, \ E = \begin{pmatrix} 0 \\ 0 \\ 1 \end{pmatrix}.$$

The state space observer, denoted as the linear extended state observer (LESO) of (4) is constructed as

$$\begin{cases} \dot{z} = Az + Bu + L(y - \hat{y}) \\ \hat{y} = Cz \end{cases} \tag{5}$$

Rewrite the differential equation (5) as

$$\begin{cases} e = z_1 - y \\ \dot{z}_1 = z_2 - \beta_1 e \\ \dot{z}_2 = z_3 - \beta_2 e + b_0 u \\ \dot{z}_3 = -\beta_3 e \end{cases} \tag{6}$$

Selecting the observer gain vector $L = [\beta_1 \ \beta_2 \ \beta_3]^T$ appropriately provides an estimate of the state of (6), $z_1 \approx x_1, z_2 \approx x_2, z_3 \approx x_3$. Here with the use of linear gains, this observer is denoted as LESO which can estimate f with bounded error if either $h = \dot{f}$ is bounded or f is bounded[11]. Moreover, to simplify the tuning process, the observer gains can be obtained using the parameterized pole placement technique as

$$L = \begin{bmatrix} 3\omega_0 & 3\omega_0^2 & \omega_0^3 \end{bmatrix}^T \tag{7}$$

where, the observer bandwidth, ω_0 is the only turning parameter. With a well-tuned observer, the observer state z_3 will track x_3 closely. Ignoring the estimation error in z_3 , the control law

$$u = (u_0 - z_3) / b_0 \tag{8}$$

Then the plant is reduced to a unity-gain double integrator,

$$\dot{x}_2 = (f - z_3) + u_0 \approx u_0 \tag{9}$$

An example of such u_0 is the common linear proportional and derivative control law

$$u_0 = k_p (\xi - z_1) - k_d z_2 \tag{10}$$

where ξ is the set point, which is constant. Note that $-k_d z_2$, instead of $k_d\left(\dot{\xi}-z_2\right)$, is used to avoid differentiation of the set point. The controller tuning is further simplified with $k_d = 2\omega_c$ and $k_p = \omega_c^2$, where ω_c is the closed loop bandwidth which is the tuning parameter.

3.2   ADRC Based Ship Track Controller Design

A linear combination of y and ψ, that is $\zeta = ky + \psi$, as a new control goal, is put forward in [2], where k is a design parameter. Its idea is that both $y \to 0$ and $\psi \to 0$ while $z \to 0$, which was proven to be correct. In this paper, the control structure borrows the idea from [2] using a new nonlinear combination of y and ψ, that is

$$\varsigma = k_1 \tanh(k_0 y) + \psi \tag{11}$$

where, $k_0 > 0$ and $k_1 > 0$ are design parameters. k_0 is used to coordinate compression and k_1 is used to adjust ship track convergence rate. Meanwhile, k_1 can restrict the maximum heading angle ψ to be used when the ship returns to the planned route. When $\varsigma \to 0$, both $y \to 0$ and $\psi \to 0$. Accordingly, the physics meaning of parameters is obvious and parameters would be tuned easily. But when ship is affected by constant wind and current, when $\psi \to 0$, y can not be convergent to zero due to leeway angle, so cross track error will always exist. In fact, the goals to implement path following is $y \to 0$, but heading angle ψ cannot be convergent to zero, and it should equal to the opposite of ship leeway angle. The ship leeway angle or drift angle is defined as

$$\beta = \arctan(v/u) \tag{12}$$

So $\psi + \beta \to 0$ is the goal that can guarantee $y \to 0$. Therefore, the following transformation can be defined

$$\varsigma = k_1 \tanh(k_0 y) + [\psi + \beta] \tag{13}$$

where, $k_0 > 0$, $k_1 > 0$. While $\varsigma \to 0$, $y \to 0$ and $\psi + \beta \to 0$. k_0 has the same meanings as above, but k_1 is used to restrict the maximum value of $(\psi + \beta)$, so the maximum heading angle to be used is $(k_1 - \beta)$ when ship travels at sea. The goal of ship tracking control is that $\varsigma \to 0$.

An ADRC controller can be designed with a $3rd$ order ESO. The structure of the ADRC ship track control is as fig.1. The reference input ξ of ADRC controller is zero, and ship control rudder angle δ is the output of ADRC controller. The input of ESO is ς, so an estimate of the state of z, $z_1 \to \varsigma$, $z_2 \to \dot{\varsigma}$, z_3 provides an estimate of unknown parts in $\dot{\varsigma}_1$. Define $e_p = \xi - z_1 = -z_1$ and $e_d = -z_2$, a PD controller $u_0 = k_p e_p + k_d e_d$ is obtained, so the final plant controller, $u = (u_0 - z_3)/b_0$ is also obtained.

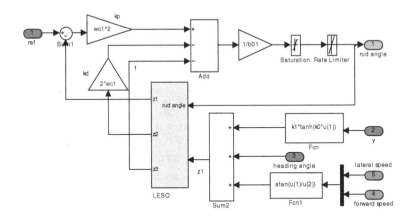

Fig. 1. The structure of ADRC ship track control

4   Simulation Study

To demonstrate the practicality of the design, simulation on ship straight line path following were performed using an underactuated surface vessel ship under Manoeuvring Mathematical Model Group (MMG), a training ship Yulong. The principal particulars are as follows: full displacement 14,635 tonnes, length of ship 126m, beam 20.8m, mean draft 8.0m, trim 0, diameter of propeller 4.6m, block coefficient 0.681. The parameters of ADRC below simulations in this paper was chosen as ω_o =0.4, ω_c =0.04 and b_o =0.478/216.

The initial states were chosen to be as following: the main engine was set to be 103.4 revolutions per minute (RPM), x=0, y=500, $\psi = 0$, $u = 7.2$ m/s. Planned route is a straight line of y=0 to the north. Disturbances of NE wind of 10m/s, and a SW current of 2 kn. The parameters of the transformation equation (14) were chosen as respectively $k_0 = 0.04$, $k_1 = \pi/9$, where $k_1 = \pi/9$ indicates $\max(\psi + \beta) = \pi/9$ before ship is stable.

By analyzing the results of Fig. 2 - Fig.3, the following conclusions can be summarized: (1) The cross track static error resulting from wind and current has been removed. The tracking control precision is high. Rudder and course response are smooth and the rudder-turning angle is small. (2) the ADRC controller has a powerful robustness to the environment disturbances and the internal dynamic uncertainties. And the ship tracking control is fast and smooth with lower energy consumption.

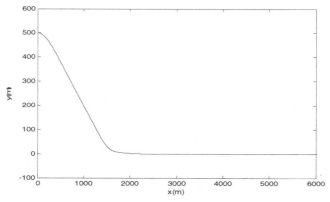

Fig. 2. Ship track of straight line following with wind and current disturbances

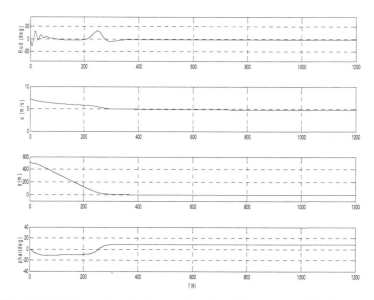

Fig. 3. Straight line path following results with wind and current disturbances

5 Conclusion

This paper has presented a novel path following control approach to underactuated
vessels under environmental disturbances of ocean current and wind. ADRC has been
applied to the control design of ship tracking utilizing the characteristic of ADRC
independence of the plant's mathematical model. The leeway angle compensation has
been added to the controller by means of designing a coordinate transformation
equation. The ADRC controller has guaranteed that the cross track error converges to
the planned path, and parameters can be tuned easily according to the ship's maneu-
verability. The high precision ship-tracking controller is robust to the ship motion
nonlinearity and the external disturbances.

References

1. Do, K.D., Pan, J.: Global waypoint tracking control of underactuated ships under relaxed assumptions. In: Proceedings of the 42nd IEEE Conference on Decision and Control, Maui, Hawaii USA, pp. 1244–1249 (2003)
2. Li, T.S., Yang, Y.S., Hong, B.G., Ren, J.S., Du, J.L.: A Robust Adaptive Nonlinear Control Approach to Ship Straight-path Tracking Design. In: Proceedings of the 2005 American Control Conference, Portland, OR, USA, pp. 4016–4021 (2005)
3. Do, K.D., Jiang, Z.P., Pan, J.: Nijmeijer. H.: A global output-feedback controller for stabilization and tracking of underactuated ODIN: a spherical underwater vehicle. Automatica 40, 117–124 (2004)
4. Do, K.D., Pan, J.: Global tracking control of underactuated ships with nonzero off-diagonal terms in their system matrices. Automatica 41, 87–95 (2005)
5. Do, K.D., Pan, J.: Underactuated ships follow smooth paths with integral actions and without velocity measurements for feedback: theory and experiments. IEEE Transactions on Control Systems Technology 14(2), 308–322 (2006)
6. Han, J.: From PID to Active Disturbance Rejection Control. IEEE Transactions on Industrial Electronics 56(3), 900–906 (2009)
7. Gao, Z.Q.: Active disturbance Rejection control: a paradigm shift in feedback control system design. In: Proceedings of the 2006 American Control Conference, Minneapolis, Minnesota, USA, pp. 2399–2405 (2006)
8. Gao, Z.Q.: Scaling and bandwidth-parameterization based controller tuning. In: Proceedings of the 2003 American Control Conference, Denver, Colrado, pp. 4989–4996 (2003)
9. Ruan, J.H., Li, Z., Zhou, F.: ADRC based ship tracking controller design and simulations. In: Proceedings of the IEEE International Conference on Automation and Logistics, pp. 1763–1768. IEEE, Piscataway (2008)
10. Liu, W.J., Sui, Q., Zhou, F.: Straight-line tracking control of ships based on ADRC. Journal of Shandong University (Engineering Science) 40(6), 48–53 (2010)
11. Yang, X.X., Huang, Y.: Capability of extended state observer for estimating uncertainties. In: Proceedings of the 2009 American Control Conference, Hyatt Regency Riverfront, St. Louis, MO, USA, pp. 3700–3705 (2009)

Canonical Correlation Analysis Neural Network for Steady-State Visual Evoked Potentials Based Brain-Computer Interfaces

Ka Fai Lao, Chi Man Wong, Feng Wan, Pui In Mak, Peng Un Mak, and Mang I Vai

Department of Electrical and Computer Engineering
Faculty of Science and Technology
University of Macau, Av. Padre Tomás Pereira Taipa, Macau, China
{da92050,fwan}@umac.mo

Abstract. Canonical correlation analysis (CCA) is a promising feature extraction technique of steady state visual evoked potential (SSVEP)-based brain computer interface (BCI). Many researches have showed that CCA performs significantly better than the traditional methods. In this paper, the neural network implementation of CCA is used for the frequency detection and classification in SSVEP-based BCI. Results showed that the neural network implementation of CCA can achieve higher classification accuracy than the method of power spectral density analysis (PSDA), minimum energy combination (MEC) and similar performance to the standard CCA method.

Keywords: Canonical Correlation Analysis (CCA), Neural Network, Brain Computer Interface (BCI), Steady-State Visual Evoked Potential (SSVEP).

1 Introduction

Non-invasive BCIs have been widely investigated in recent decades. It provides an alternative communication and control channel between human and environment through electroencephalographic (EEG) produced by brain activities [1]. This communication channel brings great benefit to patients with severe motor disabilities, which enable them to express their wishes to caregivers or operate an intelligent wheelchair by a BCI without any brain's normal pathway of peripheral nerves and muscles. A general BCI includes three working processes: signal acquisition, feature extraction and feature classification. The brain activity then is translated to a device command or message. Different types of EEG signals such as P300 evoked potential, sensorimotor mu/beta rhythms, slow cortical potential (SCP), movement-related cortical potential (MRCP) and visual evoked potential (VEP), have been paid much attention in the research of BCI [2].VEP is an evoked potential over occipital area elicited by an external visual stimulus. It can be classified into transient VEP (TVEP) and steady-state VEP (SSVEP). SSVEP is elicited by a consecutive, stable and periodic stimulus with repetition rate higher than about 4 Hz. The main characteristics of SSVEP are frequency-locked and phase-locked. SSVEP is highly spoken of among

C. Guo, Z.-G. Hou, and Z. Zeng (Eds.): ISNN 2013, Part II, LNCS 7952, pp. 276–283, 2013.

all the BCI applications because of its high signal to noise ratio (SNR), relative immunity to artifacts, high information transfer rate (ITR) and low training requirements. Basically, most BCIs based on SSVEPs utilize the frequency information of SSVEPs for identification. It means that the computer can detect which target the subject desires to select by checking the frequency information of SSVEPs [3]. Up to now, the common feature extraction methods include power spectral density analysis (PSDA), minimum energy combination (MEC) and CCA. For PSDA method, PSD is estimated by fast Fourier Transform (FFT) and its peak is detected to recognize the target stimulus. MEC method combines multiple electrode signals to cancel the noise as much as possible, and then the target stimuli is recognized according to the maximum signal power. Researches have demonstrated that CCA method has lower deviation, higher detection accuracy and higher insensitive to SNR than the tradition method [3] [4].

Although the standard CCA method has provided great performance on SSVEP-based BCI, there are still many possible improvements. Neural network (NN) implementation for BCI is one of the promising research directions. The important characteristics of NN include its self-adaptive structure, universal function approximation and the expansibility of network. In other words, NN implementation for BCI provides a more flexible and extendable environment for researchers. Moreover, some researchers have applied the neural network based approaches to achieve excellent performance on the classification accuracy [6] – [8]. Inspired by their work, we aim to apply the NN based CCA (NNCCA) models proposed by *Hsieh et al.* [9] [10] in SSVEP-based BCI to recognize SSVEPs. The performance of NNCCA is verified with the real EEG data from five healthy subjects and compared with the standard CCA, PSDA and MEC method. The preliminary offline experiment results show that NNCCA can provide the similar classification accuracy to standard CCA and the better accuracy than MEC and PSDA.

2   Methodology

2.1   CCA

CCA is a statistical technique that applied to two datasets which we believe there are some potential relationships between them. Their relationship is determined by correlation coefficient. Consider two multidimensional datasets $X \in \mathbb{R}^{p \times n}$ and $Y \in \mathbb{R}^{q \times n}$ as well as their correlation is wanted to be found, in other points of view, two inputs are wanted to be reproduced each other through linear combination of their variables:

$$U = W_1^{\mathrm{T}} X \tag{1}$$

$$V = W_2^{T} Y \tag{2}$$

$W_1 \in \mathbb{R}^{p \times 1}$ and $W_2 \in \mathbb{R}^{q \times 1}$ denote the correlation coefficient that can maximize the correlation between the canonical variates U and V. Furthermore, the canonical correlation coefficient of the variates is defined by Eq. (3):

$$\eta = \frac{cov(U, V)}{\sqrt{cov(U, U)cov(V, V)}} = \frac{W_1^T S_{XY} W_2}{\sqrt{W_1^T S_{XX} W_1 W_2^T S_{YY} W_2}} \tag{3}$$

$$F = W_1^T S_{XY} W_2 - \frac{\lambda_1}{2}(W_1^T S_{XX} W_1 - 1) - \frac{\lambda_2}{2}(W_2^T \Sigma_{YY} W_2 - 1) \tag{4}$$

S denotes the sample covariance matrix. The maximum canonical correlation corresponds to the maximum canonical correlation coefficient. To solve the maximum canonical coefficient, Lagrange Multiplier method is applied to Eq. (3) and the cost function in Eq. (4) is obtained. The maximum correlation coefficient then is solved from the cost function. According to the method presented by *Lin et al.* [3], EEG signals from multiple channels are used to calculate the canonical correlation coefficients with all stimulus frequencies in the system. In other words, the EEG signal and stimulus signal are considered as X and Y. Assume there are I stimulus frequencies and all of them are square-wave periodic, the i th stimuli signal can be decomposed into Fourier series:

$$Y_i = \begin{pmatrix} sin\,(2\pi f_i/f_s) & \cdots & sin\,(2\pi f_i n/f_s) \\ cos(\,2\pi f_i/f_s) & \cdots & cos(\,2\pi f_i n/f_s) \\ \vdots & \ddots & \vdots \\ sin\,(2\pi f_i/f_s) & \cdots & sin\,(6\pi f_i n/f_s) \\ cos(\,2\pi f_i/f_s) & \cdots & cos(\,6\pi f_i n/f_s) \end{pmatrix} \tag{5}$$

where n is the number of sample and f_s is sampling rate. Since the analysis is based on temporal and spatial information, phase different between inputs is also the influence factor of correlation. Linear combination of sine and cosine signal in Eq. (5) can match the phase between X and Y and the phase information of X is reflected on W2. Finally the frequency with the largest coefficient is identified as the stimulus frequency. CCA provides several superiorities on feature extraction of SSVEP-based BCI. First, signal preprocessing and feature extraction can be simultaneously done by CCA. Second, it provides another technique for the feature extraction. Especially most of the analysis methods are performed in frequency domain. Those superiorities provide a strong interest on this research field.

2.2   NNCCA

The core of CCA is a maximization problem. The cost function as similar to Eq. (4) thus can be used to develop the learning rule of NN and construct the NN based CCA. The NNCCA model presented by *Hsieh* [9] [10] is showed in Fig. 1. Each functional network is a three layer feed forward network which contains two hidden layers and one output layer. From the figure it can be observed that the network is combined by three parts. The double-barreled network on the left maps the input $\{x, y\}$ to $\{u, v\}$.

The neuron function of its first layer can be either performed by hyperbolic tangent function, as shown in Eq. (6) and Eq. (7), or the identity function. Second layer is usually the identity function. According to the choice of neuron function (in first layer), this network can perform either the linear (equivalent to standard CCA) or nonlinear CCA. In this study, we focus on the implementation of linear NNCCA.

$$h_k^{(x)} = \tanh\left[\left(W^{(x)}x + b^{(x)}\right)_k\right] \tag{6}$$

$$h_l^{(y)} = \tanh\left[\left(W^{(y)}x + b^{(y)}\right)_k\right] \tag{7}$$

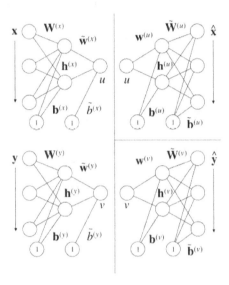

Fig. 1. NNCCA model proposed by the research of *Hsieh* [9] [10]

The networks show on the right of figure try to inversely map $\{u, v\}$ to the corresponded input $\{x', y'\}$. However, the inverse mapping is unnecessary in the implementation of SSVEP based-BCI therefore it will be ignored. The cost function of the NNCCA model is given by Eq. (8). The second, third, fourth and fifth terms of Eq. (8) are normalization constraints that force u and v to have zero mean and unit variance; the sixth term is a weight penalty whose relative magnitude is controlled by the parameter P_1. Larger values of P_1 lead to smaller weights (i.e. fewer effective model parameters), which results in a more linear model. Compare to the cost function of standard CCA in Eq. (4), Eq. (8) provides more constraints to ensure the convergence of neural network model. Due to the effect of weight penalty and the convergence precision, linear NNCCA gives slightly different canonical correlation coefficient when compared with standard CCA. The difference of them can be found in the experiment, which is discussed in the later sections.

$$J = -\eta(U,V) + \langle U \rangle^2 + \langle V \rangle^2 + \left(\langle U^2 \rangle^{\frac{1}{2}} - 1 \right)^2 + \left(\langle V^2 \rangle^{\frac{1}{2}} - 1 \right)^2$$
$$+ P_1 \left[\sum_{ki} \left(W_{ki}^{(x)} \right)^2 + \sum_{lj} \left(W_{lj}^{(y)} \right)^2 \right] \tag{8}$$

3   Offline Experiment

There were 5 healthy subjects ranging from the ages of 22 – 27 participated in this experiment. They were asked to continuously focus on the target stimulus (showed on a LCD monitor with 60 Hz refresh rate) for every 10 seconds, with 4 seconds of rest between each trial. There were 5 stimuli (flashing white squares) as shown on the LCD screen, with flashing frequency of 7.5 Hz, 8 Hz, 10Hz, 15Hz and 20Hz. Each subject carried out 9 trials for each stimulus. That is, each trial contains a 6 seconds useful EEG data and totally 45 trials were completed for each subject. All EEG signals were recorded by a g.USBamp biosignal amplifier at 600 Hz sampling rate ($f_s = 600$ Hz) from six channels PO3, PO4, POz, Oz, O1 and O2 placed on the standard position of the 10-20 international system. In order to verify whether NNCCA can perform the same functions as standard CCA, NNCCA and standard CCA are compared on the above datasets. Moreover, PSDA and MEC are also compared here.

4   Offline Experimental Result and Discussion

There are 45 trials of EEG data for each subject. Standard CCA method and linear NNCCA method are first applied to those trials. After this, 225 pairs of canonical correlation coefficients are obtained (one EEG signal is calculated with 5 stimulus signal). To eliminate the effect of individual differences, those coefficients are grouped by subject. 100 samples are randomly drawn from each group and analysis of variance (ANOVA) is used to check the differences between the results of both CCA methods. Results of ANOVA from each group of samples are 0.8824, 0.8285, 0.8961, 0.909 and 0.8709. Hence, there are no significant differences between the two methods ($p > 0.05$). On the other hands, their consistency can be directly verified through the observation of data. Fig. 2 shows the average correlation coefficient for each group. It can be found that the average canonical correlations obtained by NNCCA have a slightly larger value than standard CCA. However, the difference between them is very small (no larger than 0.005). As we have introduced in the previous part, the difference is caused by the effect of weight penalty and the convergence precision (we will just call it 'effect' in later discussion). Due to the 'effect', the canonical correlation calculated by NNCCA will not be same as the one calculated by standard CCA, but will be a similar value (can be larger or small) as the standard one. After all the average error percentage between the correlations calculated by the two methods is estimated, which is about 2 to 3%. This result indicates that linear NNCCA is a good estimator of standard CCA.

After considering the consistence of linear NNCCA and standard CCA, we then focus on the classification accuracy of linear NNCCA. Table 1 presents the classification accuracy (which successfully recognizes the SSVEP to its corresponded stimulus frequency, according to maximum canonical correlation coefficient or corresponded standard of the method) that obtains from the method of standard CCA, linear NNCCA, MEC and PSDA. The best average accuracy is given by both CCA method, where linear NNCCA method obtains 83.12% accuracy and standard CCA obtain 82.66% accuracy. The different of accuracy is induced by the 'effect'. The interesting thing is the 'effect' might coincidentally generate correct (or wrong) classification but it rarely happens. Moreover, the average computation time of linear NNCCA method for single run is around 2 seconds on an Intel Core 2 Duo E8500 CPU. The number of run depends on the complexity of the network. For instance, if the hidden layer of the network contains 5 neurons, it needs more run to obtain a convergence and true result than the 4 neurons case. In general, linear NNCCA with single neuron in each functional network (or in other words, for two data set input) can give a convergence result in about 5 to 8 runs, which means at least 10 seconds are needed to recognize SSVEP. This computation time is still too long for the online application. The computation rate is probably enhanced by the hardware that supports parallel computation. Recent work has used General-Purpose Graphic Processing Units (GPGPU) for the computation of BCI system [5], it has demonstrated that such hardware can efficiently increase the computation rate compared to the CPU system. With GPGPU receiving more and more attention, it will be the promising direction to increase the rate and enable the online application of NN implementation BCI.

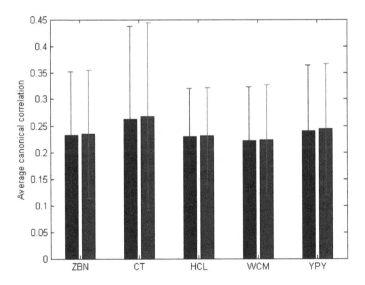

Fig. 2. Comparison of the average correlation coefficient for each group. The left bar is the average result of standard CCA and the right bar is the average result of linear NNCCA.

Table 1. The classification accuracy of feature extraction methods

Subject	CCA	NNCCA	MEC	PSDA
WCM	**91.1%**	**91.1%**	78.1%	77.8%
YPY	**93.3%**	**93.3%**	76.2%	55.8%
HCL	73.3%	**75.6%**	**97.4%**	73.3%
CT	86.7%	**88.9%**	68.3%	**88.9%**
ZBN	**68.9%**	66.7%	64.1%	55.6%
Average	82.66%	**83.12%**	76.81%	70.27%

5   Conclusion

This study attempts to implement CCA algorithm with NN for the frequency detection in SSVEP-based BCI. Results have proposed that NNCCA method can be applied to SSVEP based-BCIs as it leads to higher classification accuracy than those with the methods of PSDA, MEC, and similar performance to the standard CCA method. For future research it would be interesting to make good use of the parallel computing characteristic of NN approach. By using the parallel computing-supported hardware, such as GPGPU, the computation speed can be enhanced and the online application of NN implementation BCI can hopefully be achieved.

Acknowledgement. This work is supported in part by the Macau Science and Technology Development Fund under grant FDCT 036/2009/A and the University of Macau Research Committee under grants RG059/08-09S/FW/FST, RG080/09-10S/WF/FST and MYRG139(Y3-L2)-FST11-WF.

References

1. Wolpaw, J.R., Birbaumer, N., McFarland, D.J., Pfurtscheller, G., Vaughan, T.M.: Brain–computer interfaces for communication and control. Clinical Neurophysiology 113, 767–791 (2002)
2. Wolpaw, J.R., Boulay, C.B.: Brain-Computer Interfaces, The Frontiers Collection. Springer, Heidelberg (2010) ISBN 978-3-642-02090-2
3. Lin, Z., Zhang, C., Wu, W., Gao, X.: Frequency recognition based on canonical correlation analysis for SSVEP-Based BCIs. IEEE Transactions on Biomedical Engineering 54(6), 1172–1176 (2007)
4. Nan, W., Wong, C.M., Wang, B., Wan, F., Mak, P.U., Mak, P.I., Vai, M.I.: A comparison of minimum energy combination and canonical correlation analysis for SSVEP detection. In: 2011 5th International IEEE/EMBS Conference on Neural Engineering, NER 2011, pp. 469–472 (2011)
5. Wilson, J.A.: Using general-purpose graphic processing units for BCI systems. In: Proceedings of the Annual International Conference of the IEEE Engineering in Medicine and Biology Society, EMBS 2011, pp. 4625–4628 (2011)

6. Cecotti, H.: A time-frequency convolutional neural network for the offline classification of steady-state visual evoked potential responses. Pattern Recognition Letters 32(8), 1145–1153 (2011)
7. Coyle, D.: Neural network based auto association and time-series prediction for biosignal processing in brain-computer interfaces. IEEE Computational Intelligence Magazine 4(4), 47–59 (2009)
8. Hsu, W.-Y.: Fuzzy Hopfield neural network clustering for single-trial motor imagery EEG classification. Expert Systems with Applications 39(1), 1055–1061 (2012)
9. Hsieh, W.W.: Nonlinear canonical correlation analysis by neural networks. Neural Networks 13(10), 1095–1105 (2000)
10. Cannon, A.J., Hsieh, W.W.: Robust nonlinear canonical correlation analysis: application to seasonal climate forecasting. Nonlinear Processes in Geophysics 12, 221–232 (2008)

A Frequency Boosting Method for Motor Imagery EEG Classification in BCI-FES Rehabilitation Training System

Jianyi Liang, Hao Zhang, Ye Liu, Hang Wang, Junhua Li, and Liqing Zhang

MOE-Microsoft Key Laboratory for Intelligent Computing and Intelligent Systems, Department of Computer Science and Engineering, Shanghai Jiao Tong University, Shanghai 200240, China
{thefm.cn,juhalee.bcmi}@gmail.com, zh_chaos@sjtu.edu.cn, lyyx000@yahoo.cn, wanghanga_1@126.com, zhang-lq@cs.sjtu.edu.cn

Abstract. Common Spatial Pattern (CSP) and Support Vector Machine (SVM) are usually adopted for feature extraction and classification of two-class motor imagery. However, in a motor imagery based BCI-FES rehabilitation system, stroke patients usually are not able to conduct correct motor imagery like healthy people due to the injury of motor cortex. Therefore, motor imagery EEG of stroke patients lacks of specific discriminant features as appearances of healthy people, which significantly blocks CSP to seek the optimal projection subspace. In this paper, a method, which filters EEG into a variety of bands and improves performance through boosting principle based on a set of weak CSP-SVM classifiers, was proposed to solve the problem mentioned above and was evaluated on the EEG datasets of three stroke subjects. The proposed method outperformed the traditional CSP-SVM method in terms of classification accuracy. From data analysis, we observed that optimal spectral band for classification had been changing along with rehabilitation training, which may reveal mechanisms that dominant frequency band may be changed along with rehabilitation training and spectral power distribution may be changed in different stages of rehabilitation. In addition, this work also demonstrated the feasibility of our SJTU-BCMI BCI-FES rehabilitation training system.

Keywords: EEG, Stroke, BCI-FES Rehabilitation System, Frequency Boosting, Common Spatial Pattern, Support Vector Machine.

1 Introduction

Motor imagery based BCI-FES system is a very promising mean for rehabilitation training of strokes[1], which provides an effective training way for patients to link active motor imagery to movements of paralyzed limbs. Functional Electrical Stimulation (FES) is given to patients according to their corresponding motor imagery during training, helping patients learn external limbs controlling through simulating normal limb controlling process of heathy people[2][3].

C. Guo, Z.-G. Hou, and Z. Zeng (Eds.): ISNN 2013, Part II, LNCS 7952, pp. 284–291, 2013.

Common Spatial Pattern (CSP)is one of the most successful approaches in feature extraction of motor imagery EEG[4]. But it cannot seek the optional projection subspace when applied to stroke patients' data due to contamination of strong noise caused by irregular patterns or wrong imagery which are frequently found in the motor imagery EEG of stoke patients. In order to solve the problem, our proposed method incorporates boosting principle which is quite an effective method in dealing with series of weak learners[5]. It can improve classification performance by combining base weak classifiers, even each of them only has a performance that is slightly better than random.From data analysis, we observed that optimal spectral band for classification had been changing along with rehabilitation training, which may reveal mechanisms that dominant frequency band may be changed along with rehabilitation training and spectral power distribution may be changed in different staged of rehabilitation.

The rest of paper was organized as follows: SJTU-BCMI BCI-FES rehabilitation training system was firstly introduced in Section 2. We present the details of the frequency boosting approach in Section 3. A comparative results are given when applying our method and traditional CSP-SVM on the dataset of stroke subjects in Section 4.

2   Methodology

2.1   BCI-FES Rehabilitation Training System

Fig. 1 shows the framework of our multi-Neurofeedback BCI-FES Motor Function Rehabilitation System including data acquisition module, data server module, model training module[6][7], online classification module, online data visualization module[8] and multi-Neurofeedback module.

Raw data is recorded by 16 channels gtec EEG system with a sample rate 256Hz, among which we select medial frontal cortex and earlobe which are used as ground and reference respectively. We can adopt a variety of EEG classifiers in our framework. During our experiment, we used CSP-SVM as online classifier. EEG signals after removing artifacts are then filtered into specific subband such as 8-30Hz, detrended and converted into a format $time \times channel \times window$. CSP will be applied to calculate the optimal projection subspace. However, we noticed that the accuracy of CSP-SVM is not satisfactory. So we proposed a new algorithm described in Setion 3 to improve the performance. In our experiment each window (see details in subsection 2.2) of EEG is transformed into a 4-dimensional feature spaces which fed to SVM.

2.2   Experiment Setting

In general, the subject is required to take part in our experiment cycle 3 times per week, which consists of 3 phases: prior-training for model preparations, multi-Neurofeedback BCI task for rehabilitation and post-training for assessment. All experiments are monitored by a video camera to build tagged videos for further analysis.

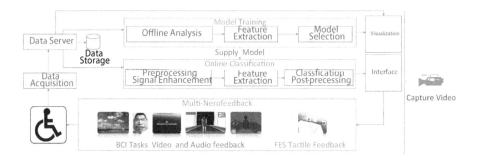

Fig. 1. The framework of multi-Neurofeedback BCI-FES Motor Function Rehabilitation System. BCI Tasks are a series of motor imagery based games giving both video and audio feedback to subjects. The whole system gives a close loop feedback to subject helping them reestablish their motor functional from stoke.

In the prior-training process, a different number of sessions ranged from 5 to 8 will be given to the subject. There are 5 minutes for subjects to relax themselves between sessions. Each session contains 15-16 trials lasting 4 seconds and balances the number of left and right motor imagery tasks. At the beginning of each trial, a bold arrow and a vocal message are given to guide subjects to concentrate on imagining movements of their corresponding part of arms. The time sequence of each trial are cut into 25 sliding windows with width of 1s and step length of 0.125s[6] for online classification. There is a 2-second interval between trials in order to help subjects adjust their mental state to avoid fatigues. The data collection of previous session is used to update model offline, and then it is used for classifying following session data online.

After the prior-training, subjects are asked to finish one or two motor imagery based games such as lifting balloons which appear at the left or right of screen randomly, balancing a ball on a beam and so on. FES is triggered and is used for stimulating one side of subjects' muscles corresponding to current motor imagery, which results in a real movement of their hands or arms. The imagination-stimulation process reconstructs the neuron feed loop between paralysed limbs and corresponding pathological brain cortex of the subject, which takes effects in the rehabilitation treatment[9]. At the end of experiment cycle, a post-training section, two sessions and 16 trials in one session, is conducted to evaluate rehabilitation efficacy.

3   Algorithm Design

CSP cannot seek the right projection subspace when EEG signals are contaminated by strong noise. This is common phenomenon while EEG signals are collected from stroke patients. Recent studies show that the brain functional compensation of damaged brain tissue may be replaced by other part. Supposing this change in spatial may cause the frequency of EEG data changed in some patterns, we try to filter the pre-processed EEG signals into a specific band which may reduce the impact of noisy and classify with SVM to produce a weak learner. Using the framework of Adaboost, we boost each weak SVM learner result by α_m to produce a form of committee whose performance will be better than any of the base classifiers. α_m is give following equation

$$\alpha_m = \ln \frac{BestAccuracy_m}{1 - BestAccuracy_m + \epsilon}. \tag{1}$$

Where $BestAccuracy_m$ is the accuracy for optimal model in round m during iteration. ϵ is to avoid infinity causing by a high accuracy around 100%.

Algorithm 1 described a boost model for two-class classification problem with data from stoke patients. A predefine $BandSet$ contains N bands which use to build model. data is also pre-filtered by band from $BandSet[n]$, detrended and splitted window into $BandWindows_1[n]$ in order to fasten our algorithm. Note that there is square root $\sqrt{\exp a_m}$ which is designed for controlling the boosting speed of incorrect classification data. $DupNum(k)$ is the number of copies for incorrect classification data k during boosting.

Algorithm 1. Frequency Boosting Model Training

1: **for** m = 1, 2, ..., M **do**
2:   **for** n = 1, 2, ..., N **do**
3:     Sample $\theta * length(BandWindows_m[i])$ data into $Sample[n]$
4:     Update $ModelSet[n]$ with
        $Band : BandSet[n]$
        $SpatialFilter$: the CSP projection matrix on data of $Sample[n]$
        SVM: SVM Model on CSP features from $Sample[n]$ by $SpatialFilter$
5:     Use $ModelSet[n].SpatialFilter$ to extract features from $BandWindows_1[n]$,
        classify with $ModelSet[n].SVM$ and calculate $Accuracy[n]$
6:   **end for**
7:   Find optimal model K by $\arg\max\limits_{k \in [1,n]} Accracy[k]$
     Let $BestModel[m] = ModelSet[K]$ and $\alpha_m = \ln \frac{Accuracy[k]}{1-Accuracy[k]+\epsilon}$
8:   $BandWindows_{m+1} = BandWindows_m$
9:   **for all** incorrect classification data k by $BestModel[m]$ in $BandWindows_1[m]$
     do
10:    $num_k = \max\left(1, Round\left(DupNum(k) * \left(\sqrt{\exp a_m} - 1\right)\right)\right)$
11:    **for** n = 1, 2, ..., N **do**
12:      $BandWindows_{m+1}[n] += [num_k$'s copies of $data_k]$
13:    **end for**
       (We boost this data num_k times more in next iteration for all bands)
14: **end for**
15: **end for**

To improve the performance of our algorithm, the band in set should have some overlap and be different in length and range to make weak learners more selective and maximize the coverage. Parameter θ is used to limit the size of training set and make sampling result randomly enough. A larger θ indicates a larger sample set which may decrease the generalization capability of model, while a extreme small θ may harm the stability during iteration.

In two-class classification problem, we use label 1 to indicate which is belonged to one class and label -1 for the other class. Algorithm 2 use the predictions of M CSP-SVM classifiers in different band with weight α_m to predict the label $y(x)$ of data x. $y[m] \in \{1, -1\}$ is the prediction of weak learner m.

Algorithm 2. Frequency Boosting Classification

1: **for all** window segment x in test *dataset* **do**
2:     **for** m = 1, 2, ..., M **do**
3:         Let data = Filter x with $BestModel[m].Band$
4:         Use $ModelSet[m].SpatialFilter$ to extract features from data
5:         Let $y[m]$ = Classification result by $ModelSet[m].SVM$
6:     **end for**
7:     $y(x) = sign(\sum_{m=1}^{M} \alpha_m y[m])$
8: **end for**

---

4   Result

Eight stroke patients from Zhejiang Taizhou Hospital participated in our study. After two months training, five patients have achieved apparently improvements while no significant improvements for the rest on three patients. We presume that these three subjects may have missed the best rehabilitation period because they suffered stroke more than eight months ago.

The algorithm is applied on the EEG datasets of three patients (out of the five patients that have achieved apparently improvements) for evaluation. The BandSet is shown in Table 1 according to the discussion in Section 3.

Table 1. Frequency Boosting parameters

Band Set (Start Hz and End Hz)	$5 - 11$ $9 - 15$	$13 - 19$ $17 - 23$	$21 - 27$ $25 - 31$	$29 - 35$		
	$5 - 13$ $9 - 17$	$13 - 21$ $17 - 25$	$21 - 29$ $25 - 33$	$29 - 37$		
	$5 - 15$ $9 - 19$	$13 - 23$ $17 - 27$	$21 - 31$ $25 - 35$	$5 - 17$		
	$9 - 21$ $13 - 25$	$17 - 29$ $21 - 33$	$25 - 37$			
Parameters	$M : 30$ $\theta : 2/3$ $\epsilon : 0.001$					

We run our algorithm on each pair of training and testing data for three times to obtain an average accuracy. For comparisons, traditional CSP-SVM method is also implemented on the dataset. We use the last session in prior-training section for testing and remaining ones for training. six weeks out of two months (three weeks per month) are chosen and the accuracy of last day in each week

Table 2. Essential information and sliding window classification accuracies

Subject	Age	Sex	Pathogenesis	week					
				1st	2nd	3rd	4th	5th	6th
1	62	female	cortex injury	48%	56%	57%	59%	65%	62%
2	71	male	basal ganglia injury	58%	60%	56%	67%	64%	67%
3	65	female	basal ganglia injury	58%	57%	74%	66%	70%	79%

has been calculated on test data. Table 2 contains the essential information and sliding window classification accuracies of three subjects.

Compared with traditional CSP-SVM method, frequency boosting gives a better accuracy in the most cases (Fig. 2). It's worthy to mention that the whole experiment also provides a powerful evidence of the feasibility of our motor imagery based BCI-FES rehabilitation system.

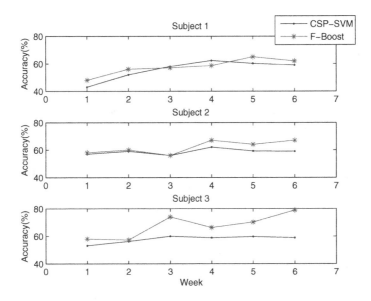

Fig. 2. Comparing between Frequency Boosting and CSP-SVM on test dataset of 3 subjects. Obviously we can find that Frequency Boosting gives a better performance in most cases.

To analyze the frequency changes, we choose the 14th and 57th day to sum up the weight of each optimal weak classification in different frequency as Fig. 3(a) and Fig. 3(b) shown. The importance of gamma-band frequency significantly increased for classification over time.

At the same time we have acquired the motor imagery EEG data of a heathy subject (25age, male), who is an engineer of our rehabilitation system. Apply

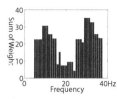 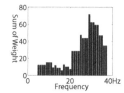 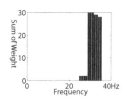

(a) 14th day of a subject, beta and gamma band contribute to classifier.

(b) 57th day of a subject, gamma band play a more important role in classifier.

(c) Use gamma band can get a nice classifier for normal subject.

Fig. 3. Comparisons of optimal spectral band for classification

frequency boosting on his EEG signals, a comparison for the weights of each bands is shown as Fig5. 3(c) (use 1 as the weight of each round in this figure). The figure of patients on the 57th day appears a more similar distribution and outline to that of the heathy subject compared with the one at the beginning of rehabilitation. It implies that the frequency of EEG changes along with brain functional compensation and the gamma band (24-37 Hz) make contribution to a high classification accuracy. This may reveal mechanisms that dominant frequency band may be changed along with rehabilitation training and spectral power distribution may be changed in different stages of rehabilitation. Oscillatory activity in the gamma-band range is related to both gestalt perception and to cognitive functions such as attention, learning, and memory[10].

Three other stroke patients were also trained with ordinary medical treatments for two months as a control group which is observed and recorded during the experiment. A much lower clinical rehabilitation parameters of the control group is observed in post assessment, which indicates that our system promotes the rehabilitation of impaired cerebral cortices and accelerates the reconstruction of the neuron feed loop of stroke patients.

5 Summary

In this paper, we proposed an adaptive Adaboost method in frequency for classifying 2-class motor imagery EEG of stroke patients. This method filtered training data with different bands and produced weak CSP-SVM classifiers for following boosted into a better one. Applying both the proposed method and traditional CSP-SVM on the same datasets of stroke subjects, we compared their classification accuracies. The simulation results proved our method outperforms the general CSP-SVM approach. By analyzing the weight of each optimal model, we provide an evidence of the band of EEG frequency changed during rehabilitation, which is also an evidence of feasibility of our BCI-FES rehabilitation system.

A shortcoming of the method is that we didn't have an auto-adapt boundary to control the number of boosting iterations to avoid over-fitting. Considering the accuracy of pervious round and the accuracy gap between suboptimal models in each round, we may improve the performance of current algorithm by given a better weight during boosting. A cross validation for choosing the best model is more convincing than a greedy way for obtaining optimal model.

For future work, we plan to focus on the EEG changes in beta band during rehabilitation to reveal the mechanism. Moreover, we plan to apply our BCI-FES system to more post-stroke cases and collect more data of stroke patients to provide generally evidence the effectiveness of our method or maybe adapt to reach a better performance.

Acknowledgement. The work was supported by the National Natural Science Foundation of China (Grant No. 90920014 and 91120305) and the NSFC-JSPS International Cooperation Program (Grant No. 61111140019).

References

1. Pfurtscheller, G., Müller-Putz, G., Pfurtscheller, J., Rupp, R.: Eeg-based asynchronous bci controls functional electrical stimulation in a tetraplegic patient. EURASIP Journal on Applied Signal Processing 2005, 3152–3155 (2005)
2. Zhao, Q., Zhang, L., Cichocki, A.: Eeg-based asynchronous bci control of a car in 3d virtual reality environments. Chinese Science Bulletin 54(1), 78–87 (2009)
3. Wolpaw, J., McFarland, D., Vaughan, T., Schalk, G.: The wadsworth center brain-computer interface (bci) research and development program. IEEE Transactions on Neural Systems and Rehabilitation Engineering 11(2), 1–4 (2003)
4. Ramoser, H., Muller-Gerking, J., Pfurtscheller, G.: Optimal spatial filtering of single trial eeg during imagined hand movement. IEEE Transactions on Rehabilitation Engineering 8(4), 441–446 (2000)
5. Freund, Y., Schapire, R., Abe, N.: A short introduction to boosting. Journal-Japanese Society For Artificial Intelligence 14(771-780), 1612 (1999)
6. Li, J., Zhang, L.: Active training paradigm for motor imagery bci. Experimental Brain Research, 1–10 (2012)
7. Li, J., Zhang, L.: Bilateral adaptation and neurofeedback for brain computer interface system. Journal of Neuroscience Methods 193(2), 373–379 (2010)
8. Hong, K., Zhang, L., Li, J., Li, J.: Multi-modal eeg online visualization and neurofeedback. In: Zhang, L., Lu, B.-L., Kwok, J. (eds.) ISNN 2010, Part II. LNCS, vol. 6064, pp. 360–367. Springer, Heidelberg (2010)
9. Meng, F., Tong, K., Chan, S., Wong, W., Lui, K., Tang, K., Gao, X., Gao, S.: Bci-fes training system design and implementation for rehabilitation of stroke patients. In: IEEE International Joint Conference on Neural Networks, IJCNN 2008 (IEEE World Congress on Computational Intelligence), pp. 4103–4106. IEEE (2008)
10. Kaiser, J., Lutzenberger, W.: Induced gamma-band activity and human brain function. The Neuroscientist 9(6), 475–484 (2003)

A Novel Ensemble Algorithm
for Tumor Classification*

Zhan-Li Sun[1],[**], Han Wang[2], Wai-Shing Lau[3] Gerald Seet[4], Danwei Wang[2],
and Kin-Man Lam[5]

[1] School of Electrical Engineering and Automation,
Anhui University, China
`zhlsun2006@126.com`
[2] School of Electrical and Electronic Engineering,
Nanyang Technological University, Singapore
[3] School of Mechanical and Systems Engineering,
Newcastle University, United Kingdom
[4] School of Mechanical and Aerospace Engineering,
Nanyang Technological University, Singapore
[5] Department of Electronic and Information Engineering,
Hong Kong Polytechnic University, China

Abstract. From the viewpoint of image processing, a spectral feature-based TLS (Tikhonov-regularized least-squares) ensemble algorithm is proposed for tumor classification using gene expression data. In the TLS model, a test sample is represented as a linear combination of atoms of an overcomplete dictionary. Two types of dictionaries, spectral feature-based eigenassays and spectral feature-based metasamples, are proposed for the TLS model. Experimental results on standard databases demonstrate the feasibility and effectiveness of the proposed method.

Keywords: Tumor classification, gene expression data, Tikhonov-regularized least-squares model.

1 Introduction

Analysis of gene expression data has become an effective auxiliary measure for disease diagnosis and treatment. In most of existing works, samples of gene expression data are treated as one-dimensional signals by means of some statistical signal processing techniques, or intelligent computation algorithms.

As a powerful data analysis technique, the sparse representation (SR) has been widely applied on signal processing. Specially, a metasample-based SRC is proposed in [1] for tumor classification. Extensive experiments on publicly

---

* This work was supported by a grant from National Science Foundation of China (No. 60905023) and a grant from National Science Foundation of Anhui Province (No. 1308085MF85).
[**] Corresponding author.

available gene expression data sets show that the proposed method can achieve a competitive recognition performance to many existing classification approaches.

Although SRC has shown an excellent performance, it's working mechanism is still not clearly revealed at present. Most researchers think that its superior classification ability is mostly due to the ℓ1-norm sparsity constraint. However, very recently, some researchers have started to doubt the function of sparsity in image classification [2] [3]. Specially, the role of collaborative representation, i.e., using the training samples from all classes to represent the query sample, is regarded as the key factor in [3].

Information in the frequency domain is useful in image classification. In [4] [5], a global feature of a scene, named "spatial envelope", is proposed by exploring the dominant spatial structure of a scene. Although the spectral feature is specially designed for scene classification, in this paper, from the viewpoint of two-dimensional signal, i.e., image, we present a spectral feature-based TLS (Tikhonov-regularized least-squares) ensemble algorithm for tumor classification using gene expression data.

To explore the spectral feature of gene expression data, the samples are transformed into two-dimensional images at first. Multi-resolution spectral images are then extracted and used as representations of gene expression data by means of a method similar to [4], thereby enlarging the size of the training set greatly. Two types of dictionaries are investigated in the TLS model: spectral feature-based eigenassays, i.e., statistically independent basis snapshots, and spectral feature-based metasamples, which are both extracted via a two-stage approach. The strategy of classifier committee learning (CCL) is designed further to combine the results obtained from different spectral features to determine the classes of the test images. Experimental results on some standard databases demonstrate the feasibility and efficiency of the proposed method.

2   Spectral Feature-Based TLS Ensemble Algorithm

2.1   Spectral Feature Image Representation

The task of tumor classification is to correctly identify the categories of new test samples given the labeled training samples. Let an $p \times n$ matrix $\mathbf{X}^0$ denote the gene expression data with p genes and n samples. The element $\mathbf{X}^0_{ij}$ of $\mathbf{X}^0$ is the gene expression level of the ith gene in the jth assay (cell sample). The ith row $\mathbf{r}_i(\mathbf{r}_i = [X^0_{i1}, \cdots, X^0_{in}])$ of $\mathbf{X}^0$ denotes the expression profile of the ith gene. And the jth column $\mathbf{c}_j(\mathbf{c}_j = [X^0_{1j}, \cdots, X^0_{pj}])$ of $\mathbf{X}^0$ is the snapshot of the jth assay.

To explore the spectral feature of gene expression data, the samples $\mathbf{c}_i$ should be transformed from one-dimensional signals into two-dimensional images $\mathbf{I}_i$. Although the images can have arbitrary sizes of rows and columns, for simplicity, all samples are reshaped as square images here. Fig. 1 shows an illustration to transform a gene sample $\mathbf{c}$ into an image $\mathbf{I}$. Sections with the same length d are drawn and put as columns of the image in sequence. When the length of the last section is not enough, a symmetry filling operation is performed at first, as shown in Fig. 1 (b).

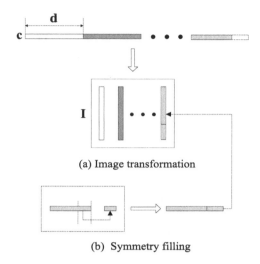

(a) Image transformation

(b) Symmetry filling

Fig. 1. An illustration to transform a gene sample into an image

After transformation, we then extract the spectral feature of these transformed images. The image is first pre-filtered by using a local normalization method of intensity variance [4]. Next, a set of Gabor filters with n_s scales and n_o orientations is applied on the Fourier transform of the prefiltered image via a dot production operation. Finally, the amplitude of the resulting image is computed as the spectral feature image. As a result, for the given N_f (i.e. $n_s \times n_o$) filters, N_f spectral feature images can be obtained for each training sample.

2.2 Dictionary Extraction of Spectral Feature Image

Dictionary extraction is a crucial component in the TLS model. The classification performance is determined by the dictionary to a large extent. Here, we present two types of spectral feature-based dictionary: metasample computed through SVD (singular value decomposition), and eigenassay obtained via ICA (independent component analysis). To utilize the statistical signal processing techniques SVD or ICA, the spectral feature image $\mathbf{F}$ is stretched as a one-dimensional signal with an inverse process shown in Fig. 1 (a).

Denote $\mathbf{X}$ the matrix that contains the spectral feature signals of $\mathbf{X}^0$ formed from one filter, in which each row is the spectral feature vector corresponding to the sample in $\mathbf{X}^0$. Suppose there are K object classes for the n samples of $\mathbf{X}^0$, and each class includes n_i samples, $\mathbf{X}$ can be divided into K data sets $\mathbf{X}_i (i = 1, \cdots, K)$, where $\mathbf{X}_i$ is a collection of the spectral feature vectors corresponding to the ith class.

1) Spectral feature-based metasamples

The matrix singular value decomposition of $\mathbf{X}_i$ can be represented as:

$$\mathbf{X}_i = \mathbf{U}_i \mathbf{S}_i \mathbf{V}_i^T, \tag{1}$$

where $\mathbf{U}_i$ and $\mathbf{V}_i$ are unitary matrices, $\mathbf{S}_i$ is a diagonal matrix with nonnegative diagonal elements in decreasing order. Denote

$$\mathbf{A}_i = \mathbf{S}_i \mathbf{V}_i^T, \tag{2}$$

(1) can be rewritten as:

$$\mathbf{X}_i = \mathbf{U}_i \mathbf{A}_i \tag{3}$$

Considering $\mathbf{A}_i$ as coefficients and the training samples of the ith class are sufficient, any samples belong to this class can be represented via the subspace spanned by the columns of $\mathbf{U}_i$. Further, the bases of all categories are collected together, denoted as $\mathbf{U} = [\mathbf{U}_1, \mathbf{U}_2, \cdots, \mathbf{U}_K]$. Regarding each column as one atom, all columns of $\mathbf{U}$ consist of a dictionary of the TLS model. The atoms formed by SVD are also called metasamples.

2) Spectral feature-based eigenassays

Assume that $\mathbf{X}_i^T$ is a linear mixture of m_i unknown statistically independent basis snapshots (eigenassays) $\mathbf{s}_j(s_j = (s_{1j}, \cdots, s_{pj})^T, j = 1, \cdots, m_i)$, the linear mixing ICA model [6] of $\mathbf{X}_i^T$ can be formulated as

$$\mathbf{X}_i^T = \mathbf{A}_i \mathbf{S}_i^T \tag{4}$$

where $\mathbf{S} = (\mathbf{s}_1, \cdots, \mathbf{s}_{m_i})$. The task of ICA algorithm is to find a linear transformation given by an $m_i \times n_i$ matrix $\mathbf{W}_i$:

$$\mathbf{Y}_i^T = \mathbf{W}_i \mathbf{X}_i^T \tag{5}$$

where $\mathbf{Y} = (\mathbf{y}_1, \cdots, \mathbf{y}_{m_i})$, so that the random variables $\mathbf{y}_j (j = 1, \cdots, m_i)$ are as independent as possible. The outputs $\mathbf{y}_j$ of ICA algorithm are the estimates of the independent eigenassay $\mathbf{s}_j$. In terms of Eqs. (4) and (5), the mixing matrix $\mathbf{A}$ can be given by

$$\mathbf{A}_i = \mathbf{W}_i^{-1} \tag{6}$$

According to Eq. (4), the ith sample $\mathbf{c}_i$ can be expressed as

$$\mathbf{c}_i = a_{i1}\mathbf{y}_1 + \cdots + a_{im_i}\mathbf{y}_{m_i} \tag{7}$$

It can be seen that $\mathbf{c}_i$ can be represented via the subspace spanned by the estimated eigenassays $\mathbf{y}_j$. Further, the eigenassays of all categories are put together, denoted as $\mathbf{Y} = [\mathbf{Y}_1, \mathbf{Y}_2, \cdots, \mathbf{Y}_K]$. Regarding each column as one atom, all columns of $\mathbf{Y}$ consist of a dictionary of the TLS model. In this paper, the well known fast and robust fixed-point algorithm (FastICA) [7] is used to extract the dictionary because of its efficiency and accuracy.

2.3 TLS Ensemble Model

Denote Φ a dictionary formed by SVD or ICA, given a test sample, its spectral feature signal $\mathbf{z}$ can be modeled as a linear representation of the atoms of Φ:

$$\mathbf{z} = \Phi\alpha + \mathbf{v} \tag{8}$$

where $\mathbf{v} \in R^p$ is the noise, and α is the coefficient vector. α can be determined by solving the least squares problem with $\ell2$-regularization, i.e., the TLS model,

$$\min_{\alpha} ||\Phi\alpha - \mathbf{z}||_2^2 + \lambda||\mathbf{z}||_2^2 \qquad (9)$$

where λ is the regularization parameter. The solution of the TLS model can be analytically derived as [8]:

$$\alpha = (\Phi^T\Phi + \lambda I)^{-1}\Phi^T\mathbf{z} \qquad (10)$$

Denote $\mathbf{P} = (\Phi^T\Phi + \lambda I)^{-1}\Phi^T$, we can see that $\mathbf{P}$ depends on Φ. As described in Section 2.2, the dictionary Φ is produced by the training samples. Therefore, Φ is independent of $\mathbf{z}$, and $\mathbf{P}$ can be regarded as a projection matrix [3]. As a result, the solutions of the test samples can be obtained in a batch model. As a comparison, the solution of the least-squares problem with $\ell1$-regularization, i.e. the SR model

$$\min_{\alpha} ||\Phi\alpha - \mathbf{z}||_2^2 + \lambda||\mathbf{z}||_1 \qquad (11)$$

is sparse but encounters a heavier computation burden. The reason for this is that the solutions of the test samples must be computed one by one, i.e. the optimization operation should be carried out for each test sample in sequence. Hence, the computation burden is heavy when the size of the test samples is large. Therefore, instead of using the SR model, the TLS model is adopted in our proposed ensemble algorithm in view of its accuracy and efficiency.

We construct one weaker classifier for each filter. As a result, for each test sample, N_f weaker classifiers are formed by means of the spectral features extracted via N_f filters. Finally, a classifier-combination strategy is adopted to determine the class label of the test sample.

3   Experimental Results

To evaluate the performance of the proposed method, a series of experiments are performed on four data sets: two two-class data sets and two multi-class data sets [9] [10] [11] [12]. As in [1], the classification accuracy obtained via a stratified 10-fold cross validation is adopted here to measure the classification performance of different methods. For the stratified 10-fold cross validation, the

Table 1. Classification results of SVD-TLS and ICA-TLS

	SVD-TLS	ICA-TLS
Colon	0.8387	0.8387
Leukemia	0.9722	0.9722
Leukemia1	0.9444	0.9444
Leukemia2	0.9722	0.9583

Table 2. Classification results of SVD-SR and ICA-SR

	SVD-SR	ICA-SR
Colon	0.8387	0.8548
Leukemia	0.9722	0.9722
Leukemia1	0.9444	0.9444
Leukemia2	0.9583	0.9583

Table 3. Classification results of MRS-SVD-TLS and MRS-ICA-TLS on four data sets

	MRS-SVD-TLS	MRS-ICA-TLS
Colon	0.9032	0.9032
Leukemia	0.9861	0.9861
Leukemia1	0.9861	0.9861
Leukemia2	0.9861	0.9861

data is divided into 10 folds at first, and then ten trials are performed. In the ten trials, each fold is used as the test samples while the remained is adopted as the training samples in sequence. The classifier performance is updated constantly, and an accumulated result is obtained after ten trials. This evaluation method can efficiently check the robustness of the algorithms.

We first investigate the classification performance of the TLS model. When the dictionary is produced by SVD (SVD-TLS) or ICA (ICA-TLS), the classification results of the TLS model are given in Table 1. We can see that the classification accuracies of these two types of dictionaries are close. Further, as a comparison, Table 2 shows the classification accuracies of the SR model when the dictionary is produced by SVD (SVD-SR) [1] or ICA (ICA-SR). The results verify again on the SR model that these two types of dictionaries have a similar classification performance. Moreover, it can be seen from Tables 1 and 2 that, for the same dictionary, the SR model and the TLS model have the nearly same classification accuracies.

We then carry out the experiments on four data sets using our proposed TLS ensemble algorithms with dictionaries produced by SVD (denoted as MRS-SVD-TLS) and ICA (denoted as MRS-ICA-TLS). Table 3 shows the classification results of MRS-SVD-TLS and MRS-ICA-TLS on two data sets. It can be seen that MRS-SVD-TLS and MRS-ICA-TLS have a close classification performance. Moreover, we can see from Tables 1, 2 and 3 that the classification accuracies of MRS-SVD-TLS and MRS-ICA-TLS are higher than those of SVD-SR, ICA-SR, SVD-TLS and ICA-TLS. The results indicate that the classification performance is improved efficiently after introducing the spectral feature and CCL. Therefore, the proposed spectral feature-based TLS ensemble classification framework is feasible and efficient for tumor classification.

4   Conclusion

From the viewpoint of image processing, in this paper we propose a spectral feature-based TLS ensemble algorithm for tumor classification using gene expression data. Experimental results on four standard data sets demonstrated the effectiveness and efficiency of the proposed algorithm.

References

1. Zheng, C.Z., Zhang, L., Ng, T.Y., Shiu, C.K., Huang, D.S.: Metasample-based sparse representation for tumor classification. IEEE/ACM Transactions on Computational Biology and Bioinformatics 8(5) (2011)
2. Rigamonti, R., Brown, M., Lepetit, V.: Are sparse representations really relevant for image classification? In: 24th IEEE Conference on Computer Vision and Pattern Recognition, pp. 1545–1552 (2011)
3. Zhang, L., Yang, M., Feng, X.: Sparse representation or collaborative representation: which helps face recognition? In: International Conference on Computer Vision, pp. 471–478 (2011)
4. Oliva, A., Torralba, A.: Modeling the shape of the scene: a holistic representation of the spatial envelope. International Journal of Computer Vision 42(3), 145–175 (2001)
5. Sun, Z.L., Rajan, D., Chia, L.T.: Scene classification using multiple features in a two-stage probabilistic classification framework. Neurocomputing 73(16-18), 2971–2979 (2010)
6. Sun, Z.L., Lam, K.M.: Depth estimation of face images based on the constrained ICA model. IEEE Transactions on Information Forensics and Security 6(2), 360–370 (2011)
7. Hyvärinen, A.: Fast and robust fixed-point algorithms for independent component analysis. IEEE Transactions on Neural Networks 10(3), 626–634 (1999)
8. Kim, S.J., Koh, K., Lustig, M., Boyd, S., Gorinevsky, D.: A method for large-scale $\ell1$-regularized least squares. IEEE Journal on Selected Topics in Signal Processing 1(4), 606–617 (2007)
9. Alon, U., et al.: Broad patterns of gene expression revealed by clustering analysis of tumor and normal colon tissues probed by oligonucleotide arrays. Proc. Nat'l Academy of Sciences USA 96, 6745–6750 (1999)
10. Golub, T.R., et al.: Molecular classification of cancer: class discovery and class prediction by gene expression monitoring. Science 286(5439), 531–537 (1999)
11. Shipp, M.A., et al.: Diffuse large B-cell Lymphoma outcome prediction by gene expression profiling and supervised machine learning. Nature Medicine 8(1), 68–74 (2002)
12. http://www.zjucadcg.cn/dengcai/Data/data.html

Reducing the Computation Time
for BCI Using Improved ICA Algorithms

Lu Huang[1,3] and Hong Wang[2,*]

[1] Sino-Dutch Biomedical and Information Engineering School, Northeastern University,
Shenyang 110819, China
[2] Northeastern University, POB 319, 110004 Shenyang, China
[3] College of Information Engineering, Dalian Ocean University, Dalian 116023, China

Abstract. P300 is a popular characteristic potential for electroencephalogram(EEG) based brain-computer interface(BCI). In P300-BCI, the extraction of P300 is a very crucial operation. Independent component analysis(ICA) technique is suitable for P300 extraction. In this paper, aiming at the current large volume of EEG data, the applications of three ICA algorithms were proposed for P300 extraction and were compared. The experiments ran on real EEG data respectively. PI and recognition accuracy were checked. The results show artificial fish swarm algorithm based ICA(AFSA_ICA) can extract P300 faster, reducing the computation time for BCI with PI remaining better.

Keywords: brain-computer interface(BCI), P300, computation time, AFSA_ICA.

1 Introduction

An electroencephalogram(EEG) based brain-computer interface(BCI) is a communication system in which messages that an individual sends to the computer do not pass through the peripheral nerve pathways but are detected through EEG activity[1]. P300 potential is often used for BCI because of its independence from training and scatheless measurement[2,3]. In a P300-BCI, the first important step is to extract the P300 potential accurately. Because of the mutual independence between spontaneous EEG, P300 and other artifacts, independent component analysis(ICA) technique is suitable for P300 extraction.

ICA is a blind source separation(BSS) technique that can extract the relevant information buried within noisy signals and allow the separation of measured signals into their fundamental underlying independent component(IC) [4]. The ICA based algorithms used in BCI need to be fast enough to meet its real-time requirement. But the fact is that the EEG data size is very huge because of the numerous electrodes and repeated measurement, and facing such a large volume of EEG data, most presented ICA based algorithms run slow and do not satisfy the real-time BCI system.

---

* Corresponding author.

C. Guo, Z.-G. Hou, and Z. Zeng (Eds.): ISNN 2013, Part II, LNCS 7952, pp. 299–304, 2013.
© Springer-Verlag Berlin Heidelberg 2013

Aiming at above-mentioned current situation, the applications of three ICA based algorithms for P300 extraction, including FastICA, PSO_ICA and AFSA_ICA, and a comparison between the implementations of them were described. Based on real EEG data, the computation time efficiencies of three algorithms were investigated with respect to iteration number and iteration time. And then, the performances of the separation were measured using performance index(PI). Our objective was to choose a better ICA based algorithm, ensuring the P300-BCI reducing the computation time with the recognition accuracy remaining unimpaired and to lay a foundation for the realization of real-time P300-BCI system.

2   Dataset

Our experiment data came from the BCI research group of Ecole Polytechnic Federale de Lausanne[5]. It contained EEG data from 4 disabled subjects and 4 healthy subjects, with measurements from 32 electrodes placed at standard sites(Standard Electrode Position Nomenclature, American Electroencephalographic Association 1990) on the scalp. There were 6 images used to be the target stimulus. In the experiment, every subject needed to finish 4 recording processes, with every process including 6 flash sequences. 6 images each flash once called a block, and 16 blocks constitute one flash sequences.

3   The ICA Based Algorithms

ICA can work well without any information about the mixing matrix. We denote the time varying observed signals(mixed signals) by $\mathbf{x}(t)=(x_1(t), x_2(t),..., x_n(t))^T$ and the source signals consisting of ICs by $\mathbf{s}(t)=(s_1(t), s_2(t),..., s_m(t))^T$, and therefore $\mathbf{x}(t)=\mathbf{A}\mathbf{s}(t)$. Then there exists a de-mixing matrix $\mathbf{w}$ such that $\mathbf{s}(t)=\mathbf{w}\mathbf{x}(t)$. The object of ICA is to find $\overline{w}$, while $\mathbf{y}(t) = \overline{w}\,\mathbf{x}(t)$ is the approximation for the source signals $\mathbf{s}(t)$.

ICA is actually an optimization problem, depending on objective function and optimization algorithm. In this paper, according to negentropy maximum criterion [6], the objective function is defined by

$$C(y) = \sum_{i=1}^{m} J(y_i) \ . \tag{1}$$

where $y_i = w_i\mathbf{x}(t)$, $J(y_i) \approx \rho\,(E\,\{G_i\,(y_i)\}\text{-}E\,\{G_i\,(v)\})^2$, ρ is a positive constant, $G_i\,(\cdot)$ is a non-quadratic function, $E\{\cdot\}$ is a mean function and v is a Gaussian variable having zero mean and unit variance. And then, three different ICA based algorithms were described according to different optimization algorithm.

3.1   FastICA

FastICA, proposed by Hyvärinen in 1999[7], is one of the more popular and referenced ICA techniques which is based on its own unique fast fixed-point iterative algorithm. Using newton iteration method and choosing an initial weight vector w, the basic form of FastICA iteration algorithm is as follows:

$$w=E\{xG(w^Tx)\}-E\{G'(w^Tx)\}w .$$

$$w= w/\| w\| . \tag{2}$$

The algorithm calculates until convergence. The shortage of FastICA is that it traps easily into the local optimum if the initial value is a poor choice.

3.2   PSO_ICA

Particle swarm algorithm(PSO) is an optimization algorithm based on swarm intelligence proposed in the mid-nineties of last century[8]. PSO algorithm, which tends to get the global optimal point by updating the speeds and positions of the particles gradually, can avoid heavy computational load and costing of time.

The updating formulas in PSO for speeds and positions of the particles are:

$$v_i(k+1)=\eta v_i(k)+c_1 rand_1(p_i(k)-x_i(k))+c_2 rand_2(p_g(k)-x_i(k)) .$$

$$x_i(k+1)=x_i(k)+v_i(k) . \tag{3}$$

where η is inertia weight, c_1 and c_2 are acceleration constants, $rand_1$ and rand2 are two random functions which values between 0 and 1, i=1,2,...,M, M is the number of particles in the swarm, k is the number of particle evolutionary steps, x_i is the current position of the ith particle, p_i indicates the best position that the ith particle have experienced and p_g means the best position that all particles have experienced.

The important problem for PSO is to look for a fitness objective function. When the negentropy of ICA technique is used to be the objective function of PSO, the PSO_ICA algorithm is built up. The steps of PSO_ICA algorithm are as follows:

1) Center and whiten the observed sigals x(t), get z(t);

2)Transform the m×n dimensional de-mixing matrix w to the 1×(m×n) dimensional particle matrix w', and initialize all the particles, including their positions $w_i(0)$ and speeds $v_i(0)$;

3) Using Eq.(1) to calculate the objective function values of all particles, specially y(t)=wz(t);

4) Select the personal best point to p_i and the best point of all particles to p_g;

5) Update the position and speed of each particle according to Eq.(3);

6) Record every global optimum, if maximum iterations or minimum error is attained, stop the evolutions, otherwise go back to the step 3.

7) Transform the 1×(m×n) dimensional particle matrix w' to the m×n dimensional de-mixing matrix $\overline{w}$, so y (t)= $\overline{w} x$ (t) is the approximation for the source signals s(t).

3.3   AFSA_ICA

Artificial fish swarm algorithm(AFSA) is a newly emerging method for swarm intelligence optimization[9]. Every artificial fish(AF) in the whole swarm has the abilities to achieve four basic actions: preying, swarming, following and moving. AFSA algorithm begins with several initial points, having the capability of parallel search and information sharing.

The moving action is defined as follow:

$$u_j = u_i + visual \cdot rand . \tag{4}$$

where visual is its vision scope, rand is a random function which values between 0 and 1, u_i is the current position of AF while u_j, the position it randomly moves to.

The preying action allows the position of AF update as follow if the food concentration of u_j is higher than u_i:

$$u_i = u_i + \frac{(u_j - u_i) + (u_{best} - u_i)}{\|(u_j - u_i) + (u_{best} - u_i)\|} \cdot step \cdot rand . \tag{5}$$

where u_{best} is the best position of all AFs, and step means the moving step length.

The center position of all AFs in the swarm is defined as u_c. The swarming action allows the position of AF update as follow if the food concentration of u_c is higher than u_i:

$$u_i = u_i + \frac{(u_c - u_i) + (u_{best} - u_i)}{\|(u_c - u_i) + (u_{best} - u_i)\|} \cdot step \cdot rand . \tag{6}$$

The best position of all AFs in the swarm is defined as u_b. The following action allows the position of AF update as follow if the food concentration of u_b is higher than u_i:

$$u_i = u_i + \frac{(u_b - u_i) + (u_{best} - u_i)}{\|(u_b - u_i) + (u_{best} - u_i)\|} \cdot step \cdot rand . \tag{7}$$

When the negentropy of ICA technique is used to be the objective function of AFSA, the AFSA_ICA algorithm is built up. The steps of AFSA_ICA algorithm are as follows:

1) Center and whiten the observed sigals $x(t)$, get $z(t)$;

2) Transform the m×n dimensional de-mixing matrix w to the 1×(m×n) dimensional AF position matrix w', and Initialize all the AF positions $w_i(0)$ and all dependent parameters ;

3) Using Eq.(1) to calculate the objective function values of all AFs, specially $y(t) = wz(t)$;

4) Evaluate every AF, and then select one of the four basic actions to execute, updating its position;

5) Record every global optimum, if maximum iterations or minimum error is attained, stop the evolutions, otherwise go back to the step 3.

6) Transform the 1×(m×n) dimensional AF position matrix w' to the m×n dimensional de-mixing matrix $\overline{w}$, so $y(t) = \overline{w} x(t)$ is the approximation for the source signals $s(t)$.

4   Results and Analysis

For the real EEG data, the obtained ICs using FastICA, PSO_ICA and AFSA_ICA respectively were averaged between 0~600ms according to the same stimulus row/column. The ICs, which got their maximum amplitude at 250~400ms after the onset of stimulus and got their top two maxima amplitudes on channel C_z, C_1 or C_2, were kept, with the others abandoned. The kept ICs were inverse operated to the scalp electrode and then 0~10Hz filtered. Finally, the peak amplitude and the wave area in the 250~400ms time window were extracted to build a 2-dimensional feature vector, and a simple liner classifier was employed to implement the classification.

Table 1. Iteration performances and PI using three algorithms

	iteration number	iteration time(s)	PI
FastICA	-	45.3	0.11
PSO_ICA	142	34.7	0.09
AFSA_ICA	127	31.2	0.08

To judge the performance of the separation of the mixed signals into ICs, we used PI of the permutation error[10]:

$$\text{PI}=\frac{1}{2M}\sum_{i=1}^{M}\left\{\left(\sum_{j=1}^{M}\frac{\left|c_{ij}\right|^2}{\max_k\left|c_{ik}\right|^2}-1\right)+\left(\sum_{j=1}^{M}\frac{\left|c_{ji}\right|^2}{\max_k\left|c_{ki}\right|^2}-1\right)\right\}. \qquad (8)$$

where $c = wA = (c_{ij})$, and M is the number of variables. PI is zero when the desired subset of ICs is perfectly separated.

The averaged iteration number and iteration time and PI for all eight subjects using three ICA based algorithms were listed in Table 1. The low PI and the small iteration

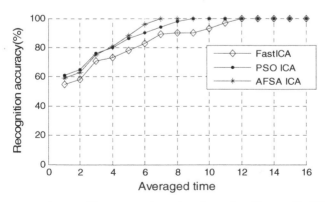

Fig. 1. The averaged recognition accuracies using three algorithms with different averaged times

number and time showed that AFSA_ICA had superior time efficiency and superior separation performance. These favorable properties could make the following classification more accurate. Fig.1 described the averaged recognition accuracies for all eight subjects using three algorithms with different averaged times. As we saw, three algorithms all achieved 100% recognition accuracy with 16 times averaged. While in the condition of less times, AFSA_ICA performed best, followed by PSO_ICA.

The experiment results show that ICA based algorithms can extract P300 satisfactorily, fitting to be the feature extraction tools in P300-BCI. Especially, AFSA_ICA ensures lower PI and the smaller iteration numbers and times, having the ability to reduce the computation time with the recognition accuracy remaining unimpaired.

5   Conclusion

The extraction of P300 potential is a very crucial operation in P300-BCI. ICA based algorithms can extract P300 accurately. As mentioned above, it worked well with just 2-dimensional feature vector built according to the extracted P300 potential using ICA based algorithms sent to the simple liner classifier. In this paper, by comparing FastICA, PSO_ICA and AFSA_ICA on real EEG dataset, we get the conclusion that AFSA_ICA, which works based on swarm intelligence optimization, can extract P300 faster. Meanwhile, the low PI of AFSA_ICA indicates its superior separation performance. All these favorable properties of AFSA_ICA make it have the ability to reduce computation time with the recognition accuracy remaining unimpaired, giving impetus to the further research for real-time BCI.

References

1. Hwang, H.J., Lim, J.H., Jung, Y.J., et al.: Development of an SSVEP-Based BCI Spelling System Adopting a QWERTY-Style LED Keyboard. J. Neurosci. Meth. 208, 59–65 (2012)
2. Salvaris, M., Cinel, C., Citi, L., et al.: Novel Protocols for P300-Based Brain–Computer Interfaces. IEEE Trans. Neural Syst. and Rehab. Eng. 20, 8–17 (2012)
3. Long, J.Y., Li, Y.Q., Yu, T.Y., Gu, Z.H.: Target Selection With Hybrid Feature for BCI-Based 2-D Cursor Control. IEEE Trans. Bio-med. Eng. 59, 132–140 (2012)
4. Hyvärinen, A., Oja, E.: Independent Component Analysis: Algorithms and Applications. Neural Networks 13, 411–430 (2000)
5. Hoffmann, U., Vesin, J.M., Ebrahimi, T., et al.: An Efficient P300-Based Brain-Computer Interface for Disabled Subjects. J. Neurosci. Meth. 167, 115–125 (2008)
6. Comon, P.: Independent Component Analysis: A New Concept? Signal Process 36, 287–314 (1994)
7. Hyvärinen, A.: Fast and Robust Fixed-Point Algorithms for Independent Component Analysis. IEEE Trans. Neural Networ. 10, 626–634 (1999)
8. Shinzawa, H., Jiang, J.H., Iwahashic, M.: Self-Modeling Curve Resolution(SMCR) by Particle Swarm Optimization(PSO). Anal. Chim. Acta. 595, 275–281 (2007)
9. Li, X.L., Shao, Z.J., Qian, J.X.: An Optimizing Method Based on Autonomous Animats: Fish-Swarm Algorithm. Systems Engineering-Theory & Practice 22, 32–38 (2002)
10. Wang, B., Lu, W.K.: A Fixed Point ICA Algorithm with Initialization Constraint. In: Proceedings of 2005 International Conference on Communications, pp. 891–894. IEEE Press, Beijing (2005)

An SSVEP-Based BCI
with Adaptive Time-Window Length

Janir Nuno da Cruz, Chi Man Wong, and Feng Wan

Department of Electrical and Computer Engineering, Faculty of Science and Technology,
University of Macau, Macao

Abstract. A crucial problem for the overall performance of steady-state visual evoked potentials (SSVEP)-based brain computer interface (BCIs) is the right choice of the time-window length since a large window results in a higher accuracy but longer detection time, making the system impractical. This paper proposes an adaptive time window length to improve the system performance based on the subject's online performance. However, since there is no known methods of assessing the online performance in real time, it is also proposed a feedback from the user, through a speller, for the system to know whether the output is correct or not and change or maintain the time-window length accordingly. The system was implemented fully online and tested in 4 subjects. The subjects have attained an average information transfer rate (ITR) of 62.09bit/min and standard deviation of 2.13bit/min with a mean accuracy of 99.00% and standard deviation of 1.15%, which represents an improvement of about 6.50% of the ITR to the fixed time-window length system.

Keywords: brain-computer interface (BCI), steady-state visual evoked potentials (SSVEP), adaptive time-window length, online performance assessment.

1 Introduction

A brain-computer interface (BCI) is a device that translates humans intentions into control signals to provide a direct communication pathway between the human brain and output devices. For instance, patients with severe motor disabilities, such as amyotrophic lateral scleroses (ALS), severe cerebral palsy, muscular dystrophies, etc., who are incapable of communicating with external environment (locked-in syndrome) [1], can express their will to other people or operate a wheelchair by BCI without any brain's normal output pathways of peripheral nerves and muscles [2].

Presently, due to its ease of implementation and its non-invasive operation, electroencephalogram (EEG) recorded along the scalp is widely used for BCI [3]. Based on the different categories of EEG signals, EEG-based BCIs are divided in many different types, such as P300, mu and beta rhythms desynchronization/synchronization (event-related desynchronization/synchronization, ERD/ERS), slow cortical potential (SCP), visual evoked potential (VEP), etc. [2]. However, nowadays, most BCI systems that are not based on SSVEP reached a low information transfer

C. Guo, Z.-G. Hou, and Z. Zeng (Eds.): ISNN 2013, Part II, LNCS 7952, pp. 305–314, 2013.
© Springer-Verlag Berlin Heidelberg 2013

rate (ITR) of around 10~35 bits/min [2][4]. It has been reported that the SSVEP scheme provides the fastest and the most reliable communication paradigm for the implementation of a non-invasive BCI system [5-7].

The recorded EEG signals can undergo considerable changes between training and feedback mode as well as during feedback itself. These variations in the signals can be due to task differences between training and feedback, variability of the recording caused by drying gel or micro movements of the electrodes, plasticity of the brain, due to experience with the task, modulation of cognitive states like attention, motivation and vigilance [12-14]. Hence, the need for an adaptive BCI based on the approach of making the computer adapt to the human's brain and not vice-versa, by constantly monitoring the system and the subject, calibrating according to the variations, thus enhancing/ optimizing the overall performance of the BCI system.

Several adaptive BCIs have been developed to overcome problems related to non-stationary in the signals and to improve the overall performance of the system. Beverina *et al.* developed two protocols used in an offline-learning phase for the system to record a set of signals from the user that should be similar to the ones it would have dealt during the online tests [9]. Volosyak adopted an online adaptation of the minimum energy combination (MEC) method and three user adaptive time segment lengths, achieving a mean ITR of 70.41 ± 25.390 bit/min and accuracy of 99.76 ± 0.583 % [10]. Valbuena *et al.* designed a faster SSVEP classification methodcalled swift BCI to detect the exact time where users start to modulate their brain signals [11].

In order to extract the features of the signals and classify them it is needed to define the time-window length that it is used to process the signal. The classification accuracy increases with the increase of the window length; however, by choosing a large window length, it will take too long to make a selection, which may make the system not practical [15][16]. Therefore, the right choice of the window length is crucial for the overall performance of the system. Accordingly, a novel adaptive time-window length based on an SSVEP-based BCI speller is introduced in this paper. The adaptive time-window length mechanism is used to increase the ITR, thus the system's performance, by selecting the appropriate time-window length based on the subject's online performance, which is assessed through a feedback from the user to the system.

2   Methods

2.1   Visual Stimulator

The proposed spellerallows to input 48 characters (26 letters: 'A-Z', 10 digits: '0-9' and 9 common used symbols) plus "Del" (delete) and "Undo". It has four pages and 16 targets in each page, 3 of the 16 buttons are reserved for turning page. The turning page buttons show the character arrangement of that page so that the users can easily find the correct position of characters. Figure 1 shows the stimuli layout of each page. The frequencies of these 16 targets range from 8Hz to 15.5Hz with increase steps of 0.5Hz. The visual stimulator was programmed in Microsoft Visual C++ 2010 and DirectX SDK 2010.

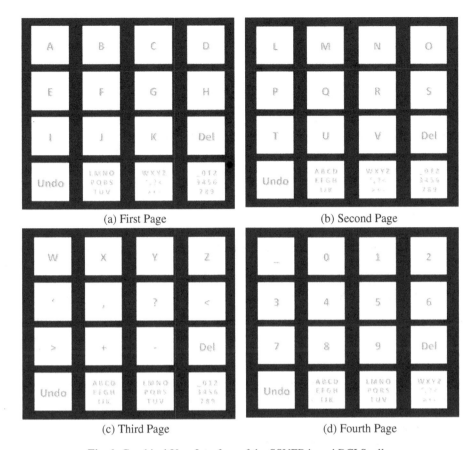

(a) First Page   (b) Second Page

(c) Third Page   (d) Fourth Page

Fig. 1. Graphical User Interface of the SSVEP-based BCI Speller

2.2 Feature Extraction

Canonical correlation analysis (CCA) as a mathematical and statistical method has been utilized widely in many different fields. It deals with the data with linear combinations, for two certain data sets, such that the correlation between the variables is maximized. Consider two multidimensional random variables X, Y and their linear combinations $x=X^T W_x$ and $y=Y^T W_y$, respectively. CCA finds the weight vectors, W_x and W_y, which maximize the correlation between x and y, as shown in (1)

$$\max_{W_x W_y} \rho(X,Y) = \frac{E[X^T Y]}{\sqrt{E[X^T X]E[Y^T Y]}} = \frac{E[W_x^T XY^T W_y]}{\sqrt{E[W_x^T XX^T W_y]E[W_x^T YY^T W_y]}} \tag{1}$$

Usually, the CCA compares the EEG signals with the preinstalled reference signals in arithmetic to calculate the CCA coefficients in SSVEP-BCI. Defining the user's command C in (2),

$$C = argmax\{\rho i\} \qquad i = 1,2,...,k \tag{2}$$

where ρ_i are the CCA coefficients. The reference signals Y_f are set as shown in (3)

$$Y_f = \begin{cases} sin(2\pi ft) \\ cos\,(2\pi ft) \\ \vdots \\ sin(2\pi N_k ft) \\ cos(2\pi N_k ft) \end{cases} \tag{3}$$

where N_k is the harmonic number, for more details refer to [8].

In our system, the sampling points are used to calculate the CCA coefficients in intervals of 0.2s. 5 correlation coefficients (CCs) are obtained in a gazing time interval. If more than two CC command labels are the same, the value is selected; otherwise the EEG is detected again. If the CC criterion is satisfied, then the coefficients are selected to calculate the arithmetic mean. If it is larger than the default threshold value, the selected CC will be chosen as the final result and the system will output a selection command. The threshold values are used as auxiliary tools to reject the influence of noise.

2.3   Adaptive Time-Window Length Mechanism

Based on our previous work [15][16], after analysing the effects of the time-window length on the classification accuracy, it was found that the classification accuracy increased with the time window length until reaching a plateau at around 5s for the CCA used in our system. However, for some subjects it is possible to decrease the length of the time-window length, without compromising the accuracy, thus, decreasing the classification time and increasing the system's overall performance. Thus, the system should be able to adaptively change the time-window length based on the subject's online performance. Nevertheless, it is difficult and there are no known methods to evaluate the online performance in real time. Hence, in this paper, it is adopted a feedback from the user through two commands ("Del" and "Undo") of the BCI Speller to inform the system whether the decision taken was correct or not and the system will increase or decrease the time-window length accordingly. Therefore making possible to achieve an online adaptive time-window length mechanism that selects the time-window length based on the performance of the subject.

First, the command "Del" (delete) is used by the user to correct a misspelling, which may be a result of the user's error. In other words, the user selects this command when he/she had selected a wrong target or the user has changed his mind about which word he/she initially wanted to write. The selection of this command doesn't affect the behavior of the system.

However, another command, "Undo", is used as a feedback from the user, indicating that the last command is wrong, even though the user was gazing at the correct target. One of the reasons for this miss detection maybe the time window length to be short for the program to make the right decision. In this case, the solution would be to increase the time window length to prevent this miss detection.

The "Undo" command allows the user to undo the last entered command, not like "Del", the undone command can be not only a character but also the action of turning a page or even the delete. The user should only the "Undo" when he/she recognizes that he/she was gazing at a certain flicker but the resulted output command was wrong.

In addition, if the system does not receive "Undo" for a long period of time, it means that the time window length (or consecutive time) is good enough. The system can then reduce the time window length a little bit, thus enhancing the system performance. In order to prevent the time-window length to become too small, thus causing errors in the detection and not allowing the user to correctly select the "Undo", to increase the time-window length, it is used two different windows (i.e., buffers) to analyze the data. A fixed one for the "Undo", which will have a higher priority and a second window for the rest. Hence, the "Undo" can be selected even if the time-window length is too short.

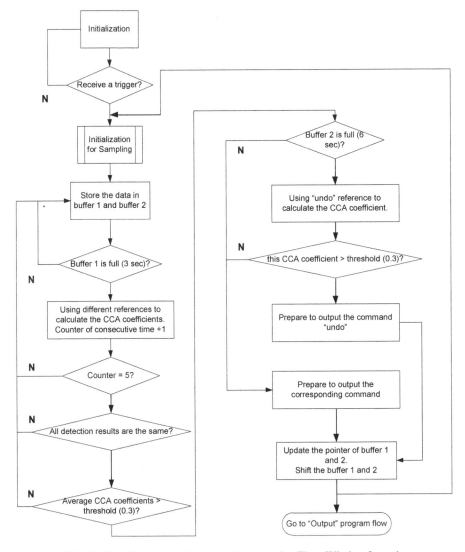

Fig. 2. Flow Chart of the Output of the Adaptive Time-Window Length

Sometimes it takes too long for the system to make a decision but eventually, it takes that decision so if the system continues to get long correct decisions it counts them as normal correct decisions and decreases the time-window length. However, for the next selections it will take even longer to select. To prevent this problem, it was added a constraint of 5 seconds for the decision to be counted as a normal right decision, which will count towards decreasing the window length.

The detailed flow-chart of the adaptive mechanism is shown in Figure 2 and Figure 3.

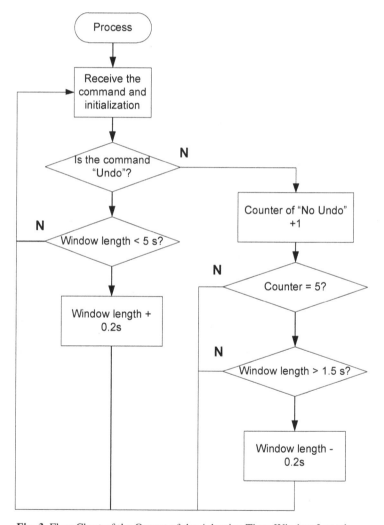

Fig. 3. Flow Chart of the Output of the Adaptive Time-Window Length

3   Results

3.1   Experimental Setup

In this experiment, an LCD monitor was used as the visual stimulator (ViewSonic 22", refresh rate 120 Hz, 1680×1050 pixel resolution). Four subjects, three males and one female (aged from 22 to 29 years old), with normal or corrected to normal vision participated in the experiment. The subjects were seated comfortably facing the monitor with a distance of about 60 cm, in a normal office circumstance. 6 standard EEG electrodes placed on PO_Z, PO_3, PO_4, O_Z, O_1, and O_2 were used as input channels. EEG signals were collected by an amplifier (g.USBamp, Guger Technologies, Graz, Austria) and were filtered by 0.5Hz to 60Hz band-pass filter and sampling rate was 256Hz.

3.2   Procedures

Before the experiments, the subjects took few minutes to get familiar with the Speller and the function of each flicker. None of the parameters were changed or adapted. The subjects were asked to input the word "SSVEP-BCI" one time in a total of 3 trials for each of the two cases: fixed time-window length and adaptive time-window length. The fixed time-window length was used as a comparison of the performance of the adaptive mechanism.

3.3   Results

The most widely used performance index of a BCI is the ITR, which measures the amount of subject information transferred per unit time in a BCI system. The ITR depends on the number of selections (N), detection speed (s) and accuracy rate (p). It provides a convenient way of measuring the performance of a BCI system without consideration of accuracy in a discrete message or a fixed time period [2][4].

$$ITR = \frac{60}{s} \times \left[log_2 N + p\, log_2 p + (1-p) log_2 \left(\frac{1-p}{N-1} \right) \right] \qquad (4)$$

First, it is important to notice that the number of flicker (N) in this system is 16, not the sum of the flicker of all the 4 pages. If a wrong command is detected, the user should correct this error. Thus, in this case the correction step is counted as a correct command classification. Hence the number of commands can increase depending of the subject's performance. In case of an incorrectly classified selection command, the wrongly spelled letter should be corrected. This results in one additional command to select the "Del" or "Undo".

The classification accuracy (p) is calculated as the number of correct command classifications divided by the total number of classified commands. Since all the 16 frequencies can be (erroneously) classified, it could be assumed that all choices are equally probable.

Table 1 shows the contrast between the two time-window length types (adaptive and fixed) average measurable values for each subject as well as the mean (Mean) and standard deviation (S.D.) for the whole experiment. The detection time is measured between two consecutive detections and gives a good comparison between the two systems besides the usual ITR.

Table 1. Adaptive Time-Window Length vs Fixed Time-Window Length

Subject	Time-Window Length Type	Accuracy (%)	Detection Time (s)	ITR (bit/min)
S1	Fixed	100.00	4.42	53.92
	Adaptive	98.00	4.05	59.47
S2	Fixed	100.00	3.89	60.89
	Adaptive	100.00	3.76	64.32
S3	Fixed	100.00	3.92	60.78
	Adaptive	100.00	3.80	63.21
S4	Fixed	100.00	4.11	57.83
	Adaptive	98.00	3.73	61.34
Mean	Fixed	100.00	4.09	58.36
	Adaptive	99.00	3.85	62.09
S.D.	Fixed	0.00	0.24	3.28
	Adaptive	1.15	0.13	2.13

It can be seen from the Table 1, the overall ITR of the adaptive system is 62.09bit/min and a standard deviation of 2.13bit/min with a mean accuracy of 99.00% and a standard deviation of 1.15%, which represents an improvement of about 6.50%of the ITR to the fixed time-window length system. All the subjects experienced an increase of the ITR, obtaining improvements of 10.29%,5.63%,4.00% and 6.07% for the subjects 1, 2, 3 and 4, respectively. This increase of the ITR is a result of the decrease of the time-window length of an average of 5.87% for each subject. However, for clearer conclusions it is needed to test it to a larger population and also observe the performance of the adaptive system in subjects that cannot attain good control of the SSVEP-based BCIs to confirm that by increasing the time-window length it is possible to increase their performance. Moreover, the word used is short, a subject, with 100% accuracy, only needs 14 selections to write "SSVEP-BCI". Hence, it is needed to test the system for more complex sentences to see its performance in a long run.

It is worth mentioning that the system is fully online and that offline calibration of the stimuli frequencies and the threshold of the CCA coefficients could improve the system. However, insisting in a fully online system the results presented in this study are satisfactory and very promising.

4   Conclusions

In this paper, an adaptive time-window length BCI system is proposed. The motivation of this work is to present a mechanism that adaptively selects the appropriate time-window length based on the subject's online performance to enhance the overall system performance.Since to evaluate the online performance is difficult, it was considered a feedback from the user to system to allow the system to know whether the output was correct or not and then make a decision to change the size of the time-window length or not. This project builds up on the existing system in our laboratory and uses an SSVEP-based BCI Speller as the core of the proposed algorithm. Even though this is an on-going project and we only present some preliminary results, these results are promising since the proposed mechanism allowed an increase of around to 6.50% of the ITR without compromising the accuracy.

In further work, in order to reasonably increase or decrease the time-window length the system should also take into account the time interval to calculate the CCA coefficients, the number of correlation coefficients and the detection time. Besides, a more efficient and practical way of assessing the real time online performance is to be studied.

Acknowledgement. This work is supported in part by the Macau Science and Technology Development Fund under grant FDCT 036/2009/A and the University of Macau Research Committee under grants RG059/08-09S/FW/FST, RG080/09-10S/WF/FST and MYRG139(Y3-L2)-FST11-WF.

References

1. Pfurtscheller, G., Neuper, C., Guger, C., Harkam, W., Ramoser, H., Schlogl, A., Obermaier, B., Pregenzer, M.: Current trends in Graz brain- computer interface (BCI) research. IEEE Trans. Rehabil. Eng. 8, 216–219 (2000)
2. Wolpaw, J.R., Birbaumer, N., McFarland, D.J., Pfurtscheller, G., Vaughan, T.M.: Brain-computer interfaces for communication and control. Clin. Neurophysiol. 113(6), 767–791 (2002)
3. van Gerven, M., Farquhar, J., Schaefer, R., Vlek, R., Geuze, J., Nijholt, A., Ramsey, N., Haselager, P., Vuurpijl, L., Gielen, S., Desain, P.: The brain-computer interface cycle. J. Neural Eng. (6) (2009)
4. MacFarland, D.J., Krusienski, D.J., Wolpaw, J.R.: Brain-computer interface signal processing at the Wadsworth Center: mu and sensorimotor beta rhythms. Prog. Brain Res. 411(9), 159 (2006)
5. Vialatte, F.-B., Maurice, M., Dauwels, J., Cichocki, A.: Steadystate visually evoked potentials: Focus on essential paradigms and future perspectives. Prog. Neurobiol. 90, 418–438 (2010)
6. Wang, Y., Gao, X., Hong, B., Jia, C., Gao, S.: Brain-Computer Interfaces Based on Visual Evoked Potentials: Feasibility of Practical System Designs. IEEE Eng. Med. Biol. Mag., 64–71 (2008)
7. Zhu, D., Bieger, J., Molina, G.G., Aarts, R.M.: A survey of stimulation methods used in SSVEP-based BCIs. Comput. Intell. Neurosci. (2010)
8. Bin, G., Gao, X., Yan, Z., Hong, B., Gao, S.: An online multi-channel SSVEP-based brain-computer interface using a canonical correlation analysis method. J. Neural Eng. 6(4) (2009)

9. Beverina, F., Palmas, G., Silvoni, S., Piccione, F., Giove, S.: User adaptive BCIs: SSVEP and P300 based interfaces. PsychNol. J 1(4), 331–354 (2003)

10. Volosyak, I.: SSVEP based Bremen-BCI – boosting information transfer rates. J. Neural Eng. 8(3) (2011)

11. Valbuena, D., Volosyak, I., Gräser, A.: sBCI: Fast Detection of Steady-State Visual Evoked Potentials. In: Proc. Int. Conf. IEEE Eng. Med. Biol. Soc (EMBC 2010), pp. 3966–3969 (2010)

12. Millan, J.R., Mourino, J.: Asynchronous BCI and Local Neural Classifiers: An Overview of the Adaptive Brain Interface Project. IEEE Trans. Neural Syst. and Rehabil. Eng. 11(2) (2003)

13. Shenoy, P., Krauledat, M., Blankertz, B., Rao, R., Muller, K.R.: Towards adaptive classification for BCI. J. Neural Eng. 3, 13 (2006)

14. Girouard, A., Solovey, E.T., Hirshfield, L.M.: Adaptive Brain-Computer Interface. In: Brain Computer Interfaces, pp. 221–237. Springer (2010)

15. Wong, C.M.: Phase Information Enhanced Steady-State Visual Evoked Potential-based Brain-Computer Interface. Unpublished Masters Thesis. University of Macau (2011)

16. Cao, T., Wang, X.: Idle State Detection in SSVEP based BCI System: Algorithm, Implementation and Application, Unpublished Bachelor Thesis. University of Macau (2011)

A Novel Geodesic Distance
Based Clustering Approach
to Delineating Boundaries of Touching Cells

Xu Chen[1], Yanqiao Zhu[1], Fuhai Li[2], Zeyi Zheng[3], Eric Chang[3], Jinwen Ma[1,⋆],
and Stephen T.C. Wong[2]

[1] Department of Information Science, School of Mathematical Sciences,
Peking University, Beijing, 100871, China
[2] NCI Center for Modeling Cancer Development, Department of Systems Medicine
and Bioengineering, The Methodist Hospital Research Institute,
Weill Cornell Medical College of Cornell University, Houston, TX, 77030, U.S.A.
[3] Department of Molecular and Cellular Biology,The Breast Center, Baylor
College of Medicine, One Baylor Plaza, Houston, TX, 77030, U.S.A.
jwma@math.pku.edu.cn

Abstract. In this paper, we propose a novel geodesic distance based
clustering approach for delineating boundaries of touching cells. In specific, the Riemannian metric is firstly adopted to integrate the spatial
distance and intensity variation. Then the distance between any two
given pixels under this metric is computed as the geodesic distance in a
propagational way, and the K-means-like algorithm is deployed in clustering based on the propagational distance. The proposed method was
validated to segment the touching Madin-Darby Canine Kidney (MDCK)
epithelial cell images for measuring their N-Ras protein expression patterns inside individual cells. The experimental results and comparisons
demonstrate the advantages of the proposed method in massive cell segmentation and robustness to the initial seeds selection, varying intensity
contrasts and high cell densities in microscopy images.

Keywords: Cell Segmentation, Riemannian Metric, Distance
Propagation, Clustering Analysis.

1 Introduction

High content screening (HCS) has been popular for discovering novel drugs and
targets by investigating the morphological changes of interested proteins inside
individual cells [1–3]. Along with the advance of automated image acquisition
equipments, on one hand, researchers have the access to large scale cell images
with multiple fluorescent markers, whereas, on the other hand, great challenges
have been posed on the automatic quantification of individual cell morphology
due to their complex appearances, uneven intensity, low signal-to-noise ratio

---

⋆ Corresponding author.

C. Guo, Z.-G. Hou, and Z. Zeng (Eds.): ISNN 2013, Part II, LNCS 7952, pp. 315–322, 2013.
© Springer-Verlag Berlin Heidelberg 2013

(SNR), and cell touching. Automated and accurate detection and segmentation of touching cells is crucial for biological studies such as cell morphological analysis, cell tracking and cell phase identification [4, 5]. However, it remains an open problem due to the aforementioned challenges.

A set of approaches of automatic cell segmentation in various kinds of images have been proposed [6]. In general, the widely used segmentation approaches, for example,watershed [5, 7] and deformable models, such as the snake model, and level set methods [8, 9], are sensitive to initializations(e.g., seeds selection), and intensity variation inside cells. For example, the watershed algorithm is frequently criticized for the over-segmentation problem, which requires further processing. Active contour or level set based deformable models are very sensitive to boundary initializations. In addition, the computational cost is heavy for these approaches. Therefore they are unsuitable for processing large scale image data sets in high content screening studies. Clustering methods and statistical approaches have also been employed for cell segmentation [10–12]. This kind of methods are efficient but usually provide incomplete segmentations because they operate on the single pixel level and neglect the fact that cells are continuous regions. Especially, these methods have limited ability in delineating boundaries of touching cells in cell membrane segmentation [6]. Moreover, large intensity variations both inside the cell membrane and across the whole cell have made it a much more challenging problem to delineate cell boundaries, which usually lead to biased segmentation results. To segment touching cells and delineate the dividing boundaries, Jones et al. [13] proposed a new distance metric defined on image manifolds which combined image gradients and inter-pixel distance. This metric has demonstrated its effectiveness, but segmentation results heavily depend on the initial selection of seeds. Centers of nuclei are adopted as seeds in CellProfiler which implements the above idea [14]. However, this often fails due to morphological variations of cells, and also requires accurate detection results of cell nuclei, which takes extra computational cost and introduces errors from the nuclear segmentation procedure. In addition, the nuclei image might not be available in some HCS studies.

To improve the boundary delineation results of touching cells, in this paper, we propose a geodesic distance based clustering method using the manifold distance. Instead of requiring a seed point (e.g., detected cell nuclear center) for each cell, seed points are iteratively updated in our method and will finally converge to cell centers. Therefore, our method is robust to the initialization of seeds. Moreover, it avoids the incomplete segmentation problem of clustering methods mentioned before thanks to the spatial regularization term in the defined distance metric.

The rest of this paper is organized as follows. In Section 2, we present the geodesic distance based clustering method. Section 3 shows experimental results on segmenting cell images with green fluorescent protein (GFP) tagged N-Ras protein. Finally, a brief discussion of our findings is provided in Section 4.

2   Methodology

2.1   The Riemannian Metric

We first introduce the Riemannian metric defined at each pixel in an image $\mathcal{I}$ [13]. Specifically, $G(\cdot)$ is defined as

$$G(\cdot) = \frac{\nabla g(\cdot) \nabla g^T(\cdot) + \lambda \mathrm{E}}{1 + \lambda}, \tag{1}$$

where E is the 2×2 identity matrix, $\lambda \geq 0$ is a regularization parameter and ∇ is a gradient operator. The function g is introduced to reduce the effect of noise and usually set to be a weighted averaging function in certain neighborhood. The infinitesimal distance at each pixel is then calculated by

$$\| dx \|_{G(\cdot)}^2 \equiv dx^T G(\cdot) dx = \frac{\left(dx^T \nabla g(\cdot) \right)^2 + \lambda dx^T dx}{1 + \lambda}. \tag{2}$$

The item $\left(dx^T \nabla g(\cdot) \right)^2$ increases greatly along directions which are parallel to large gradients in the image and captures the boundary information. The operator $G(\cdot)$ becomes Euclidean as λ increases to infinity and when $\lambda = 0$, the operator considers only gradient information. More details can be found in [13].

2.2   The Geodesic Distance Based Clustering

Under the proposed Riemannian metric, the geodesic distance between any two pixels can be obtained and utilized in cell segmentation. Specifically, the geodesic distance from any pixel to another is calculated along the shortest path via Dijkstra's algorithm [13]. In general, the distance calculation propagates from a smaller neighborhood of the starting pixel to a larger one, and finally reaches the ending pixel. Similarly, the nonlinear dimensionality reduction method ISOMAP also adopted the geodesic distance to discover the true structure of data [15].

Based on the geodesic distance introduced above, we propose a clustering method to delineate cell boundaries. Inspired by the K-means algorithm [16], which finds the optimal clustering result by iteratively updating cluster centers and data's belongingness until converged, we try to update cell seeds in a similar way. Specifically, the algorithm can be abstracted as four steps:

1. Set initial seeds;
2. Compute pixels' geodesic distances to each seed based on the Riemannian metric and assign each pixel to the cell with the nearest distance;
3. Update each seed to the central point of its current enclosing cell;
4. Repeat step 2 and step 3 until the algorithm has converged or the maximum number of iterations is reached.

Seeds will gradually converge to cells' spatial centers. Thus the algorithm is robust to the initial seeds selection. Moreover, since the algorithm is still based on the Riemannian metric, it can accurately delineate touching cell boundaries in images with high noise, low contrast, and large intensity variations.

After the algorithm has converged, cell boundaries are easy to draw according to the current clustering result.

3   Experimental Results

We applied the proposed algorithm on epithelial cells images from dog kidney (MDCK) that were used to study the effects of drug compounds on regulating Ras protein levels, which was determined by measuring the fluorescence intensity of a GFP-tagged N-Ras reporter. Thus the accurate segmentation of cells was crucial for the computation of GFP-tagged Ras level changes between the controlled and drug-treated cells, as shown in green around cell boundaries in images. The obtained images were with high noise, low contrast and varying densities and morphologies, as illustrated in Fig. 1 and Fig. 2.

Images were first compressed and resized to 256×256 from 512×512, and then transformed into gray scale images as the input images. Function g was taken as the Gaussian low pass filter in 3×3 neighborhood and λ was set to 0.8 in our experiments.

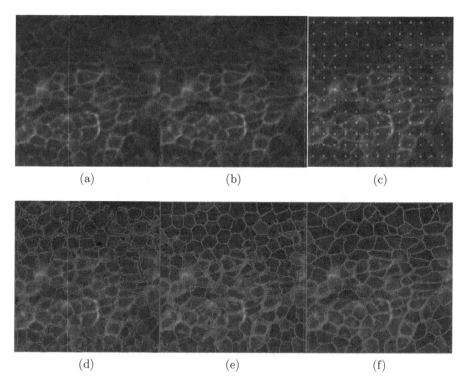

Fig. 1. (a) The original cell image, (b) the corresponding gray scale image, (c) initial seeds for the proposed method and K-means method, (d), (e), (f), segmentation results of CellProfiler, K-means clustering and the proposed method, respectively. Red lines were the detected cell boundaries.

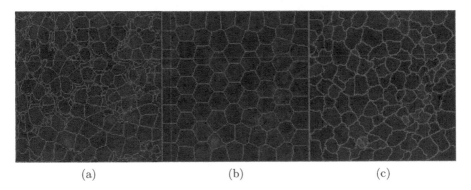

(a)   (b)   (c)

Fig. 2. Segmentation results on a low contrast image, (a) CellProfiler, (b) K-means, (c) the proposed method. Red lines were the detected cell boundaries.

3.1 Adaptive Initial Seeds Selection

We proposed a simple adaptive method to initialize seeds for the algorithm using local intensity variations. An approximately cell-sized square window was first constructed. Then the window was slided in a zigzag way with certain sliding step to scan the whole image. A seed was placed at the window center if and only if there was no seed in the current window and that the average intensity inside the window deviated from those of its already scanned neighboring window-sized regions by a pre-set threshold value. In the following experiments, the size of window was set to 20×20, the sliding step was set to 10, and the intensity difference threshold was set to 0.5.

Fig. 1 (c) illustrated the seeds selection result on a sample image. Selected seeds captured the variation of intensity in the image. However, they didn't seem to be good initial seeds in the way that they deviated from cell centers a lot and several of them were even outside cells. We would show in the following part that the proposed method could provide accurate delineation of cell boundaries even under such an initialization and thus verify its robustness to seed initializations.

3.2 Clustering Results and Analysis

Fig. 3 demonstrated the evolution of cell seeds and corresponding distance maps on a subregion of the image shown in Fig. 1 (a). Seeds quickly converged to centers of corresponding cells. Cell morphologies were greatly recovered by the distance map at the final iteration and true cell boundaries were then easy to tell according to the clustering result.

Fig. 1 (f) showed the result of applying the proposed method on a MDCK image. For comparison, results of CellProfiler and K-means algorithm were also provided, as shown in Fig. 1 (d) and (e), respectively. For the K-means algorithm, seeds initialization was the same as the proposed method and the feature vector for each pixel was taken as $(i, j, g(i, j))$, where i and j were the row and

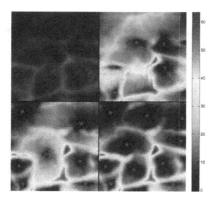

Fig. 3. The original image (*up left*) and the corresponding maps of distances to nearest seeds at the initial state (*up right*), iteration 5 (*down left*) and iteration 15 (*down right*). The red * denoted the corresponding seed positions.

column number of the pixel in the image, and $g(i, j)$ was the intensity in the transformed gray scale image. The introduction of spatial coordinates in the feature vector was inspired by the superpixel analysis to pursue integrate local clustering [17].

CellProfiler generated a lot of false boundaries, especially in the low contrast regions of the image. The result of K-means included some debris and more importantly, the detected boundaries fluctuated a lot. While the proposed method accurately delineated the smooth boundaries of cells. Moreover, K-means algorithm took more than 100 iterations to converge. But our method converged within 15 iterations, which saved a lot of time. Experiment results of the proposed method on images with extreme low contrast and varying cell densities were also satisfying, as illustrated in Fig. 2.

To quantitatively evaluate the segmentation results, comparisons with manual segmentation results were made. The distance between the automatic segmentation results and manual segmentation results were calculated and the cumulative distributions were presented in Fig. 4 [13]. 20 cell images were experimented with and the average result was taken. The percentages of detected boundary pixels in the accuracy of within 2 pixels from the manually delineated boundaries were 83.2%, 77.1% and 63.6% for the proposed method, CellProfiler and K-means algorithm, respectively. Therefore, our method performed consistently better than CellProfiler and K-means algorithm.

Considering the width and fuzziness of cell boundaries, we could get that our method succeeded in delineating cell boundaries reliably. Since our seeds selections were not good at initialization, as illustrated in Fig. 1 (c), the proposed method was robust to the initialization of seeds. In addition, it overcame the challenges of low contrast, high intensity variation and varying cell density.

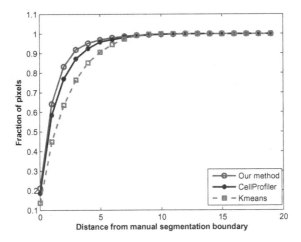

Fig. 4. Cumulative distributions of distance between automatically delineated boundaries and the manual segmenation result

4   Conclusions and Discussions

In this paper, we have proposed a novel geodesic distance based clustering algorithm that could well address the challenge of reliable segmentation of dense and overlapping cells for discovering novel drugs regulating the spatial patterns of the N-Ras protein inside single cells. Experimental results show that the proposed algorithm is effective and robust. Compared with traditional methods, the main advantages of our proposed method include multiple cell segmentation, distinct and smooth cell boundaries, high cell detection rates, robustness to initial seeds selection and efficiency. Moreover, our method is successful in finding the dividing boundaries between touching cells, even in very low contrast images.

Although the proposed method is robust to the initialization of seeds , the availability of prior information, such as nuclei locations, may lead to faster convergence and a better segmentation result. In addition, a fusion procedure may be necessary in dealing with cells with irregular sizes, which are prone to be segmented into several parts. Besides, the optimization of regularization parameter λ would need further investigation, though our studies indicates that a reference of proper λ for a gray scale image is in interval $[0.2, 5]$.

Acknowledgement. We would like to thank Ahmad Hammoudi for helpful discussion about the local clustering. This work was supported by the Natural Science Foundation of China for grant 61171138 and NIH R01 CA121225, NIH U54 CA149196.

References

1. Li, F., Zhou, X., Ma, J., Wong, S.: Multiple nuclei tracking using integer programming for quantitative cancer cell cycle analysis. IEEE Transactions on Medical Imaging 29(1), 96–105 (2010)
2. Li, F., Zhou, X., Ma, J., Wong, S.: An automated feedback system with the hybrid model of scoring and classification for solving over-segmentation problems in rnai high content screening. Journal of Microscopy 226(2), 121–132 (2007)
3. Wang, M., Zhou, X., Li, F., Huckins, J., King, R., Wong, S.: Novel cell segmentation and online svm for cell cycle phase identification in automated microscopy. Bioinformatics 24(1), 94–101 (2008)
4. Ficarra, E., Cataldo, S., Acquaviva, A., Macii, E.: Automated segmentation of cells with ihc membrane staining. IEEE Transactions on Biomedical Engineering 58(5), 1421–1429 (2011)
5. Zhou, X., Li, F., Yan, J., Wong, S.: A novel cell segmentation method and cell phase identification using markov model. IEEE Transactions on Information Technology in Biomedicine 13(2), 152–157 (2009)
6. Meijering, E.: Cell segmentation: 50 years down the road [life sciences]. IEEE Signal Processing Magazine 29(5), 140–145 (2012)
7. Cheng, E., Challa, S., Chakravorty, R.: Microscopic cell segmentation and dead cell detection based on cfse and pi images by using distance and watershed transforms. In: Digital Image Computing: Techniques and Applications, DICTA 2009, pp. 32–39. IEEE (2009)
8. Kass, M., Witkin, A., Terzopoulos, D.: Snakes: Active contour models. International Journal of Computer Vision 1(4), 321–331 (1988)
9. Chan, T., Vese, L.: Active contours without edges. IEEE Transactions on Image Processing 10(2), 266–277 (2001)
10. Bak, E., Najarian, K., Brockway, J.: Efficient segmentation framework of cell images in noise environments. In: 26th Annual International Conference of the IEEE Engineering in Medicine and Biology Society, IEMBS 2004, vol. 1, pp. 1802–1805. IEEE (2004)
11. Muda, T., Salam, R.: Blood cell image segmentation using hybrid k-means and median-cut algorithms. In: 2011 IEEE International Conference on Control System, Computing and Engineering, ICCSCE, pp. 237–243. IEEE (2011)
12. Palacios, G., Beltran, J.: Cell nuclei segmentation combining multiresolution analysis, clustering methods and colour spaces. In: International Machine Vision and Image Processing Conference, IMVIP 2007, pp. 91–97. IEEE (2007)
13. Jones, T.R., Carpenter, A., Golland, P.: Voronoi-based segmentation of cells on image manifolds. In: Liu, Y., Jiang, T.-Z., Zhang, C. (eds.) CVBIA 2005. LNCS, vol. 3765, pp. 535–543. Springer, Heidelberg (2005)
14. Lamprecht, M., Sabatini, D., Carpenter, A., et al.: Cellprofiler: free, versatile software for automated biological image analysis. Biotechniques 42(1), 71 (2007)
15. Tenenbaum, J., De Silva, V., Langford, J.: A global geometric framework for nonlinear dimensionality reduction. Science 290(5500), 2319–2323 (2000)
16. MacQueen, J.: Some methods for classification and analysis of multivariate observations. In: Proceedings of the Fifth Berkeley Symposium on Mathematical Statistics and Probability, California, USA, vol. 1, pp. 281–297 (1967)
17. Achanta, R., Shaji, A., Smith, K., Lucchi, A., Fua, P., Süsstrunk, S.: Slic superpixels. EPFL Technical Report no. 149300 (June 2010)

Seizure Detection in Clinical EEG
Based on Entropies and EMD

Qingfang Meng[1,2], Shanshan Chen[1,2], Weidong Zhou[3], and Xinghai Yang[1,2]

[1] School of Information Science and Engineering, University of Jinan, Jinan 250022, China
[2] Shandong Provincial Key laboratory of Network Based Intelligent Computing,
Jinan 250022, China
[3] School of Information Science and Engineering, Shandong University, Jinan 250100, China
ise_mengqf@ujn.edu.cn

Abstract. Considering the EEG signals are nonlinear and nonstationary, the nonlinear dynamical methods have been widely applied to analyze the EEG signals. Directly extracted the approximate entropy and sample entropy as features are efficient methods to analysis the EEG signals of epileptic parents. To detect the epilepsy seizure signals from epileptic EEG, choose an appropriate threshold value as the discrimination criteria is simplest. The experiment indicated the approximate entropy provide a higher accuracy in distinguishing the epileptic seizure signals from the EEG than sample entropy. To improve the accuracy of sample entropy, empirical mode decomposition (EMD) is used to decompose EEG into multiple frequency subbands, and then calculate sample entropy for each component. The results show that the accuracy is up to 91%, which could be used to discriminate epileptic seizure signals from epileptic EEG.

Keywords: epileptic EEG, approximate entropy, sample entropy, empirical mode decomposition (EMD).

1 Introduction

Epilepsy is a sudden neurological disorder characterized by the presence of recurring seizures. For the epileptic patients, the nerve cells in the brain usually discharge excessive electrical impulses. So the electrical activities of the brain during epilepsy seizure period are different from the normal. The EEG (Electroencephalogram) contains important information about the conditions and functions of the brain, epilepsy can be assessed by the EEG. EEG is fundamental for diagnosing epilepsy disease, and useful for both physiological research and medical applications.

Because the brain is a complex nonlinear dynamical system, the scalp EEG signals are complex, nonlinear and nonstationary. The nonlinear dynamical methods have been widely applied to analyze EEG signals [1-4]. Reference [1] analysis blanket dimension and fractal intercept features, and applies fractal intercept to epileptic EEG detection. Reference [2] proposes a method using subband nonlinear parameters and genetic algorithm for automatic seizure detection in EEG. Reference [3] analyses

C. Guo, Z.-G. Hou, and Z. Zeng (Eds.): ISNN 2013, Part II, LNCS 7952, pp. 323–330, 2013.

seizure detection features and their combinations using a probability-based scalp EEG seizure detection framework. Reference [4] analysis EEG signals based on approximate entropy (ApEn) and discrete wavelet transform (DWT).

Approximate entropy and sample entropy, which are studied in the past two decades [5-13], are believed to be effective methods to analysis nonlinear and nonstationary signals. In addition to the advantage of being less dependent on the length of data set, direct feature-based seizure detection is easier to understand and simplest. Fast computation of sample entropy and approximate entropy has been developed to speed up the computation for short data length [13]. So the two methods have been proved to apply to the nonlinear signals, such as EEG signals [6, 7, 8], heart rate variability signal [9, 10], Measurement of cardiac synchrony [11], schizophrenic [12].

Studies have showed that individual EEG frequency subbands can provide more information than the entire EEG [2]. The empirical mode decomposition (EMD), introduced by huang et al. [14], has good performance in analyzing nonstationary signals [15, 16]. EMD decomposes a raw signal into a set of complete and almost orthogonal components called intrinsic mode functions (IMFs). IMFs represent the natural oscillatory modes embedded in the raw signal. Owing to the benefit of self-adaptive capacity, it is widely used in physiological signals researches [17, 18]. In this work, each EEG signal is decomposed into three subbands by EMD.

In this paper, firstly extract the approximate entropy (ApEn) and sample entropy (SampEn) as feature values to detect the epileptic signals. Considering the excessive discharge during seizure, we proposed the method of combination of EMD and sample entropy to distinguishing the epilepsy seizure EEG from EEG signals.

2   Seizure Detection in Clinical EEG Based on Approximate Entropy, Sample Entropy and EMD Methods

Approximate entropy and sample entropy are similar regularity or complexity measurements of a time series. The epileptic seizure detection is directly based on the two entropies. They are less affected by data length and noise. Given an appropriate embedding dimension m and tolerance r, they are easy to calculate approximate entropy and sample entropy values.

The Approximate entropy algorithm can be defined as follows. For N points time series {u (i): $1 \leq i \leq N$}, the following vector sequence can be formed.

$$X_i^m = \left\{ u(i), u(i+1), \ldots, u(i+m-1) \right\}, 1 \leq i \leq N-m+1 \tag{1}$$

Here X_i^m represents m consecutive u values, commencing with the ith point. The distance d_{ij}^m between two vectors X_i^m and X_j^m is defined as:

$$d_{ij}^m = d[X_i^m, X_j^m] = \max_{k=0 \sim m-1} \left| u(i+k) - u(j+k) \right|, i, j = 1, \cdots, N-m+1, i \neq \mathrm{j} \tag{2}$$

Given tolerance r, and then define

$$C_r^m(i) = \frac{1}{N-m+1} \sum_{j=1, j\neq i}^{N-m} \Theta\left(d_{ij}^m - r\right) \tag{3}$$

$$\Phi^m(r) = \frac{1}{N-m+1} \sum_{i=1}^{N-m} \ln(C_r^m(i)) \tag{4}$$

Here Θ is the Heaviside function

$$\Theta(z) = \begin{cases} 1, z \leq 0 \\ 0, z > 0 \end{cases} \tag{5}$$

Then, calculate $(m+1)$ dimensional embedding vectors X_i^{m+1} and $\Phi^{m+1}(r)$, exactly the same way using X_i^{m+1}. Finally, the estimate approximate entropy of the time series (obtained in the limit of N→∞) is defined by:

$$ApEn(m, r, N) = \Phi^m(r) - \Phi^{m+1}(r) \tag{6}$$

For a N points time series {u (i): 1≤ i≤ N}, similar to the definition of approximate entropy, the sample entropy as following:
Firstly, form the following vector sequence:

$$X_i^m = \left\{u(i), u(i+1), \ldots, u(i+m-1)\right\}, 1 \leq i \leq N-m+1 \tag{7}$$

The distance d_{ij}^m between X_i^m and X_j^m is defined as

$$d_{ij}^m = d[X_i^m, X_j^m] = \max_{k=0 \sim m-1} \left|u(i+k) - u(j+k)\right| \tag{8}$$

Given tolerance r, and then define

$$B_r^m(i) = \frac{1}{N-m-1} \sum_{j=1, j\neq i}^{N-m} \Theta\left(d_{ij}^m - r\right) \tag{9}$$

$$B_r^m = \frac{1}{N-m} \sum_{i=1}^{N-m} B_r^m(i) \tag{10}$$

Similarly, form the vector sequence { X_i^{m+1} } and get the function A_r^m

$$A_r^m(i) = \frac{1}{N-m-1} \sum_{j=1, j\neq i}^{N-m} \Theta\left(d_{ij}^m - r\right) \tag{11}$$

$$A_r^m = \frac{1}{N-m} \sum_{i=1}^{N-m} A_r^m(i)$$ (12)

For finite data sets, the sample entropy can be estimated form the formula

$$SampEn = -\ln(A_r^m / B_r^m)$$ (13)

EMD is an adaptive and efficient method applied to analysis non-stationary signals.

In this paper, each EEG signal is decomposed into a series of intrinsic mode function (IMF) by EMD. Each IMF must satisfy two conditions: in the whole data set, the number of extreme and the number of zero crossings must either equal or differ at most by one; and at any point, the mean value of the upper envelope and lower envelope is zero. The EMD algorithm for the signal $x(t)$ can be summarized as follows:

(1) Identify the local maxima and minima of the original data $x(t)$, then connect respectively by a cubic spline line to produce the upper and lower envelops: U_{max} and U_{min}.

(2) Obtain the mean value of corresponding data point

$$m_1 = \frac{U_{max} + U_{min}}{2}$$ (14)

(3) Define the difference between $x(t)$ and m_1 as the first component

$$h_1 = x(t) - m_1$$ (15)

(4) Regard h_1 as new $x(t)$ and repeat the operation above until h_1 satisfies the IMF conditions, then obtain the first-order IMF, designate it as $c_1 = h_1$

(5) Defined the residue r_1 as $x(t)$ minus c_1

$$r_1 = x(t) - c_1$$ (16)

(6) Taking the residue r_1 as a new data and repeating (1) ~ (5) and the second IMF component is obtained. If c_1 or r_1 is smaller than a predetermined value, or r_1 becomes a monotone function, the sifting process is stopped, or else repeated as the last step. Thus, a series of IMF can be obtained. The signal $x(t)$ can be expressed as

$$x(t) = \sum_{i=1}^{m} c_i + r_m \tag{17}$$

EMD decomposes each EEG signal into $(m+1)$ frequency components. Here, m intrinsic mode functions (IMFs) represent the different higher frequency components of the original signals, while r_m corresponding to the lower frequency residue.

3   Experiment Results and Analysis

The data used in this research were obtained from the six clinical diagnosed epilepsy cases in the Qilu Hospital of Shandong University. All EEG signals were sampled at a rate of 128 Hz. For the purpose of comparison, we select 200 episodes of epileptic data during a seizure free interval and 200 episodes of data during seizure, each episode of 1024 points.

Fig.1. showed approximate entropy and sample entropy values of the epileptic EEG, the classification based on the features directly. Displayed in Fig.1.(a) are the Approximate Entropy analysis results of interictal EEG and epileptic EEG episodes. It could be found that the mean entropy value of interictal EEG was greater than that of ictal EEG. We chose a constant threshold value 0.25 as the discrimination criteria, represented by the dotted line in the picture. The classification results are displayed in the first row of Table 1.

Similarity, the best threshold of sample entropy was determined as 0.75 which is shown in Fig.2.(b). The classification results using sample entropy method are displayed in the second row of table 1.

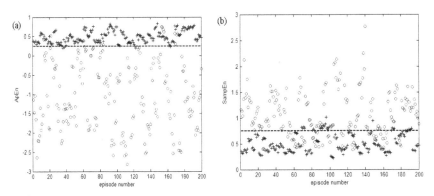

Fig. 1. The ApEn (a) and SampEn (b) of epileptic EEG ('+'represents interictal EEG, 'o' represents ictal EEG, the dotted horizontal line is the threshold)

However, look at the Fig. 1(b), the sample entropy values have many overlapping. In order to improve the classification accuracy of sample entropy, we use the EMD method to decompose every original signal into frequency subband components, and then make sample entropy value for every subband. The original EEG and three IMF components are showed in Fig.2, and the sample entropy values of the IMF components arc displayed in Fig. 3.

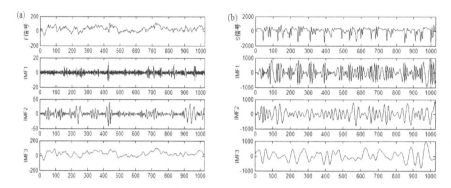

Fig. 2. Examples of a segment of EEG signals: (a) represent interictal EEG episodes; (b) represent ictal EEG episodes

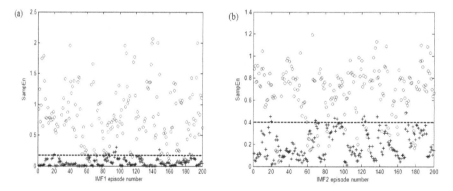

Fig. 3. Sample entropy of IMF1 (a) and IMF2 (b) ('+'represents interictal EEG, 'o' represents ictal EEG)

Table 1. The classification of ictal EEG and interictal EEG using ApEn and SampEn Analysis

	thres-hold	ictal EEG		Interictal EEG		Accu-racy(%)
		Sensitivity	Specificity	Sensitivity	Specificity	
ApEn	0.25	93.50	97.50	97.50	93.50	95.50
SampEn	0.75	73.50	85.50	85.50	73.50	79.50

Table 2. The classification of ictal EEG and interictal EEG based on frequency subband components using SampEn Analysis

	Threshold	ictal EEG		interictal EEG		Accuracy (%)
		Sensitivity	Specificity	Sensitivity	Specificity	
IMF1	0.17	88.50	95.00	95.00	88.50	91.75
IMF2	0.40	83.50	95.00	95.00	83.50	89.00

For IMF1, determine the best threshold as 0.17 which is shown in Fig. 4(a). The classification results are displayed in the first row of table 2. For IMF2, as is showed in the Fig. 4(b), we choose the threshold 0.4. The classification results are displayed in the second row of table 2. It could be found that the accuracy of two frequency subband components is 91.75%, 89.0%, respectively. The accuracy is higher than using sample entropy directly

4   Conclusions

Nonlinear features have good discriminatory power for the analysis the EEG signal. In the paper, we extract approximate entropy and sample entropy as the non-linear features could distinguish epileptic EEG from interictal EEG. Only calculating a feature from the EEG signals and then threshold the feature to make detections is typically more transparent and easier to understand, but it is stricter to the feature. We propose a new method that combined sample entropy with EMD, the classification accuracy of the sample entropy of high frenquence subband compnoment is improved. As a result, for scalp EEG, the approximate entropy and subband sample entropy as the classification features could be used to discriminate ictal EEG from interictal EEG for seizure detection.

Acknowledgments. Project supported by the National Natural Science Foundation of China (Grant No. 61201428, 61070130), the Natural Science Foundation of Shandong Province, China (Grant No. ZR2010FQ020), the Shandong Distinguished Middle-aged and Young Scientist Encourage and Reward Foundation, China (Grant No. BS2009SW003), and the China Postdoctoral Science Foundation (Grant No. 20100470081), the Shandong Province Higher Educational Science and Technology Program (Grant NO. J11LF02).

References

1. Wang, Y., Zhou, W.D., Li, S.F., Yuan, Q., Geng, S.J.: Fractal Intercept Analysis of EEG and Application for Seizure Detection. Chinese Journal of Biomedical Engineering 30, 562–566 (2011)
2. Kai, C.H., Sung, N.Y.: Detection of Seizures in EEG using Subband Nonlinear Parameters and Genetic Algorithm. Computers in Biology and Medicine 40, 823–830 (2010)

3. Kuhlmann, L., Burkitt, A.N., Cook, M.J., Fuller, K., Grayden, D.B., Seiderer, L., Maieels, I.M.: Seizure Detection Using Seizure Probability Estimation: Comparison of Features Used to Detect Seizures. Ann. Biomed. Eng. 37, 2129–2145 (2009)
4. Ocak, H.: Automatic detection of epileptic seizures in EEG using discrete wavelet transform and approximate entropy. Expert Systems with Applications 36, 2027–2036 (2009)
5. Pincus, S.M.: Approximate entropy as a measure of system complexity. Proc. Natl. Acad. Sci. USA 88, 2297–2301 (1991)
6. Kannatha, N., Choob, M.L., Acharya, U.R., Raudys, P.K.: Entropies for detection of epilepsy in EEG. Computer Methods and Programs in Biomedicine 80, 187–194 (2005)
7. Sabeti, M., Katebi, S., Boostani, R.: Entropy and complexity measures for EEG signal classification of schizophrenic and control participants. Artificial Intelligence in Medicine 47, 263–274 (2009)
8. Achary, U.R., Molinari, F., Sree, S.V., Chattopadhyay, S., Ng, K.H., Suri, J.S.: Automated diagnosis of epileptic EEG using entropies. Biomedical Signal Processing and Control 7, 401–408 (2012)
9. Zhuang, J.J., Ning, X.B., Du, S.D., Wang, Z.Z., Huo, C.Y., Yang, X., Fan, A.H.: Short-term nonlinear heart rate variability forecast of ventricular tachycardia. Chinese Science Bulletin 53, 520–527 (2008)
10. Huang, X.L., Cui, S.Z., Ning, X.B., Bian, C.H.: Multiscale Base-scale Entropy Analysis of Heart Rate Variability. Phys. Sin. 58, 8160–8165 (2009)
11. Cullen, J., Saleem, A., Swindell, R., Burt, P., Moore, C.: Measurement of cardiac synchrony using Approximate Entropy applied to nuclear medicine scans. Biomedical Signal Processing and Control 5, 32–36 (2010)
12. Sabeti, M., Katebi, S., Boostani, R.: Entropy and complexity measures for EEG signal classification of schizophrenic and control participants. Artificial Intelligence in Medicine 47, 263–274 (2009)
13. Pan, Y.H., Wang, Y.H., Liang, S.F., Lee, K.T.: Fast computation of sample entropy and approximate entropy in biomedicine. Comput. Methods Programs Biomed. 104, 382–396 (2011)
14. Huang, N., Shen, Z., Long, S.R., Wu, M.C., Shin, H.H., Zheng, Q., Yen, N.C., Tung, C.C., Liu, H.H.: The empirical mode decomposition and the Hilbert spectrum for nonlinear nonstationary time series analysis. Proc. R. Soc. London A 454, 903–995 (1998)
15. Li, S.F., Zhou, W.D., Cai, D.M., Liu, K., Zhao, J.L.: EEG Signal Classification Based on EMD and SVM. Journal of Biomedical Engineering 28, 891–894 (2011)
16. Pachori, R.B., Bajaj, V.: Analysis of normal and epileptic seizure EEG signals using empirical mode decomposition. Computer Methods and Programs in Biomedicine 104, 373–381 (2011)
17. Tafreshi, A.K., Nasrabadi, A.M., Omidvarnia, A.H.: Epileptic Seizure Detection Using Empirical Mode Decomposition. In: 8th IEEE International Symposium on Signal Processing and Information Technology, Sarajevo, Bosnia, pp. 238–242 (2008)
18. Braun, S., Feldman, M.: Decomposition of nonstationary signals into varying time scales: Some aspects of the EMD and HVD methods. Mechanical Systems and Signal Processing 25, 2608–2630 (2011)

A New Hybrid Intelligent System
for Fast Neural Network Training

Anantaporn Hanskunatai

Department of Computer Science, King Mongkut's Institute of Technology Ladkrabang,
Ladkrabang, Bangkok, 10520, Thailand
ksananta@kmitl.ac.th

Abstract. A major drawback of artificial neural network is long training time depending on a number of training data. Thus, the contribution of this work is to present the intelligent hybrid system for faster training on neural network. The concept of the proposed method is applying DBSCAN for removing noise and outliers then selecting the represented instances to form a smaller training set for further model training. The experimental results indicate that the proposed method can dramatically reduce a size of training set while the predictive performance of the classifiers are better or almost the same as models trained with original training sets.

Keywords: data preprocessing, data reduction, data cleaning, DBSCAN, neural network, fast training.

1 Introduction

Neural network [1] is a machine learning algorithm that has been successful on data mining applications. Many real world data are modeled with neural network such as stock market [2-4], river flow forecast [5], and rainfall-runoff [6]. The advantages of neural network are high tolerance of noise, has ability to classify samples which it never been trained, and almost provides a better prediction performance than others learning algorithms especially for continuous-valued inputs and outputs. However, an important problem of neural network is long training time. If training data are more complex or larger, the training time is longer than a normal dataset. For this reason, the aim of this paper is to present an efficient method for reducing the size of training data by applying DBSCAN for removing outliers or noise and selecting represented data to generate a new smaller training dataset while the predictive performance of the trained models using preprocessed data are better or almost the same as original datasets.

The rest of this paper is organized as follows; Section 2 briefly introduces data preprocessing and overviews some researches using data preprocessing before construct a neural network model, Section 3 provides a basic idea of the DBSCAN algorithm, Section 4 describes the proposed hybrid system, Section 5 explains the experiment and gives the experimental results. Finally, the conclusion of this work is summarized in Section 6.

C. Guo, Z.-G. Hou, and Z. Zeng (Eds.): ISNN 2013, Part II, LNCS 7952, pp. 331–340, 2013.

2   Data Preprocessing in Neural Network

Data preprocess is an important task in data mining. This process can improve the performances of a classifier both in time complexity for model construction and model accurate. For basically, there are four major tasks of data preparation [7], which are data cleaning, data integration, data reduction, and data transformation.

Data cleaning is a useful step in data preprocessing. The tasks of this step are handling with missing values, detecting and removing outliers, and smoothing noise. Data integration relate to integrating multiple sources of data. This process may cause data inconsistency and data redundancy, thus in this step, redundant data must be detected and removed. Data reduction is the most frequently used in real world applications. It attempts to reduce the size of data by using two techniques: dimensionality reduction or numerosity reduction. For dimensionality reduction, it tries to reduce dimensions or attributes of data whereas numerosity reduction attempts to select some instances as the data representation of a training set. Finally, data transformation is used to transform data into a suitable format of specific mining algorithm such as nearest-neighbor, neural networks or clustering. The popular technique of data transformation is data normalization which attributes are scaled to a smaller range.

Backpropagation (BP) neural network is a popular model for various data mining applications. Most of the researches on neural network applications must use data preprocessing step before training a neural network model. In [8] a neural network model was applied for flow stress prediction with plane strain compression (PSC) data. In data preprocessing, Yang used data cleaning approach to remove noise based on data rationalizing and then tried to smooth out the remaining noise with mean average and median average filtering techniques.

For data reduction technique, there are many neural network application researches applied this step to improve a training set. In [9] used numerosity method by selecting minimal subset of the training set that would correctly classify the remaining samples from the training set by using the 1-NN algorithm. Other works applying numerosity method are [10] and [11]. At the first work [10], the k-mean clustering technique was used to select the representation of data to generate a new training set for financial time series prediction. For the second paper [11], there are two steps of data reduction. At the first step, dimensionality reduction was used to select a subset of attributes based on functional dependency, after that a numerosity technique was applied to remove the duplicated instances in a training set.

The examples of works using data transformation before training a neural network model are [12] and [13]. In [12], a neural network was applied to forecast the river flow located in Italy. The authors investigated the effects of data transformation on model performance. The techniques used as data transformation in this paper are wavelet transforms and data discretization. The results showed that the neural networks trained with data preprocessing can improve the model performance especially for using the data discretization technique that was obtained the best results. In [13], the application on fault detection and prediction in Boiler of power plant was addressed based on neural network. The proposed preprocessing process are data reduction which using an average method, data elimination; remove noise and missing values, data transformation which using a min-max normalization technique, and data clustering.

3   The DBSCAN Algorithm

DBSCAN (Density-Based Spatial Clustering of Applications with Noise) is a data clustering algorithm which uses density-based approach to find a group of data. This technique is different from partitioning or hierarchical clustering that it can find an arbitrary shape of clusters such as "S", oval, interlocking, or spiral shape. In addition, DBSCAN can detect noise or outlier points and discard them to form clusters.

There are two user-specified parameters in DBSCAN: *Eps* and *MinPts*. *Eps* is the maximum radius of a neighborhood, and *MinPts* is the minimum number of points in an *Eps*-neighborhood of that point. In DBSCAN, all points (objects or samples in a training set) are assigned into three types: core point, border point, or noise point. The core point is in the interior of a density-based cluster. A point is a core point if the *Eps*- neighborhood of the point contains at least *Minpts* points. The border point is a neighborhood of a core point, but it is not a core point. The noise point is a point that is neither a core point nor a border point. The pseudocode of the DBSCAN algorithm [7] is shown in Fig. 1.

Algorithm DBSCAN:
Input:
   - *D*: a dataset containing *n* objects
   - *Eps*: the radius parameter
   - *MinPts*: the neighborhood density threshold
Output: A set of density-based clusters
Method:
   mark all objects as **unvisited**;
   do
         randomly select an unvisited object *p*;
         mark *p* as **visited**;
         if the *Eps*-neighborhood of *p* has at least *MinPts* objects
               create a new cluster *C*, and add *p* to *C*;
               let *N* be the set of objects in the *Eps*-neighborhood of *p*;
               for each point *p′* in *N*
                     if *p′* is unvisited
                           mark *p′* as **visited**;
                           if the *Eps*-neighborhood of *p′* has at least *MinPts* points,
                              add those points to *N*;
                        if *p′* is not yet a member of any cluster, add *p′* to *C*;
               end for
               output *C*;
         else mark *p* as noise;
      until no object is unvisited

Fig. 1. The DBSCAN algorithm

4   The Proposed Hybrid System

This paper applies DBSCAN for data cleaning and data reduction. The prominent characteristic of DBSCAN is that it can discover outliers or noise. In addition, this method can find an arbitrary shape of clusters depending on the density of data without determine k value (where k is the number of the clusters) in advances. For these reasons, DBSCAN is a suitable technique to detect the outliers in a training dataset, and select the represented data to form a new training set.

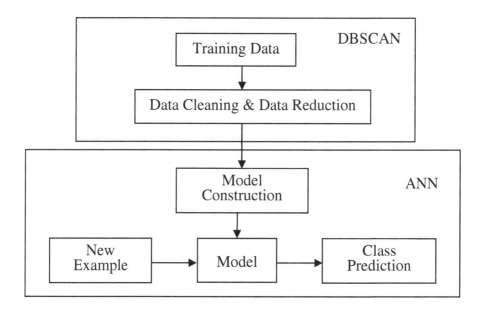

Fig. 2. An overall process of the proposed method

The proposed hybrid system as shown in Fig. 2 is a combination between DBSCAN and artificial neural network (ANN). The proposed method consists of two parts. The first part is to preprocess data using DBSCAN and the second step is to construct a classification model based on ANN. At data preprocessing phase, data cleaning and data reduction techniques are used to handle with all training data. In data cleaning, this step tries to detect and remove outliers using DBSCAN. At the same time, the output from DBSCAN can identify all border points of each cluster. Thus in data reduction, these border points are selected to generate a new training set which has a smaller size than the original training set. At model construction phase, ANN is trained based on the new training set from the previous step. Finally, the trained model is used to classify a new example.

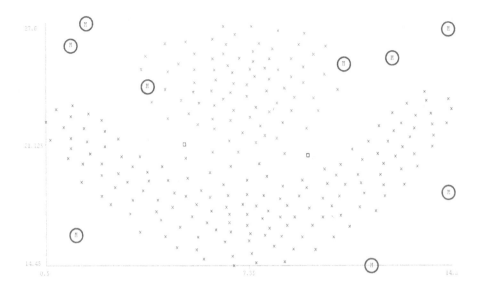

Fig. 3. Outlier detection by DBSCAN

Fig. 3 and 4 illustrate the results from running DBSCAN. In Fig. 3, data points marked with the circle are outliers detected by DBSCAN. The results from DBASCAN provided two clusters, and the border points of each cluster are shown in Fig. 4.

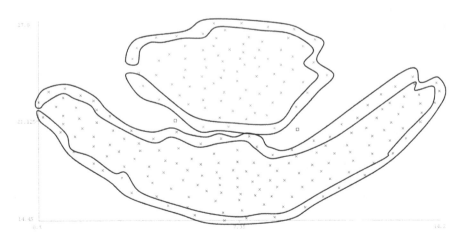

Fig. 4. Border points from DBSCAN

5   Experiments

The datasets using in the experiments for evaluating the performance of the proposed model are from http://cs.joensuu.fi/sipu/datasets. There are three datasets which are Jain, Flame, and Aggregation. The dataset characteristics are shown in Table1 and data visualization of all datasets are shown in Fig. 5.

Table 1. The characteristics of three datasets

Dataset	# Attribute	# Instance	# Class
Flame	2	240	2
Jain	2	373	2
Aggregation	2	788	7

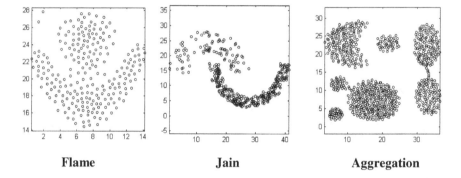

Flame   **Jain**   **Aggregation**

Fig. 5. Shape of datasets

In the experiments, several models of neural networks are constructed depending on these parameters: learning rate and generation. Learning rate is set to 0.01, 0.05, 0.1, and 0.5 respectively and the generation (epoch) is set to 100, 200, 300, 400 and 500. The topology of all neural network models is set to three layers. The number of input node of each model varies by the number of attribute in the dataset, and the number of output node is equal to the number of class. Finally, the number of hidden node is $(a+b)/2$ where a and b are the number of input and output node. Each model is evaluated by 5-fold cross-validation technique. In addition, to run the original dataset, noisy data are also added to all training sets with 5%, 10%, and 15%. The experimental results are shown in Table 2 and Table 3.

In Table 2, the accuracies of two methods are compared. The ANN column means the neural network models trained with the original dataset while the DBSCAN +ANN column represents the proposed models trained with data preprocessing step based on DBSCAN. The results in Table 2 show that most of the proposed models produced better predictive performance than the ANN model with no data preprocessing step. Only Jain dataset with no noise of the ANN model can perform a higher accuracy than the proposed method. And the result of two methods has the same accuracy for Aggregation dataset with 5% of noise.

Table 2. The predictive performance of two methods

Dataset	Add noise	ANN	DBSCAN+ ANN
Flame	0%	98.72	99.15
	5%	98.70	99.15
	10%	98.74	99.14
	15%	98.31	98.33
Jain	0%	95.70	94.90
	5%	95.20	95.47
	10%	96.18	97.00
	15%	96.51	96.78
Aggregation	0%	99.62	99.74
	5%	99.87	99.87
	10%	99.37	99.62
	15%	98.22	99.49

Table 3. Running time of each model

Dataset	Add noise	Running Time				
		ANN1	ANN2	Reducing time1 (%)	DBSCAN +ANN	Reducing time2 (%)
Flame	0%	0.104	0.05	51.92	0.0864	16.92
	5%	0.098	0.048	51.02	0.0844	13.88
	10%	0.106	0.050	52.83	0.0664	37.36
	15%	0.114	0.044	61.40	0.0870	23.68
Jain	0%	0.146	0.044	69.86	0.0908	37.81
	5%	0.170	0.068	60.00	0.1148	32.47
	10%	0.162	0.054	66.67	0.1008	37.78
	15%	0.166	0.052	68.67	0.0988	40.48
Aggregation	0%	1.966	0.372	81.08	0.4344	77.90
	5%	2.050	0.404	80.29	0.4664	77.25
	10%	2.078	0.378	81.81	0.4404	78.81
	15%	2.056	0.376	81.71	0.4384	78.68

In Table 3, the ANN1 column represents the training time of neural network using the original training dataset and the ANN2 column indicates the training time of neural network with a reduced size of training set using DBSCAN. The DBSCAN+ANN column represents the training time of ANN2 including with preprocessing time (DBSCAN running time in the preprocessing step). The "Reducing time1" column shows the percentage of time reduction of ANN2 when compared with ANN1 whereas the "Reducing time2" column indicates the percentage of time reduction of DBSCAN+ANN when compared with ANN1. Considering only the training time, the model trained with data preprocessing by using DBSCAN are extremely faster than the original training set for all datasets. The proposed method in data preprocessing can reduce the training time more than 50%, particularly on Aggregation datasets which can reduce the training time up to 80%. In addition, the total time of the proposed method (including data preprocessing time) is also faster than ANN1 for all datasets and the time reduction of DBSCAN+ANN is up to 78%.

The results in Table 4 show the percentage of data reduction using DBSCAN. The trend of the percentage of data reduction is increasing according to the volume of noise in the training sets except for Aggregation dataset. The experimental results in Table 4 indicate that using DBSCAN in the preprocessing step can reduce a huge number of instances yielding in faster training time while the predictive performance of the proposed models are greater than or almost the same as original neural network models.

Table 4. Percentage of data reduction

Dataset	Add noise	Data reduction (%)
Flame	0%	68.15
	5%	68.67
	10%	70.84
	15%	75.44
Jain	0%	75.54
	5%	77.89
	10%	77.78
	15%	79.43
Aggregation	0%	82.96
	5%	79.96
	10%	82.13
	15%	82.96

6   Conclusion

This work proposes the intelligent hybrid system for an efficient of neural network training. The main idea of the proposed method is to apply DBSCAN in data preprocessing step for removing outliers or noise and then to selecting the represented data from all training set. In the experiments, three standard datasets were used to run with the proposed system. For model predictive performance, 5-fold cross-validation technique was used to evaluate the model. The experimental results show that the new set of training generated by DBSCAN can reduce the size of the original dataset up to 82%. The sizes of the new training sets yield dramatically decrease in neural network training time. It can reduce training time up to 80% and up to 78% in total time (including data preprocessing step). In addition, the predictive performance of the proposed model is better or almost equal to the model trained with the original dataset.

References

1. Haykin, S.: Neural Networks and Learning Machines. Prentice-Hall, M.New Jersey (2008)
2. Lian, Y., Jiang, J.: Neural Network and its Application in the Stock Market Forecast. Journal Inner Mongolia University 17(5), 406–407 (2002)
3. Chaigusin, S., Chirathamjaree, C., Clayden, J.: The Use of Neural Networks in the Prediction of the Stock Exchange of Thailand (SET) index. In: The International Conference on Computational Intelligence for Modeling Control & Automation, pp. 670–673 (2008)
4. Sutheebanjard, P., Premchaiswadi, W.: Stock Exchange of Thailand Index Prediction using Back Propagation Neural Networks. In: The Second International Conference on Computer and Network Technology, pp. 377–380 (2010)
5. Baratti, R., Cannas, B., Fanni, A., Pintus, M., Sechi, G.M., Toreno, N.: River Flow Forecast for Reservoir Management through Neural Networks. Neurocomputing 55, 421–437 (2003)
6. Cannas, B., Montisci, A., Fanni, A., See, L., Sechi, G.M.: Comparing Artificial Neural Networks and Support Vector Machines for Modeling Rainfall-runoff. In: The 6th International Conference on Hydroinformatics. World Scientific Publishing Company (2004)
7. Han, J., Kamber, M., Pei, J.: Data Mining Concepts and Techniques, 3rd edn. Morgan Kaufmann, United States of America (2011)
8. Yang, Y.Y., Linkens, D.A.: Modeling of the Flow Stress Using BP Network. In: The Second International Conference on Intelligent Processing and Manufacturing of Materials, vol. 2, pp. 755–761 (1999)
9. Nedeljkovic, V., Milosavljevic, M.: On the Influence of the Training Set Data Preprocessing on Neural Networks Training. In: 11th IAPR International Conference on Pattern Recognition Methodology and Systems, vol. 2, pp. 33–36 (1992)
10. Zhang, J., Yang, Y.: BP Neural Network Model Based on the K-means Clustering to Predict the Share Price. In: The Fifth International Joint Conference on Computational Sciences and Optimization, pp. 181–184 (2012)

11. Wong, M.T., Geva, S., Orlowski, M.: Patten Recognition from Neural Network with Functional Dependency Preprocessing. IEEE Speech and Image Technologies for Computing and Telecommunications 1, 387–390 (1997)
12. Cannas, B., Fanni, A., See, L., Sias, G.: Data Preprocessing for River Flow Forecasting Using Neural Networks: Wavelet Transforms and data partitioning. Physics and Chemistry of the Earth 31, 1164–1171 (2006)
13. Rakhshani, E., Sariri, I., Rouzbehi, K.: Application of Data Mining on Fault Detection and Prediction in Boiler of Power Plant Using Artificial Neural Network. In: International Conference on Power Engineering, Energy and Electrical Drives, pp. 473–478 (2009)

EDA-Based Multi-objective Optimization Using Preference Order Ranking and Multivariate Gaussian Copula

Ying Gao, Lingxi Peng, Fufang Li, MiaoLiu, and Xiao Hu

Department of Computer Science and Technology, Guangzhou University
Guangzhou, 510006, P.R. of China
falcongao@sina.com.cn

Abstract. Estimation of distribution algorithms (EDAs) are a class of evolutionary optimization algorithms based on probability distribution model. This article extends the basic EDAs for tackling multi-objective optimization problems by incorporating multivariate Gaussian copulas for constructing probability distribution model, and using the concept of preference order. In the algorithm, the multivariate Gaussian copula is used to construct probability distribution model in EDAs. By estimating Kendall's τ and using the relationship of correlation matrix and Kendall's τ, correlation matrix R in Gaussian copula are firstly estimated from the current population, and then is used to generate offsprings. Preference order is used to identify the best individuals in order to guide the search process. The population with the current population and current offsprings population is sorted based on preference order, and the best individuals are selected to form the next population. The algorithm is tested to compare with NSGA-II, GDE, MOEP and MOPSO based on convergence metric and diversity metric using a set of benchmark functions. The experimental results show that the algorithm is effective on the benchmark functions.

Keywords: EDA, Multi-objective optimization, Pareto optimal, Preference order ranking, Multivariate Gaussian copula.

1 Introduction

In various fields of science and technology, optimization problems have two or more objectives that we wish to optimize simultaneously. These are called *multi-objective optimization problems* (MOPs), and their solution involves the design of algorithms different from those adopted for dealing with single-objective optimization problems. In the absence of preference information, there does not exist a unique or straightforward way to determine if a solution is better than other in multi-objective optimization. The notion of optimality most commonly adopted is the one called *Pareto optimality* which leads to trade-offs among the objectives in MOPs. The solution of MOPs is usually a set of acceptable trade-off optimal solutions. The solution set is called the *Pareto optimal set*. Compared with traditional algorithms,

C. Guo, Z.-G. Hou, and Z. Zeng (Eds.): ISNN 2013, Part II, LNCS 7952, pp. 341–350, 2013.

evolutionary algorithms are more suitable for solving MOPs. In the last decade, many evolutionary algorithms have been widely developed to solve MOPs, such as Non-dominated Sorting Genetic Algorithm (NSGA-II)[1], and Strength Pareto Evolutionary Algorithm (SPEA2) [2], Pareto Archived Evolution Strategy (PAES)[3], Multi-objective optimization with artificial weed colonies[4], Generalized Differential Evolution(GDE) [5], Multi-Objective Evolutionary Algorithm based on Decomposition (MOEAD)[6], Multi-Objective Evolutionary Programming (MOEP)[7], Multi-Objective Particle Swarm Optimization(MOPSO) [8], etc.

Recently, a class of novel evolutionary algorithm, called estimation of distribution algorithm (EDA) [9], has become a favorite topic in the field of evolutionary computation. With EDA an entirely new paradigm of evolutionary computation has been introduced, which is a combination of statistical learning theory and stochastic optimization algorithm without using conventional evolutionary operators such as crossover and mutation. EDA tries to estimate the probabilistic distribution of an entire population or to describe its evolutionary trend directly from a macroscopic point of view. It has been proven that EDA has some special characteristics of concise concept, good global searching ability and been successfully extended to multi-objective optimization problems[10-11]. The performance of an EDA highly depends on how well it estimates and samples the probability distribution. A wide variety of EDAs using probabilistic graphical modeling techniques[12-14] to estimate and sample the probability distribution have been proposed and are the subject of active research. However, EDA using probabilistic graphical modeling techniques generally spend too much time on the learning about the probability distribution of the promising individuals.

Since the introduction of copulas by Sklar as a tool for constructing multivariate distributions they have become more popular[15-16]. According to copula theory, a joint probability distribution can be decomposed into n marginal probability distributions and a copula function. So, the joint probability distribution of multivariate can be constructed utilizing a copula function and the marginal probability distributions of every variable. Gaussian copula is a member of the Elliptical copulas family which has attracted particular interest since they have a number of properties which make them simple to analyze. It can be applied to constitute the probabilistic model in multi-objective EDAs. In this paper, EDA is extended to multi-objective optimization problems by using preference order ranking and multivariate Gaussian copula. The algorithm employs multivariate Gaussian copulas to construct probability distribution model in EDA. By estimating Kendall's τ and using the relationship of correlation matrix and Kendall's τ, correlation matrix R in Gaussian copula are firstly estimated from the current population, and then is used to generate offsprings. And instead of Pareto dominance, preference order is used to identify the best individuals in order to guide the search process. The population with the current population and current offsprings population is sorted based on preference order, and the best individuals are selected to form the next population. The proposed algorithm is tested to compare with NSGA-II[1], GDE[5], MOEP[7] and MOPSO[8] using a set of benchmark functions. Convergence metric, diversity metric are used to evaluate the performance of the algorithm. The experimental results show that the algorithm is effective on the benchmark functions.

2   Background

2.1   Multi-objective Optimization and Pareto Dominance

The multi-objective optimization problem can be formally defined as the problem of finding all $\mathbf{x} = (x_1, x_2, \cdots, x_n)$ which satisfy the m inequality constraints:

$$g_i(\mathbf{x}) \leq 0, \quad i = 1, 2, ..., m$$

p equality constraints: $h_i(\mathbf{x}) = 0, \quad i = 1, 2, ..., p$

and optimize the vector function: $\min \mathbf{f}(\mathbf{x}) = \left(f_1(\mathbf{x}), f_2(\mathbf{x}), \cdots, f_k(\mathbf{x}) \right)$

where $\mathbf{x} = (x_1, x_2, \cdots, x_n)$ is an n-dimensional decision variable vector. The constraints define the feasible region Ω and any vector $\mathbf{x}$ in the feasible region Ω is called a feasible solution. $\Pi = \{ \mathbf{y} | \mathbf{y} = \mathbf{f}(\mathbf{x}), \mathbf{x} \in \Omega \}$ is referred to as objective space. The concept of optimum commonly adopted in multi-objective optimization is Pareto dominance.

A feasible solution $\mathbf{x}_1 \in \Omega$ is said to strictly dominate another feasible solution $\mathbf{x}_2 \in \Omega$, denoted by $\mathbf{x}_1 \prec \mathbf{x}_2$, if and only if (iff) $\forall i \in \{1, 2, \cdots, k\}: f_i(\mathbf{x}_1) \leq f_i(\mathbf{x}_2)$ and $\exists j \in \{1, 2, \cdots, k\}: f_j(\mathbf{x}_1) < f_j(\mathbf{x}_2)$.

A feasible solution $\mathbf{x} \in \Omega$ is said to be Pareto optimal with respect to Ω iff there is no other feasible solution that dominates $\mathbf{x}$ in Ω. The set of all Pareto optimal solutions in the feasible region Ω is called Pareto optimal set and the corresponding set of objective vector is called Pareto optimal front. An illustrative example of a multi-objective minimization problem with two objectives, f_1 and f_2, that are plotted in the objective space Π mapped from the feasible region Ω is shown in Figure 1. The bold part in the feasible region Ω indicates the Pareto optimal set. The bold curve in the objective space Π indicates the Pareto front.

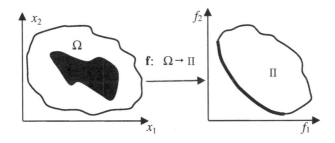

Fig. 1. An illustrative example

2.2   Preference Order

Among the multi-objective methods, the majority of research is concentrated on Pareto-based approaches. Pareto-based methods select a part of individuals based on the Pareto dominance notion as leaders. These methods use different approaches to select non-dominated individuals as leaders which are maintained in an external archive[17]. Preference order[18] is a generalization of Pareto optimality. It provides a way to designate some Pareto solutions superior to others when the size of the non-dominated solution set is very large. An individual $\mathbf{x}$ is considered efficient of order k if it is not Pareto-dominated by any other individual for any of the k-element subspaces where are considered only k objectives at a time. Efficiency of order M for a MOP with exactly M objectives simply corresponds to the original Pareto optimality definition. If $\mathbf{x}$ is efficient of order k, then it is efficient of order $k+1$. Analogously, if $\mathbf{x}$ is not efficient of order k, then it is not efficient of order $k-1$. Efficiency of order can be used to reduce the number of points in a non-dominated set by retaining only those regarded as the "best compromise" [19].

2.3   Multivariate Gaussian Copula

A copula is a distribution function with known marginals. Any continuous Multivariate joint distribution of n random variables $x_1, x_2, ..., x_n$

$F(x_1, x_2, ..., x_n) = \text{Prob}\{X_1 \leq x_1, X_2 \leq x_2, ..., X_n \leq x_n\}$, can be represented by a copula C as a function of the marginal distribution $F_{X_i}(x_i) = \text{Prob}\{X_i \leq x_i\}, i = 1, 2, \cdots, n$;   i.e.

$$F(x_1, x_2, ..., x_n) = C(F_1(x_1), F_2(x_2), ..., F_n(x_n)) \overset{\Delta}{=} C(u_1, u_2, ..., u_n) \qquad (1)$$

Where $u_i = F_{X_i}(x_i), i = 1, 2, \cdots, n$ and $C(u_1, u_2, ..., u_n)$ is the associated copula function. Furthermore, application of the chain rule shows that the corresponding density function $f(x_1, x_2, ..., x_n)$ can be decomposed as

$$f(x_1, x_2, ..., x_n) = \frac{\partial^n C(u_1, u_2, ..., u_n)}{\partial u_1 \partial u_2 ... \partial u_n} \overset{\Delta}{=} c(u_1, u_2, ..., u_n) \cdot f_1(x_1) \cdot f_2(x_2) \cdot ... \cdot f_n(x_n) \qquad (2)$$

From the above it may be seen that the joint density function is the product of the marginals $f_i(x_i), i = 1, 2, \cdots, n$ and copula densities function $c(u_1, u_2, ..., u_n)$.

Let $\mathbf{R} = (r_{i,j}), i = 1, 2, \cdots, n, j = 1, 2, \cdots, n$ be a symmetric, positive definite matrix with unit diagonal entries. The *multivariate Gaussian copula* is defined as

$$C(u_1, u_2, ..., u_n; \mathbf{R}) = \Phi_{\mathbf{R}}(\varphi^{-1}(u_1), \varphi^{-1}(u_2), \cdots, \varphi^{-1}(u_n)) \qquad (3)$$

Where $\Phi_{\mathbf{R}}$ denotes the standardized multivariate normal distribution with correlation matrix $\mathbf{R} = (r_{i,j}), i = 1, 2, \cdots, n, j = 1, 2, \cdots, n$. $\varphi^{-1}(x)$ denotes the inverse of the univariate standard normal distribution $\varphi(x)$.

The corresponding density is

$$c(u_1, u_2, ..., u_n; \mathbf{R}) = \frac{1}{|\mathbf{R}|^{1/2}} \exp\left(-\frac{1}{2}\omega^T(\mathbf{R}^{-1} - \mathbf{I})\omega\right) \qquad (4)$$

with $\omega = (\varphi^{-1}(u_1), \varphi^{-1}(u_2), \cdots, \varphi^{-1}(u_n))^T$

3   The Proposed Algorithm

3.1   Scheme of Algorithm

The proposed algorithm is designed based on EDA. EDA estimate a probability distribution over the search space, and then sample the offspring individuals from this distribution. The main calculation procedure of the EDA is that (1) firstly, the M selected individuals are selected from the population in the previous generation. (2) Secondly, the probabilistic model is estimated from the genetic information of the selected individuals. (3) A new population with size N is then sampled by using the estimated probabilistic model. (4) Finally, the new population is evaluated. (5) Steps (1)-(4) are iterated until stopping criterion is reached.

EDAs are undoubtedly a powerful search tool for solving single objective optimization problems. However, the original scheme has to be modified for solving multi-objective optimization problems. As we saw in Section 2, the solution set of a problem with multiple objectives does not consist of a single solution. Instead, in multi-objective optimization, we aim to find Pareto optimal set. In the proposed algorithm, besides using *multivariate Gaussian copula* for constructing probability distribution model in the multi-objective EDA, we have used the preference order ranking for selecting the next population. The algorithm steps are as follows:

$t=0$, Initialization
1) Generate initial population P(0)with size N;
2) Estimate the correlation matrix of Gaussian copula from P(0);
3) Sample Gaussian copula to generate $P'(0)$ with size N;

WHILE $t < T$ DO
1) $S(t) \leftarrow P(t) \cup P'(t)$
2) Rank $S(t)$ using preference order ranking algorithm;
3) Select M best individuals from $S(t)$ to form $Q(t)$;
4) Estimate the correlation matrix of Gaussian copula from $Q(t)$;
5) Sample Gaussian copula to generate $P'(t)$ with size N;
6) $t \leftarrow t + 1$
END WHILE
Output the obtained Pareto optimal set;

3.2   Initialization

The population size N, M and the maximum number of iterations T are initialized according to the problem concerned. The individuals in P(0) are initialized randomly. The individuals in $P'(0)$ are generated by sampling Gaussian copula.

3.3   Preference Order Ranking Algorithm

According to preference order properties, the order of efficiency of an individual $\mathbf{x}$ is the minimum k value for which $\mathbf{x}$ is efficient. Formally:

$$order(\mathbf{x}) = \min_{k=1}^{M} \left(k | \text{isEfficient}(\mathbf{x},k) \right)$$

Where isEfficient($\mathbf{x}$, k) is to be true if $\mathbf{x}$ is efficient of order k. The order or efficiency can be used as the rank of individuals. The smaller the order of efficiency, the better an individual is. The resulting ranking scheme is as follows[20]:

1). Identify the Pareto non-dominated solutions of P and group them into the subset $Q^{(1)}$, which is given rank 1.
2). Assign to the individuals of $Q^{(1)}$ a rank according to preference order and the worst given rank is w.
3). Identify the Pareto non-dominated individuals of $P \backslash Q^{(1)}$ and group them into the subset $Q^{(2)}$, which will be given rank $w+s$.
4). Iterate 2) and 3) until $P \backslash Q^{(s)}$, where $Q^{(s)}$ is the subset that contains the worst individuals.

3.4   Estimating Correlation Matrix of Gaussian Copula

A simple method[16] based on Kendall's tau for estimating the correlation matrix $\mathbf{R}$ in Gaussian copula. The method consists of constructing an empirical estimate of Kendall's *tau* for each bivariate margin of the copula and then using relationship (5)

$$\tau(\mathbf{x}_k, \mathbf{x}_m) = \frac{2}{\pi} \arcsin(r_{k,m}), (k, m = 1, ..., N) \tag{5}$$

To infer an estimate of the relevant element of $\mathbf{R}$. More specifically we estimate $\tau(\mathbf{x}_k, \mathbf{x}_m)$ by calculating the standard sample c coefficient

$$\hat{\tau}(\mathbf{x}_k, \mathbf{x}_m) = \frac{2}{n(n-1)} \sum_{1 \le i < j \le n} \text{sign}[(x_{i,k} - x_{j,k})(x_{i,m} - x_{j,m})] \tag{6}$$

From the original data samples $\mathbf{x}_i = (x_{1,i}, x_{2,i}, ..., x_{n,i}), (i = 1, ..., N)$, and write the jth component of the ith sample as $x_{i,j}$; this yields an unbiased and consistent.

An estimator of $r_{k,m}$ is then given by $\hat{r}_{k,m} = \sin(\frac{\pi}{2} \hat{\tau}(\mathbf{x}_k, \mathbf{x}_m))$. In order to obtain an estimator of the entire matrix $\mathbf{R}$ we can collect all pairwise estimates $\hat{r}_{k,m}$ and then construct the estimator

$$\hat{\mathbf{R}} = \left(\sin(\frac{\pi}{2} \hat{\tau}(\mathbf{x}_k, \mathbf{x}_m)) \right), k = 1, \cdots N, m = 1, \cdots, N \tag{7}$$

3.5   Sampling Gaussian Copula

To simulate a random vector $(u_1, u_2,..., u_n)$ distributed according to a Gaussian copula with correlation matrix $\mathbf{R}$, we start with the canonical representation:

$$C(u_1, u_2,..., u_n; \mathbf{R}) = \mathbf{\Phi}_\mathbf{R}(\varphi^{-1}(u_1), \varphi^{-1}(u_2), \cdots, \varphi^{-1}(u_n))$$

Where $\mathbf{\Phi}_\mathbf{R}$ is the c.d.f. of an n-variate normal distribution function with linear correlation matrix $\mathbf{R}$ and $\varphi(x)$ is the univariate standard normal c.d.f.. A multivariate random numbers $\mathbf{x} = (x_1, x_2,..., x_n)$ with distribution function $F_\mathbf{x}$ defined by assigning marginals $F_1, F_2, \cdots, F_n$ and a Gaussian copula function $\mathbf{\Phi}_\mathbf{R}$ can be generated as follows[16].

(1) Find the Cholesky decomposition of $\mathbf{R}$, so that $\mathbf{AA}^T = \mathbf{R}$, with $\mathbf{A}$ lower triangular;
(2) Generate a sample of n independent random variables $(z_1, z_2,..., z_n)$ from $N(0, 1)$;
(3) Set $\mathbf{Y} = \mathbf{AZ}$ with $\mathbf{Z} = (z_1, z_2,..., z_n)^T$ and $\mathbf{Y} = (y_1, y_2,..., y_n)^T$; and finally,
(4) Return $\mathbf{x} = (x_1, x_2,..., x_n) = (F_1(y_1), F_2(y_2),..., F_n(y_n))$.

4   Experimental Results

To validate efficiency of the proposed algorithm(EDAMOPGC), the performance of the algorithm is compared with that of some multi-objective optimization algorithms. The two common metrics used to compare are (1) convergence metric, (2) divergence metric. For these metrics we need to know the true Pareto front for a problem. In our experiments we use 1000 uniformly spaced Pareto optimal solutions as the approximation of the true Pareto front. An introduction of these metrics is given here:

Convergence metric γ was proposed by Deb[1], measures the distance between the obtained non-dominated front NF and optimal Pareto front PF. Mathematically, it may be defined as:

$$\gamma = \frac{\sqrt{\sum_{i=1}^{N} d_i^2}}{N} \tag{8}$$

Where N is the number of non-dominated solutions found by the algorithm being analyzed and d_i is the minimum Euclidean distance (measured in the objective space) between the ith solution of NF and the solutions in PF.

Diversity metric Δ was proposed by Deb[1]. It measures the extent of spread achieved among the obtained solutions and is defined as:

$$\Delta = \frac{d_f + d_i + \sum_{i=1}^{s-1}|d_i - \overline{d}|}{d_f + d_i + (s-1)\overline{d}} \quad \text{where} \quad \overline{d} = \frac{\sum_{i=1}^{s-1} d_i}{s-1} \tag{9}$$

Where s is the number of members in the set of non-dominated solution found so far, and the parameter d_f and d_i are the Euclidean distances between the extreme solutions and the boundary solutions of the obtained non-dominated set. The parameter $\overline{d}$ is the average of all distances $d_i (i = 1,...,s-1)$, assuming that there are s solutions on the best non-dominated front.

Five test problems as examples are used to compare with the performance of NSGAII and MOPSO. Test problems are chosen from a number of significant studies in this area, including Schaer's study (SCH1 and SCH2) [21] and Zitzler's test set (ZDT1, ZDT2, ZDT4) [22].

The smaller the value of these metrics is, the better the performance of the algorithm is. The initial population was generated from a uniform distribution in the ranges specified below. Population size N=100. All experiments were repeated for 40 runs. The maximum number of iterations is set to 5000 in each running. The solutions accepted after iteration processes are used to calculate the performance metrics. The results of the performance metrics are shown in Tables1. Tables1 listed the mean and variances of the convergence and diversity metrics obtained using NSGA-II[1], GDE[5], MOEP[7], MOPSO[8] and the proposed algorithm on the multi-objective

Table 1. Mean and variances of two metrics for the multi-objective benchmark functions

Problem	Algorithm	Diversity Δ (mean ± variances)	Convergence γ (mean ± variances)
SCH1	NSGA-II	0.5792 ± 0.0113	0.0043 ± 0.0076
	GDE	0.7204 ± 0.0674	0.0072 ± 0.0039
	MOEP	0.9561 ± 0.0357	0.0095 ± 0.0014
	MOPSO	0.4915 ± 0.0134	0.0063 ± 0.0047
	EDAMOPGC	0.5006 ± 0.0108	0.0041 ± 0.0052
SCH2	NSGA-II	0.6932 ± 0.0129	0.0068 ± 0.0017
	GDE	0.7102 ± 0.0262	0.0064 ± 0.0026
	MOEP	0.5912 ± 0.0274	0.0078 ± 0.0013
	MOPSO	0.6018 ± 0.0136	0.0032 ± 0.0038
	EDAMOPGC	0.5843 ± 0.0203	0.0029 ± 0.0024
ZDT1	NSGA-II	0.5403 ± 0.0149	0.2033 ± 0.0069
	GDE	0.7283 ± 0.0171	0.3029 ± 0.0053
	MOEP	0.4807 ± 0.0476	0.4692 ± 0.0046
	MOPSO	0.3092 ± 0.0263	0.2015 ± 0.0017
	EDAMOPGC	0.3011 ± 0.0285	0.2029 ± 0.0002
ZDT2	NSGA-II	0.5341 ± 0.0146	0.1266 ± 0.0064
	GDE	0.6813 ± 0.0743	0.3705 ± 0.0039
	MOEP	0.6237 ± 0.0521	0.4709 ± 0.0024
	MOPSO	0.2948 ± 0.0326	0.3206 ± 0.0018
	EDAMOPGC	0.2209 ± 0.0372	0.2021 ± 0.0006
ZDT4	NSGA-II	0.6725 ± 0.0321	0.2062 ± 0.0029
	GDE	0.6254 ± 0.0637	0.3205 ± 0.0016
	MOEP	0.7318 ± 0.0183	0.4228 ± 0.0077
	MOPSO	0.5034 ± 0.0141	0.2153 ± 0.0058
	EDAMOPGC	0.5102 ± 0.0036	0.1924 ± 0.0029

benchmark functions SCH1, SCH2, ZDT1, ZDT2 and ZDT4. It can be known from Tables1, in both aspects of convergence and distribution of solutions, the proposed algorithms is effective on the benchmark functions.

5   Conclusions

We proposed an EDA-based multi-objective optimization algorithm using preference order ranking and multivariate Gaussian copula. The algorithm applies the multivariate Gaussian copulas to construct probability distribution model in EDA, and the new individuals are generated according to the probability distribution model by sampling Gaussian copula. The preference order ranking is incorporated into the algorithm for forming the next population. The proposed algorithm is tested to compare with NSGA-II, GDE, MOEP and MOPSO using a set of multi-objective benchmark functions SCH1, SCH2, ZDT1, ZDT2 and ZDT4. Both convergence and diversity metrics are used to evaluate the performance of the algorithm. The experimental results show that the algorithm is effective in two metrics. Future work for our investigation will focus on the design of algorithms with other copulas.

Acknowledgments. This work is supported by the Scientific and Technological Innovation Projects of Department of Education of Guangdong Province, P.R.C. Grant No.2012KJCX0082 and Science and Technology Projects of Guangdong Province, P.R.C. under Grant No.2011B090400623, 2012B091100337, Guangzhou Science and Technology Projects under Grant No.12C42011563, 11A11020499. Authors of the paper express great acknowledgment for these supports.

References

1. Deb, K., Agrawal, S., Pratap, A., Meyarivan, T.: A fast and elitist multi-objective genetic algorithm: NSGA-II. IEEE Transactions on Evolutionary Computation 6(2), 182–197 (2002)
2. Zitzler, E., Laumanns, M., Thiele, L.: SPEA2: Improving the strength pareto evolutionary algorithm for multi-objective optimization. In: Proceedings of the EUROGEN 2001 Conference, pp. 95–100 (2001)
3. Knowles, J.D., Corne, D.W.: The Pareto archived evolution strategy: A new baseline algorithm for Pareto multi-objective optimization. In: Proc. of the 2003 IEEE Conf. on Evolutionary Computation, pp. 98–105 (1999)
4. Kundu, D., Suresh, K., Ghosh, S., Das, S., Panigrahi, B.K., Das, S.: Multi-objective optimization with artificial weed colonies. Information Sciences 181(12), 2441–2454 (2011)
5. Kukkonen, S., Lampinen, J.: Performance assessment of generalized differentia evolution with a given set of constrained multi-objective test problems. In: Proceeding of Congress on Evolutionary Computation, CEC 2009, pp. 1943–1950 (2009)
6. Zhang, Q., Liu, W., Li, H.: The performance of a new version of MOEA/D on CEC 2009 unconstrained mop test instances. In: Proceeding of Congress on Evolutionary Computation, CEC 2009, pp. 203–208 (2009)

7. Qu, B.Y., Suganthan, P.N.: Multi-objective evolutionary programming without non-domination sorting is up to twenty times faster. In: Proceeding of Congress on Evolutionary Computation, CEC 2009, pp. 2934–2939 (2009)

8. Coello Coello, C.A., Pulido, G.T., Lechuga, M.S.: Handling multiple objectives with particle swarm optimization. IEEE Transactions on Evolutionary Computation 8(3), 256–279 (2004)

9. Larranaga, P., Lozano, J.A.: Estimation of Distribution Algorithms: A New Tool for Evolutionary Computation. Kluwer Academic Publishers, Dordrecht (2002)

10. Pelikan, M., Sastry, K., Goldberg, D.E.: Multi-objective Estimation of Distribution Algorithms. Studies in Computational Intelligence 33, 223–248 (2006)

11. Sastry, K., Pelikan, M., Goldberg, D.E.: Limits of scalability of multiobjective estimation of distribution algorithms. In: Proceedings of the Congress on Evolutionary Computation, pp. 2217–2224 (2005)

12. Etxeberria, R., Larrañaga, P.: Global optimization using Bayesian networks. In: Ochoa, A., Soto, M.R., Santana, R. (eds.) Proceedings of the Second Symposium on Artificial Intelligence, CIMAF 1999, Havana, Cuba, pp. 151–173 (1999)

13. Khan, N.: Bayesian optimization algorithms for multi-objective and hierarchically difficult problems. Master's thesis, University of Illinois at Urbana-Champaign, Urbana, IL (2003)

14. Shakya, S., McCall, J.: Optimisation by Estimation of Distribution with DEUM framework based on Markov Random Fields. International Journal of Automation and Computing 4, 262–272 (2007)

15. Nelsen, R.B.: An Introduction to copula. Springer, New York (1998)

16. Cherubini, U., Luciano, E., Vecchiato, W.: Copula Methods in Finance. John Wiley & Sons Ltd., England (2004)

17. Deb, K.: Multi-Objective Optimization Using Evolutionary Algorithms. John Wiley & Sons (2001)

18. Das, I.: A preference ordering among various Pareto optimal alternatives. Structural and Multidisciplinary Optimization 18(1), 30–35 (1999)

19. di Pierro, F., Khu, S.T., Savi, D.A.: An investigation on preference order ranking scheme for multiobjective evolutionary optimization. IEEE Transactions on Evolutionary Computation 11(1), 17–45 (2007)

20. Jaimes, A.L., Quintero, L.V.S., Coello, C.A.C.: Ranking methods in many-objective evolutionary algorithms. In: Chiong, R. (ed.) Nature-Inspired Algorithms for Optimisation. SCI, vol. 193, pp. 413–434. Springer, Heidelberg (2009)

21. Schaffer, J.D.: Multiple Objective Optimization with Vector Evaluated Genetic Algorithms. In: 1st International Conference on Genetic Algorithms, pp. 93–100 (1985)

22. Zitzler, E., Deb, K., Thiele, L.: Comparison of multi-objective evolutionary algorithms: Empirical results. Evolutionary Computation 8, 173–195 (2000)

Resource Scheduling of Cloud with QoS Constraints

Yan Wang, Jinkuan Wang, Cuirong Wang, and Xin Song

College of Information Science and Engineering
Northeastern University, Shenyang, 110004, China
holdwangyan@gmail.com

Abstract. According to the dynamic, distribution and complexity of cloud computing, resource scheduling effectively with users' QoS demand and achieving maximum benefit is the unprecedented challenge. To solve the above problem, we propose to use genetic algorithm: design for the crossover operator and build a cloud resource optimization scheduling model that promised to address user needs while optimizing resource allocation. With the experiments, this paper verifies the superiority of models made in this paper. The results show that the use of genetic algorithm to optimize cloud resource scheduling has the rationality and feasibility. Meanwhile, using the genetic algorithm is useful for effectively scheduling of cloud resource meeting the users' QoS.

Keywords: cloud computing, QoS constraints, resource scheduling.

1 Introduction

In just a few years, cloud computing has become a very popular paradigm and a business success story. Having the characteristics of providing resource promptly, fewer inputs, less interaction with service providers and providing users with a powerful resource pool to meet their demand for resources, cloud computing resource allocation and scheduling has become the focus of cloud computing research.

Nowadays, a lot of methods of cloud resource allocation are introduced. In [1-3], economics, game theory, auction model are used for analysis of cloud resource allocation problem. In [4,5] the interests of the cloud resource providers and the pricing of cloud resources are improved. However, due to the complexity of the cloud computing environment, one aspect of optimization and improvement does not satisfy the cloud resource scheduling requirements, some algorithms consider scheduling performance improvements, but ignore the load balancing [6,7].Therefore, scheduling algorithms and models about the cloud resource load balancing are proposed in [8,9], and solutions of load balancing about cloud resource in business application are included in [10], but, the balance in these papers doesn't fully consider the needs of users' QoS.

The emphasis in cloud computing is on-demand service model. When allocating resource to users' requests, many kinds of resources are proposed. While providing multiple resources to meet users' needs, such as to satisfy the user for certain service proposed response time, energy consumption and cost, it needs integration, collaboration and optimize the execution in the cloud computing.

C. Guo, Z.-G. Hou, and Z. Zeng (Eds.): ISNN 2013, Part II, LNCS 7952, pp. 351–358, 2013.

In this paper, with the superiority of genetic algorithm in solving the multi-objective optimization, we introduce the resource scheduling policy and implementation process for cloud resource satisfying the QoS constraints. In addition, we redesign the operator which affects the genetic algorithm convergence, and establish the scheduling model for cloud resource optimization scheduling. Finally, experiments are carried out to verify the design of genetic algorithm in the convergence speed and efficiency, which has certain advantages comparing to the basic genetic algorithm.

The remainder of this paper is organized as follows: section 2 describes the model structure and the design of the objective function, chromosome population initialization, the choice of the fitness function, and genetic manipulation of design. Evaluations are given in section 3 and conclusion is in the final section.

2   Model Structure

2.1   Description of the Problem

For cloud resource scheduling problem, we give the following assumptions:

1. The users' requirements are divided into multiple sub-tasks, the granularity of sub-tasks is uniform, and these subtasks are independent task.

2. Using the batch mode, that can give reasonable task mapping strategy taking advantage of the ample resources.

3. The time of each user's requirements in the standard resources to run is known.

2.2   Objective Function

For the emphasis of cloud service, we propose the standard for evaluation scheduling is as follows:

$$Q = \{R, E, C\} \tag{1}$$

Here, R, E, C indicate response time, energy consumption, cost, respectively.

The created model is based on the goal that the effectiveness has the maximum value while various users' target should be achieved after the cloud resources scheduling, as in order to evaluate a variety of resources, we need to establish the objective function, first, we should define a 0-1 variable.

1. 0-1 Variable

$$D_{ij} = \begin{cases} 1 & resource\ \ i\ is\ assigned\ to\ user\ j \\ & i = 1, 2, ..., n \quad j = 1, 2, ..., m \\ 0 & else \end{cases} \tag{2}$$

When the value is 1 in formula (2), it indicates the resource i is assigned to the user j, and 0 otherwise.

2. Design of Objective Function

To ensure the response time, energy consumption and cost, set each objective function as follows:

$$R = \min(\sum_{i=1}^{n}\sum_{j=1}^{m} R_{ij} D_{ij}) \quad i=1,2,...,n \quad j=1,2,...,m \tag{3}$$

$$E = \max(\sum_{i=1}^{n}\sum_{j=1}^{m} E_{ij} D_{ij}) \quad i=1,2,...,n \quad j=1,2,...,m \tag{4}$$

$$C = \min(\sum_{i=1}^{n}\sum_{j=1}^{m} C_{ij} D_{ij}) \quad i=1,2,...,n \quad j=1,2,...,m \tag{5}$$

The constraint condition is:

$$n \le \sum_{i=1}^{n}\sum_{j=1}^{m} D_{ij} \le 2n \quad i=1,2,...,n \; j=1,2,...,m \tag{6}$$

In the expressions of (3) - (5), R_{ij} indicates the response time that the resource i is assigned to the user j, E_{ij} represents the power consumption when the resource i providing service for the user j, C_{ij} represents the costs of services provided by the resource i for user j.

We use the single goal as the goal of genetic algorithm optimization function, and preferences of users to be considered for each target, using linear weighted way to construct the objective function, as follows:

$$G = \lambda_1 R + \lambda_2 E + \lambda_3 C \tag{7}$$

$\lambda_i \, (i=1,2,3)$ as a weighted value, and $\lambda_1 + \lambda_2 + \lambda_3 = 1$.

2.3   Chromosome

The design of chromosomes includes chromosome encoding and decoding. Based on the comparison of binary encoding and floating-point encoding, as well as the characteristics of the cloud resources scheduling, this paper uses binary encoding.

1. Encoding Process

Assumptions:

a. Resource pool has n resources available for scheduling, there are m users.($m < n$).

b. Each resource can be assigned to any user for service, and each user can also accept two resources to provide service.

c. In one scheduling, each resource can be assigned only once.

Table 1. Encoding corresponding

user 1	user 2	user 3	...	user m
$u_{11}u_{12}...u_{1n}$	$u_{21}u_{22}...u_{2n}$	$u_{31}u_{32}...u_{3n}$	...	$u_{m1}u_{m2}...u_{mn}$

In table 1, u_{mn} indicates resource n is assigned to user m, for example, eight available resources can be allocated to the two users, randomly generated chromosome is encoded as: {10100000, 00010000}. This indicates resource 1 and resource 3 are allocated to user 1, and resource 4 is allocated to user 2.

2. Decoding Process

Every eight bits set to one, indicates one user, and 1 as the starting number, then, {10100000, 00010000} means that user 1 shares resource 1 and resource 3 meanwhile, user 2 shares resource 4.

2.4   Populations Initialization and the Selection of the Fitness Function

1. Population Initialize

The initial population is the search space of iterative evolution, also, it is generated by the random function. Taking into account the search speed and convergence of the algorithm, we use a randomly generated initial population M. In each iteration, the population size is not changed.

2. Selection of the Fitness Function

In this article, the goal of the objective function is to make the benefits maximum. The fitness function is as follows:

$$F = \begin{cases} G - T_{\min} & ,G > T_{\min} \\ 0 & \end{cases} \tag{8}$$

$T_{\min}$ is the estimated value of lower limit to the objective function

2.5   Genetic Manipulation Design

1. Select the Operating

Select or copy operation is to determine which individual can enter the next generation. Roulette selection method is chosen in this article. Selection formulas:

$$P_i = \frac{f_i}{\sum_{k=1}^{M} f_k} \tag{9}$$

f_i is the fitness of individuals in one population, P_i is the probability for individual i to be selected.

Selecting step is as follows:

a. In the tth generation, $\sum_{k=1}^{M} f_k$ and P_i is calculated by the formula (9).

b. Generates a random number of [0,1] and $s = rand() \sum_{k=1}^{M} f_k$.

c. Find the smallest of the k in $\sum_{k=1}^{M} f_k \geq s$,and the kth individual is selected.

d. Repeat the steps b and c for N times, the obtained N individuals become a new generation population, that is $t = t + 1$.

2. Crossover

We design probability of crossover considering the relevance. Definition: suppose tth population expressed as P^t, size of population is M, individual i in tth population is expressed as X_i^t, its fitness value is f_i^t, the relevance to the tth population is expressed as:

$$D_t = \sqrt{\frac{\sum_{i=1}^{M}\sum_{j=1}^{M}\left(f_i^t - f_j^t\right)^2}{M(M-1)}} . \tag{10}$$

The relevance reflects the degree of dissimilarity of the individuals in a population. The value of D_t is larger and the individuals are more dissimilar, which indicates in order to save search time, should reduce the crossover operator, otherwise increase the crossover operator. Accordingly, the crossover operator of the tth population can be expressed as follows:

$$P_c^t = \begin{cases} e^{\left(1+\frac{1}{D_t+1}\right)} \\ 0.8, \quad D_t = 0 \end{cases} . \tag{11}$$

When D_t is 0, P_c^t is the traditional genetic algorithm fixed operator value of 0.8. With the increase of D_t, the crossover operator is reduced accordingly, because the relevance of the population and the diversity is strong at this time, increasing the crossover probability may cause search time consuming, also is easy to destroy the excellent individuals. Contrary, to increase the crossover operator can enhance the diversity and improve the ability of the evolution of populations.

3. Mutation Operation

In this paper using basic bit mutation operator: that is a locus of original genetic value of 0, the mutation operation to 1. Conversely, if the original genetic value is 1, the mutation operator turns to 0.

3   Experiments and Results Analysis

3.1   Parameter Setting

In this experiment, assuming the number of cloud resources is eight, the number of users is three, the initial population is randomly generated, the number of iterations is 50, and selection operator is automatically given in accordance with the roulette. In this paper, the crossover probability model is given by equation (11), and the mutation probability is 0.001. The conventional genetic algorithm crossover probability is 0.8, and mutation probability is 0.001. The weight for $G = \lambda_1 R + \lambda_2 E + \lambda_3 C$ is 0.3, 0.3, 0.4.

Evaluation as shown in Table 2, the initial population may produce populations, as shown in Table 3:

Table 2. Evaluation

	user 1	user 2	user 3
response time (s)	0.02	0.05	0.01
energy consumption (kwh)	0.06	0.07	0.05
cost ($/h)	1.1	1.4	1.7

Table 3. The initial population

user 1	user 2	user 3
$u_{11}u_{12}...u_{18}$	$u_{21}u_{22}...u_{28}$	$u_{31}u_{32}...u_{38}$
00100000	00001000	01000000
01100000	00000010	00100000
...	...	...
00000001	11000000	00100001
00000100	00010000	10000000
01000000	00101000	00000001

3.2   Results Analysis

As shown in Fig. 1 and Fig. 2, at the beginning, fitness improves quickly, because it is far from the ideal value. From Fig. 2, at about the 35th generation, there is a big leap

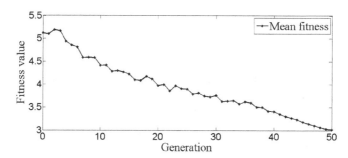

Fig. 1. Changes in Fitness with Traditional Genetic Algorithm (population50)

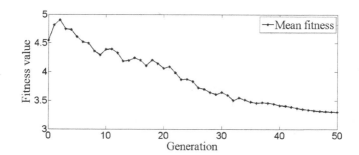

Fig. 2. Changes in Fitness with Modified Genetic Algorithm (population50)

degree. At about 40 generations, the improvement of the fitness in the modified algorithm in this paper is relatively slow. After the 45th generation, it stabilizes and reaches the solution relative. Then, in Fig. 1, the fitness of traditional genetic algorithm reaches stability after the 50 generation, from which it could indicate the modified algorithm is feasible, also, it can quickly improve calculation speed.

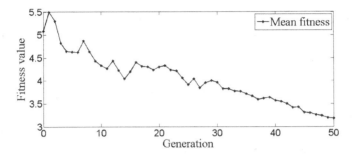

Fig. 3. Changes in Fitness with Traditional Genetic Algorithm (population20)

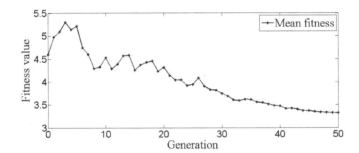

Fig. 4. Changes in Fitness with Modified Genetic Algorithm (population20)

As shown in Fig. 3 and Fig. 4, obviously, when the population is small scale (the number of population is 20 in Fig. 3 and Fig. 4 and that is 50 in Fig. 1 and Fig. 2), the difference of the two algorithms is not very large, about at the 45th generation, they begin to stabilize. That indicates the modified algorithm in this paper with crossover operator changed and some improvement is rational, especially is suitable for large scale. Respect to the traditional genetic algorithm, the genetic algorithm proposed in this paper can shorten the overall execution time of the scheduling.

4   Conclusion

In this paper, considering the influence of genetic algorithm of multi-objective optimization, we solve the scheduling problem of cloud resources. We have described a resource scheduling model with genetic algorithm for QoS constrained cloud resource meanwhile design the factor-crossover operator. Experimental results show that the algorithm have a better search speed, showing certain superiority. In next work, we consider the workflow and random optimization of resource allocation.

References

1. Wei, G., Vasilakos, A.V., Zheng, Y.: A Game-Theoretic Method of Fair Resource Allocation for Cloud Computing Services. J. Supercomputing 54(2), 252–269 (2010)
2. An, B., Vasilakos, A.V.: Evolutionary Stable Resource Pricing Strategies. In: Proceedings of ACM SIGCOMM 2009, pp. 17–21. ACM Press, New York (2009)
3. An, B., Lesser, V., Irwin, D.: Automated Negotiation with Decommitment for Dynamic Resource Allocation in Cloud Computing. In: 9th International Conference on Autonomous Agents and Multi-Agent Systems, pp. 981–988. ACM Press, New York (2010)
4. Mihailescu, M., Teo, Y.M.: Strategy-Proof Dynamic Resource Pricing of Multiple Resource Types on Federated Clouds. In: Hsu, C.-H., Yang, L.T., Park, J.H., Yeo, S.-S. (eds.) ICA3PP 2010, Part I. LNCS, vol. 6081, pp. 337–350. Springer, Heidelberg (2010)
5. Gao, H.Q., Xing, Y.: Research on Cloud Resource Management Model Based on Economics. J. Computer Engineering and Design 31(19), 4139–4142 (2010)
6. Zhang, Y.X., Yao, Y.P.: A Dynamic Partitioning Algorithm Based on Approximate Local Search for Optimistic Parallel Discrete Event Simulation. J. Computers 33(5), 813–821 (2010)
7. Kong, X.H., Ye, B., Xu, W.B.: Ant Colony Optimization for Multi-objective Grid Scheduling Algorithm. J. Computer Engineering and Applications 43(30), 88–90 (2007)
8. Chang, H., Tang, X.: A Load-Balance Based Resource-Scheduling Algorithm under Cloud Computing Environment. In: Luo, X., Cao, Y., Yang, B., Liu, J., Ye, F. (eds.) ICWL 2010. LNCS, vol. 6537, pp. 85–90. Springer, Heidelberg (2011)
9. Fang, Y., Wang, F., Ge, J.: A Task Scheduling Algorithm Based on Load Balancing in Cloud Computing. In: Wang, F.L., Gong, Z., Luo, X., Lei, J. (eds.) Web Information Systems and Mining. LNCS, vol. 6318, pp. 271–277. Springer, Heidelberg (2010)
10. Tejaswi, R.: Windows azure platform. Apress, New York (2009)

Hybird Evolutionary Algorithms for Artificial Neural Network Training in Rainfalll Forecasting

Linli Jiang[1] and Jiansheng Wu[1,2]

[1] Department of Mathematics and Computer, Liuzhou Teacher College,
Liuzhou, 545004, Guangxi, China
jll200@163.com
[2] School of information Engineering, Wuhan University of Technology
Wuhan, 430070, Hubei, China
wjsh2002168@163.com

Abstract. This paper investigates the effectiveness of the Genetic Algorithm (GA) and Simulated Annealing algorithm (SA) training artificial neural network weights and biases for rainfall forecasting, namely GAS–ANN. Firstly, a hybrid GA and SA method is used to train the begining connection weights and thresholds of ANN. Secondly, the back propagation algorithm is used to search around the global optimum. Finally, a numerical example of monthly rainfall data in a catchment located in a subtropical monsoon climate in Linzhou of China, is used to elucidate the forecasting performance of the proposed GASA–ANN model. The forecasting results indicate that the proposed model yields more accurate forecasting results than the autoregressive integrated moving average (ARIMA), back–propagation neural network (BP–NN) and pure Genetic Algorithm training Artificial Neural Network model (GA–ANN). Therefore, the GASA–ANN model is a promising alternative for rainfall forecasting.

Keywords: Rainfall Forecasting, Artificial Neural Network, Genetic Algorithm, Simulated Annealing.

1   Introduction

Accurate and timely rainfall forecasting has been a difficult subject in hydrology due to the complexity of the physical processes involved and the variability of rainfall in space and time [1, 2]. Recurrently, Artificial neural network (ANN) have being used in rainfall modeling [3–5]. There were still certain issues that need attention, such as optimal network structure, choice of training algorithm and choice of input training subset from spatial rainfall information. ANN model might be doing well in predicting past incidents, but unable to predict future events [6, 7]. Therefore, different operators with different experience and knowledge may obtain different results for the same issue. That is, the network is intelligent but capricious, which will greatly limit applications of ANN in actually rainfall forecasting [8].

C. Guo, Z.-G. Hou, and Z. Zeng (Eds.): ISNN 2013, Part II, LNCS 7952, pp. 359–366, 2013.

Genetic Algorithm (GA) has been widely used in the last few years for training and/or automatically designing neural networks. However, GA is diffcult to solve optimization problems in a larger range of optimization problems because it is easily trapped in local optimum in searching optimal solution [9]. Simulated annealing (SA) has some institution to be able to escape from local minima and reach to the global minimum [10]. However, SA costs more computation time. In order to overcome these drawbacks from BP, GA and SA, it is necessary to find some effective approach and improvement to avoid misleading to the local optimum and to search optimum objective function efficiently. This paper propose a novel and specialized hybrid optimization strategy by incorporating SA into GA to train an optimal begining connection weights and thresholds of ANN for rainfall modelling, then the BP algorithm is used to search around the global optimum, namely HGASA–ANN.

The rest of this study is organized as follows. Section 2 describes the proposed HGASA–ANN, ideas and procedures. For further illustration, this work employs the method set up a prediction model for rainfall forecasting in Section 3. Discussions are presented in Section 4 and conclusions are drawn in the final Section.

2   The Developed HGASA–ANN Approach

2.1   ANN Rainfall Forecasting Model

The most popular type of ANN, i.e. multilayer feed-forward neural network (MFNN), the three-layer feed-forward neural network (TFNN), is shown in Fig. 1. In general, the backpropagation method is used for training multilayer perception neural network.

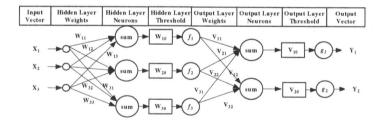

Fig. 1. The structure of a three–layer feed–forward artificial neural network

Referring to Fig. 1, each neuron in the network operates by taking the sum of its weighted inputs and passing the result through a nonlinear activation function (transfer function). Each hidden neuron's output is calculated using Equation (1), while the output neuron's output is calculated using Equation (2)

$$f_i = tanh(\sum_{i=1}^{n} x_i W_{ji} + W_{j0}) \tag{1}$$

$$Y_i = \sum_{i=1}^{m} f_j V_{kj} + V_{k0} \tag{2}$$

where x_i is the value of the input variable, W_{ji} and V_{kj} are connection weights between the input and hidden neuron, and between the hidden neuron and output neuron, W_{j0} and W_{k0} are the threshold (or bias) for the ith and kth neuron, respectively, and i, j and k are the number of neurons for the layers, respectively.

2.2 Methods and Hybrid GASA with ANN

Suppose we are given training data $(x_i, y_i)_{i=1}^{n}$, where $x_i \in R^n$ is the input vector; y_i is the output value and n is the total number of data dimension. The fitness function is defined as follows:

$$f_{fitness} = 1/[1 + \sqrt{\sum_{i=1}^{n}(y_i - \hat{y}_i)^2 / \sum_{i=1}^{n}(y_i)^2}\;] \tag{3}$$

In this paper, a real code genetic algorithm encoding strategy is used vector encoding strategy, what we have to do is just to set each neuron's connection weights and bias to its correspondent gene segments. Each chromosome has four genes, which represent four parameters W_{ji}, V_{kj}, W_{j0} and V_{k0}. For example, for the TFNN with the structure of 3–4–1, the corresponding encoding style for each chromosome can be represented as:

$$\text{generation}(i) = [w_{11}w_{12}w_{13}w_{21}w_{22}w_{23}w_{31}w_{32}w_{33}w_{10}w_{20}w_{30}v_{11}v_{21}v_{31}v_{10}] \tag{4}$$

$$\text{population} = [\text{generation}(1); \text{generation}(2); \cdots ; \text{generation}(M)] \tag{5}$$

where M is the number of the total population, $i = 1, ..., M$.

In this encoding strategy, every chromosome is encoded for a matrix. We also take the TFNN with the structure of 3–3–1 for an example, the encoding strategy can be written as

$$\text{generation}(:, :, :, :, i) = [W, W_0, V, V_0] \tag{6}$$

$$W = \begin{bmatrix} w_{11} & w_{12} & w_{13} \\ w_{21} & w_{22} & w_{23} \\ w_{31} & w_{32} & w_{33} \end{bmatrix} \quad W_0 = \begin{bmatrix} w_{10} \\ w_{20} \\ w_{30} \end{bmatrix} \quad V = \begin{bmatrix} v_{11} \\ v_{21} \\ v_{31} \end{bmatrix} \quad V_0 = \begin{bmatrix} v_{10} \end{bmatrix} \tag{7}$$

where W is the hidden layer weight matrix, V is the output layer weight matrix, while W_0 is the hidden layer bias matrix and of V_0 is the output layer bias matrix.

The hybrid GASA training ANN process consists of two stages: firstly employing GASA to search for optimal or approximate optimal connection weights and thresholds for the network, then using the BP to adjust the final weights. Fig.2 shows flowchart of the proposed algorithm. The major steps of the proposed algorithm are as follows:

1. Generate initial population. The connection weights and thresholds are encoded as float string, randomly generated within $[-1, 1]$. Input training data and calculate the fitness of each chromosome according to Equation (4).
2. Perform GA process. The selection operator of genetic algorithm is implemented by using the roulette–wheel algorithm to determine which population members are chosen as parents that will create offspring for the next generation.
3. The crossover of connection weights and thresholds are operated with probability p_c at float string according to Equations (8) and (9)

$$x_i^{t+1} = \alpha x_i^t + (1 - \alpha)x_{i+1}^t \qquad (8)$$

$$x_{i+1}^{t+1} = (1 - \alpha)x_i^t + \alpha x_{i+1}^t \qquad (9)$$

where x_i^{t+1} stands for the real values of the ith individual of the $(t+1)$th generation, x_i^t and x_{i+1}^t are a pair of individuals before crossover, x_i^{t+1} and x_{i+1}^{t+1} are a pair of individuals after crossover, α is taken as random value within $[0, 1]$.
4. The mutation of connection weights and thresholds are operated with probability p_m at float string according to Equation 10

$$x_i^{t+1} = x_i^t + \beta \qquad (10)$$

where x_i^t stands for the real values of the ith individual of the tth generation, x_i^t is individual before mutation, x_i^{t+1} is individual after mutation, β is taken as random value within $[0, 1]$.
5. Stop condition. If the number of generation is equal to a given scale, then the best chromosomes are presented as a solution, otherwise go to the step 1 of the SA part. GA will deliver its best individual to SA for further processing.
6. Perform SA operators. Generate initial current state. Receive values of the weights and thresholds from GAs. The values of forecasting error, fitness, shown as Equation (3), is defined as the system state (E). Here, the initial state (E_0) is obtained.
7. Provisional state. The existing system state is denoted by S_{old}, Make a random move to change the existing system state to a provisional state, namled S_{new}. Another set of weights and thresholds are generated in this stage S_{new}.
8. Metropolis criterion tests. The probability of accepting the new state is given by the following probability function

$$P = \begin{cases} 1 & if \quad E(S_{new}) > E(S_{old}) \\ exp(\frac{E(S_{new}) > E(S_{old})}{kT}) & if \quad E(S_{new}) > E(S_{old}) \end{cases} \qquad (11)$$

T is the thermal equilibrium temperature, k represents the Boltzmann constant. If the provisional state is accepted, then set the provisional state as the current state.
9. Temperature reduction. After the new system state is obtained, reduce the temperature. The new temperature reduction is obtained by the Equation (12)

$$T_{i+1} = \alpha T_i \qquad (12)$$

where T_i is i-th temperature stage and determines the gradient of cooling,α is set at 0.8 in this paper. If the pre-determined temperature is reached, then stop the algorithm and the latest state is an approximate optimal solution. Otherwise, go to step 8.
10. Perform GA process. Once the termination condition is met, output the final solution, obtain the appropriate connection weights and thresholds. Input validation data, compute fitness for all the individuals by a fitness function.
11. Input testing data and Output forecasting results by the GASA–ANN.

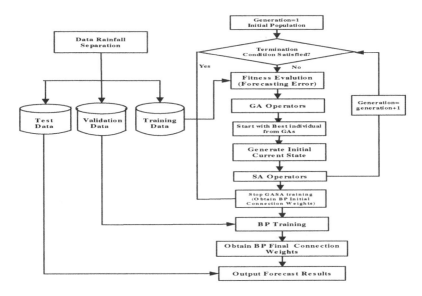

Fig. 2. Flowchart of GASA–ANN algorithm

3   Experimental Results and Discussion

The platform adopted to develop the GASA–ANN approach is a PC with the following features: Intel Celeron M 1.86 GHz CPU, 1.5 GB RAM, a Windows XP operating system and the MATLAB development environment. GASA–ANN algorithm parameters are set as follows: the iteration times are 100; the population size is 20; the crossover probability is 0.5; the mutation probability is 0.05; the initial temperature is 5000; the termination temperature is 0.9.

3.1   Study Area and Data

The data was collected from 18 stations of the Liuzhou Meteorology Administration rain gauge networks for the period from 1991 to 2009, a total of 25 years. Thus the training data set contains 252 data points, whose training set is 144 (1991–2002), validation set is 72 (2003-2008), and testing set is 36 (2007–2009).

3.2   Criteria for Evaluating Model Performance

Three different types of standard statistical performance evaluation criteria were employed to evaluate the performance of various models developed in this paper. These are average absolute relative error (AARE), root mean square error (RMSE), and the Relative Coefficient (RC) which be found in many paper [11].

The input vector is represented by rainfall and runoff values for the preceding 6 monthly rainfall (i.e., $t-1$, $t-2$, $t-3$, $t-4$). Accordingly, the output vector represents the expected rainfall value for mothly $\hat{y}_t$. Through calculating auto-correlation function, partial correlation function and AIC, The AR model is as follows:

$$x_t = 139.8 + 0.28x_{t-1} + 0.07x_{t-2} - 0.08x_{t-3} - 0.09x_{t-4} \tag{13}$$

For the purpose of comparison by the same six input variables, we have also built other three rainfall forecasting models: multi–layer perceptron neural network model based on the back–propagation learning algorithm (BP–NN), and pure genetic algorithm evolutionary neural network method(GA–ANN) [12] proposed by Irani et al. In this paper data, the best NN architecture is: 4–6–1 (6 input units, 6 hidden neurons, 1 output neuron), and the best ANN parameters is chosen as a benchmark model for comparison by the trial–and–error method with the minimum testing root mean square error. Before training and testing all source data are normalized into the range between and 1, by using the maximum and minimum values of the variable over the whole data sets.

3.3   Analysis of the Results

The well-trained models, AR(4), BP–ANN, GA–ANN and GASA–ANN, are applied to forecast the monthly rainfall in Liuzhou. Fig.3 shows the actual values and the forecast values obtained using various forecasting models at 36 testing samples. From the Fig. 3, the output of the GASA–ANN model, simulated with testing data, shows a good agreement with the target. We can find that the GASA–ANN methods have better results than the methods only using AR(4), BP–ANN or GA–ANN in accuracy. Especially GASA–ANN, it has the smallest error rates and standard deviation. Table 1 illustrates the fitting, validation and

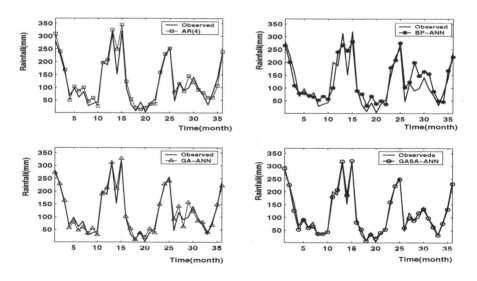

Fig. 3. Comparison between observed and predicted of model in testing samples

testing accuracy and efficiency of the model in terms of various evaluation indices for training, validation and testing samples, respectively.

From the Table 1, we can generally see that learning ability of GASA–ANN outperforms the other three models under the same network input. As a consequence, poor performance indices in terms of AARE, RMSE and PRC can be observed in AR(4) model than other three model. Table 1 also shows that the performance of GASA–ANN is the best in case study for training samples, validation samples and testing samples. The more important factor to measure performance of a method is to check its forecasting ability of testing samples in order for actual rainfall application. Table 1 indicates that GASA–NN not only has the smallest AARE, but also the smallest RMSE. This means that GASA–ANN is the most stable method. In addition, the RC of GASA–ANN is the maximum in all models. The results indicate that the deviations between observed and forecasting value are very small, are capable to capture the average change tendency of the mothly rainfall data.

Table 1. Performance statistics of the five models for rainfall fitting and forecasting

Model	AR(4)	BP–ANN	GA–ANN	GASA–ANN
Index	**Training data (from 1991 to 2002)**			
AARE(%)	47.29(%)	27.04(%)	27.07(%)	19.40(%)
RMSE	27.68	16.47	16.35	12.33
RC	0.958	0.966	0.9284	0.9736
	Validation data (from 2003 to 2006)			
AARE(%)	46.89 (%)	31.25(%)	29.85(%)	19.82(%)
RMSE	27.10	20.84	17.93	17.25
RC	0.947	0.930	0.942	0.9629
	Testing data (from 2007 to 2009)			
AARE(%)	49.58(%)	65.25(%)	31.26(%)	20.01(%)
RMSE	25.70	41.33	22.73	13.05
RC	0.967	0.889	0.972	0.989

4   Conclusion

In this paper, we use a novel technique for the automatic training of ANN begining connection weights and thresholds by evolving to the optimal GSAS algorithm for rainfall forecasting. The fundamental idea for this hybrid algorithm is that at the beginning stage of searching for the optimum, the GASA is employed to accelerate the training speed and helps to avoid trapping into local minimum. When the fitness function value has not changed for some generations, or value changed is smaller than a predefined number, the searching process is switched to gradient descending searching according to this heuristic knowledge. The new hybrid GASA–ANN approach is compared with the BP–ANN and GA–ANN methods with a set of benchmark mathematical functions. The improved hybrid GASA–ANN method is shown to outperform both individual optimization methods for rainfall forecasting.

Acknowledgments. This work was supported in part by the Natural Science Foundation of China under Grant No.11161029, and in part by the Education Department of Guangxi under Grant No.201203YB186.

References

1. Wu, J., Liu, M.Z., Jin, L.: A Hybrid Support Vector Regression Approach for Rainfall Forecasting Using Particle Swarm Optimization and Projection Pursuit Technology. International Journal of Computational Intelligence and Applications 9(3), 87–104 (2010)
2. Wu, J., Jin, L.: Study on the Meteorological Prediction Model Using the Learning Algorithm of Neural Networks Ensemble Based on PSO agorithm. Journal of Tropical Meteorology 15(1), 83–88 (2009)
3. Wu, J.: Prediction of Rainfall Time Series Using Modular RBF Neural Network Model Coupled with SSA and PLS. In: Pan, J.-S., Chen, S.-M., Nguyen, N.T. (eds.) ACIIDS 2012, Part II. LNCS, vol. 7197, pp. 509–518. Springer, Heidelberg (2012)
4. Nourani, V., Kisi, Ö., Komasi, M.: Two Hybrid Artificial Intelligence Approaches for Modeling Rainfall–runoff Process. Journal of Hydrology 402, 41–59 (2011)
5. Wu, J.: An Effective Hybrid Semi–Parametric Regression Strategy for Rainfall Forecasting Combining Linear and Nonlinear Regression. International Journal of Applied Evolutionary Computation 2(4), 50–65 (2011)
6. Yu, J., Wang, S., Xi, L.: Evolving Artificial Neural Networks Using an Improved PSO and DPSO. Neurocomputing 71(4–6), 1054–1060 (2008)
7. Kiranyaz, S., Ince, T., Yildirim, A., Gabbouja, M.: Evolutionary Artificial Neural Networks by Multi–dimensional Particle Swarm Optimization. Neural Networks 22, 1448–1462 (2009)
8. Wu, J.: A Semiparametric Regression Ensemble Model for Rainfall Forecasting Based on RBF Neural Network. In: Wang, F.L., Deng, H., Gao, Y., Lei, J. (eds.) AICI 2010, Part II. LNCS, vol. 6320, pp. 284–292. Springer, Heidelberg (2010)
9. Rogers, A., Prü-Bennett, A.: Genetic Drift in Genetic Algorithm Selection Schemes. IEEE Transaction Evoling of Computation 3(4), 298–303 (1999)
10. Eglese, R.W.: Simulated annealing: A Tool for Operation Research. European Journal of Operational Research 46, 271–281 (1990)
11. Wang, K., Yang, J., Shi, G., Wang, Q.: An Expanded Training Set Based Validation Method to Avoid Over Fitting for Neural Network Classifier. In: Fourth International Conference on Natural Computation, vol. 3, pp. 83–87 (2008)
12. Irani, R., Nasimi, R.: Evolving Neural Network Using Real Coded Genetic Algorithm for Permeability Estimation of The Reservoir. Expert Systems with Applications 38, 9862–9866 (2011)

Artificial Fish Swarm Optimization Algorithm Based on Mixed Crossover Strategy[*]

Li-yan Zhuang and Jing-qing Jiang

College of Computer Science and Technology, Inner Mongolia University for Nationalities,
Tongliao 028043, P.R. China
zhuangliyan88@163.com

Abstract. The nonlinear constrained optimization problems have been widely used in many fields, such as engineering optimization and artificial intelligence. According to the deficiency of artificial fish swarm algorithm (AFSA), that the artificial fishes walk around aimlessly and randomly or gather in non-global optimal points, a hybrid algorithm-artificial fish swarm optimization algorithm based on mixed crossover strategy is presented. By improving the artificial fish's behaviors, the genetic operation of mixed crossover strategy is used as a local search strategy of AFSA. So the efficiency of local convergence of AFSA is improved, and the algorithm's running efficiency and solution quality are improved obviously. Based on test verification for typical functions, it is shown that the hybrid algorithm has some better performance such as fast convergence and high precision.

1   Introduction

A great deal of the nonlinear constrained optimization problems have been often met in many fields, such as engineering optimization and artificial intelligence. Some of these problems belong to the NP-hard problem, and it is the normal way the constrained problem is transformed into the unconstrained problem. Intelligent algorithm has developed by analyzing biological evolutionary theory in recent years, such as genetic algorithm, artificial fish swarm algorithm and so on, which has been widely used in the optimization field because of its unique optimization mechanism, generalization and flexibility [1-2].

Artificial fish swarm algorithm (ASFA) is an optimizing method based on animal autonomy, which is proposed by simulating fish behavior, and it is a concrete application of swarm intelligence theory. The main feature of ASFA is that it is not need to learn special information of problems, but need to compare the advantages and disadvantages of problems, and the global optimal is finally emerged out by local optimizing behavior of every artificial fish individual [3]. The basic ASFA has characteristics of seizing search direction and avoiding the problem of the local optimal to some extent, but when the artificial fishes walk around aimlessly and randomly or

[*] This work is supported by Scientific Foundation of Educational Department of Inner Mongolia Autonomous Region (No.NJZY11208)

C. Guo, Z.-G. Hou, and Z. Zeng (Eds.): ISNN 2013, Part II, LNCS 7952, pp. 367–374, 2013.

gather in non-global optimal points, or when the region of optimizing is large or tends to be flat, the convergence to the global optimum is slowed down and searching performance is deteriorated and even it can falls into the local minimum. The algorithm has generally fast convergence speed in initial of optimizing, but that is often relatively slow in late of optimizing. The solution attained is the satisfactory solution, which has the low precision [3-4].

In order to escape more easily the local optimal for the basic AFSA, and to improve searching efficiency and precision, the genetic operation of mixed crossover strategy and the Gaussian mutation operation are introduced into the basic ASFA in the article. The artificial fishes can effectively escape from the local optimal through rational allocation of parameters, and the diversity of artificial fish can be kept. Meanwhile local fields can be further searched and the searching efficiency is accelerated, so that the swarm converges quickly to the global optimum and the solution of high precision is attained.

2   Treatment of Nonlinear Constrained Optimization Problems

The nonlinear constrained optimization problem is mainly considered in the article, which is described as following [5]:

$$\min_{x \in D \subset [L,U]} f(x)$$

$$st. \ g_i(x) \leq 0 , \ i = 1,2,...,m \tag{1}$$

where $[L, U]$ is n dimension vector field of the space domain Rn and $[L , U] = \{x = (x_1, x_2 , ..., x_n)| \ l_i \leq x_i \leq u_i, \ i =1,2, ..., n\}$. The set $D = \{x| \ x \in [L,U], \ g_i(x) \leq 0, \ i =1,2, ..., m\}$ is the feasible region of solution. If there exists in $x^* \in D$ and makes $f(x^*) \leq f(x)$ that is tenable to any $x \in D$, then x^* is called as the global optimal solution and $f(x^*)$ is called as the global optimal.

For nonlinear constrained optimization problems as formula (1), it is transformed into multi-objective optimization problem which has only two objectives [5].

$$\min\{ f_1(x), f_2(x)\} \tag{2}$$

In which $f1(x)$ is the objective function of the original problem (1) and $f_2(x) = max \ (0, g_i(x), i = 1, 2,..., m)$. Obviously, the minimization of the first objective function means to find x^* which can make the objective function of the original problem to reach the minimum, while the second objective function is to select the maximum between zero and the constrained function value that the violation of the constraints is maximum in all constraints. So as long as $f_2(x) = 0$, all constraints of the original problem must be less than zero, that is the minimization process of $f_2(x)$, in essence, is to try to find out the point x^*, which meets all constraints. Therefore, the simultaneous minimizing of two objectives is to find the point meets all constraints and makes $f_1(x)$ to reach the minimum, namely the optimal solution of the original nonlinear constrained optimization problems. The relation of the model (1) and the model (2) is as follow [6]:

Theorem 1. The necessary and sufficient condition x^* is the optimal solution: x^* is efficient solution of model (2), that is

$$x^* \in D, \exists f_1(x^*) = \min_{x \in D} f_1(x), \ x \in D \tag{3}$$

Proof 1. See reference [6].

3   Artificial Fish Swarm Optimization Algorithm Based on Mixed Crossover Strategy

3.1   Artificial Fish Swarm Algorithm

Artificial fish swarm algorithm is a class of stochastic optimization algorithm based on swarm intelligence, and its mathematical model is described as follow [7]: supposed in a objective searching space of n-dimension, there has N artificial fishes which composed of a swarm, and the state of every artificial fish can be expressed as the vector $X = (x_1, x_2, ..., x_n)$, in which x_i ($i = 1, 2, ..., n$) is the variable of optimization. The food concentration of the current location of artificial fish is expressed as $y = f(x)$, in which y is the objective function. The distances between individuals of artificial fish are expressed as $d_{ij} = \|X_i - X_j\|$. The feeling range of artificial fish and the crowded degree factor are expressed respectively as visual and σ. The step artificial fish moves and the maximum tentative number artificial fish preys each time are expressed respectively as step and *try_number*.

In the process of iteration for each time, artificial fishes are renewed by preying behavior, gathering behavior, following behavior and so on, so the optimizing is realized. The concrete behaviors are described as follows [3]:

①preying behavior: Supposed the current state of artificial fish is X_i, and a state X_j is selected randomly within its visual range (namely $d_{ij} <= Visual$). If the food concentration of this state is more than that of the current state, then the artificial fish forwards a step to the direction. Conversely, a state X_j is reselected randomly, and judges whether it meets the forward conditions. After repeated *try_number* times, if the forward conditions do not still be met, then artificial fish moves randomly a step.

②gathering behavior: Supposed the current state of artificial fish is Xi, the number of partners within its visual range is n_f. If n_f/N is less than δ, which shows that the central position of partners is not too crowded, meanwhile, the food concentration of the central position is more than that of the current state, then the artificial fish forwards a step to the central position, otherwise, preying behavior is carried.

③following behavior: Supposed the current state of artificial fish is X_i, an optimal partner within its visual range is X_{max}. If the number of partners within its visual range is n_f and meets $n_f/N < \delta$, meanwhile, the food concentration of X_{max} position is more than that of the current state, then artificial fish forwards a step to X_{max} position, otherwise, preying behavior is carried.

④stochastic behavior: Supposed the current state of artificial fish is X_i, and a state X_j is selected randomly within its visual range, in order to enlarge search range.

⑤moving strategy: The current environment of artificial fish do be evaluated, that is gathering behavior and following behavior are simulated and executed, then the behavior that has high the food concentration value is executed, and the default behavior way is preying behavior.

⑥constraint behavior: In the process of optimizing, the corresponding constraint conditions need to be added, in order to adjust the situation that the solved solution can be not feasible solution because of some operations, such as gathering behavior, stochastic behavior and so on.

Finally, artificial fishes aggregate around the local optimal, and around the optimal fields of the better value can aggregate generally more artificial fishes.

3.2   Genetic Operation of Mixed Crossover Strategy

Mixed strategies are defined as follows [8]: In a normal game $G = \{S_1,\ldots,S_n; u_1,\ldots,u_n\}$ with n players, set the strategy space S_i for player i is $S_i = \{s_{i1},\ldots,s_{ik}\}$. The player i selects a strategy randomly in the available k strategies based on a probability distribution $p_i = (p_{i1},\ldots,p_{ik})$. The strategies for every player obtained by this way are called mixed strategies, where $j = 1,\ldots,k$, $0 \le p_{ij} \le 1$ and $p_{i1} + \ldots + p_{ik} = 1$. Meanwhile, the original strategy is called pure strategy.

Crossover operator is the most important genetic operator in GA, so the research on crossover operators reflects the research progress on GA. A new crossover operator, mixed crossover strategy, is proposed by mixing the four different crossover operators, including one-point, two-point, uniform and uniform two-point crossover [9]. Every crossover operator is called as pure crossover strategy in mixed crossover strategy, and has its own probability distribution for crossover. The probability distribution of every crossover strategy is adjusted by strengthen or weaken in current population so that the mixed strategy is adjusted. This algorithm implements the choice of crossover strategy automatically. The performance of the algorithm becomes more stable and effective.

The crossover strategy is described as follows [9]:

1) Initialization:

Using pure crossover strategy h, $h \in \{1,2,3,4\}$, which expressed as the crossover strategy of one-point, two-point, uniform, and uniform two point crossover strategy, respectively. The probability distribution of mixed strategy vector ρ is initialized.

2) Crossover operation based on mixed strategy

(1) In every generation g, selects a crossover strategy h according to mixed strategies vector ρ, and uses crossover strategy h to cross to individuals, then generates offspring.

(2) Calculate offspring fitness value, and rank according to fitness.

(3) The method of adjustment of mixed strategy in offspring generation is as follow:

If the number of offspring individuals, which fitness value greater than the parents fitness, is more than half of the number of parent after crossover then strengthens the crossover strategy, that is

if $\forall l \ne h$ then

$$\rho_1^{(k+1)} = \rho_1^{(k)} - \rho_1^{(k)} \times \gamma$$

else $\rho_h^{(k+1)} = \rho_h^{(k)} + (1 - \rho_h^{(k)}) \times \gamma$

Else weakens the crossover strategy, that is

if $\forall l \neq h$ then $\rho_1^{(k+1)} = \rho_1^{(k)} + \dfrac{1}{(no\_strategy - 1)} \times \rho_1^{(k)} \times \gamma$

else $\rho_h^{(k+1)} = \rho_h^{(k)} - \rho_h^{(k)} \times \gamma$

where $\gamma \in (0,1)$ uses to adjust the probability distribution of mixed strategies, "no_strategy" is the number of mixed pure crossover strategy, where $\gamma = 1/2$.

3.3   Artificial Fish Swarm Optimization Algorithm Based on Mixed Crossover Strategy (MCSG-AFSA)

Artificial fish swarm optimization algorithm based on mixed crossover strategy (MCSG-AFSA) is proposed by introducing the genetic operation of mixed crossover strategy in basic ASFA. The algorithm can make effectively artificial fishes to get rid of the limitation of the local optimal, and the local field is further searched, so the searching efficiency is accelerated. Meanwhile the diversity of population is increased by introducing mutation operator [10], so the swarm converges quickly to the global optimal and the solution of high precision is attained finally.

The algorithm flow of MCSG-AFSA algorithm as follow [1]:

(1) Initialization:

The iteration number of the initial bulletin board *Beststep* is equal to 0 when the state of the optimal artificial fish does continuously not change or change little, and set the initial iteration number *Num* is equal to 0 and the maximal threshold of the times that the optimal do not continuously change *maxbest* is equal to 5. *n* artificial fishes are generated randomly in feasible region of control variable, so the initial fish swarm is formed.

(2) The initialization of genetic parameters:

The initial probability of mixed strategy vector is set at 0.25, and the crossover probability and the maximum times of running are set.

(3) The initial value of the bulletin board assignment:

The formula (1) is selected to calculate function value *y* of the current state of individual artificial fish for the initial fish swarm. Then the size of *y* is compared, the minimum of *y* is selected into the bulletin board, and the fish is assigned to the bulletin board.

(4) Behavior selection:

Every artificial fish simulates following behavior and gathering behavior, and the behavior that *y* value is less is executed actually after behavior selected (the default behavior is preying behavior).

(5) Renew the bulletin board:

After every behavior for every artificial fish, *y* of its own and that of the bulletin board are tested. If the former is better than the latter, then *y* of the bulletin board is substituted by that of its own and *Beststep* is set as 0. Otherwise, the behavior of artificial fish that the minimum *y* is calculated by formula (2) is executed actually, and the default behavior is preying behavior. Then the step (5) is executed.

(6)The condition judgment that mixed crossover strategy and mutation operator of genetic algorithm introduced:

Judge whether the value of *Beststep* reaches *Maxbest*. If the value of *Beststep* reaches *Maxbest*, then mixed crossover strategy and mutation algorithm of the step (7) are executed, otherwise, the algorithm is transformed to execute the step (8).

(7) Mixed crossover strategy and mutation operation:

All artificial fishes in swarm execute operations as follows, except the optimal individual of the bulletin board.

①mixed crossover operation:

According to the crossover probability P_c , the corresponding individuals are selected from artificial fish swarm, and execute the genetic operation of mixed crossover strategy in 3.2 section, so function values *y* of new individuals are calculated out and every *y* is compared with the optimal of the bulletin board. If the current *y* is better than the optimal of the bulletin board, so y of the bulletin board is substituted by the current *y*.

②mutation operation:

The corresponding individuals are selected randomly from artificial fishes swarm according to the mutation probability P_m, and execute Gaussian mutation. Then function values of new individuals are calculated out and every *y* is compared with the optimal of the bulletin board. If the current *y* is better than the optimal of the bulletin board, so y of the bulletin board is substituted by the current *y*.

③ *Beststep* is set as 0.

(8) The terminal condition judgment:

Judge whether Num reaches the maximum iteration number *Maxnumber* or whether the optimal reaches within satisfactory error. If the both do not meet, then *Num* is set as *Num*+1, *Beststep* is set as *Beststep*+1, and the algorithm is transformed to execute the step (4). Otherwise, the algorithm is transformed to execute the step (9).

(9) The algorithm terminal and the optimal output, namely the state and function value of artificial fish in the bulletin board.

4   Optimization Test

Five group typical functions are selected to test verification in the paper, and the running result of the algorithm in the paper is compared with that of other algorithms.

Based on test verification for typical functions, it is shown that the hybrid algorithm has some better performance such as fast convergence and high precision, as shown in table1.

Table 1. The Comparison the running result of test functions between MCSG-AFSA and other algorithms

Test function	F1	F2	F3	F4	F5
The optimal	0.25000	-6961.814	13.59084	1.00000	3.791340
MCSG-AFSA	0.25000	-6961.814	13.59084	1.00000	3.791340
Other algorithms [see reference1]	0.25000	-6961.814	13.59085	0.95825	3.791340
The average result of reference[1]	0.25004	-6961.352	13.59393	0.99999	3.791297
The average result of MCSG-AFSA	0.25001	-6961.669	13.59065	0.99999	3.791331

The results of table show that MCSG-AFSA is better than other algorithms obviously, and embodies the certain superiority. MCSG-AFSA is programmed to realize on Matlab7.0 language platform. Test functions F1-F5 are shown as follows:

F1:
$$\min \ f(x) = 100\,(x_2 - x_1)^2 + (1 - x_1)^2$$
$$s.t.\begin{cases} g_1(x) = -x_1 - x_2{}^2 \le 0 \\ g_2(x) = -x_1{}^2 - x_2 \le 0 \\ -0.5 \le x_1 \le 0.5, -1 \le x_2 \le 1 \end{cases}$$

F2:
$$\min \ f(x) = (x_1 - 10)^3 + (x_2 - 20)^3$$
$$s.t.\begin{cases} g_1(x) = 100 - (x_1 - 5)^2 - (x_2 - 5)^2 \le 0 \\ g_2(x) = -82.81 + (x_1 - 6)^2 + (x_2 - 5)^2 \le 0 \\ 13 \le x_1 \le 100, 0 \le x_2 \le 100 \end{cases}$$

F3:
$$\min f(x) = (x_1{}^2 + x_2 - 11)^2 + (x_1 + x_2{}^2 - 7)^2$$
$$s.t.\begin{cases} g_1(x) = 4.84 - x_1{}^2 - (x_2 - 2.5)^2 \le 0 \\ g_2(x) = -4.84 + (x_1 - 0.05)^2 + (x_2 - 2.5)^2 \le 0 \\ 0 \le x_1 \le 6, 0 \le x_2 \le 6 \end{cases}$$

F4:
$$\min \ f(x) = \sin^3(2\pi x_1)\sin^3(2\pi x_2)$$
$$s.t.\begin{cases} g_1(x) = x_1{}^2 + x_2 + 1 \le 0 \\ g_2(x) = 1 - x_1 + (x_2 - 4)^2 \le 0 \\ 0 \le x_1 \le 10, 0 \le x_2 \le 10 \end{cases}$$

F5:
$$\min \ f(x) = x_1{}^2 + x_2{}^2$$
$$s.t.\begin{cases} g_1(x) = x_1 + x_2 - 1 \le 0 \\ g_2(x) = x_1 + x_2{}^2 - 1 \le 0 \\ g_3(x) = x_1{}^2 + x_2{}^2 - 9 \le 0 \end{cases}$$

5   Conclusion

According to the deficiency of AFSA, that the artificial fishes walk around aimlessly and randomly or gather in non-global optimal points, artificial fish swarm optimization algorithm based on mixed crossover strategy is presented, in order to solve the nonlinear constrained optimization problems. The genetic operation of mixed crossover strategy and the Gaussian mutation operation are used in the algorithm, which can jump out the local optimization and avoid the limitation of early maturity when the optimal that the algorithm solves does continuously not change or change little. Based on test verification for typical functions, it is shown that new algorithm has some better performance such as fast convergence and high precision, and can solve very well this kind of optimization problems.

However, in order to improve the precision and the convergence speed of this kind of problems, AFSA will be fused with other algorithms, and some concepts will be introduced into AFSA such as multi-population, synergetic algorithm and so on, and how to use these to solve actual problems, so these will to be done in the next step research work.

References

1. Liu, B., Zhou, Y.-Q.: Artificial fish swarm optimization algorithm based on genetic algorithm. Computer Engineering and Design 29(22), 5827–5829 (2008)
2. Wang, H.-Y., Zhang, Y.-G.: An Improved Artificial Fish-Swarm Algorithm of Solving Clustering Analysis Problem. Computer Technology and Development 20(3), 84–87 (2010)
3. Qu, D.-L., He, D.-X.: Artificial Fish Swarm Algorithm Based on hybrid mutation operators. Computer Engineering and Applications 44(35), 50–52 (2008)
4. Qu, D.-L., He, D.-X.: Bi-group artificial fish-school algorithm based on simplex method. Computer Applications 28(8), 2103–2104 (2008)
5. Liu, C.-A.: New Particle Swarm Optimization Algorithm for the Solution to Nonlinear Constrained Programming Problem. Journal of Chongqing Institute of Technology 20(11), 118–120 (2006)
6. Wang, Y., Liu, D., Cheung, Y.-M.: Preference bi-objective evolutionary algorithm for constrained optimization. In: Hao, Y., Liu, J., Wang, Y.-P., Cheung, Y.-m., Yin, H., Jiao, L., Ma, J., Jiao, Y.-C. (eds.) CIS 2005. LNCS (LNAI), vol. 3801, pp. 184–191. Springer, Heidelberg (2005)
7. Yao, X.-G., Zhou, Y.-Q., Li, Y.-M.: Hybrid algorithm with artificial fish swarm algorithm and PSO. Application Research of Computers 27(6), 2084–2086 (2010)
8. Fan, R.-G., Han, M.-C.: Game Theory, pp. 4–20. Wuhan University press, Wuhan (2006)
9. Zhuang, L.-y., Dong, H.-b., Jiang, J.-Q., Song, C.-Y.: A Genetic Algorithm Using a Mixed Crossover Strategy. In: Sun, F., Zhang, J., Tan, Y., Cao, J., Yu, W. (eds.) ISNN 2008, Part I. LNCS, vol. 5263, pp. 854–863. Springer, Heidelberg (2008)
10. Huang, H.-J., Zhou, Y.-Q.: Hybrid artificial fish swarm algorithm based on mutation operator. Computer Engineering and Applications 45(33), 28–30 (2009)

Bias-Guided Random Walk
for Network-Based Data Classification

Thiago Henrique Cupertino and Liang Zhao

Institute of Mathematics and Computer Science
University of São Paulo, São Carlos, Brazil
{thiagohc,zhao}@icmc.usp.br

Abstract. This paper presents a new network-based classification technique using limiting probabilities from random walk theory. Instead of using a traditional heuristic to classify data relying on physical features such as similarity or density distribution, it uses a concept called ease of access. By means of an underlying network, in which nodes represent states for the random walk process, unlabeled instances are classified with the label of the most easily reached class. The limiting probabilities are used as a measure for the ease of access by taking into account the biases provided by an unlabeled instance in a specific adjacency matrix weight composition. In this way, the technique allows data classification from a different viewpoint. Simulation results suggest that the proposed scheme is competitive with current and well-known classification algorithms.

Keywords: network-based learning, data classification, supervised learning, random walk, limiting probabilities.

1 Introduction

Supervised machine learning comprises the construction of a model by using information extracted from a training data set. The constructed model defines decision borders that are used to classify unlabeled data [1]. An unlabeled instance is classified depending on its relative position to the decision borders. Due to its importance in various real applications, many classification techniques have been developed, such as Neural Networks, k-Nearest Neighbors (kNN), Linear Discriminate Analysis (LDA), Naive-Bayes Method, Support Vector Machines (SVM) and Decision Tree [1, 2, 3, 4, 5]. These traditional classification techniques divide the data space according to physical features (similarity, distance, or distribution) of the training data. In this way, many intrinsic and semantic relations among data items are ignored as, for example, topological structure and pattern formation.

On the other hand, the usage of an underlying network can take into account these previously mentioned relationships among data. In the machine learning domain, many recent works have applied random walk processes to perform semi-supervised learning. In this learning paradigm, just a few data compose

C. Guo, Z.-G. Hou, and Z. Zeng (Eds.): ISNN 2013, Part II, LNCS 7952, pp. 375–384, 2013.

the training data set, and so the classification processes makes use of the information provided by the unlabeled data, which is most commonly represented by nodes in a network. Many of these works share the regularization framework, differing only in the particular choice of the loss function and the regularizer [6, 7, 8, 9, 10, 11]. In these works, the concept of relationship among data is the measurement of how easy labels propagate or how easy a random walker reaches target nodes on a network structure. In these techniques, the underlying link structure is responsible for giving the probabilities or weights between two neighboring nodes to support the label propagation or the walker transition between two linked nodes. Similarly, random walks have also been extensively applied to unsupervised learning, in which there is no labeled data, such as community detection and data clustering [12, 13, 14, 15, 16]. However, very few efforts have been done for network-based supervised learning [17, 18, 19, 20], and thus this work is also a contribution to the use of network-based techniques in the supervised learning field.

Here, we propose a new network-based classification heuristic which consider the *ease of access* of unlabeled instances to each class. Differently from previous works, the proposed technique uses the dynamical process measure called random walk limiting probabilities. Limiting probabilities are applied to random walk processes to measure the limiting state transitions through an underlying network [21]. In the proposed scheme, the training data set is used to construct the network, in which instances (nodes) represent the states a random walker visits during the process. An unlabeled instance is considered belonging to the class that is most easily reached, that is, the limiting transition probability for a random walker to that class, after the insertion of the unlabeled instance bias in the underlying adjacency weight matrix, is large. As a consequence of the dynamical processes, both local and global relationships among nodes are taken into account.

This paper is organized as follows: section 2 describes the model for the supervised classification technique. In section 3, simulation results and comparisons are presented. Finally, section 4 concludes the paper.

2   Supervised Inductive Classification Model

In this section, the technique is derived for a supervised inductive classification model. To be classified, an unlabeled instance is first inserted into the network of training data as a virtual state. The concept of virtual state means that the probability of belonging to this state is not considered, that is, the random walker can visit this virtual state, but only the information extracted directly from the training data is used for classification. The insertion of an unlabeled instance as a virtual state is carried out by a specific weight composition and aims to provide a bias to the classification process by enhancing the probabilities of the unlabeled instance's network neighborhood. Therefore, the bias prioritizes near classes in the state space (classes with the large transition probability) by adopting the assumption that close instances belong to the same class. The mathematical formulation is given next.

Training Phase. We consider it is given a labeled data set $\mathcal{X}^{(l)} = \{\mathbf{x}_i^{(l)}, i = 1, \ldots, n\}$ containing only labeled instances, where each instance is described by q attributes $\mathbf{x}_i = \{x_{i1}, x_{i2}, \ldots, x_{iq}\}$. A weighted undirected network $\mathcal{N} = \{\mathcal{V}, \mathcal{E}\}$ without self-loops is constructed, in which data instances are represented by nodes, $\mathcal{V} = \mathcal{X}^{(l)}$, and similarities among instances are represented by weights of the edges, $\mathcal{E} = [\mathcal{W}_{ij}], i, j = 1, \ldots, n$. The network similarity matrix $\mathcal{W} = \{w_{ij}\}$ can be calculated by using any distance function. w_{ij} is the similarity between the pair of instances $\mathbf{x}_i^{(l)}$ and $\mathbf{x}_j^{(l)}$.

Classification Phase. To perform the classification of an unlabeled instance $\mathbf{x}^{(u)}$, the set of nodes $\mathcal{V}$ is considered as a state space set, meaning that each node is a possible state for a random walker. The transition probabilities among states are given by a normalized transition matrix P, whose construction is further explained.

First, the unlabeled instance $\mathbf{x}^{(u)}$ is inserted into the network $\mathcal{N}$. To do so, a similarity vector $\mathcal{S} = [\mathcal{S}_i], i = 1, \ldots, n$, between $\mathbf{x}^{(u)}$ and all other nodes $\mathbf{x}_i^{(l)}$ is calculated by using a distance function. Next, a new asymmetric and $n \times n$ modified similarity matrix $\hat{\mathcal{W}}$ is constructed by composing the matrix of weight biases $\hat{\mathcal{S}}$:

$$\hat{\mathcal{W}} = \mathcal{W} + \epsilon \hat{\mathcal{S}}, \tag{1}$$

where ϵ is a non-negative parameter and $\hat{\mathcal{S}}$ is an $n \times n$ matrix composition:

$$\hat{\mathcal{S}} = \begin{bmatrix} \mathcal{S}^T \\ \mathcal{S}^T \\ \vdots \\ \mathcal{S}^T \end{bmatrix},$$

where $\mathcal{S}^T$ is the transpose of vector $\mathcal{S}$.

In Eq. 1, it can be observed that the weight biases of the virtual state $\mathbf{x}^{(u)}$, encoded in matrix $\hat{\mathcal{S}}$, are applied over all edges $\mathcal{W}$ of network $\mathcal{N}$, that is, the weight of each edge is linearly added up with the corresponding weight bias. The idea behind this operation is that the distance between any pair of nodes is reduced because of the new route introduced by the insertion of the unlabeled data instance. The higher the proximity between the unlabeled instance and a node, say node j, the more strengthened the connections from all other nodes to node j are. The parameter ϵ controls the influence of weight bias provided by matrix $\hat{\mathcal{S}}$ on the original network. The larger is the value of parameter ϵ, the greater will be the influence of the bias weights provided by the virtual state.

After the bias composition, the virtual state $\mathbf{x}^{(u)}$ is effectively inserted into network $\mathcal{N}$ as a virtual state. To do so, an $(n + 1) \times (n + 1)$ adjacency matrix $\mathcal{A} = \{a_{ij}\}$ is constructed:

$$\mathcal{A} = \begin{bmatrix} \hat{\mathcal{W}} & \mathcal{S} \\ \mathcal{S}^T & 0 \end{bmatrix}. \tag{2}$$

In this formulation, without loss of generality, the virtual state is inserted as the last entry $(n + 1)$ of matrix $\mathcal{A}$.

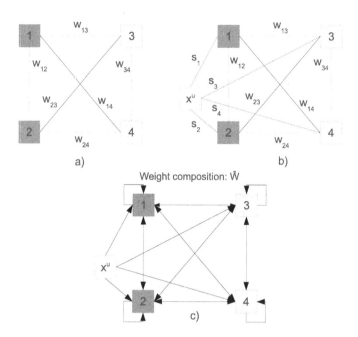

Fig. 1. Illustration for supervised inductive classification of unlabeled instance $\mathbf{x}^u$. a) An undirected and complete network $\mathcal{N}$ is formed by using 4 training instances of 2 classes: green and yellow; b) the similarities $\mathcal{S}$ are calculated for the unlabeled instance $\mathbf{x}^u$; c) modified network $\mathcal{N}$, directed with self-loops, after bias composition (Eq. 1) and insertion of the unlabeled virtual state $\mathbf{x}^u$ (Eq. 2).

The two steps described above can be easily understood by using the toy example depicted in Fig. 1. In this example, a network is formed by 4 labeled instances belonging to 2 distinct classes, *blue* and *red*, each one containing 2 representative instances (Fig. 1a). In the initial network, the links are undirected and the similarity matrix is symmetric. Next, giving an unlabeled instance $\mathbf{x}^u$ to be classified, the similarity vector $\mathcal{S}$ between $\mathbf{x}$ and all other nodes is computed (Fig. 1b). After this computation, the weight biases of $\mathbf{x}^u$ are added up to the original similarity matrix to form a biased similarity matrix for the same network (Eq. 1). After that, the unlabeled instance $\mathbf{x}^u$ is effectively inserted into the network (Eq. 2). It can be seen from Fig. 1c that the network becomes directed and with self-loops (except the virtual state, that has no self-loop) and the biased weight matrix is no more symmetric.

After the virtual state insertion, we are able to compute the entries of the transition matrix $\mathcal{P} = [\mathcal{P}_{ij}]$ by scaling the entries of matrix $\mathcal{A}$:

$$p_{ij} = a_{ij} / \sum_{j=1}^{n+1} a_{ij}.$$

Algorithm 1. Classification Algorithm

Input:
$\mathcal{X}^{(l)}$: training data set
$\mathbf{x}^{(u)}$: unlabeled instance
Parameters:
t : number of largest probabilities to be selected
ϵ : bias weighting
Output:
c : estimated class ($c \in \{1, \ldots, C\}$) for $\mathbf{x}^{(u)}$
Training:
1. $\mathcal{N}$ = Create a network from $\mathcal{X}^{(l)}$
Classification:
2. $\hat{\mathcal{W}}$ = Compose bias weights into $\mathcal{N}$ (Eq. 1)
3. $\mathcal{A}$ = Insert the virtual state $\mathbf{x}^{(u)}$ into $\mathcal{N}$ (Eq. 2)
4. $\mathbf{p}^{\infty}$ = Compute limiting probabilities (Eq. 3)
5. $\mathcal{T}$ = Select the t largest limiting probabilities from $\mathbf{p}^{\infty}$
6. c = Assign $\mathbf{x}^{(u)}$ the most representative class in $\mathcal{T}$

With the above matrix $\mathcal{P}$ at hand, the limiting probabilities can be calculated. This calculation is performed by two ways: by finding the eigenvector corresponding to the unit eigenvalue of matrix $\mathcal{P}$ or, as a faster manner, iterating the system

$$\mathbf{p}_{i+1} = \mathcal{P}\mathbf{p}_i, \tag{3}$$

to the stationary state, where $\mathbf{p}$ is an $(n + 1) \times (n + 1)$ normalized vector. It results in the following vector:

$$\mathbf{p}^{\infty} = [p_1 \, p_2 \ldots p_{n+1}]$$

where each column represents a state, and each entry p_i represents the probability of $\mathbf{x}^{(u)}$ belonging to the class of state i.

As the final step, the classification of $\mathbf{x}^{(u)}$ is accomplished by assigning it the most representative label from the set of states. To achieve that, a set $\mathcal{T}$ containing the t states with the largest limiting probabilities are selected in a descend order and the most representative class in $\mathcal{T}$ is associated to $\mathbf{x}^{(u)}$.

In a concise form, the proposed supervised inductive classification process can be summarized by Algorithm 1.

3   Results for Real Data Sets and Comparisons

We present classification results using the proposed supervised technique, as well as a comparative study against some current classifiers. In the experiments, 15 data sets were selected from the UCI machine learning repository [22]. Table 1 shows the metadata for all data sets. As can be seen in this table, the selection was made to encompass diversity on data domains as well as to consider different number of classes, attributes and class sizes. They vary from 3 to 15, 4 to 91

Table 1. Information of all data sets used in simulations

Domain	Instances	Attributes	Classes
Zoo	101	16	7
Hayes-Hoth	132	5	3
Iris	150	4	3
Teaching	151	5	3
Wine	178	13	3
Image	210	19	7
Glass	214	9	6
E. Coli	336	8	8
Libras	360	91	15
Balance	625	4	3
Vehicle	846	18	4
Vowel	990	13	11
Yeast	1484	8	10
Wine Q. (Red)	1599	12	6
Segment	2310	19	7

and 101 to 1599, respectively. The Euclidean distance was used in all simulations as a distance measurement. Eventual categorical attributes, in data sets such as Balance and Zoo, were treated as numerical. As a data preparation, each instance vector was normalized to have a magnitude of 1. Individual cases were normalized by dividing each attribute of the instance by the square root of the sum of the squares of the individual attributes. Thus, an instance $\mathbf{x}_i = (x_{i1}, x_{i2}, \ldots, x_{ip})$ was normalized by dividing each attribute x_{ij} by $\sum_{j=1}^{p} x_{ij}^2$.

The parameter optimization for the proposed technique was done as follows. It was created a complete network $\mathcal{N}$ (Step 1 of Alg. 1) by using the Euclidean measurement as a distance function. Parameter t (Step 5 of Alg. 1) ranged from 1 to the number of instances in the largest class of the training set. Parameter ϵ (Eq. 1) was evaluated by the grid method with the values $\{0, 0.1, 0.2, \ldots, 10\}$.

The influence of parameter ϵ in the classification accuracy was evaluated. As stated before in Sec. 2, ϵ is responsible for weighting the biases provided by the virtual state. Figure 2 depicts the accuracy in function of parameter ϵ for eight selected data sets. The results were averaged over 50 simulations. Each simulation was performed by using a 10-fold stratified cross-validation process [23]. In this process, the data set is split in 10 disjoint sets and, in each run, 9 sets are used as training data and 1 set is used as the test data, resulting in a total of 10 runs. Therefore, 50×10 runs were executed. It can be seen on Fig. 2 that next to value 0 - where the biases influence are reduced because of a small weight - the classification accuracies are poor. On the other hand, as ϵ becomes larger, the accuracies increase and stabilize before it approaches 10. In this later case, the biases play a main role due to the large weight applied to them (Eq. 1). These scenarios configure a convergent behavior for parameter ϵ and can help in the simulations by restricting the search space.

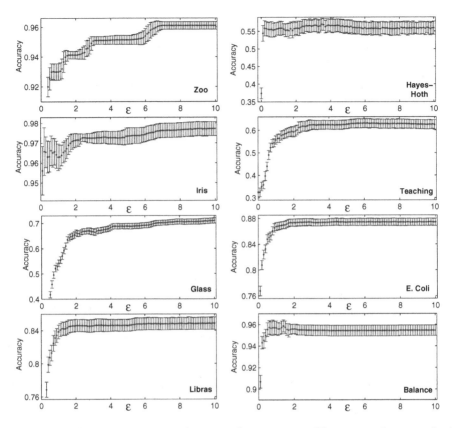

Fig. 2. Classification accuracy in function of parameter ϵ. The curves show standard deviations for each point in error bars. 50 simulations were averaged.

The proposed technique was compared to other 4 well-known multi-class classification algorithms: Weighted kNN (WkNN), Decision Tree C4.5 [24], Multi-Class SVM (MSVM) [25] and the network-based k-Associated Optimal Graph (kAOG) [17]. For the parametric algorithms (all of the algorithms except kAOG), a repeated cross-validation [23] was done in order to optimize their respective parameters. For the MSVM algorithm, we used the *one-against-one* multi-class version, in which $C(C-1)/2$ binary classifiers distinguishes between every pair of classes by using a voting scheme. To avoid ties, the output of each MSVM corresponded to the real valued decision functions. For reducing the parameter search space in MSVM model selection, the only kernel in consideration was the radial basis function, $K(\mathbf{x}_i, \mathbf{x}_j) = e^{-\gamma \|\mathbf{x}_i - \mathbf{x}_j\|^2}$, and the stopping criterion for the optimization method was defined as the Karush-Kuhn-Tucker violation to be less than 10^{-3}, the same condition used in [26]. In the WkNN, the classification process was performed by using the sum of the weights between the instance to be labeled and its k-nearest neighbors. Specifically, the weight between two instances $\mathbf{x}_i$ and $\mathbf{x}_j$ is defined by $1/\mu(\mathbf{x}_i, \mathbf{x}_j)$, where μ is a

Table 2. Classification accuracy (%) followed by standard deviation and rank for each algorithm. Each result shows the adjusted parameters for model selection. Best result for each data set is in bold face.

Data set	Proposed (ϵ, t)	kAOG	WkNN (k)	C4.5 (cf, m)	MSVM (cp, γ)
Zoo	96.2 ± 0.3 (7.8, 1)	**97.0 ± 5.2**	96.2 ± 5.8 (1)	95.8 ± 5.3 (0.1, 0)	96.3 ± 6.4 ($2^1, 2^1$)
Hayes-Hoth	**57.0 ± 1.7** (3.5, 1)	55.7 ± 12.6	56.8 ± 13.2 (3)	46.7 ± 10.5 (1, 0)	45.4 ± 13.1 ($2^{12}, 2^{13}$)
Iris	**98.1 ± 0.2** (3.1, 29)	97.4 ± 3.2	97.9 ± 3.3 (19)	95.0 ± 5.8 (0.25, 2)	97.0 ± 4.6 ($2^{-2}, 2^3$)
Teaching	**63.1 ± 2.0** (6.0, 1)	62.5 ± 11.6	63.0 ± 12.3 (9)	58.2 ± 14.9 (1, 0)	52.5 ± 7.9 ($2^6, 2^3$)
Wine	84.6 ± 1.8 (9.4, 1)	83.5 ± 8.5	84.1 ± 8.2 (1)	91.7 ± 6.7 (0.5, 1)	**94.4 ± 5.8** ($2^{11}, 2^2$)
Image	75.5 ± 1.0 (1.8, 1)	75.3 ± 7.5	75.4 ± 8.2 (3)	80.7 ± 7.6 (0.8, 3)	**86.7 ± 7.4** ($2^{10}, 2^{-3}$)
Glass	**72.5 ± 1.2** (10.0, 2)	72.5 ± 8.1	71.8 ± 9.0 (1)	66.9 ± 9.4 (0.1, 3)	69.5 ± 5.6 ($2^{10}, 2^4$)
E. Coli	**87.5 ± 0.6** (4.4, 9)	85.8 ± 6.4	87.4 ± 5.4 (9)	83.6 ± 6.1 (0.25, 3)	86.7 ± 8.2 ($2^{12}, 2^{-9}$)
Libras	84.9 ± 0.7 (9.0, 1)	85.4 ± 5.3	84.8 ± 5.4 (1)	71.6 ± 7.5 (0.8, 1)	**86.6 ± 5.0** ($2^7, 2^2$)
Balance	96.3 ± 0.5 (4.1, 9)	94.9 ± 2.5	96.7 ± 2.1 (11)	89.6 ± 3.7 (0.5, 1)	**98.2 ± 0.9** ($2^7, 2^0$)
Vehicle	67.6 ± 0.7 (10, 4)	67.9 ± 4.4	67.6 ± 4.1 (5)	70.7 ± 3.5 (0.5, 2)	**84.4 ± 3.4** ($2^10, 2^3$)
Vowel	97.5 ± 0.9 (10.0, 1)	**98.9 ± 0.7**	98.8 ± 0.9 (11)	78.6 ± 4.3 (0.5, 0)	97.5 ± 1.9 ($2^7, 2^0$)
Yeast	59.4 ± 0.4 (10.0, 12)	53.6 ± 3.8	**60.9 ± 3.6** (16)	55.8 ± 3.6 (0.1, 5)	58.9 ± 4.8 ($2^{11}, 2^0$)
Wine Q. Red	61.0 ± 0.3 (10.0, 1)	61.8 ± 3.6	**64.0 ± 3.8** (19)	59.8 ± 2.4 (1, 0)	60.4 ± 3.2 ($2^9, 2^1$)
Segment	93.1 ± 0.5 (10.0, 1)	93.7 ± 1.5	93.6 ± 1.4 (5)	95.4 ± 1.2 (1, 0)	**96.6 ± 1.2** ($2^11, 2^0$)
Avg. Rank	**2.67 ± 1.50**	3.33 ± 1.80	2.80 ± 1.26	5.00 ± 1.96	3.13 ± 2.20

distance measurement. The number of neighbors k ranged from 1 to the number of instances in the largest class of the training set. For the C4.5 algorithm, two parameters were adjusted, the confidence factor which assumes the values $cf \in \{0, 0.1, 0.25, 0.5, 0.8, 1\}$, where smaller values incur more pruning (1 is for no pruning), and the minimum number of instances that a set must have in order to be further partitioned is $m \in \{0, 1, 2, 3, 4, 5, 10, 15, 20, 50\}$.

Table 2 presents the classification accuracy on the test set followed by the standard deviation. Best results are in bold face. At the last line, the average rank for each algorithm is shown. The calculation procedure for the rank measurement is as follows: i) for each data set, the algorithms were ranked according to their average performance, that is, the best algorithm was ranked as first, the second best was ranked as second, and so on; and ii) for each algorithm, the average rank was based on the rank values on all the data sets. It can be seen that the proposed technique achieved the best average ranking amongst all simulated techniques. Moreover, it is worth emphasizing that our technique exhibited the smallest deviation values over all results.

4   Conclusions

This paper has presented a new network-based classification technique which applies limiting probabilities from random walk theory. In the supervised model, these probabilities represent the ease of access an unlabeled instance has to the classes in the training set in an underlying network. Unlabeled nodes are classified with the label of the class most easily reached. Simulations have suggested that the proposed technique is competitive with the current well-known classification techniques. As future works, mathematical models could be studied to describe and shed more light to the proposed technique. We hope this research contributes to the network-based learning area, specially to development of new supervised classification heuristics.

Acknowledgments. The authors would like to acknowledge the São Paulo State Research Foundation (FAPESP) and the Brazilian National Council for Scientific and Technological Development (CNPq) for the financial support given to this research.

References

[1] Duda, R.O., Stork, D.G., Hart, P.E.: Pattern classification, 2nd edn. Wiley-Interscience (2000)

[2] Friedman, N., Geiger, D., Goldszmidt, M.: Bayesian network classifiers. Machine Learning 29, 131–163 (1997)

[3] Burges, C.J.C.: A tutorial on support vector machines for pattern recognition. Data Mining and Knowledge Discovery 2(2), 121–167 (1998)

[4] Haykin, S.: Neural Networks: A Comprehensive Foundation, 2nd edn. Prentice Hall, Englewood Cliffs (1998)

[5] Tan, P.N., Steinbach, M., Kumar, V.: Introduction to Data Mining, 1st edn. Addison-Wesley (2005)

[6] Blum, A., Chawla, S.: Learning from labeled and unlabeled data using graph mincuts. In: Proceedings of the Eighteenth International Conference on Machine Learning, San Francisco, pp. 19–26. Morgan Kaufmann (2001)

[7] Zhu, X., Ghahramani, Z., Lafferty, J.: Semi-supervised learning using gaussian fields and harmonic functions. In: Proc. 20th Int. Conf. Mach. Learn., pp. 912–919 (2003)

[8] Belkin, M., Matveeva, I., Niyogi, P.: Regularization and semi-supervised learning on large graphs. In: Shawe-Taylor, J., Singer, Y. (eds.) COLT 2004. LNCS (LNAI), vol. 3120, pp. 624–638. Springer, Heidelberg (2004)

[9] Belkin, M., Niyogi, P., Sindhwani, V.: On manifold regularization. In: Proc. 10th Int. Workshop Artif. Intell. Stat., pp. 17–24 (2005)

[10] Zhou, D., Schölkopf, B.: Adaptive computation and machine learning. In: Discrete Regularization, pp. 237–250. MIT Press, Cambridge (2006)

[11] Silva, T., Zhao, L.: Network-based stochastic semisupervised learning. IEEE Transactions on Neural Networks and Learning Systems 23(3), 451–466 (2012)

[12] Zheng, X., Lin, X.: Automatic determination of intrinsic cluster number family in spectral clustering using random walk on graph. In: 2004 International Conference on Image Processing, ICIP 2004, vol. 5, pp. 3471–3474 (October 2004)

[13] Alamgir, M., von Luxburg, U.: Multi-agent random walks for local clustering on graphs. In: 2010 IEEE 10th International Conference on Data Mining, ICDM, pp. 18–27 (December 2010)

[14] Cai, B., Wang, H., Zheng, H., Wang, H.: An improved random walk based clustering algorithm for community detection in complex networks. In: 2011 IEEE International Conference on Systems, Man, and Cybernetics, SMC, pp. 2162–2167 (October 2011)

[15] Breve, F., Zhao, L., Quiles, M., Pedrycz, W., Liu, J.: Particle competition and cooperation in networks for semi-supervised learning. IEEE Transactions on Knowledge and Data Engineering 24(9), 1686–1698 (2012)

[16] Silva, T., Zhao, L.: Stochastic competitive learning in complex networks. IEEE Transactions on Neural Networks and Learning Systems 23(3), 385–398 (2012)

[17] Bertini, J.R., Zhao, L., Motta, R., de Andrade Lopes, A.: A nonparametric classification method based on k-associated graphs. Information Sciences 181, 5435–5456 (2011)

[18] Cupertino, T.H., Silva, T., Zhao, L.: Classification of multiple observation sets via network modularity. Neural Computing and Applications (Print) (2012)

[19] Cupertino, T.H., Zhao, L.: Using katz centrality to classify multiple pattern transformations. In: Proceedings of the 2012 Brazilian Symposium on Neural Networks, pp. 1–6 (2012)

[20] Silva, T., Zhao, L.: Network-based high level data classification. IEEE Transactions on Neural Networks and Learning Systems 23(6), 954–970 (2012)

[21] Gallager, R.G.: Discrete Stochastic Processes, 1st edn. Springer (1996)

[22] Frank, A., Asuncion, A.: UCI machine learning repository (2010)

[23] Kim, J.H.: Estimating classification error rate: Repeated cross-validation, repeated hold-out and bootstrap. Computational Statistics and Data Analysis 53, 3735–3745 (2009)

[24] Quinlan, J.R.: C4.5: Programs for Machine Learning, 1st edn. Morgan Kaufman Publishers (1993)

[25] Vapnik, V.: The Nature of Statistical Learning Theory, 1st edn. Springer (1999)

[26] Hsu, C.W., Lin, C.J.: A comparison of methods for multiclass support vector machines. IEEE Transactions on Neural Networks 13, 415–425 (2002)

A Robust Multi-criteria Recommendation Approach with Preference-Based Similarity and Support Vector Machine

Jun Fan and Linli Xu

School of Computer Science and Technology
University of Science and Technology of China,
230027, Hefei, Anhui
floydfan@mail.ustc.edu.cn, linlixu@ustc.edu.cn

Abstract. In the next generation of recommender systems, multi-criteria recommendation could be regarded as one of the most important branches. Compared with traditional recommender systems with usually one single rating, multi-criteria recommender systems have several ratings from different aspects, and generally describe users' interests more accurately. However, owing to the cost of ratings, multi-criteria recommender systems meet more severe data sparsity problem than traditional single criteria recommender systems.

In this paper, We design a new approach to compute the similarity between users, which tackles the challenge posed by data sparsity that one cannot obtain the similarity between users with no common rated items. With a new method of data preprocessing, the features of items are combined to eliminate the effect of noise and evaluation scale. We model the aggregation function using support vector regression which is more accurate and robust than linear regression. The experiments demonstrate that our method produces a better performance, while providing more powerful suitability on sparse and noisy datasets.

Keywords: recommendation, multi-criteria, support vector regression, sparsity, preference.

1 Introduction

While the Internet gives users easy access to a lot of resources, information overload has become a big challenge to us. There are a huge amount of movies, books, CDs, articles published every day, hour and minute, it is impossible for an individual to find interesting items just by himself. Recommender systems can tackle this problem by helping users to find information or items that they may like the best.

The problem of recommendation has been identified as the way to help individuals in a community to find information or items that are most likely to be interesting to them or to be relevant to their needs[1][2][3]. A utility function f: $Users \times Items \rightarrow R$ is defined to measure the interests of users to items, where R

C. Guo, Z.-G. Hou, and Z. Zeng (Eds.): ISNN 2013, Part II, LNCS 7952, pp. 385–394, 2013.
© Springer-Verlag Berlin Heidelberg 2013

is typically represented by non-negative integers or real numbers within a certain range. For each user, we want to find items that maximize his utility. Unfortunately, utility function is just defined on some subset of $Users \times Items$ space. Therefore, the crucial problem of recommendation systems is to extrapolate the missing values of the utility function to the whole $Users \times Items$ space.

In most existing recommender systems, utility function is usually considered to be a single rating, named as overall rating. However, a single rating is sometimes not sufficient to illustrate the users' preference. For example, many fans of "Transformers" like the movie because of the scenes, but others chase it for the story. So drilling down the overall rating to find out the explicit preference in each aspect can help to understand users' interests more accurately.

Therefore, with growing number of real-world applications, extending recommendation techniques to incorporate multi-criteria ratings has been regarded as one of the most important issues for the next generation of recommender systems[4]. There already exist some applications of multi-criteria recommender systems: TripAdvisor [1] provides five criteria for evaluating hotels from "Overall", "Food", "Service", "Value" and "Atmosphere"; DianPing.Com [2] provides four criteria for restaurants from "Overall", "Taste", "Environment" and "Service". The utility function of multi-criteria recommendation can be defined as follows:

$$f : Users \times Items \rightarrow R_0 \times R_1 \times \cdots \times R_k \tag{1}$$

where R_0 is the overall rating, R_i $(i = 1, 2, \cdots, k)$ is the rating of each individual criterion, named sub-criterion in this paper. In addition to the overall rating, sub-criteria ratings provide more information about users' preference over different aspects of items. Recommender systems could benefit from leveraging this additional information since it could potentially increase the recommendation accuracy.

With the development of e-commerce in service domain, multi-criteria recommendations get more and more attention. Adomavicius and Kwon[5] propose a framework named as aggregation-function-based approach, which assumes the overall rating serves as an aggregation function of the sub-criteria ratings. They also extend the traditional collaborative filtering(CF)[6] approaches to multi-criteria recommender systems with distance-based similarity[5]. Tang and McCalla[7] think that the users' similarity can be represented as a weighted sum of users' similarities over individual criteria. Liu et al.[8] consider the overall rating can be represented as a linear combination of ratings of sub-criteria. But these methods mentioned above all involve decomposing the multi-criteria problems into several single criterion problems, which ignore the relationship among criteria. Xin et al.[9] propose an algorithm named CMAP, which combines quality-based and relevance-based algorithms together. Li et al.[10] propose a novel approach to improve a traditional CF algorithm by utilizing the MSVD[11](Multilinear Single Value Decomposition) technique.

Despite the significant research progress in multi-criteria recommendations, the existing techniques draw lessons from CF which calculates similarity between

---

[1] www.tripadvisor.com
[2] www.dianping.com

users by integrating the ratings of the common rated items. That makes data sparsity more severe, since most users have no common rated items. In most existing recommender systems, such as Amazon, BestBuy, there are a great many of categories of items. It is quite inaccurate to use the rating of an old lady assigns to ipad to estimate the rating of an enthusiastic apple fan. So we need to verify the neighbors of the target user by checking whether they have common shopping records in the past. Due to difficulties to find a set of uniform criteria for different kinds of items, in existing multi-criteria recommender systems, usually the items are in the same category, such as movies or restaurants. Any user in these systems can give a relatively accurate evaluation to items in this unique category. Therefore, we design a preference-based similarity, by which we can obtain similarities between any users.

In this paper, our approach falls into the framework of aggregation-function-based approach. We construct a preference-based similarity formulation, by which we can obtain the similarity value between any users without decomposing the multi-criteria problems. Then we design a set of methods to pre-process the original ratings to get the user preference. Finally, support vector regression(SVR)[12] is used to model the aggregation function for each user, which is more accurate and robust than linear regression(LR) employed in [8].

The rest of the paper is organized as follows: our approach is detailed in section 2. Relevant experiments and results are included in section 3. Section 4 contains the conclusions of the paper and the future work.

2   Multi-criteria Recommendation with Preference-Based Similarity and SVR

2.1   Framework

The framework of our approach is shown in figure 1, which falls into aggregation-function-based approach framework. In aggregation-function-based approach[5], it is assumed that the overall rating can be represented by an aggregation function of sub-criteria ratings. i.e.

$$r_0 = f(r_1, ..., r_k) \tag{2}$$

The aggregation function which is chosen based on domain expertise or machine learning techniques, has three scopes: total, user-based, item-based. For the total scope, different users share the same aggregation function which is learned from the entire dataset; for user-based or item-based, a separate aggregation function is learned for each user or item.

In this paper, aggregation functions are learned for every user, it is more individualised than total scope. We state our approach as following steps:

1. Predict sub-criteria ratings r_1, r_2, ..., r_k;
2. Learn an aggregation function f for each user;
3. Synthesize overall ratings $r_0 = f(r_1, ..., r_k)$.

Then we can recommend items to users with r_0 as traditional recommendation systems.

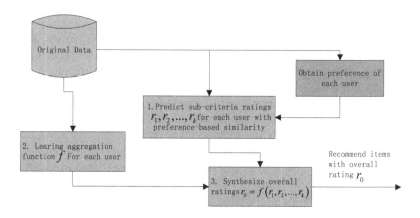

Fig. 1. The Framework of Our Approach

2.2   Predict Sub-criteria Ratings

In this section, we first formulate the users' preference, and then propose preference-based similarity with multi-dimensional distance metrics. Lastly, we design a formulation to generate the sub-criteria ratings r_1, r_2, ..., r_k of the target user by integrating the information of his neighbors.

Obtain Preference. It is noticed that original ratings cannot express users' evaluations and preference exactly. Due to different evaluation scales, a strict user may give 3/5 points to an item which he considers good, while a loose user may give 4/5 points to an item which he considers not good. Also we cannot get the conclusion that one user is strict just because his average rating is lower than the public, since it is possible that all items he rated are low quality. So we combine the features of items and evaluation scales to preprocess the original data. Here, we use a simple method: the average ratings, to represent the features of items.

$$f_{ic} = \frac{1}{|U|} \sum_{u \in U} r^c_{ui}, c \in [1, k] \tag{3}$$

where f_{ic} is the feature of c-th criterion of item i, U is the set of users who rated the item i and r^c_{ui} is the rating of c-th criterion that user u assigns to item i. In order to reflect one user's preference over each criterion from the rating he assigns to an item, features of this item should be combined to transform the original ratings. So we give the following formulation, where rt^c_{ui} represents the transformed rating of c-th criterion that user u assigns to i, f_{ij} represents the feature of j-th criterion of item i.

$$rt^c_{ui} = r^c_{ui} * ((\sum_{j=1}^{k} f_{ij})/f_{ic}), \ c \in [1, k] \tag{4}$$

We still need to normalize the ratings to eliminate the effects of evaluation scale. User preference can be measured by averaging normalized ratings of all items that he rated. p_{uc} is the preference of user u over c criterion, I is the set of items that user u rated, and $\widetilde{rt}^c_{ui}$ is the normalized rating of c criterion that user u give to item i.

$$\widetilde{rt}^c_{ui} = rt^c_{ui} / \sqrt{\sum_{j=1}^{k} (rt^j_{ui})^2}, \; c \in [1, k] \tag{5}$$

$$p_{uc} = \frac{1}{|I|} \sum_{i \in I} \widetilde{rt}^c_{ui}, c \in [1, k] \tag{6}$$

Preference-Based Similarity. Due to the reasons discussed in section 1, we can compute the similarities between users with preference instead of ratings of common rated items. The similarity formula can be defined as

$$sim(u, v) = \frac{1}{1 + dist(\overrightarrow{p_u}, \overrightarrow{p_v})} \tag{7}$$

$\overrightarrow{p_u}$ and $\overrightarrow{p_v}$, which are calculated by Eq.6, represent the preference vectors of user u and v, and $dist(\overrightarrow{p_u}, \overrightarrow{p_v})$ is the euclidean distance between preference vectors $\overrightarrow{p_u}$ and $\overrightarrow{p_v}$.

We are delighted to find that operands are just two k-dimensions vectors in (6), in contrast to $|I| * k$-dimensions vectors, where I represents the common rated items of two users. More importantly, we can obtain the similarities of users without common rated items, so we could integrate more information to predict the unknown ratings.

Preference Order Based Cluster. In traditional CF, the neighbors of the target user are identified by similarity of them. Here, we assume neighbors also should share similar preference over each criterion. For the target user who takes "Taste" as the most important criterion and "Service" as the least important criterion, any users have the same opinion and high similarity with him should be considered as his neighbors. It is noticed that the criterion highly valued by one user gets a small number in his preference. Therefore, we cluster the users by their similarity and the order of the criteria in preference. However, users whose preferences include k criteria, will be divided into factorial of k clusters, there will be not enough data to utilize in each cluster. So, we can cluster users according to the order of their significant criteria. Here, we choose top-$n(n \leq k)$ important criteria as the significant criteria, and cluster users into factorial of n clusters according to the order of significant criteria.

Predict Ratings of Sub-criteria. When we get the neighbors of user u, we can predict the ratings of the sub-criteria that u gives to an item i based on the ratings of i rated by neighbors of u. The similarity-based weighted sum approaches are often used in CF. However, using these methods, the ratings we obtained are between the lowest and highest ratings given by neighbors. Thereby, for one user whose rating is lower or higher than his neighbors, the rating predicted is inaccurate. We need to take evaluation scale into consideration. So given a user u and his neighbors $U = \{u_1, u_2, ..., u_n\}$, we predict the sub-criteria rating of u as follows:

$$C_{ui}^{\;c} \underset{c=1,...,k}{=} \frac{1}{N} (\sum_{v \in U} \frac{\overline{u^c}}{\overline{v^c}} * C_{vi}^c) \tag{8}$$

$\overline{u^c}$ and $\overline{v^c}$ are the average ratings of c th criterion of user u and v. C_{vi}^c represents the rating of item i on criteria c rated by user v. N is the size of user u's neighbors.

2.3   Model Aggregation Function

In order to synthesize the overall rating, we should model an aggregation function for each user in advance. Aggregation function is considered as linear function of sub-criteria rating in [8]. It is well known that LR is vulnerable to singular points. In recommender systems, there always exist a large amount of noise and random factors due to diverse collection of ratings. Sometimes users may give random ratings for bad mood or saving time; the rivals may give abnormal ratings deliberatively. So we model aggregation functions by another approach: SVR.

Support Vector Regression. SVR is a supervised model used for regression analysis. It's aimed at maximizing the margin which is the distance of training samples that are closest to separating hyperplane. More detailed explanations can be found in [13]. Here we present the final formulation of the model.

$$\max \quad -\frac{1}{2}\sum_{i,j=1}^{l}\left(a_i - a_i^*\right)\left(a_j - a_j^*\right)K\left(x_i, x_j\right)$$
$$-\varepsilon\sum_{i=1}^{l}\left(a_i + a_i^*\right) + \sum_{i=1}^{l}y_i\left(a_i - a_i^*\right) \qquad (9)$$
$$s.t. \quad \sum_{i=1}^{l}\left(a_i - a_i^*\right) = 0, a_i, a_i^* \in [0, C]$$

a_i^*, a_i are named dual variables, which represent the weights of every sample, $K\left(x_i, x\right)$ is the kernel function. Only a few training samples, namely Support Vectors, obtain non-zero weights, determine the prediction. This basic nature of SVR makes it not sensitive to singular points and tackle the disadvantage of overfitting. We could obtain dual variables from the solution of Quadratic Programming.

We can get the prediction by the following function, where b is a constant threshold.

$$f\left(x\right) = \sum_{i=1}^{N}(a_i^* - a_i)K\left(x_i, x\right) + b \qquad (10)$$

where $f\left(x\right)$ is the aggregation function we want to model for each user. At last, we can synthesize the overall ratings with sub-criteria ratings and aggregation functions, then recommend items to users according to the overall ratings.

3   Experimental Results

In order to evaluate the performance of our approach, we conduct a set of experiments where we compare the proposed methods with some existing techniques.

All experiments were implemented based on PC with Windows 7, with AMD Phenom X4 CPU, 3.20GHz and 4GB RAM.

3.1 Dataset

The ratings dataset which is an example of service domain is collected from DianPing.Com. There are three sub-criteria ratings, from "Taste", "Environment", "Service", and an overall rating. These three sub-criteria ratings range from 0 to 4, with 4 as the excellent, and overall rating ranges from 1 to 5. There are 422,284 records in our dataset, with 5,484 restaurants and 6,503 users. We discard the users who have rated less than 20 restaurants and the restaurants which have less than 20 users' ratings, and the records with all sub-criteria ratings equal to zero. After cleaning, there are 163,057 records left with 3,560 restaurants and 2,939 users. The sparsity level of data is 1.56%. 80% records are selected randomly as the training set, and 20% records as the testing set.

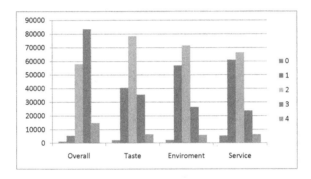

Fig. 2. Distribution of Restaurant Ratings in Dataset

The number of ratings of each criterion in our dataset is ploted in Fig. 2. It shows the distribution of restaurant ratings in our dataset. In order to bring the overall rating in accordance with sub-criteria ratings, we rescale the overall rating range from 1-5 to 0-4. We can observe that the data distribution is very concentrated, so it is conclude that the results could be very good.

3.2 Evaluation Metric

Mean Absolute Error(MAE)[14] is one of the most commonly used statistical accuracy metric. It measures the accuracy by the average absolute deviation between predicted ratings and true ratings.

$$MAE = \frac{\sum_{i=1}^{n} |p_i - q_i|}{n} \tag{11}$$

where p_i is the predicted rating of testing sample i, q_i is the true rating of testing sample i, n is the size of testing set. The lower the MAE is, the more accurate predictions are[15].

3.3   Performance of Four Methods

In the framework of our algorithm, the first step is to predict the ratings of sub-criteria with Preference-Based Similarity(PBS). In order to validate the effective of our new similarity formulation, we show not only the results of overall rating prediction, but the results of sub-criteria ratings prediction. We choose the traditional user-based CF as the baseline named Single in Fig. 3, by decomposing multi-criteria problem into 4 single criteria problems and solving them separately. ExtendedCF[4] is the method extended from traditional CF with euclidean-distance-based similarity. PBS-Cluster and PBS-NoCluster correspond to two variations of our method detailed in Section 2.2, where PBS-NoCluster does not cluster users by preference order, while PBS-Cluster involves clustering users with preference order.

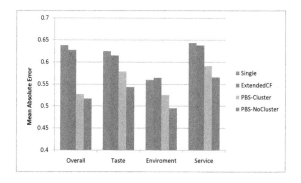

Fig. 3. MAEs of different approaches

As is shown in Fig. 3, our approaches outperform the Single and ExtendedCF. With PBS-Cluster, the accuracy improves around 10% on sub-criteria ratings prediction and 16% on overall rating prediction, compared with Single and Extended CF. It is proved that our framework with preference-based similarity is very effective. An interesting observation is that the method PBS-Cluster performs poorer than PBS-NoCluster. This is because after clustering, we can only utilize the information in same cluster, which worsens the data sparsity problem. However, PBS-Cluster has superiority in time cost over PBS-NoCluster, as is shown in table 1.

Table 1. Time Cost

Method	PBS-NoCluster	PBS-Cluster	ExtendedCF
Time	4783s	411s	2269s

The results demonstrate that with clustering, PBS-Cluster provides a more than ten times speed than the method PBS-NoCluser, which indicates a tradeoff between accuracy and time-complexity. For time-consuming problem, PBS-Cluster method may be a better choice.

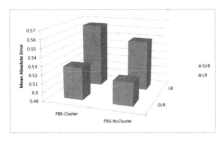
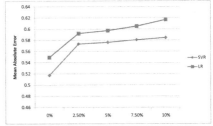

Fig. 4. Accuracy of SVR and LR   **Fig. 5.** Robustness of SVR and LR

3.4 SVR versus LR

As we discussed in section 2.3, there always exists a large amount of noise and random factors in multi-criteria recommender systems. So we use the more robust method SVR to learn the aggregation functions instead of LR which is used in [8]. The result is shown in Figure 4. According to the result, the aggregation function learned by SVR have a better performance, the accuracy is improved by 5%-6% compared with aggregation function learned by LR.

In order to check the robustness of SVR, we add outliers to training set, and conduct another set of experiments based on the ratings of sub-criteria predicted by PBS-NoCluster. we add some records as outliers, of which sub-criteria ratings generated from [0,2] and overall rating generated from [4,5] randomly, or vice verse. We notice that, the performance of the both two algorithms get worse as the number of outliers increase. But compared with LR, SVR performs more stably. The comparison of LR and SVR is shown in figure 5. The X axis represents the percentage of outliers in training set.

4 Conclusion and Future work

In this paper, we are motivated by three facts: original ratings cannot express preference of users exactly; multi-criteria recommender systems usually recommend items in the same category; the data is sparse and noisy in multi-criteria recommender systems. So we propose a new method to preprocess the original data, to alleviate the issues of evaluation scale and ignorance of items feature. Similarity is calculated by users' preference instead of ratings of common rated items, which tackles the data sparsity problem. To speed up the predicting process we cluster the users with their preference order, at some cost of accuracy. In order to improve the accuracy and robustness, we learn the aggregation function by SVR instead of LR. In the end, we prove that our methods provide a good performance compared with other methods. Thus, from a technical standpoint, we believe that the work in this paper will be helpful in improving recommender system in real applications.

Compared with ratings, reviews are not affected by evaluation scales, can express users' feelings more straightforwardly and objectively. Therefore, more information can be obtained by mining and summarizing customer reviews[16]. In our future work, we plan to exploit reviews to improve the performance of the proposed technique.

Acknowledgments. Research supported by the National Natural Science Foundation of China (No. 61003135) and the Fundamental Research Funds for the Central Universities (WK0110000022).

References

1. Adomavicius, G., Tuzhilin, A.: Toward the next generation of recommender systems: A survey of the state-of-the-art and possible extensions. IEEE Transactions on Knowledge and Data Engineering 17(6), 734–749 (2005)
2. Konstan, J.: Introduction to recommender systems: Algorithms and evaluation. ACM Transactions on Information Systems (TOIS) 22(1), 1–4 (2004)
3. Resnick, P., Iacovou, N., Suchak, M., Bergstrom, P., Riedl, J.: Grouplens: an open architecture for collaborative filtering of netnews. In: Proceedings of the 1994 ACM Conference on Computer Supported Cooperative Work, pp. 175–186. ACM (1994)
4. Adomavicius, G., Manouselis, N., Kwon, Y.: Multi-criteria recommender systems. In: Recommender Systems Handbook, pp. 769–803 (2011)
5. Adomavicius, G., Kwon, Y.: New recommendation techniques for multicriteria rating systems. IEEE Intelligent Systems 22(3), 48–55 (2007)
6. Breese, J., Heckerman, D., Kadie, C.: Empirical analysis of predictive algorithms for collaborative filtering. In: Proceedings of the Fourteenth Conference on Uncertainty in Artificial Intelligence, pp. 43–52. Morgan Kaufmann Publishers Inc. (1998)
7. Tang, T., McCalla, G.: The pedagogical value of papers: a collaborative-filtering based paper recommender. Journal of Digital Information 10(2) (2009)
8. Liu, L., Mehandjiev, N., Xu, D.: Multi-criteria service recommendation based on user criteria preferences. In: Proceedings of the Fifth ACM Conference on Recommender Systems, pp. 77–84. ACM (2011)
9. Xin, X., Lyu, M., King, I.: Cmap: effective fusion of quality and relevance for multi-criteria recommendation. In: Proceedings of the Fourth ACM International Conference on Web Search and Data Mining, pp. 455–464. ACM (2011)
10. Li, Q., Wang, C., Geng, G.: Improving personalized services in mobile commerce by a novel multicriteria rating approach. In: Proceeding of the 17th International Conference on World Wide Web, pp. 1235–1236 (2008)
11. De Lathauwer, L., De Moor, B., Vandewalle, J.: A multilinear singular value decomposition. SIAM Journal on Matrix Analysis and Applications 21(4), 1253–1278 (2000)
12. Vapnik, V., Golowich, S., Smola, A.: Support vector method for function approximation, regression estimation, and signal processing. Advances in Neural Information Processing Systems, 281–287 (1997)
13. Joachims, T.: Text categorization with support vector machines: Learning with many relevant features. Machine Learning: ECML 1998, 137–142 (1998)
14. Willmott, C., Matsuura, K.: Advantages of the mean absolute error (mae) over the root mean square error (rmse) in assessing average model performance. Climate Research 30(1), 79 (2005)
15. Sarwar, B., Karypis, G., Konstan, J., Riedl, J.: Item-based collaborative filtering recommendation algorithms. In: Proceedings of the 10th International Conference on World Wide Web, pp. 285–295. ACM (2001)
16. Hu, M., Liu, B.: Mining and summarizing customer reviews. In: Proceedings of the Tenth ACM SIGKDD International Conference on Knowledge Discovery and Data Mining, pp. 168–177. ACM (2004)

Semi-Supervised Learning Using Random Walk Limiting Probabilities

Thiago Henrique Cupertino and Liang Zhao

Institute of Mathematics and Computer Science
University of São Paulo, São Carlos, Brazil
{thiagohc,zhao}@icmc.usp.br

Abstract. The semi-supervised learning paradigm allows that a large amount of unlabeled data be classified using just a few labeled data. To account for the minimal *a priori* label knowledge, the information provided by the unlabeled data is also used in the classification process. This paper describes a semi-supervised technique that uses random walk limiting probabilities to propagate label information. Each label is propagated through a network of unlabeled instances via a biased random walk. The probability of a vertex receiving a label is expressed in terms of the limiting conditions of the walk process. Simulations show that the proposed technique is competitive with benchmarked techniques.

Keywords: network-based learning, semi-supervised learning, random walk, limiting probabilities, stationary distribution.

1 Introduction

Semi-supervised learning (SSL) is a machine learning paradigm that overcomes the problem originated when labeling a training data set becomes expensive and very time-consuming. The main idea behind this paradigm is to classify data using just a few labeled instances and the information provided by many unlabeled instances [1]. This is practicable due to three SSL assumptions: manifold, smoothness and cluster. The manifold assumption states that the high-dimensional data lies on a low-dimensional manifold whose properties ensure more accurate density estimation and more appropriate similarity measures. The smoothness assumption states that if two points are close to each other in a high density region, then their correspondent labels should be close to each other as well. Finally, the cluster assumption states that if two points are in the same cluster, then they are likely to be of the same class (or, in other words, to have the same label). In this way, the SSL approach can provide high classification accuracies using less human effort and exploiting the unlabeled massive group of data.

Random walk theory has been applied in many machine learning problems. In image analysis, for example, a random walk process can be executed through pixels represented by network vertices. Texture discrimination and edge detection were performed by comparing *boundary distributions* of such process [2, 3].

C. Guo, Z.-G. Hou, and Z. Zeng (Eds.): ISNN 2013, Part II, LNCS 7952, pp. 395–404, 2013.

As an alternative way to perform edge detection and image segmentation, other measurements were derived. In [4], the *first time passage* probability is computed when a random walker passes through a labeled vertex (pixel) after starting from an unlabeled vertex. A similar approach applied to content-based image retrieval can be found in [5]. An agglomerative network-based classification method was introduced in [6]. In this work, the hierarchy is based on *life-time* of restricted random walks, in which the steps of a random walker are limited by a distance function of antecedent steps. The time approach was also used in [7], in which classification is achieved by comparing *commute times* to labeled points of different classes. In [8], an unlabeled instance is classified as the class which maximizes a *posterior probability* by considering the time a walker starting in a vertex of the same class reaches the unlabeled instance.

Despite many concepts of random walk processes were already applied to classification and correlate tasks, *limiting probabilities* has never been directly applied, to our knowledge, as is done in this work. To account for the usage of the information provided by the unlabeled instances in the SSL task, the random walk process takes into account the whole network, which is composed by the unlabeled instances. The few labeled instances are inserted into the network by a specific weight composition responsible for creating a bias to the classification process. This bias is taken into account when the limiting probabilities are calculated. As simulation results showed, this is an effective measurement to capture intrinsic relations among labeled and unlabeled vertices in a network.

2   Background: Random Walks and Limiting Probabilities

Random walks can be understood in terms of Markov chains [9]. Consider a stochastic process Ω with a finite state space Γ. For each $n \in \mathbb{N} = \{0, 1, 2, \ldots\}$, $\Omega_n \in \Omega$ is an element from Γ. Then, the stochastic process $\Omega = \{\Omega_n\}$ is called a Markov chain if probability $P(\Omega_{n+1} = i | \Omega_0, \ldots, \Omega_n) = P(\Omega_{n+1} = i | \Omega_n)$, $i \in \Gamma$, that is, the process is independent of past states provided that the current state Ω_n is known. In this work, a time-homogeneous chain is considered: when $P(\Omega_{n+1} = j | \Omega_n = i) = p_{ij}$ is independent of n [10]. The probabilities p_{ij} can be arranged into a Markov matrix $\mathcal{P} = \{p_{ij}\}$.

The probability for a random walk starting at state i_0 to end at a state i_m is given by the probability of the chain $P(i_0, i_m) = P(i_0, i_1)P(i_1, i_2) \ldots P(i_{m-1}, i_m)$. If an infinite number of transitions ($m \to \infty$) is considered, then a limiting probability (stationary state) needs to be calculated. It can be shown that the limiting probability $P^{\infty}(i_m) = \lim_{n \to +\infty} P^n(i_0, i_m)$ exists given a recurrent, non-null and aperiodic state i_m. Therefore, considering $|\Gamma| = q$, one is able to calculate the Markov matrix. The limiting probability of the final state i_m is independently of the initial state i_0 [10].

3   The Proposed Semi-Supervised Technique

Given a dataset $\mathcal{X} = \{\mathbf{x}_i, i = 1, \ldots, r\}$, the objective is to classify the subset of unlabeled instances $\mathcal{X} \supset \mathcal{X}^{(u)} = \{\mathbf{x}_i^{(u)}, i = 1, \ldots, n\}$ using the subset composed of just a few labeled instances $\mathcal{X} \supset \mathcal{X}^{(l)} = \{\mathbf{x}_i^{(l)}, \ i = 1, \ldots, m\}, \ l \in \{1, 2, \ldots, c\}$ ($\mathcal{X}^{(l)} \cap \mathcal{X}^{(u)} = \emptyset$, $\mathcal{X}^{(l)} \cup \mathcal{X}^{(u)} = \mathcal{X}$). To characterize a SSL task, $m \ll n$.

First, an undirected network $\mathcal{N} = \{\mathcal{V}, \mathcal{E}\}$ without self-loops is created. In this network, instances are represented by vertices, $\mathcal{V} = \mathcal{X}^{(u)}$, and similarities among instances are represented by edges, $\mathcal{E} = [\mathcal{W}_{ij}], \ i, j = 1, \ldots, n$. The network similarity matrix $\mathcal{W} = \{w_{ij}\}$ is calculated by using some sort of distance function as, for example, the Euclidean distance. w_{ij} is the similarity between a pair of instances $\mathbf{x}_i^{(u)}$ and $\mathbf{x}_j^{(u)}$. $w_{ij} = 0$ means that there is no link between $\mathbf{x}_i^{(u)}$ and $\mathbf{x}_j^{(u)}$.

In the next step, an labeled instance $\mathbf{x}_j^{(l)}$ is inserted into the network $\mathcal{N}$. To do so, the similarities $\mathcal{S}_j = [s_{j1}, s_{j2}, \ldots, s_{jn}]^T$, between $\mathbf{x}_j^{(l)}$ and all other vertices $\mathbf{x}_i^{(u)} \in \mathcal{V}$ are calculated by using a distance function, and a new asymmetric $n \times n$ modified similarity matrix $\hat{\mathcal{W}}_j$ is constructed by composing it with the matrix of weight biases $\hat{\mathcal{S}}_j$:

$$\hat{\mathcal{W}}_j = \mathcal{W} + \epsilon \hat{\mathcal{S}}_j, \tag{1}$$

where ϵ is a non-negative parameter and $\hat{\mathcal{S}}_j$ is the following $n \times n$ matrix composition:

$$\hat{\mathcal{S}}_j = \begin{bmatrix} \mathcal{S}_j^T \\ \mathcal{S}_j^T \\ \vdots \\ \mathcal{S}_j^T \end{bmatrix}, \tag{2}$$

where $\mathcal{S}_j^T$ is the transpose of vector $\mathcal{S}_j$.

In Eq. 1, it can be observed that the weight biases of $\mathbf{x}_j^{(l)}$, encoded in matrix $\hat{\mathcal{S}}$, are applied over all edges w_{ij} of network $\mathcal{N}$, that is, the weight of each edge is linearly added up with the corresponding weight bias. The idea behind this operation is that the distance between any pair of vertices is reduced because of the new route introduced by the insertion of the labeled data instance. The higher the proximity between the labeled instance and a vertex, say vertex i, the more strengthened the connections from all other vertices to vertex i are. The parameter ϵ controls the influence of weight bias provided by matrix $\hat{\mathcal{S}}$ on the original network. The larger is the value of parameter ϵ, the greater will be the influence of the bias weights provided by $\mathbf{x}_j^{(l)}$.

After the bias composition, $\mathbf{x}_j^{(l)}$ is effectively inserted into network $\mathcal{N}$. To do so, an $(n+1) \times (n+1)$ adjacency matrix $\mathcal{A}_j = \{a_{ij}\}$ is constructed:

$$\mathcal{A}_j = \begin{bmatrix} \hat{\mathcal{W}} & \mathcal{S}_j \\ \mathcal{S}_j^T & 0 \end{bmatrix}. \tag{3}$$

In this formulation, without loss of generality, $\mathbf{x}_j^{(l)}$ is inserted as the last entry $(n+1)$ of matrix $\mathcal{A}_j$.

The two steps described above can be easily understood by using the toy example depicted in Fig. 1. In this example, a network is formed by 4 unlabeled

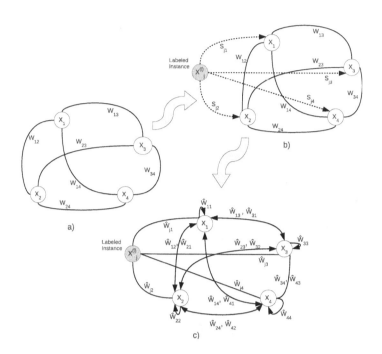

Fig. 1. Composition process for the modified similarity matrix $\hat{\mathcal{W}}_j$. a) An undirected and complete network $\mathcal{N}$ is formed by using 4 unlabeled instances; b) similarities $\mathcal{S}_j$ are calculated for the labeled instance $\mathbf{x}_j^{(l)}$; c) modified network $\mathcal{A}_j$, directed with self-loops, after bias composition (Eq. 1) and insertion of the labeled virtual state $\mathbf{x}_j^{(l)}$ (Eq. 3).

instances (Fig. 1a). In this initial network, the links are undirected and the similarity matrix is symmetric. Next, the similarity vector $\mathcal{S}_j$ between $\mathbf{x}_j^{(l)}$ and all other vertices is computed (Fig. 1b). After this computation, the weight biases of $\mathbf{x}_j^{(l)}$ are added up to the original similarity matrix to form a biased similarity matrix for the same network (Eq. 1). After that, $\mathbf{x}_j^{(l)}$ is effectively inserted into the network (Eq. 3). It can be seen from Fig. 1c that the network becomes directed and with self-loops (except for $\mathbf{x}_j^{(l)}$, that has no self-loops) and the biased weight matrix is no more symmetric.

After the insertion of $\mathbf{x}_j^{(l)}$, we are able to compute the entries of the transition Markov matrix $\mathcal{P}_j = [p_{ik}]$ by scaling the entries of matrix $\mathcal{A}_j$:

$$p_{ik} = a_{ik} / \sum_{k=1}^{n+1} a_{ik}. \tag{4}$$

With the above matrix $\mathcal{P}_j$ at hand, the limiting probabilities are calculated. This calculation can be performed by two ways: finding the eigenvector corresponding to the unit eigenvalue of matrix $\mathcal{P}_j$ or, in a faster way, iterating the system

$$\mathbf{p}_j(t+1) = \mathcal{P}_j \mathbf{p}_j(t), \tag{5}$$

to the stationary state, where $\mathbf{p}_j$ is an $(n+1) \times (n+1)$ normalized vector. It results in the following vector:

$$\mathbf{p}_j^\infty = [p_1 \, p_2 \ldots p_{n+1}] \tag{6}$$

where each column represents an unlabeled vertex, and each entry p_i represents the probability that $\mathbf{x}_i^{(u)}$ belongs to the class l of the labeled vertex $\mathbf{x}_j^{(l)}$ (the probability p_{n+1} is ignored as it represents $\mathbf{x}_j^{(l)}$, which is inserted into $\mathcal{N}$ by means of Eq. 3).

Finally, the classification is completed by repeating the above steps (Eq. 1 through 6) for all labeled instances $\mathbf{x}_j^{(l)} \in \mathcal{X}^{(l)}$. The classification of $\mathbf{x}_i^{(u)}$ is accomplished by assigning it the most representative label, that is, after averaging all $\mathbf{p}_j^\infty$ for each label (each $\mathbf{x}_j^{(l)}$), the following matrix is constructed:

$$\mathbf{p}^\infty = \begin{bmatrix} p_{11} & p_{21} & \cdots & p_{n1} \\ p_{12} & p_{22} & \cdots & p_{n2} \\ \vdots & \vdots & \ddots & \vdots \\ p_{1c} & p_{2c} & \cdots & p_{nc} \end{bmatrix}, \tag{7}$$

where p_{il}, $l = 1, 2, \ldots, c$, is the averaged probability that instance $\mathbf{x}_i^{(u)}$ belongs to class l. Then, the label $p_{max} = argmax\{p_{il}\}$, corresponding to the largest probability value, is assigned to $\mathbf{x}_i^{(u)}$.

In a concise form, the proposed semi-supervised transductive classification can be summarized by Algorithm 1.

4   Illustrative Toy Example

In this subsection, simulation results on a toy example are presented. It was used a toy data set that captures different class characteristics, such as different shapes and densities, to illustrate the behavior of the semi-supervised technique. The toy example in Fig. 2 encompasses a challenging classification task. This data set is composed of 3 different class distributions (from left to right): Gaussian, Highleyman and Lithuanian. The data was generated by using the PRTools toolbox [11]. Each class has 500 instances, totaling 2500 instances for the entire data set. In addition, each class comprises 10 labeled instances, representing (2%) of the entire data. Figure 2b shows that the proposed technique satisfactorily detected the 5 classes.

Algorithm 1. The proposed semi-supervised technique

Input:
c : number of classes
$\mathcal{X}^{(u)}$: unlabeled dataset
$\mathcal{X}^{(l)}$: labeled dataset
Parameters:
ϵ : bias weighting
Output:
Estimated class ($l \in \{1, \ldots, c\}$) for $\mathbf{x}_i^{(u)} \in \mathcal{X}^{(u)}$
Training:
1. $\mathcal{N}$ = Create a network from $\mathcal{X}^{(u)}$
Classification:
for each $\mathbf{x}_j^{(l)} \in \mathcal{X}^{(l)}$ do
   2. $\hat{\mathcal{W}}_j$ = Compose bias weights into $\mathcal{N}$ (Eq. 1)
   3. $\mathcal{A}_j$ = Insert the virtual state $\mathbf{x}_j^{(l)}$ into $\mathcal{N}$ (Eq. 3)
   4. $\mathbf{p}_j^{\infty}$ = Compute limiting probabilities (Eq. 5)
end for
5. $\mathbf{p}^{\infty}$ = Averaged $\mathbf{p}_j^{\infty}$
6. Assign $\mathbf{x}_i^{(u)}$ the most representative class in $\mathbf{p}^{\infty}$

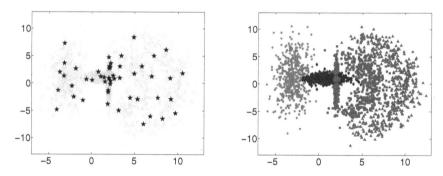

(a) Black-filled stars represent labeled in- (b) Classification achieved by the proposed
stances.   technique.

Fig. 2. Mix of different cluster shapes for semi-supervised classification. This artificial
data set is composed of 2500 instances divided into 5 balanced and distinct clusters
shapes: Gaussian, Highleyman and Lithuanian. Each cluster contains 10 labeled instances.

5 Benchmark Data Sets

The proposed semi-supervised technique was tested and compared using 7 benchmark data sets. Table 1 shows a brief description of them. Three artificial sets
(*g241c*, *g241d* and *Digit1*) were created in order to encompass some of the semi-
supervised assumptions: manifold, smoothness and cluster. The other four data

Table 1. Meta-data of the datasets composing the SSL benchmark

Data Set	Classes	Dimension	Points	Type
g241c	2	241	1500	artificial
g241d	2	241	1500	artificial
Digit1	2	241	1500	artificial
USPS	2	241	1500	unbalanced
COIL	6	241	1500	
BCI	2	117	400	
Text	2	11960	1500	sparse discrete

Table 2. References for semi-supervised learning techniques used for comparisons

Abbreviation	Technique	Ref.
MVU + 1-NN	Maximum Variance Unfolding	[14]
LEM + 1-NN	Laplacian Eigenmaps	[15]
QC + CMR	Quadratic Criterion and Class Mass Reg.	[16]
Discrete Reg.	Discrete Regularization	[17]
TSVM	Transductive Support Vector Machines	[18]
SGT	Spectral Graph Transducer	[19]
Cluster-Kernel	Cluster Kernels	[20]
Data-Dep. Reg.	Data-Dependent Regularization	[21]
LDS	Low-Density Separation	[18]
Laplacian RLS	Laplacian Regularized Least Squares	[22]
CHM (normed)	Conditional Harmonic Mixing	[23]
LGC	Local and Global Consistency	[12]
LP	Label Propagation	[24]
LNP	Linear Neighborhood Propagation	[13]

sets (*USPS*, *COIL*, *BCI* and *Text*) were derived from real data. The benchmarks were developed to evaluate the power of different algorithms as neutral as possible [1]. For each data set, 24 independent splits of labeled data for the training set are available. 12 splits contain 10 labeled instances for each data set and the other 12 splits contain 100 labeled instances. For each split, at least 1 instance of each class is labeled. A more detailed explanation of each data set can be found in [1]. The proposed technique was compared to 16 well-known and established SSL techniques. Table 2 shows a brief description and the related references for them. All simulation results were extracted from the reference [1], where it can be found values for parameter optimization and model selection in order to minimize test errors. For LGC, LP and LNP, σ was selected from the set $\{0, 1, \ldots, 100\}$ and α was fixed to $\alpha = 0.99$ (the same setup done in [12] and [13]). For the LNP, k was evaluated for the values in $\{1, 2, \ldots, 100\}$. The configuration and parameter optimization for the proposed technique was done as follows. For the network construction (Step 1 of Alg. 1), the k-nearest neighbor technique was used: each vertex was linked with its k most similar neighbors. Parameter k was evaluated for the values in $\{1, 2, \ldots, 100\}$ and parameter ϵ was evaluated for the values in $\{0, 0.1, 0.2, \ldots, 10\}$. In all simulations, no data preprocessing was performed by the techniques and the Euclidean distance was used as a distance function.

Tables 3 and 4 shows the simulations results for 10 and 100 labeled instances, respectively. For 10 labeled instances, the proposed technique achieved an average rank of 5.86 (2nd place) and, for 100 labeled instances, an average rank of 9.29 (11th place). Overall, it achieved an average rank of 7.57 (5th place) - preceded by LP (7.21), LDS (6.93), Laplacian RLS (5.43) and SGT (5.33). Interestingly, it achieved a very high position (2nd) in the case of only 10 labeled instances, a very challenging semi-supervised task in which as only as a small portion of 0.67% of the dataset is labeled. Hence, concerning the 17 techniques and the 7 benchmark datasets, we conclude that the proposed technique is at least comparable to the best known semi-supervised techniques.

Table 3. Classification error rate (%) and the corresponding average rank of each technique. Best results are in bold face. Data sets with 10 labeled points.

	g241c	g241d	Digit1	USPS	COIL	BCI	Text	**Avg. Rank**
1-NN	47.88	46.72	13.65	16.66	63.36	49.00	38.12	9.57
SVM	47.32	46.66	30.60	20.03	68.36	49.85	45.37	14.00
MVU + 1-NN	47.15	45.56	14.42	23.34	62.62	47.95	45.32	9.86
LEM + 1-NN	44.05	43.22	23.47	19.82	65.91	48.74	39.44	10.00
QC + CMR	39.96	46.55	9.80	13.61	59.63	50.36	40.79	7.71
Discrete Reg.	49.59	49.05	12.64	16.07	63.38	49.51	40.37	10.57
TSVM	24.71	50.08	17.77	25.20	67.50	49.15	31.21	10.71
SGT	**22.76**	**18.64**	8.92	25.36	N/A	49.59	29.02	6.17
Cluster-Kernel	48.28	42.05	18.73	19.41	67.32	48.31	42.72	10.86
Data-Dep. Reg.	41.25	45.89	12.49	17.96	63.65	50.21	N/A	9.83
LDS	28.85	50.63	15.63	17.57	61.90	49.27	**27.15**	8.29
Laplacian RLS	43.95	45.68	**5.44**	18.99	**54.54**	48.97	33.68	6.00
CHM (normed)	39.03	43.01	14.86	20.53	N/A	46.90	N/A	7.20
LGC	45.82	44.09	9.89	**9.03**	63.45	47.09	45.50	7.29
LP	42.61	41.93	11.31	14.83	55.82	**46.37**	49.53	**5.57**
LNP	47.82	46.24	8.58	17.87	55.50	47.65	41.06	7.14
Proposed Method	40.30	41.74	13.94	19.98	59.40	46.69	34.32	5.86

Table 4. Classification error (%) and the corresponding average rank of each technique. Best results are in bold face. Data sets with 100 labeled points.

	g241c	g241d	Digit1	USPS	COIL	BCI	Text	**Avg. Rank**
1-NN	43.93	42.45	3.89	5.81	17.35	48.67	30.11	12.57
SVM	23.11	24.64	5.53	9.75	22.93	34.31	26.45	9.29
MVU + 1-NN	43.01	38.20	2.83	6.50	28.71	47.89	32.83	11.71
LEM + 1-NN	40.28	37.49	6.12	7.64	23.27	44.83	30.77	12.00
QC + CMR	22.05	28.20	3.15	6.36	10.03	46.22	25.71	7.43
Discrete Reg.	43.65	41.65	2.77	4.68	**9.61**	47.67	24.00	8.14
TSVM	18.46	22.42	6.15	9.77	25.80	33.25	24.52	8.71
SGT	17.41	9.11	2.61	6.80	N/A	45.03	**23.09**	**4.50**
Cluster-Kernel	**13.49**	**4.95**	3.79	9.68	21.99	35.17	24.38	6.71
Data-Dep. Reg.	20.31	32.82	**2.44**	5.10	11.46	47.47	N/A	6.83
LDS	18.04	23.74	3.46	4.96	13.72	43.97	23.15	5.43
Laplacian RLS	24.36	26.46	2.92	4.68	11.92	**31.36**	23.57	4.86
CHM (normed)	24.82	25.67	3.79	7.65	N/A	36.03	N/A	8.80
LGC	41.64	40.08	2.72	**3.68**	45.55	43.50	46.83	9.86
LP	30.39	29.22	3.05	6.98	11.14	42.69	40.79	8.86
LNP	44.13	38.30	3.27	17.22	11.01	46.22	38.48	12.14
Proposed Method	27.92	26.19	3.57	9.59	20.01	44.11	25.54	9.43

6 Conclusions

This paper has presented a new network-based semi-supervised classification technique. The set of unlabeled instances compose a network in which vertices represent the state space for a random walker. Via a specific matrix composition, each labeled instance is inserted into the network to provide a bias to the classification process. This label bias is propagated through the unlabeled vertices of the network by means of the limiting probabilities. The local and global topology are taken into account by the random walk process and thus both clustering and smoothness SSL assumptions are satisfied, making the usage of the unlabeled instances information effective. Simulations have showed that the proposed technique is capable of detecting classes that present different shapes and distribution. The technique has also been demonstrated to be competitive with some well-known semi-supervised techniques using benchmark data sets.

Acknowledgments. The authors would like to acknowledge the São Paulo State Research Foundation (FAPESP) and the Brazilian National Council for Scientific and Technological Development (CNPq) for the financial support given to this research.

References

[1] Chapelle, O., Schölkopf, B., Zien, A. (eds.): Semi-Supervised Learning. MIT Press, Cambridge (2006)

[2] Wechsler, H., Kidode, M.: A random walk procedure for texture discrimination. IEEE Transactions on Pattern Analysis and Machine Intelligence 1(3), 272–280 (1979)

[3] Wechsler, H., Citron, T.: Feature extraction for texture classification. Pattern Recognition 12(5), 301–311 (1980)

[4] Grady, L.: Random walks for image segmentation. IEEE Transactions on Pattern Analysis and Machine Intelligence 28(11), 1768–1783 (2006)

[5] Bul, S.R., Rabbi, M., Pelillo, M.: Content-based image retrieval with relevance feedback using random walks. Pattern Recognition 44(9), 2109–2122 (2011)

[6] Schll, J., Schll-Paschinger, E.: Classification by restricted random walks. Pattern Recognition 36(6), 1279–1290 (2003)

[7] Zhou, D., Schölkopf, B.: Learning from labeled and unlabeled data using random walks. In: Rasmussen, C.E., Bülthoff, H.H., Schölkopf, B., Giese, M.A. (eds.) DAGM 2004. LNCS, vol. 3175, pp. 237–244. Springer, Heidelberg (2004)

[8] Szummer, M., Jaakkola, T.: Partially labeled classification with markov random walks. In: Advances in Neural Information Processing Systems, pp. 945–952. MIT Press (2001)

[9] Gallager, R.G.: Discrete Stochastic Processes, 1st edn. Springer (1996)

[10] Çinlar, E.: Introduction to stochastic processes. Prentice Hall (1975)

[11] Duin, R., Juszczak, P., Paclik, P., Pekalska, E., de Ridder, D., Tax, D., Verzakov, S.: Prtools4.1, a matlab toolbox for pattern recognition (2007)

[12] Zhou, D., Bousquet, O., Lal, T.N., Weston, J., Schlkopf, B.: Learning with local and global consistency. In: Advances in Neural Information Processing Systems 16, vol. 16, pp. 321–328. MIT Press (2004)

[13] Wang, F., Zhang, C.: Label propagation through linear neighborhoods. IEEE Transactions on Knowledge and Data Engineering 20(1), 55–67 (2008)

[14] Sun, J., Boyd, S., Xiao, L., Diaconis, P.: The fastest mixing markov process on a graph and a connection to a maximum variance unfolding problem. SIAM Review 48(4), 681–699 (2006)

[15] Belkin, M., Niyogi, P.: Laplacian eigenmaps for dimensionality reduction and data representation. Neural Computation 15(6), 1373–1396 (2003)

[16] Delalleau, O., Bengio, Y., Roux, N.L.: Nonparametric function induction in semi-supervised learning. In: Workshop Artif. Intell. Stat. (2005)

[17] Zhou, D., Schölkopf, B.: Adaptive computation and machine learning. In: Discrete Regularization, pp. 237–250. MIT Press, Cambridge (2006)

[18] Chapelle, O., Zien, A.: Semi–supervised classification by low density separation. In: Proceedings of the International Workshop on Artificial Intelligence and Statistics, pp. 57–64 (2005)

[19] Joachims, T.: Transductive learning via spectral graph partitioning. In: Proc. Int. Conf. Mach. Learn., pp. 290–297 (2003)

[20] Chapelle, O., Weston, J., Schölkopf, B.: Cluster Kernels for Semi-Supervised Learning. In: NIPS 2002, vol. 15, pp. 585–592. MIT Press, Cambridge (2003)

[21] Corduneanu, A., Jaakkola, T.: Adaptive computation and machine learning. In: Data-dependent Regularization, pp. 163–190. MIT Press, Cambridge (2006)

[22] Sindhwani, V., Niyogi, P.: Beyond the point cloud: from transductive to semi-supervised learning. In: Proc. 22nd Int. Conf. Mach. Learn., pp. 824–831 (2005)

[23] Burges, C.J.C., Platt, J.C.: Adaptive computation and machine learning. In: Semi-supervised Learning with Conditional Harmonic Mixing, pp. 251–273. MIT Press, Cambridge (2006)

[24] Zhu, X., Ghahramani, Z.: Learning from labeled and unlabeled data with label propagation. Technical Report CMU-CALD-02-107, School Comput. Sci., Carnegie Mellon Univ., Pittsburgh, PA, Tech. Rep (2002)

A Purity Measure
Based Transductive Learning Algorithm

João Roberto Bertini Junior and Liang Zhao

Instituto de Ciências Matemáticas e de Computação, Universidade de São Paulo
Avenida Trabalhador são-carlense, 400, São Carlos, SP, Brasil
{bertini,zhao}@icmc.usp.br

Abstract. The increasing on the human ability to gather data has led to an increasing effort on labeling them to be used in specific applications such as classification and regression. Therefore, automatic labeling methods such as semi-supervised transdutive learning algorithms are of a major concern on the machine learning and data mining community nowadays. This paper proposes a graph-based algorithm which uses the purity measure to help spreading the labels throughout the graph. The purity measure determines how intertwined are different subspaces of data regarding its classes. As high values of purity indicate low mixture among patterns of different classes, its maximization helps finding well-separated connected subgraphs; which facilitates the label spreading process. Results on benchmark data sets comparing to state-of-the-art methods show the potential of the proposed algorithm.

Keywords: Graph-based Transduction, Purity Measure, KNN Mutual Graph, Semi-supervised Learning.

1 Introduction

Semi-supervised learning concerns the problem of automatic classification with the restriction that only a small portion of the available data patterns present labels [4], [14]. With the increasing capacity of collecting data, it also has increased the demand of labeled data in order to develop applications such as classification and prediction. Regular labeling data process though, is generally associated to some drawbacks, such as time and monetary costs, not to mention the reliability on a manual process [3]. Facing to this scenario, transductive learning methods provide an automatic way to spread the labels from a very small portion of the data to the whole data set. As a consequence, the expert needs to classify only a small portion of the data, which decreases the aforementioned costs and enhances reliability.

Transductive learning is a branch of semi-supervised learning which concerns the task of labeling a finite set of data; differently from inductive learning algorithms, whose aim is to determine a prediction function from the available data (labeled and unlabeled) defined on the entire data space [7]. Among other transductive models such as generative [10], co-training [3] and low-density separation

C. Guo, Z.-G. Hou, and Z. Zeng (Eds.): ISNN 2013, Part II, LNCS 7952, pp. 405–412, 2013.

[12], the techniques based on graphs have been receiving a special attention over the last years, see [6], [8], [13], [11]. In graph-based methods, each vertex of the graph represents a data pattern and the edges stand for some relation of similarity between vertices. In general, the higher the similarity among data, the higher the probability of connection between them. Therefore, similar patterns tend to be heavily linked together while non-similar patterns may hardly be connected. As a result, the graph obtained from data represents arbitrary shapes of data classes and it can be viewed as a natural structure to spread the labels. However, most of the graph-based techniques present cubic order of complexity, which limits its applications to small and middle sized data sets [14].

In this paper, we propose a new graph-based transductive algorithm, which bases on the purity measure to aid spreading the labels throughout the graph. The so called Purity Measure based Transductive Learning Algorithm (PMTLA) resembles the KAOGSS algorithm [2], which is a semi-supervised inductive algorithm. Therefore, the KAOGSS builds a classifier, named K-associated optimal graph, which is a graph structure that, allied to the purity measure, is used to classify new unlabeled data. Here we propose a version proper to cope with transductive tasks, whose objective is to classify the already existing patterns. Similar to the K-associated optimal graph construction, the PMTLA algorithm does not depend on the parameter K, since it also constructs a KNN graph incrementally; i.e. K, initially set to 1, is increased by one until all vertices are labeled. The labels are spread as the components are being merged with the increasing of K. Purity is then used along with a Monte Carlo method to best separate the components with more than one class.

The remainder of the paper is organized as follow. Section 2 presents an overview on the purity measure. Section 3 details the proposed graph-based transducive algorithm. Section 4 presents some experimental results on an important semi-supervised learning benchmark. At last, Section 5 concludes the paper and gives prospective development directions regarding the algorithm proposed in this paper.

2   The Purity Measure

This section presents an overview on the purity measure, further details can be found in Refs. [1], [9]. Prior to introduce the purity measure, let $G = (V, E)$ be a graph, with a set of vertices V and a set of edges E. A maximal connected subgraph of G is called component, an isolated vertex is also a component. Consider also a vector-based data set $X = \{\mathbf{x}_1, \ldots, \mathbf{x}_N\}$ where a data pattern $\mathbf{x}_i = (x_{i1}, x_{i2}, \ldots, x_{ip}, c_i)$ described by p attributes and a class label c_i, where $c_i \in \Omega = \{\omega_1, \omega_2, \ldots, \omega_M\}$ and M is the number of classes in the problem. A K-associated graph is a directed graph built from a vector-based data set by abstracting each pattern to a vertex and connecting each vertex v_i to the set $\Delta_{v_i,K} = \{v_j | v_j \in \Lambda_{v_i,K} \wedge c_i = c_j\}$, where $\Lambda_{v_i,K}$ is the set of K nearest neighbors of v_i. The resulting K-associated graph can be viewed as a set of disjoint components, $G = \{C_1, \ldots, C_\alpha, \ldots, C_R\}$, where each component C_α is composed by vertices of a single class. The number of components R varies in the range $N \geq R \geq M$ accordingly to the magnitude of K; where N is the number of patterns in the training set and M the number of classes.

The total number of edges among the vertices of a component C_α is proportional to K and can be at most equal to KN_α, where N_α is the number of vertices in component C_α. This maximum value is only achieved if all vertices in the neighborhood of any vertex of the component have the same class. Likewise, nearby vertices of other classes decrease the number of connections of the given component. Thus, a measure of "purity" can be defined as: Given a component C_α of a K-associated graph with number of vertices $N_\alpha > 1$, given also the average degree D_α, the purity measure of the component C_α, denoted as Φ_α, is defined by Eq. (1).

$$\Phi_\alpha = \frac{D_\alpha}{2K}. \tag{1}$$

In this way, $\Phi_\alpha = 1$, if and only if, for every v_i in the component C_α, all the K neighbors have the same class label of v_i. On the other hand, if there exist noise or two or more classes are mixed together, vertices in this region are unable to make their K connections due to the existence of vertices of other classes in the neighborhood of some vertices. Indeed, the more mixing the components are, the lower their average degrees D_α and consequently their respective purities. This paper take the definition of purity less severe; basically to conform to the KNN mutual graph, purity might not reach high values as in the K-associated graph but it is sufficient to the aimed applications, once here we do not compare components formed with different K.

3   The Proposed Purity Measure Based Transductive Learning Algorithm

The problem addressed here regards the classification of a set of known unlabeled patterns in a given data set. Considering the constraint that there exist too few labeled data to employ a regular supervised method. In such scenario, it is necessary to consider a transductive semi-supervised method in order to spread the labels from the labeled to the unlabeled patterns. Hence, consider the data set $X = \{(\mathbf{x}_1, c_1), \ldots, (\mathbf{x}_L, c_L), \mathbf{x}_{L+1}, \ldots, \mathbf{x}_N\}$ with L labeled patterns $(\mathbf{x}_i, c_i)$ and $N - L$ unlabeled patterns $\mathbf{x}$ (or $(\mathbf{x}_j, \emptyset)$).

The Purity Measure Transductive Learning Algorithm (PMTLA) proposed here, relies on iteratively building a mutual KNN graph while spreading the labels within those formed components where there exist a labeled vertex. Similar to its counterpart, the inductive semi-supervised K-associated algorithm, which constructs a sequence of graphs to end up with a optimal graph proper to perform classification. The PMTLA algorithm, however, builds one graph by increasing K and adding edges correspondingly, up to the point where all vertices have label. The PMTLA algorithm consider a mutual KNN graph, defined as a undirected graph $G = (V, E)$ which consists of a set of labeled vertices $V = \{v_1, \ldots, v_N\}$

matching the given vector-based data set X; i.e. vertex v_i represent pattern $\mathbf{x}_i$ with its corresponding label c_i or without label ($c_i = \emptyset$). The set of edges E is formed incrementally as K increases along time. Therefore, at every value of K, and for each pattern $\mathbf{x}_i$ it is verified if its Kth nearest neighbor, represented as $\mathbf{x}_i^{(K)}$, not only have $\mathbf{x}_i$ as one of its K nearest neighbors, but also present the same label. Then, supposing v_j represents $\mathbf{x}_i^{(K)}$, the $\{v_i, v_j\}$ connection is performed. In other words, v_i connects to those vertices which also have v_i as a K nearest neighbor and belong to the same class of v_i or do not have label. If v_i itself does not have a class label, it connects to all its K mutual nearest neighbors without considering their classes.

As a consequence of connecting unlabeled vertices to labeled vertices, components with more than one class may be formed. Since each component ideally represents a local group of data, thus, it should not have more than one class. Therefore, to overcome this problem, it is necessary to split those components with vertices belonging to different classes. Intuitively, several methods can be employed for such task. For instance, a cut procedure can be addressed to separate components into well-connected clusters of vertices, such as the min-cut [5] or the cut procedure for multi-class problems proposed in Ref. [2]. Basically, it is possible to consider cutting a component before or after spreading the labels. By cutting before spreading, it can be said that the process relies strongly on the graph structure, which can be thought to be the natural order to proceed. However, depending on the cutting criterion, components with no class can be isolated and even last through various interactions of K without being labeled. On the other hand, by spreading the labels before cutting, one can guarantee that no vertex will be left unlabeled. In this case, the problem is how to spread the labels consistently to the graph structure. The PMTLA employ the purity measure to help finding a suitable configuration of labels for the unlabeled vertices. As high values of purity indicate low mixture of data patterns regarding their classes, the idea is that, by maximizing the sum of the purities of the formed components, we can obtain components to represent well-separated smooth groups of data patterns. The details on the proposed algorithm is given in the Algorithm 1.

In the algorithm, the graph evolves by increasing K while considering connecting each of the vertices to its Kth nearest neighbor. To establish the connections, at a given value of K, the Kth nearest neighbor of each pattern $\mathbf{x}_i$, noted as $\mathbf{x}_i^{(K)}$, is found. Suppose $\mathbf{x}_i^{(K)}$ correspond to pattern $\mathbf{x}_j$, then the connection $\{v_i, v_j\}$ is established if and only if $\mathbf{x}_j$ also has $\mathbf{x}_i$ as one of its K nearest neighbors and they do not belong to different classes, i.e. they can have the same label or at least one may be unlabeled (see line 7 of Algorithm 1). After performing all possible connections for a given K, the function $findComponents()$ returns a set of disjoint components, represented as $G^{(K)} = \{C_1, \dots, C_\alpha, \dots, C_R\}$. The algorithm proceeds by considering a component C_α at a time. If there exist labeled vertices in C_α and if they belong to the same class (verified by function $hasOneClass()$), then this label is spread throughout the whole component by the $spreadLabel()$ function. Those vertices will no longer have its label changed.

Algorithm 1 . Purity Measure based Transductive Learning Algorithm - PMTLA

Input: $X = \{(\mathbf{x}_1, c_1), \ldots, (\mathbf{x}_L, c_L), \mathbf{x}_{L+1}, \ldots, \mathbf{x}_N\}$; L - Number of labeled patterns;

Symbols: $G^{(K)}$ - KNN Mutual graph, $\mathbf{x}_i^{(K)}$ - Kth nearest neighbor of pattern $\mathbf{x}_i$;
   $itMax$ - Maximum of steps for the algorithm to test labels configurations;
   $\bar{C}$ - A subset of disjoint components; N_α^u - number of unlabeled vertices in
   component; C_α; $\Lambda_{\mathbf{x}_j, K}$ - Set of K nearest neighbors of pattern; $\mathbf{x}_j$

1: $K \Leftarrow 1, C \Leftarrow \emptyset$
2: $G^{(K)} = (V, E); V = \{v_1, \ldots, v_N\}, E \Leftarrow \emptyset$
3: $purity \Leftarrow 0, higherPurity \Leftarrow 0$
4: **repeat**
5:    **for all** $\mathbf{x}_i \in X$ **do**
6:       Find $\mathbf{x}_j$ such that $\mathbf{x}_j = \mathbf{x}_i^{(K)}$ {Kth nearest neighbor of $\mathbf{x}_i$}
7:       **if** ($\mathbf{x}_i \in \Lambda_{\mathbf{x}_j, K}$ **and** ($c_j = c_i$ **or** $c_i = \emptyset$ **or** $c_j = \emptyset$)) **then**
8:          $E \Leftarrow E \cup \{v_i, v_j\}$
9:       **end if**
10:    **end for**
11:    $G^{(K)} \Leftarrow findComponents(V, E)$ {Graph is composed of disjoint components}

12:    **for all** $C_\alpha \in G^{(K)}$ **do**
13:       **if** $hasOneClass(C_\alpha)$ **then**
14:          $C_\alpha \Leftarrow spreadLabel(C_\alpha)$
15:          $L \Leftarrow L + N_\alpha^u$
16:       **else if** $hasMoreThanOneClass(C_\alpha)$ **then**
17:          **for** $it = 1$ **to** $itMax$ **do**
18:             $C_\alpha \Leftarrow spreadRandomLabel(C_\alpha)$
19:             $sum \Leftarrow 0$
20:             **for all** $v_i \in C_\alpha$ **do**
21:                **for all** $v_j \in \Lambda_{v_i, K}$ **do**
22:                   **if** $c_i = c_j$ **then**
23:                      $sum \Leftarrow sum + 1$
24:                   **end if**
25:                **end for**
26:             **end for**
27:             $purity \Leftarrow (sum/N_\alpha)/2K$
28:             **if** $purity > higherPurity$ **then**
29:                $higherPurity \Leftarrow purity$
30:                $C_\alpha^{aux} \Leftarrow C_\alpha$ {C_α^{aux} stores best label configuration}
31:             **end if**
32:          **end for**
33:          $\bar{C} \Leftarrow separateComponent(C_\alpha^{aux})$
34:          $G^{(K)} = G^{(K)} - C_\alpha \bigcup \bar{C}$
35:          $L \Leftarrow L + N_\alpha^u$
36:       **end if**
37:    **end for**
38:    $K \Leftarrow K + 1$
39: **until** $L = N$ {Ends when all patterns are labeled}
40: **Output:** All patterns in X labeled.

Figure 1 shows an example of the label propagation process. Figure 1(a) shows the original data patterns, note that only two vertices presents labels. Then the algorithm starts to connect the vertices, Fig. 1(b) stands for $K = 2$, at this stage there are three components and each of then presents one class (or no class), therefore the label is spread without problem.

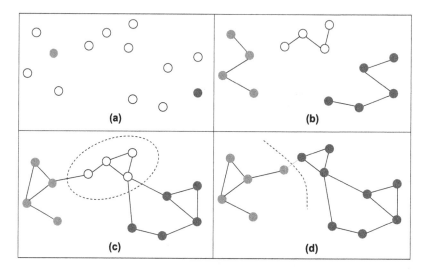

Fig. 1. Label spreading and component division process examples. Figure (a) Raw data vertices (b) PMTLA at $K = 2$ (c) PMTLA at $K = 3$, a component with more than one class arises (d) A possible configuration of labels and the component separation.

Although direct connections between vertices of different classes are not allowed, components presenting vertices of more than one class might appear as a result of an unlabeled component eventually merge with components of different classes (such as in Fig. 1(c)). If this is the case, the algorithm sets random labels to the unlabeled vertices (in a multiclass situation, of course, only labels that are presented in the component are considered). Once random labels are set it is necessary to verify if it consists of a valid configuration. This is necessary to guarantee the smoothness of the classes distributions. A configuration of labels is accepted only if for every recent labeled vertex there exist a path, with vertices of the same label, connecting it to some vertex that has been labeled in a previous step, i.e. in a previous K. Consider that the function $spreadRandomLabel()$ returns only valid configurations, then for each obtained valid configuration of labels, the purity measure is estimated for each of the formed components as stated in lines 20 to 27. The final configuration is the one with the highest sum of purity obtained after some steps ($itMax = 100$ in the experiments). The components are then separated by cutting off the connections between vertices of different classes, as shown in Fig. 1(d). Afterwards, the new obtained components from C_α, noted as $\bar{C}$, substitutes the old component C_α in the graph (lines

33-34 of the Algorithm). At last, K is increased by one and the whole process takes place again, until all vertices are labeled.

4 Experiments Results

This section reports the obtained results on the benchmark[1] introduced by Chapelle et al. [4] and compare to 13 state-of-the-art algorithms. Further details on the comparison algorithms as well as their corresponding references can be found in Ref. [4]. The results were obtained through a 12-fold cross validation process. Each fold considered 100 labeled patterns whose indices was set as the same as used in Ref. [4]. Table 1 shows the test errors percentages for the proposed algorithm along with the 13 algorithms under comparison.

Table 1. Comparison results concerning the proposed algorithm (PMTLA) and 13 state-of-the-art algorithms. The results are the test error percentage averaged over a 12-fold cross-validation process considering 100 labeled patterns.

Algorithm	g241c	g241d	Digit1	USPS	COIL	BCI	Text
PMTLA	43.13	43.50	1.94	5.77	9.98	34.12	25.52
1-NN	43.93	42.45	3.89	5.81	17.35	48.67	30.11
SVM	23.11	24.64	5.53	9.75	22.93	34.31	26.45
MVU + 1-NN	43.01	38.20	2.83	6.50	28.71	47.89	32.83
LEM + 1-NN	40.28	37.49	6.12	7.64	23.27	44.83	30.77
QC + CMN	22.05	28.20	3.15	6.36	10.03	46.22	25.71
Discrete Reg.	43.65	41.65	2.77	4.68	9.61	47.67	24.00
TSVM	18.46	22.42	6.15	9.77	25.80	33.25	24.52
SGT	17.41	9.11	2.61	6.80	-	45.03	23.09
Cluster-Kernel	13.49	4.95	3.79	9.68	21.99	35.17	24.38
Data-Dep. Reg.	20.31	32.82	2.44	5.10	11.46	47.47	-
LDS	18.04	23.74	3.46	4.96	13.72	43.97	23.15
Laplacian RLS	24.36	26.46	2.92	4.68	11.92	31.36	23.57
CHM (normed)	24.82	25.67	3.79	7.65	-	36.03	–

As can be seen from Table 1, no algorithm is the best concerning all the tested data sets. According to Chapelle et al. [4], the domains g241c, g241d and Text are cluster-like data sets, and the rest are manifold-like data sets, regarding their respective data distribution. Though, as one can expect the proposed graph-based algorithm PMTLA had good results when considering the manifold-like data sets. Moreover, it had ranked in the first place in the Digit1 domain and in the second place in COIL. Regarding the cluster-like data sets, the PMTLA algorithm performed at about average in the Text domain, but it presented bad results for g241c and g241d when compared to others algorithms. Summarizing, evidences have showed that the PMTLA algorithm is a very sound choice for some domains, in special manifold-like domains.

---

[1] The benchmark data sets can be found in
http://olivier.chapelle.cc/ssl-book/benchmarks.html

5 Conclusions

This paper presents a new graph-based transductive algorithm based on the purity measure, named PMTLA. Results on expressive benchmark has show the potential of the algorithm, in special when considering manifold-like applications domains, such as image classification. What needs to be emphasised is that the PMTLA algorithm presents performance comparable to the state-of-the-art semi-supervised algorithms, with low computational cost compared to regular graph-based approaches. Some future works include employing other method to construct the graph, extend purity to more levels of neighborhood in order to better address cluster-like domains, develop mechanisms to detect possible initially incorrectly labeled patterns and consider more data sets and less labeled data for future experiments.

References

1. Bertini Jr., J.R., Zhao, L., Motta, R., Lopes, A.A.: A Nonparametric Classification Method based on K-associated Graphs. Inform. Sciences 181, 5435–5456 (2011)
2. Bertini Jr., J.R., Lopes, A.A., Zhao, L.: Partially Labeled Data Stream Classification with the Semi-supervised K-associated Graph. J. Brazil. Comp. Soc. 18, 299–310 (2012)
3. Blum, A., Mitchell, T.: Combining Labeled and Unlabeled Data with Co-training. In: Proc. 11th Annual Conf. on Computational Learning Theory, pp. 92–100 (1998)
4. Chapelle, O., Zien, A., Schölkopf, B. (eds.): Semi-supervised Learning. MIT Press (2006)
5. Cormen, T., Leiserson, C., Rivest, R., Stein, C.: Introduction to Algorithms. MIT Press (2009)
6. Culp, M., Michailidis, G.: Graph-based Semisupervised Learning. IEEE T. Pattern Anal. 30, 174–179 (2008)
7. Delalleau, O., Bengio, Y., Roux, N.: Efficient Non-parametric Function Induction in Semi-supervised Learning. In: Proc. 10th Int. Workshop on Artificial Intelligence and Statistics, Society for Artificial Intelligence and Statistics, pp. 96–103 (2005)
8. Joachims, T.: Transductive Learning via Spectral Graph Partitioning. In: Proc. 20th Int. Conf. on Machine Learning, pp. 290–297 (2003)
9. Lopes, A.A., Bertini Jr., J.R., Motta, R., Zhao, L.: Classification based on the Optimal K-associated Network. In: Zhou, J. (ed.) Complex 2009. LNICST, vol. 4, pp. 1167–1177. Springer, Heidelberg (2009)
10. Nigam, K., Mccallum, A.K., Thrun, S., Mitchell, T.: Text Classification from Labeled and Unlabeled Documents using EM. Machine Learning 39, 103–134 (2000)
11. Silva, T.C., Zhao, L.: Network-based Stochastic Semisupervised Learning. IEEE T. Neural Networ. 23, 451–466 (2012)
12. Vapnik, V.N.: Statistical Learning Theory. Wiley, New York (2008)
13. Zhu, X., Ghahramani, Z., Lafferty, J.: Semi-supervised Learning using Gaussian fields and Harmonic Functions. In: Proc. 20th Int. Conf. on Machine Learning, pp. 912–919 (2003)
14. Zhu, X.: Semi-Supervised Learning Literature Survey. Technical Report 1530, Computer-Science, University of Wisconsin-Madison (2008)

Novel Class Detection within Classification for Data Streams[*]

Yuqing Miao, Liangpei Qiu, Hong Chen,
Jingxin Zhang, and Yimin Wen

School of Computer Science and Engineering,
Guilin University of Electronic Technology, Guilin, China
miaoyuqing@guet.edu.cn, {qlp_1018,chenhong0327}@126.com

Abstract. Traditional data stream classification techniques are not capable of recognizing new classes emerged in data stream. Recently, an ensemble classification framework focused on the new challenge. But the novel class detection technique is limited to the numeric data in the framework. And, both the lower process speed and the larger model size of base classifier trouble the framework. In this paper, a novel class instance detection technique is proposed to deal with mixed attribute data and the VFDTc is adopted as base classifier to speed up the process and reduce the model size. Experimental results showed that the algorithm outperformed the previous one in both classification accuracy and processing speed.

Keywords: Data Stream, Novel Class Detection, Classification, Data Mining.

1   Introduction

Data stream classification task is an important research field of data stream mining. There are four major problems in data stream classification. (1) Limited training data, infinite storage and infinite running time. (2) Upper accuracy and real-time result. (3) The potential concept-drift. (4) Novel classes emerged. Most of the existing solutions just pay close attention to the first three problems, such as single model incremental learning algorithm and ensemble techniques (e.g. SEA, WCE and KBS et al.). The capability of detecting novel classes is not be involved in them.

Spinosa et al. [1] described a clustering approach to detect both concept-drift and novel class. It assumed only one existing class. Masud et al. [2] presented an ensemble classification framework called MineClass, which could handle any number of existing classes. It handled concept-drift by using WCE (Weighted Classifier Ensemble), and detected novel classes by applying k-means clustering. WCE used C4.5 decision tree and K-NN as base classifier. In their follow-up work [3], the novel class detection

---

[*] The work described in this paper is supported by Foundation of Guangxi Key Laboratory of Trustworthy Software, China (kx201116) and Educational Commission of Guangxi Province, China(201204LX122).

C. Guo, Z.-G. Hou, and Z. Zeng (Eds.): ISNN 2013, Part II, LNCS 7952, pp. 413–420, 2013.

technique was enhanced by using adaptive decision boundary and Silhouette Gini coefficient. They detected multiple novel classes by constructing a graph and identifying the connected components in the graph.

But to learn C4.5 model requires scanning and sorting data for many times, and the model expends lots of storage. The traditional k-means deals with numerical attribute only. It is sensitive to initial centers and "noise". The method may lead to lower accuracy of novel detection and classification. In order to improve the accuracy and efficiency of classification, we adopted VFDTc (an extension of Very Fast Decision Tree) [4-5] as base learner, and proposed k-prototypes++ algorithm to deal with the mixed attribute data and initial centers. All these elements were assembled to an improved framework DTNC (Detecting Novel Classes in Data Streams).

2   Novel Class Detection within Classification Framework

MineClass and MCM [3] have a similar framework. Data stream is divided into equal sized chunks. The latest unlabeled chunk is the input of WCE with M classification models $E=\{E_1, E_2, ..., E_m\}$. Data will be estimated whether it belongs to existing class first, if yes, WCE will vote and determine the label, if no, it will be used to detect and generate novel class. If the chunk is classified, a new classifier E_0 can be trained to update the ensemble. The concepts of stream are real-time updated in models.

2.1   Novel Class Detection

Definition 1. if at least one of the ensemble models $E_i \in E$ has been trained with the instances of class c, c is an **existing class**. Otherwise, c is a **novel class** [2].

The data of the same class should be closer to each other (cohesion) and farther apart from the data belonging to any other classes (separation) [3]. The instances close to each other can reasonably be supposed to belong to the same class. Introduce the notion of used space to denote feature space occupied by instances of existing classes, and the unused space to denote feature space not be occupied. Novel class instances must fall in the unused space with strong cohesion. Thus, the two basic principles to detect novel class are keeping track of the used space and finding strong cohesion among the instances that fall into the unused space.

2.2   The Used Spaces Construction

MineClass and MCM built K clusters with the training data chunk, and save the useful information as pseudopoints ("p-points" for short) to describe the used space. Each p-point ψ_i corresponds to a hypersphere $RE(\psi_i)$ in the feature space. The union of the regions covered by all p-points is the union of used space, which forms a decision boundary $B(E_j) = \cup_{\psi_i \in \psi_j} RE(\psi_i)$, for a classifier E_j. The decision boundary of the ensemble is the union of all models' decision boundary.

If instance falls into the decision boundary, then it is classified using the ensemble of models. Otherwise, it is an F-outlier. F-outliers are temporarily stored in a buffer *buf* to observe whether they also satisfy the cohesion property. It is done by computing a metric called q-Neighborhood Silhouette Coefficient (q-NSC). which is a unified measure of cohesion and separation.

2.3 The Novel Class Detection Approach

Algorithm 1 outlines the novel class detection approach. K_0 clusters and F-p-points are built with F-outliers. For each F-p-point h, q-NSC(h) is computed with the weighted average distance from h to the n-nearest p-points of h, and the weighted average distance from h to the r-nearest F-p-points of h(including itself). The q-NSC of h is the approximate average of the q-NSC value of each instance x in h. If the q-NSC(x) value is negative, x is regarded as an existing class instance.

The q-NSC value may be different in different classifier, maximum value is saved. If tp, the total number of instances having positive q-NSC value, is greater than the threshold q, then the corresponding classifier votes in favor of a novel class. If all the classifiers vote for novel class, then we find the positive novel class instances.

Nscore(x) for all instances x with positive q-NSC is declared to measure the cohesion with other F-outliers, and the separation from the existing class instances. $G(s)$ is the discrete Gini Coefficient of Nscore. If Nscore(x) > $G(s)$, instance x is stored in a buffer N-List. The last step is to detect multiple novel classes by clustering with N-List, constructing a graph and identifying the connected components in the graph. Finally, instances will be labeled with component number.

Algorithm 1.   Detect-Novel(E,Buf)

Input: Current ensemble classifiers $E=\{E_1, E_2, ..., E_m\}$, F-outlier instances set Buf
Output: The novel class instances identified, if found
1: $K_0\leftarrow(K*|Buf|/S)$ // K is the number of p-points per chunk
2: $H\leftarrow K$-means(Buf, K_0)
3: **for** each classifier $E_i\epsilon E$ **do** $tp\leftarrow 0$
4:   **for** each $h\epsilon H$ **do** $h.sc\leftarrow q$-NSC(h)
5:     **if** $h.sc>0$ **then** tp += $h.w$
6:       **for** each instance $x\epsilon h$ **do** $x.sc \leftarrow$ max($x.sc, h.sc$)
7:     **end if**
8:   **end for**
9:   **if** $tp > q$ **then** $vote$++
10: **end for**
11: **if** $vote == m$
12:   **for** all x **if** $x.sc > 0$ **do** $X\leftarrow x$
13:   **for** all $x\epsilon X$ **do** $x.ns\leftarrow Nscore(x)$
14:     **if** $x.ns > G(s)$ **then** $N\_list\leftarrow N$-list$\cup x$
15:   **end for**
16:   Detect-Multinovel(N_list)
17: **end if**

---

3   Using VFDTc as Base Learner

Domingos and Hulten [5] put forward one of the most successful algorithms for mining data streams named VFDT. It is an incremental, online, and any-time decision tree learning algorithm. The main innovation of VFDT is using the Hoeffding bounds to decide the necessary examples to be observed before installing a split-test at leaf.

VFDT uses a subset of all training data and only once scan to split leaves and produce an approximate classification model. Compute is reduced rapidly, and a similar accuracy is provided. It has good noise immunity and multifarious concept processing performance. The memory of model is independent of the processing sample size, occupied with the leaf number, the number of attributes and attribute values and labels only. But the ability to deal with numerical attributes and concept-drift in data stream are not been considered in VFDT. VFDTc proposed two major extensions to VFDT: a condition of the form $attr_i \leq cut\_point$ in split-test for numeric attribute, and naive Bayes classifier applied in tree leaves.

4   Improving the Cluster Algorithm for Novel Class Detection

K-means is a clustering technique partitioning n points set χ in $\mathbb{R}^d$ into k homogenous clusters efficiently. It aims at choosing k centers so as to minimize the sum of dissimilarity measure between each point and the closest center. But its use is limited to numeric data, and the empirical speed and simplicity come to the price of accuracy. K-means++ [6] proposed a way to initialize k-means by choosing starting centers. K-prototypes [7] presented a measure similar to the squared distances to deal with both numeric and categorical attributes. We will combine the two technique, the new algorithm is named as k-prototypes++.

4.1   Seeding the Initial Centers

The k-means++ chooses a point c as an initial center with probability proportional to c's contribution to the overall potential [7]. Let $D(x)$ denote the shortest distance from a data point x to the closest center, which is already chosen. The algorithm 2 is the summary of this seeding technique. It is $O(\log k)$-competitive with the optimal clustering. The seeding technique of k-means++ is $O(\log k)$-competitive with the optimal clustering. It is as simple and fast as k-means. More introduce and discussion for the seeding technique in [7].

Algorithm 2. Seed the initial centers for cluster algorithm

Input: a set of n data points χ in $\mathbb{R}^d$, an integer k
Output: a set of k initial centers C
1: Choose an initial center $x_1 \rightarrow C$ uniformly at random from χ
2: **for** each point $x \in \chi$ **do** compute the distance $D(x)$ from x to the closest center
3: choose the next center $x' \rightarrow C$, with probability $D(x')^2 \Big/ \sum_{x \in \chi} D(x)^2$
4: **repeat** step 2 and step 3 **until** $|C| == k$

4.2   Dealing with Both Numeric and Categorical Attributes

K-modes extends *k*-means to categorical domains. It uses a simple matching dissimilarity measure for categorical points and replaces means of clusters by modes found by a frequency-based method. *K*-prototypes based on *k*-modes and *k*-means extending their domains with mixed numeric and categorical values whilst preserving efficiency.

Let $\chi = \{X_1, X_2, \ldots, X_n\}$ denote a set of n data points and $X_i = [x_{i1}, x_{i2}, \ldots, x_{im}]$ be a point x represented by m attribute values combined with the first r numeric attribute values and the last m-r categorical attribute values. The mixed dissimilarity measure is:

$$d_p(X, Y) = \sum_{j=1}^{r} (x_j - y_j)^2 + \gamma \sum_{s=r+1}^{m} \delta(x_s, y_s) \tag{1}$$

Where $\delta(x_s, y_s) = \begin{cases} 0 & (x_s = y_s) \\ 1 & (x_s \neq y_s) \end{cases}$. The first part is the Euclidean distance measure,

and the second is the simple matching dissimilarity measure. The weight γ is used to avoid favouring type of attribute. [8] showed that when γ ranged in (1.5, 2.5), clustering performance was favorable. Here the value of γ defaults to 2.

5   Experiments

Weka is a famous data mining tool implemented in Java. MOA based on Weka implements algorithms and running experiments for online learning from data streams. Experiments were done in Weka and MOA. The parameter settings are as follows, unless mentioned: K (number of p-points per chunk) = 50, S (chunk size) = 2000, L (ensemble size) = 6, q (minimum number of F-outliers to declare a novel class) = 50. These values of parameters are turned to achieve an overall satisfactory performance. About the setting' discussion of these parameters see [9].

5.1   Clustering Algorithm Performance Comparison

The *credit approval dataset* is chosen to test the *k*-prototypes++ algorithm in Weka. It has 690 instances described by 6 numeric, 9 categorical attributes and 2 classes. As to *k*-means, we removed all categorical attributes. Obviously, as to all attributes of data points are numeric, *k*-prototypes++ is the same as *k*-means. In the case of all attributes categorical, *k*-prototypes++ is equal to *k*-modes with the seeding technique.

Table 1. Summary of three clustering algorithms performance comparison

The algorithm	Error Rate			Iterations			Time(sec × 0.01)		
	μ	σ	Cv	μ	σ	Cv	μ	σ	Cv
k-means	37.97	**0.20**	**0.01**	9.56	3.50	0.37	**2.5**	0.8	32.7
k-prototypes	25.88	6.80	0.26	5.27	1.72	0.33	2.6	0.8	30.2
k-prototypes++	**22.06**	3.55	0.16	**4.57**	**1.34**	**0.29**	2.7	**0.5**	**19.5**

Table 1 shows the result of three algorithms applied to the dataset for 100 times in three aspects, namely, error rate, iterations and running time. Each aspect is represented by μ mean value, σ standard deviation and C_v coefficient of variance. C_v is a normalized measure of dispersion around the mean. K-prototypes++ outperforms k-means and k-prototypes at error rate and iterations. K-means deals with the 6 numeric attributes only, hence it takes the least time. K-prototypes and k-prototypes++ obtains a higher dispersion around the mean and an unstable accuracy. Contrast with k-prototypes, the seeding technique of k-prototypes++ increases the time, but decreases the error rate and iterations.

5.2   Modeling Performance Comparison with C4.5 or VFDTc as Base Learner

RandomTreeGenerator in MOA can generate a stream favouring decision tree learners. The generator has parameters to control the number of classes and attributes, we used the default settings. WCE ensemble classification model with different base learner was evaluated by the generated data stream in MOA.

Fig. 1(a) describes the curves about the time learning ensemble classification model when $M = 6$ and S ranges from 1K to 10K. When S comes to 10K, the modeling time with VFDTc learner is one third of C4.5. Fig. 1(b) shows the curves about the ensemble size with M range from 5 to 120 and $S = 4000$. When M comes to 120, the ensemble model size with VFDTc learner is one-eleventh of C4.5. The faster model is learned and the smaller the model size is, the less depth is needed to pass by the instance, the greater advantage is achieved in classification of data stream.

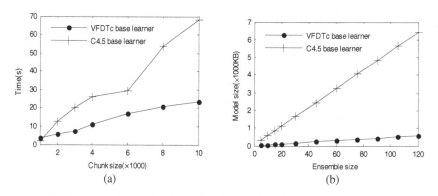

Fig. 1. The modeling time (a) and model size (b) with different parameters

5.3   Performance Comparison of Classification and Novel Detection

KDDCup 99 network intrusion detection: The dataset is extracted from a simulating intrusions network at MIT Lincoln Labs. The 10% version (about 49.4W records) consists of 34 numeric attributes, 8 categorical attributes and 23 different classes. Different classes appear and disappear frequently, making the new class detection very challenging.

Forest cover dataset (UCI repository): The dataset (about58.1W records) is extracted from USFS RIS containing geospatial descriptions of different types of forests. Each record consists of 12 numeric attributes and, 42 categorical attributes and 7 classes. The dataset was arranged so that in each chunk at most 3 and least 2 classes co-occur, and new class appears randomly.

Because the categorical attributes cannot be processed by k-means, we combine the k-means and k-modes for MineClass or MCM. The performance metrics are defined first. N_e = % of novel class instances misclassified as existing class, E_n = % of existing class instances falsely identified as novel class, *ERR* = % Total misclassification error (including N_e, E_n and misclassifications within the existing classes themselves) [3].

Fig. 2(a,b) show the total error for each approach throughout the stream in different datasets. Each approach gets a high error rate in the early chunks because of insufficient training of model, and gradually tends to stable. On the whole, the improved framework DTNC, which assembles k-prototypes++ and VFDTc, achieves a lower error rate than MineClass and MCM throughout two data streams.

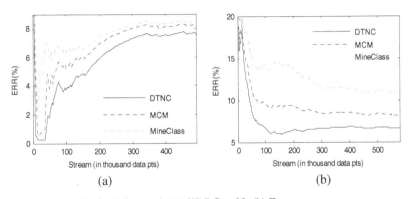

(a)   (b)

Fig. 2. ERR rates in (a) KDDCup 99, (b) Forest cover

Table 2. Summary of performance about each approach in KDD and forest

Dataset	Approach	Time(s)	N_e	E_n	ERR
KDD	MineClass	632.49	9.42	3.57	8.39
	MCM	610.06	8.75	3.01	8.21
	DTNC	**540.54**	**8.43**	**2.12**	**7.66**
Forest Cover	MineClass	698.42	13.26	5.18	10.70
	MCM	671.94	10.01	4.33	8.15
	DTNC	**460.74**	**7.30**	**2.54**	**6.60**

Table 2 shows the overall performance about each approach in KDD and forest dataset with running time (training time and classification time), N_e, E_n, and *ERR*. Generally, DTNC is 1.17 and 1.52 times faster than MineClass in KDD and forest Cover dataset. Also, DTNC is 1.13 and 1.45 times faster than MCM in KDD and forest

Cover dataset. The classification accuracy of DTNC is better than MineClass and MCM in KDD and forest Cover dataset. It is the result of the better base learner and the more precise novel detection.

6   Conclusion

Traditional data stream classification techniques are not capable of recognizing new classes emerged in concept-drift data stream. Recently, an ensemble classification framework focused on the new question. But, the framework is limited by numeric data, lower process speed, higher error rate and larger model size. We applied VFDTc as the base learner of ensemble model to accelerate the classification of data stream and decrease the model size. We also proposed k-prototypes++, a modified clustering method, to improve the technique of novel class detection by combining the seeding technique and mixed attribute dissimilarity measure. Experimental results on several datasets show that our approach obtains much better performance than existing techniques. In the future, we want to improve the dissimilarity measure to obtain stable result in k-prototypes++ and reduce the complexity of novel detection.

References

1. Spinosa, E.J., de Leon, A.P., de Carvalho, F., Gama, J.: OLINDDA: A cluster-based approach for detecting novelty and concept drift in data streams. In: Proc. of the Annual ACM Symposium of Applied Computing, pp. 448–452. ACM, New York (2007)
2. Masud, M.M., Gao, J., Khan, L., Han, J., Thuraisingham, B.: Integrating novel class detection with classification for concept-drifting data streams. In: Buntine, W., Grobelnik, M., Mladenić, D., Shawe-Taylor, J. (eds.) ECML PKDD 2009, Part II. LNCS (LNAI), vol. 5782, pp. 79–94. Springer, Heidelberg (2009)
3. Masud, M.M., Chen, Q., Khan, L.: Classification and adaptive novel class detection of feature-evolving data streams. IEEE Trans. Knowl. Data Eng. 99 (2012) (preprints), http://doi.ieeecomputersociety.org/10.1109/TKDE.2012.109
4. Gama, J., Rocha, R., Medas, P.: Accurate decision trees for mining high-speed data streams. In: Proceedings of the Nineth ACM SIGKDD International Conference on Knowledge Discovery and Data Mining, pp. 523–528. ACM, New York (2003)
5. Domingos, P., Hulten, G.: Mining high-speed data streams. In: Proceeding of the Sixth ACM SIGKDD International Conference on Knowledge Discovery and Data Mining, pp. 71–80. ACM, New York (2000)
6. Arthur, D., Vassilvitskii, S.: k-means++: The advantages of careful seeding. In: SODA 2007, pp. 1027–1035. Society for Industrial and Applied Mathematics (2007)
7. Huang, Z.: Extensions to the k-Means Algorithm for Clustering Large Data Sets with Categorical Values. Data Mining and Knowledge Discovery 2(3), 283–304 (1998)
8. Wang, J., Zhu, Y.: Research on the weighting exponent in fuzzy K-Prototypes algorithm. Computer Application 25(02), 348–351 (2005)
9. Masud, M.M., Gao, J., Khan, L., Han, J., Thuraisingham, B.M.: Classification and novel class detection in concept-drifting data streams under time constraints. IEEE Trans. Knowl. Data Eng. 23(6), 859–874 (2011)

Spatial Pyramid Formulation
in Weakly Supervised Manner

Yawei Yue[1], ZhongTao Yue[2], Xiaolin Wang[1], and Guangyun Ni[1]

[1] School of Computer Science and Technology, Shandong University
250001 Jinan, China
[2] School of Electronic Information and Electrical Engineering,
Shanghai JiaoTong University 200240 Shanghai, China
yue123161@gmail.com

Abstract. Spatial pyramid match scheme (SPM) is an important scheme in local feature based image classification which effectively adopts geometric structure information into image classification. Most previous approach formulized Spatial Pyramid in unsupervised manner by hierarchical splitting images into separate bins. We found that weak supervised information exists in this process totally unused. We cannot use information directly, because those information corresponding to the combination of all bins, thus we can use those weakly supervised information for bins selection. In this paper, we proposed to select those bins with better discriminative properties .The discriminative property can be well defined from neighborhood entropy. We incorporate local sensitive hash for fast neighborhood identification. We set those bins with higher neighborhood entropy weight zero. Analysis shows that our approach can down weight those non-discriminative bins, in contrast highlighting those discriminative bins. Experiments show that our approach can improve the performance of spatial pyramid match, especially for those categories with complex background. We also proof that under our scheme, result kernel matrix can still preserve positive semi-definite, which can guarantee that our algorithm will coverage.

Keywords: Spatial Pyramid Match, Local Sensitive Hash, Weakly Supervised.

1   Introduction

Image Classification is one of the most important and fundamental problems in computer vision area. Image classification mainly focuses about determining which category given image may belong based on classifiers trained on labeled datasets.[1]

Based on features used in classification, image classification can be generally divided into two categories: global feature based and local feature based methods. Global feature based methods employ features like color, texture to describe image. And classify images based on those descriptions. One major weakness of global feature based image classification called Semantic gap, which means that visual similarity cannot infer semantic similarity. One reason for Semantic Gap is that global

C. Guo, Z.-G. Hou, and Z. Zeng (Eds.): ISNN 2013, Part II, LNCS 7952, pp. 421–428, 2013.

features only concern about global statistics in image and discard local structures in image.

Local feature based image classification methods represent image as a collection of local features. Local features can be feature representation of local patch features after image segmentation, or can be collection of local consistency features. SIFT is the most popular two local features widely used in various computer vision applications. SIFT is based on theory of scale space, aims at generate scale invariance features.

Traditional classification approaches focus on vector representation, where single training vector corresponds to single label. In local feature setting, a collection of local features correspond to single label, where traditional classifiers don't work. We either need classifiers deal with classify collections of local features, or we need a transform from collection representation to vector representation. Bag of visual words (BOVW) model is a realization of second ideas. BOVW aims at transform from a collection of local features into the frequency of visual words. Visual words are usually generated by clustering local features. Local feature can be quantified by finding nearest visual word of it. Then image can be described by frequency of visual words. We can use traditional multi class classifier for image classification.

BOVW well captures local features of image, thus it totally ignores geometric relationships between local features. So BOVW model can be further improved by taking geometric constrains into consideration. Spatial pyramid match kernel capture geometric information between local features by partition the image into increasing fine bins. Final histograms will be generated by concatenate histogram of local bins. BOVW model is only an approximation of global geometric representation while keep the computation efficiency without consideration about the discriminative ability of bins.

In image, many bins contain little discriminative information, thus damn the performance of classification. In this paper, we further improve spatial pyramid match by identify those bins and remove the effect of those bins. We measure the information of an instance by calculating information entropy or other information measure of its neighborhood. We identify the neighborhood of an instance by local sensitive hash.

The structure of our paper as follows: Section 2 gives a brief introduction about related works .Section 3 proposes our method and gives out our proof about coverage of our algorithm. Section 4 describes our experiments.

2   Related Works

Bag of visual words model is an extension of bag of words model in computer vision, which represent images as a collection of unordered vision words. Bag of visual words model represent image as a collection of visual words generated by clustering. In Bag of visual words model spatial information is discarded. Similarity between different images can be calculated by distance between histograms. Thus, BOVW represents images as a bag of unordered visual word, totally neglects geometric information widely exist in original image.

2.1   Spatial Pyramid Match

Spatial pyramid match [3] is an extension of pyramid match[2] considering geometric information. Geometric information is approximated by split image into bins hierarchically, histogram is calculated on each region, and final histogram is formed by concatenate histograms of all bins. Informally, pyramid matching works by place a sequence of increasing coarser grids over the feature space and taking a weighted sum of the number of matches that occur at each level of resolution. Consider a pyramid represented as a sequence of grids at resolution $0,1,....,L$, such that grid at level l has 2^l cells along each dimensions. Let H_X^l and H_Y^l denote the histograms of X and Y at resolution l , so that H_X^l and H_Y^l denote the histograms of X and Y at resolution l , so that H_X^l and H_Y^l denote the numbers of local features fall into the $i-th$ cell of the grid. Then the number of match at level l is given by the histogram intersection kernel $I(H_X^l,H_Y^l) = \sum_{i=1}^{D} \min(H_X^l,H_Y^l)$ .Note from the definition that, the number of match at level l include the finer level match ,the number of new match can be defined as $I^l - I^{l+1}$ for $l = 0,1,...L-1$,and we further gives weight associate level l as $1/(2^{L-l})$,the final pyramid match kernel can be defined as

$$k^L(X,Y) = (1/2^L)I^0 + \sum_{i=1}^{L}(1/2^{L-l+1})I^l \qquad (1)$$

Illustration of Spatial pyramid match can be illustrate as in Fig 1

2.2   Analysis

Spatial pyramid match incorporate geometric information into image classification process. Thus it gives equal weight to each bin without considering the difference between bins. Those bins have strong tendency to appear in small number of categories should be highlighted. The tendency can be reflected by distribution of neighborhood labels. We measure the tendency with Information Entropy of neighborhood label distribution. The purer the neighborhood labels, the lower the neighborhood label entropy.

Thus, in image classification, labels are provided on images level. No specific label is provided on bins level. Image label is just a weak supervise information for particular bin. But those weakly supervised information can still reflect the categories tendency of given bins. Traditional spatial pyramid process is totally unsupervised, which is a waste of useful information.

So we need a framework which well incorporates geometric information and weakly supervised information from image labels.

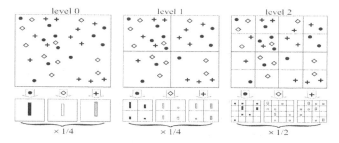

Fig. 1. Illustration of spatial pyramid match kernel. Match in finer grid will receive more weight, while coarser match will get less weight. Final kernel will be obtained by calculating the sum of weighted match over every level[3].

In this paper, we combine geometric information and weak supervised information through neighborhood information entropy, which reflect category tendency of given bins, details of our framework is given in Section 3.

3   Our Approach

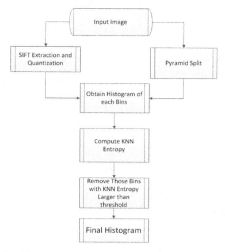

Fig. 2. Neighborhood is identified through local sensitive hash. Weakly supervised information is incorporated in Neighborhood Information Entropy calculation process. Final histograms is generated by remove those non-informative bins and reweighted each histograms through pyramid weighting scheme.

3.1   Neighborhood Identification

Naïve method for neighborhood identification is k nearest neighbor algorithm. Thus for large scale dataset, computation cost for k-nn can be extremely unbearable. Tree based approximate nearest neighbor algorithm such as kd-tree,rp-tree organize data through tree based space partition. Queries will traverse from root node to leaf for nearest neighbor search. Tree based structure has been proved to be approximate for nearest neighbor search. But for high dimensional data ,tree based structure will not approximate.

Local sensitive hash [5][6] become popular for large scale high dimensional approximate nearest neighbor search. Similarity between instances can be measured by hamming distance which can be obtained by bitwise operation. In this paper, we only concern about the neighborhood of given instance. To avoid boundary effect, we adopt several hash tables and select neighbors from candidates returned by each hash table.

Assume after hash, we adopt k neighbors for each query, the conditional distribution of query given its neighbor nn_i can be given by normalized distance of query and nn_i its neighbor

$$p(q \mid nn_i) = dist(q, nn_i) / \sum_{j=1}^{k} dist(q, nn_j) \qquad (2)$$

3.2   Neighborhood Entropy

In information theory, Entropy[4] is used as a measure of information contained in data. The purer the data, the smaller code we need for data description. In classification, information entropy can be used as a measure of complexity of data. Information entropy can be defined as in eq 3

$$I = -\sum_{i} p(w_i) \log_2 p(w_i) \qquad (3)$$

Where $p(w_i)$ represent the frequency of instances belong to class w_i .It is easy to verify that if all instances belong to the same class, the value of I achieve its minimum zero. If all classes have the same frequency, the information entropy achieves its maximum.

With neighborhood in previous step, we can calculate neighborhood information entropy through eq 4

$$I_{NE}(q) = -\sum_{c \in C} (\sum_{nn \in c} p(q \mid nn)) \log_2 (\sum_{nn \in c} p(q \mid nn)) \qquad (4)$$

3.3   Non-informative Bins Removal

After neighborhood identification and neighborhood entropy computation, we need to identify non-informative bins, and setting those histograms to zero. We intuitively set threshold to remove 10% of bins.

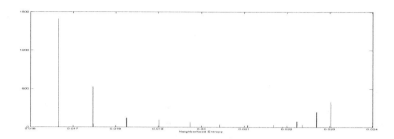

Fig. 3. Distribution of neighborhood entropy at pyramid level 2,we can learn from the image that , most instances have a pure neighborhood, while few have a complex neighborhood.

3.4   Analysis

Through remove those non-informative bins, histograms corresponding to those bins will be down weighted or totally removal. Different from original spatial pyramid match is that our method treats different regions with different weights according to their neighborhood information entropy. Define $Ind_i^l = \{0,1\}$ represents whether bin i at layer l is informative. Weight of bins can be calculated through

$$w_i = \sum_{l=0}^{\#Layers} 2^{l-\#Layers} * Ind_j^l \ (where \ Bin_i \subseteq Bin_j) \tag{5}$$

We can easily find those patches never appear as non-informative bins will get biggest weight in final histogram. Those often appear in non-informative bins will get less weight. Thus traditional spatial pyramid match treat those bins equally.

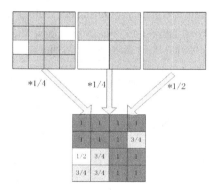

Fig. 4. Illustration of non-informative bins removal based bins weighting

From the theory of support vector machine, we know that kernels we used for support vector machine training and testing must preserve positive semi-definite. Traditional pyramid match kernel has been proved to be positive semi-definite, we will prove that our kernel also satisfy the positive semi-definite condition.

Theorem 1. With non-informative bins removal, spatial pyramid kernel can still preserve positive semi-definite (PSD).

Proof. 1 Basic Kernels we use such as Histogram Intersection Kernels is PSD

2 We don't remove any bins in layer 0, so its kernel matrix is PSD

3 For layers $\neq 0$, assume layer l .bin's index b , assume indices of images preserve bin b in layer l are $I_{lb}(P)$,indices of images remove bin b in layer l are $I_{lb}(R)$,$|I_{lb}(P)|+|I_{lb}(R)|=|I|$

4. Kernel matrix for bin b in layer l can be transform into

$$K_{lb} = \begin{pmatrix} K_{lb(P)} & 0 \\ 0 & K_{lb(R)} \end{pmatrix}$$ Where $K_{lb(R)}$ is zero matrix of $|I_{lb}(R)|\times|I_{lb}(R)|$

5.for any $|I|$ dimensional vector z

$$z \bullet K_{lb} \bullet z^T = zp \bullet K_{lb(P)} \bullet zp^T + zr \bullet K_{lb(R)} \bullet zr^T$$ Where zp is vector of $|I_{lb}(P)|$ and
$$= zp \bullet K_{lb(P)} \bullet zp^T$$

zr is vector with dimension $|I_{lb}(R)|$

From line 1, we can learn that $zp \bullet K_{lb(P)} \bullet zp^T >= 0$, so kernel matrix of layer l bin b is PSD.

6, Final Kernel matrix is a weighted addition of single kernel matrix will preserve PSD .So we can be guarantee that our algorithm will coverage.

4   Experiments

We do image classification on two commonly used datasets, Caltech 101[7] and Scene 15 which was also used in original spatial pyramid match paper.

We remove 10% bins in layer 2 and layer 3 of pyramid .We use chi-square kernel for histogram comparison. With multi class SVM for classification, we get comparable result precision on two datasets.

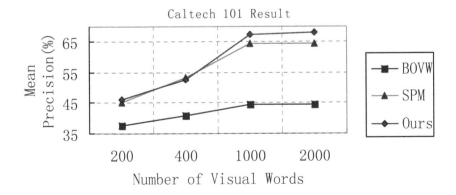

Fig. 5. Mean Precision of Image Classification

For scene-15 we training a vocabulary with length 400,for that larger vocabulary will not increase performance so much .Mean precision of BOVW,SPM and Our Methods are 73.8%,80.3%and82.1%.

5   Conclusion

In this paper, we propose a new method which incorporates weakly supervised information into spatial pyramid scheme. We give out our scheme and proof that under our scheme result kernel matrix can still preserve Positive Semi-definite which can guarantee our algorithm will coverage. Similar algorithms can be found in text classification areas, stop words such as 'the','a', is removed for better performance. Different from stop words removal, we do the removal on bins level instead of words level.

References

1. Duda, R.O., Hart, P.E., Stork, D.G.: Pattern Classification, ch. 8, 2nd edn., pp. 394–413. Wiley, U.K. (2001)
2. Grauman, K.: Matching Sets of Features for Efficient Retrieval and Recognition. Ph.D. Thesis, MIT (2006)
3. Lazebnik, S.: Local, Semi-Local and Global Models for Texture, Object and Scene Recognition. Ph.D. Dissertation, University of Illinois at Urbana-Champaign (May 2006)
4. Shannon, C.E.: A Mathematical Theory of Communication. Bell System Technical Journal 27, 379–423 & 623–656 (1948)
5. Andoni, A., Indyk, P.: Pătraşcu, M.: On the optimality of the dimensionality reduction method. Manuscript (2006)
6. Gionis, A., Indyk, P., Motwani, R.: Similarity search in high dimensions via hashing. In: Proceedings of the 25th International Conference on Very Large Data Bases, VLDB (1999)
7. Fei-Fei, L., Perona, P.: A Bayesian hierarchical model for learning natural scene categories. In: CVPR (2005)
8. Cai, H., Yan, F., Mikolajczyk, K.: Learning weights for codebook in image classification and retrieval. In: Proc. CVPR (2010)

Ship Maneuvering Modeling
Based on Fuzzy Rules Extraction and Optimization

Yiming Bai, Tieshan Li*, and Xiaori Gao

Navigational College, Dalian Maritime University, Dalian 116026, China
{hs_bym,tieshanli}@126.com

Abstract. This paper aims to verify the capability of fuzzy inference system in establishing time series model for ship manoeuvrability. The traditional modeling approaches are usually based on a unified framework. Due to the presence of outliers or noises in ship sailing records, it is difficult in achieving satisfactory performance directly from data. In this paper, we propose a combined time series modeling method by the use of data mining technique and fuzzy system theory. Data mining concepts are introduced to improve the fuzzy rule extraction algorithm to make the resulting fuzzy inference system more robust with respect to the noises or outliers. A ship 20°/20° zig-zag test is simulated. The data point records in time series are obtained from an actual manoeuvring test. With comprehensive robustness analysis, our fuzzy inference system using data mining technology is proved to be a robust and accurate tool for ship manoeuvring simulation.

Keywords: Fuzzy modeling, data mining, optimization, robustness, time series.

1 Introduction

The ship maneuvering model characterizes the ship dynamic behaviors. It is vital to construct an accurate model. There are two kinds of ship maneuvering models: the hydrodynamic model and the response model. Most of the ship manoeuvrability predictors use the hydrodynamic model, such as the Abkowitz model [1], the MMG model [2], etc. Nowadays, it is popular and effective to build a response model of ship maneuvering motion based on the full-scale trials or the actual ship tests [3] [4]. To apply this method, standard maneuvers are conducted to obtain system inputs (rudder angle, propeller revolution, etc) and outputs (ship velocity, heading, yaw rate, location of ship center, etc). Given the input-output data, the ship maneuvering models can be determined by using system identification. The key point for using above two kinds of models is to accurately determine the hydrodynamic derivatives. Although the existing

* This work was supported in part by the National Natural Science Foundation of China (Nos.51179019, 60874056), the Natural Science Foundation of Liaoning Province (No. 20102012) and the Program for Liaoning Excellent Talents in University (LNET) (Grant No.LR2012016), the Fundamental Research Funds for the Central Universities (No. 3132013005), and the Applied Basic Research Program of Ministry of Transport of China.

C. Guo, Z.-G. Hou, and Z. Zeng (Eds.): ISNN 2013, Part II, LNCS 7952, pp. 429–435, 2013.
© Springer-Verlag Berlin Heidelberg 2013

models work well for some cases, they have a limitation in that they can only be used to characterize the ship manoeuvrability in a unified framework. This drawback often constructs a mismatched model. Therefore, it is significant to explore a data-driven approach to model the ship maneuvering characteristics with a structure adaptation mechanism.

On the other hand, ship maneuvering characteristics are often affected by the complex flow, wind and wave. Lots of unavoidable noise and outliers are mixed in ship time series records. In order to deal with this kind of data, we will establish a robust fuzzy inference system with data mining concepts. The introduction of data mining concepts enhances the fuzzy inference system to be flexible to data description and prediction [5]-[9]. Our improved algorithm employs the degree of support and confidence defined by association rule to mine fuzzy rules and reduce the rule number. Acquisition of time series records in real ship zig-zag test

To identify the time series model of ship manoeuvrability with fuzzy inference systems, we need to collect some experimental data and then employ our improved data mining algorithm to extract a set of fuzzy rules. In this paper, the time series data is an actual zig-zag test record from the Training Ship Yukun of Dalian Maritime University, China. Assume that we are given M data points of this time series. These data points can be easily rearranged into M-m input-output data pairs:

$$
\left.
\begin{array}{l}
\{x(t+(M-m)*\Delta), \ldots, x(t+(M-1)*\Delta); \quad x(t+M*\Delta)\} \\
\qquad\qquad \ldots \\
\{x(t), \ldots, x(t+m*\Delta); \quad x(t+(m+1)*\Delta)\}
\end{array}
\right\} \qquad (1)
$$

Then, the time series modeling problem could be transformed into train a fuzzy inference system with these desired M-m data pairs to match the mapping function F_m:

$$\hat{x}(k+p*\Delta)=F_m\{x(k-(m-1)*\Delta), \ldots, x(k)\} \qquad (2)$$

2   Extraction Fuzzy Rules for Ship Manoeuvring Model

The M-m sampled data pairs of ship heading variation could be simplified as:

$$(X(n), y(n)), \qquad n=1,2,\ldots,N \qquad (3)$$

where $X(n)=(x_1(n), x_2(n), \ldots, x_m(n))\in R^m$, and $x_{m+1}(n)\in R$; N is the total number of the data pairs. Then, our proposed approach consists of the following steps:

Step 1 Partition the input-output spaces into fuzzy regions
As both the input and output variables are the ship heading variation, the domain intervals of $x_1, x_2, \ldots, x_m$ and y could be set with the same range: $[x^-, x^+]$. Then the domain interval $[x^-, x^+]$ is divided into p regions. The lengths of these regions can be equal or unequal. And each fuzzy region is a fuzzy set denoted by an linguist label $A_k\in\{A_1, A_2, \ldots, A_p\}$.

Step 2 Convert ordinary records into fuzzy records
Each dimension of input-output data pair (i.e. each data point in the time series) could be represented by an expression of the linguist label and its membership function value. For instance, A_K ($k \in \{1, 2, 3, 4, 5\}$) is one of the linguistic labels associated with a triangular fuzzy membership function. And the triangular membership function is defined as:

$$\left. \begin{array}{l} \mu(x; x_c, \sigma_1) = \dfrac{\sigma_1 - (x_c - x)}{\sigma}, x \in [x_c - \sigma_1, x_c] \\[2mm] \mu(x; x_c, \sigma_2) = \dfrac{\sigma_2 - (x - x_c)}{\sigma_2}, x \in [x_c, x_c + \sigma_2] \\[2mm] \mu(x; x_c, \sigma_1, \sigma_2) = 0, x \in R - [x_c - \sigma_1, x_c + \sigma_2] \end{array} \right\} \qquad (4)$$

where σ_1, σ_2 and the center location x_c are determined in step 1.

Step 3 Generating fuzzy rule base with the degree of support
In order to emphasize and to clarify the basic ideas of our improved approach, a simple two-input one-output case is chosen here. The divisions of input and output spaces are show in fig. 1. And the fuzzy IF-THEN rules are described as:

$$\text{Rule i: IF } x_1 \text{ is } A^{l_1} \text{ and } x_2 \text{ is } A^{l_2}, \text{ THEN } y \text{ is } A^{l_3}, \qquad (5)$$

where $l_1 \in [1,2,...,p]$, $l_2 \in [1,2,...,p]$ and $l_3 \in [1,2,...,p]$ denote the current division. The task of this step is to determine the fuzzy rule's THEN part which could suit the IF part in the most reasonable way by the use of the degree of support. The support degree of rule *i* is calculated as follow:

$$\sup_{B^{l_3}}(Ri) = \frac{\sum_{p=1}^{N} \mu_{(x_1^p)}^{A_1^{l_1}} \mu_{(x_2^p)}^{A_2^{l_2}} \mu_{(y^p)}^{B^{l_3}}}{\sum_{p=1}^{N} \mu_{(x_1^p)}^{A_1^{l_1}} \mu_{(x_2^p)}^{A_2^{l_2}}} \qquad (6)$$

where M-m is the total number of the input-output data pairs, l_1 and l_2 are fixed according to the grid into which the data pairs are put, and l_3 is a variable belonging to a set of $[1,2,...,p]$. We choose the linguistic label B^k as the fuzzy rule's THEN part, whose $\sup_{B^k}(Ri)$ $k \in [1,2,...,p_3]$ is the maximum value among the p values for rule i. When the data pairs have traversed through all the input grids, a "complete" fuzzy rule base is generated. (In some grids $\omega = \sum \mu_{(x_1^p)}^{A_1^{l_1}} \mu_{(x_2^p)}^{A_2^{l_2}}$ is zero, then no rule will be generated.) The flow chart of the traversal algorithm is shown in fig.2.

Step 4 Reduce the rule number with the degree of confidence
A "complete" rule base is generated from the above three steps. In some cases, the input variable is a high-dimensional variable and a huge number of rules are obtained in the rule base. Then the degree of confidence should be used to remove some trivial rules from the rule base:

$$conf(Ri) = \frac{Sup(Ri)}{\dfrac{1}{N} \sum_{p=1}^{N} \mu_{(x_1^p)}^{A_1^{l_1}} \mu_{(x_2^p)}^{A_2^{l_2}}} \qquad (7)$$

The degree of confidence is an index of the density of data pairs. If few data pairs are covered in some of the input grid, the degree of confidence for these rules is very low.

Although this reduction may result in a worse approximation performance, it is very necessary to make such a tradeoff between the complexity and the accuracy in practice.

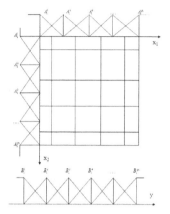

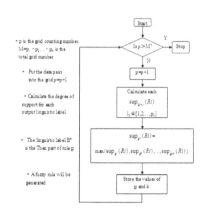

Fig. 1. Divisions of the input and output spaces into fuzzy regions and the corresponding membership functions

Fig. 2. Tthe flow chart of traversal algorithm

Step 5 The centroid defuzzification strategy
We use the following centroid defuzzification formula to determine the output y for the given input X:

$$y = \frac{\sum_{i=1}^{R} \omega^j y_c^i}{\sum_{i=1}^{R} \omega^j} \tag{8}$$

R is the total number of fuzzy rules. ω is the combination of the antecedents in the ith fuzzy rule using product operation. y_c^i denotes the center value of output fuzzy region i.

Thus, a fuzzy inference system could be built by the simple five steps. Our new method is straightforward: it is a one-pass build-up procedure which does not require time-consuming training.

Remark: The Robustness of Fuzzy Inference System

The success of fuzzy modeling relies heavily on the quality of fuzzy rule base [10] [11]. To identify the fuzzy basis function model, Wang and Mendel proposed a simple and practical algorithm for the extraction of fuzzy rules directly from numerical data in paper [12][13].

A further study of these algorithms revealed that there is further opportunity to improve the robustness of the fuzzy rule base [14]. To make the fuzzy system more robust against input noise or outliers in modeling tasks, we improved the algorithm for the extraction of fuzzy rules in section 3. The sampling output values were replaced by its membership function values in computing the weighted average output values for each fuzzy rule. Since the membership function values are regularized into an interval of [0, 1], the noisy error could be easily averaged with enough sample data.

3   Simulation of the Manoeuvrability Characteristics of Ships

In the actual experiment, the training ship was sailing at 15 knots. In order to obtain an accurate record, the heading angle deviations are taken record every 8 seconds by three

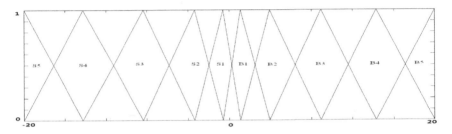

Fig. 3. The refined membership function

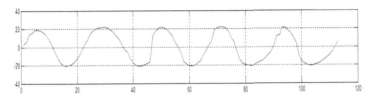

(a)The curve of observing heading angle deviation

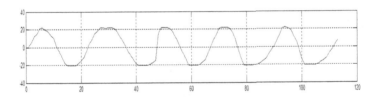

(b)The curve of forecasting heading angle deviation

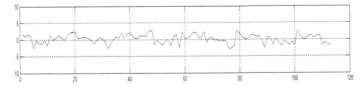

(c)The Prediction error of DM data mining method

Fig. 4. The curves of heading angle deviation forecasting

test clerks. The heading angle deviations values took the mean values of these three records. According to the test record, the domain intervals of heading angle variation is [-20°, 20°]. And the m in expression (1) is set as 4. I.e. the sample data points are rearranged into a four-input and single-out pattern. And 10 fuzzy sets are employed to divide this range interval equally. By optimizing with the partition refinement strategy[15][16], the locations of the fuzzy membership functions are adjusted. The final membership function structure is shown in Fig.3. The modeling errors are less than 3° in all test data points. And the forecasting curves in fig.4 certify that our fuzzy inference system is a suitable tool for robust time series modeling of ship manoeuvrability.

4   Conclusions

This paper develops a framework of time series modeling for ship manoeuvrability. Firstly, we employ an improved data-mining algorithm for the extraction of fuzzy rules. Then, the membership functions are optimized with the sample I/O data pairs. And the adjusting process is operated in an autonomous way. At last, a real ship zig-zag test record is introduced in simulation test. The results prove that our robust fuzzy modeling algorithm is effective in dealing with noisy records in practice.

References

1. Abkowitz, M.A.: Measurement of hydrodynamic characteristic from ship maneuvering trials by system identification. Transactions of Society of Naval Architects and Marine Engineers 88, 283–318 (1980)
2. Ogawa, A., Kasai, H.: On the mathematical model of manoeuvring motion of ship. International Ship-building Progress 25(292), 306–319 (1978)
3. Zhang, X.G., Zou, Z.J.: Identification of Abkowitz model for ship manoeuvring motion using support vector regression. Journal of Hydrodynamics 23, 353–360 (2011)
4. Huang, L.W., Zhang, D.: Ship maneuvering simulation based on real-time numerical simulation of flow. In: 2010 International Conference on Computer and Communication Technologies in Agriculture Engineering (2010)
5. Weng, Y.J., Zhu, Z.Y.: Research on time series data mining based on linguistic concept tree technique. In: Proceedings of the IEEE International Conference on Systems, Man and Cybernetics, vol. 2, pp. 1429–1434 (2003)
6. Au, W.H., Chan, K.C.C.: Mining fuzzy rules for time series classification. In: Proceedings of 2004 IEEE International Conference on Fuzzy Systems, vol. 1, pp. 239–244 (2004)
7. Last, M., Klein, Y., Kandel, A.: Knowledge discovery in time series databases. IEEE Transactions on Systems, Man, and Cybernetics, Part B: Cybernetics 31(1), 160–169 (2001)
8. Batyrshin, I.Z., Sheremetov, L.B.: Perception-based approach to time series data mining. Applied Soft Computing Journal 8(3), 1211–1221 (2008)
9. Chen, C.H., Hong, T.P., Tseng, V.S.: Fuzzy data mining for time-series data. Applied Soft Computing Journal 12(1), 536–542 (2012)
10. Rojas, I., Pomares, H., Ortega, J., Prieto, A.: Self-organized fuzzy system generation from training examples. IEEE Trans. Fuzzy Syst. 8, 23–26 (2000)

11. Wang, L.X.: A Course in Fuzzy Systems and Control, USA (1997)
12. Wang, L.X., Mendel, J.M.: Generating fuzzy rules by learning from examples. IEEE Trans. Syst., Man, Cybern. 22, 1414–1427 (1992)
13. Wang, L.X.: The WM Method Completed: A Flexible Fuzzy System Approach to Data Mining. IEEE Transaction on Fuzzy Systems 11(6), 768–782 (2003)
14. Wang, Y.F., Wang, D.H., Chai, T.Y.: Extraction of Fuzzy Rules with Completeness and Robustness. Acta Automatica Sinica 36(9), 1337–1342 (2010)
15. Guillaume, S.: Designing fuzzy inference systems from data: An interpretability oriented review. IEEE Trans. Fuzzy Syst. 9, 426–443 (2001)
16. Bai, Y.M., Meng, X.Y., Han, X.J.: Partition Refined Fuzzy Inference Systems from Data. American Journal of Engineering and Technology Research 11(12) (2011)

Learning to Create an Extensible Event Ontology Model from Social-Media Streams

Chung-Hong Lee[1], Chih-Hung Wu[1], Hsin-Chang Yang[2], and Wei-Shiang Wen[1]

[1] Dept of Electrical Engineering,
National Kaohsiung University of Applied Sciences, Kaohsiung, Taiwan
leechung@mail.ee.kuas.edu.tw, williamwu.tw@gmail.com,
weishiang@dml.ee.kuas.edu.tw
[2] Dept of Information Management, National University of Kaohsiung, Kaohsiung, Taiwan
yanghc@nuk.edu.tw

Abstract. In this work we utilize the social messages to construct an extensible event ontology model for learning the experiences and knowledge to cope with emerging real-world events. We develop a platform combining several text mining and social analysis algorithms to cooperate with our stream mining approach to detecting large-scale disastrous events from social messages, in order to achieve the aim of automatically constructing event ontology for emergency response First, we employ the developed event detection technique on Twitter social-messages to monitor the occurrence of emerging events, and record the development and evolution of detected events. Furthermore, we store the messages associated with the detected events in a repository. Through the developed algorithms for analyzing the content of social messages and ontology construction the event ontology can be established, allowing for developing relevant applications for prediction of possible evolution and impact evaluation of the events in the future immediately, in order to achieve the goals for early warning of disasters and risk management.

Keywords: Text Mining, Information Retrieval, Data Mining, Machine Learning, Social Mining.

1 Introduction

For emergency response of disastrous events, the decision making process particularly concerns the real-time information need for situational awareness, and a well-established and suitable standard operating procedure the people can follow. Event related social messages, produced by local people and their friends, truthfully depict the actual disastrous process and detailed information of the event development. The corpus of such information contains fruitful learning materials for establishing standard operating procedures to the domain of emergency response. One of the valuable applications might be an event ontology extracted from the social messages regarding the events. Motivated by this, in this work we make use the potentials of social messages for learning the experiences to cope with emerging real-world events.

By analyzing the publicly available social streams, most of the large accidents or disasters could be immediately detected for investigation [1, 2]. Our approach identifies

C. Guo, Z.-G. Hou, and Z. Zeng (Eds.): ISNN 2013, Part II, LNCS 7952, pp. 436–444, 2013.

and extracts candidate terms and event ontologies from social media streams. It is worth mentioning that, the focus of this study is not to create a new structure of ontology from scratch; instead, we expected to design a way to support constructing the entities of extended event-ontology. As a result, our work started with the development of a real-time event detection system which is able to detect emerging events from tweets. Subsequently, we incorporate the system with an event-ontology construction module to mining temporal, spatial, topical and potential relational entities to dynamically create the ontology upon the collected social data sets. Through the developed algorithms for analyzing the content of social messages and ontology construction the event ontology can be established, allowing for developing relevant applications for prediction of possible evolution and impact evaluation of the events in the future immediately, in order to achieve the goals for early warning of disasters and risk management.

2   Related Work

Several ontology-based approaches have been developed to support crisis management and decision making [3]. In Ye's work, three categories of ontologies including static ontology, dynamic ontology, and social ontology [4], were developed to tackle challenging issues in different perspectives for crisis contagion management in financial institutions. Kang proposed a knowledge level model (KLM), which integrates top-level and domain-level ontologies, for systemic risk management in financial institutions [5]. Wang [6] proposed a conceptual modeling approach which was developed to detect real time quality problems and support quality information sharing between agents in the whole milk supply-chain. Still, lots of developed ontology applications mainly rely on people with domain expertise to construct the ontology manually. In modern dynamic environments, how to keep up with real-time information and quickly formulate the data sets into a structured/semi-structured format (i.e. event-ontology form) from unstructured data like social text streams has been becoming a challenging application issue. In general, ontology construction could be categorized in three ways including manual, semi-automatic and automatic ontology constructions [7] as follows:

(i) Manual Ontology Construction

Ye [3] proposed three types of ontology, including static ontology, dynamic ontology and social ontology, for crisis management in financial institutions. Ye [5] proposed a knowledge level model which integrates domain level ontology, top level ontology and problem solving model for systemic risk management in financial institutions. Wang [6] proposed an approach of manual ontology construction which based on conceptual modeling for supply chain management.

(ii) Semi-Automatic Ontology Construction

Wei [8] proposed an approach that utilizes HTML parser to crawl the information about agriculture from web, and then utilizes webpage wrapper to capture the information and translate these data to structure data for storing in database. Subsequently, the top taxonomy and defined the attribute of these data are constructed by experts. Finally, the architecture of top-down hierarchical ontology could be constructed based on top taxonomy with OWL language. Maedche [9] proposed an architecture and system, named Text-To-Onto, to extract entities and relations from text which places several natural language processing and information extracting approaches within a coherent framework. [10] is a completely re-design and re-engineering of KAON-TextToOnto combining machine learning approaches with basic linguistic processing, such as tokenization or lemmatizing and shallow parsing.

(iii) Automatic Ontology Construction

Dahab [11] proposed a semantic pattern-based approach to construct ontology automatically from natural domain documents. Navigli [12] proposed an approach, OntoLearn, that extracts a domain terminology from available documents. Gacitua [13] provided a framework, OntoLancs, for ontology learning form text to support and evaluate the usefulness and accuracy of different techniques as well as the possible combinations of techniques into specific processes. Narayan [14] utilized twitter API to collect tweets and categorized events from these messages with keyword query. Finally, there are five attributes which include name, time, location, type and url for construction the social event ontology. However, the ontology-construction work mentioned above mainly using news, literature, and web information. In this work, we propose a novel method to construct event ontology by using user generated content from social media.

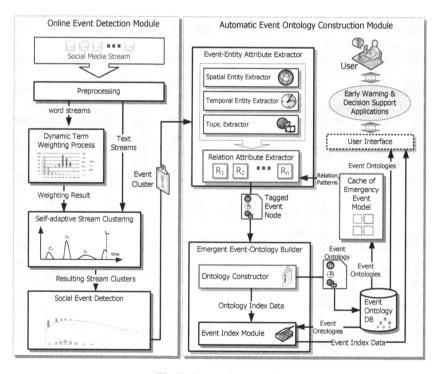

Fig. 1. System framework

3   System Framework

In this section, the system framework for event mining and construction process of event ontology is described. As shown in Fig. 1, in this work we develop a self-adaptive stream clustering algorithm which combines several technical modules, including a *Dynamic Term Weighting* process, a *Social Event Detection* technique, an *Event-Entity Attribute Extractor* and an *Emergent Event-Ontology Builder*. More detailed description of our developed model is described as follows.

(i) Dynamic Term Weighting Process

To process texts with a chronological order, a fundamental problem we concerned is how to find the significant features in collected streams. The design of weighting process of real-time message should be constantly updated. Here we apply the term weighting scheme BursT which was proposed in [1, 2, 15]. The experimental results indicate that has a better performance in weighting words of messages than incremental TFIDF & TFPDF of the word w at time t in BursT will be constituted as BS (Burst Score) and TI (Term Importance) as equation (1) [16]:

$$BursT_{w,t} = BS_{w,t} \times TI_{w,t} \tag{1}$$

(ii) Self-adaptive Stream Clustering

This study uses a self-adaptive window to extract event clusters and then calculates the message intensity within such clusters to estimate the evolving stage and detect the potential incident in real-time. Each cluster will be decided to survive or phase out by checking its decay factor. As same as the message intensity, we not only consider the evolvement of a cluster but also the number of messages in it. Even with a high similarity, a new message would not be joined into the cluster which is going to fade away. Here a time-decay function $d(\cdot)$ is implemented to determinate the lifecycle of an event cluster and to erase slack messages which is denoted as equation (2) [16]:

$$d\left(t_{c_i}, t\acute{\ }_{c_i}, t_{m_r}, n\right) = \eta n^{-\alpha} e^{\frac{-(t\acute{\ }_{c_i} - t_{m_r})}{1 + (t_{c_i} - t_{m_r})}} \tag{2}$$

Where $\alpha > 1$ is the power factor and $\eta = 1 + \ln n$ controls the slope and optimizes our scheme according to the number of real-time posts. The similarity between message m_r and selected cluster c_i would be determined by a threshold θ. The m_r is considered dissimilar to c_i and cannot be joined to c_i accurately if as equation (3) [16]:

$$sim(c_i, m_r)d(\cdot) < \theta \tag{3}$$

(iii) Social Event Detection

The key to catch the hot event is to detect spread messages fast and efficiently, and two major processes need to be accomplished including: 1) monitoring the temporal variability of a term and the intensity of online message, 2) tracing the evolvement of an incident according to its density and life-cycle. Here, we first extract the event threat by tracing the evolvement of event clusters. Let $C_i^{y \leq \varepsilon_{max}} = \{c_1^y, c_2^y, \dots, c_k^y\}$ be the set of clusters of the collected event thread y and $\forall y \ C_i^y \neq \emptyset$, and

$$\exists c_t^{'} = \begin{cases} c_{i+1}^y, if \ sim(c_i^y, c_t^{'}) \geq \theta^c \\ c_1^{y+1}, if \ sim(c_i^y, c_t^{'}) < \theta^c \end{cases} \tag{4}$$

$$sim(c_i^y, c_t') = \max_{y=1,\cdots,\mathcal{E}} \sum_i cos(c_i^y, c_t') \tag{5}$$

where θ^c is the threshold determining the extracted cluster c_t belongs to a existed thread or a new event.

(iv) Event-Entity Attribute Extractor

The clustered online event provides precious information such as the distribution of temporal evolution, spatial factor (region or geoName), message density, and word bursting score. Such information facilitates to construct necessary entities of event ontology. van Hage [17] illustrates an example of a simple event ontology model. There are four core classes: sem:Event (what happens), sem:Actor (who or what participated), sem:Place (where), sem:Time (when). These entities can be fulfilled by combining clues of online event collected form our online detection process.

● Temporal Feature Extractor

The temporal features were extracted from clustered events. The starting time could be set as the creating time point of first cluster, and vice versa the ending time is the last updating time point of the last cluster. Moreover, compared to traditional approaches, our method can also record the density distribution of messages which can support to discover the potential relations among different events, e.g. a sub-event.

● Spatial Named Entity Extractor

Users of microblogs often use a limited vocabulary in the messages, including names of objects are used in singular and plural; a smaller number of words are used to produce many of the compound terms and phrases. For instance, in the event of Mar.11 2011 Japan Earthquake, we can find out that the terms tokyo and japan were used more intensively than other spatial words in the messages, and the importance of each word could be represented as the bursty score. Here, the candidate of spatial entity ($E_{spatial}$) for a selected event can be obtained by calculating the product of message number and bursty score of each keyword. Then we can get the most possible candidate of $E_{spatial}$.

$$E_{spatial}^{evt_i} = \max_{w \in S}(\sum_{t \in TF} MS_{w_{evt_i}^t} \times BS_{w_{evt_i}^t}) \tag{6}$$

where S is collected terminology of spatial entity, e.g. GeoNames geographical database[18], TF is the time frame of event evt_i, $MS_{w_{evt_i}^t}$ is the number of messages containing a specific geographical name and in the event evt_i at time t.

● Topic Extractor

Similar to spatial entity, the most possible topic words of event evt_i can be selected by weighting their number and bursty score as:

$$W_{evt_i}^{topic} = \max_{w \in D}(\sum_{t \in TF} MS_{w_{evt_i}^t} \times BS_{w_{evt_i}^t}) \qquad (7)$$

where D is the collection of domain specific terminology, e.g. natural disaster, and the candidate topic entity can be constructed as the combination of spatial entity and topic word:

$$E_{topic}^{evt_i} = W_{evt_i}^{topic} + E_{spatial}^{evt_i} \qquad (8)$$

● Relation Attribute Extractor

Relation attribute extractor mainly extracts the possible relationships among real-time events and the past events, by analyzing the temporal and spatial relationships from all related event content which have been stored in the event ontology database. We generalized some specific attributes which can be extracted from real-time events, by using the relation entity defined by some important ontology, such as sub-events, dependent events, continues event, derived events and novel events.

(v) Emergent Event-Ontology Builder

The entity attributes of real-time events will be automatically tagged by the *ontology constructor*. These attributes are being encapsulated into appropriate structured document to fit various ontology forms. To allow users quickly locate and track the historical information associated with some real-time events, a module is designed to map keywords of existing events by an index table. This is the key for implementing a social search system. In real-time event detection phase, our system encapsulates the bursty scores which correspond to the event-message clusters. The event index module can perform sorting events via the intensity of keywords.

4   Experimental Result

In the experiment, the test data set were collected through Twitter Stream API. We analyze the message corpus collected from Oct 26, 2012 to Nov 2nd, 2012. There were totally about more than 2 million tweets utilized for this study. During the period, the *Hurricane Sandy* event is selected for investigation as our case study. Fig. 2 illustrates the spatio-temporal impacts of the case event. In the beginning, the spatial tags of tweets were wildly spread around large regions and getting concentration as the storm made more damages day by day.

Subsequently, we also use the Japan earthquake event to verify our proposed event-entity extraction method. The earthquake event was detected at 14:03:28, and then the tsunami event follows the earthquake at 14:28:12. Fig. 3 illustrates the partial result of entity extraction for "311 Japan earthquake" event. We can found that where and when the tsunami happened by detecting messages posted from social media stream, and then the topic was drifted from earthquake to tsunami. Hence, the proposed event-entity extraction method can extract the correct entity of event topic accurately.

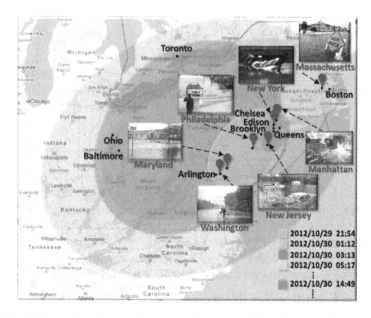

Fig. 2. Illustration of the spatio-temporal impacts on *"Hurricane Sandy"* event

eID	BS	StartTime	EndTime	KW1	KW1%	KW2	KW2%	L1	L1%	L2	L2%
36155789:129987529	2.630036446	Fri Mar 11 14:28:12	Fri Mar 11 14:42:49	tsunami	6.15%	earthquake	3.41%	japan	78.18%	tokyo	5.45%
36160938:129987617	3.661421143	Fri Mar 11 14:42:57	Fri Mar 11 14:47:38	tsunami	13.93%	earthquake	5.25%	japan	61.42%	philippines	4.72%
36162756:129987649	7.433504033	Fri Mar 11 14:48:11	Fri Mar 11 14:53:33	tsunami	14.21%	earthquake	4.43%	japan	59.41%	philippines	6.93%
36164546:129987682	12.08302991	Fri Mar 11 14:53:44	Fri Mar 11 14:59:58	tsunami	9.71%	earthquake	5.13%	japan	73.08%	tokyo	1.10%
36166848:129987722	11.8933249	Fri Mar 11 15:00:20	Fri Mar 11 15:03:11	tsunami	19.50%	earthquake	2.37%	japan	50%	taiwan	7.79%
36167933:129987739	6.509132447	Fri Mar 11 15:03:15	Fri Mar 11 15:04:15	tsunami	23.98%	earthquake	5.24%	japan	50.62%	indonesia	7.41%
36168613:129987750	22.18908585	Fri Mar 11 15:05:06	Fri Mar 11 15:07:46	tsunami	17.35%	earthquake	4.25%	japan	66.36%	taiwan	3.64%
36169576:129987766	31.39012775	Fri Mar 11 15:07:47	Fri Mar 11 15:10:25	tsunami	18.43%	earthquake	3.91%	japan	54.63%	taiwan	6.71%
36170540:129987782	21.50148077	Fri Mar 11 15:10:28	Fri Mar 11 15:12:24	tsunami	21.90%	earthquake	2.44%	japan	52.07%	philippines	4.73%
36171262:129987795	37.64577136	Fri Mar 11 15:12:30	Fri Mar 11 15:14:48	tsunami	30.80%	earthquake	2.68%	japan	48.75%	indonesia	6.45%
36172102:129987808	16.62280969	Fri Mar 11 15:14:49	Fri Mar 11 15:15:37	tsunami	31.55%	earthquake	2.49%	japan	46.32%	indonesia	7.37%
36172387:129987813	41.9865764	Fri Mar 11 15:15:38	Fri Mar 11 15:17:52	tsunami	33.41%	earthquake	2.61%	japan	51.60%	indonesia	4.80%
36173310:129987829	36.43095551	Fri Mar 11 15:18:11	Fri Mar 11 15:19:36	tsunami	34.18%	earthquake	2.90%	japan	47.69%	indonesia	6.15%
36173848:129987837	38.42318533	Fri Mar 11 15:19:36	Fri Mar 11 15:20:59	tsunami	30.04%	earthquake	2.91%	japan	62.50%	indonesia	7.14%
36174412:129987846	55.62022589	Fri Mar 11 15:21:02	Fri Mar 11 15:23:07	tsunami	33.93%	earthquake	2.33%	japan	47.46%	indonesia	8.14%

Fig. 3. Illustration of the result of entity extraction on "311 Japan earthquake" event

5   Conclusion

Social-media messages contain fruitful experiences and knowledge regarding how to relieve victims of some disaster for the applications of emergency response. One of the valuable applications might be an event ontology extracted from the social messages regarding the events. The view is taken, therefore in this work we make use the potentials of social messages to develop a novel ontology system for emergency response. Our work started with the development of a real-time event detection system which is able to detect emerging events from tweets. A self-adaptive stream clustering method is implemented to detect online events. Subsequently, we incorporate the system with an event-ontology construction module to mining temporal, spatial, topical and potential relational entities to dynamically create the ontology upon the collected social data sets. Through the developed algorithms for analyzing the content of social messages and ontology construction the event ontology can be established, allowing for developing relevant applications for prediction of possible evolution and impact evaluation of the events in the future immediately, in order to achieve the goal for risk management.

References

1. Lee, C.H.: Mining spatio-temporal information on microblogging streams using a density-based online clustering method. Expert Systems with Applications 39(10), 9623–9641 (2012)
2. Lee, C.H., et al.: Computing Event Relatedness Based on a Novel Evaluation of Social-Media Streams. In: Proceedings of the 2012 International Workshop on Social Computing, Network, and Services (SocialComNet-2012), June 26-28, vol. 164, pp. 697–707 (2012)
3. Ye, K., et al.: Ontologies for crisis contagion management in financial institutions. Journal of Information Science 35(5), 548–562 (2009)
4. Jurisica, I., et al.: Ontologies for Knowledge Management: An Information Systems Perspective. Knowledge and Information Systems 6(4), 380–401 (2004)
5. Ye, K., et al.: Knowledge level modeling for systemic risk management in financial institutions. Expert Systems with Applications 38(4), 3528–3538 (2011)
6. Wang, S., et al.: A conceptual modeling approach to quality management in the context of diary supply chain. In: Proceedings of the 2010 2nd International Conference on Information Science and Engineering, ICISE (2010)
7. Chen, R.C., Liang, J.Y., Pan, R.H.: Using recursive ART network to construction domain ontology based on term frequency and inverse document frequency. Expert Systems with Applications 34(1), 488–501 (2008)
8. Wei, Y.Y., et al.: From Web Resources to Agricultural Ontology: a Method for Semi-Automatic Construction. Journal of Integrative Agriculture 11(5), 775–783 (2012)
9. Maedche, A., Staab, S.: Ontology Learning for the Semantic Web. IEEE Intelligent Systems 16(2), 72–79 (2001)
10. Cimiano, P., Volker, J.: Text2Onto - A Framework for Ontology Learning and Data-driven Change Discovery. In: Proceedings of the 10th International Conference on Applications of Natural Language to Information Systems, Alicante, Spain, pp. 227–238
11. Dahab, M.Y., Hassan, H.A., Rafea, A.: TextOntoEx: Automatic ontology construction from natural English text. Expert Systems with Applications 34(2), 1474–1480 (2008)
12. Navigli, R., Velardi, P.: Learning Domain Ontologies from Document Warehouses and Dedicated Web Sites. Computational Linguistics 30(2), 151–179 (2004)
13. Gacitua, R., Sawyer, P., Rayson, P.: A flexible framework to experiment with ontology learning techniques. Knowledge-Based Systems 21(3), 192–199 (2008)
14. Narayan, S., et al.: Population and Enrichment of Event Ontology using Twitter. In: Proceedings of the 1 st Workshop on Semantic Personalized Information Management (2010)

15. Lee, C.H., Wu, C.H., Chien, T.F., Burs, T.: A Dynamic Term Weighting Scheme for Mining Microblogging Messages. In: Proceedings of the 8th International Symposium on Neural Networks (ISNN-2011), Guilin, China, May 29-June 1, pp. 548–557 (2011)

16. Lee, C.H., et al.: Being aware of the world: An Early-Warning System Framework by Detecting Real-time Social-Media Streams. In: Proceedings of the 2012 International Conference on Innovations in Electronics and Energy Engineering, Singapore, December 14-15, pp. 226–230 (2012)

17. van Hage, W.R., et al.: Design and use of the Simple Event Model (SEM). Journal of Web Semantics 9(2), 128–136 (2011)

18. GeoNames, http://www.geonames.org/

A Revised Inference for Correlated Topic Model

Tomonari Masada[1] and Atsuhiro Takasu[2]

[1] Nagasaki University, 1-14 Bunkyo-machi, Nagasaki-shi, Nagasaki, 852-8521 Japan
masada@nagasaki-u.ac.jp
[2] National Institute of Informatics, 2-1-2 Hitotsubashi,
Chiyoda-ku, Tokyo, 101-8430 Japan
takasu@nii.ac.jp

Abstract. In this paper, we provide a revised inference for correlated topic model (CTM) [3]. CTM is proposed by Blei et al. for modeling correlations among latent topics more expressively than latent Dirichlet allocation (LDA) [2] and has been attracting attention of researchers. However, we have found that the variational inference of the original paper is unstable due to almost-singularity of the covariance matrix when the number of topics is large. This means that we may be reluctant to use CTM for analyzing a large document set, which may cover a rich diversity of topics. Therefore, we revise the inference and improve its quality. First, we modify the formula for updating the covariance matrix in a manner that enables us to recover the original inference by adjusting a parameter. Second, we regularize posterior parameters for reducing a side effect caused by the formula modification. While our method is based on a heuristic intuition, an experiment conducted on large document sets showed that it worked effectively in terms of perplexity.

Keywords: Topic models, variational inference, covariance matrix.

1   Introduction

Topic modeling is one of the dominant trends of recent data mining research. This approach uses latent variables for modeling topical diversity in document data and represents the data in a lower-dimensional "topic" space. This line of thinking reminds us that multilayer undirected graphical models also extract a lower-dimensional data representation. In fact, neural network researchers also propose a topic model, e.g. by using restricted Boltzmann machines [12] or by a hybridization [13], where we can find an affinity between both approaches.

However, many topic models, including latent Dirichlet allocation (LDA) proposed by the inaugural paper of Blei et al. [2], do not consider *correlations* among latent topics explicitly. It is natural to assume that latent topics interwoven into a semantic content of each document are correlated to some extent. For example, articles on worldwide energy consumption may often mention geopolitical conflicts among countries. Therefore, Blei et al. [3] have proposed correlated topic model (CTM) to offer an established way to model correlations among latent topics. While we know that recent sophisticated approaches can also model topic correlations [11,4], we here concentrate on CTM.

C. Guo, Z.-G. Hou, and Z. Zeng (Eds.): ISNN 2013, Part II, LNCS 7952, pp. 445–454, 2013.

In CTM, per-document topic distributions are obtained from a corpus-wide logistic normal distribution, where a covariance matrix Σ models correlations among latent topics. More precisely, we draw a K-dimensional vector $\boldsymbol{m}_d$ for document d from a Gaussian distribution $\mathcal{N}(\boldsymbol{\mu}, \boldsymbol{\Sigma})$, where K is the number of topics, and obtain a topic distribution $\boldsymbol{\theta}_d$ as $\theta_{dk} = \exp(m_{dk})/\sum_{k'} \exp(m_{dk'})$. This construction makes CTM more expressive than LDA.

In the variational inference for CTM proposed in the original paper [3], the covariance matrix Σ is estimated by maximizing the variational lower bound of the log evidence. Further, Σ needs to be inverted in every iteration. Therefore, Σ should be kept less singular all through the iterations of the inference. However, our preliminary experiment conducted on large document sets has shown that the inference often gets unstable due to the fact that Σ is almost singular. When we apply CTM to a large data set, we would like to set the number of topics to a large value, say 300. Under that situation, the original inference gives a result disastrous in terms of perplexity [2]. Therefore, we have revised the inference. Through trying several heuristic methods for making the covariance matrix less singular, we have found an effective one, which is given in this paper.

However, after revising the inference, we have found another problem. It is known that variational inference for topic models is likely to give worse perplexity than inference by sampling [1], though it is not known whether there are any efficient sampling methods for CTM like collapsed Gibbs sampling (CGS) for LDA [6]. Therefore, for achieving as good perplexity as possible, we can take the following strategy: run CGS for LDA first and then run the variational inference for CTM. Fortunately, we can initialize the parameters of CTM based on a result of CGS for LDA. Therefore, we can start the variational inference for CTM with the parameter values giving a good perplexity. However, a preliminary experiment has shown that our revised inference is likely to make the initial perplexity, which is achieved by CGS for LDA, worse as the inference proceeds. Therefore, we regularize some of the variational posterior parameters so that they are kept close to their initial values inherited from CGS for LDA.

In sum, our method consists of the following two features: 1) Keep the covariance matrix less singular; and 2) Keep the values of some posterior parameters close to their values initialized based on a result of CGS for LDA. With respect to the former, we modify the update formula of the covariance matrix by taking an approach similar to shrinkage estimation. With respect to the latter, we add a regularization term so that the parameter values do not deviate significantly from their initial values. The rest of the paper is organized as follows. Section 2 describes the details of the original variational inference for CTM. Section 3 presents our proposal. Section 4 contains the results of an experiment conducted over large document sets. Section 5 concludes the paper with discussions.

2   Correlated Topic Models

With correlated topic model (CTM) [3], we can explicitly model correlations among latent topics. The variety of correlations that can be modeled in CTM is richer than that in LDA, because Dirichlet prior distribution used in LDA is not that powerful in modeling correlations among drawn multinomial probabilities. To make the paper self-contained, we describe CTM in detail. In the following, we denote the number of topics, different words, and documents by K, W, and D, respectively, and identify each entity with its index number.

As in latent Dirichlet allocation (LDA) [2], we represent each document d as a mixture of K latent topics by using a multinomial distribution Multi($\boldsymbol{\theta}_d$) defined over topics, also in CTM. The multinomial parameters $\boldsymbol{\theta}_d = (\theta_{d1}, \ldots, \theta_{dK})$ satisfy $\sum_k \theta_{dk} = 1$, where θ_{dk} is a probability that topic k is expressed by a word token in document d. Topic k is in turn represented by a multinomial distribution Multi($\boldsymbol{\phi}_k$) defined over words. The parameters $\boldsymbol{\phi}_k = (\phi_{k1}, \ldots, \phi_{kW})$ satisfy $\sum_w \phi_{kw} = 1$, where ϕ_{kw} is a probability that a token of word w expresses topic k. Both in LDA and in CTM, we assume that $\boldsymbol{\phi}_k$s are drawn from a corpus-wide Dirichlet prior distribution Dir($\boldsymbol{\beta}$), where $\boldsymbol{\beta} = (\beta_1, \ldots, \beta_W)$ is a set of its hyperparameters. However, with respect to per-document topic multinomial distributions Multi($\boldsymbol{\theta}_d$), LDA and CTM show a difference.

In LDA, per-document topic distributions are drawn from a corpus-wide Dirichlet distribution Dir($\boldsymbol{\alpha}$). However, Dirichlet distribution can only model a restricted variety of correlations among topics. Therefore, CTM adopts logistic-normal distribution as the prior for $\boldsymbol{\theta}_d$s. We below describe CTM generatively.

1. For each topic $k \in \{1, \ldots, K\}$, draw parameters $\boldsymbol{\phi}_k = (\phi_{k1}, \ldots, \phi_{kW})$ of a multinomial distribution Multi($\boldsymbol{\phi}_k$) defined over words $\{1, \ldots, W\}$ from the corpus-wide Dirichlet prior distribution Dir($\boldsymbol{\beta}$).
2. For each document $d \in \{1, \ldots, D\}$,
   (a) Draw a K-dimensional vector $\boldsymbol{m}_d = (m_{d1}, \ldots, m_{dK})$ from the corpus-wide K-dimensional Gaussian distribution $\mathcal{N}(\boldsymbol{\mu}, \boldsymbol{\Sigma})$.
   (b) Obtain a topic distribution $\boldsymbol{\theta}_d = (\theta_{d1}, \ldots, \theta_{dK})$ as $\theta_{dk} = \frac{\exp(m_{dk})}{\sum_{k'} \exp(m_{dk'})}$.
   (c) Let n_d be the length (i.e., the number of word tokens) of document d. For the ith word token in document d, where $i \in \{1, \ldots, n_d\}$,
      i. Draw a topic from the topic multinomial distribution Multi($\boldsymbol{\theta}_d$). Let the drawn topic be the value of a latent variable z_{di}, which gives the topic to which the ith word token in document d is assigned.
      ii. Draw a word from the word multinomial distribution Multi($\boldsymbol{\phi}_{z_{di}}$). Let the drawn word be the value of an observed variable x_{di}, which gives the word appearing as the ith word token of document d.

Based on this description, we obtain the full joint distribution of CTM as follows:

$$p(\boldsymbol{x}, \boldsymbol{z}, \boldsymbol{\phi}, \boldsymbol{m} | \boldsymbol{\beta}, \boldsymbol{\mu}, \boldsymbol{\Sigma}) = p(\boldsymbol{\phi}|\boldsymbol{\beta})p(\boldsymbol{m}|\boldsymbol{\mu}, \boldsymbol{\Sigma})p(\boldsymbol{z}|\boldsymbol{m})p(\boldsymbol{x}|\boldsymbol{\phi}, \boldsymbol{z})$$

$$= \prod_k \frac{\Gamma(\sum_w \beta_w)}{\prod_w \Gamma(\beta_w)} \phi_{kw}^{\beta_w - 1} \cdot \prod_d \frac{1}{(2\pi)^{K/2}|\boldsymbol{\Sigma}|^{1/2}} \exp\left\{ -\frac{1}{2}(\boldsymbol{m}_d - \boldsymbol{\mu})^T \boldsymbol{\Sigma}^{-1}(\boldsymbol{m}_d - \boldsymbol{\mu}) \right\}$$

$$\cdot \prod_d \prod_i \frac{\exp(m_{dz_{di}})}{\sum_k \exp(m_{dk})} \phi_{z_{di} x_{di}} . \tag{1}$$

The variational inference proposed in the original paper [3] approximates the posterior $p(\boldsymbol{z}, \boldsymbol{\phi}, \boldsymbol{m} | \boldsymbol{x}, \boldsymbol{\beta}, \boldsymbol{\mu}, \boldsymbol{\Sigma})$ by a factorized variational posterior $q(\boldsymbol{z})q(\boldsymbol{\phi})q(\boldsymbol{m})$.

Consequently, a lower bound of the log evidence $\ln p(\boldsymbol{x}|\boldsymbol{\beta}, \boldsymbol{\mu}, \boldsymbol{\Sigma})$ is obtained by applying Jensen's inequality as follows:

$$\ln p(\boldsymbol{x}|\boldsymbol{\beta}, \boldsymbol{\mu}, \boldsymbol{\Sigma}) \geq \int \sum_{\boldsymbol{z}} q(\boldsymbol{z})q(\boldsymbol{m}) \ln p(\boldsymbol{z}|\boldsymbol{m})d\boldsymbol{m} + \int q(\boldsymbol{\phi}) \ln p(\boldsymbol{\phi}|\boldsymbol{\beta})d\boldsymbol{\phi}$$

$$+ \int \sum_{\boldsymbol{z}} q(\boldsymbol{z})q(\boldsymbol{\phi}) \ln p(\boldsymbol{x}|\boldsymbol{\phi}, \boldsymbol{z})d\boldsymbol{\phi} + \int q(\boldsymbol{m}) \ln p(\boldsymbol{m}|\boldsymbol{\mu}, \boldsymbol{\Sigma})d\boldsymbol{m}$$

$$- \sum_{\boldsymbol{z}} q(\boldsymbol{z}) \ln q(\boldsymbol{z}) - \int q(\boldsymbol{\phi}) \ln q(\boldsymbol{\phi})d\boldsymbol{\phi} - \int q(\boldsymbol{m}) \ln q(\boldsymbol{m})d\boldsymbol{m} . \tag{2}$$

We denote the right hand side of Eq. (2) by $\mathcal{L}$, which needs to be maximized. With respect to each variational posterior distribution, we assume the followings.

- $q(\boldsymbol{z})$ is factorized as $\prod_d \prod_i \gamma_{diz_{di}}$, where γ_{dik} is an approximated probability that the ith word token in document d expresses topic k.
- $q(\boldsymbol{\phi})$ is factorized as $\prod_k q(\boldsymbol{\phi}_k|\boldsymbol{\zeta}_k)$, where $q(\boldsymbol{\phi}_k|\boldsymbol{\zeta}_k)$ is the density of an approximated Dirichlet posterior whose parameters are $\boldsymbol{\zeta}_k = (\zeta_{k1}, \dots, \zeta_{kW})$.
- $q(\boldsymbol{m})$ is factorized as $\prod_d \prod_k q(m_{dk}|r_{dk}, s_{dk})$, where $q(m_{dk}|r_{dk}, s_{dk})$ is a density of a univariate Gaussian distribution whose mean and standard deviation parameters are r_{dk} and s_{dk}, respectively.

The variational parameters γ_{dik}s, ζ_{kw}s, and ν_ds can be updated by a closed formula: $\gamma_{dik} \propto \exp(r_{dk}) \cdot \frac{\exp \Psi(\zeta_{kx_{di}})}{\exp \Psi(\sum_w \zeta_{kw})}$; $\zeta_{kw} = \beta_w + \sum_d \sum_i \sum_k \gamma_{dik}$; and $\nu_d = \sum_k \exp(r_{dk} + s_{dk}^2/2)$, where ν_ds are the parameters introduced to make the inference tractable. Details of the derivation are referred to the original paper [3].

However, the parameters r_{dk}s and s_{dk}s cannot be updated by any closed formulas. In this paper, we use L-BFGS [10,7] for maximizing the relevant terms in $\mathcal{L}$ and update these parameters. The target functions are given below.

$$\mathcal{L}(r_{dk}) = n_{dk}r_{dk} - \frac{n_d}{\nu_d} \exp(r_{dk} + s_{dk}^2/2)$$

$$+ \frac{1}{2}r_{dk}^2(\boldsymbol{\Sigma}^{-1})_{kk} - r_{dk} \sum_{k'}(r_{dk'} - \mu_{k'})(\boldsymbol{\Sigma}^{-1})_{kk'} \tag{3}$$

$$\mathcal{L}(s_{dk}) = -\frac{n_d}{\nu_d} \exp(r_{dk} + s_{dk}^2/2) - \frac{1}{2}s_{dk}^2(\boldsymbol{\Sigma}^{-1})_{kk} + \ln s_{dk} \tag{4}$$

Since the parameters $r_{d1}, \dots, r_{dK}$ and $s_{d1}, \dots, s_{dK}$ for a fixed d are dependent on each other, we update them not separately but in concert by using L-BFGS.

The mean parameter $\boldsymbol{\mu}$ of the K-dimensional Gaussian distribution $\mathcal{N}(\boldsymbol{\mu}, \boldsymbol{\Sigma})$, which models the correlation among latent topics, can be updated as $\boldsymbol{\mu} = \frac{1}{D}\sum_d \boldsymbol{r}_d$. The covariance matrix $\boldsymbol{\Sigma}$ can be updated as

$$\boldsymbol{\Sigma} = \frac{1}{D}\sum_d \boldsymbol{S}_d + \frac{1}{D}\sum_d (\boldsymbol{r}_d - \boldsymbol{\mu})(\boldsymbol{r}_d - \boldsymbol{\mu})^T , \tag{5}$$

where S_d is a $K \times K$ diagonal matrix whose kth diagonal entry is s_{dk}^2. However, Eq. (5) is likely to give an almost singular matrix when K is large. This is a serious problem, because we need the inverse of Σ in Eqs. (3) and (4). When we apply CTM to a large document set, we would like to set the number of topics K to a large number, say 300. This type of situation is likely to make Σ almost singular and thus to make the entire inference unstable. Therefore, we propose a revised inference for CTM.

3   A Revised Inference for CTM

Our proposal aims to achieve a stable inference for CTM by making the covariance matrix Σ far from singular. The proposal has the following two features:

- We modify the update formula Eq. (5) in a manner that we can recover the original inference by adjusting a parameter called *interpolation parameter*.
- We initialize the parameters of CTM by using a result of collapsed Gibbs sampling (CGS) for LDA and regularize some parameters lest they deviate substantially from their initial values. The strength of regularization can be adjusted by a parameter called *regularization parameter*.

3.1   Covariance Matrix Update

First, we discuss a revised update of Σ. We focus on the $K \times K$ matrix appearing as the second term of the right hand side of Eq. (5), i.e., $\hat{T} = \frac{1}{D} \sum_d (r_d - \mu)(r_d - \mu)^T$. We can view $\hat{T}$ as the maximum likelihood estimator of the covariance matrix T of a Gaussian distribution $\mathcal{N}(\mu, T)$, from which $r_1, \ldots, r_D$ are drawn. Chen et al. [5] give the following matrix as a "naive but most well-conditioned estimate" for T: $\hat{F} = \frac{\mathrm{Tr}(\hat{T})}{K} I$, where $\mathrm{Tr}(\cdot)$ is the trace of a matrix and I is the identity matrix. $\hat{F}$ is a diagonal matrix whose diagonal entries are all equal to the average of the diagonal elements of $\hat{T}$. Our approach uses $\hat{F}$ in place of the first term of the right hand side of Eq. (5) and update Σ as $\Sigma = \frac{\mathrm{Tr}(\hat{T})}{K} I + \hat{T}$. Further, we obtain a linear interpolation of this equation and Eq. (5) by introducing an *interpolation parameter* π as follows:

$$\Sigma = \left\{ (1 - \pi) \cdot \frac{1}{D} \sum_d S_d + \pi \cdot \frac{\mathrm{Tr}(\hat{T})}{K} I \right\} + \frac{1}{D} \sum_d (r_d - \mu)(r_d - \mu)^T . \quad (6)$$

When $\pi = 0$, we can recover the original inference. It can be said that we use $\hat{F}$ to conduct a shrinkage operation on the average of the covariance matrices $S_1, \ldots, S_D$ of the variational Gaussian posteriors $q(m_d | r_d, S_d)$, $d = 1, \ldots, D$.

3.2   Variational Mean Regularization

Second, we discuss a regularization of parameters. We update the variational means, i.e., r_{dk}s, in a regularized manner, because r_{dk}s are likely to deviate from their initial values when we use Eq. (6), in place of Eq. (5), for updating Σ.

Table 1. Document set specifications

	# docs	# words	# training tokens (# test tokens)
CORA[1]	36,183	8,542	2,127,005 (235,566)
MOVREV[2]	27,859	18,616	8,145,228 (905,427)
TDT4[3]	96,246	15,153	15,070,250 (1,674,304)
NSF[4]	128,181	19,066	12,284,568 (1,366,399)
MEDLINE[5]	2,495,210	134,615	225,844,065 (25,097,215)

The probability of topic k in document d is calculated as $\frac{\exp(m_{dk})}{\sum_{k'} \exp(m_{dk'})}$, and r_{dk} is of the same dimension with m_{dk}. Therefore, based on a result of CGS for LDA, we can initialize r_{dk} as follows: $r_{dk} = \log(n_{dk} + \alpha_k)$, where n_{dk} is the number of the word tokens in document d that are assigned to topic k. α_k is a Dirichlet hyperparameter corresponding to topic k and is updated by Minka's method [8,1] in our experiment. To keep r_{dk}s close to their initial values, we add a regularization term to the right hand side of Eq. (3) as follows:

$$\mathcal{L}_{\text{reg}}(r_{dk}) = n_{dk} r_{dk} - \frac{n_d}{\nu_d} \exp(r_{dk} + s_{dk}^2/2)$$
$$+ \frac{1}{2} r_{dk}^2 (\boldsymbol{\Sigma}^{-1})_{kk} - r_{dk} \sum_{k'} (r_{dk'} - \mu_{k'})(\boldsymbol{\Sigma}^{-1})_{kk'}$$
$$- \pi\rho\{r_{dk} - \log(n_{dk} + \alpha_k)\}^2 . \tag{7}$$

We call ρ *regularization parameter*, which determines the strength of the regularization and takes a non-negative value. We maximize $\mathcal{L}_{\text{reg}}(r_{dk})$ in Eq. (7) in place of $\mathcal{L}(r_{dk})$ in Eq. (3). Note that ρ is multiplied by the interpolation parameter π. Therefore, even when $\rho > 0$, the inference is reduced to the original one as long as $\pi = 0$. When ρ takes a large positive value, r_{dk}s are kept close to their initial values. This means that we keep staying close to the result of CGS for LDA. When we applied this regularization to our revised inference in our experiment, we always set ρ to 1.0, because other settings gave no interesting differences. Therefore, we have only one parameter π to be adjusted by hand.

Consequently, $\pi = 0.0$ means that we use the original inference for CTM and, at the same time, do not regularize r_{dk}s. Further, $\pi = 1.0$ means that we use the revised inference for CTM in its full capacity and, at the same time, regularize r_{dk}s with strength 1.0. However, we needed to directly apply the regularization to the original inference in our experiment for comparison. Therefore, in this case, we set ρ to 1.0 after eliminating π from Eq. (7).

---

[1] http://people.cs.umass.edu/~mccallum/data.html
[2] http://www.cs.cornell.edu/people/pabo/movie-review-data/
polarity_html.zip
[3] http://projects.ldc.upenn.edu/TDT4/
[4] http://archive.ics.uci.edu/ml/datasets/NSF+Research+
Award+Abstracts+1990-2003
[5] This is the set of the abstracts extracted from the XML files whose names range from medline12n0600.xml to medline12n0699.xml.

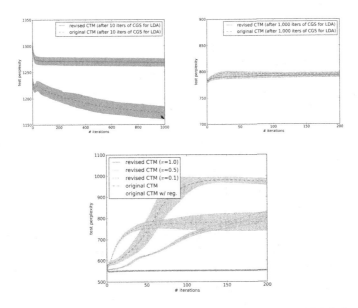

Fig. 1. Comparing the revised inference with the original one on CORA document set

4   Comparison Experiment

We compared our revised inference with the original one on the document sets in Table 1. For each document set, we conducted a series of appropriate preprocessings, e.g. changing text case to lower case, stemming, removing high and low frequency words, etc. The evaluation measure is *test perplexity* [2], which represents predictive power of topic models and is defined as $\exp(-\frac{\sum_d \sum_i \log \sum_k \lambda_{dk} \zeta_{k x_{di}}}{N_{test}})$, where N_{test} is the number of the word tokens used for calculating perplexity, called test word tokens. λ_{dk} is defined as $\sum_{i=1}^{n_d} \gamma_{dik}$. We randomly select 90% word tokens from every document for running inference and use the rest for calculating test perplexity. Note that smaller perplexity is better.

Fig. 1 gives test perplexities of the revised and the original inferences conducted on CORA document set under various settings. The horizontal axis represents number of iterations, and the vertical axis represents perplexity. For each setting, we repeated inference 10 times and calculated a mean and a standard deviation of the corresponding 10 perplexities. The mean and the standard deviation, which is depicted by the error bar, are given at every iteration.

First, we clarify when the original inference gave a better perplexity. The top left panel of Fig. 1 presents the result obtained when the number of topics was 30, a fairly small number, and we ran only 10 iterations of CGS for LDA before initiating the revised or the original inference for CTM. The number of iterations of the inference for CTM was set to 1,000, because CGS for LDA iterates only 10 times and could not reduce perplexity enough. The interpolation parameter π was set to 1.0 for the revised inference and to 0.0 for the original inference.

Since 10 iterations of CGS for LDA led to a perplexity nearly equal to 1,320, both of the line graphs, each corresponding to the revised inference (green solid line) and the original one (red dashed line), start from the perplexity of around 1,320. Obviously, the original inference is significantly better. It can be said that the original inference works without any modification when we set the number of topics to be relatively small and start the variational inference for CTM after a small number of iterations of CGS for LDA.

Second, we increased the number of iterations of CGS for LDA. The top right panel of Fig. 1 gives the result when we ran 1,000 iterations of CGS for LDA and then ran 200 iterations of the revised or the original inference for CTM. π was set to 1.0 for the revised inference and to 0.0 for the original inference. 1,000 iterations of CGS reduced perplexity to around 780. In this manner, CGS for LDA reduced perplexity significantly before we initiated the inference for CTM. Consequently, both of the revised and the original inferences could not improve the perplexity achieved by CGS for LDA,[6] though they behave a little differently. However, it is clear that the original inference gave almost the same perplexity with the revised inference, and thus that the revised inference could show no advantage also in this case.

Third, we increased the number of topics to 300. The bottom panel of Fig. 1 gives the corresponding result, where we ran 1,000 iterations of CGS for LDA and ran 200 iterations of the revised or the original inference for CTM. This panel contains five line graphs, each corresponding to the following cases: 1) $\pi = 1.0$ (green solid line), 2) $\pi = 0.5$ (blue dotted line), 3) $\pi = 0.1$ (magenta dash-dot line), 4) the original inference (red dashed line), and 5) the original inference with regularization (black pixel marker). The regularization parameter ρ was set to 1.0 for the revised inferences. We also applied our regularization to the original inference by setting $\pi\rho = 1.0$ in Eq. (7) and, at the same time, by setting $\pi = 0.0$ in Eq. (6). As this panel shows, the perplexities for the cases $\pi = 1.0$ and $\pi = 0.5$ gave almost the same perplexity, which is a little better than that achieved by CGS for LDA. However, the case $\pi = 0.1$ and the original inference gave significantly worse perplexities than those two cases. Further, even when we apply the regularization to the original inference, the improvement was not remarkable.[7] That is, our regularization did not work for the original inference. At the same time, this result also shows that the regularization with $\pi\rho = 1.0$ is not so strong to forcibly keep r_{dk}s almost the same with their initial values.

In sum, the original inference for CTM works if we set the number of topics to be fairly small. However, note that the bottom panel in Fig. 1 contains the best perplexity among the three panels in this figure. Therefore, it is better to choose the revised inference when we would like to make topic models fully show their predictive power. In contrast, if we are under a situation where perplexity hardly matters for some reasons, the original inference may find relevance.

---

[6] Needless to say, if we do not conduct the inference for CTM, we just have a result of CGS for LDA and cannot consider correlations among topics.

[7] When we conduct the revised inference without regularization, we consistently observed worse perplexities, though we do not present the corresponding results here.

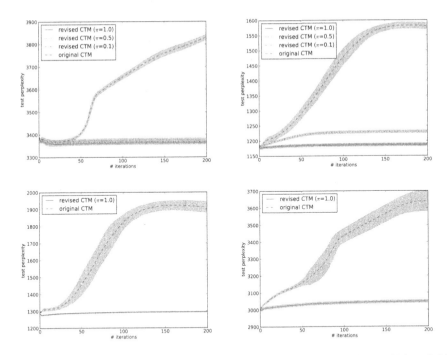

Fig. 2. Comparing the revised inference with the original one on MOVREV (top left), TDT4 (top roght), NSF (bottom left), and MEDLINE (bottom right) document sets

We obtained similar results also for the other document sets in Table 1. Fig. 2 contains all results. As in case of the bottom panel of Fig. 1, we ran 1,000 iterations of CGS for LDA and then started 200 iterations of the revised or the original inference for CTM. For MOVREV and TDT4 document sets, we tested the following four settings: $\pi = 1.0, 0.5, 0.1$, and 0.0. The regularization is applied only to the revised inference with $\rho = 1.0$ in Eq. (7). As the top left panel shows, for MOVREV document set, the three settings $\pi = 1.0, 0.5$, and 0.1 gave almost the same perplexity, which is slightly better than the perplexity achieved by CGS for LDA. In contrast, the original inference led to a far larger perplexity. For TDT4, we obtained almost the same perplexity when $\pi = 1.0$ and 0.5, and the two other cases resulted in worse perplexities, as the top left panel depicts. The bottom left and the bottom right panels present the results for NSF and MEDLINE document sets, respectively. For these relatively large document sets, we only provide the results for the two extreme cases, i.e., the case $\pi = 1.0$ and the original inference. Obviously, the revised inference achieved a better perplexity by a large margin.

In the end, we add comments on implementation. With respect to matrix inversion, we used `dgetrf_()` and `dgetri_()` in CLAPACK. As a byproduct of `dgetrf_()`, we can calculate the determinant of the covariance matrix. When we ran the original inference, the absolute value of the determinant went to a value close to zero in case the number of topics was relatively large. Consequently, $\mathcal{L}$, i.e., the lower bound of the log evidence, decreased by a large amount in earlier iterations, though $\mathcal{L}$ needed to be maximized. While we knew that there was an implementation by the authors of the original paper [3], it consumed main memory by a considerable factor when compared with our implementation and thus could not load any document sets in a reasonable time. Therefore, we evaluated the original inference by setting π to 0.0 in our own implementation.

5   Conclusion

In this paper, we provide a revised inference for CTM. We modify diagonal elements of the covariance matrix lest it be almost singular and regularize variational mean parameters lest they deviate from their initial values. However, as long as the model is described by using a $K \times K$ covariance matrix, it seems difficult to completely avoid the problem of matrix singularity, because K needs to be larger for a topic analysis over larger document sets. Therefore, recent proposals [11,4] may find their relevance. However, CTM has an advantage in its simplicity and efficiency of inference. We tried to make such CTM to be utilized in a wider range of situations. An important future work is to compare our revised inference with the original one in terms of other evaluation measures [9].

References

1. Asuncion, A., Welling, M., Smyth, P., Teh, Y.W.: On smoothing and inference for topic models. In: Proc. of UAI 2009, pp. 27–34 (2009)
2. Blei, D.M., Ng, A.Y., Jordan, M.I.: Latent Dirichlet allocation. JMLR 3, 993–1022 (2003)
3. Blei, D., Lafferty, J.: Correlated topic models. NIPS 18, 147–154 (2006)
4. Chen, C., Ding, N., Buntine, W.: Dependent hierarchical normalized random measures for dynamic topic modeling. In: Proc. of ICML 2012 (2012)
5. Chen, Y., Wiesel, A., Eldar, Y.C., Hero III, A.O.: Shrinkage algorithms for MMSE covariance estimation. IEEE Trans. on Sign. Proc. 58(10), 5016–5029 (2010)
6. Griffiths, T.L., Steyvers, M.: Finding scientific topics. PNAS 101(suppl. 1), 5228–5235 (2004)
7. Liu, D.C., Nocedal, J.: On the limited memory method for large scale optimization. Math. Programming B 45(3), 503–528 (1989)
8. Minka, T.P.: Estimating a Dirichlet distribution (2000), http://research.microsoft.com/en-us/um/people/minka/papers/dirichlet/
9. Newman, D., Karimi, S., Cavedon, L.: External Evaluation of Topic Models. In: Proc. of ADCS 2009, pp. 11–18 (2009)
10. Nocedal, J.: Updating quasi-Newton matrices with limited storage. Math. Comp. 35, 773–782 (1980)
11. Paisley, J., Wang, C., Blei, D.: The discrete infinite logistic normal distribution for mixed-membership modeling. In: Proc. of AISTATS 2011, pp. 74–82 (2011)
12. Salakhutdinov, R., Hinton, G.: Replicated Softmax: an Undirected Topic Model. NIPS 23, 1607–1614 (2009)
13. Wan, L., Zhu, L., Fergus, R.: A Hybrid Neural Network-Latent Topic Model. In: Proc. of AISTATS 2012, pp. 1287–1294 (2012)

Adaptive Fault Estimation
of Coupling Connections for Synchronization
# of Complex Interconnected Networks

Zhanshan Wang, Chao Cai, Junyi Wang, and Huaguang Zhang

School of Information Science and Engineering, Northeastern University, Shenyang,
Liaoning, 110004, People's Republic of China
zhanshan_wang@163.com

Abstract. Adaptive fault estimation problems of coupling connections
for a class of complex networks have been studied in the concept of
drive-response synchronization. By constructing a suitable response networks and designing the adaptive control laws and adaptive estimator of
coupling connections, the coupling connections in drive networks can be
estimated and correspondingly monitored online, which can be as indica-
tors to judge whether a fault connection occurs or not. Meanwhile, when
the actuator fault occurs, a passive fault tolerant controller is designed
to guarantee the synchronization between drive and response networks.

Keywords: Complex interconnected networks, fault estimation,
synchronization, adaptive coupling.

1 Introduction

Recently, complex networks have become an important part in our daily life.
Analysis and control of the behaviors of complex networks consisting of a large
number of dynamical nodes have attracted wide attention in different fields in
the past decade. In particular, special attention has been focused on the control and synchronization of large scale complex dynamical networks with certain
types of topology.[3–8]. Neuronal systems are the prototypic examples where
synchronization plays an important role in their functions. Neurons in a population get into synchrony and this process causes binding, cognition, information
processing, and computing in the brain [9]. Various brain disorders have been
linked to the abnormal levels of synchronization in the brain [10, 11]. Recently,
arrays of coupled neural networks or complex interconnected neural networks
have attracted much attention of researchers in different research fields. They
can exhibit many interesting phenomena, such as spatio-temporal chaos [12].
Moreover, experiment and theoretical analysis have revealed that a mammalian
brain not only displays in its storage of associative memories, but also modulate oscillatory neuronal synchronization by selective perceive attention [13, 14].
Therefore, the study of synchronization of complex interconnected neural networks is an important step for both understanding brain science and designing
complex neural networks for practical use [15, 16].

C. Guo, Z.-G. Hou, and Z. Zeng (Eds.): ISNN 2013, Part II, LNCS 7952, pp. 455–462, 2013.

The most significant difference between a dynamical network and an isolated dynamical system is that the function of a network is implemented through the connections between the nodes systems. Through connections the nodes systems in a network can impact each other to achieve the desired objectives. However, the network may fail occasionally, for example, traffic jam makes the vehicles change their routes; the extinction of some species of animals These kinds of accidents are sure to destroy parts of the connections or even all of the connections in the network. There are also some other situations when the networks become disconnected on purpose for economical reasons or others. Network failure refers to the case when partial or all connections in the network become disconnected; as a result, the objective of the network cannot be fulfilled if the network is not recovered. There is no doubt that network failures will affect the synchronization behavior of the complex dynamical networks. In general, there are two ways to solve this problems. One is to use the adaptive adjusting methods to update the coupling connections adaptively. Another way is to use the external control law to compensate the fault effect of the coupling connections. Unfortunately there are few works on this topic to our knowledge. Therefore, in this paper, fault synchronization of complex dynamical networks with network failures is studied.

Fault tolerant control has been an active research topic during the past three decades for increasing the safety and reliability of complex dynamical systems [17–20]. In the literature, faults normally occur at two places: the actuators and sensors. Actuator faults are faults that act on the system, resulting in the deviation of the process variables. The result is the command (control) signal sent to this device has no effect. Fault tolerant control includes the controller design and parameter updating laws in order to guarantee the proper dynamics of the controlled systems during both normal and faulty conditions [21–24]. For the controlled synchronization of complex interconnected neural networks with delay, fault tolerant synchronization is essential to achieve the synchronization no matter whether a fault happens or not. Based on above discussion, we will study the adaptive fault estimation of coupling connections and then using the estimated information to be fault indicator to produce alarm. Then a fault tolerant synchronization control law is designed for the case of actuator fault.

2   Problem Description and Preliminaries

In this paper we will discuss the following complex networks with N identical nodes,

$$\dot{x}_i(t) = f(x_i(t)) + \sum_{j=1}^{N} c_{1ij}(t)\Gamma_1 x_j(t) + \sum_{j=1}^{N} c_{2ij}(t)\Gamma_2 x_j(t - \tau), \qquad (1)$$

where $x_i(t) = (x_{i1}(t), \cdots, x_{in}(t))^T \in \Re^n$ is the state vector of the ith node, $f(x_i(t)) = (f_1(x_{i1}(t)), f_2(x_{i2}(t)), \cdots, f_n(x_{in})(t))^T$, $f_i(\cdot)$ is a smooth vector field, node system is $\dot{s}(t) = f(s(t))$, Γ_1 and Γ_2 are the inner-coupling connections,

$C_1 = (c_{1ij})_{N \times N}$ and $C_2 = (c_{2ij})_{N \times N}$ are the outer-coupling connections, which are easily effected by accidental faults, $i, j = 1, \cdots, N$.

In order to estimate the outer-coupling connections, we construct a response system as follows,

$$\dot{\hat{x}}_i(t) = f(\hat{x}_i(t)) + \sum_{j=1}^{N} \hat{c}_{1ij}(t) \Gamma_1 \hat{x}_j(t) + \sum_{j=1}^{N} \hat{c}_{2ij}(t) \Gamma_2 \hat{x}_j(t - \tau) + u_i(t), \quad (2)$$

where $\hat{x}_i(t) = (\hat{x}_{i1}(t), \cdots, \hat{x}_{in}(t))^T \in \Re^n$ is the response system state vector of the ith node, $u_i(t)$ is the control input to be designed, $\hat{C}_1 = (\hat{c}_{1ij})_{N \times N}$ and $\hat{C}_2 = (\hat{c}_{2ij})_{N \times N}$ are the estimations of outer-coupling connections C_1 and C_2, respectively. $\Gamma_1 = \text{diag}(\gamma_{11}, \gamma_{12}, \cdots, \gamma_{1n})$, $\Gamma_2 = \text{diag}(\gamma_{21}, \gamma_{22}, \cdots, \gamma_{2n})$, $i, j = 1, \cdots, N$.

Assume that $\tilde{x}_i(t) = \hat{x}_i(t) - x_i(t)$, $\tilde{c}_{1ij}(t) = \hat{c}_{1ij}(t) - c_{1ij}$, and $\tilde{c}_{2ij}(t) = \hat{c}_{2ij}(t) - c_{2ij}$, then we have the following error system,

$$\dot{\tilde{x}}_i(t) = f(\hat{x}_i(t)) - f(x_i(t)) + \sum_{j=1}^{N} \tilde{c}_{1ij}(t) \Gamma_1 \hat{x}_j(t) + \sum_{j=1}^{N} c_{1ij}(t) \Gamma_1 \tilde{x}_j(t)$$

$$+ \sum_{j=1}^{N} \tilde{c}_{2ij}(t) \Gamma_2 \hat{x}_j(t - \tau) + \sum_{j=1}^{N} c_{2ij}(t) \Gamma_2 \tilde{x}_j(t - \tau) + u_i(t), \quad (3)$$

where $i, j = 1, \cdots, N$.

In order to derive the main result, the following preliminaries are required.

Assumption 3. The activation function $f_i(x_i(t))$ is bounded and continuous, which satisfies $|f_i(x_i(t))| \leq G_i^b$, $G_i^b > 0$ is a positive constant,

$$0 \leq \frac{f_i(\eta) - f_i(v)}{\eta - v} \leq \delta_i, \quad (4)$$

for any $\eta \neq v, \eta, v \in \Re$, and $\delta_i > 0$, $i = 1, \cdots, n$. Let $\Delta = \text{diag}(\delta_1, \cdots, \delta_n)$.

Lemma 1. (see [1, 2]): For a dynamic system $\dot{X}(t) = f(X(t))$, $f(\cdot) : \Re^n \to \Re^n$, is a C^1 continuous map. If function $V(X(t)) : \Re^n \to \Re^+$ exists, for each $X(t) \in \Re^n$, $\dot{V}(X(t)) \leq 0$ and definition $E := \{X(t) \in \Re^n : \dot{V}(X(t)) = 0\}$. Given that B_0 is the largest invariant set in E, then each bounded solution will converge to B_0 when $t \to \infty$.

3   Main Results

In this section, we will establish some conditions to guarantee the fault tolerant synchronization for the drive network (1) and response network (2).

Theorem 1. Suppose that Assumption 3 holds. Response complex networks (2) can be fault tolerantly synchronized to the drive complex networks (1) under the following control law and coupling updating laws,

$$u_i(t) = - d_i(t)\tilde{x}_i(t), \tag{5}$$
$$\dot{d}_i(t) = k_i \tilde{x}_i^T(t) P \tilde{x}_i(t),$$

$$\dot{\tilde{c}}_{1ij}(t) = - \tilde{x}_i^T(t) P \Gamma_1 \hat{x}_j(t), \tag{6}$$
$$\dot{\tilde{c}}_{2ij}(t) = - \tilde{x}_i^T(t) P \Gamma_2 \hat{x}_j(t - \tau). \tag{7}$$

Moreover, the coupling connections C_1 and C_2 of complex networks (1) can be estimated asymptotically by the coupling adjusting laws (6) and (7), where P is a given positive diagonal matrix, k_i is a known positive constant, $i = 1, \cdots, N$.

Proof. Let us consider the Lyapunov functional $V(t) = V_1(t) + V_2(t)$, where

$$V_1(t) = \frac{1}{2} \sum_{i=1}^{N} \tilde{x}_i^T(t) P \tilde{x}_i(t) + \frac{1}{2} \sum_{i=1}^{N} \sum_{j=1}^{N} \tilde{c}_{1ij}^2(t) + \frac{1}{2} \sum_{i=1}^{N} \sum_{j=1}^{N} \tilde{c}_{2ij}^2(t), \tag{8}$$

$$V_2(t) = \frac{1}{2} \sum_{i=1}^{N} \frac{1}{k_i}(d_i(t) - d^*)^2 + \sum_{m=1}^{n} p_m \int_{t-\tau}^{t} e_m^T(s) M e_m(s) ds, \tag{9}$$

where $P = \text{diag}(p_1, \cdots, p_n)$ is a positive diagonal matrix, d^* is a sufficiently large positive constant, M is a positive definite symmetric matrix, and $e_m(t) = (\tilde{x}_{1m}(t), \tilde{x}_{2m}(t)), \cdots, \tilde{x}_{Nm}(t))^T \in \Re^N$, $m = 1, \cdots, n$.

The derivative of $V_1(t)$ is as follows

$$\dot{V}_1(t) = \sum_{i=1}^{N} \tilde{x}_i^T(t) P \Big[f(\hat{x}_i(t)) - f(x_i(t)) + \sum_{j=1}^{N} \tilde{c}_{1ij}(t) \Gamma_1 \hat{x}_j(t) + \sum_{j=1}^{N} c_{1ij} \Gamma_1 \tilde{x}_j(t)$$

$$+ \sum_{j=1}^{N} \tilde{c}_{2ij}(t) \Gamma_2 \hat{x}_j(t-\tau) + \sum_{j=1}^{N} c_{2ij} \Gamma_2 \tilde{x}_j(t-\tau) + u_i(t) \Big]$$

$$+ \sum_{i=1}^{N} \sum_{j=1}^{N} \tilde{c}_{1ij}(t) \dot{\tilde{c}}_{1ij}(t) + \sum_{i=1}^{N} \sum_{j=1}^{N} \tilde{c}_{2ij}(t) \dot{\tilde{c}}_{2ij}(t)$$

$$\leq \sum_{i=1}^{N} \tilde{x}_i^T(t) P \Big[\Delta \tilde{x}_i(t) + \sum_{j=1}^{N} c_{1ij} \Gamma_1 \tilde{x}_j(t) + \sum_{j=1}^{N} c_{2ij} \Gamma_2 \tilde{x}_j(t-\tau) - d_i(t) \tilde{x}_i(t) \Big]$$

$$+ \sum_{i=1}^{N} \sum_{j=1}^{N} \Big[\tilde{x}_i^T(t) P \tilde{c}_{1ij}(t) \Gamma_1 \hat{x}_j(t) + \tilde{c}_{1ij}(t) \dot{\tilde{c}}_{1ij}(t) \Big]$$

$$\sum_{i=1}^{N} \sum_{j=1}^{N} \Big[\tilde{x}_i^T(t) P \tilde{c}_{2ij}(t) \Gamma_2 \hat{x}_j(t-\tau) + \tilde{c}_{2ij}(t) \dot{\tilde{c}}_{2ij}(t) \Big]$$

$$= \sum_{m=1}^{n} p_m \delta_m e_m^T(t) e_m(t) + \sum_{m=1}^{n} p_m \gamma_{1m} e_m^T C_1 e_m(t) + \sum_{m=1}^{n} p_m \gamma_{2m} e_m^T C_2 e_m(t - \tau)$$

$$- \sum_{i=1}^{N} \tilde{x}_i^T(t) P d_i(t) \tilde{x}_i(t) \Big]. \tag{10}$$

The derivative of $V_2(t)$ is as follows,

$$\dot{V}_2(t) = \sum_{i=1}^{N} \frac{1}{k_i} (d_i(t) - d^*) \dot{d}_i(t) + \sum_{m=1}^{n} p_m [e_m^T(t) M e_m(t) - e_m^T(t - \tau) M e_m(t - \tau)]. \tag{11}$$

Combining (10) and (11), it yield,

$$\dot{V}(t) \le \sum_{m=1}^{n} p_m (e_m^T(t) \quad e_m^T(t - \tau)) \Xi (e_m^T(t) \quad e_m^T(t - \tau))^T, \tag{12}$$

where

$$\Xi = \begin{bmatrix} \delta_m I + 0.5 \gamma_{1m}(C_1 + C_1^T) + M - D^* & 0.5 \gamma_{2m} C_2 \\ 0.5 \gamma_{2m} C_2^T & -M \end{bmatrix},$$

$$D^* = \mathrm{diag}(d^*, \cdots, d^*).$$

According to Schur Complement, Ξ is negative definite if and only if $\delta_m I + 0.5 \gamma_{1m}(C_1 + C_1^T) + M - D^* + 0.25 \gamma_{2m}^2 C_2 M^{-1} C_2^T < 0$. Obviously, if d^* is sufficiently large positive constant, Ξ is negative definite. Therefore, we have $\dot{V}(t) \le 0$ for $e_m(t) \ne 0$ and $e_m(t - \tau) \ne 0$ or $e_m(t) = 0$ and $e_m(t - \tau) \ne 0$. $\dot{V}(t) = 0$ if and only if $e_m(t) = 0$. That is, the set $B_0 = \{\tilde{x}(t) = 0, \tilde{C}_1 = 0, \tilde{C}_2 = 0, d_i(t) = d^*\}$ is the largest invariant set of the set $E = \{\dot{V}(t) = 0\}$ for the error system (3). According to Lemma 1, starting with arbitrary initial values, the trajectories of the error system (3) asymptotically converges to the set B_0. Therefore, one gets $\lim_{t \to \infty} \hat{C}_1 = C_1$ and $\lim_{t \to \infty} \hat{C}_2 = C_2$. As a result, the coupling connections C_1 and C_2 have been estimated by the adaptive updating laws. Thus, the proof is complete.

Let $K = \mathrm{diag}(K_1, K_2, \cdots, K_n)$, $K_i = 1$ represents the normal case while $K_i = 0$ represents the fault case, $i = 1, \cdots, n$. When actuator fault K occurs in the control channel, that is, $u_i(t) = -d_i(t) K \tilde{x}_i(t)$, we have the following result.

Theorem 2. Suppose that Assumption 3 holds. Response complex networks (2) can be fault tolerantly synchronized to the drive complex networks (1) under the actuator fault,

1) for given K_m, if there exists positive definite symmetric matrix M such that the following conditions hold, $m = 1, \cdots, n$,

$$\begin{bmatrix} \delta_m I + 0.5 \gamma_{1m}(C_1 + C_1^T) + M - K_m D^* & 0.5 \gamma_{2m} C_2 \\ 0.5 \gamma_{2m} C_2^T & -M \end{bmatrix} < 0, \tag{13}$$

2) the following control laws and coupling updating laws are chosen,

$$u_i(t) = -d_i(t)K\tilde{x}_i(t), \tag{14}$$
$$\dot{d}_i(t) -k_i\tilde{x}_i^T(t)PK\tilde{x}_i(t),$$

$$\dot{\hat{c}}_{1ij}(t) = -\tilde{x}_i^T(t)P\Gamma_1\hat{x}_j(t), \tag{15}$$
$$\dot{\hat{c}}_{2ij}(t) = -\tilde{x}_i^T(t)P\Gamma_2\hat{x}_j(t-\tau). \tag{16}$$

Moreover, the coupling connections C_1 and C_2 of complex networks (1) can be estimated asymptotically by the coupling adjusting laws (15) and (16), where P is a given positive diagonal matrix, k_i is a known positive constant, $i = 1, \cdots, N$.

For the case of no delayed coupling, we can have the following corollary.

Corollary 1. Suppose that Assumption 3 holds. For the case of no delayed coupling, response complex networks (2) can be fault tolerantly synchronized to the drive complex networks (1) under the following control law and coupling updating laws,

$$u_i(t) = -d_i(t)\tilde{x}_i(t), \tag{17}$$
$$\dot{d}_i(t) =k_i\tilde{x}_i^T(t)P\tilde{x}_i(t),$$
$$\dot{\hat{c}}_{1ij}(t) = -\tilde{x}_i^T(t)P\Gamma_1\hat{x}_j(t), \tag{18}$$

Moreover, the coupling connections C_1 of complex networks (1) can be estimated asymptotically by the coupling adjusting laws (18), where P is a given positive diagonal matrix, k_i is a known positive constant, $i = 1, \cdots, N$.

Remark 1. From the proof procedure of Theorem 1, it is obvious that the parameter d^* always exist if their magnitudes are large enough. Therefore, the adopted adaptive laws can guarantee the synchronization of the whole nodes.

Remark 2. Similar research on the estimation of coupling connections have been found in [25–28]. From the view point of topology identification, the authors have discussed the estimation problems of the coupling connections. In this paper, we stand at the point of reliable or fault tolerant synchronization of complex networks under parameter fault and actuator faults. In the fault case, some measures should be take to prevent the performance deterioration. For example, fault detection should first be conducted and an alarm should be evoked. In this point, estimation problems of the coupling connections can be converted to the fault detection problems. More importantly, if some active fault tolerant control laws are adopted, the estimation of the coupling connections may be used to compensate the incomplete action. Therefore, the background and the motivation of the present research are different from those in [25–28], which may lead to different application of the related results.

Remark 3. In the simulations of [25, 27], the initial conditions of the drive system and response system are in the complex-value space. This selection contradicts with the elementary requirements of the complex networks, which belong to the real space. While in the present simulation, the initial conditions are chosen randomly in [0, 1], which belong to the real space.

Remark 4. Theorem 2 presents a passive fault tolerant synchronization scheme for the case of actuator fault. This scheme can determine which control channel of node network is important and then some preventive measure can he taken.

4 Conclusions

For a class of complex interconnected networks with delayed coupling, a fault tolerant control scheme is proposed to guarantee the synchronization when the coupling connections are fault. The adaptive estimation laws of coupling connections are designed, which can monitor the status of coupling connection on line, which can act as an indicator to make an alarm of fault. At the same time, for the case of actuator fault, an adaptive fault tolerant controller is design, which is on the basis of adaptive parameter estimation.

Acknowledgements. This work was supported by the National Natural Science Foundation of China under Grants 61074073 and 61034005, Program for New Century Excellent Talents in University of China (NCET-10-0306), and the Fundamental Research Funds for the Central Universities under grants N110504001 and N100104102.

References

1. Hassan, K.: Nonlinear Systems, 3rd edn. Prentice-Hall, Englewood Cliffs (2002)
2. Khalil, H.: Nonlinear Systems, 2nd edn. Prentice-Hall, Englewood Cliffs (1996)
3. Strogatz, S.: Exploring complex networks. Nature 410, 268–276 (2001)
4. Watts, D., Strogatz, S.: Collective dynamics of small-world networks. Nature 393, 440–443 (1998)
5. Pecora, L., Carroll, T.: Master stability functions for synchronized coupled systems. Phys. Rev. Lett. 80, 2109–2119 (1998)
6. Wu, C., Chua, L.: Synchronization in an array of linearly coupled dynamical systems. IEEE Trans. Circuits Syst., I: Fundam. Theory Appl. 42, 430–447 (1995)
7. Barahona, M., Pecora, L.: Synchronization in small-world systems. Phys. Rev. Lett. 89, 054101 (2002)
8. Arenas, A., Diaz-Guilera, A., Kurths, J., Moreno, Y., Zhou, C.: Synchronization in complex networks. Phys. Rep. 469, 93–178 (2008)
9. Fries, P.: Neuronal gamma-band synchronization as a fundamental process in cortical computation. Annual Review of Neuroscience 32, 209–224 (2009)
10. Uhlhaas, P., Singer, W.: Neuronal dynamics and neuropsychiatric disorders: toward a translational paradigm for dysfunctional large-scale networks. Neuron 75, 963–980 (2012)

11. Uhlhaas, P., Roux, F., Rodriguez, E.: Neural synchrony and the development of cortical networks. Trends in Cognitive Sciences 14, 72–80 (2010)

12. Zheleznyak, A., Chua, L.: Coexistence of low-and high-dimensional spatio-temporal chaos in a chain of dissipatively coupled Chua's circuits. Int. J. Bifur. Chaos. 4, 639–672 (1994)

13. Fries, P., Reynolds, J., Rorie, J., Desimore, R.: Modulation of oscillatory neuronal synchronization by selective visual attention. Science 291, 1560–1563 (2001)

14. Steinmetz, P., Roy, A., Fitzgerald, P., Hsiao, S., Johnson, K., Niebar, E.: Attention modulate synchronized neuronal firing in primate somatosensory cortex. Nature 404, 487–490 (2000)

15. Hoppensteadt, F., Izhikevich, E.: Pattern recognition via synchronization in phase-locked loop neural networks. IEEE Trans. Neural Netw. 11, 734–738 (2000)

16. Chen, G., Zhou, J., Liu, Z.: Global synchronization of coupled delayed neural networks and applications to chaos CNN models. Int. J. Bifurc. Chaos. 14, 2229–2240 (2004)

17. Gao, Z., Jiang, B., Shi, P., Liu, J., Xu, Y.: Passive fault tolerant control design for near space hypersonic vehicle dynamical system. Circuits Systems and Signal Processing 31, 565–581 (2012)

18. Xu, Y., Jiang, B., Tao, G., Gao, Z.: Fault accommodation for near space hypersonic vehicle with actuator fault. International Journal of Innovative Computing, Information and Control 7, 2187–2200 (2011)

19. Shumsky, A., Zhirabok, A., Jiang, B.: Fault accommodation in nonlinear and linear dynamic systems: fault decoupling based approach. International Journal of Innovative Computing, Information and Control 7, 4535–4550 (2011)

20. Gayaka, S., Yao, B.: Accommodation of unknown actuator faults using output feedback-based adaptive robust control. Int. J. Adapt. Control Signal Process. 25, 965–982 (2011)

21. Du, D., Jiang, B., Shi, P.: Sensor fault estimation and compensation for time-delay switched systems. International Journal of Systems Science 43, 629–640 (2012)

22. Gao, Z., Jiang, B., Shi, P., Qian, M., Lin, J.: Active fault tolerant control design for reusable launch vehicle using adaptive sliding mode technique. Journal of the Franklin Institute 349, 1543–1560 (2012)

23. Jiang, B., Gao, Z., Shi, P., Xu, Y.: Adaptive fault-tolerant tracking control of near-space vehicle using TakagiCSugeno fuzzy models. IEEE Transactions on Fuzzy Systems 18, 1000–1007 (2010)

24. Jiang, B., Staroswiecki, M., Cocquempot, V.: Fault accommodation for nonlinear dynamic systems. IEEE Transactions on Automatic Control 51, 1578–1583 (2006)

25. Zhou, J., Lu, J.: Topology identification of weighted complex dynamical networks. Physica A 386, 481–491 (2007)

26. Guo, W., Chen, S., Sun, W.: Topology identification of the complex networks with non-delayed and delayed coupling. Physics Letters A 373, 3724–3729 (2009)

27. Xu, Y., Zhou, W., Fang, J., Sun, W.: Topology identification and adaptive synchronization of uncertain complex networks with adaptive double scaling functions. Commun. Nonlinear Sci. Numer. Simulat. 16, 3337–3343 (2011)

28. Zhao, J., Li, Q., Lu, J., Jiang, Z.: Topology identification of complex dynamical networks. Chaos 20, 023119 (2010)

Time Series Fault Prediction in Semiconductor Equipment Using Recurrent Neural Network

Javeria Muhammad Nawaz, Muhammad Zeeshan Arshad, and Sang Jeen Hong*

Department of Electronic Engineering, Myongji University, Korea
{javeriamnawaz,mzeeshanarshad}@gmail.com, samhong@mju.ac.kr

Abstract. This paper presents a model of Elman recurrent neural network (ERNN) for time series fault prediction in semiconductor etch equipment. ERNN maintains a copy of previous state of the input in its context units, as well as the current state of the input. Derivative dynamic time warping (DDTW) method is also discussed for the synchronization of time series data set acquired from plasma etcher. For each parameter of the data, the best ERNN structure was selected and trained using Levenberg Marquardt to generate one-step-ahead prediction for 10 experimental runs. The faulty experimental runs were successfully distinguished from healthy experimental runs with one missed alarm out of ten experimental runs.

Keywords: Time series prediction, recurrent neural network, derivative dynamic time warping.

1 Introduction

As semiconductor device technology moves toward a few tens of nanometer scale, no more process margin is allowed in manufacturing environment. Small amount of process shift or drift can jeopardize the product quality, and it may increase the quantity of wafer scrap. In order to improve production yields, it is essential to detect process and tool faults in a timely manner. On time detection of any suspicious run in the semiconductor processes is crucial to guarantee productivity and reliability [1].

Semiconductor manufacturing processes usually show complex interactions with multiple input signals and multiple output variables. Previous research suggested that neural networks (NNs) have emerged as a powerful means of obtaining quantitative models for input–output mapping, and NNs have been successfully and widely applied for modeling semiconductor processes [2-8]. In this paper, we employed Elman's recurrent neural network (ERNN) to detect fault using a healthy data set collected from an industrial plasma etcher. The ERNN is trained using Levenberg Marquardt back-propagation (LMBP) algorithm.

Section 2 describes data collection and preprocessing and Section 3 provides background information of the employed ERNN. In Section 4, the implementation of ERNNs for semiconductor equipment data has been presented. Finally, discussion and summary are followed in Section 5.

---

* Corresponding author.

C. Guo, Z.-G. Hou, and Z. Zeng (Eds.): ISNN 2013, Part II, LNCS 7952, pp. 463–472, 2013.

Table 1. Experimental runs and the type of perturbation added

Run No.	Added Perturbation
1	None
2	-0.5mT from base pressure
3	+0.5mT from base pressure
4	-1% MFC conversion shift
5	+1% MFC conversion shift
6	Source RF Cable: loss simulation
7	None
8	Bias RF Cable: power delivered
9	None
10	Added chamber leak by 1.3 mT/min

2   Semiconductor Equipment Data

2.1   Data Acquisition and Preprocessing

The data for time series modeling was acquired from Applied Materials' DPS-II Centura dielectric etcher. It consists of 10 repeated normal, herein called 'healthy,' runs and 10 experimental runs with some arbitrary perturbations as shown in Table 1. Among the experimental runs, no perturbation was added to Run No. 1, 7 and 9.

Table 2. List of parameters selected through PCA for fault detection

Par.	Par. Name	Par.	Par. Name
1	Throttle Gate Valve Current	9	RF Probe Phase
2	RF Source Forward	10	E-chuck Voltage
3	RF Matcher Current 1	11	Flow Splitter- Flow 1
4	RF Matcher Current 2	12	Flow Splitter – Total Flow
5	RF Bias Forward	13	Gas Flow – 12
6	RF Bias Shunt	14	RF Probe V_{p-p}
7	RF Probe Voltage	15	RF Probe DC Bias
8	RF Probe Current		

Data Reduction: The original data set contains 55 parameters, collected from the semiconductor manufacturing equipment hardware through System V Interface Definition (SVID) with 10 Hz data collection frequency. Key success for correct fault detection and classification (FDC) depends on selecting useful parameters that containing effective information on equipment status. To alleviate the concern, principle component analysis (PCA) was employed for the selection of statistically significant parameters [9]. The parameters which carry the most useful information were selected by choosing the top 15 most statistically significant parameters presented in Table 2.

Data Normalization and Synchronization: The data acquired consists of different parameters each having a different range of values including negative values as well, depending on the measured property and its units. In addition, 10 Hz of data acquisition speed over 55 parameters requires precise data synchronization to avoid Type I error and Type II error from established FDC models. Min-max normalization (using Equation 1) was initially performed, and derivative dynamic time warping (DDTW) was employed for the time series data synchronization.

$$x_{i'} = \frac{x_i - \min(\mathbf{X})}{\max(\mathbf{X}) - \min(\mathbf{X})} \tag{1}$$

Where x_i is the un-scaled value from matrix X to be scaled, $\min(X)$ is the minimum value in X, $\max(X)$ is the maximum value in X and $x_{i'}$ is the resulting normalized value of x_i. When the multi-parameter time series data set is not synchronized, established model may suffer from Type I and Type II errors induced by time delay from equipment hardware system. We make use of the Derivative Dynamic Time Warping (DDTW) algorithm [10] to synchronize the data runs. DDTW is modified form of Dynamic Time Warping (DTW) which was originally developed in 1960s by Bellman [11] for recognition of speech. The use of this technique has been extended for the purpose of similarity measurement [12].

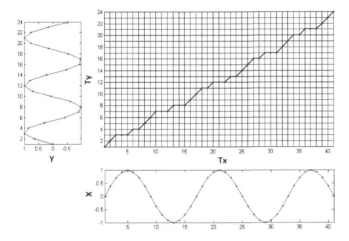

Fig. 1. The warping path is shown as an example for two sample time sequences, X and Y with number of data samples $N=24$ and $M=41$ respectively

The DDTW algorithm compares two independent time sequences in order to find the right alignment, a path consisting of a set of coordinates across the map is computed which is called the warping path W as shown in Fig. 1. Kassidas showed the use of DTW for synchronization of batch. In this research a modified form of this methodology was used for our application [13]. We fixed a reference run from the healthy data set and synchronized all other runs from healthy and test data set to this reference. Hence we obtained the synchronized runs each having total data samples equal to those of the reference run.

3   Elman's Recurrent Neural Networks (ERNN)

In this study we consider Elman's recurrent neural network (ERNN) that consists of two-layer back propagation network with feedback connections from the hidden nodes to additional nodes, called the "context units" which contain the previous states of the hidden nodes [14].

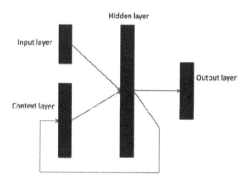

Fig. 2. A general structure of an Elman Recurrent Neural Network

The structure of ERNN is represented in Fig. 2 which represents connections from hidden layer to the context layer with a fixed weight (w=1), maintaining a copy of the previous states of the hidden neurons that serves as a short term memory as input for the current state [15-16]. The number of context layer neurons is equal to the number of hidden layer neurons, considering the fact that there is one neuron in the former for each corresponding neuron in the later.

In order to train ERNN, Levenberg Marquardt (LM) back-propagation algorithm has been selected due to its better performance on time series data [17]. The LM algorithm has the following weight update equation for an input training pattern k:

$$\mathbf{w}(k+1) = \mathbf{w}(k) - (\mathbf{J}^T\mathbf{J} + \lambda\mathbf{I})^{-1}\mathbf{J}^T E\,(k) \tag{2}$$

where $\mathbf{J}$ is the Jacobian matrix, $\mathbf{E}$ is a vector of network errors, λ is the Marquardt's parameter, δ is the weight update and $\mathbf{I}$ is the identity matrix. The Jacobian matrix contains first derivatives of the network errors with respect to the weights and biases and can be computed through a standard back-propagation technique [18-19]. LM algorithm can be implemented by the following steps [20-21];

1) The inputs are presented to the network and the corresponding output of the network is computed using the initial values of weights:

2) The network error E_k for each corresponding input training pattern k can be calculated by

$$E_k = \frac{1}{2}\sum_{j\in Y}\left(y_{kj} - d_{kj}\right)^2 \tag{3}$$

Where p is the number of training patterns, y_{kj} is network output, d_{kj} is the target output, and Y is the set of output neurons:

3) The new weights $w(k+1)$ for the network are then computed using equation (1):

4) Using the new weights, the new output of the network and resulting new network error $E(k+1)$ is calculated:

5) If $E(k+1)$ is less than $E(k)$, then the value of λ is reduced. Otherwise, if $E(k+1)$ is greater than $E(k)$, is increased and process repeats from Step 3. $E(k)$ is increased or decreased by a fixed adjustment parameter β:

6) Unless the network error has reached a pre-set goal E_x, the algorithm keeps iterating from Step 3.

Once the data was preprocessed, normalized and synchronized, ERNN was applied by selecting most suitable structure and training algorithm which are described in the following paragraphs.

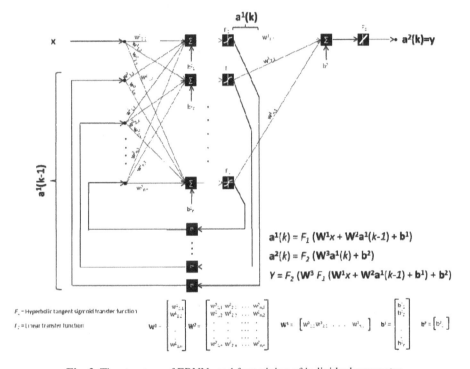

$$a^1(k) = F_1 (W^1 x + W^2 a^1(k-1) + b^1)$$
$$a^2(k) = F_2 (W^3 a^1(k) + b^2)$$
$$Y = F_2 (W^3 F_1 (W^1 x + W^2 a^1(k-1) + b^1) + b^2)$$

Fig. 3. The structure of ERNN used for training of individual parameter

Structure: The ERNN used for modeling semiconductor equipment data is shown in the Fig.3. For the ERNN with n number of hidden layer neurons, the output of the hidden neuron layer can be expressed as

$$\mathbf{a}^1(k) = F_1 (\mathbf{W}^1\mathbf{x} + \mathbf{W}^2\mathbf{a}^1(k-1)+ \mathbf{b}^1) \qquad (4)$$

where F_1 is a hyperbolic tangent sigmoid transfer function for the hidden layer, $\mathbf{x}$ is the data input, $\mathbf{a}^1(k-1)$ is the delayed feedback from the hidden layer, $\mathbf{w}^1$ is the weight matrix of input to hidden layer synapses, $\mathbf{w}^2$ is the weight matrix of context to hidden layer synapses, and $\mathbf{b}^1$ is the bias matrix for hidden layer. The output, $\mathbf{a}^2(k)$ can be expressed as

$$a^2(k) = F_2 (W^3 a^1(k) + b^2) \tag{5}$$

where F_2 is a linear transfer function for the hidden layer, w^3 is the weight matrix of hidden to output layer synapses, and b^2 is the bias matrix for the output layer. Using (4), the output of ERNN, Y can be expressed as

$$Y = a^2(k) = F_2 (W^3 a^1(k) + b^2) \tag{6}$$

or,

$$Y = F_2 (W^3 F_1 (W^1 x + W^2 a^1(k-1) + b^1) + b^2) \tag{7}$$

As evident from the implemented ERNN structure shown in Fig. 3, the input and output layers have one neuron each. This is because a separate one-input, one-output ERNN was constructed for each of the 15 parameters of the equipment data.

Table 3. Comparison of candidate ERNN structures for each parameter

	Par. 1	Par. 2	Par. 3	Par. 4	Par. 5	Par. 6	Par. 7	Par. 8	Par. 9	Par. 10	Par. 11	Par. 12	Par. 13	Par. 14	Par. 15
(1-4-1)	0.178	0.189	0.154	0.306	0.186	0.123	0.149	0.039	0.131	0.020	0.092	*0.028*	0.023	0.079	0.049
(1-6-1)	0.016	0.075	0.044	0.063	0.186	*0.019*	0.119	0.064	*0.036*	0.023	0.038	0.069	0.015	*0.054*	0.045
(1-8-1)	0.012	0.099	0.063	0.038	*0.057*	0.101	0.078	0.050	0.044	0.030	*0.015*	0.040	0.018	0.056	*0.040*
(1-10-1)	0.176	*0.043*	*0.037*	*0.021*	0.114	0.094	0.089	0.261	0.092	0.037	0.227	0.057	0.171	0.135	0.219
(1-12-1)	*0.007*	0.064	0.039	0.085	0.067	0.025	*0.022*	*0.033*	0.050	*0.008*	0.048	0.094	*0.006*	0.063	0.062

The structure for each of the ERNNs, in terms of the number of hidden layer neurons, $n_1, n_2 \dots n_{15}$, was individually defined for each parameter. The number of hidden neurons is varied to choose the best architecture that suits our tool dataset. The candidate structures evaluated for the selection of the best ERNN model are (1-4-1), (1-6-1), (1-8-1), (1-10-1) and (1-12-1). Each of the parameters was trained using the

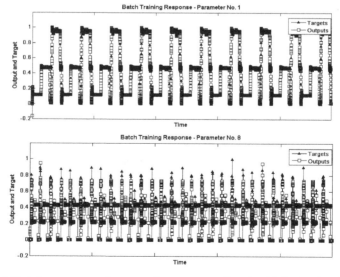

Fig. 4. Training response of two sample parameters, Par. 1 and 8, by batch training

candidate models and the best one was selected considering the least value of root mean square error (RMSE). For the purpose of this comparison, the maximum number of epochs was fixed at 6 and the performance of each model was compared after being trained for 6 epochs. The result is presented in Table 3. For each parameter, the selected model has been highlighted in the table.

Network Training: The data for each parameter from the healthy data set was used to train each corresponding ERNN model selected for that parameter. The input series was set as the combined sequence of each parameter, from all healthy runs, while the target was set as the same input shifted ahead by one sample in time, so that the network is trained to predict the next data sample, for a given input sample. For example, for a data sample x_i presented at the input, x_{i+1} is set as the target.

The networks were trained using previously explained, Levenberg Marquardt algorithm but there were two different well known approaches followed in order to train the network; batch training and incremental training. In batch training, the weights are updated after all the training data input has been presented to the network. On the other hand, in incremental training the weights are updated each time an input is presented [22-23].

For both types of approaches, the training stopping criteria included the error goal of RMSE=0.01. Also the maximum number of epochs was fixed at 40. Fig. 5 shows the network training response for Par. 1 and Par. 8 for batch training algorithms.

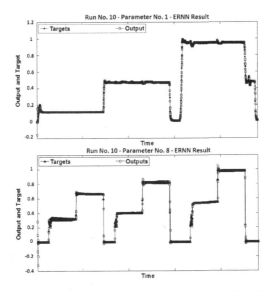

Fig. 5. ERNN result on two sample parameters from Run No. 10

The results for the two kinds of training have shown that batch training has improved the training performance by 52.7% as compared to incremental training for the same training data and conditions. Hence batch training was used for the subsequent steps.

4   Application to Equipment Fault Detection

Once the networks were trained for each parameter, they were applied on the test runs from the experimental data set, in order to give a one-step-ahead prediction. Fig. 6 shows the test response of networks, for Par. 1 and Par. 8 by batch training.

Table 4. ERNN results in terms of RMSE for each parameter of each experimental run

	Par. 1	Par. 2	Par. 3	Par. 4	Par. 5	Par. 6	Par. 7	Par. 8	Par. 9	Par. 10	Par. 11	Par. 12	Par. 13	Par. 14	Par. 15
Run No. 1	0.0168	0.0612	0.0634	0.0636	0.0536	0.0292	0.0413	0.0378	0.0555	0.0257	0.0390	0.0355	0.0457	0.0505	0.0504
Run No. 2	0.0169	0.0584	0.0614	0.0623	0.0559	0.0287	0.0420	0.0378	0.0587	0.0267	0.0417	0.0353	0.0458	0.0527	0.0614
Run No. 3	0.0164	0.0611	0.0598	0.0631	0.0530	0.0284	0.0411	0.0382	0.0582	0.0274	0.0403	0.0354	0.0462	0.0520	0.0555
Run No. 4	0.0169	0.0637	0.0615	0.0621	0.0546	0.0288	0.0414	0.0369	0.0588	0.0280	0.0405	0.0354	0.0459	0.0509	0.0535
Run No. 5	0.0173	0.0611	0.0656	0.0628	0.0554	0.0286	0.0411	0.0375	0.0575	0.0290	0.0420	0.0353	0.0457	0.0513	0.0561
Run No. 6	0.0167	0.0601	0.0576	0.0628	0.0562	0.0282	0.0435	0.0370	0.0586	0.0264	0.0403	0.0356	0.0460	0.3088	0.0538
Run No. 7	0.0166	0.0635	0.0617	0.0628	0.0610	0.0289	0.0420	0.0385	0.0601	0.0288	0.0404	0.0354	0.0458	0.0509	0.0603
Run No. 8	0.0168	0.0652	0.0604	0.0637	0.0479	0.0293	0.0428	0.0408	0.0558	0.0289	0.0404	0.0353	0.0456	0.0548	0.0587
Run No. 9	0.0171	0.0654	0.0608	0.0616	0.0512	0.0289	0.0408	0.0374	0.0571	0.0279	0.0404	0.0353	0.0458	0.0516	0.0511
Run No. 10	0.0170	0.0644	0.0645	0.0643	0.0537	0.0296	0.0452	0.0432	0.0602	0.0285	0.0390	0.0355	0.0457	0.0511	0.0601

Using the residuals for each of the parameters for test runs, the computed RMSE is summarized in Table 4. For each parameter in Table 2, the batches with highest fault probability have been highlighted. Observing the values closely we can notice the primary and secondary effects of the perturbations mentioned in Table 1. For example in Run No. 6 'Loss in source RF cable' was simulated. This fault is clearly indicated by the high value of RMSE for Run No. 6 in Par. 7: RF Probe Voltage and Par. 14: RF Probe Peak to Peak Voltage. Similarly Run No. 4 and 5 which had been induced with 'Conversion shift in mass flow controllers,' have high value of fault probability for Par. 11: Flow Splitter – Flow 1.

In order to find the combined probability of fault in each run, the RMSE values for each parameter in Table 4 were normalized and combined to form a run-wise result represented in the chart in Fig. 7(a). The perturbed runs, Run No. 2-6, 8 and 10 have resulted in high values of error providing evidence for fault in these runs. Also considering the low error values for the unperturbed runs, Run No. 1 and 9, they can be regarded as controlled runs.

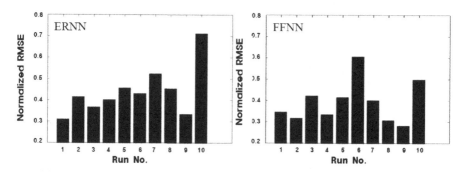

Fig. 6. (a) ERNN combined RMSE and (b) FFNN combined RMSE

Table 5. FFNN results in terms of RMSE for each parameter of each experimental run

	Par. 1	Par. 2	Par. 3	Par. 4	Par. 5	Par. 6	Par. 7	Par. 8	Par. 9	Par. 10	Par. 11	Par. 12	Par. 13	Par. 14	Par. 15
Run No. 1	0.012164	0.046638	0.045821	0.045935	0.048850	0.018638	0.034899	0.028124	0.041252	0.018446	0.029584	0.029569	0.033831	0.042641	0.060228
Run No. 2	0.012383	0.046629	0.045804	0.045396	0.041669	0.017746	0.035778	0.027933	0.042443	0.017183	0.029513	0.029498	0.033464	0.042940	0.062637
Run No. 3	0.012382	0.046642	0.045977	0.045936	0.041948	0.023554	0.035935	0.028146	0.045274	0.020438	0.029589	0.029563	0.033019	0.042457	0.060020
Run No. 4	0.011696	0.046653	0.044525	0.044780	0.131238	0.019713	0.035306	0.027603	0.043109	0.024776	0.029417	0.029369	0.032949	0.042669	0.060184
Run No. 5	0.012072	0.046653	0.045322	0.045066	0.043678	0.019323	0.035664	0.027567	0.043096	0.021684	0.029731	0.029673	0.032570	0.042778	0.060613
Run No. 6	0.012004	0.046626	0.049030	0.069089	0.119201	0.028630	0.062278	0.027210	0.050671	0.017303	0.029548	0.029534	0.031981	0.045115	0.061118
Run No. 7	0.012498	0.046642	0.046583	0.045686	0.047469	0.018020	0.035678	0.027481	0.045826	0.021229	0.029550	0.029548	0.032572	0.042386	0.061991
Run No. 8	0.012136	0.046630	0.046518	0.046030	0.038603	0.018790	0.035881	0.028925	0.045121	0.021349	0.029573	0.029574	0.032656	0.038418	0.057644
Run No. 9	0.012021	0.046651	0.044988	0.045488	0.046205	0.017672	0.034867	0.027704	0.042186	0.019251	0.029553	0.029532	0.032430	0.042269	0.059698
Run No. 10	0.012347	0.046661	0.045708	0.046100	0.053719	0.018795	0.038008	0.029811	0.046009	0.021193	0.029582	0.029553	0.031162	0.043778	0.064028

However, Run No. 7 which was performed without adding any perturbation was detected as a faulty run. It could be argued that although no perturbation was added to Run No. 7 intentionally, yet, a fault may have occurred during the process, generating variations in its data when compared to the model runs.

In order to compare the performance of ERNN, feed forward neural network (FFNN) was also applied with similar criteria and specifications to make a fair comparison. The RMSE results for FFNN are shown in Table 5 and the combined probability results are represented in Fig 7(b). The results show that FFNN could not distinguish well between faulty and healthy runs and using the same standard as for ERNN, the accuracy was found to be 60%.

5   Summary

The ERNN has been applied on semiconductor etch data. It has been showed that ERNNs are useful for detecting faults in semiconductor etch equipment condition data. Assuming that the controlled run (Run No. 7) was actually flawless, this misdetection of controlled run as a fault brings down the accuracy of this technique to 90%. Currently, we are investigating uncertain information inference with Bayesian networks, and the ERNN prediction will be employed for the priori fault probability for real-time fault detection and classification in semiconductor manufacturing equipment.

Acknowledgement. This work was supported by Small and Medium Business Administration (Cont. No. S2063463), and authors are grateful to Mr. Woo Young Lee and Bio Lim at TES for his technical discussion and support.

References

1. Kerdprasop, K., Kerdprasop, N.: Feature Selection and Boosting Techniques to Improve Fault Detection Accuracy in the Semiconductor Manufacturing Process. In: Int. Multi. Conf. Engineers and Computer Scientists 2011, pp. 398–403. IAENG, Hong Kong (2011)
2. Stokes, D., May, G.S.: Real-Time Control of Reactive Ion Etching Using Neural Networks. IEEE Trans. Semi. Manufac. 13, 469–480 (2000)
3. Hong, S.J., May, G.S.: Neural Network Based Time Series Modeling of Optical Emission Spectroscopy Data for Fault Detection in Reactive Ion Etching. In: SPIE Conf. Adv. Microelectronic Manufacturing, vol. 5041, pp. 1–8 (2003)

4. Hong, S.J., May, G.S., Park, D.-C.: Neural Network Modeling of Reactive Ion Etching Using Optical Emission Spectroscopy Data. IEEE Trans. Semi. Manufac. 16, 598–608 (2003)
5. May, G.S.: Manufacturing ICs the Neural Way. IEEE Spectrum 31, 47–51 (1994)
6. Baker, M., Himmel, C., May, G.S.: Time Series Modeling of Reactive Ion Etching Using Neural Networks. IEEE Trans. Semi. Manufac. 8, 62–71 (1995)
7. Nadi, F., Agogino, A., Hodges, D.: Use of Influence Diagrams and Neural Networks in Modeling Semiconductor Manufacturing Processes. IEEE Trans. Semi. Manufac. 4, 52–58 (1991)
8. Hong, S.J., Lim, W.Y., Cheong, T.S., May, G.S.: Fault Detection and Classification in Plasma Etch Equipment for Semiconductor Manufacturing e-Diagnostics. IEEE Trans. Semi. Manufac. 25, 83–93 (2012)
9. Pearson, K.: Principal Components Analysis. The London, Edinburgh and Dublin Philosophical Magazine and Journal 6, 559 (1901)
10. Keogh, E.J., Pazzani, M.J.: Derivative dynamic time warping. In: First SIAM International Conference on Data Mining (SDM 2001), Chicago, USA (2001)
11. Bellman, R., Kalaba, R.: On adaptive control processes. IEEE Trans. Automat. Contr. 4, 1–9 (1959)
12. Senin, P.: Dynamic time warping algorithm review. University of Hawaii, Honolulu (2008)
13. Kassidas, A., MacGregor, J.F., Taylor, P.A.: Synchronization of batch trajectories using dynamic time warping. AIChE Journal 44, 864–875 (1998)
14. Elman, J.L.: Finding Structure in Time. Cognitive Science 14, 179–211 (1990)
15. Mankar, V.R., Ghatol, A.A.: Design of Adaptive Filter Using Jordan/Elman Neural Network in a Typical EMG Signal Noise Removal. Advances in Artificial Neural Systems (2009)
16. Swingler, K.: Applying Neural Networks, A Practical Guide. Morgan Kaufmann, San Francisco (1996)
17. Singh, E.: Comparative Economic Forecasting with Neural Networks Forecasting Aggregate Business Sales from S&P 500 and Interest Rates. Pace University, New York (2007)
18. Marquardt, D.W.: An Algorithm for Least-Squares Estimation of Nonlinear Parameters. Journal of the Society for Industrial and Applied Mathematics 11, 431–441 (1963)
19. Levenberg, K.: A method for the solution of certain non-linear problems in least squares. Quart. Appl. Math 2, 164–168 (1944)
20. Dohnal, J.: Using of Levenberg-Marquardt method in identification by Neural Networks. In: Student EEICT 2004, pp. 361–365 (2004)
21. Hagan, M.T., Menhaj, M.B.: Training Feed forward Networks with the Marquardt Algorithm. IEEE Trans. on Neural Networks 5, 989–993 (1994)
22. Du, K.L., Swamy, M.N., Swamy, S.: Neural Networks in a soft computing Framework. Springer, London (2006)
23. Si, Nelson, Runger: Artificial Neural Network Models for Data Mining. In: The Handbook of Data Mining. Lawrence Erlbaum Associates Publishers, New Jersey (2003)

Relearning Probability Neural Network for Monitoring Human Behaviors by Using Wireless Sensor Networks

Ming Jiang[1] and Sen Qiu[2]

[1] School of Electronic Engineering, Dongguan University of Technology,
Guangdong, Dongguan, China
[2] School of Control Science and Engineering, Dalian University of Technology
Dalian, Liaoning, China
{onlyjiang,qantasy}@gmail.com

Abstract. Human behaviors monitoring by using wireless sensor networks has gained tremendous interest in recent years from researchers in many areas. To distinguish behaviors from on-body sensor signals, many classification methods have been tried, but most of them lack the relearning ability, which is quite important for long-term monitoring applications. In this paper, a relearning probabilistic neural network is proposed. The experimental results showed that the proposed method achieved good recognition performance, as well as the relearning ability.

Keywords: Behavior monitoring, classification method, relearning ability, probabilistic neural network.

1 Introduction

So far, studies on using wireless sensor networks to monitor human behaviors have grown up rapidly in many health-related applications [1]. To recognize human behaviors from on-body sensor data, a range of classification methods have been tried [2], including hidden Markov models (HMMs) [3][4], decision trees [5], k-nearest neighbor (kNN) [6], artificial neural networks (ANNs) [7][8] and support vector machines (SVMs) [9][10]. The primary limitation of these methods is the lack of relearning ability, which is quite important for long-term monitoring applications.

To distinguish human behaviors reliably, a classifier should be easily updated from new training data. First, because the characteristics of human behaviors are various, even for the same person, sensor data collected from a short time period is usually not enough to train a classifier. Second, annotation of training instances is a hard work and easy to lead label errors, so the trust of one training dataset is dangerous.

In this study, a relearning method based on probabilistic neural network (PNN) is proposed. The method has good relearning features, including: 1) it can be easily updated from new training data to improve its accuracy; 2) it can improve the robustness against label errors. The experimental results from realistic sensor data have shown that the proposed method obtained good recognition performance, as well as the relearning ability.

C. Guo, Z.-G. Hou, and Z. Zeng (Eds.): ISNN 2013, Part II, LNCS 7952, pp. 473–480, 2013.

2   Theories

2.1   Brief Introduction of PNN

PNN [11] is a model based on Bayes strategy [12] for making decision, and Parzen's theorem [13] for estimating probability density functions (PDFs). Bayes strategy identifies an unknown instance x into class a, if inequality (1) holds,

$$p_a f_a(x) > p_{a'} f_{a'}(x), \ \forall a' \neq a \tag{1}$$

where p_a is the prior probability of class a, and f_a is the PDF. Parzen's theorem is a common tool to estimate a univariate PDF from training instances. Cacoullos [14] suggests that the unknown PDF can be estimated according to (2),

$$f_a(x) = \frac{1}{(2\pi)^{\frac{p}{2}} \sigma^p q_a} \sum_{i=1}^{q_a} \exp\left(-\frac{\|x - c_{a,i}\|^2}{2\sigma^2} \right) \tag{2}$$

where q_a is the number of classification patterns of class a, σ is the smoothing parameter, p is the dimension of the measurement space, and $c_{a,i}$ is the ith classification pattern.

PNN usually consists of four layers. The input layer just distributes the input to the pattern layer. The neurons in pattern layer are used to store classification patterns. A common way to find classification patterns from training instances is by using clustering algorithm. The center vectors of clusters could be used as classification patterns [15]. The output of a pattern neuron is a Gaussian function of the input, as the exponential part in (2). In the summation layer, the number of neurons is equal to the number of classes. By averaging the outputs from pattern neurons belonging to a same class, a summation neuron calculates the likelihood of an input being classified into each class. There is only one neuron in the decision layer, deciding which class an input vector should be classified into. Detailed introduction of PNN can be found in the paper of Specht [11].

2.2   Brief introduction of FCM

Fuzzy c-means (FCM) [16] is a clustering method, which partitions a set of instances S into q clusters. The ith ($i = 1, \cdots, q$) cluster is supposed to have a center vector c_i. For each $s \in S$, there is a membership $u_i(s)$ ($\in [0,1]$) indicating with what degree s belongs to the ith cluster. The objective function J of FCM is defined as follows:

$$J = \sum_{i=1}^{q} \sum_{s \in S} (u_i(s))^m \|s - c_i\|^2 \tag{3}$$

$$c_i = \sum_{s \in S} (u_i(s))^m s \bigg/ \sum_{s \in S} (u_i(s))^m \tag{4}$$

$$u_i(s) = \left(\sum_{k=1}^{q} (\|s - c_i\| / \|s - c_k\|)^{2/m-1} \right)^{-1} \tag{5}$$

where m is the fuzzy degree member. FCM is briefly described as follows:

Step 1: Choose appropriate integer q and threshold value ε. Initialize each $u_i(s)$ by a random value within interval $[0,1]$.

Step 2: Compute c_i according to (4).

Step 3: Compute $u_i(s)$ according to (5).

Step 4: Compute the J according to (5). If the difference between two adjacent values of J is less than ε, terminate. Otherwise, go to step 2.

3   The Proposed Method

3.1   Establish a New PNN

Let the first training dataset be represented by Ω. Suppose there are A behaviors in Ω. The proposed method first divides Ω into A subsets. The subset Ω_a $(a = 1, \cdots, A)$ consists of instances belonging to behavior a. FCM is then carried out to Ω_a to generate q_a clusters. Each center vector $c_{a,i}$ $(i = 1, \cdots, q_a)$ is stored in the pattern layer. Given an unlabelled instance x, the unknown PDF $f_a(x)$ is calculated according to (6).

$$f_a(x) = \sum_{i=1}^{q_a} r_{a,i} \frac{1}{(2\pi)^{p/2} \sigma^p} \exp\left(-\frac{\|x - c_{a,i}\|^2}{2\sigma^2} \right) \tag{6}$$

The weighting coefficient $r_{a,i}$, which is calculated according to (7), is used to reduce the impact of label errors in training instances, where $n_{a,i}$ denotes the number of instances in Ω_a belonging to ith cluster.

$$r_{a,i} = n_{a,i} \bigg/ \sum_{j=1}^{q_a} n_{a,j} \tag{7}$$

The decision neuron makes a final decision according to (1).

After established a new PNN, the reconstruction error of ith cluster of ath behavior is calculated according to (8),

$$v_{a,i} = \sum_{x \in \Omega_{a,i}} \|x - \hat{x}\|^2 \tag{8}$$

where $\Omega_{a,i} = \left\{ i = \arg\max_j \|x - c_{a,j}\| \,|\, x \in \Omega_a \right\}$, and $\hat{x}$ is calculated according to (9).

$$\hat{x} = \sum_{i=1}^{q_a} (u_{a,i}(x))^m c_{a,i} \bigg/ \sum_{i=1}^{q_a} (u_{a,i}(x))^m \tag{9}$$

The distance between x and $\hat{x}$ constitutes a viable measure of the quality of clusters [17]. The reconstruction error of each cluster is used for updating process.

3.2   Update a PNN from New Training Data

Let the new training data be represented by Ω^{new}. In this section, only consider that there is no new behaviors in Ω^{new}. The proposed method first divides Ω^{new} into A subsets Ω_a^{new}. For behavior a, the reconstruction error of existing q_a clusters are re-calculated according to (8) and (9), but Ω_a is replaced by Ω_a^{new} at this time. Let the new reconstruction error of ith cluster of ath behavior be denoted by $\hat{v}_{a,i}$.

If $\tau_1 < \max\{\hat{v}_{a,i}\} - \max\{v_{a,i}\}$, where $0 \le \tau_1$, the proposed method splits i_0th $\left(i_0 = \arg\max_i\{\hat{v}_{a,i}\}\right)$ cluster into two clusters to lower $\max\{\hat{v}_{a,i}\}$. Then, the corresponding pattern neuron is replaced by two new pattern neurons. The calculation of two new center vectors $c_{a,h}^{new}$ ($h = 1, 2$) are carried out iteratively according to (10) and (11),

$$c_{a,h}^{new} = \sum_{x\in\Omega_{a,i_0}^{new}} \left(u_{a,h}^{new}(x)\right)^m x \Bigg/ \sum_{x\in\Omega_{a,i_0}^{new}} \left(u_{a,h}^{new}(x)\right)^m \tag{10}$$

$$u_{a,h}^{new}(x) = u_{a,i_0}(x)\left[\sum_{j=1}^{2}\left(\left\|x - c_{a,h}^{new}\right\| \Big/ \left\|x - c_{a,j}^{new}\right\|\right)^{\frac{1}{m-1}}\right]^{-1} \tag{11}$$

where $\Omega_{a,i_0}^{new} = \left\{x\in\Omega_a^{new}\Big|i_0 = \arg\max_j\left\|x - c_{a,j}\right\|\right\}$. The splitting process is repeated until $\max\{\hat{v}_{a,i}\} - \max\{v_{a,i}\} \le \tau_1$.

If $\max\{\hat{v}_{a,i}\} - \max\{v_{a,i}\} < \tau_2$, where $\tau_2 < \tau_1$, the proposed method merges i_1th and i_2th clusters, where $(i_1, i_2) = \arg\min_{(j,k)}\{\|c_{a,j} - c_{a,k}\| \mid j \ne k\}$. Then, the corresponding two pattern neurons are replaced by a new pattern neuron. The new center vector c_a^{new} is calculated according to (12).

$$c_a^{new} = \frac{\displaystyle\sum_{x\in\Omega_{i_1}^{new}\cup\Omega_{i_2}^{new}} \left(u_{a,i_1}(x) + u_{a,i_2}(x)\right)^m x}{\displaystyle\sum_{x\in\Omega_{i_1}^{new}\cup\Omega_{i_2}^{new}} \left(u_{a,i_1}(x) + u_{a,i_2}(x)\right)^m} \tag{12}$$

The merging process is repeated until $\tau_2 < \max\{\hat{v}_{a,i}\} - \max\{v_{a,i}\} < \tau_1$.

After splitting and merging processes, the weighting coefficient of each pattern neuron is re-calculated according to (7), where $n_{a,i} = n_{a,i} + n_{a,i}^{new}$.

4   Experiment and Results

4.1   Experiment Design and Data Collection

The experiment data used in this study is from Lab of Intelligent System in Dalian University of Technology [18]. The experiment platform consisted of five signal collection nodes placed at five body locations (lower left forearm, lower right forearm, waist, left ankle, and right ankle), as well as a signal reception node attaching to a computer. A tri-axial accelerometer (ADXL330) was integrated in a signal collection node, which could measure acceleration with a minimum full-scale range of ±3g. Accelerometry was regarded as a practical method of objectively monitoring human movements [19]. The frequency of wireless transmission of acceleration signals was 50Hz which achieved a good trade-off between sampling density and packet loss rate.

Four subjects (4 males, ages from 25-28) took part in this experiment. Each subject was asked to perform a range of daily behaviors during a two-hour period, including standing (A1), sitting on a couch (A2), lying on a bed (A3), walking around (A4), going upstairs (A5), going downstairs (A6), sweeping the floor (A7), wiping a table (A8), putting on a vest (A9), and taking off a vest (A10).

4.2   Pre-processing of Sensor Data

In the pre-processing step, raw signals are first cut into small segments. The length of sliding window was 256 samples (about 5.12 second), and there is also a 50% overlap between adjacent windows. Five kinds of features were extracted from each window to characterize corresponding behavior, including mean, variance, correlation, energy and entropy. These features have shown their efficiency of representing the characteristics of human behaviors [5]. Extracted features from a window form a feature vector. All the feature vectors along with their labels are used as training/testing instances.

4.3   Simulation 1: Update an Existing PNN

The collected eight-hour acceleration dataset was divided into sixteen half-hour subsets. Each subset included all ten behaviors. The first dataset was used to establish a new PNN, and the other fifteen datasets were used to update it. To establish a new PNN, the initial number of clusters, iteration-stopping threshold and fuzzy parameter used in FCM were selected to be 20, 0.5 and 2. To update a PNN, the thresholds used in splitting and merging processes were selected to be 2.8 and 0.4 on the optimal result among a range of trials. The detection accuracy of the proposed method was shown in Figure 1. Five other classification methods, including HMMs, SVMs, MLP, kNN and C4.5, were also carried out for comparison. At each step, the new subset and all previously used subsets were combined together to re-train the above classifiers. The detection results of the five classification methods were also shown in Figure 1.

At the first testing step, the accuracy rates of HMMs, SVMs, MLP, kNN, C4.5 and the proposed method were 80.72%, 90.21%, 90.82%, 80.52%, 81.29% and 80.15% respectively. At the sixteenth step, the accuracy rates were 89.24%, 93.45%, 91.01%, 90.48%, 86.89% and 93.14% respectively. The improvement of the proposed method was the largest. The experiment results showed that the proposed method achieved good performance of "updating from new training data to improve the accuracy".

4.4   Simulation 2: Involve Label Errors in Training Data

The eight-hour acceleration dataset was divided into sixteen half-hour subsets. Two simulations were carried out. In the first simulation, 80% of instances in the sixth subset

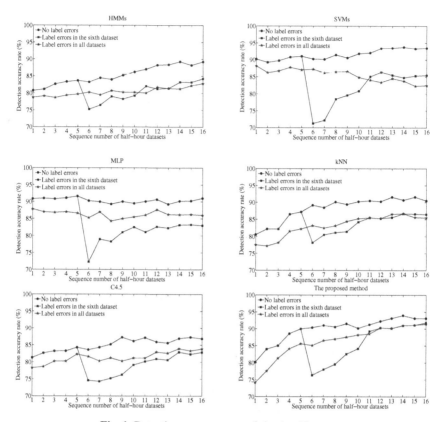

Fig. 1. Detection accuracy rate of six classifiers

were mislabelled on purpose. In the second simulation, the label errors were equally distributed into all sixteen subsets, and each subset had 5% of mislabelled training instances. The calculation results of two simulations were also shown in Figure 1.

In the first simulation, the detection accuracy of all classifiers dropped sharply at the sixth testing. At the sixteenth testing step, the accuracy rates of HMMs, SVMs, MLP, kNN, C4.5 and the proposed method decreased by 5.38%, 8.19%, 8.29%, 4.64%, 4.30% and 1.81 % respectively. In the second simulation, there was no sharp reduction of accuracy rate, but the accuracy of each classification method was decreased at all the testing. The average drop percent of HMMs, SVMs, MLP, kNN, C4.5 and the proposed method were 4.95%, 6.06%, 4.07%, 5.06%, 3.69% and 3.67% respectively. The results of both simulations showed that the proposed method gained the most impressive performance of both recovery speed and recovery quality against label errors.

5   Conclusion

In this paper, a relearning method is proposed for monitoring human behaviors. The proposed method is based on PNN and FCM algorithm. The experimental results have shown that the proposed method may obtain a satisfactory performance for long-term monitoring applications.

Acknowledgements. The authors express their sincere gratitude to Lab of Intelligent System in Dalian University of Technology for providing the experimental data used in this study, especially to Prof. Zhelong Wang for his invaluable guidance and assistance throughout this research.

References

1. Atallah, L., Yang, G.Z.: The use of pervasive sensing for behaviour profiling-a survey. Pervasive and Mobile Computing 5(5), 447–464 (2009)
2. Preece, S., Goulermas, J., Kenney, L., Howard, D., Meijer, K., Crompton, R.: Activity identification using body-mounted sensors-a review of classification techniques. Physiological Measurement 30, R1–R33 (2009)
3. Jamie, A.W., Lukowicz, P., Troster, G., Starner, T.E.: Activity recognition of assembly tasks using body-worn microphones and accelerometers. IEEE Transactions on Pattern Analysis and Machine Intelligence 28(10), 1553–1567 (2006)
4. Nguyen, N.T., Phung, D.Q., Venkatesh, S., Bui, H.: Learning and detecting activities from movement trajectories using the hierarchical hidden Markov models. IEEE Computer Society Conference on Computer Vision and Pattern Recognition 2, 955–960 (2005)
5. Bao, L., Intille, S.S.: Activity recognition from user-annotated acceleration data. In: Ferscha, A., Mattern, F. (eds.) PERVASIVE 2004. LNCS, vol. 3001, pp. 1–17. Springer, Heidelberg (2004)
6. Zhang, T., Wang, J., Liu, P., Hou, J.: Fall detection by embedding an accelerometer in cellphane and using KFD algorithm. International Journal of Computer Science and Network Security 6(10), 277–284 (2006)
7. Wang, N., Ambikairajah, E., Lovell, H.N., Celler, G.B.: Accelerometry based classification of walking patterns using time-frequency analysis. In: Proceedings of the 29th Annual International Conference of the IEEE Engineering in Medicine and Biology Society, pp. 4899–4902 (2007)
8. Sun, Z.L., Mao, X.C., Tian, W.F., Zhang, X.F.: Activity classification and dead reckoning for pedestrian navigation with wearable sensors. Measurement Science and Technology 20, 1–10 (2009)
9. Zhang, T., Wang, J., Xu, L., Liu, P.: Using wearable sensor and NMF algorithm to realize ambulatory fall detection. In: Jiao, L., Wang, L., Gao, X.-b., Liu, J., Wu, F. (eds.) ICNC 2006. LNCS, vol. 4222, pp. 488–491. Springer, Heidelberg (2006)
10. Yin, J., Yang, Q., Pan, J.J.: Sensor-based abnormal human-activity detection. IEEE Transactions on Knowledge and Data Engineering 20(8), 1082–1090 (2008)
11. Specht, D.F.: Probabilistic neural networks. Neural Networks 3, 109–118 (1990)
12. Mood, A.M., Graybill, F.A.: Introduction to the theory of statistics. Macmillan, New York (1962)
13. Parzen, E.: On estimation of probability density function and mode. Annals of Mathematical Statistics 33, 1065–1076 (1962)
14. Cacoullos, T.: Estimation of a multivariate density. Annals of the Institute of Statistical Mathematics 18(2), 179–189 (1966)

15. Musavi, M.T., Ahmed, W., Chan, K.H., Hummels, D.M., Kalantri, K.: A probabilistic model for evaluation of neural network classifiers. Pattern Recognition 25, 1241–1251 (1992)
16. Dunn, J.C.: Some recent investigations of a new fuzzy partition algorithm and its application to pattern classification problems. Cybernetics and Systems 4(2), 1–15 (1974)
17. Pedrycz, W.: A dynamic data granulation through adjustable fuzzy clustering. Pattern Recognition Letters 29, 2059–2066 (2008)
18. http://ei.dlut.edu.cn/lis
19. Mathie, M.J., Coster, A.C.F., Lovell, N.H., Celler, B.G.: Accelerometry: providing an integrated, practical method for long-term, ambulatory monitoring of human movement. Physiological Measurement 25, R1–R20 (2004)

Different ZFs Leading to Various ZNN Models Illustrated via Online Solution of Time-Varying Underdetermined Systems of Linear Equations with Robotic Application

Yunong Zhang*, Ying Wang, Long Jin, Bingguo Mu, and Huicheng Zheng

School of Information Science and Technology,
Sun Yat-sen University, Guangzhou 510006, China
zhynong@mail.sysu.edu.cn

Abstract. Recently, by following Zhang et al.'s design method, a special class of recurrent neural network (RNN), termed Zhang neural network (ZNN), has been proposed, generalized and investigated for solving time-varying problems. In the design procedure of ZNN models, choosing a suitable kind of error function [i.e., the so-called Zhang function (ZF) used in the methodology] plays an important role, and different ZFs may lead to various ZNN models. Besides, differing from other error functions such as nonnegative energy functions associated with the conventional gradient-based neural network (GNN), the ZF can be positive, zero, negative, bounded, or unbounded even including lower-unbounded. In this paper, different newly-designed ZNN models are proposed, developed and investigated to solve the problem of time-varying underdetermined systems of linear equations (TVUSLE) based on different ZFs. Computer-simulation results (including the robotic application of the newly-designed ZNN models) show that the effectiveness of the proposed ZNN models is well verified for solving such time-varying problems.

Keywords: Recurrent neural network (RNN), Zhang function (ZF), time-varying, underdetermined system of linear equations.

1 Introduction and Problem Formulation

Recently, a special class of recurrent neural network, Zhang neural network (ZNN), has been proposed, generalized and investigated for a number of time-varying problems solving. In the last decade, ZNN method/models have been applied to effectively solving many time-varying problems. In the design procedure of ZNN models, choosing a suitable Zhang function (ZF) is very important. Usually, different ZFs lead to various ZNN models by employing the Zhang et al.'s design method. According to such reasoning, we are motivated to propose and investigate two types of ZFs to construct the ZNN models for time-varying underdetermined system of linear equations (TVUSLE) solving in

---

* Corresponding author.

C. Guo, Z.-G. Hou, and Z. Zeng (Eds.): ISNN 2013, Part II, LNCS 7952, pp. 481–488, 2013.

this paper. Now, let us consider the following problem formulation of TVUSLE with smoothly time-varying vector $b(t) \in R^m$ known or measurable for any time instant $t \in [0, +\infty)$:

$$A(t)x(t) = b(t), \tag{1}$$

where coefficient matrix $A(t) \in R^{m \times n}$ (with $m < n$) with full rank are smoothly time-varying and known or measurable. The unknown vector $x(t) \in R^n$ is to be found in an error-free and real-time manner. Note that (1) can be viewed as a general time-varying system of m real-valued time-varying linear equations and n real-valued time-varying variables. For further discussion, we employ the right Moore-Penrose inverse $A^+ = A^T(AA^T)^{-1}$ of full-rank coefficient matrix $A(t)$ with the consideration of $m < n$ in this paper.

2   ZNN Solvers

In this section, by employing two different ZFs [denoted by $e(t)$], the specific steps of constructing the ZNN models with different activation function arrays $F(\cdot)$ are established. According to Zhang et al.'s design method, any monotonically-increasing odd activation function $f(\cdot)$ can be used to construct the ZNN models [3]. Note that $f(\cdot)$ denotes the element of vector-to-vector activation function array $F(\cdot)$. For the convenience of further discussion, the following three activation functions are used in this paper.

1) linear activation function (LAF) $f(e_i) = e_i$;
2) power-sigmoid activation function (PSIAF)

$$f(e_i) = \begin{cases} e_i^p, & \text{if } |e_i| \geq 1 \\ \frac{1+\exp(-\zeta)}{1-\exp(-\zeta)} \cdot \frac{1-\exp(-\zeta e_i)}{1+\exp(-\zeta e_i)}, & \text{otherwise} \end{cases}$$

with the suitable design parameters $\zeta > 2$ and $p \geq 3$;
3) power-sum activation function (PSUAF) $f(e_i) = \sum_{k=1}^{N} e_i^{2k-1}$.

Note that e_i denotes the ith element of $e(t)$ in the research.

2.1   ZNN Models Derived from ZF I

In this subsection, we present the ZNN models which are derived from a usually-used ZF, and the corresponding design steps are established as follows.

Step 1 (Choose a suitable ZF). It follows from our previous work [1] that the following usual error-monitoring function (which is termed ZF I) is used:

$$e(t) := A(t)x(t) - b(t) \in R^m. \tag{2}$$

Step 2 (Use Zhang design formula). The following ZNN design formula is then adopted [1,2,3]:

$$\dot{e}(t) = \frac{de(t)}{dt} := -\gamma F(e(t)), \tag{3}$$

where design parameter $\gamma > 0$, being the reciprocal of a capacitance parameter in the analogue-circuit implementation, should be large enough and is used to scale the convergence rate of the network. As mentioned at the beginning of this section, $F(\cdot) : R^m \to R^m$ denotes the activation function array used now in (3).

Step 3 (Generate ZNN models). Expanding Zhang formula (3) yields firstly the following implicit dynamic equation of a ZNN model solving TVUSLE (1):

$$A(t)\dot{x}(t) = \dot{b}(t) - \dot{A}(t)x(t) - \gamma F(A(t)x(t) - b(t)), \qquad (4)$$

and then the following explicit dynamic equation of the ZNN model:

$$\dot{x}(t) = A^+(t)\dot{b}(t) - A^+(t)\dot{A}(t)x(t) - \gamma A^+(t)F(A(t)x(t) - b(t)), \qquad (5)$$

where $x(t)$, starting from initial condition $x(0) \in R^n$, is the neural state corresponding to theoretical time-varying solution $x^*(t)$ of (1). Besides, if LAF is applied directly to (4), then we have the following implicit linear ZNN model:

$$A(t)\dot{x}(t) = \dot{b}(t) - \dot{A}(t)x(t) - \gamma(A(t)x(t) - b(t)). \qquad (6)$$

2.2 ZNN Models Derived from ZF II

In this subsection, a new ZF is proposed and investigated for constructing new ZNN models. For such a new ZF, the right Moore-Penrose inverse $A^+(t)$ of time-varying coefficient $A(t)$ is employed, and its time-derivative $dA^+(t)/dt$ (i.e., $\dot{A}^+(t)$ for short) is approximated in the following proposition.

Proposition. For time-varying matrix $A(t) \in R^{m \times n}$ (with $m < n$) and its time-varying right Moore-Penrose inverse $A^+(t)$, we approximately have $\dot{A}^+(t) = -A^+(t)\dot{A}(t)A^+(t)$.

Proof. Since $A(t)A^+(t) = I$ with $I \in R^{m \times m}$ being the identity matrix, we have $d(A(t)A^+(t))/dt = dI/dt = 0$. Expanding the left hand side of the equation, we obtain

$$\frac{dA(t)}{dt}A^+(t) + A(t)\frac{dA^+(t)}{dt} = 0,$$

$$A(t)\frac{dA^+(t)}{dt} = -\frac{dA(t)}{dt}A^+(t) = -\dot{A}(t)A^+(t).$$

It follows approximately that $dA^+(t)/dt = -A^+(t)\dot{A}(t)A^+(t)$, which, via a simpler notation, is $\dot{A}^+(t) = -A^+(t)\dot{A}(t)A^+(t)$. The proof is now complete. $\square$

Based on the above result and the several similar steps mentioned in the last subsection, new ZNN models can be established. That is, firstly the following new error-monitoring function (which is termed ZF II) is designed by exploiting the time-varying Moore-Penrose inverse $A^+(t)$:

$$e(t) := x(t) - A^+(t)b(t) \in R^n. \qquad (7)$$

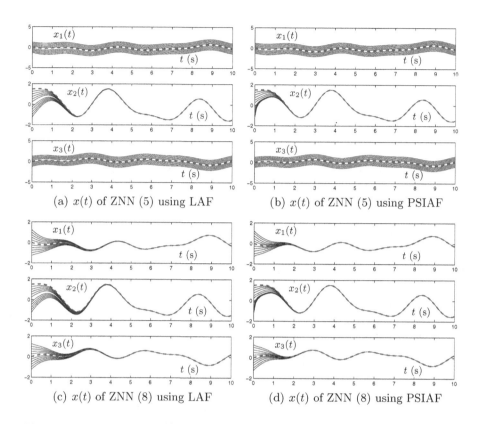

(a) $x(t)$ of ZNN (5) using LAF

(b) $x(t)$ of ZNN (5) using PSIAF

(c) $x(t)$ of ZNN (8) using LAF

(d) $x(t)$ of ZNN (8) using PSIAF

Fig. 1. State trajectories $x(t)$ of ZNN (5) and (8) using different activation functions

Then, by expanding Zhang formula (3), the explicit dynamic equation of a new ZNN model for TVUSLE (1) solving is obtained as follows:

$$\dot{x}(t) = A^+(t)\dot{b}(t) - A^+(t)\dot{A}(t)A^+(t)b(t) - \gamma F(x(t) - A^+(t)b(t)). \qquad (8)$$

Besides, by using the LAF array and $x(t) = A^+(t)b(t)$, another new explicit linear ZNN model is generalized from (8) for solving the same TVUSLE (1):

$$\dot{x}(t) = A^+(t)\dot{b}(t) - A^+(t)\dot{A}(t)x(t) - \gamma(x(t) - A^+(t)b(t)). \qquad (9)$$

In addition, from (9), we have the following implicit linear ZNN model:

$$A(t)\dot{x}(t) = \dot{b}(t) - \dot{A}(t)x(t) - \gamma(A(t)x(t) - b(t)). \qquad (10)$$

Evidently, derived from different ZFs, ZNN models (6) and (10) are the same. Thus, comparing ZNN models (4)-(6) and (8)-(10), we draw a conclusion that different ZFs lead to various ZNN models (including the different and same).

3   Simulations and Verifications

In this section, two illustrative examples are presented and analyzed for comparisons among the ZNN models for the online solution of TVUSLE (1).

Example 1. The time-varying coefficients of TVUSLE (1) are as follows:

$$A(t) = \begin{bmatrix} \sin(0.6t) & \cos(0.6t) & -\sin(0.6t) \\ -\cos(0.6t) & \sin(0.6t) & \cos(0.6t) \end{bmatrix}, \; b(t) = \begin{bmatrix} 1.5\cos(t) \\ \sin(2t) \end{bmatrix}.$$

The corresponding MATLAB-simulation results are shown in Fig. 1. In the time period $t \in [0, 10]$s, state trajectories of the elements $x_1(t)$, $x_2(t)$ and $x_3(t)$ (denoted by blue solid curves) synthesized by ZNN models (5) and (8) with $\gamma = 1$ are illustrated. Evidently, starting from ten randomly-generated initial states $x(0) \in [-2, 2]^3$, some of the simulated state trajectories synthesized by ZNN model (5) [e.g., $x_1(t)$ of Fig. 1(a) and (b)] do not converge to the trajectories of the referenced theoretical solution $x^*(t) = A^+(t)b(t)$ (denoted by red dash-dotted curves), but run in parallel with the theoretical-solution trajectories. The reason is that there are multiple time-varying solutions satisfying the TVUSLE (1) with different initial states $x(0)$ used. In contrast, other simulated state trajectories of ZNN model (5) and all simulated state trajectories of ZNN model (8), starting from randomly-generated initial states $x(0) \in [-2, 2]^3$, relatively fast converge to the trajectories of the referenced theoretical solution $x^*(t) = A^+(t)b(t)$, as shown in the subplots of Fig. 1. Furthermore, comparing Fig. 1 (a) and (b) or Fig. 1 (c) and (d), we see that, for a same ZNN model, the state trajectories synthesized by the model using PSIAF converge slightly faster than those synthesized by the model using LAF.

Moreover, Fig. 2 depicts the residual errors $\|A(t)x(t) - b(t)\|_2$ of ZNN models (5) and (8) with $\gamma = 1$ used, where $\| \cdot \|_2$ denotes the 2-norm of a vector. As seen from the figure, the residual errors $\|A(t)x(t) - b(t)\|_2$ of the proposed ZNN model (5) using LAF and the proposed ZNN model (8) using PSIAF fast

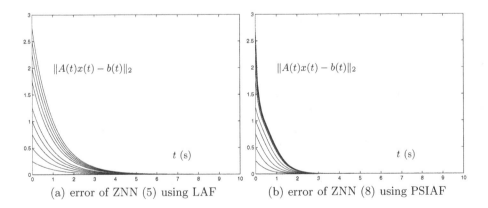

(a) error of ZNN (5) using LAF   (b) error of ZNN (8) using PSIAF

Fig. 2. Residual errors of ZNN (5) and (8) using different activation functions

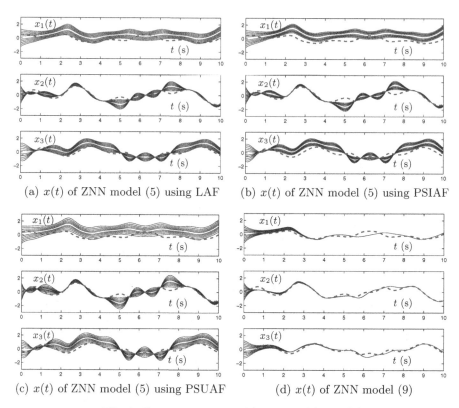

(a) $x(t)$ of ZNN model (5) using LAF   (b) $x(t)$ of ZNN model (5) using PSIAF

(c) $x(t)$ of ZNN model (5) using PSUAF   (d) $x(t)$ of ZNN model (9)

Fig. 3. State trajectories $x(t)$ of ZNN (5) and (9)

converge to zero. Note that, due to results similarity and space limitation, the corresponding figures of ZNN model (5) using PSIAF and ZNN model (8) using LAF are omitted.

Example 2. The time-varying coefficients of TVUSLE (1) are now

$$A(t) = \begin{bmatrix} \sin(t) & \cos(2t) & \sin(2t) \\ \cos(2t) & \sin(2t) & -\cos(2t) \end{bmatrix}, \ b(t) = \begin{bmatrix} \sin(0.5t) \\ \cos(t) \end{bmatrix}.$$

The corresponding MATLAB-simulation results are shown in Fig. 3, where phenomena are similar to those of the previous example. Note that, corresponding to $x(t)$ in the figure, the residual errors of the ZNN models all converge to zero, whose figures are omitted due to results similarity and space limitation.

In summary, the above computer-simulation results of the two examples substantiate the effectiveness of the proposed ZNN models on solving TVUSLE (1).

4   Application to Robotic Motion Planning

In this section, the proposed ZNN models (5), (8) and (9) activated by different functions are applied to the inverse-kinematics motion planning of a four-link

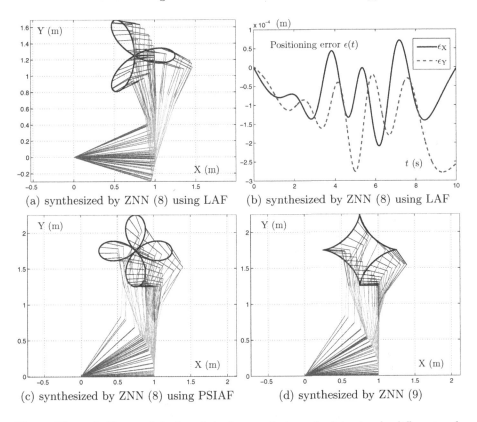

(a) synthesized by ZNN (8) using LAF

(b) synthesized by ZNN (8) using LAF

(c) synthesized by ZNN (8) using PSIAF

(d) synthesized by ZNN (9)

Fig. 4. The end-effector of the four-link planar robot manipulator tracks different end-effector paths synthesized by the proposed ZNN models with $\gamma = 10^4$

planar robot manipulator. For such a robot, its joint-angle vector is generally written as $\theta = [\theta_1, \theta_2, \theta_3, \theta_4]^T \in R^4$. Besides, according to [4,5], the relation between the end-effector velocity $\dot{r}(t) \in R^2$ and the joint velocity $\dot{\theta}(t)$ is

$$J(\theta(t))\dot{\theta}(t) = \dot{r}(t), \tag{11}$$

where $J(\theta) \in R^{2 \times 4}$ is the Jacobian matrix defined as $J(\theta) = \partial f(\theta)/\partial \theta$ [with $f(\theta)$ denoting the forward-kinematics mapping]. Evidently, when the velocity-level expression (11) equals zero, the path-tracking task can be achieved. In other words, the motion planning of the four-link planar robot manipulator can be achieved by solving (11) via the proposed ZNN models. The simulation results of tracking different end-effector paths via the proposed ZNN models using different activation functions are shown in Fig. 4. Specifically, Fig. 4(a) shows the robot motion trajectories synthesized by ZNN model (8) using LAF when the end-effector of the four-link planar robot manipulator tracks a three-leafed rose path; and Fig. 4(b) presents the corresponding positioning error $\epsilon(t) = r_d(t) - f(\theta)$, with $r_d(t)$ as well as ϵ_X and ϵ_Y denoting respectively the desired end-effector path as well as the X- and Y-axis components of $\epsilon(t)$. Note that the maximal positioning error synthesized by ZNN model (8) using LAF is of order 10^{-4}m, and that, due to space limitation, the figures about the positioning errors synthesized by ZNN (8) using PSIAF and ZNN (9) are omitted.

In summary, this application to the inverse-kinematics control of the four-link planar robot manipulator demonstrates (again) the effectiveness of the proposed ZNN method/models on solving the TVUSLE and its related robotic problem.

5   Conclusions

In this paper, by following Zhang et al.'s design method and defining two Zhang functions, six ZNN models (with two models happening to be the same) have been proposed and investigated for solving the TVUSLE. In addition, two non-linear activation functions have been applied to the ZNN models for faster convergence speed in solving the TVUSLE problems. Computer-simulation results have further demonstrated the efficacy of the proposed ZNN models for the on-line solution of TVUSLE. Besides, the proposed ZNN models have been applied to the inverse-kinematics motion planning of a robot manipulator with good performance shown.

Acknowledgments. This work is supported by the National Natural Science Foundation of China (under Grants 61075121 and 60935001), the Specialized Research Fund for the Doctoral Program of Institutions of Higher Education of China (with project number 20100171110045) and also by the National Innovation Training Program for University Students (under Grant 201210558042). Besides, the corresponding author, Yunong, would like to thank the coauthors of this paper by sharing the following thoughts: "We see links, in addition to differences", "Around success, there are many traps called failure", and "The river said to the streams: You and I will forget all our transient ways when we reach the heart of the sea" (Gibran).

References

1. Zhang, Y., Peng, H.: Zhang Neural Network for Linear Time-Varying Equation Solving and its Robotic Application. In: 6th International Conference on Machine Learning and Cybernetics, pp. 3543–3548. IEEE Press, New York (2007)
2. Zhang, Y., Ruan, G., Li, K., Yang, Y.: Robustness Analysis of the Zhang Neural Network for Online Time-Varying Quadratic Optimization. J. Phys. A: Math. Theor. 43, 245202 (2010)
3. Zhang, Y., Yang, Y., Tan, N., Cai, B.: Zhang Neural Network Solving for Time-Varying Full-Rank Matrix Moore-Penrose Inverse. Comput. 92, 97–121 (2011)
4. Zhang, Y., Wang, J.: Obstacle Avoidance for Kinematically Redundant Manipulators Using a Dual Neural Network. IEEE Trans. Syst. Man. Cybern. B: Cybern. 34, 752–759 (2004)
5. Hou, Z., Cheng, L., Tan, M.: Multicriteria Optimization for Coordination of Redundant Robots Using a Dual Neural Network. IEEE Trans. Syst. Man. Cybern. B: Cybern. 40, 1075–1087 (2010)

Estimation of Indicated Torque
for Performance Monitoring in a Diesel Engine

María de Lourdes Arredondo[1], Yu Tang[2,*], Angel Luís Rodríguez[2],
Saúl Santillán[2], and Rafael Chávez[2]

[1] Universidad Autónoma de Querétaro, Querétaro, México
[2] Universidad Nacional Autónoma de México, Querétaro, México
tang@unam.mx

Abstract. In this paper we presents an observer based on artificial neuro-fuzzy networks approach to estimate the indicated torque of a diesel engine from crank shaft angular position and velocity measurements. These variables can be measured by low-cost sensors, since the indicated torque is an important signal for monitoring and/or control of a diesel engine; however, it is not practical to measure it, due to it is not easily measured and need expensive sensors. A model of average value of a diesel engine is used in the simulation to test the estimator of the indicated torque, these results are presented. This estimator may be useful in the implementation of control strategies or diagnostic where the indicated torque measurements are required.

Keywords: Diesel engines, indicated torque, neuro-fuzzy networks.

1 Introduction

The indicated torque in a diesel engine is very important for monitoring and/or control of combustion in a vehicle in order to achieve the designed performance, lead to lower fuel consumption and have a better comfort. The indicated torque could be measured indirectly by measuring the pressure in the cylinders using high performance pressure sensors in each cylinder. However, high performance pressure sensors have high cost, beside that can be an uncertainty in error when measure indirectly. An alternative option is therefore to estimate indicated torque using low cost sensors based on a model. It has been developed and tested new models for estimating indicated torque [3, 4 and 12].

On the other hand, today diesel engines are typically equipped with control and post-treatment devices [6, 10 and 13] in order to comply with the environmental requirements currently imposed. These entire devices make the indicated torque affected by a large number of variables, it is impossible to consider the interaction of all these variables when building a model, especially when wanted to keep it simple. A viable option would be to develop an observer to estimate the indicated torque using available signals that are easy to measure. In this work we present the design of

---

* Corresponding author.

C. Guo, Z.-G. Hou, and Z. Zeng (Eds.): ISNN 2013, Part II, LNCS 7952, pp. 489–499, 2013.

the indicated torque estimator based in a recurrent neuro-fuzzy networks and evaluate it using computer simulations. First, we use a mean value model [6, 10 and 13] to describe the dynamics of the system in crank angle domain, which is used later to evaluate the indicated torque estimator. Then we presented a technique based on neuro-fuzzy networks for diesel engine indicated torque estimation. The network is a recurrent neural network, obtained by adding a feedback in the input network. Simulations results are presented in order to compare estimator whit the real behavior.

2   Model of Production of Torque and Crankshaft Dynamics

In this part, we describe the dynamics of the combustion torque, also referred as the indicated torque. A nonlinear dynamic crankshaft rotational equation can be derived from the Lagrangian or Newtonian equations.

$$M_{comb} - M_{mass} - M^*{}_{load} = 0 \qquad (1)$$

where combustion torque M_{comb} is the indicated torque in a diesel engine, $M^*{}_{load} = M_{load} + M_F$ refers to extended load torque (load torque more friction torque) and load torque and friction torque can be determined if we know the crankshaft speed [7]. The mass torque M_{mass} is derived from the kinetic energy E_{mass} of the masses moving in the engine as described in Figure 1.

$$E_{mass} = \int_0^{2\pi} M_{mass} d\alpha = \frac{1}{2} J(\alpha)\,\dot{\alpha}^2 \qquad (2)$$

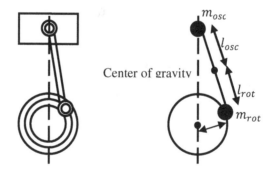

Fig. 1. Mass Model (source: Chauvin, 2004)

The mass torque M_{mass} can be expressed as

$$\frac{dE_{mass}}{dt} = M_{mass}\dot{\alpha} = \left(J\ddot{\alpha} + \frac{1}{2}\frac{dJ}{d\alpha}\dot{\alpha}^2\right)\dot{\alpha} \qquad (3)$$

Solving the equation (3) for M_{mass}, (1) can be expressed as

$$J(\alpha)\ddot{\alpha} = M_{comb}(\alpha) - M^*{}_{load}(\alpha) - f(\alpha)\dot{\alpha}^2 \qquad (4)$$

This depends of the crankshaft velocity α, over a engine cycle. We built our simulation model based on empirical data available in the literature [7,11,14]. The indicated torque response being observed, the inputs to the simulation are the fuel command and the load torque, the engine speed response is used as one of the inputs to the network, due to this variable is observable as shown in (4).

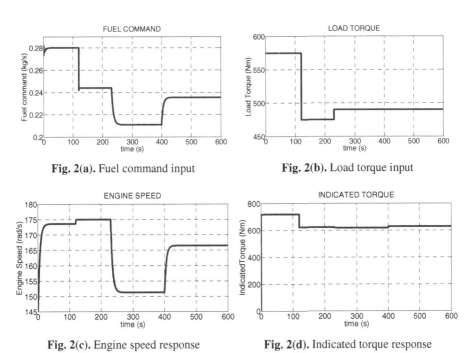

Fig. 2(a). Fuel command input   **Fig. 2(b).** Load torque input

Fig. 2(c). Engine speed response   **Fig. 2(d).** Indicated torque response

3   Indicated Torque Estimator Design

In this section, a neuro fuzzy network is used as an observer which calculates the response of the indicated torque, using only two signals, the engine craft speed and the mass flow of fuel, which are available to measure. The network is of type perception with three layers, one input layer, one hidden layer and one output layer. It has been used the feed-forward back-propagation algorithm to train the networks using the data obtain in the simulation. This algorithm uses a sequential method based on the gradient, to adjust the weights of the neurons with the purpose of minimize the error and approximate the real response. There are n_r fuzzy IF-THEN rules for vector mapping linguistic input $X = [x_1, \ldots, x_n] \in R^n$ to a linguistic output variable. Consider a fuzzy logic system given by n_r fuzzy IF-THEN rules

$$R^r: if \ x_1 \ is \ A_1{}^r(x_1) \ and \ \ldots and \ x_{n_i} \ is \ A_{n_i}{}^r\left(x_{n_i}\right)$$
$$then \ y_1 is \ b_1{}^r and \ \ldots and \ y_{n_o} is \ b_{n_o}{}^r \tag{5}$$

where R^r denotes the rth rule, $1 \leq r \leq n_r$, $x = (x_1, ..., x_{n_i})^T \in X \subset \mathbb{R}^{n_i}$ and $y = (y_1, ..., y_{n_o})^T \in X \subset \mathbb{R}^{n_o}$ are the input (n_i inputs) and the output (n_o outputs) of the neuro-fuzzy observer, $b^r{}_j$ is the fuzzy singleton for the jth output in the rth rule, and $A_1{}^r(x_1) ... A_{n_i}{}^r(x_{n_i})$ are fuzzy sets with Gaussian membership functions

$$A_j^r(x_j) = exp\left\{ -\left(\frac{x_j - c_j^r}{\sigma_j^r}\right)^2 \right\} \tag{6}$$

with c_j^r and σ_j^r design parameters. Using the product-inference rule, singleton fuzzifier, center average defuzzifier, and Gaussian membership function given by the n_r fuzzy IF–THEN rules, the kth output is given by

$$y_k = \sum_{r=1}^{n_r} w^r(x) b_k^r, k = 1, ..., n_o \tag{7}$$

$$w^r(x) = \frac{\prod_{j=1}^{n_i} A_j^r(x_j)}{\sum_{j=1}^{n_r} \prod_{i=1}^{n_i} A_i^r(x_i)}, r = 1, ..., n_r \tag{8}$$

where $b^r(x) = [b^1(x), ..., b^r(x)]^T$ is the vector that contain the weight b where $\mu_{B_i} = 1$. Therefore, the equation (7) can be seen as the internal product of two vectors

$$y(x) = \begin{bmatrix} y_1(x) \\ y_2(x) \\ \cdots \\ y_{n_o}(x) \end{bmatrix} = \begin{bmatrix} b_1^1 & b_1^2 & \cdots & b_1^{n_r} \\ b_2^1 & b_2^2 & \cdots & b_2^{n_r} \\ \cdots & \cdots & \cdots & \cdots \\ b_{n_o}^1 & b_{n_o}^2 & \cdots & b_{n_o}^{n_r} \end{bmatrix} \begin{bmatrix} w_1(x) \\ w_2(x) \\ w^{n_r}(x) \end{bmatrix} = BW(x) \tag{9}$$

The equation (9) is known as a universal approximator in the sense that given any real continuous function $f(x)$ in a compact set $X \in \mathbb{R}^n$ and any $\rho > 0$ there exist a system (9) such that

$$\underset{x \in X}{sup} \|y(x) - f(x)\| < \rho \tag{10}$$

where $\|\cdot\|$ denotes Euclidean norm or its induced matrix norm. In light of this result, the function $f(x)$ can be rewritten as

$$f(x) = BW(x) + \tilde{f}(x), \qquad \forall x \in X \subset \mathbb{R}^n \tag{11}$$

where the $\tilde{f}(x)$ approximation error depends on some factor like the numbers of rules, membership functions or design parameters. In the practice, the weights may be unknown. Several methods based the gradient of an error function are available to estimate it.

3.1   Recurrent Neuro-fuzzy Observer

Figure 3 shows the model of identification for diesel engine using the recurrent neuro-fuzzy observer in its process of training, which uses the variables of the input fuel command m_f and the response of engine speed ω as inputs to the observer, the dynamics network consists in add one input that fed back from the output for minimizing the convergence error to zero; the difference between the indicated torque response of the engine T_i and the output of the network $\hat{T}_i$.

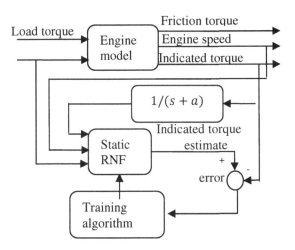

Fig. 3. Training process of recurrent neuro-fuzzy observer

The neuronal network adopted in this section has three inputs, one hidden layer with 32 neurons and one output layer. The transfer functions used at the hidden are Gaussian. The extra input just has two levels, low or high, and then it have 32 fuzzy IF-THEN rules that are given by

$$R^r : if\ m_f\ is\ A_1^r(m_f)\ and\ \omega\ is\ A_2^r(\omega)\ and\ T_i'\ is\ A_{n_i}^r(T_i')\ then\ \hat{T}_i = \hat{\vartheta}_i \tag{12}$$

with T_i' defined by

$$T_i' = \frac{1}{s+a}w^T\vartheta = \frac{1}{s+a}\hat{T}_i\ ; a > 0 \tag{13}$$

where R^r denotes r-th rule, $1 \le r \le 32$, $x = (m_f, \omega, T_i')^T \in X \subset \mathbb{R}^{n_i}$ are the inputs ($n_i = 3$ inputs) and the $y = (\hat{T}_i)^T \in Y \subset \mathbb{R}^{n_o}$ is the output ($n_o = 1$ output), b_j^r is the value for the j-th output in the r-th rule, a is a constant.

$A_1^r(m_f)$, $A_2^r(\omega)$ and $A_{n_i}^r(T_i')$ are fuzzy sets with Gaussian membership functions given by (6), where c_j^r and σ_j^r are design parameters of the levels established for each function and denotes like previous network as the levels established for the first two inputs are the same and each parameter has been established and tuned according to the error so that the difference between T_i and $\hat{T}_i$ is minimized. For each input it defines four levels: low, medium low, medium high and high. Each parameter has been established given the results of engine model, and tuned according to the error so that the difference between T_i and $\hat{T}_i$ is minimized. Table 1 lists the centers and widths for each input variable. The Indicated torque premium T_i' only has two levels, low and high.

Table 1. Design parameters

	Level	center	width		Level	center	width
	Low	168	4		Low	230	11
Engine	Medium low	173.5	1.5	Fuel	Medium low	244	3
speed	Medium high	176.5	1.5	rate	Medium high	253	6
	High	184	6		High	283	24

then from the equation (7) the output $\hat{T}_i$ can be rewritten as following

$$\hat{T}_{ik} = w_k(m_f, \omega) \cdot \vartheta_k \tag{14}$$

and the final output $\hat{T}_i$ can be calculated as the sum of each kth output and it can be rewritten as the internal product of the two vectors as (9)

$$\hat{T}_i = \begin{bmatrix} w_1, \cdots w_{n_r} \end{bmatrix} \begin{bmatrix} \vartheta_1 \\ \vdots \\ \vartheta_{n_r} \end{bmatrix} = w^T \vartheta \tag{15}$$

The final output $\hat{T}_i$ is calculated as (14), and is fed back to the network; consequently, the neuronal model is a dynamic model. Figure 4 shows the neuronal network architecture used in this section. Like the previous network this network is trained so that the difference between T_i and $\hat{T}_i$ is as small as possible, and as static network all antecedent parameters have been established and only the consequence parameters are trained on-line.

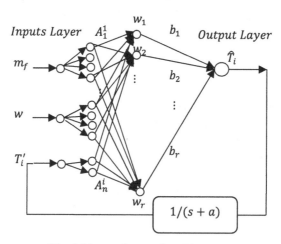

Fig. 4. Neuronal network architecture

The network is trained so that the difference is as small as possible, all antecedent parameters have been established and only the consequence parameters are trained on-line. Several methods based the gradient error of an error function are available to estimated it [15]. During the training, the data of the engine are collected in the laboratory, where the combustion m_f, the craft speed w and the indicated torque T_i are measured. The error is given by

$$e_T = \hat{T}_i - T_i \qquad (16)$$

The singleton vector is updated by

$$\dot{\vartheta} = -\gamma w e_T \qquad (17)$$

where γ is a constant define like design parameter. Figure 5 shows the compact set of system (16).

Modeling error (15) has been used to train the fuzzy neural networks the equation (14) such that $\hat{T}_i$ can approximate to T_i, according to the theory of function approximation of fuzzy logic [13] and neural network [2] identification process can be represented as

$$\hat{T}_i = f(m_f, \omega) = w(m_f, \omega) \cdot \vartheta + \varepsilon \qquad (18)$$

where ε is the approximation error, and ϑ is the vector that contains the weights b that can minimize ε.

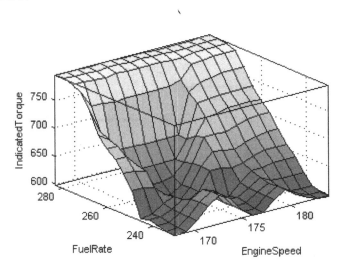

Fig. 5. Compact set of system

Theorem 1. *If it has take that the neuronal fuzzy networks the equation (15), the next descent algorithm of gradient with a proportion of learning of variant time can be the identification of error is bounded.*

$$\vartheta_{(k+1)} = \vartheta_k - \eta_k e_k W^T \tag{19}$$

where $\eta_k = \frac{\eta}{1+\|w[X(k)]\|^2}$, $0 < \eta \le 1$. *The standard identification error satisfies the following average performance*

$$e_N^2(k) = \frac{e^2(k)}{1 + \underset{k}{\text{máx}}(\|w[X(k)]\|^2)}$$

$$J = \lim_{T \to \infty} \sup \frac{1}{T} \sum_{k=1}^{T} e_N^2(k) \le \bar{\mu}$$

where $\bar{\mu} = \max_k [\mu^2(k)]$.

We have used the inputs m_f and w, data collected of the performance engine, and we use the fuzzy set as Table 2 as well as the number of fuzzy set for each input variable. In this simulation we let $t = 600s$, the same time as engine model, so the observer can be training on-line with the data engine. We use the equation (18) to update the weights for online identification. From Theorem 1 we have a necessary condition for a stable learning of η, with $\eta \le 1$. In this case, we find that learning is stable with $\eta = 1$ and if $\eta \ge 1.65$ the learning process becomes unstable. The result of identification is show in Figure 6(a) and quadratic mean error is expressed in Figure 6(b).

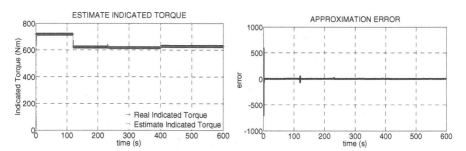

Fig. 6(a). Approximation of Indicated Torque of engine using recurrent network

Fig. 6(b). Mean square error

After training there are 16 fuzzy rules that are validated by Indicated Torque response. Table 2 shows the rules used in this section.

Table 2. Rules for recurrent network

Engine speed	Fuel rate	Indicated torque prima	Indicated torque
Low	Low	Low	Low
Low	Low	High	Medium high
Low	Medium low	Low	Medium low
Low	Medium low	High	Medium high
Low	Medium high	Low	Medium low
Low	Medium high	High	Medium high
Low	High	Low	Medium high
Low	High	High	Medium high
Medium low	Low	Low	Low
Medium low	Low	High	Medium low
Medium low	Medium low	Low	Low
Medium low	Medium low	High	Medium low
Medium low	Medium high	Low	Medium low
Medium low	Medium high	High	Medium high
Medium low	High	Low	Medium high
Medium low	High	High	High
Medium high	Low	Low	Medium high
Medium high	Low	High	High
Medium high	Medium low	Low	Low
Medium high	Medium low	High	Medium low
Medium high	Medium high	Low	Medium high
Medium high	Medium high	High	High
Medium high	High	Low	Medium high
Medium high	High	High	High
High	Low	Low	Medium low
High	Low	High	Medium high
High	Medium low	Low	Medium high
High	Medium low	High	High
High	Medium high	Low	Medium high
High	Medium high	High	High
High	High	Low	Medium high
High	High	High	High

4   Neuronal Networks Validation

Figures 6(a) show the response of observer output, respectively, these responses are the results of training process for the observers; then, we had use the final data training for validation, in order to show the capacity of the observers to predict the indicated torque. Figure 7 shows the output recurrent neuro-fuzzy observer using the data training.

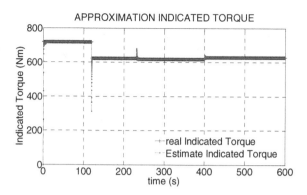

Fig. 7. Recurrent neuro-fuzzy observer (red) using the data training

The results are in good agreement with the output of the indicated torque in the model, calculated in the first section. The static network has a behavior similar to the simulation in the model, however would require more neurons to achieve performance closer to the exact values, when these are required. This required a more complex structure and faster computational algorithm. The recurrent network gets faster to the behavior of the output variable of the indicated torque, this has more neurons and consequently their algorithm is larger. The simulations results are compared with8 results generate by experiments [14], and then the neuronal network used in this work can help like a tool of identification.

5   Conclusion

First, the model used for simulation of the interest variable can be use like a mean value model for a turbocharged diesel engine, the simulation results are in good agreement with the experimental data. The model can be expanded to cover other variables when this are need, or the state variables can be change in the sense of expand the no linear of the model, however, increasing the number of variables can lead to uncertainty in the model when not have the data of the constants. Finally, we suggested a neural fuzzy network for identification that are capable of predicting the behavior of indicated torque that can be applied to the engine's control and that can be easily integrated in a control then. The weights and design parameter can be replaced in the neural network for a particular engine.

Acknowledgements. This work was supported in part by CONACyT under grant 129800 and by PAPIIT 116412.

References

1. Dobner, D.J.: Int. J. of Vehicle Design, Technological Advances in Vehicle Design Series, SP4, Application of Control Theory in the Automotive Industry, pp. 54–74 (1982)
2. Haykin, S.: Neural Networks, A Comprehensive Foundation, 2nd edn. Prentice Hall U.S. (1999)
3. Heywood, J.B.: International Combustion Engine Fundamentals. McGraw-Hill (1988)

4. Horlock, J.H., Winterbone, D.E.: The Thermodynamics and Gas Dynamics of Internal-Combustion Engines, vol. II. Clarendon Press (1986)
5. Ledger, J.D., Benson, R.S., Whitehouse, N.D.: Dynamic modelling of a turbocharged diesel engine. SAE Transactions, paper number 710177, pp. 1-12 (1971)
6. Amman, M., Fekete, N.P., Guzzella, N., Glattfelder, A.H.: Model based control of the VGT and EGR in a turbocharged common rail diesel engine: Theory and passenger car implementation. Presented at the SAE World Congr. Exhib., Detroit, MI, 2003-01-0357 (2003)
7. Kao, M., Moskwa, J.J.: Turbocharged Diesel Engine Modeling for Nonlinear Engine Control and State Estimation. Transactions of the ASME 117 (March 1995)
8. Moskwa, J.J., Hedrick, J.K.: Modeling and Validation of Automotive Engines for Control Algorithm Development. ASME Journal of Dynamic System, Measurement, and Control (June 1992)
9. Palmer, C.A.: Dynamics Simulation of a Solid Fueled Gas Turbine System. Department of Mechanical Engineering, University of Wisconsin Madison, Ph.D. thesis (1991)
10. Mital, R., Li, J., Huang, S.C., Stroia, B.J., Yu, R.C., Howden, K.C.: Diesel exhaust emissions control for light duty vehicles. Presented at the SAE World Congr. Exhib., Detriot, MI, 2003-01-0041 (March 2003)
11. Omran, R., Younes, R., Champoussin, J.-C.: Optimal Control of a Variable Geometry Turbocharged Diesel Engine Using Neural Networks: Applications on the ETC Test Cycle. IEEE Transactions on Control Systems Technology 17(2) (March 2009)
12. Watson, N., Janota, M.S.: Turbocharging the Internal Combustion Engine. Wiley, New York (1982)
13. Leu, Y.G., Lee, T.T., Wang, W.Y.: Observer-based adaptive fuzzy-neural control for unknown nonlinear dynamical systems. IEEE Trans. Syst., Man, Cybern. B 29, 583–591 (1999)
14. Yanakiev, D., Kanellakopoulos, I.: Engine and Transmission Modeling for Heavy-Duty Vehicles, Institute of Transportation Studies University of California, Berkeley, Path Technical Note 95-6 (August 1995)
15. Tang, Y., Vélez-Díaz, D.: Robust Fuzzy Control of Mechanical Systems. IEEE Transactions on Fuzzy Systems 11(3), 411 (2003)

Blind Single Channel Identification Based on Signal Intermittency and Second-Order Statistics

Tiemin Mei

School of Information Science and Engineering,
Shenyang Ligong University, Shenyang 110168 China
`meitiemin@163.com`

Abstract. For intermittent channel output signal, namely, the active periods followed by nonactive periods alternatively, the blind single-input single-output (SISO) system identification problem can be transformed into a blind multichannel identification problem. It is possible and feasible to blindly identify the channel using only second-order statistics from the channel output signal. A two-stage approach is proposed in this paper. At the first stage, two or more segments of channel input signal are estimated from the single channel observation; at the second stage, the channel impulse response is identified by exploiting the estimated channel input signal segments and their corresponding channel output signal segments. Simulations show that the proposed approach works well.

Keywords: Blind channel identification, SISO system, Signal intermittency, Second-order statistics.

1 Introduction

In many cases, we need to identify a channel so as to equalize it to cancel the inter-symbol interference [1], to reshape it to cancel the reverberation effect [2], to estimate the environmental scale [3] [4], or to improve the recognition percentage in human-machine interface applications. Blind channel identification means that the channel parameters are estimated by exploiting only the received signals. For non-minimum phase and finite impulse response (FIR) SISO systems, second-order statistics have been proved to be insufficient for blind channel identification without further assumptions on the input signal, such as cyclostationarity [5]. For multichannel or multirate models, the situation is different, second order statistics are sufficient for the channel identification. Mathematically, multichannel and multirate models are equivalent to each other [1].

For comparison, we review the multichannel model [1] [6] [7]:

$$x_i(n) = \sum_{k=0}^{L-1} h_i(k)s(n-k), \quad (i = 1, 2), \tag{1}$$

C. Guo, Z.-G. Hou, and Z. Zeng (Eds.): ISNN 2013, Part II, LNCS 7952, pp. 500–505, 2013.
© Springer-Verlag Berlin Heidelberg 2013

where $x_i(n)$ and $h_i(n)$ $(i = 1, 2)$ are the multichannel output signals and impulse responses, respectively; $s(n)$ is the single-channel input signal; L is the length of $h_i(n)$. If the input signal is unknown and the channel impulse responses are co-prime to each other, the cross-relation between the outputs of different channels can be exploited to estimate the channel impulse responses by following the fact that,

$$x_1(n) * h_2(n) = x_2(n) * h_1(n), \tag{2}$$

where '*' is the convolution operator. For the estimation of $h_i(n)$ $(i = 1, 2)$, the corresponding LMS algorithm is as follows.

Let $e(n) = \mathbf{x}_1^T(n)\mathbf{h}_2 - \mathbf{x}_2^T(n)\mathbf{h}_1$, where $\mathbf{x}_i(n) = [x_i(n), x_i(n-1), \dots, x_i(n-L+1)]^T$ $(i = 1, 2)$ and $\mathbf{h}_i = [h_i(0), h_i(1), \dots, h_i(L-1)]^T$ $(i = 1, 2)$, the conditioned objective function is defined as

$$J(\mathbf{h}) = E[e^2(n)] = \mathbf{h}^T \mathbf{R} \mathbf{h}, \quad \text{subject to } \|\mathbf{h}\| = 1, \tag{3}$$

where $\mathbf{h} = [\mathbf{h}_2^T, \mathbf{h}_1^T]^T$, $\mathbf{x}(n) = [\mathbf{x}_1^T(n), -\mathbf{x}_2^T(n)]^T$, $\mathbf{R} = E[\mathbf{x}(n)\mathbf{x}^T(n)]$, and $\|\mathbf{h}\|$ is the 2-norm of $\mathbf{h}$.

The corresponding LMS algorithm is

$$\mathbf{h}^{n+1} = \mathbf{h}^n - \mu e(n)\mathbf{x}(n), \quad \text{subject to } \|\mathbf{h}\| = 1. \tag{4}$$

In this paper, we try to tackle the SISO case. If the channel output signal is of the property of intermittency, i.e., the active and nonactive intervals of the channel output signals are alternatively following each other, then this problem turns out to be mathematically a multichannel identification problem.

2   Problem Formulation

Let $x(n)$ be the output of the unknown nonminimum FIR time-invariant system $h(n)$, and let $s(n)$ be the unknown input signal. Then the input-output relation is as follows,

$$x(n) = \sum_{k=0}^{L-1} h(k)s(n-k) + v(n), \tag{5}$$

where $v(n)$ is the received noise, but for simplicity, it will not be considered in the following deduction.

We suppose that the channel input signal $s(n)$ is intermittent in time, and further, we suppose that there are at least two active segments, which are denoted by $s_1(n)$ and $s_2(n)$, the periods of the nonactive segments before and after these two active segments are longer than the duration of the impulse response $h(n)$. The corresponding segments of the channel output signal, which are denoted by $x_1(n)$ and $x_2(n)$, will be complete convolutions of the channel input signal segments and the impulse response of the system,

$$x_i(n) = \sum_{k=0}^{L-1} h(k)s_i(n-k) \quad (i = 1, 2). \tag{6}$$

If we exchange the positions of $h(n)$ and $s_i(n)$ in (6), then we have

$$x_i(n) = \sum_{k=0}^{L_i-1} s_i(k)h(n-k) \quad (i=1,2). \tag{7}$$

Comparing (7) with (1), they are mathematically equivalent if we replace $h_i(n)$ and $s(n)$ in (1) with $s_i(n)$ and $h(n)$, respectively. So the identification of $s_1(n)$ and $s_2(n)$ is equivalently a multichannel identification problem, all approaches proposed for tackling the multichannel problem can be used to estimate the channel input signal segments $s_1(n)$ and $s_2(n)$.

The cross-relation between different segments of the channel output signal is as follows,

$$x_1(n) * s_2(n) = x_2(n) * s_1(n). \tag{8}$$

This relation is the basis for $s_1(n)$ and $s_2(n)$ estimation.

After the estimation of $s_1(n)$ and $s_2(n)$, the channel impulse response $h(n)$ will be easily identified with the observed channel output signal segments $x_1(n)$ and $x_2(n)$ and the estimated channel input signal segments $s_1(n)$ and $s_2(n)$.

3   Algorithm Development

The two-stage algorithm will be formulated in the following two subsections.

3.1   The Identification of Channel Input Signal Segments

We suppose that the lengths of the input signal segments are L_1 and L_2, respectively. Let

$$\mathbf{s}_1 = [s_1(0), s_1(1), ..., s_1(L_1-1)]^{\mathrm{T}},$$

$$\mathbf{s}_2 = [s_2(0), s_2(1), ..., s_2(L_2-1)]^{\mathrm{T}}$$

and $\mathbf{s} = [\mathbf{s}_2^{\mathrm{T}}, \mathbf{s}_1^{\mathrm{T}}]^{\mathrm{T}}$. As shown in Fig. 1, the error signal is

$$e(n) = \mathbf{x}^{\mathrm{T}}(n)\mathbf{s}, \tag{9}$$

where $\mathbf{x}(n) = [x_1(n), x_1(n-1), ..., x_1(n-L_1+1), -x_2(n), -x_2(n-1), ..., -x_2(n-L_2+1)]^{\mathrm{T}}$.

To avoid the trivial solution $\mathbf{s} = 0$, the conditioned least mean squares objective function is defined as

$$J(\mathbf{s}) = \mathrm{E}[e^2(n)] = \mathbf{s}^{\mathrm{T}}\mathbf{R}\mathbf{s}, \text{subject to MAX}\{|\mathbf{s}|\} = 1, \tag{10}$$

where $\mathbf{R} = \mathrm{E}[\mathbf{x}(n)\mathbf{x}^{\mathrm{T}}(n)]$ is the correlation matrix of $\mathbf{x}(n)$; MAX$\{|\mathbf{s}|\}$ represents the maximum absolute component of vector $\mathbf{s}$.

Let $\frac{\mathrm{d}J(\mathbf{s})}{\mathrm{d}\mathbf{s}} = 0$, then we have

$$\mathbf{R}\mathbf{s} = 0, \text{ subject to MAX}\{|\mathbf{s}|\} = 1. \tag{11}$$

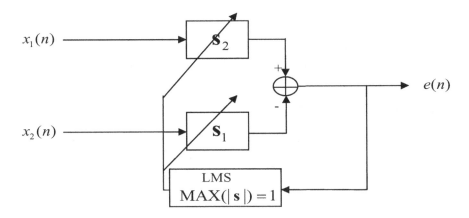

Fig. 1. The estimation of the channel input signal segments $s_1(n)$ and $s_2(n)$ from the observed intermittent channel output signal segments $x_1(n)$ and $x_2(n)$ in SISO system identification

The solution of equation (11) is the least mean square estimation of the channel input signal segments $\mathbf{s}_i$ ($i = 1, 2$) if $s_1(n)$, $s_2(n)$ are coprime to each other in Z domain.

But often, similar to that of (4), we have the online learning algorithm:

$$\mathbf{s}^{n+1} = \mathbf{s}^n - \mu e(n)\mathbf{x}(n), \text{subject to MAX}\{|\mathbf{s}|\} = 1. \tag{12}$$

In practice, the channel output signal segments are used repeatedly until the algorithm converges. In addition, the vector $\mathbf{s}$ is normalized to MAX$\{|\mathbf{s}|\} = 1$ in each iteration.

3.2 The Channel Identification

After the channel input signal segments $s_1(n)$ and $s_2(n)$ being estimated, $h(n)$ can be easily estimated according to the relationship (6). The LMS algorithm is

$$\mathbf{h}_i^{n+1} = \mathbf{h}_i^n + \mu e_i(n)\mathbf{s}_i(n) \quad (i = 1, 2), \tag{13}$$

where $e_i(n) = x_i(n) - \mathbf{s}_i^T(n)\mathbf{h}_i^n$; $\mathbf{h}_i = [h_i(0), h_i(1), ..., h_i(L-1)]^T$; $\mathbf{s}_i(n) = [s_i(n), s_i(n-1), ..., s_i(n-L+1)]^T$. Lastly, the average $\mathbf{h} = (\mathbf{h}_1 + \mathbf{h}_2)/2$ is taken as the single channel estimation.

Alternatively, we can estimate the $h(n)$ directly in frequency domain. When the Fourier transform is applied to both sides of equation (6),we have

$$X_i(e^{j\omega}) = H(e^{j\omega})S_i(e^{j\omega}) \quad (i = 1, 2). \tag{14}$$

This yields

$$h_i(n) = \text{FT}^{-1}\left[\frac{X_i(e^{j\omega})}{S_i(e^{j\omega})}\right] \quad (i = 1, 2), \tag{15}$$

where $\mathrm{FT}^{-1}[.]$ is the inverse Fourier transform operator. In the same way, we take $\mathbf{h} = (\mathbf{h}_1 + \mathbf{h}_2)/2$ as the single channel estimate.

The LMS algorithm (13) gives a better estimation than the direct frequency-domain algorithm (15), especially when observation noise exists.

4   Simulation Results

In the experiment, the two input signal segments are two Gaussian white noise processes of lengths $L_1, L_2 = 1250$. The original channel impulse response $h(n)$ is generated with the image method [8] and of length $L = 1000$. The channel input and output signals are shown in Fig. 2.

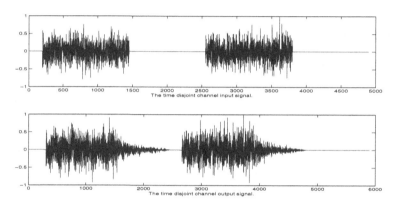

Fig. 2. The time intermittent channel input and output signals. Top: the channel input signal: the two active segments are $s_1(n)$ and $s_2(n)$, respectively; Bottom: the channel output signal: the two active segments are $x_1(n)$ and $x_2(n)$, respectively.

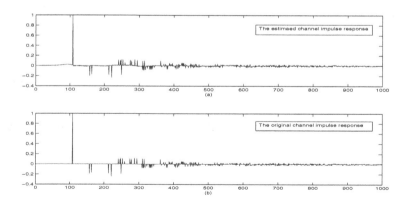

Fig. 3. The normalized original and the estimated channel impulse response, $\mathrm{SNR}_h = 17.4\mathrm{dB}$: (a) the estimated channel impulse response; (b) the original channel impulse response

The estimation result is shown in Fig. 3. The signal-to-noise ratio (SNR) of the estimated channel impulse response is as $\text{SNR}_h = 10 \log \frac{\sum_{n=0}^{L} h^2(n)}{\sum_{n=0}^{L} (\hat{h}(n) - h(n))^2}$, where $\hat{h}(n)$ represents the estimated channel impulse response. In this experiment, $\text{SNR}_h = 17.4$ dB.

The SNR of the estimated input signal is as $\text{SNR}_{s_i} = 10 \log \frac{\sum_{n=0}^{L_i} s_i^2(n)}{\sum_{n=0}^{L_i} (\hat{s}_i(n) - s_i(n))^2}$, where $\hat{s}_i(n)$ represents the estimated channel input signal. In this experiment, $\text{SNR}_{s_1} = 18.3$ dB; $\text{SNR}_{s_2} = 19.9$ dB, respectively.

5 Conclusion

Theoretic analysis and simulation results show that blind identification of SISO systems is possible on the basis of second-order statistics if the observed channel output signal is intermittent in time.

References

1. Xu, G., Liu, H., Tong, L., Kailath, T.: A least-square approach to blind channel identification. IEEE Trans. on Signal Processing 43(12), 2982–2992 (1995)
2. Mei, T., Mertins, A., Kallinger, M.: Room impulse response shortening with infinity-norm optimization. In: Proc. IEEE Int. Conf. Acoust. Speech and Signal Processing, pp. 3745–3748 (2009)
3. Shabtai, N., Rafaely, B., Zigel, Y.: Room Volume Classification from Reverberant Speech. In: Proceedings of International Workshop on Acoustic Echo and Noise Control (IWAENC 2010), Tel Aviv (2010)
4. Filos, J., Habets, E., Naylor, P.A.: A Two-step Approach to Blindly Infer Room Geometries. In: Proceedings of International Workshop on Acoustic Echo and Noise Control (IWAENC 2010), Tel Aviv (2010)
5. Tong, L., Xu, G., Kailath, T.: Blind Identification and Equalization Based on Second-Order Statistics: A Time Domain Approach. IEEE Trans. on Information Theory 40(2), 340–348 (1994)
6. Huang, Y., Benesty, J.: Adaptive Multi-channel Least Mean Square and Newton Algorithms for Blind Channel Identification. Signal Processing 82, 1127–1138 (2002)
7. Huang, Y., Benesty, J.: A Class of Frequency-Domain Adaptive Approaches to Blind Multichannel Identification. IEEE Trans. on Signal Processing 51(1), 11–24 (2003)
8. Allen, J.B., Berkley, D.A.: Image method for efficiently simulating small-room acoustics. Journal of the Acoustical Society of America 65(4), 943–950 (1979)

Constructing Surrogate Model for Optimum Concrete Mixtures Using Neural Network

Lin Wang[1], Bo Yang[1,2,*], and Na Zhang[3]

[1] Shandong Provincial Key Laboratory of Network Based Intelligent Computing,
University of Jinan, Jinan, 250022, China
[2] School of Informatics, Linyi University, Linyi, 276000, China
[3] Information Department, China United Network Communications Co. Ltd. Shandong branch,
Jinan, 250101, China
yangbo@ujn.edu.cn

Abstract. The determination of concrete mix ratio is known as the concrete mix design which involves many theories and practice knowledge and must satisfy some requirements. In order to get high performance concrete, the mix design should be tuned using optimization. However, a lot of concrete experiments are needed to correct models which are very time-consuming and expensive. In this paper, a neural network surrogate model based method is proposed to optimize concrete mix design. This approach focuses on the optimization of compressive strength. Experimental results manifest that the optimum design which achieves high compressive strength can be found by employing the novel approach.

Keywords: Neural Network, Concrete, Surrogate Model, Genetic Algorithm.

1 Introduction

Concrete mix ratio refers to the mass ratio of components in concrete per unit of volume. The determination of concrete mix ratio is known as the concrete mix design. Commonly, concrete mixture ratio design involves many theories and practice knowledge. Concrete mix design must satisfy the following four basic requirements. It should meet the compressive strength requirements of structural design, should meet the requirements of concrete construction and ease of use, should meet the requirements for durability in particular environment, and should save the cost.

The optimization of concrete mix design, which searches for the best concrete mix ratio using optimization method, was proposed by Cannon J P and Krishna Murti G R[1]. They adopted the simplex method as optimizer. The regression and multi objective optimization were also introduced and achieved good results in practice. In recent years, the combination of artificial neural network and nonlinear optimization has become a new pathway for concrete mix design. I-Cheng Yeh [2] presented a novel Computer Aided Design tool for undertaking the design of concrete mixes incorporating Super Plasticizer, fly ash, and slag to attempt to insure that the resulting concrete product will not only be economical and strong enough for the intended purpose, but will have some

---

* Corresponding author.

C. Guo, Z.-G. Hou, and Z. Zeng (Eds.): ISNN 2013, Part II, LNCS 7952, pp. 506–513, 2013.

assurance of adequate workability characteristics as well. However, in order to build a satisfactory computer aided design tool, a lot of real experiments are needed to correct models. These experiments are very time-consuming and expensive.

When an outcome of interest cannot be easily directly measured or calculated, the surrogate model can be used instead. Surrogate model refers to a simple mathematical model built from a small number of samples. It is used to take place of original experiment or simulation. However, its precision is very close to original one. It has become a hot issue in the study of simulation and optimization. The calculation results of surrogate model are very close to the results of experiment or to the results of simulation. Therefore, in optimization problems, this simple surrogate model can be used in place of the complex and time-consuming experiment or simulation. Furthermore, it is updated during the process of optimization to improve accuracy. The surrogate model reduces computational complexity and computational time, and increases the efficiency of design optimization. It has been widely used in agricultural[3], manufacturing[4], airplane[5], investment and emergency[6].

The difficulties faced by time and cost in real experiments and the progress achieved by surrogate model promoted us to explore constructing surrogate model for optimum concrete mixtures using neural network. This work focuses on the optimization of compressive strength.

This paper is arranged as follows. Section 2 describes the details of how to implement constructing surrogate model for optimum concrete mixtures using neural network. Section 3 outlines and discusses our experimental results followed by conclusions in Section 4.

2 Methodology

In concrete mix design, the amount of each component (such as cement, fly ash, slag, water, superplasticizer, coarse aggregate and fine aggregate) produces different effects on the final compressive strength. It is a complex process with multicomponent factors.

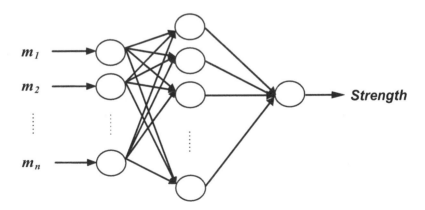

Fig. 1. The feedforward neural network surrogate model. m_i represents the weight of ith component

Conventionally, in order to get the best mix ratio, scientists have to do a great deal of experimentations to find the best solution. It requires huge amounts of money to finish these experiments. Furthermore, it will also take a long time to wait for the slow hydration until predefined ages.

Therefore, the surrogate model is adopted to accelerate the searching process for concrete mix design. Surrogate models are integrated methods which involve experimental design and approximate method. Approximate method is the core. We use surrogate model instead of time-consuming and expensive concrete strength test, and use optimization algorithm to search for the best design in the space of constructed model. In addition, the searched best design in each searching and its measured strength will be fed back to the data and further enhance the model.

Many surrogate models including response surface method[7], Kriging method[8], and neural networks[9,10] etc. have been proposed. Among these techniques, neural network model has been proven to be a practical approach with lots of success stories in several tasks. Feedforward neural networks have ability to approximate any nonlinear functions, it is very suitable for them to play the role of surrogate model instead of concrete experiments. In the input layer, each neuron corresponds to the weight of a kind of component. The number of neurons in input layer is set to the number of components n. In output layer, there is only one neuron, the compressive strength. Therefore, one feedforward neural network is enough to approximate the concrete experiment. Fig. 1 illustrates the adopted feedforward neural network with n input neurons and one output neuron.

In this research, neural network can be viewed as a simple model of mapping relationship which transforms the mix design to compressive strength in place of concrete experiments. It is constructed according to a given data set which collected by experiments and is trained using Levenberg-Marquardt algorithm (LM). Once the model has been built, it can calculates the approximate compressive strength by inputting new mix ratio.

The task to find the best solution is in charge of genetic algorithm which is a procedure inspired in biologic evolution[11]. It was formally introduced by John Holland in the United States in the 1970s and is performed in computers. It has ability to find global best solutions for problems that have been difficult to solve with classical approaches.

In the initial stage, we need only provide a small amount of measured data. When finished a searching, the optimal solution is tested by concrete experiments. Then, the results are added into data set for the next modeling and searching. The flow chart of searching algorithm is shown in Fig. 2.

The problem of concrete mixture optimization should subject to several constraints related to absolute volume, component contents, and component ratios. Absolute volume method assumed that the volume of mixture of concrete is equal to the sum of absolute volume of its components. That is to say, the sum of volume of components is equals to 1000 litre.

$$\sum_{i=1}^{n} \frac{m_i}{d_i} = 1000 \qquad (1)$$

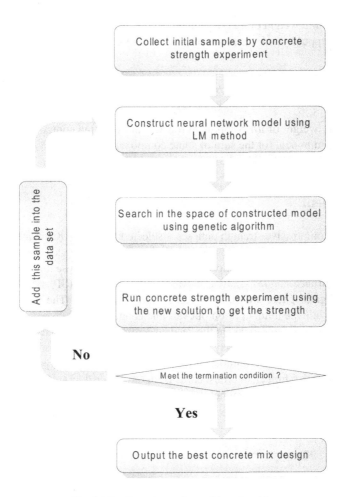

Fig. 2. The flow chart of searching algorithm

where m_i is the unit weight of ith component, d_i is specific gravity of this component, n is the number of components. However, in practice, it is hard to achieve this objective. We can make the sum approximate 1000 by setting lower bound V^{lower} and upper bound V^{upper} around 1000.

$$V^{lower} \leq \sum_{i=1}^{n} \frac{m_i}{d_i} \leq V^{upper} \tag{2}$$

The weight of each component must be kept within bounds.

$$M_i^{lower} \leq m_i \leq M_i^{upper} \tag{3}$$

where m_i is the unit weight of ith component, M_i^{lower} is lower bound of this component, and M_i^{upper} is upper bound. Furthermore, the ratio of components should also be kept within bounds.

$$R^{lower} \leq \frac{m_i}{m_j} \leq R^{upper} \qquad (4)$$

where m_i is the weight of ith component (or the sum of some components), m_j is the weight of jth component (or the sum of some components). R^{lower} and R^{upper} represent the lower bound and upper bound of this ratio, respectively.

3   Experiment

This concrete compressive strength data set is selected from the UCI machine learning repository, which are available from the web site: http://archive.ics.uci.edu/ml/. This data set is a highly nonlinear function of age and ingredients. These ingredients include cement, blast furnace slag (Slag), fly ash, water, superplasticizer (SP), coarse aggregate, and fine aggregate. There are 1030 instances and 9 attributes. The number of input variables is 8 while the number of output variable is one.

However, since we only focus on the 28 days strength which is the standard, the attribute age is deleted and the 28 days samples are extracted. Therefore, the number of input variables is 7 while the number of instances is 425. A three-layered feedforward neural network with 5 hidden neurons learns from the whole data set to replace strength experiment. Then, another neural network, which is the same as the above mentioned network, is used as surrogate model. To better illustrate the performance of proposed

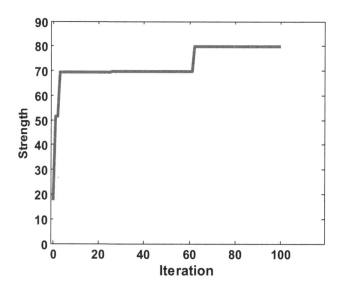

Fig. 3. The evolution process of compressive strength of the best mix ratio design for concrete

method, only 30 samples whose strength are lowest in the whole data set are extracted as the initial set. The highest compressive strength in the subset is 17.34MP. The maximum iteration is set to 100. In each iteration, genetic algorithm is used for searching. The population size is set to 20. Maximum generation is set to 100. The constraints of concrete design are shown in Table 1.

Table 1. The concrete design constraints

	Lower Bound	Upper Bound
Cement (*kg*)	102	540
Slag (*kg*)	0	360
Fly Ash (*kg*)	0	200
Water (*kg*)	122	247
SP (*kg*)	0	32
Coarse Aggregate (*kg*)	801	1145
Fine Aggregate (*kg*)	594	993
Volume (*L*)	949	1046
Water / Cement	0.27	1.88
Water / Binder	0.24	0.9
Water / Solid	0.05	0.13

Fig. 3 shows the evolution process of compressive strength of the best mix ratio design for concrete. It can be observed in the figure that we can get good results by a few times of strength experiments. The best concrete mix design is achieved after 62 times of compressive strength experiments. Taking the initial strength experiments into account, only 92 times of compressive strength experiments are needed. The proposed method greatly reduces cost and what is more important, the time. The surrogate model in design optimization can be considered as a guide to the the best location. Actually, it is the predictor which is enhanced during iteration. Table 2 shows the best design of concrete mix ratio after searching.

Table 2. The best design of concrete mix ratio

Cement	Slag	Fly Ash	Water	Sp	Coarse Aggregate	Fine Aggregate
467	50	59	235	30	873	654

The evolution of landscape are shown in Fig. 4. The weight of first five components are fixed to $(475, 118, 0, 181, 9)$ while coarse aggregate and fine aggregate increases from 500 to 1000. It can be observed that the landscape generated at each iteration is gradually closed to the landscape generated from all of samples.

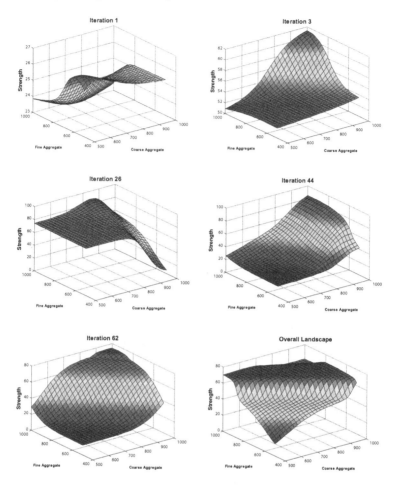

Fig. 4. The evolution of landscape on coarse aggregate and fine aggregate. The weight of first five components are fixed to $(475, 118, 0, 181, 9)$.

4   Conclusion

In this paper, we proposed a surrogate model based method to optimize concrete mix design using neural network and focused on the optimization of compressive strength. In the initial stage, a small amount of measured data are feeded to the method. When finished a circle of searching, the optimal solution is tested by concrete experiments. Then, the results are added into data set for the next modeling and searching. The proposed method reduces computational complexity and computational time, and increases the efficiency of design optimization.

To evaluate the performance of the proposed approach, the concrete compressive strength data set is selected from the UCI machine learning repository. Experimental results manifest that the better design which achieves high compressive strength can be found by employing the novel approach.

Acknowledgement. This work was supported by National Key Technology Research and Development Program of the Ministry of Science and Technology under Grant 2012BAF12B07-3. National Natural Science Foundation of China under Grant No. 61173078, No. 61203105, No. 61173079, No. 61070130, No. 60903176. Provincial Natural Science Foundation for Outstanding Young Scholars of Shandong under Grant No. JQ200820. Shandong Provincial Natural Science Foundation, China, under Grant No. ZR2010FM047, No. ZR2012FQ016, No. ZR2012FM010. Program for New Century Excellent Talents in University under Grant No. NCET-10-0863.

References

1. Cannon, J.P., Krishna Murti, G.R.: Concrete optimized mix proportioning (COMPO). Cement and Concrete Research 1, 353–366 (1971)
2. Yeh, I.-C.: Computer-aided design for optimum concrete mixtures. Cement & Concrete Composites 29, 193–202 (2007)
3. Starbird, S.A.: A Metamodel Specification for a Tomato Processing Plan. European Journal of Operational Research 41, 229–240 (1990)
4. Durieux, S., Pierreval, H.: Regression Metamodeling for the Design of auto mated Manufacturing System Composed of Parallel Machines sharing a Mate rial handling Resource. Int. J. Production Economics 89, 1–30 (2004)
5. Jimmy, M., Tai, C., Mavris, D.N., Schrage, D.P.: Applicati on of a Response Surface Method to the Design of Tipjet Driven Stopped Rotor/Wing Concepts. In: 1st AIAA Aircraft Engineering, Technology and Operations Congress, Los Angeles. CA, pp. 19–21 (1995)
6. Kilmer, R.A., Smith, A.E., Shuman, L.J.: An Emergency Department Simulation and a Neural Network Metamodel. Journal of the Society for Health Systems 5, 63–79 (1997)
7. Hosder, S., Watson, L.T., Grossman, B.: Polynomial Response Surface Approximations for the Multidisciplinary Design Optimization of a High Speed Civil Transport. Optimization and Engineering 2, 431–452 (2001)
8. Dubourg, V., Sudret, B., Bourinet, J.-M.: Reliability-based design optimization using kriging surrogates and subset simulation. Structural and Multidisciplinary Optimization 44, 673–690 (2011)
9. Sreekanth, J., Datta, B.: Multi-objective management of saltwater intrusion in coastal aquifers using genetic programming and modular neural network based surrogate models 393, 245–256 (2010)
10. Dun-wei, G., Jie, R., Xiao-Yan, S.: Neural network surrogate models of interactive genetic algorithms with individual' sinterval fitness. Control and Decision 24, 1522–1530 (2009)
11. Holland, J.H.: Adaptation in Natural and Artificial Systems. University of Michigan Press, Ann Arbor (1975)

Flickr Group Recommendation Based on User-Generated Tags and Social Relations via Topic Model

Nan Zheng and Hongyun Bao

The State Key Laboratory of Management and Control for Complex Systems,
Institute of Automation, Chinese Academy of Sciences, Beijing 100190, China
{nan.zheng,hongyun.bao}@ia.ac.cn

Abstract. The boom of Flickr, a photo-sharing social tagging system, leads to a dramatical increasing of online social interactions. For example, it offers millions of groups for users to join in order to share photos and keep relations. However, the rapidly increasing amount of groups hampers users' participation, thus it is necessary to suggest groups according to users' preferences. By analyzing user-generated tags, one can explore users' potential interests, and discover the latent topics of the corresponding groups. Furthermore, users' behaviors are affected by their friends. Based on these intuitions, we propose a topic-based group recommendation model to predict users' potential interests and conduct group recommendations based on tags and social relations. The proposed model provides a way to fuse tag information and social network structure to predict users' future interests accurately. The experimental results on a real-world dataset demonstrate the effectiveness of the proposed model.

Keywords: Group recommendation, topic model, social tagging.

1   Introduction

With the dramatic development of Web 2.0, social media become more and more important as a way for users to access to valuable information, express individual opinions, share one's experiences as well as keep in touch with others. Flickr as a typical representative of social media websites is an online photo-sharing tagging system with over 6 billion images[1]. On Flickr, social actions consist of tagging photos, building contacts and joining interested groups, etc. Groups are self-organized community formation based on common thematic interests to collect similar photos and users of common interests. Joining groups facilitates flexibility in managing self photos, making them more accessible to the public and seeking photos and users of similar interests. However, huge amounts of groups hamper users' participation, it is necessary to recommend appropriate groups for users to help their social interactions.

The user-generated tags, one of the main factors in Flickr, provide large scale meaningful data for photo retrieval and management. Meanwhile, a user's list of tags

---

[1] http://news.softpedia.com/news/Flickr-Boasts-6-Billion-Photo-Uploads-215380.shtml

C. Guo, Z.-G. Hou, and Z. Zeng (Eds.): ISNN 2013, Part II, LNCS 7952, pp. 514–523, 2013.

can be considered as a description of the interests he/she holds [1]. As a group consists of photos associated with tags, the aggregation of tags for photos in a group pool provides descriptions to the group. Thus, tags are good indications of both users' preferences and the hidden topics of the corresponding groups.

Flickr allows users to designate others as "contacts", and delivers the latest photos that their contacts upload instantly. By adding contacts, users can track others whose photos are of interest to them, and also keep relation with their family and friends on Flickr. In recent years, some research has tended to focus on recommendation based on social relations, and their work demonstrated the effectiveness of friendship relations in representing users' interests [2, 3]. In this paper, we aim to predict a user's interests by analyzing both tag and contact information.

In particular, we simultaneously model the interests of users and the hidden topics of groups by considering tag usage patterns through a generative model. Based on the model, semantic associations between users and groups can be captured through the same hidden topics. Then groups can be recommended to users according to the semantic associations. On the other hand, the preferences of users' contacts can also be calculated according to the hidden topics, and then group recommendations can be conducted according to contact preferences. A hybrid algorithm is then proposed to provide final group recommendations. To evaluate the effectiveness of the proposed method, we perform experiments on a real-world dataset crawled from Flickr. Experimental results verify the usefulness of the proposed model which indicate the importance of both tag and contact information when predicting users' preferences.

2   Related Work

With the number of social tagging grows, personalized recommendation has gradually become a critical technique for users to efficiently find resources in folksonomies. For example, traditional user-based and item-based collaborative filterings (CF) have been applied in CiteULike for recommending scientific articles to users, and the results showed that user-based CF performed better than item-based CF in CiteULike [4]. To estimate the latent topics across the communities through the number of user's posts, Kang and Yu [5] applied Latent Dirichlet Allocation (LDA) for community recommendations, where each user was treated as a document, and each community the user joins was treated as a word in the document. Their research inspires us to emphasize topic models in group recommendations; however, they ignored the usage of tags, which is one of the most important factors in folksonomies. More recently, Krestel and Fankhauser [6] investigated language models and LDA for personalized tag recommendation and demonstrated their method outperformed state-of-the-art tag recommendation algorithms. J schke et al. proposed *FolkRank*, a modification of PageRank, to deal with undirected triadic hyperedges in social tagging systems instead of directed binary edges [7]. As the users, tags and resources can naturally be represented as a three-mode tensor, tensor decompositions have been applied to solve recommendation problems in social tagging systems. For example, Symeonidis et al. presented a unified model based on Higher-Order Singular Value Decomposition (HOSVD), which could recommend resources, tags and users simultaneously [8].

In our previous work [9], we proposed a Non-negative CANDECOMP/ PARAFAC (NNCP) decomposition-based recommendation model to discover latent associations among the three entities in social tagging systems and demonstrated its effectiveness in recommending Flickr groups. In a recent work, a generalized latent factor model was built to model user behaviors in social tagging system and showed the user-comment-item and user-tag-item relations could be mutually inferred based on common latent factors and thus improved prediction performance [10]. This line of research indicates the importance and necessity to integrate tag information when conducting personalized recommendations in social tagging systems.

As related people in a social network influence each other to become more similar, recent research have paid great attentions to recommendations by considering social relations. For example, Ma et al. [11] studied the relationship between the trust network and the user-item matrix and proposed a probabilistic factor analysis framework that integrated the user's characteristics and the user's trusted friends' recommendations. Jamali et al. [12] proposed SocialMF model by incorporating trust propagation into a matrix factorization for movie recommendations. Guy et al. proposed a probability method for recommending social media items, and showed a combination of directly used tags and tags applied by others was effective in representing the user's topics of interest [13]. Konstas, et al. adopted Random Walk with Restarts to represent social networks by considering both the social annotations and friendships [14]. They evaluated their approach using data from Last.fm and gained better performance in item recommendations. Yang et al. established a joint friendship-interest propagation model and showed that the interest and friendship information were highly relevant and mutually helpful [3]. These work indicate that adding friendship can discover more reliable semantic associations. Suggesting groups to users was recently studied in [15] by exploring image visual contents and users' group participation activity through a joint topic model. However, the approach neglects the friendship relation in social media, which is a focus in our work. Overall, in this paper, we provide Flickr group recommendations by considering both tag and contact information via a topic model.

3   Proposed Approach

In this section, we illustrate how to fuse tag and contact information in our model to suggest groups for Flickr users. In particular, we propose a computational approach by following three steps. First, we extract hidden topics of users and groups simultaneously by tag usage patterns via a probabilistic generative model. Second, based on the same hidden topics, groups can be suggested according to user's own preferences. Third, by considering users' contacts, groups can be recommended according to contacts' topic distributions. Finally, a linear fusion function is applied to combine user's own preferences and contacts' preferences to conduct group recommendations.

3.1   Hidden Topic Extraction

On Flickr, users participate in groups according to their own interests, and the aggregation of users provides descriptions to the group. Users also annotate photos with tags and share photos in groups, and the aggregation of tags for photos in a group pool provides descriptions to the group. Formally, each group g contains a Ug-vector of users $\mathbf{u}_g=\{u_{1g}, ..., u_{ig}, ...u_{Ugg}\}$, where each u_{ig} takes the value of 1 indicating user i has joint group g, and a Ng-vector of tags $\mathbf{t}_g=\{t_{1g}, ..., t_{ig}, ...,t_{Ngg}\}$, where t_{ig} is the times of occurrence of tag i in group g. All the unique tags in the collection constitute a vocabulary of size V, and all the unique users in the collection constitute a set of users of size U. A collection of G groups is defined by $\mathbf{G}=\{(\mathbf{t}_1,\mathbf{u}_1), ..., (\mathbf{t}_G,\mathbf{u}_G)\}$. Given this collection of $\mathbf{G}$, we aim to extract the hidden topics automatically to model both the themes of groups and the interests of users.

The author-topic (AT) model [16], a generative model for documents that extends LDA to include authorship information, has been proved to be effective in simultaneously modeling the content of documents and the interests of authors. Here, we extract the hidden topics in group participation activity for Flickr users based on AT model. Each user is associated with a mixture of different topics, and each topic is represented by a probabilistic distribution over tags. A group with multiple users is modeled as a distribution over topics each of which is a mixture of the distribution associated with the users. Let $z_1, ..., z_T$ be T topics to be extracted, the generative process for a corpus $\mathbf{G}$ is given below:

For each user $u=1, ..., U$:
    Draw a multinomial distribution over topics, θ_u, from Dirichlet prior α;
For each topic $z=1, ..., T$:
    Draw a multinomial distribution over tags, ϕ_z, from Dirichlet prior β;
For each group $g=1, ..., G$:
    For each tag token $i=1, ..., Ng$ in group g:
        Sample a user u_i uniformly from group g's members $\mathbf{u}_g$;
        Sample a topic z_i from multinomial distribution θ_u conditioned on u_i;
        Sample a tag t_i from multinomial topic distribution ϕ_z conditioned on z_i.

The user-topic distribution θ and the topic-tag distribution ϕ are two main variables of interest. They are estimated using the Gibbs sampling method [17], a Markov chain Monte Carlo algorithm to sample from the posterior distribution over parameters. The probability of assigning the current tag token t_i to each topic z_i with group member u_i, conditioned on all other variables, is calculated as:

$$P(z_i = j, u_i = k \mid t_i = m, \mathbf{z}_{-i}, \mathbf{u}_{-i}) \propto \frac{C_{mj}^{VT} + \beta}{\sum\limits_{m'=1}^{V} C_{m'j}^{VT} + V\beta} \frac{C_{kj}^{UT} + \alpha}{\sum\limits_{j'=1}^{T} C_{kj'}^{UT} + T\alpha} \tag{1}$$

where $z_i = j$ and $u_i = k$ represent the assignments of t_i to topic j and user k respectively, $t_i = m$ represents the observation that t_i is the mth word in the lexicon, and $\mathbf{z}_{-i}$, $\mathbf{u}_{-i}$ represent all topic and user assignments not including t_i. C^{VT} is the tag-topic matrix of counts with dimensions $V \times T$, and C_{mj}^{VT} is the number of times tag m being assigned to topic j, excluding the current instance. Similarly, C^{UT} is the user-topic matrix of counts with dimensions $U \times T$, where C_{kj}^{UT} is the number of times user k is assigned to topic j, excluding the current instance. From the count matrices C^{VT} and C^{UT}, we can easily estimate the topic-tag distributions ϕ and user-topic distributions θ by:

$$P(t_i = m \mid z_i = j) = \phi_{mj} = \frac{C^{VT}_{mj} + \beta}{\sum\limits_{m'=1}^{V} C^{VT}_{m'j} + V\beta} \tag{2}$$

$$P(z_i = j \mid u_i = k) = \theta_{kj} = \frac{C^{UT}_{kj} + \alpha}{\sum\limits_{j'=1}^{T} C^{UT}_{kj'} + T\alpha} \tag{3}$$

where ϕ_{mj} is the probability of using tag m in topic j, and θ_{kj} is the probability of using topic j by user k. A group with multiple users is modeled as a distribution over topics each of which is a mixture of the distribution associated with the corresponding users. The probability of using topic j by group g is given by:

$$P(z_i = j \mid g_i = g) = \sum\limits_{k \in \mathbf{u}_g} P(z_i = j \mid u_i = k). \tag{4}$$

3.2   Group Recommendations Based on User's Own Preferences

The hidden topics govern the associations between users and groups through tag usage patterns. For user k, $[P(z_i=1|u_i=k), \ldots, P(z_i=T|u_i=k)]$ provides an additive linear combination of factors which indicates the latent preferences of user k. The higher weight user k is assigned to a factor, the more interests user k has in the relevant topic. For group g, $[P(z_i=1|g_i=g), \ldots, P(z_i=T|g_i=g)]$ provides an additive linear combination of factors which indicates the topics of group g. The higher weight a group is assigned to a factor, the more related the group is with the relevant topic. Consequently, groups can be recommended according to the captured associations as follows:

$$P_u(g_i = g \mid u_i = k) = \sum\limits_{j=1}^{T} P(g_i = g \mid z_i = j)P(z_i = j \mid u_i = k)$$

$$= \sum\limits_{j=1}^{T} \frac{P(z_i = j \mid g_i = g)P(g_i = g)}{P(z_i = j)} P(z_i = j \mid u_i = k) \tag{5}$$

$$\propto \sum\limits_{j=1}^{T} \frac{P(z_i = j \mid g_i = g)}{P(z_i = j)} P(z_i = j \mid u_i = k)$$

A high score indicates that user k has a great probability to join group g in the future. Therefore, for user k, groups can be recommended according to the scores.

3.3   Group Recommendations Based on Contacts' Preferences

If a user is interested in others' photo streams, he/she may add others as contacts, and Flickr will deliver the latest photos that their contacts upload instantly. The contact relation is asymmetric. That is, if user u_1 designates u_2 as a contact, u_1 can see the photo stream of u_2, but not vice versa. By adding contacts, users can track others whose photos are of interest to them. To some extent, user k's list of contacts F_k is a good indication of the user's latent preferences. Thus, user k's topic distributions can be calculated by aggregating his/her contacts' distributions over topics:

$$P_f(z_i = j \mid u_i = k) = \frac{1}{|F_k|} \sum_{h \in F_k} P(z_i = j \mid u_i = h) \tag{6}$$

where $P_f(z_i = j \mid u_i = k)$ is the probability of using topic j by user k according to his/her contacts' topic distributions, and $|F_k|$ is the total number of user k's contacts. Groups can be recommended according to the captured associations as follows:

$$P_f(g_i = g \mid u_i = k) = \sum_{j=1}^{T} P(g_i = g \mid z_i = j) P_f(z_i = j \mid u_i = k)$$

$$\propto \sum_{j=1}^{T} \frac{P(z_i = j \mid g_i = g)}{P(z_i = j)} P_f(z_i = j \mid u_i = k) \tag{7}$$

3.4 Combining User's Own and Contacts' Preferences

As $P_u(g_i = g \mid u_i = k)$ and $P_f(g_i = g \mid u_i = k)$ both constitute probability distributions, we can combine these two by straightforward linear interpolation:

$$P(g_i = g \mid u_i = k) = \lambda \cdot P_u(g_i = g \mid u_i = k) + (1 - \lambda) \cdot P_f(g_i = g \mid u_i = k) \tag{8}$$

where parameter λ is introduced to adjust the significance of the two parts, and the impact of λ will be examined experimentally in section 4.

4   Experiments

4.1   Dataset

We use Flickr API to gather our dataset with an arbitrary user as the seed then crawling by contacts. For each user, we collect all his/her public photos with photos' tags and groups as well as his/her contact information before Sep 28th, 2012. After removing the users who have annotated with less than 10 tags and joint less than 5 groups, we end up with 193 users, 16,499 tags, 663 groups and 159,341 {user, tag, group} triples. The total number of contact relation among 193 users is 3465.

In our experiments, the hyper-parameters α and β are set at 50/T and 0.01, which have been widely adopted in the literature [17]. The number of topics T is set at 25 on the basis of human judgment of meaningful topic plus measured models with the lower the best. Gibbs sampling is repeated for 50 iterations.

4.2   Evaluation Metric

In this paper, we choose the top-k recommendations metric [18] as the evaluation metric, which measures the position of the correct group in the top-k recommendation list. We randomly select one joined group and k-1 unjoined groups for each user to form the test set and the remaining joined groups form the training set. In the recommendation result, there are k possible ranks for the joined group and the best result is that no unjoined groups appear before the joined one. We have repeated the procedure of training/test partition 20 times with different random seeds. The reported numbers are the mean performance averaged over the 20 runs. In addition, we set k to be 201 in the experiments.

4.3   Results and Discussions

Topic Extraction Results. Before performing group recommendations, we first discuss if our model could reasonably identify the latent user-topic and topic-tag distributions. Table 1 illustrates 3 different topics out of 25, discovered from the 50th iteration of a particular Gibbs sampler run. The tags associated with a topic are quite intuitive and precise in the sense of conveying a semantic context of a specific kind of photos. Also, by analyzing the users' photo streams on Flickr, we find that the users associated with each topic are quite representative. For example, the tags from TOPIC 1 express the topic related to *portrait*, and by examining the photos of users in TOPIC 1, we find the users uploaded a certain amount of photos related to *portrait* and shared many photos annotated with high probability tags in TOPIC 1. As shown, our model can successfully discover the latent topics.

Table 1. Three example topics extracted from the Flickr dataset. Each topic is shown with the top 10 tags and users that have highest probability conditioned on that topic.

TOPIC 1		TOPIC 2		TOPIC 3	
Tags	Prob.	Tags	Prob.	Tags	Prob.
film	0.07999	playa	0.04534	nikon	0.04638
portrait	0.05961	beach	0.04397	d3000	0.03625
pretty	0.05111	mar	0.04305	light	0.02585
girl	0.03298	sea	0.03540	sky	0.02010
female	0.03197	verde	0.02695	bn	0.01855
model	0.02579	horizon	0.01542	camera	0.01838
hair	0.01694	yellow	0.01439	bw	0.01675
cute	0.01403	rocks	0.01428	white	0.01563
skin	0.01290	sky	0.01256	luz	0.01495
fashion	0.01284	fly	0.01222	colours	0.01477
Users	Prob.	Users	Prob.	Users	Prob.
43903625@N02	0.38157	81974018@N00	0.35598	26199251@N05	0.20541
20989422@N03	0.11086	38822352@N07	0.04422	46497476@N06	0.20482
32749946@N04	0.11080	36766265@N03	0.03943	53088693@N08	0.05504
41418433@N06	0.09954	29233834@N00	0.03876	44194025@N06	0.04939
42319899@N06	0.08095	22840406@N05	0.03698	46788954@N06	0.03995
53088693@N08	0.03678	34771988@N03	0.02640	49941824@N02	0.02849
22272810@N06	0.02628	38432971@N07	0.02562	38512579@N08	0.02166
46893049@N07	0.02469	48339866@N05	0.02328	49557553@N06	0.02040
29363336@N03	0.01455	41384867@N08	0.02038	26362046@N05	0.01913
29233834@N00	0.01056	23770762@N06	0.01816	49259990@N04	0.01652

Parameter Settings. Parameter λ in Equation 8 should be decided in the experiments. We range it from 0 to 1 by an interval of 0.1, where λ=1 and λ=0 practically switch off the group recommendation results based on user's own topic distributions and his/her contacts' topic distributions. Table 2 shows the percentage of cumulative quantity of ranks for the test joined groups at the first N position @N by different parameter settings. We can find that model with both user's own and contacts' topic distributions achieves a higher performance than model with one's own (λ=1) or one's contacts (λ=0) preferences alone. Comparing with the two parts, model with personal interests gets a higher performance than model with only friends' interests, which indicates tags can express one's preferences more accurately than social relations in our dataset. The best precision achieves when λ=0.7, so we set λ=0.7 in the following experiments.

Table 2. The variation of top-k recommendation results of different λ

	λ=0	λ=0.1	λ=0.2	λ=0.3	λ=0.4	λ=0.5	λ=0.6	λ=0.7	λ=0.8	λ=0.9	λ=1
@1	0.0375	0.0373	0.0368	0.0398	0.0418	0.0423	0.0470	**0.0557**	0.0522	0.0490	0.0470
@5	0.1025	0.1080	0.1097	0.1139	0.1035	0.1199	0.1097	**0.1363**	0.1281	0.1199	0.1159
@10	0.1736	0.1731	0.1654	0.1587	0.1806	0.1881	0.1990	**0.2100**	0.2005	0.1990	0.1821
@15	0.2199	0.2199	0.2090	0.2005	0.2266	0.2333	0.2450	0.2567	**0.2650**	0.2450	0.2374
@20	0.2529	0.2587	0.2498	0.2413	0.2654	0.2721	0.2851	**0.2980**	0.2851	0.2721	0.2682

Group Recommendation Results. In this sub-section, to demonstrate the usefulness and effectiveness of our proposed model, we compare it with three other models:

1) User-based (UB) CF [4]: This is a memory-based CF that first finds the neighborhood of a target user by similar group-participation patterns. Then all groups that neighbors have participated but have not been joined by the target user are sorted and considered to be possible recommendations. We use the jaccard similarity metric to determine similarity between users. The neighborhood size is set to be 25.
2) Non-negative Matrix Factorization (NMF) model [19]: This is a model-based CF, which predicts unobserved user-group pairs by the non-negative low-rank factors learned from the observed data in the user-group matrix. We set number of topics to be 25 and times of iteration to be 50.
3) Non-negative CANDECOMP/ PARAFAC (NNCP) decomposition model [9]: This model-based CF first represents interactions among users, tags and groups into a three-mode tensor, and then discovers the latent factors via tensor decomposition, and finally recommends relevant groups to users based on the latent factors. This method has been applied to Flickr group recommendations, and shown to achieve the best result. We set number of factors to be 25 and times of iteration to be 50.

Figure 1 shows the comparison results, in which we can clearly see the competitive results of our approach compared with other methods. More specially, UB gets the worst performance, due to the low coverage of memory-based models under sparse data. However, NMF, which models users' preferences using the latent factors, performs much better than UB. The manner can extract more discriminative and informative features and remove useless information for relation mining, therefore better reflect users' preferences than UB. As shown, NNCP achieves better performances than NMF, which indicates that adding tags can improve the quality of recommendations. Our model is more capable to model the problem and gains the best results. Because both semantic tags and social relations are involved in mining associations between users and groups, which may address the problem of sparsity and deal with the *cold-start* problem to some extent.

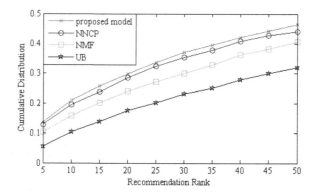

Fig. 1. Top-50 ranks of recommendation performances

5   Conclusions

In this paper, we propose a new approach to predict users' interests to offer better personalized services in Flickr. The proposed model makes use of collaborative information, user-generated tags and social relations in an integrated framework. The experimental results on a real-world dataset show that our approach is more effective compared with the state-of-the-art group recommendation methods. Our future work aims at proposing a unified topic model which combines both collaborative information and social interactions.

Acknowledgments. This research is supported by the Early Career Development Award of SKLMCCS, Youth Innovation Promotion Association of CAS and the NSFC project 61072084.

References

[1] Zheng, N., Li, Q.: A recommender system based on tag and time information for social tagging systems. Expert Systems with Applications 38(4), 4575–4587 (2011)
[2] Ye, M., Liu, X., Lee, W.-C.: Exploring social influence for recommendation - A probabilistic generative model approach. In: International Conference on Research & Development on Information Retrieval (SIGIR 2012), pp. 671–680 (2012)

[3] Yang, S.H., Long, B., Smola, A., Sadagopan, N., Zheng, Z., Zha, H.: Like like alike - Joint friendship and interest propagation in social networks. In: Proceedings of the 20th International Conference on World Wide Web (WWW), pp. 537–546. ACM, Hyderabad (2011)

[4] Bogers, T., van den Bosch, A.: Recommending scientific articles using citeulike. In: Proceedings of the 2008 ACM Conference on Recommender Systems, pp. 287–290 (2008)

[5] Kang, Y., Yu, N.: Soft-constraint based online LDA for community recommendation. In: Qiu, G., Lam, K.M., Kiya, H., Xue, X.-Y., Kuo, C.-C.J., Lew, M.S. (eds.) PCM 2010, Part II. LNCS, vol. 6298, pp. 494–505. Springer, Heidelberg (2010)

[6] Krestel, R., Fankhauser, P.: Personalized topic-based tag recommendation. Neurocomputing 76, 61–70 (2012)

[7] Jäschke, R., Marinho, L., Hotho, A., Schmidt-Thieme, L., Stumme, G.: Tag recommendations in folksonomies. In: Kok, J.N., Koronacki, J., Lopez de Mantaras, R., Matwin, S., Mladenič, D., Skowron, A. (eds.) PKDD 2007. LNCS (LNAI), vol. 4702, pp. 506–514. Springer, Heidelberg (2007)

[8] Symeonidis, P., Nanopoulos, A., Manolopoulos, Y.: A unified framework for providing recommendations in social tagging systems based on ternary semantic analysis. IEEE Transactions on Knowledge and Data Engineering 22(2), 179–192 (2010)

[9] Zheng, N., Li, Q., Liao, S., Zhang, L.: Which photo groups should I choose? A comparative study of recommendation algorithms in Flickr. Journal of Information Science 36(6), 733–750 (2010)

[10] Yin, D., Guo, S., Chidlovskii, B., Davison, B.D., Archambeau, C., Bouchard, G.: Connecting comments and tags: improved modeling of social tagging systems. In: Proceedings of the Sixth ACM International Conference on Web Search and Data Mining, pp. 547–556 (2013)

[11] Ma, H., King, I., Lyu, M.R.: Learning to recommend with social trust ensemble. In: Proceedings of the 32nd International ACM SIGIR Conference on Research and Development in Information Retrieval, pp. 203–210 (2009)

[12] Jamali, M., Ester, M.: A matrix factorization technique with trust propagation for recommendation in social networks. In: Proceedings of the 4th ACM Conference on Recommender Systems, pp. 135–142 (2010)

[13] Guy, I., Zwerdling, N., Ronen, I., Carmel, D., Uziel, E.: Social media recommendation based on people and tags. In: Proceeding of the 33rd International ACM SIGIR Conference on Research and Development in Information Retrieval, pp. 194–201 (2010)

[14] Konstas, I., Stathopoulos, V., Jose, J.: On social networks and collaborative recommendation. In: Proceedings of the 32nd International ACM SIGIR Conference on Research and Development in Information Retrieval, pp. 195–202 (2009)

[15] Wang, J., Zhao, Z., Zhou, J., Wang, H., Cui, B., Qi, G.: Recommending Flickr groups with social topic model. Information Retrieval 15(3-4), 278–295 (2012)

[16] Rosen-Zvi, M., Griffiths, T., Steyvers, M., Smyth, P.: The author-topic model for authors and documents. In: Proceedings of the 20th Conference on Uncertainty in Artificial Intelligence, pp. 487–494. AUAI Press (2004)

[17] Steyvers, M., Griffiths, T.: Probabilistic topic models. Latent Semantic Analysis: A Road to Meaning, Handbook of Latent Semantic Analysis 427(7), 424–440 (2007)

[18] Koren, Y.: Factorization meets the neighborhood: a multifaceted collaborative filtering model. In: Proceeding of the 14th ACM SIGKDD, pp. 426–434 (2008)

[19] Zhang, S., Wang, W., Ford, J., Makedon, F.: Learning from incomplete ratings using non-negative matrix factorization. In: 6th SIAM Conference on Data Mining, pp. 548–552 (2006)

Measure Method of Fuzzy Inclusion Relation in Granular Computing

Wenyong Zhou, Chunhua Liu, and Hongbing Liu

School of Computer and Information Technology,
Xinyang Normal University, 464000 Xinyang China
xynuzwy@126.com

Abstract. For granular computing in discrete space, the inclusion relation between two granules is partially ordered. How to measure the fuzzy inclusion degree is one of the key issues. We proposed a fuzzy inclusion relation between two hyperbox granules using an inclusion measure function based on a linear positive valuation function induced by the longest diagonal a hyperbox granule. The fuzzy algebraic system is formed by the granule set and fuzzy inclusion relation between two granules, and used to guide the design of algorithms in granular computing.

Keywords: Hyperbox granule, fuzzy inclusion relation, positive valuation function, granular computing.

1   Introduction

Granular Computing is a novel intelligent computation theory and method based on partition of the problem apace[1]. Recently, GrC is widely used in pattern recognition, information systems and other areas. GrC fusion of research fields of artificial intelligent, such as rough set, fuzzy set. Representation of granules, relation between two granules, and operation between two granules are mainly researched in GrC. Professor Yao explored the motivation for people to study granular computing from a philosophical point of view, the granular represented as a hierarchy structure[2,3]. In the quotient space model proposed by Zhang Bo, a granule is represented as an equivalence class[4]. T. Y. Lin Professor represented the granule as the point's neighborhood[5]. The relationship between two granule is an important research field in granular computing, and the base of forming granular computing, the different relation can form the different granular computing model. Granular computing model based on equivalence relation is a major granular computing model. Quotient space model and the traditional rough set are the partition of space based on equivalence relation. Cover model based on compatibility relation is another important granule computing model, which partitions the space by compatible relation. Professors Zhang Bo and Zhang Ling researched granularity model according to fuzzy compatible relation [6]. Professor Zadeh proposed organization process, the granulation process and the causal relationship reflects the interchange and interconnection among the granular space [7], moreover, it also presents the capability of the transformation from one granule space to another one without difficulty [6].

C. Guo, Z.-G. Hou, and Z. Zeng (Eds.): ISNN 2013, Part II, LNCS 7952, pp. 524–532, 2013.

In this paper, the measure method of fuzzy inclusion relation between two hyperbox granules fuzzy granules is proposed. Firstly, the granule is represented as the vector including the beginning point and the end one. Secondly, a linear positive valuation function is introduced to form the fuzzy inclusion function. Thirdly, the theorems are proposed by the fuzzy inclusion function.

2   Lattice Algebraic Systems

For classification problems

$$S = \{(\mathbf{x}_i, lab_i) \mid \mathbf{x}_i \in R^N, lab_i \in \{1, 2, ..., N\}, i = 1, 2, ..., \ell\}$$

is the training set composed by ℓ input variables of N dimensional and ℓ output variable of one dimensional. Two points $\mathbf{x} = (x_1, x_2, ..., x_N)$ and $\mathbf{y} = (y_1, y_2, ..., y_N)$ are used to represent granule $G = (\mathbf{x}, \mathbf{y})^T$, which $\mathbf{x} \preceq \mathbf{y}$, the granule G is the hyperbox including the beginning point $\mathbf{x}$ and end of $\mathbf{y}$. Such as the two-dimensional space, $G = (0.1, 0.2, 0.4, 0.6)^T$ denotes the beginning point $(0.1, 0.2)$, the end point $(0.4, 0.6)$, and the hyperbox granules with length 0.4 and width 0.3. In particular, in the one-dimensional space hyperbox granule is in the form of the interval. If the beginning point and the end point are the same one, the hyperbox granule is called the atomic hyperbox granule.

For the hyperbox granules $G_1 = (\mathbf{x}_1, \mathbf{y}_1)$ and $G_2 = (\mathbf{x}_2, \mathbf{y}_2)$, Design join operator $\vee$

$$G_1 \vee G_2 = (\mathbf{x}_1 \wedge \mathbf{x}_2, \mathbf{y}_1 \vee \mathbf{y}_2) \tag{1}$$

Where

$$\mathbf{x}_1 \wedge \mathbf{x}_2 = (x_{11} \wedge x_{21}, x_{12} \wedge x_{22}, ..., x_{1N} \wedge x_{2N})$$
$$\mathbf{y}_1 \vee \mathbf{y}_2 = (y_{11} \vee y_{21}, y_{12} \vee y_{22}, ..., y_{1N} \vee y_{2N})$$

are the operators between the vectors. For join operator, two smaller hyperbox granules are united into one larger hyperbox granule. Join operator and inclusion have the following relationship:

$$G_1 \subseteq G_1 \vee G_2, \quad G_2 \subseteq G_1 \vee G_2$$

Design meet operator $\wedge$

$$G_1 \wedge G_2 = \begin{cases} (\mathbf{x}_1 \vee \mathbf{x}_2, \mathbf{y}_1 \wedge \mathbf{y}_2) & \text{if } \mathbf{x}_1 \vee \mathbf{x}_2 \preceq \mathbf{y}_1 \wedge \mathbf{y}_2 \\ \varnothing & \text{otherwise} \end{cases} \tag{2}$$

The lager granule size of the two hyperbox granules is divided into smaller hyperbox granules. Meet operation and inclusion have the following relationship:

$$G_1 \wedge G_2 \subseteq G_1, G_1 \wedge G_2 \subseteq G_2$$

Definition 1. The inclusion relation between two hyperbox granules $G_1 = (\mathbf{x}_1, \mathbf{y}_1)$ and $G_2 = (\mathbf{x}_2, \mathbf{y}_2)$ is $G_1 \subseteq G_2 \Leftrightarrow \mathbf{x}_2 \preceq \mathbf{x}_1 \ \& \ \mathbf{y}_1 \preceq \mathbf{y}_2$

Theorem 1. Inclusion relation between two hyperbox granules is a partial order relation.

For granule space induced by N dimensional space, hyperbox granules with length $2N$ $G_1 = (\mathbf{x}_1, \mathbf{y}_1)$ and $G_2 = (\mathbf{x}_2, \mathbf{y}_2)$

where $\mathbf{x}_1 = (x_{11}, x_{12}, ..., x_{1N})$, $\mathbf{y}_1 = (y_{11}, y_{12}, ..., y_{1N})$.   If $G_1 \subseteq G_2$,   then $(\mathbf{x}_1, \mathbf{y}_1) \preceq (\mathbf{x}_2, \mathbf{y}_2)$, namely

$$G_1 \subseteq G_2 \Leftrightarrow x_{21} \le x_{11} \ \& \ x_{22} \le x_{12} \ \& ... \& \ x_{2N} \le x_{1N} \\ \& \ y_{11} \le y_{21} \ \& \ y_{12} \le y_{22} \ \& ... \& \ y_{1N} \le y_{2N}$$

(3)

For a length $2N$ vector space, relation between the two vectors is

$$(\mathbf{x}_1, \mathbf{y}_1) \preceq (\mathbf{x}_2, \mathbf{y}_2) \Leftrightarrow x_{11} \le x_{21} \ \& \ x_{12} \le x_{22} \ \& ... \& \ x_{1N} \le x_{2N} \\ \& \ y_{11} \le y_{21} \ \& \ y_{12} \le y_{22} \ \& ... \& \ y_{1N} \le y_{2N}$$

(4)

Relations (3) and (4) are partial order relations. it is very obvious that the two partial order relations is inconsistent. How to eliminate this inconsistency, the isomorphic function $\theta(G)$ is introduced to eliminate the inconsistency between (3) and (4).

Definition 2. For $G = (\mathbf{x}, \mathbf{y})$, $\theta(G) = (\theta(\mathbf{x}), \mathbf{y})$, if $\theta(G_1 \subseteq G_2) \Leftrightarrow (\theta(\mathbf{x}_1), \mathbf{y}_1) \preceq (\theta(\mathbf{x}_2), \mathbf{y}_2)$, then $\theta(G)$ is called the isomorphic function between the granule space and vector space.

Because $\langle R^N, \le \rangle$ and $\langle R^N, \ge \rangle$ are the dual, $\mathbf{x}_1 \succeq \mathbf{x}_2$ such that $\theta(\mathbf{x}_1) \preceq \theta(\mathbf{x}_2)$, in order to ensure $G_1 \subseteq G_2$, $\theta(\cdot)$ is an isomorphic mapping between lattice $\langle R^N, \le \rangle$ and the dual lattice $\langle R^N, \ge \rangle$.

If $\theta(G)$ is a decreasing function in the one-dimensional, then formula (3) and (4) has the consistency of partial order relation under $\theta(G)$. Combining operator (1) and (2), the following relation can be outlined that is based on between the hyperbox granule inclusion relations and hyperbox granules operator.

$$G_1 \subseteq G_2 \Leftrightarrow G_1 = G_1 \wedge G_2, G_2 = G_1 \vee G_2$$

(5)

For classification problem S of N-dimensional space, GS is the hyperbox granule set, the algebraic system $\langle GS, \subseteq, \vee, \wedge \rangle$ can be formed by the inclusion function, join operator, meet operator, and can be expressed as $\langle GS, \subseteq \rangle$. The following theorem is easy to be proved.

Theorem 2. $\langle GS, \subseteq \rangle$ is a lattice.

3   Fuzzy Inclusion Function between Two Hyperbox Granules

For classification problems the training set S, the hyperbox granule is a subset of S, namely the hyperbox granule is a set. The traditional inclusion relation between two sets can not reflect the inclusion relation between two hyperbox granules. In Fig.1, from (a) to (b), the granularity of two hyperbox granules have not changed, because their relative position is different, the inclusion relation is different. The inclusion degree $G_1 \subseteq G_2$ in (b) is greater than that in (a). Therefore, the inclusion relation between two hyperbox granules is fuzzy, and the fuzzy inclusion relation must be discussed.

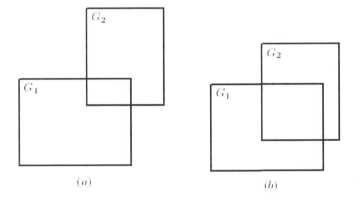

(a) (b)

Fig. 1. The fuzzy inclusion relation between two granules

We can construct the fuzzy inclusion function between hyperbox granules.

$$\sigma : GS \times GS \rightarrow [0,1]$$

The fuzzy inclusion degree function satisfies the following four conditions

(1) $G \in GS, G \neq \varnothing, \sigma(G,\varnothing) = 0$

(2) $\sigma(G,G) = 1$

(3) $G_1 \subseteq G_2 \Rightarrow \sigma(G,G_1) \leq \sigma(G,G_2)$

(4) $G_1 \wedge G_2 \subset G_1 \Rightarrow \sigma(G_1,G_2) < 1$

The fuzzy inclusion degree is a scalar, while the hyperbox granule of the N dimensional space is the vector of length $2N$. In order to construct the fuzzy inclusion function, the mapping between the high dimensional space and one-dimensional space is formed to reflect the size of hyperbox granules, namely granularity. References[8] defines a linear positive valuation function $\upsilon(\mathbf{x})$ to measure the granularity of hyperbox granule

$$\upsilon(\mathbf{x}) = \sum_{i=1}^{N} x_i \qquad (6)$$

Where $\upsilon(\mathbf{x})$ satisfies the nonnegative, equality and inequality properties, namely $\upsilon(\mathbf{x}) \geq 0$ (for $\mathbf{x} \geq \mathbf{0}$), $\upsilon(\mathbf{x} \wedge \mathbf{y}) + \upsilon(\mathbf{x} \vee \mathbf{y}) = \upsilon(\mathbf{x}) + \upsilon(\mathbf{y})$, and $\mathbf{x} \prec \mathbf{y} \Leftrightarrow \upsilon(\mathbf{x}) < \upsilon(\mathbf{y})$.

This paper constructs a nonlinear positive evaluation function as follows

$$\upsilon(\mathbf{x}) = \sum_{i=1}^{N} \sin\left(\frac{\pi x_i}{2}\right) \tag{7}$$

Because the space $[0,1]$ and R are isomorphic, we can map R^N to $[0,1]^N$. In the one-dimensional space, $\langle [0,1], \leq \rangle$ is a lattice, the greatest lower bound is 0, the minimum upper bound of 1. We know by Theorem 1, the algebra system $\langle G([0,1]), \subseteq \rangle$ induced by the algebra system $\langle [0,1], \leq \rangle$ is the lattice.

Theorem 3. Let $\theta(x) = 1 - x$ is the isomorphic mapping from lattice $\langle [0,1], \geq \rangle$ to its dual lattice $\langle [0,1], \leq \rangle$, then the function

$$\upsilon([a,b]) = \upsilon(\theta(a)) + \upsilon(b) \tag{8}$$

is a mapping of the granule set $G([0,1])$ to one-dimensional space R, where $\upsilon(\cdot)$ is the function (7).

Proof:

(1) nonnegative property. One dimensional hyperbox granule $G = [a,b]$, for $0 \leq a \leq b \leq 1$, $0 \leq \theta(a) = 1 - a \leq 1$

$$\upsilon(G) = \upsilon([a,b]) = \upsilon(\theta(a)) + \upsilon(b) = \sin\left(\frac{\pi(1-a)}{2}\right) + \sin\left(\frac{\pi b}{2}\right) \geq 0$$

(2) equality property

In hyperbox granule space $G([0,1])$, $G_1, G_2 \in G([0,1])$ is hyperbox granules, Let $G_1 = [a,b], G_2 = [c,d]$

$$\upsilon(G_1 \wedge G_2) = \upsilon([a,b] \wedge [c,d]) = \upsilon([a \vee c, b \wedge d]) = \upsilon(\theta(a \vee c)) + \upsilon(b \wedge d)$$

By $\theta(x) = 1 - x$ is a linear decreasing function

$$a \leq c \Rightarrow \begin{cases} a \vee c = c \Rightarrow \theta(a \vee c) = \theta(c) \\ \theta(c) \leq \theta(a) \Rightarrow \theta(a) \wedge \theta(c) = \theta(c) \end{cases} \Rightarrow \theta(a \vee c) = \theta(a) \wedge \theta(c)$$

$$a \leq c \Rightarrow \begin{cases} a \wedge c = a \Rightarrow \theta(a \wedge c) = \theta(a) \\ \theta(c) \leq \theta(a) \Rightarrow \theta(a) \vee \theta(c) = \theta(a) \end{cases} \Rightarrow \theta(a \wedge c) = \theta(a) \vee \theta(c)$$

$$c \leq a \Rightarrow \begin{cases} a \vee c = a \Rightarrow \theta(a \vee c) = \theta(a) \\ \theta(a) \leq \theta(c) \Rightarrow \theta(a) \wedge \theta(c) = \theta(a) \end{cases} \Rightarrow \theta(a \vee c) = \theta(a) \wedge \theta(c)$$

$$c \leq a \Rightarrow \begin{cases} a \wedge c = c \Rightarrow \theta(a \wedge c) = \theta(a) \\ \theta(a) \leq \theta(c) \Rightarrow \theta(a) \wedge \theta(c) = \theta(a) \end{cases} \Rightarrow \theta(a \wedge c) = \theta(a) \vee \theta(c)$$

That is

$$v([a,b]\wedge[c,d]) = v([a\vee c,b\wedge d]) = v(\theta(a\vee c))+v(b\wedge d)$$
$$= v(\theta(a))+v(\theta(c))-v(\theta(a)\vee\theta(c))+v(b)+v(d)-v(b\vee d)$$
$$= v([a,b])+v([c,d])-v(\theta(a)\vee\theta(c))-v(b\vee d)$$

The same method
$$v(G_1\vee G_2) = v([a,b]\vee[c,d]) = v([a\wedge c,b\vee d]) = v(\theta(a\wedge c))+v(b\vee d)$$
$$= v(\theta(a)\vee\theta(c))+v(b)+v(d)$$

The two formulae
$$v([a,b]\wedge[c,d])+v([a,b]\vee[c,d]) = v([a,b])+v([c,d])$$

That is
$$v(G_1\wedge G_2)+v(G_1\vee G_2) = v(G_1)+v(G_2)$$

That is function $v([a,b]) = v(\theta(a))+v(b)$ satisfies the equality.

(3) inequality property

If $G_1\subset G_2$ (that is $[a,b]\subset[c,d]$),there are $c<a$ and $b\leq d$ 、 $c\leq a$ and $b<d$ 、 $c<a$ and $b<d$ three cases, by $\theta(x)=1-x$ know of ,when $c\leq a$, $\theta(a)\leq\theta(c)$,when $c<a$, $\theta(a)<\theta(c)$

If $c<a$ and $b\leq d$,then $v([a,b]) = v(\theta(a))+v(b) < v(\theta(c))+v(d) = v([c,d])$
If $c\leq a$ and $b<d$,then $v([a,b]) = v(\theta(a))+v(b) < v(\theta(c))+v(d) = v([c,d])$
If $c<a$ and $b<d$,then $v([a,b]) = v(\theta(a))+v(b) < v(\theta(c))+v(d) = v([c,d])$
That is $v(G_1) < v(G_2)$, function satisfies the equality.

Theorem 4. Only discussed the partial order relation between the space of the one-dimensional and one-dimensional space induced the hyperbox granule, and the algebra system $\langle G([0,1]),k\rangle$ is form as following by the fuzzy inclusion function induced by the nonlinear positive valuation function.

Theorem 5. $\langle G([0,1]),k\rangle,\langle G([0,1]),s\rangle$ are the fuzzy lattice, in which

$$k([a,b],[c,d]) = \frac{v([c,d])}{v([a,b]\vee[c,d])}$$
$$s([a,b],[c,d]) = \frac{v([a,b]\wedge[c,d])}{v([a,b])}$$

(9)

are the fuzzy inclusion function between two hyperbox granules.

Proof: Verify $k([a,b],[c,d])$ satisfies four conditions of fuzzy inclusion function
(1) $\forall[a,b]\in G([0,1])$, $O = [1,0]$

$$\begin{aligned}
v([1,0]) &= v(\theta(1)) + v(0)\\
&= v(1-1) + v(0)\\
&= v(0) + v(0)\\
&= \sin 0 + \sin 0\\
&= 0
\end{aligned}$$

Namely $k([a,b],O) = 0$

(2) $\forall [a,b] \in G([0,1])$, $[a,b] \vee [a,b] = [a,b]$, $v([a,b] \vee [a,b]) = v([a,b])$

Namely $k([a,b],[a,b]) = 1$

(3) $\forall [x,y],[a,b],[c,d] \in G([0,1]),[a,b] \preceq [c,d]$

$$[x,y] \vee [a,b] \preceq [x,y] \vee [c,d]$$
$$[x,y] \wedge [a,b] \preceq [x,y] \wedge [c,d]$$

Known by the positive evaluation function inequality
$$v([x,y] \vee [a,b]) \leq v([x,y] \vee [c,d])$$
$$v([x,y] \wedge [a,b]) \leq v([x,y] \wedge [c,d])$$

Namely
$$v([x,y] \vee [c,d]) - v([x,y] \vee [a,b])$$
$$\leq v([x,y] \vee [c,d]) - v([x,y] \vee [a,b]) + v([x,y] \wedge [c,d]) - v([x,y] \wedge [a,b])$$

Known by positive evaluation function equation
$$v([x,y] \vee [c,d]) - v([x,y] \vee [a,b]) + v([x,y] \wedge [c,d]) - v([x,y] \wedge [a,b])$$
$$= v([x,y]) + v([c,d]) - v([x,y] \wedge [c,d]) - (v([x,y]) + v([a,b]) - v([x,y] \wedge [a,b]))$$
$$+ v([x,y] \wedge [c,d]) - v([x,y] \wedge [a,b])$$
$$= v([c,d]) - v([a,b])$$

$v([x,y] \vee [c,d]) - v([x,y] \vee [a,b]) \leq v([c,d]) - v([a,b])$ can be transformed into
$$v([x,y] \vee [c,d]) \leq v([c,d]) - v([a,b]) + v([x,y] \vee [a,b])$$

Namely
$$\frac{v([a,b])v([x,y] \vee [c,d])}{v([c,d])} \leq \frac{v([a,b])(v([c,d]) - v([a,b]) + v([x,y] \vee [a,b]))}{v([c,d])}$$
$$= \frac{v([a,b])(v([c,d]) - v([a,b])) + v([a,b])v([x,y] \vee [a,b]))}{v([c,d])}$$
$$\leq \frac{v([a,b] \vee [x,y])(v([c,d]) - v([a,b])) + v([a,b])v([x,y] \vee [a,b]))}{v([c,d])}$$
$$= \frac{v([a,b] \vee [x,y])(v([c,d]) - v([a,b]) + v([a,b]))}{v([c,d])}$$
$$= v([a,b] \vee [x,y])$$

Therefore $\dfrac{v([a,b])}{v([a,b] \vee [x,y])} \leq \dfrac{v([c,d])}{v([c,d] \vee [x,y])}$

Namely $k([x,y],[a,b]) \leq k([x,y],[c,d])$

(4) $\forall [a,b],[c,d] \in G([0,1]), [a,b] \wedge [c,d] \preceq [a,b]$,have $[a \vee c, b \wedge d] \preceq [a,b]$

$\preceq$ is a partial order relation $c \leq a \vee c, b \wedge d \leq d$

$$k([a,b],[c,d]) = \frac{\upsilon([c,d])}{\upsilon([a,b] \vee [c,d])}$$

$$= \frac{\upsilon(\theta(c)) + \upsilon(d)}{\upsilon([a \wedge c, b \vee d])} = \frac{\upsilon(\theta(c)) + \upsilon(d)}{\upsilon(\theta(a \wedge c)) + \upsilon(b \vee d)}$$

By decrease of $\theta(x)$ and increase of $\upsilon(x)$

$$\theta(c) \leq \theta(a \wedge c)$$
$$\upsilon(\theta(c)) \leq \upsilon(\theta(a \wedge c))$$
$$\upsilon(d) \leq \upsilon(b \vee d)$$

That is $k([a,b],[c,d]) \leq 1$

Discussed above is the granular set lattice in the one-dimensional space ,for the N dimensional space ,the following theorem can be obtained.

Theorem 6. The classification problems in N dimensional space straining set S induced the hyperbox granule set and the fuzzy inclusion relations based on hyperbox granules construct a algebraic system, the algebraic system is a fuzzy lattice, that $\langle GS, k \rangle$ and $\langle GS, s \rangle$ are fuzzy lattice,

$$k(G_1, G_2) = \frac{\upsilon(G_2)}{\upsilon(G_1 \vee G_2)}, s(G_1, G_2) = \frac{\upsilon(G_1 \wedge G_2)}{\upsilon(G_1)} \tag{10}$$

4   Conclusions

In the paper, we proposed a novel fuzzy inclusion function by the nonlinear valuation function compounded by the function with the form of sin. The fuzzy inclusion function reflects the fuzziness of inclusion relations between two hyperbox granules. The fuzzy inclusion functions are proved by the four properties. How to design the classification algorithms based on the fuzzy inclusion function is the future work.

Acknowledgement. This work was in part supported by Natural Science Foundation of Henan Province and Education Department of Henan Province (112300410197, 102102210241, 13B520267).

References

1. Skowron, A., Stepaniuk, J.: Information Granular: Towards Foundation of Granular Computing. International Journal of Intelligent System 16(1), 57–85 (2001)
2. Yao, Y.Y.: Granular Computing: Past, Present and Future. In: Proceedings of IEEE International Conference on Granular Computing, Hangzhou, China, pp. 80–85 (2008)

3. Yao, Y.Y.: Interpreting Concept Learning in Cognitive Informatics and Granular Computing. IEEE Transactions on Systems, Man, and Cybernetics, Part B: Cybernetics 39(4), 855–866 (2009)
4. Zhang, L., Zhang, B.: Theory and Applications of Problem Solving, 2nd edn. Tsinghua University Press, Beijing (2007) (in Chinese)
5. Lin, T.Y.: Granular Computing on Binary Relations II : Rough Set Representations and Belief Functions. A. Skowron, L. Polkowski. Rough Sets in Knowledge Discovery (1998)
6. Zhang, L., Zhang, B.: Fuzzy Tolerance Quotient Spaces and Fuzzy Subsets. Scientia Sinica (Informationis) 41(1), 1–11 (2011) (in Chinese)
7. Zadeh, L.A.: Towards a Theory of Fuzzy Information Granulation and Its Centrality in Human Reasoning and Fuzzy Logic. Fuzzy Sets and Systems 19, 111–127 (1997)
8. Kaburlasos, V.G., Athanasiadis, I.N., Mitkas, P.A.: Fuzzy Lattice Reasoning (FLR) Classifier And Its Application For Ambient Ozone Estimation. International Journal of Approximate Reasoning 45, 152–188 (2007)

Zhang-Gradient Controllers of Z0G0, Z1G0 and Z1G1 Types for Output Tracking of Time-Varying Linear Systems with Control-Singularity Conquered Finally

Yunong Zhang*, Jinrong Liu, Yonghua Yin, Feiheng Luo, and Jianhao Deng

School of Information Science and Technology,
Sun Yat-sen University, Guangzhou 510006, China
zhynong@mail.sysu.edu.cn

Abstract. Recently, Zhang dynamics (ZD) and gradient dynamics (GD) have been used frequently to solve various kinds of online problems. In this paper, the output tracking of time-varying linear (TVL) systems is considered. Then, for such a problem, three different types of tracking controllers (i.e., Z0G0, Z1G0 and Z1G1 controllers) are designed by exploiting the ZD and GD methods. Simulation results on different TVL systems show that such three types of controllers can be feasible and effective for the output-tracking problem solving. Especially, the Z1G1 controller is capable of conquering the control-singularity of systems.

Keywords: Zhang dynamics (ZD), Gradient dynamics (GD), Time-varying linear system, Output tracking, Control-singularity.

1 Introduction and Problem Formulation

In order to solve online problems efficiently and effectively, a novel type of dynamic method termed Zhang dynamics (ZD) has recently been proposed [1–4]. Meanwhile, another type of dynamic method termed gradient dynamics (GD), which is intrinsically feasible and efficient to solve time-invariant problems, has been generalized to solve time-varying problems as well [5].

Output tracking control of time-invariant or time-varying linear systems are extensively encountered in many fields and they have attracted a lot of interest of researchers [6, 7]. Most of the control theory, including the output-tracking control, is devoted to the study of time-invariant linear (TIL) systems, and the key reason is of course that the TIL systems are simpler. However, it is known that in reality almost nothing is time-invariant. As a matter of fact, general time-varying linear (TVL) systems are normally too hard to analyze and study due to the difficulties (e.g., control-singularity) existing in TVL systems. Hence, the output-tracking control of TVL systems is an issue worthy of research.

Recent studies have shown that ZD and GD are two types of powerful methods for online problems solving [1–5]. Inspired by that, in this paper, the

---

* Corresponding author.

C. Guo, Z.-G. Hou, and Z. Zeng (Eds.): ISNN 2013, Part II, LNCS 7952, pp. 533–540, 2013.

output-tracking control of TVL systems is considered and three different types of tracking controllers (i.e., Z0G0, Z1G0 and Z1G1 controllers) are designed by exploiting the ZD and GD methods.

The description of a general TVL system is given as

$$\begin{cases} \dot{x}(t) = A(t)x(t) + B(t)u(t), \\ y(t) = C(t)x(t) + D(t)u(t), \end{cases} \tag{1}$$

where $x(t) \in R^n$ is the system state, $u(t) \in R^m$ is the control input, $y(t) \in R^l$ is the system output, and $A(t) \in R^{n \times n}$, $B(t) \in R^{n \times m}$, $C(t) \in R^{l \times n}$ and $D(t) \in R^{l \times m}$ are system matrices. For tracking the desired output trajectory $y_d(t) \in R^l$, the output-tracking error of system (1) is defined as

$$e(t) := y(t) - y_d(t) \in R^l. \tag{2}$$

Note that the conventional output tracking control problem [6, 7] is described as follows: given the desired output trajectory $y_d(t)$, a control input in the general form of $u(t)$ [or termed, $u(t)$-form] is to be designed such that it can drive the system output $y(t)$ to track the desired output trajectory $y_d(t)$ as close as possible [i.e., $e(t)$ asymptotically approaching zero]. In addition, the output-tracking error $e(t)$ is actually nonzero in the computer simulation due to various kinds of errors (e.g., truncation errors and round-off errors).

2   Zhang-Gradient Controllers Design

Corresponding to the TVL system (1) mentioned above, in this section, three different types of tracking controllers (i.e., Z0G0, Z1G0 and Z1G1 controllers), termed Zhang-gradient controllers, are designed and depicted in detail for the output tracking control problem solving.

2.1   Z0G0 Controller

From the TVL system (1), the output-tracking error $e(t)$ and the basic thought of output-tracking control, it follows that

$$e(t) = y(t) - y_d(t) = C(t)x(t) + D(t)u(t) - y_d(t) = 0,$$

which can be rewritten as

$$D(t)u(t) = y_d(t) - C(t)x(t).$$

Then, the (conventional) Z0G0 controller in the $u(t)$-form is designed as below:

$$u(t) = D^+(t)\left(y_d(t) - C(t)x(t)\right), \tag{3}$$

where $D^+(t)$ denotes the pseudoinverse of $D(t)$ [8]. Note that, if $D(t) \equiv 0$, this Z0G0 controller becomes inapplicable in this case.

2.2   Z1G0 Controller

For more complex and general situations to be handled, the following Zhang function (or termed, Zhangian) is constructed as

$$z(t) := e(t) = y(t) - y_d(t) = C(t)x(t) + D(t)u(t) - y_d(t) \qquad (4)$$

so that the tracking process can be monitored and controlled. Note that the Zhangian $z(t)$ can generally be matrix-, vector-, or scalar-valued, in addition to being positive, zero, negative, bounded or unbounded (even including lower-unbounded). Via ZD method [1–4], the following ZD design formula is used:

$$\dot{z}(t) = \frac{\mathrm{d}z(t)}{\mathrm{d}t} := -\gamma z(t), \qquad (5)$$

where the design parameter $\gamma > 0 \in R$ is used to scale the exponential convergence rate of the ZD solution.

It follows from (4) and (5) that

$$\dot{C}(t)x(t) + C(t)\dot{x}(t) + \dot{D}(t)u(t) + D(t)\dot{u}(t) - \dot{y}_d(t)$$
$$= -\gamma(C(t)x(t) + D(t)u(t) - y_d(t)).$$

Manipulating the above equation, we thus obtain

$$\dot{C}(t)x(t) + C(t)A(t)x(t) + C(t)B(t)u(t) + \dot{D}(t)u(t) + D(t)\dot{u}(t)$$
$$-\dot{y}_d(t) + \gamma(C(t)x(t) + D(t)u(t) - y_d(t)) = 0, \qquad (6)$$

which can be rewritten as

$$D(t)\dot{u}(t) = -(\dot{C}(t)x(t) + C(t)A(t)x(t) + C(t)B(t)u(t) + \dot{D}(t)u(t)$$
$$-\dot{y}_d(t) + \gamma(C(t)x(t) + D(t)u(t) - y_d(t))).$$

Then the Z1G0 controller in the $\dot{u}(t)$-form, which is quite different from the conventional $u(t)$-form, is designed as below:

$$\dot{u}(t) = -D^+(t)(\dot{C}(t)x(t) + C(t)A(t)x(t) + C(t)B(t)u(t) + \dot{D}(t)u(t)$$
$$-\dot{y}_d(t) + \gamma(C(t)x(t) + D(t)u(t) - y_d(t))). \qquad (7)$$

2.3   Z1G1 Controller

From (6), we can further define

$$h := \dot{C}(t)x(t) + C(t)A(t)x(t) + C(t)B(t)u(t) + \dot{D}(t)u(t) + D(t)\dot{u}(t)$$
$$-\dot{y}_d(t) + \gamma(C(t)x(t) + D(t)u(t) - y_d(t)), \qquad (8)$$

which should be zero theoretically. Based on the GD method [5], a norm-based scalar-valued nonnegative energy function is defined as $\varepsilon := \|h\|_2^2$, where $\|\cdot\|_2$ denotes the two-norm of a vector.

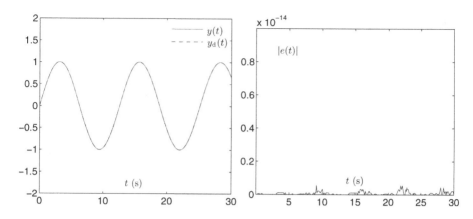

Fig. 1. Output-tracking performance of the TVL system (10) equipped with the Z0G0 controller for the desired output trajectory $y_d(t) = \sin(0.5t)$

Then, by the GD method [5], we can adopt the GD design formula $\dot{v}(t) := -\lambda \partial \varepsilon / \partial v(t)$, where $v(t) := \dot{u}(t)$, and the design parameter $\lambda > 0 \in R$ is used to scale the convergence rate of the GD solution.

Finally, substituting (8) into the GD design formula yields the following Z1G1 controller in the $\ddot{u}(t)$-form:

$$
\begin{aligned}
\ddot{u}(t) = -2\lambda D^{\mathrm{T}}(t)(\dot{C}(t)x(t) + C(t)A(t)x(t) + C(t)B(t)u(t) + \dot{D}(t)u(t) \\
+ D(t)\dot{u}(t) - \dot{y}_d(t) + \gamma(C(t)x(t) + D(t)u(t) - y_d(t))),
\end{aligned}
\tag{9}
$$

where superscript ^{T} denotes the transpose of a matrix/vector.

3   Simulations and Verifications

In order to verify the effectiveness of Z0G0, Z1G0 and Z1G1 controllers, this section compares the performance of three controllers for two TVL systems. Such two systems are different, as the control-singularity does not exist in the first system, while it exists in the second one.

3.1   Example 1: TVL System without Control-Singularity

In Example 1, the following TVL system is considered:

$$
\begin{cases}
\dot{x}(t) = \begin{bmatrix} -2 & 1 & 0 \\ 1 & -1-t & 0 \\ 0 & 0 & -0.1t^2 \end{bmatrix} x(t) + \begin{bmatrix} 1 & t & 1 \\ \cos t & 3 & 2 \\ t & t & 0 \end{bmatrix} u(t), \\
y(t) = \begin{bmatrix} \sin t & t^2 & t \end{bmatrix} x(t) + \begin{bmatrix} t+1 & \sin t & \cos t \end{bmatrix} u(t),
\end{cases}
\tag{10}
$$

where control-singularity (or say, zero-crossing) does not exist for the coefficient before $u(t)$. In the simulation, for controllers of Z1G0 and Z1G1 types, design

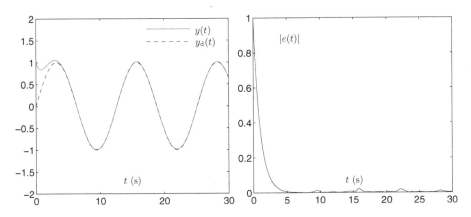

Fig. 2. Output-tracking performance of the TVL system (10) equipped with the Z1G0 controller for the desired output trajectory $y_d(t) = \sin(0.5t)$

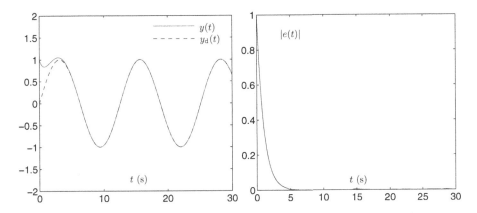

Fig. 3. Output-tracking performance of the TVL system (10) equipped with the Z1G1 controller for the desired output trajectory $y_d(t) = \sin(0.5t)$

parameters $\gamma = 1$ and $\lambda = 1000$. Besides, the desired output trajectory and the running time are $y_d(t) = \sin(0.5t)$ and 30 seconds, respectively. It is worth mentioning that the initial values [e.g., $x_1(0)$, $x_2(0)$ and $x_3(0)$] are set to 0.5. Corresponding to the Z0G0, Z1G0 and Z1G1 controllers, the simulation results on the tracking control of system (10) are shown in Figs. 1-3, respectively.

Specifically, Figs. 1-3 show that the outputs of the TVL system (10) equipped with the three controllers are capable of tracking the desired output trajectory $y_d(t) = \sin(0.5t)$. Besides, compared with the output-tracking errors in Figs. 1 and 3, the output-tracking error in Fig. 2 is larger and decreases slower. This illustrates that the tracking performance of the TVL system (10) equipped with the Z0G0 or Z1G1 controller is better than that of the TVL system (10) equipped with the Z1G0 controller.

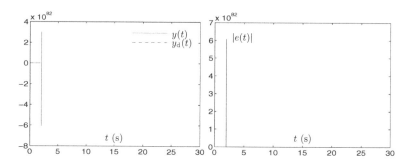

Fig. 4. Output-tracking performance of the TVL system (11) equipped with the (conventional) Z0G0 controller for the desired output trajectory $y_d(t) = 10\sin(2t) + 0.5t$

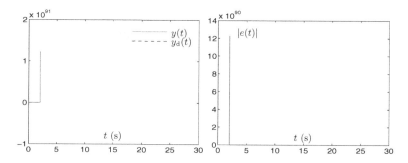

Fig. 5. Output-tracking performance of the TVL system (11) equipped with the Z1G0 controller for the desired output trajectory $y_d(t) = 10\sin(2t) + 0.5t$

3.2   Example 2: TVL System with Control-Singularity

In Example 2, another TVL system is considered as follows:

$$\begin{cases} \dot{x}(t) = \begin{bmatrix} -(t-10)^2 & 1 \\ -1 & -(t-20)^4 \end{bmatrix} x(t) + \begin{bmatrix} 1 \\ 2 \end{bmatrix} u(t), \\ y(t) = \begin{bmatrix} 1 & 5 \end{bmatrix} x(t) + \begin{bmatrix} t-2 \end{bmatrix} u(t). \end{cases} \tag{11}$$

Similarly, for controllers (specifically, Z1G0 and Z1G1 controllers), design parameters $\gamma = 1$ and $\lambda = 1000$ are set, and the running time is 30 seconds. In addition, all initial values [e.g., $x_1(0)$ and $x_2(0)$] are set to be 0.5. For further verification, a different desired output trajectory $y_d(t) = 10\sin(2t) + 0.5t$ is intentionally selected in the simulation. Note that, for the conventional controller design of the system, there exists a singularity (i.e., zero-crossing) problem when t approaches 2 seconds, which hinders the Z0G0 and Z1G0 controllers in this example. The corresponding simulation results are shown in Figs. 4-6.

Specifically, Fig. 4 shows that the output-tracking process of the closed-loop system with the Z0G0 controller stops when t is near 2 seconds. In detail, as time t approaches 2 seconds, the magnitude of the closed-loop system output equipped

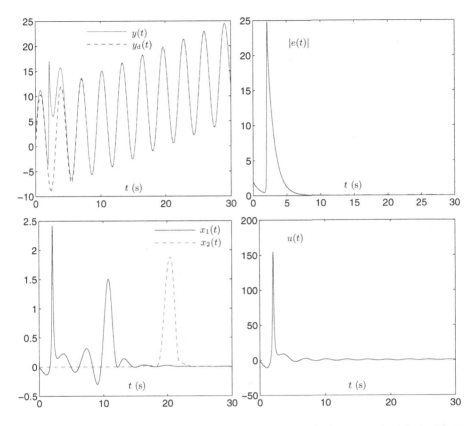

Fig. 6. Output-tracking performance of the TVL system (11) equipped with the Z1G1 controller for the desired output trajectory $y_d(t) = 10\sin(2t) + 0.5t$

with the (conventional) Z0G0 controller becomes extremely large, which leads to system crash. As observed from Fig. 5, the output-tracking process of the closed-loop system with the Z1G0 controller stops also at around $t = 2$ seconds, and the crash of this system is similar to that of the closed-loop system with the Z0G0 controller. The reason is that, when time t approaches 2 seconds, the magnitude of the term $D^+(t)$ of the Z0G0 or Z1G0 controller tends to be extremely large, which makes the control input $u(t)$ or $\dot{u}(t)$ become too large to implement, and consequently the closed-loop system crashes with its output being out of control.

By contrast, as observed from Fig. 6, the output of the closed-loop system with the Z1G1 controller is capable of tracking the desired output trajectory $y_d(t) = 10\sin(2t) + 0.5t$ successfully. More specifically, as time t approaches 2 seconds, the control input becomes large and the system state start to fluctuate, which drives the output to deviate the trajectory, but after 2 seconds (i.e., $t > 2$ seconds), the closed-loop system can adjust itself and the output is able to track the desired output trajectory automatically again. This well illustrates that the

Z1G1 controller conquers the singularity problem successfully while the other two controllers (i.e., Z0G0 and Z1G0 controllers) fail.

In summary, the simulation results of Examples 1 and 2 demonstrate that the Z0G0, Z1G0 and Z1G1 controllers are effective for the output-tracking control of TVL systems. More importantly, the Z1G1 controller shows its superiority for conquering the control-singularity of the system.

4   Conclusions

In this paper, the output tracking of TVL systems has been formulated and investigated. By exploiting and combining the ZD and GD methods, three types of tracking controllers (i.e., Z0G0, Z1G0 and Z1G1 controllers), termed Zhang-Gradient controllers, have been designed to solve such a tracking-control problem. Illustrative and comparative simulation results have been presented and the results have well verified the effectiveness of these three controllers for the output tracking of TVL systems. In particular, the controller of Z1G1 type has successfully conquered the troublesome control-singularity problem.

Acknowledgments. This work is supported by the National Natural Science Foundation of China (under Grants 61075121 and 60935001), the Specialized Research Fund for the Doctoral Program of Institutions of Higher Education of China (with project number 20100171110045) and also by the National Innovation Training Program for University Students (under Grant 201210558042).

References

1. Zhang, Y., Ma, W., Cai, B.: From Zhang Neural Network to Newton Iteration for Matrix Inversion. IEEE Trans. Circuits Syst. 56, 1405–1415 (2009)
2. Zhang, Y., Yi, C., Guo, D., Zheng, J.: Comparison on Zhang Neural Dynamics and Gradient-Based Neural Dynamics for Online Solution of Nonlinear Time-Varying Equation. Neural Comput. Appli. 20, 1–7 (2011)
3. Zhang, Y., Yi, C., Ma, W.: Simulation and Verification of Zhang Neural Network for Online Time-Varying Matrix Inversion. Simul. Model. Practice Theory 17, 1603–1617 (2009)
4. Zhang, Y., Li, Z.: Zhang Neural Network for Online Solution of Time-Varying Convex Quadratic Program Subject to Time-Varying Linear-Equality Constraints. Phys. Lett. A 373, 1639–1643 (2009)
5. Zhang, Y., Ke, Z., Xu, P., Yi, C.: Time-Varying Square Roots Finding via Zhang Dynamics versus Gradient Dynamics and The Former's Link and New Explanation to Newton-Raphson Iteration. Inf. Process. Lett. 110, 1103–1109 (2010)
6. Chen, M.: A Tracking Controller for Linear Time-Varying Systems. J. Dyn. Syst., Meas. Control 120, 111–116 (1998)
7. Zheng, D.: Linear System Theory. Tsinghua University Press, Beijing (2002)
8. Zhang, Y., Yang, Y., Tang, N., Cai, B.: Zhang Neural Network Solving for Time-Varying Full-Rank Matrix Moore-Penrose Inverse. Comput. 92, 97–121 (2011)

Utilizing Social and Behavioral Neighbors for Personalized Recommendation

Gang Xu, Linli Xu, and Le Wu

School of Computer Science and Technology
University of Science and Technology of China
{gangxu,wule}@mail.ustc.edu.cn, linlixu@ustc.edu.cn

Abstract. As a successful and effective technique, recommendation systems have been widely studied. Recently, with the popularity of social networks, some researchers have proposed the social recommendation, which considers the social relations between users besides the rating data. However, in real world scenarios, both the social relations and ratings are very sparse, how to combine them together to improve the performance becomes a critical issue. To that end, in this paper, we propose a unified three-stage recommendation framework named **R**andom **W**alk **N**eighborhood-aware **M**atrix **F**actorization(RWNMF), which can effectively integrate the social and rating data together and alleviate the sparsity problem. Specifically, we first perform random walk on social graph to find potential neighbors of each user, then select behavioral neighbors based on the rating data. Lastly, both the social neighbors and behavioral neighbors can be incorporated into traditional SocialMF, leading to more accurate recommendations. Experimental results on Epinions and Flixster datasets demonstrate our approach outperforms the state-of-the-art algorithms.

Keywords: Recommendation Systems, Social Network, Matrix Factorization.

1 Introduction

Recommendation systems, which aim at helping users overcome the problem of information overload, have been popular in recent years. Among all the techniques for building recommendation systems, Collaborative Filtering has received significant attention and empirical success. Typically in collaborative filtering, there is a set of users and a set of items respectively. Users' ratings for the items are expressed in a two-dimensional user-item rating matrix, where only some of the elements are observed, and the goal of CF is to predict the missing values.

In general, there are two categories of successful approaches in CF: Neighborhood methods(also namely K Nearest Neighbor, or KNN) [2,12] and Matrix Factorization [5,10,11]. Neighborhood methods mainly include user-oriented KNN(UKNN) [2] and item-oriented KNN(IKNN) [12] methods. UKNN assumes

C. Guo, Z.-G. Hou, and Z. Zeng (Eds.): ISNN 2013, Part II, LNCS 7952, pp. 541–551, 2013.

that similar users will have similar preference on the same item. While, IKNN assumes that a user will have similar preference on similar items. On the other hand, matrix factorization exploits the sparse user-item rating matrix to learn the characteristics of users and items in a low dimensional feature space, and then use the resulting feature vectors to do further rating prediction. In recent years, researchers have tried lots of sophisticated techniques to improve the recommendation accuracy, but the results are still unsatisfactory due to the extreme sparsity of the rating matrix. For example, in the Netflix Prize, the sparsity of the data is less than 1%, while in real-world recommendation systems, the sparsity of data is even worse.

Recently, with the popularity of social networks, it is easy to find users' social friends from the social based websites (for example, facebook.com). Based on the intuition that a user's preference is similar to her friends, a lot of work has incorporated the social information into traditional CF, named social recommendation. Social recommendations are based on the effects of social selection and social influence that have long since been assumed by sociologists. Social selection means that people tend to contact with people with similar attributes, and due to social influence the related people in a social network influence each other to become more similar [13]. Recent work on social recommendation includes [3,6,7].

Among these methods, SocialMF [3] has received a lot of attention. SocialMF is a matrix factorization method, in which the feature vector of each user is dependent on the feature vectors of her direct neighbors in the social network. This allows SocialMF to handle the transitivity of trust and trust propagation. The Epinions dataset published by the authors in [8] show that, lots of users have no more than ten social friends, and it is too rough to treat each neighbor equally. Therefore in this paper we propose a method to increase the number of social friends. At the same time, relying only on social friends is far from satisfactory. In order to integrate the other information, other neighbors should aslo be exploited, such as the behavioral neighbors, which refer to the neighbors whose behavior is similar to the users. It has been shown that the social interaction and behavioral similarity graphs have little overlap [1], sharing fewer than 15% of their edges. In order to improve the performance of recommendation, the similarity of the user-item rating matrix can be exploited to obtain behavioral neighbors due to the fact that the user-item rating matrix provides an independent source of information.

Following the work in SocialMF [3], we propose a fusion of social and behavioral neighbors for more accurate recommendation. The framework is composed of three steps: **1)Discovery of Social Neighbors**: we perform random walk on these users to discover their potential social neighbors since most users have only a few direct friends on the social network. **2)Discovery of Behavioral Neighbors**: user-item rating matrix is used to calculate the users(items)' behavioral neighbors and the neighbors are recorded. **3)Neighborhood Fusion**: in this step, we fuse the social neighbors and the behavioral neighbors to form the neighbors set for each user in the recommendation systems, and the behavioral neighbors of the items are also incorporated into the method. After that we can incorporate the more accurate neighbors into the SoicalMF framework. Experiments on two large real world datasets demonstrate the effectiveness of our method, especially when the social relations are very sparse.

The remainder of this paper is organized as follows. In Section 2 we provide an overview of social recommendation. Section 3 presents our method. The experimental results are reported in Section 4, followed by conclusions in Section 5.

2   Social Recommendation Method

In a social recommendation service systems, there is a user set U, an item set I, a trust network among users and some ratings of users. The task of recommendation systems is to predict the missing rating $r_{u,i}$ for the user u on the item i, given the user u and the item i. The predicted rating is denoted by $\hat{r}_{u,i}$. Generally, users rate only on a small fraction of the items, and $r_{u,i}$ is unknown for most pairs of (u, i).

Because SocialMF[3] is the state-of-the-art way in social recommendation, we focus on introducing the SocialMF method. In SocialMF, the mechanism of trust propagation is introduced into the framework since trust propagation has been shown to be a crucial factor in social network analysis and social recommendation. Based on the assumption that the behavior of the user u is affected by her direct neighbors N_u, the latent feature vector of the user u is dependent on the latent feature vectors of all her direct neighbors. This is formulated as in SocialMF:

$$\hat{U}_u = \frac{\sum_{v \in N_u} T_{u,v} U_v}{\sum_{v \in N_u} T_{u,v}} = \frac{\sum_{v \in N_u} T_{u,v} U_v}{|N_u|} \tag{1}$$

where u and v denote two users, $\hat{U}_u$ is the estimated latent feature vector of the user u , U_u and U_v denote the feature vectors of the user u and v respectively. Moreover, N_u represents the direct trusted neighbors of the user u, and $T_{u,v}$ is the weight that quantifies how much the user u trusts the user v. Since all none-zero values of $T_{u,v}$ are 1 in binary social networks, we normalize each row of the trust matrix so that $\sum_{v \in N_u} T_{u,v} = 1$ [3].

In general, the performance of the existing method will be limited only considering social neighbors, so we put forward our method, which can not only find the potential social neighbors but also integrate the information of the behavioral neighbors.

3   Random Walk Neighborhood-Aware Matrix Factorization

3.1   Discovery of Social Neighbors in Social Graph

As is revealed in the real world datasets, most users have only a few direct friends. For example, there are respectively 36% and 48% users who have no more than

10 social neighbors in the Epinions and Flixster datasets, which makes it hard to discover users' social neighbors based on their direct relationships. To alleviate this problem, we perform Random Walk with Restart (RWR) [9] to discover their potential friends and select the top K social neighbors based on the stationary distribution of each user.

To get the users' top K potential social neighbors, each time we cast a random walker which starts from each user node. Firstly the random walker goes back to the starting user with the same probability and chooses randomly among the available edges every time. Assuming that the random walker which starts from user node p, we use the probability that it reaches a user node q to denote the trust weight that quantifies how much the user p trusts the user q, which is expressed as a element of the trust weight vector $u_p(q)$. For further details, please refer to [9]. The detailed algorithm for discovery of social neighbors can be summarized as Algorithm 1.

Algorithm 1. Random Walk with Restart.

Input:
> The number of users, n;
> The social network graph, $G_{n \times n}$;
> The number of random walk steps, m;
> The n-dimensional trust weight vector $\boldsymbol{u_q}$, the n-dimensional restart vector $\boldsymbol{v_q}$ for each user;

Output:
> The n-dimensional trust weight vector $\boldsymbol{u_q}$ for each user
> 1: Make sure $v_q(q) = 1$ and $v_q(i) = 0, i \neq q$ in $\boldsymbol{v_q}$;
> 2: Normalize the adjacency matrix $\boldsymbol{A}$ by column, make the out-degrees sum to 1;
> 3: Initialize $\boldsymbol{u_q} = \boldsymbol{v_q}$;
> 4: While(iterations $< m$);
> $\boldsymbol{u_q} = (1 - \alpha)A\boldsymbol{u_q} + \alpha\boldsymbol{v_q}$;
> 5: Select the top K neighbors;
> 6: Normalize the trust weights of the neighbors, make them sum to 1.

3.2   Discovery of Behavioral Neighbors

Besides social information, user behavior can also be used to find similar users. For example, if two users have seen similar movies, they may be similar. Since the adjusted cosine similarity can achieve better performance than cosine similarity and pure cosine similarity [12], we choose the the adjusted cosine similarity to obtain the similarity between users or items. The metric equation of similarity between users could be expressed as:

$$sim(i, j) = \frac{\sum_{v \in IC_{i,j}} (r_{i,v} - \overline{r_v})(r_{j,v} - \overline{r_v})}{\sqrt{\sum_{v \in IC_i} (r_{i,v} - \overline{r_v})^2}\sqrt{\sum_{v \in IC_j} (r_{j,v} - \overline{r_v})^2}} \quad (2)$$

where $IC_{i,j}$ is the set of common items who users i and j have both rated, and $\bar{r}_v$ denotes the average of ratings item v received, $sim(i, j)$ denotes the similarity between user i and j. The metric equation of similarity between items can be obtained similarly.

After obtaining similarity between users, the K nearest neighbors of the user i, M_{U_i} can be easily selected. Identical procedures can be used to characterize the similarity of each pair of item j and k, and the K nearest neighbors of the item j is expressed as M_{V_j}. For notational convenience in later sections, we normalize the similarities between each user i and her neighbors M_{U_i}, and those between each item j and its neighbors M_{V_j} to ensure $\sum_{k \in M_{U_i}} S_{i,k} = 1$ and $\sum_{l \in M_{V_j}} W_{j,l} = 1$.

3.3   Fusion of the Social Neighborhood and Behavioral Neighborhood

In order to define the model more realistically, the feature vectors of the users should reflect both of the two factors of social neighbors and behavioral neighbors as stated in Section 1. Similarly, the latent feature vectors of the items should rely on the weighted average of their behavioral neighbors. Based on this intuition, the following equations are formulated as:

$$\hat{U}_i = \alpha \sum_{l \in N_{U_i}} T_{i,l} U_l + (1 - \alpha) \sum_{k \in M_{U_i}} S_{i,k} U_k + \theta_P,$$

$$\theta_P \sim N(0, \sigma_P^2 \mathbf{I}) \tag{3}$$

$$\hat{V}_j = \sum_{l \in M_{V_j}} W_{j,l} V_l + \theta_Q,$$

$$\theta_Q \sim N(0, \sigma_Q^2 \mathbf{I}) \tag{4}$$

The terms in the two equations above can be classified into two parts:

1) In Eq. (3), the first term and the second term together characterize the group feature of a user, which is the weighted average of the fusion of her social neighbors and behavioral neighbors. They are smoothed by the parameter α. In Eq. (4), the first term characterizes the group feature of an item, which is the weighted average of its behavioral neighbors.

2) The third term in Eq. (3) and the second term in Eq. (4) emphasize the uniqueness of each user(item)'s feature vectors, which could diverge from her(its) neighbors to some extent. The divergence is controlled by the variance parameters σ_P^2 and σ_Q^2.

Based on Eq. (3) and Eq. (4), the latent feature of user and item can be reformulated to the following:

$$p(U|T, S, \sigma_U^2, \sigma_P^2) \propto p(U|\sigma_U^2) * p(U|T, S, \sigma_P^2)$$

$$= \prod_{i=1}^{M} N(U_i|0, \sigma_U^2 \mathbf{I}) * \prod_{i=1}^{M} N(U_i|\alpha \sum_{l \in N_{U_i}} T_{i,l} U_l + (1 - \alpha) \sum_{k \in M_{U_i}} S_{i,k} U_k, \sigma_P^2 \mathbf{I}) \tag{5}$$

$$p(V|W, \sigma_V^2, \sigma_Q^2) \propto p(V|\sigma_V^2) * p(V|W, \sigma_Q^2)$$

$$= \prod_{j=1}^{N} N(V_j|0, \sigma_V^2 \mathbf{I}) * \prod_{j=1}^{N} N(V_j|\sum_{l \in M_{V_j}} W_{j,l} V_l, \sigma_Q^2 \mathbf{I}) \tag{6}$$

where we add a zero-mean Gaussian prior as in Eq. (5) and Eq. (6) to avoid over-fitting and the regularization is controlled by the variance parameters σ_U^2 and σ_V^2.

Therefore, the conditional probability equation of observed rating is as follows:

$$p(R|U, V, \sigma_R^2) = \prod_{i=1}^{M} \prod_{j=1}^{N} [N(R_{i,j}|U_i^T V_j), \sigma_R^2]^{I_{i,j}^R} \tag{7}$$

Then through Bayesian inference, the log of the posterior distribution over the user and item latent feature is given as:

$$p(U, V|R, T, S, W, \sigma_R^2, \sigma_U^2, \sigma_V^2, \sigma_P^2, \sigma_Q^2)$$
$$\propto p(R|U, V, \sigma_R^2) * p(U|T, S, \sigma_U^2, \sigma_P^2) * p(V|W, \sigma_V^2, \sigma_Q^2) \tag{8}$$

Keeping the parameters fixed, maximizing the log posterior in Eq. (8) is equivalent to minimizing the following sum-of-squared cost function:

$$E = \frac{1}{2} \sum_{i=1}^{M} \sum_{j=1}^{N} I_{i,j}^R (R_{i,j} - U_i^T V_j)^2$$
$$+ \frac{\lambda_U}{2} \sum_{i=1}^{M} \|U_i\|_F^2 + \frac{\lambda_V}{2} \sum_{j=1}^{N} \|V_j\|_F^2$$
$$+ \frac{\lambda_P}{2} \sum_{i=1}^{M} \left\| U_i - \alpha \sum_{l \in N_{U_i}} T_{i,l} U_l - (1 - \alpha) \sum_{k \in M_{U_i}} S_{i,k} U_k \right\|_F^2$$
$$+ \frac{\lambda_Q}{2} \sum_{j=1}^{N} \left\| V_j - \sum_{l \in M_{V_j}} W_{j,l} V_l \right\|_F^2 \tag{9}$$

In the equation above, $\lambda_U = \sigma_R^2/\sigma_U^2$, $\lambda_V = \sigma_R^2/\sigma_V^2$, $\lambda_P = \sigma_R^2/\sigma_P^2$ and $\lambda_Q = \sigma_R^2/\sigma_Q^2$ are the parameters to smooth the five terms in the objective function, which naturally fuses social and behavioral neighborhood information with matrix factorization in social recommendation. The parameters λ_U and λ_V control the strength of fitting on the training data. The parameter λ_P controls how much the user neighbors influence the user feature vectors, while the parameter λ_Q controls how much the item neighbors influence the item feature vectors.

Then a local minimum of the objective function Eq. (9) can be found by performing gradient descent on U_i and V_j for all users u and all items i given the derivatives below.

$$\frac{\partial E}{\partial U_i} = \sum_{j=1}^{N} I_{i,j}^R V_j (U_i^T V_j - R_{i,j}) + \lambda_U U_i$$
$$+ \lambda_P (U_i - \alpha \sum_{l \in N_{U_i}} T_{i,l} U_l - (1 - \alpha) \sum_{k \in M_{U_i}} S_{i,k} U_k)$$

$$- \alpha \lambda_P \sum_{\{l|i\in N_{U_l}\}} T_{l,i}(U_l - \alpha \sum_{w\in N_{U_l}} T_{l,w}U_w - (1-\alpha) \sum_{k\in M_{U_l}} S_{l,k}U_k)$$

$$- (1-\alpha)\lambda_P \sum_{\{l|i\in M_{U_l}\}} S_{l,i}(U_l - \alpha \sum_{w\in N_{U_l}} T_{l,w}U_w - (1-\alpha) \sum_{k\in M_{U_l}} S_{l,k}U_k) \quad (10)$$

$$\frac{\partial E}{\partial V_j} = \sum_{i=1}^{M} I_{i,j}^R U_i(U_i^T V_j - R_{i,j}) + \lambda_V V_j$$

$$+ \lambda_Q(V_j - \sum_{l\in M_{V_j}} W_{j,l}V_l)$$

$$- \lambda_Q \sum_{\{l|j\in M_{V_l}\}} W_{l,j}(V_l - \sum_{k\in M_{V_l}} S_{l,k}V_k) \quad (11)$$

To reduce the model complexity, λ_U is set equal to λ_V in our experiments. In each iteration, we update U and V based on the latent feature vectors from the previous iteration.

4 Experiments

We conduct several experiments on two real world datasets and demonstrate the following: **1)** the comparison of effectiveness between RWNMF and other state-of-the-art methods; **2)** the influence of parameter settings in RWNMF; **3)** the performance of the RWNMF on the cold-start users.

4.1 Datasets

We briefly introduce the two public datasets used in our experiments: the Epinions dataset [1], and the Flixster dataset [2]. Both datasets contain user-item rating data and user relation data. We remove users who rate less than 5 times, and also remove items that have less than 5 rating scores. Then we leave only the corresponding relation data of these users in the social relation data. Overall ratings of the Epinions dataset range from 1 to 5. Possible ratings values in the Flixster dataset are 10 discrete numbers in the range [0.5,5] with step size 0.5. More statistics of two datasets are listed in Table 1.

4.2 Baselines and Evaluation Metric

We compare our model with the following methods. We choose these models as baselines because they are proven to be the state-of-art methods for social recommendation which shows good results on the social rating datasets.

PMF: This method is the baseline matrix factorization method proposed in [11], which only uses rating matrix to conduct recommendation.

---

[1] www.trustlet.org/wiki/Downloaded_Epinions_dataset
[2] www.cs.sfu.ca/~sja25/personal/datasets/

Table 1. General statistics of the Epinions and Flixster

Statistics	Epinions	Flixster
Sparsity of rating records	0.10%	0.45%
# of users	20397	69367
# of items	21901	25678
# of rating records	446892	8000204
# of relation records	350012	970288
Average ratings per user	21.9	115.33
Average ratings per item	20.4	311.55
Average trust times per user	22.02	15.65
Average trusted times per user	18.38	15.36

STE: This method is proposed in [6], which combines the users' tastes and their trusted friends' interests together by a linear combination of the basic matrix factorization approach and the social recommendation approach.

SocialMF: This is a method proposed in [3], which takes the trust propagation for recommendation in social network into account. On the other hand, this can be seen as a special case of RWNMF when both the behavioral neighbors of the user and the behavioral neighbors of the item are unknown, specifically, when we set $\alpha=1$, $\lambda_Q=0$.

In the experiments of our proposed method RWNMF, we set $\lambda_U = \lambda_V=0.1$, $K=30$. Discovery of social neighbors is done on the users who have no more than 30 social neighbors and select the top 30 social neighbors. The evaluation metric we use in our experiments is Root Mean Squared Error(RMSE) [4], which is widely used to measure the performance of rating prediction accuracy in CF and Social Recommendation. For each dataset, we take 80% of the data as the training data and the remaining 20% as the testing data.

4.3   Experimental Results

Overall Performance. In this section, we compare the performance of our model with the baseline methods. The dimensionality of the latent feature vectors D in matrix factorization methods is set to 5, 10, 20, 30 respectively. As to the other parameters, we perform parameter selection in advance for each method and use the best settings found in all the experiments for fairness.

Table 2. RMSE comparisons for different latent feature dimension D

Model	D=5		D=10		D=20		D=30	
	Epinions	Flixster	Epinions	Flixster	Epinions	Flixster	Epinions	Flixster
PMF	1.1064	0.9085	1.1034	0.9047	1.1014	0.9023	1.0971	0.9015
STE	1.0961	0.8901	1.0912	0.8885	1.0872	0.8847	1.0815	0.8824
SocialMF	1.0784	0.8810	1.0735	0.8792	1.0729	0.8744	1.0704	0.8726
RWNMF	**1.0487**	**0.8512**	**1.0439**	**0.8459**	**1.0414**	**0.8426**	**1.0402**	**0.8414**

Table 2 reports the RMSE values of all the algorithms under different settings of latent feature dimension D respectively. From the comparison, we have the following conclusion: our method RWNMF performs the best comparing to other methods. Generally, RWNMF improves the RMSE of the best results of PMF, STE and SocialMF about 5.91%, 4.40%, 3.20%.

Different Parameter Setting. In our model, parameters λ_P and λ_Q control the divergence of the active users' and items' latent features from the fusion of their behavioral neighbors and social neighbors or their behavioral neighbors set respectively.

Figure 1 compares RMSE of our model with different range of regularization parameters on two datasets. For convience, we set $\lambda_P = \lambda_Q = \lambda$, and λ is set to 0, 0.01, 0.05, 0.1, 0.5, 1, 5, 10, 20, 30 respectively. And we set D=30. It is easy to observe that the value of λ impacts the recommendation results significantly, which demonstrates that incorporating neighborhood information into social recommendation improves the recommendation accuracy. Clearly in Figure 1, the best regularization parameters setting for both datasets is $\lambda_P = \lambda_Q = 0.1$.

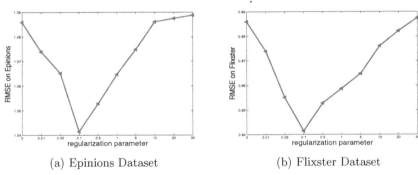

(a) Epinions Dataset   (b) Flixster Dataset

Fig. 1. Impact of regularization parameters on two different data sets

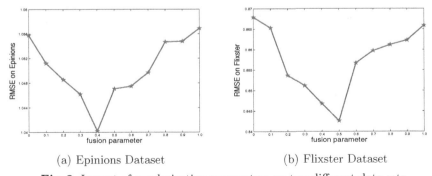

(a) Epinions Dataset   (b) Flixster Dataset

Fig. 2. Impact of regularization parameters on two different data sets

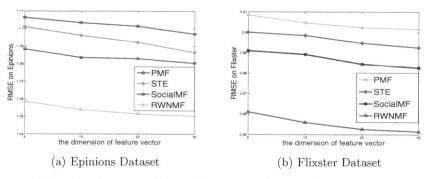

(a) Epinions Dataset   (b) Flixster Dataset

Fig. 3. Performance of the cold-start users in two different data sets

Figure 2 compares RMSE score of our model with different range of parameter α on two datasets. Parameter α controls extent of the influence of the social neighbors and similarity neighbors on the behavior of users. We set $\lambda_P = \lambda_Q = 0.1$, and we set D=30. In fact the fusion of social neighbors and the behavioral neighbors achieves better prediction due to a combination of two different factors.

4.4   Performance on the Cold-Start Users

We remove 90% of the user relationship in the user relation data. Figure 3 compares the RMSE of our model with the other methods on the cold-start users with different range of the dimension of the feature vectors on the Epinions and Flixster datasets. For convience, the dimension of the feature vectors is set to 5, 10, 20, 30 respectively. As shown in Figure 3, our method still achieve good results owing to the discovery of social neighbors and the neighbors fusion.

5   Conclusion

In this paper, we present a novel approach for recommendation in social networks. Through the random walk in the first step of RWNMF, it is effective to find the potential social neighbors, and they could be adopted in the following matrix factorization framework. Next it is necessary to find the behavioral neighbors of the users and items. Lastly we take into consideration the fact that the latent feature of each user is dependent on the feature vectors of her social neighbors and behavioral neighbors and the latent feature of each item is dependent on the feature vectors of its behavioral neighbors. This allows RWNMF to exploit the information of both the social network and the similarity network to improve the recommendation accuracy. Experiments on the public real world datasets of Epinions and Flixster demonstrate that RWNMF outperforms the existing methods for social recommendation.

Acknowledgments. Research supported by the National Natural Science Foundation of China (No. 61003135).

References

1. Crandall, D., Cosley, D., Huttenlocher, D., Kleinberg, J., Suri, S.: Feedback effects between similarity and social influence in online communities. In: Proceeding of ACM SIGKDD (2008)
2. Herlocker, J., Konstan, J., Borchers, A., Riedl, J.: An algorithmic framework for performing collaborative filtering. In: Proceedings of ACM SIGIR (1999)
3. Jamali, M., Ester, M.: A matrix factorization technique with trust propagation for recommendation in social networks. In: Proceedings of ACM RecSys (2010)
4. Koren, Y.: Factorization meets the neighborhood: a multifaceted collaborative filtering model. In: Proceedings of ACM SIGKDD (2008)
5. Koren, Y., Bell, R., Volinsky, C.: Matrix factorization techniques for recommender systems. Computer (2009)
6. Ma, H., King, I., Lyu, M.: Learning to recommend with social trust ensemble. In: Proceedings of ACM SIGIR (2009)
7. Ma, H., Yang, H., Lyu, M.R., King, I.: Sorec: social recommendation using probabilistic matrix factorization. In: Proceedings of ACM CIKM (2008)
8. Massa, P., Avesani, P.: Trust-aware recommender systems. In: Proceedings of ACM RecSys (2007)
9. Pan, J., Yang, H., Faloutsos, C., Duygulu, P.: Automatic multimedia cross-modal correlation discovery. In: Proceedings of ACM SIGKDD (2004)
10. Paterek, A.: Improving regularized singular value decomposition for collaborative filtering. In: Proceedings of KDD Cup and Workshop (2007)
11. Salakhutdinov, R., Mnih, A.: Probabilistic matrix factorization. In: Advances in Neural Information Processing Systems (2008)
12. Sarwar, B., Karypis, G., Konstan, J., Riedl, J.: Item-based collaborative filtering recommendation algorithms. In: Proceedings of ACM WWW (2001)
13. Wasserman, S., Faust, K.: Social network analysis: Methods and applications, vol. 8. Cambridge University Press (1994)

Artificial Fish Swarm Algorithm for Two-Dimensional Non-Guillotine Cutting Stock Problem[*]

Lanying Bao[1], Jing-qing Jiang[2,3,**], Chuyi Song[1], Linghui Zhao[1], and Jingying Gao[1]

[1] College of Mathematics, Inner Mongolia University for Nationalities,
Tongliao 028043, China
[2] College of Computer Science and Technology, Inner Mongolia University for Nationalities,
Tongliao 028043, China
[3] Department of Computer Science, University of Missouri, Columbia 65201, USA
{Jingqing Jiang,jiangjingqing}@yahoo.com.cn

Abstract. In this paper we present Artificial Fish Swarm Algorithm (AFSA) applying to a two-dimensional non-guillotine cutting stock problem. Meanwhile, we use a converting approach which is similar to the Bottom Left (BL) algorithm to map the cutting pattern to the actual layout. Finally, we implement Artificial Fish Swarm Algorithm on several test problems. The simulated results show that the performance of Artificial Fish Swarm Algorithm is better than that of Particle Swarm Optimization Algorithm.

Keywords: Artificial Fish Swarm Algorithm; Cutting Stock Problem; Bottom Left Algorithm.

1 Introduction

The two-dimensional cutting stock problem can be stated as the problem of fitting rectangular pieces of predetermined sizes onto a large but finite rectangular plate (the stock plate); or equivalently, the problem of cutting small rectangular pieces of predetermined sizes from a large rectangular plate. Non-guillotine cut means that the cuts need not go from one edge of a rectangle to the opposite edge in a straight line. The aim is to minimize the unused area. Because of the need to reduce the costs of raw material and the need to avoid material wastage, the cutting stock problem is of great interest in various production processes in the glass, steal, wood, paper and textile industries. This problem belongs to the class of combinatorial optimization problems, so a solution among all possible solutions has to be found, which optimizes or quasi-optimizes a criterion function subject to a set of constraints.

Gilmore and Gomory [1, 2] used linear programming to solve such kind of a problem exactly. But because of the complexity of the cutting stock problem, the exact algorithm only fits the case in which fewer pieces are being cut. These algorithms fail if there are more pieces to be cut. A problem of cutting more than 20 pieces would cause some difficulty. Recently, with the extended application of various optimization methods, some methods such as heuristic searching algorithms

---

[*] Supported by the National Natural Science Funds of China under Grant No. 61163034.
[**] Corresponding author.

[3], evolutionary algorithms [4-6], swarm intelligence algorithm [7-9] and simulated annealing approach [10] have been applied to this kind of problems. Artificial Fish Swarm Algorithm (AFSA) was presented by Li in 2002 [11]. It is a kind of the swarm intelligence optimization methods based on fish behaviors. It has been used to deal with some combinatorial optimization problems [12-13], but it is rarely applied to cutting stock problem. In this paper, we will consider applying Artificial Fish Swarm Algorithm to a two-dimensional cutting stock problem.

2   Artificial Fish Swarm Algorithm

The details of the artificial fish behavior are as follows [11]:

An artificial fish denotes a vector $X_i = (p_{i1}, p_{i2},..., p_{in})$ $i = 1,2,...,N$, where N denotes the size of fish swarm, $p_{ik}(k = 1,2,...,n)$ is the variable that needs to be optimized, n denotes the dimension of the problem space. The food consistence of the position which an artificial fish locate is denoted by $Y = f(X)$. The distance between two artificial fishes is denoted by $d_{i,j} = \|X_i - X_j\|$. The other properties of an artificial fish are the visual field ($visual$), the crowded factor (δ), the largest number of every trying ($trynumber$).

Behavior of an artificial fish is described as follows:

In a water field, a fish could find the most nutrition position by itself or following other fishes. The most nutrition position has more fishes. The Artificial Fish Swarm Algorithm implements the optimization by simulating preying behavior, swarming behavior and following behavior.

(1) Preying behavior: Set the current position of the ith artificial fish is X_i. Select a position X_j in its visual field randomly. If $Y_i < Y_j$ (forward condition) then move a step to this direction. Otherwise randomly select a position X_j in its visual field randomly again, and then judge whether it satisfies the forward condition. Try $trynumber$ times. If the forward condition is not yet satisfied then perform random moving behavior.

(2) Swarming behavior: Explore the partner number n_f in an artificial fish's neighborhood. If $n_f/N < \delta, (0 < \delta < 1)$, it shows that the center of the partners has more foods and not crowded. In this case, if $Y_i < Y_j$ then the artificial fish moves a step to the center X_c. Otherwise perform random preying behavior.

(3) Following behavior: Explore the optimal neighbor X_{max} in its neighborhood. If $Y_i < Y_{max}$ and the partner number n_f in the neighborhood of X_{max} satisfies $n_f/N < \delta, (0 < \delta < 1)$, it shows that the position X_{max} have more food and not crowded. Then the artificial fish move a step to the position X_{max}. Otherwise performs preying behavior.

(4) Random behavior: The artificial fish selects a random position in its visual field and then moves a step to this direction. It is a default behavior of preying behavior. It provides a random swimming in preying behavior when the *trynumber* is small. It also increases the variety of the swarm so that the artificial fish could jump out of the local optimal value.

(5) Bulletin board: Bulletin board is used to record the optimal artificial fish. In the optimization process, each artificial fish compares its position with the bulletin board after a moving. If its position is better than the bulletin board, the bulletin board is rewritten by this artificial fish. So the historical optimal position is written in bulletin board.

3   Stock Cutting Algorithm

3.1   Some Constrains

For the sake of simplicity, the following assumptions are adopted [5]:

(1) All the pieces have fixed orientation, i.e. a piece with length l_k and width w_k is different from the piece with length w_k and width l_k if l_k does not equal to w_k.

(2) The pieces must be placed into the stock plate orthogonally, thus the sides of the small rectangles are parallel to the stock plate. In other words, there is no rotation when placing the pieces into the plate.

(3) The length and width of each piece does not exceed the corresponding dimension of the stock plate.

(4) All cuts on the stock plate are infinitesimally thin, i.e. the edges of the pieces do not occupy any area.

(5) Each piece may be positioned at any place in the stock plate and in proximity to any other piece in the plate, i.e. there is no restriction that two pieces can not be put together.

3.2   Encoding Mechanism

To describe a layout of pieces on a stock plate, we must first specify the piece list and the stock plate available for the placement of the pieces. The stock plate to be used is a rectangle with specified dimension. The pieces cutting from the stock plate are also rectangular in shape. We describe the stock plate and the pieces in a free coordinates. The left bottom corner of the stock plate is placed at the origin. Each piece is denoted by a four-dimension vector $p_k = (x_k, y_k, l_k, w_k)$, where (x_k, y_k) is the position of the left bottom corner of the piece on the stock plate, l_k and w_k are the length and width of the piece, respectively. The pieces are generated within the following ranges:

$$0 \leq x_k, \ l_k \leq Length$$

$$0 \leq y_k, w_k \leq Width$$

where *Length* and *Width* are the stock plate length and width.

Each fish in the swarm denotes a cutting pattern. A cutting pattern is consisted of a set of pieces p_k.

3.3   Actual Layouts

In order to place the pieces on the stock plate, we should convert the cutting pattern to an actual layout. In this paper, we use a converting approach which is similar to the Bottom Left (BL) algorithm [5]. It is called the Coordinate-based Bottom Left Algorithm (CBL) [8]. In the BL algorithm, the piece is firstly put at the right upper corner of the stock plate and then it is moved to the left. In this paper, in order to use the coordinates of the piece's left bottom corner, we place the piece according to the coordinate then move it. We sort the pieces on the x_k to reduce the probability of overlapping. The steps of the CBL are as follows:

Step 1. Sort the pieces in ascending order of x_k.
Step 2. Place the pieces on the stock plate according to the coordinates of the left bottom orderly. For each piece, if it is placed entirely on the plate and does not overlap the pieces that have been placed on the plate, then try to move it down (as long as it is not blocked by another piece). Then move it to the left and then repeat to move it down and left again, until it cannot be moved. The finial position is where the rectangular piece stays. When placing a piece if it overlaps any of the pieces which have been placed entirely on the stock place or if it cannot be placed onto the stock plate completely, we do not place it on the stock plate temporarily. If all of the pieces have been tried to place, go to step 3.
Step 3. For the pieces that have not been placed on the stock plate, we put each piece on the right upper corner of the stock plate, then try to move it down and to the left repeatedly until it cannot be moved. If the piece is placed entirely on the plate, the final position is where the piece stays. Otherwise the piece will be discarded.

4   AFSA for Cutting Stock Problem

The cutting algorithm deals with the fishes. Each fish denotes a cutting pattern. $X_i = (p_{i1}, p_{i2}, ..., p_{in})$ is a fish, where $i = 1,2,...,N$. N denotes the size of fish swarm. p_{ik} is a piece denoted by a four-dimension vector $p_{ik} = (x_{ik}, y_{ik}, l_{ik}, w_{ik})$, where $k = 1,2,...,n$. n is the number of pieces that should be cut from the given stock plate. (x_{ik}, y_{ik}) is the position of the left bottom corner of the piece on the stock plate, l_{ik} and w_{ik} are the length and width of the piece, respectively. The distance between a fish X_i and a fish X_j is denoted by

$$d_{i,j} = \left\| X_i - X_j \right\| = \frac{1}{n} \sum_{k=1}^{n} \left\| p_{ik} - p_{jk} \right\| = \frac{1}{n} \sum_{k=1}^{n} \sqrt{(x_{ik} - x_{jk})^2 + (y_{ik} - y_{jk})^2}$$

In AFSA, the food consistence $f(X)$ of the position which an artificial fish X located is taken as the ratio of the summed areas of the pieces completely placed on the stock plate to the total area of the stock plate, i.e.

$$f(X) = \frac{s(X)}{T}$$

where X denotes a cutting pattern, $s(X)$ is the summed area of the pieces which area placed completely on the stock plate using the CBL corresponding to the cutting pattern X, and $T = Length*Width$ denotes the designated area of the stock plate.

The steps of Artificial Fish Swarm Algorithm for cutting stock problem are as follows:

Step 1. Initialize the left bottom coordinates of each piece in each artificial fish randomly.

Step 2. Calculate the food consistence of each fish and record the best fish on the bulletin board.

Step 3. For each artificial fish, perform one of the three behaviors: Perform following behavior. If the new position is better than the bulletin board, update bulletin board and go to step 4. Otherwise perform swarming behavior. If the new position after swarming is better than the bulletin board, update bulletin board and go to step 4. Otherwise perform preying behavior. If the new position after preying is better than the bulletin board, then update the bulletin board.

Step 4: If the maximum number of iterations is reached or the designated fitness is achieved, the process is stopped. Otherwise go to step 3.

5   Simulation Results

Five cutting stock problems are used to examine Artificial Fish Swarm Algorithm. Each of the five test problems has its own optimal solution of zero trim loss. Thus we can estimate easily the performance of the algorithm. The number of the rectangular pieces in each stock plate ranges from 10 to 30. The population size of the swarm N is taken as 60, and the parameters in artificial fish swarm algorithm are taken as $\delta = 0.3$, $trynumber = 10$, $visual = 20,50,20,30,50$ for the five test problems respectively. The algorithms are written in C and run in a DELL Optiplex 380 computer. The maximum number of iterations is taken as 1000. For comparison, Particle Swarm Optimization (PSO) is used to the same five cutting stock problem. Fig. 1 shows one of the five test problems, where 30 pieces will be cut from the stock plate. Fig. 2 shows the actual layout generated using Artificial Fish Swarm Algorithm. Twenty-eight pieces are cut from the stock pieces. The shade in the stock plate is the lost area.

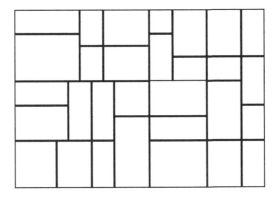

Fig. 1. A test problem with 30 pieces to be cut from the stock plate

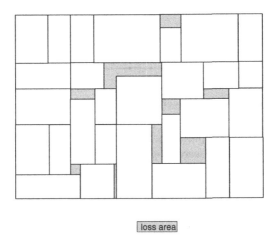

loss area

Fig. 2. Cutting results using Artificial Fish Swarm Algorithm

Table 1 shows the simulation results for the five test cases of the two-dimensional cutting problems. Table 2 shows the size of the five test cases. From the simulation results it can be seen that the performance of Artificial Fish Swarm Algorithm is better than that of Particle Swarm Optimization in dealing with the cases with more pieces to be cut.

Table 1. Simulated results for the five test cases

No.	1	2	3	4	5
Trim loss of PSO (%)	0	0	7.5	6.3	7.1
Trim loss of AFSA(%)	0	0	0	3	6.6

Table 2. The size of the five test cases

No.	Number of pieces	Size of stock plate	Sizes of pieces
1	10	100×80	$(30 \times 40) \times 2, (10 \times 40), (20 \times 40) \times 3,$ $(20 \times 50), \quad (20 \times 30), \quad (30 \times 20) \times 2$
2	10	40×20	$(20 \times 4), (16 \times 4), (10 \times 6) \times 2, (16 \times 6),$ $(4 \times 10), (10 \times 5) \times 2, (10 \times 10), (20 \times 10)$
3	15	40×20	$(5 \times 6) \times 4, (10 \times 3), (12 \times 5) \times 2, (10 \times 7) \times 2$ $(20 \times 4) \times 2, (10 \times 5) \times 3, (8 \times 10)$
4	20	40×40	$(11 \times 15) \times 4, (17 \times 5) \times 4, (12 \times 4) \times 4,$ $(3 \times 9) \times 4, (5 \times 15) \times 4$
5	30	65×45	$(17 \times 6), (6 \times 9) \times 3, (6 \times 12) \times 5, (9 \times 12) \times 4$ $(9 \times 6) \times 2, (6 \times 15) \times 2, (17 \times 12), (12 \times 9) \times 2$ $(11 \times 12), (9 \times 15), \quad (9 \times 18), (14 \times 9), (14 \times 6)$ $(9 \times 9), (15 \times 9), (15 \times 6), (15 \times 12), (6 \times 6),$

6   Conclusions and Discussions

From the results of the test problems it can be seen that Artificial Fish Swarm Algorithm has better ability to search for the global optimum for the two-dimensional cutting stock problem. When the number of the pieces is smaller, both Particle Swarm Optimization and Artificial Fish Swarm Algorithm work well. With larger number of pieces, the performance of Artificial Fish Swarm Algorithm is better than the PSO's.

References

1. Gilmore, P.C., Gomory, R.E.: A Linear Programming Approach to the Cutting Stock Problem. Operations Research 9, 849–859 (1961)
2. Gilmore, P.C., Gomory, R.E.: Multistage Cutting Stock Problems of Two and More Dimensions. Operations Research 13, 94–120 (1965)
3. Huang, W., Chen, D., Xu, R.: A New Heuristic Algorithm for Rectangle Packing. Computers & Operations Research 34, 3270–3280 (2007)
4. Jokobs, S.: On Genetic Algorithms for the Packing of Polygons. European Journal of Operational Research 88, 165–181 (1996)
5. Leung, T.W., Yung, C.H., Troutt Marvin, D.: Applications of Genetic Search and Simulated Annealing to the Two-dimensional Non-guillotine Cutting Stock Problem. Computer and Industrial Engineering 40, 201–214 (2001)
6. Gonçalves, J.F.: A Hybrid Genetic Algorithm-Heuristic for a Two-dimensional Orthogonal Packing Problem. European Journal of Operational Research 183, 1212–1229 (2007)
7. Levine, J., Ducatelle, F.: Ant Colony Optimization and Local Search for Bin Packing and Cutting Stock Problems. Journal of the Operational Research Society 55, 705–716 (2004)

8. Jiang, J.Q., Liang, Y.C., Shi, X.H., Lee, H.P.: A Hybrid Algorithm Based on PSO and SA and Its Application for Two-Dimensional Non-guillotine Cutting Stock Problem. In: Bubak, M., van Albada, G.D., Sloot, P.M.A., Dongarra, J. (eds.) ICCS 2004. LNCS, vol. 3037, pp. 666–669. Springer, Heidelberg (2004)
9. Shen, X., Li, Y., Yang, J., Yu, L.: A Heuristic Particle Swarm Optimization for Cutting Stock Problem Based on Cutting Pattern. In: Shi, Y., van Albada, G.D., Dongarra, J., Sloot, P.M.A. (eds.) ICCS 2007, Part IV. LNCS, vol. 4490, pp. 1175–1178. Springer, Heidelberg (2007)
10. Lai, K.K., Chan, W.M.: Developing a Simulated Annealing Algorithm for the Cutting Stock Problem. Computer and Industrial Engineering 33, 115–127 (1997)
11. Li, X., Shao, Z., Qian, J.: An Optimizing Method Based on Autonomous Animats: Fish-swarm Algorithm. Systems Engineering-theory & Practice 22, 32–38 (2002)
12. Lei, J.: Application Research of Artificial Fish Swarm Algorithm of in Combinatorial Optimization Problems. Xi'an University of Technology (2010)
13. Wei, Y.: One-Dimensional Cutting Stock Problem Based on Artificial Fish Swarm Algorithm. South China University of Technology (2010)

Utility-Driven Share Scheduling Algorithm in Hadoop

Cong Wan, Cuirong Wang, Ying Yuan, Haiming Wang, and Xin Song

College of Information Science and Engineering,
Northeastern University, Shenyang 110044, China
10000Cong@163.com

Abstract. Job scheduling in hadoop is a hot topic, however, current research mainly focuses on the time optimization in scheduling. With the trend of providing hadoop as a service to the public or specified groups, more factors should be considered, such as time and cost. To solve this problem, we present a utility-driven share scheduling algorithm. Considering time and cost, algorithm offers a global optimization scheduling scheme according to the workload of the job. Furthermore, we present a model that can estimate job execute time by cost. Finally, we implement the algorithm and experiment it in a hadoop cluster.

Keywords: scheduling, hadoop, utility, QoS.

1 Introduction

Recently years, with the development of network applications, The explosive growth of personal data and enterprise data caused many effective data storage and data processing problems. Just increasing the number of computer in the cluster does not fundamentally solve these problems. Google released GFS [1], BigTable [2] and MapReduce [3] to solve this problem.

MapReduce is a distributed programming model focus on large data sets on clusters of computers. Apache Hadoop [4] is a software framework that supports data-intensive distributed applications, which implements MapReduce and GFS. Companies, such as Facebook and Yahoo, are using Hadoop and a variety of applications around the Hadoop platform has been developed. Companies rely more and more on big data when making their decisions. Amazon, Cloudera, and IBM have announced their Hadoop-as-a-Service offerings, while Microsoft promises to do the same job recently. Amazon was the first company to provide Hadoop-as-a-Service. Liking many other IaaS service coming from Amazon, this service provides the minimum hardware and software necessary to run analytics on big data.

Commercialization Hadoop need to pay more and more attention to user satisfaction. Previous studies on Hadoop scheduling only use job running time as optimization objective. User satisfaction not only depends on the job running time, but also related to other factors, such as cost.

In this paper, we present a utility-driven share scheduling algorithm in Hadoop. We have set up a model which could estimate the job running time by the users bid and

C. Guo, Z.-G. Hou, and Z. Zeng (Eds.): ISNN 2013, Part II, LNCS 7952, pp. 560–568, 2013.

job workload. We designed an algorithm which maximum the whole user satisfaction by adjusting user fees.

2   Relate Works

There are some schedulers released with hadoop. FIFO is the default scheduler in hadoop. It select job according to the priority of the job and then the arrival time. Capability scheduler [5] supports multi queues which are allocated a fraction of the total resource capability. Fair scheduler [6] allows jobs to obtain the resource fairly with the passage of time.

Many researchers present there schedulers. Thomas Sandholm and Kevin Lai [7] present the Dynamic Priority (DP) parallel task scheduler for hadoop. It allows users to control their allocated capacity by adjusting their spending over time. But there work was not related to user satisfaction. Xicheng Dong et al. [8] present a mixed real-time and non-real-time job scheduler in hadoop. They try to meet users' QoS demands, but only consider the time factor. Jord`a Polo et al. [9] present a model to estimates individual job completion times given a particular resource allocation, and uses these estimates as to meeting job's performance goal. Kamal Kc and Kemafor Anyanwu [10] present a scheduler to meet the deadlines of jobs in hadoop. Hsin-Han You [11] and his team proposed the Load-Aware scheduler for heterogeneous environments with dynamic loading which could modify slot number according to the environment.

3   Design

3.1   Scheduling in Hadoop

MapReduce in hadoop consists of four processes: JobTracker, TaskTracker, JobClient and Task. A hadoop MapReduce job usually submitted by JobClient to the JobTracker, then JobTracker received the request and added it into the job queue. TaskTracker send a heartbeat to JobTracker regularly to ask for job. If the job queue is not empty, JobTracker selected a suitable task from a suitable job send to TaskTracker .All scheduling in hadoop are on the task level.

3.2   Problem Formulation

Time is the primary optimization objective of any scheduling problems, so our scheduling algorithm should take it into account. Any system providing service should consider problems such as priority, resource allocation and how to measure system workload and user satisfaction. Economic-based approach is an effective mean in the scheduling and price is a commonly used indicator. So we believe that users mainly focus on job running time and cost in a hadoop as a service environment. we assume that the user is not just concerned about price, but about his satisfaction.

We define a utility function described as equations (1) to describe the users' satisfaction.

$$U = U_0 - \alpha * T - \beta * Cost \quad . \tag{1}$$

U_0 represents the maximum satisfaction, which is a pre-defined. α and β represent the weight of time and cost which determined by the user.

First, we formulate the system. We defined job as Job_i , it related to the following parameters:

AT_i : job arriving time.

MW_i : map workload. we define the workload of MapReduce WordCount programme process a map task which have 64MB data as 1 unit.

RW_i : reduce workload.

RN_i : number of reduce task.

MN_i : number of map task.

α_i : the weight of time.

β_i : the weight of cost.

$FinMN_i$: the number of completed map.

P_i : the price of unit time.

smT_i : single map time.

$rate_i$: the proportion of resources should be allocated to Job_i . It can be calculated as equations (2).

$$rate_i = \frac{P_i}{\sum_{i=1}^{n} P_i} \quad . \tag{2}$$

$am_i = rate_i * SlotM$: assigned map slot number.

T_i : running time.

The cost of user include cost of the map phase and cost of reduce phase. In the map phase, each maptask must be paied and we assume each maptask spend the same time. In the reduce phase, the number of reduce task is defined by user, but cost can be caluculate by workload. The equation (3) describes the cost calculation method.

$$Cost_i = smT_i * MN_i * P_i + RW_i * P_i \quad . \tag{3}$$

There are some parameters about cluster. $CurrT$ represent the current time. $SlotM$ represent the number of map slot in the cluster. $SlotR$ represent the number of reduce slot in the cluster. P_{max} and P_{min} represent the system defined max and min price.

Our algorithm is a share scheduling algorithm which allocated to each Job a certain percentage of the resources and the proportion of resources is decided by the bid. The time of map stage time can be determined by the number of map task and allocated resources. It can be calculated using the equation (4).

$$
\begin{aligned}
mT_i &= CurrT - AT_i + smT_i * \frac{MN_i - FinMN_i}{am_i} \\
&= CurrT - AT_i + smT_i * \frac{MN_i}{\dfrac{P_i}{\sum_{i=1}^{n} P_i} * SlotM} \\
&= CurrT - AT_i + \frac{smT_i * (MN_i - FinMN_i)}{SlotM} * \frac{\sum_{i=1}^{n} P_i}{P_i}
\end{aligned}
\tag{4}
$$

The time of reduce phase can by calculated as equation (5).

$$
rT_i = \frac{RW_i}{RN_i}
\tag{5}
$$

The job running time can by calculated as equation (6).

$$
\begin{aligned}
T_i &= mT_i + rT_i \\
&= CurrT - AT_i + \frac{smT_i * (MN_i - FinMN_i)}{SlotM} * \frac{\sum_{i=1}^{n} P_i}{P_i} + \frac{RW_i}{RN_i}
\end{aligned}
\tag{6}
$$

We can use the equation (7) to calculate the utility of the job.

$$
\begin{aligned}
U_i &= U_0 - \alpha_i * T_i - \beta_i * Cost_i \\
&= U_0 - \alpha_i * (CurrT - AT_i + \frac{smT_i * (MN_i - FinMN_i)}{SlotM} * \frac{\sum_{i=1}^{n} P_i}{P_i} + \frac{RW_i}{RN_i}) - \beta_i * (smT_i * MN_i * P_i + RW_i * P_i)
\end{aligned}
\tag{7}
$$

For simplicity, we define three coefficients for each job which are described as equation (8) (9) (10).

$$a_i = -\frac{\alpha_i * smT_i * (MN_i - FinMN_i)}{SlotM} \quad .$$ (8)

$$b_i = -\beta_i * (smT_i * MN_i + RW_i)$$ (9)

$$c_i = U_o - \frac{\alpha_i * RW_i}{RN_i} - \alpha_i * (CurrT - AT_i)$$ (10)

Finally, the utility is defined as equation (11)

$$U_i = a_i * \frac{\sum_{i=1}^{n} P_i}{P_i} + b_i * P_i + c_i \quad .$$ (11)

The utility of the system can be calculated using equation (12) when the number of jobs is n.

$$U = \sum_{i=1}^{n} U_i$$

$$= \sum_{i=1}^{n} (a_i * \frac{\sum P_i}{P_i} + b_i * P_i + c_i) \quad .$$ (12)

$$= \sum_{i=1}^{n} \sum_{j=1}^{n} a_i * \frac{P_j}{P_i} + \sum_{i=1}^{n} b_i * P_i + \sum_{i=1}^{n} c_i$$

Our Scheduling goal is to maximize U by calculating the P_i. So, the optimization model defined as equation (13).

$$MAX \sum_{i=1}^{n} \sum_{j=1}^{n} a_i * \frac{P_j}{P_i} + \sum_{i=1}^{n} b_i * P_i + \sum_{i=1}^{n} c_i \quad .$$ (13)

$$s.t. \ \forall i, p_{max} > p_i > p_{min}$$

3.3   Genetic Algorithm

We use genetic algorithm to solve the optimization problem. The implementation of genetic algorithm can be detailed as following steps.

1. Encoding. Suppose there are n job, we use a 5*n length binary code to represent decision variables $p_1, p_2 ...... p_n$. When decoding we split the binary code into n binary series with five bit, the convert them to decimal code and plus 1.

2. Fitness. Since the optimization goal is to seek maximum value, so we defined the individual fitness to the corresponding utility function value, if the value is less than 0 then fitness is 0.

3. Selection. Using selection operator based on fitness ratio.

4. Crossover. Using one point crossover operator.

5. Mutation. Using bit mutation operator.

3.4   Scheduling Algorithm

The scheduling algorithm consists of three functions: jobadded and assigntask. They would be called when a job added and response the heartbeat request of TaskTracker.

When a new job arrives, first of all, we calculated a set of decision variables $P = \{ p_1, p_2 ... p_n \}$ using the utility as the optimization objective. Then, if the new utility greater than before, we recalculated the resource should be allocated to each job depending on P. Otherwise, the new job will be rejected.

When idle slot appears, calculate the hungry degree which is defined as the number of assigned slot divided by the number of required slot of each job and allocate the slot to the hungriest job. Algorithm pseudo-code is shown below.

Pseudo-code of function assigntask

```
void assigntask(newslot){
  int selected
  float minHungry = Float.maxFloat
  For each job_i in job queue{
    int desire stands for the desired slot
    int due stands for slot should be allocated
    int alc stands for slot already allocated
    if( job_i is running and desire>0){
      float hungry=alc/min(desire,due)
      if(hungry< minHungry)
      selected=i
    }
  }
  Assign new slot to job selected
}
```

4   Performance Evaluation

We set up Hadoop on a computer cluster which has 1 NameNode and 4 DataNodes. Both NameNode and 4 DataNodes are composed of 2.93GHz Intel Core2 Duo CPU and 2GB memory, All nodes were connected by a gigabit switch. In terms of software, we implement the application using jdk-6u26-windows-i586 and eclipse-SDK-3.6.1. The operating system is Ubuntu10.04 and the Hadoop version is Hadoop-1.0.3.

We configure 4 map slot and 2 reduce slot in each tasktracker. Genetic algorithm settings: Crossover probability is 0.35, mutation probability is 0.05, iteration time is 5000 and initial population size is 10.

We executed 3 real job to experiment out scheduling algorithm. The parameters of the three jobs are shown in Table 1. We defined workload of one map task in wordcount as workload unit. In this experiment, we first submit a wordcount job and a sencondarySort job 30 seconds later and then a TeraSort job 50 seconds later. The maximum satisfaction was defined as 100.

Table 1. Information of jobs

Job name	Job type	Map task number	Reduce task number	Map workload	Reduce workload	α	β
Job1	wordcount	58	1	58	2.48	0.1	0.9
Job2	secondarySort	59	1	29	11.96	0.9	0.1
Job3	TeraSort	58	1	43.07	16.5	0.5	0.5

Fig. 1 shows the number of slot allocated to the three job changes over time. In the beginning, Job1 as the only job got all slots. There is no idle slot when Job2 was submitted until the running map tasks were finished one by one at the 45th second. Job1 and Job2 were assigned part of slots. Slots were re-allocated because Job3 was submitted at the 50th second. Job2 was finished at the 253th second and the idle slots were assigned to the other jobs. Job3 was finished at the 287th second and Job1 got all the slots.

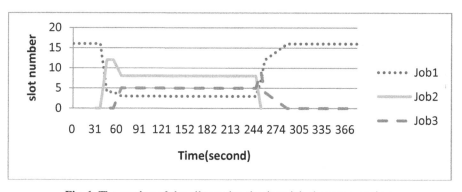

Fig. 1. The number of slot allocated to the three job changes over time

Table 3 shows the running time, cost and utility of the three jobs.

Table 2. Information of jobs

Job name	cost	Map running time(second)	Running time(second)	utility
Job1	60.48	380	485	44.5
Job2	188.96	223	921	62.6
Job3	128.5	237	743	27.25

5   Conclusion and Future Work

In this paper, we presented a scheduling algorithm in hadoop environment. Assume that the user focus on satisfaction, rather than just time or cost. The proportion of the allocation of resources is determined by the user bid which determined by the system. The user's preferences will affect their charges. Algorithm charge more to reduce the job running time; to the contrary, extend the running time to reduce costs to the users. Price also will float with the utilization of the resources which making the system a high satisfaction.

Experimental shows that more resources are allocated to jobs emphasizing on time than cost which improve the satisfaction of the entire system by adjusting the proportion of resources.

In future work, we will consider the satisfaction both system and individual job and solve it using game theory.

Acknowledgments. This work was supported by The Central University Fundamental Research Foundation, under Grant. N110323009, N110323007.

References

1. Ghemawat, S., Gobioff, H., Shan-Tak Leung: The Google file system. Operating Systems Review, Vol. 37(5), pp. 29-43(2003)
2. Chang, F., et al.: Bigtable: a distributed storage system for structured data. ACM Transactions on Computer Systems, Vol. 26(2), pp. 4-30(2008)
3. Dean, J., Ghemawat, S.: MapReduce: simplified data processing on large clusters. Commun. ACM 51(1), pp. 107-113(2008)
4. Hadoop: Open source implementation of MapReduce, http://hadoop.apache.org/
5. Hadoop CapacityScheduler, http://hadoop.apache.org/common/docs/current/capacity_ scheduler . html.
6. Hadoop FairScheduler, http://hadoop.apache.org/common/docs/current/fair_scheduler.html.
7. Thomas, S., Kevin L.: Dynamic Proportional Share Scheduling in Hadoop. Lecture Notes in Computer Science, Vol. 6253, pp. 110-131(2010)

8. Xicheng, D., Ying, W., Huaming, L.: Scheduling Mixed Real-time and Non-real-time Applications in MapReduce Environment. In: 17th IEEE International Conference on Parallel and Distributed Systems (ICPADS), pp. 9-16, Tainan(2011)

9. Polo, J., et al.: Performance-Driven Task Co-Scheduling for MapReduce Environments. In: 2010 IEEE/IFIP Network Operations and Management Symposium - NOMS, pp. 373-380(2010)

10. Kc, K., Anyanwu, K.: Scheduling Hadoop Jobs to Meet Deadlines. In: Proceedings of the 2010 IEEE 2nd International Conference on Cloud Computing Technology and Science (CloudCom 2010), pp. 388-392(2010)

11. You, H., Yang, C., Huang, J.: A load-aware scheduler for MapReduce framework in heterogeneous cloud environments. In: Proceedings of the ACM Symposium on Applied Computing, pp. 127-132(2011)

Research on Fault Diagnosis
of Transmission Line Based on SIFT Feature

Shujia Yan[1], Lijun Jin[1,*], Zhe Zhang[2], and Wenhao Zhang[1]

[1] College of Electronic and Information Engineering, Tongji University,
Shanghai, 201804, China
{144yanshujia,jinlij,xiwen}@tongji.edu.cn
[2] Tongyuan Architectural Design Institute, Tongji Architectural Design (GROUP) CO., LTD,
Shanghai, 200092, China
zz4193@163.com

Abstract. Recent interest in line-tracking methods using UAV has been introduced in the research of pattern recognition and diagnosis of transmission system. A fault diagnosis method for transmission line based on Scale Invariant Feature Transform (SIFT) is proposed in this paper, which recognizes fault images by comparing aerial images with model images. The reliability and efficiency of the system is effectively improved by pro-calculating local scale-invariant features of models. The research can provide a new method for predictive maintenance of the transmission line.

Keywords: Transmission line, scale invariant feature transform (SIFT), feature matching, fault diagnosis.

1   Introduction[1]

Taking into account China's vast territory and geographical structure, development of regional economies is imbalance. Energy bases in the mainland are far away from load centers loaded along the coast and especially in South China, which gradually formed a complex topology structure of transmission system. The running state directly affects the security and stability of power grid. Because of long exposure to the nature, transmission lines constantly suffer from mechanical tension, aging of materials, artificial damage, natural disasters and other factors, which produce some damages as collapse of tower, broken of strand, abrasion and corrosion. Table 1 shows the statistics of fault types in Sichuan province during 2005-2010[1]. It will eventually lead to severe accidents without timely maintenance and replacement, which cause widespread blackouts and huge economic losses. Therefore, conducting regular transmission line inspection and timely detecting equipment's defects that endanger the safe operation are important to power system.

---

[1] This work was supported by the National Natural Science Foundation of China under Grant 51177109.
* Corresponding author.

C. Guo, Z.-G. Hou, and Z. Zeng (Eds.): ISNN 2013, Part II, LNCS 7952, pp. 569–577, 2013.

Table 1. Statistics of fault types in Sichuan province during 2005-2010

Year	Icing	Fire	Birds	Pollution Flasfover	Product Quality	Artificial Damage	Unknown Causes	Others	Total
2005	2	3	1	3	2	0	1	0	12
2006	1	3	5	0	0	2	2	0	13
2007	7	0	4	0	0	0	0	0	11
2008	29	2	1	0	1	3	0	0	36
2009	1	4	4	2	0	1	1	5	18
2010	4	2	3	1	2	2	0	0	14
Total	44	14	18	6	5	8	4	5	104
Ratio	42	13	17	6	5	8	4	5	100

Currently, the inspection method widely used in transmission system is artificial method. This method has disadvantages of heavy workload, high risk, great administration costs, low detecting precision and poor reliability, etc[2]. In the 1950's, some developed western countries started to use helicopters for rush repairs of overhead transmission lines. After decades of research and exploration, helicopters' application in overhead transmission lines has expanded to inspection, maintenance of live wires, and line construction. Recent research on Unmanned Aerial Vehicles (UAVs) inspection system becomes a hot topic about smart grid[3,4].

The new transmission system inspection method with helicopter or UAV is a good way to solve the problems in artificial inspection, but meanwhile bring some new challenges. There are a huge number of aerial images need to be selected after each inspection, which will spend a lot of labor power. Thus how to identify fault images from the whole set of aerial images becomes the key and difficult point in the smart inspection system. Utal state university published a white paper *Aerial Surveillance System for Overhead Power Line Inspection* in 2003, in which the feasibility and technical requirements of transmission line automatic inspection based on aerial images were discussed. This paper also pushed forward the application of image processing in transmission system inspection[5].

Transmission system inspection based on image processing is an emerging industry, still in its infancy at the present. The research on identification of transmission equipment and fault localization has rarely involved at home and abroad. Li et al.[6] presented an algorithm of automatic extraction of 550kV transmission lines from in cluttered backgrounds. The potential power line pixels are acquired by Ratio operator, and then part Radon transform is used to acquire and link the segments. A PCNN filter is developed to remove background noise from the images prior to the Hough transform being employed to detect straight lines in [7]. The image segmentation is applied to S component in HSI space by Huang and Zhang[8] using the maximum entropy threshold method based on genetic algorithm.

In 1999, Lowe[9] proposed a new feature descriptor *Scale Invariant Feature Transform* (SIFT) based on image gradient distribution. The features are invariant to image scale and rotation, and provide robust matching across a substantial range of affine distortion. This paper presents a new method for aerial images recognition and fault diagnosis of transmission system based on SIFT features matching, which provides a basis for follow-up maintenance.

2 Fault Diagnosis for Transmission System

Camera shake in the course of aerial photography causes distortion, blurring and anamorphous of aerial images. And illumination variation can lead to brightness variance problems as well. The image features extracted by conventional feature extraction algorithms will be changed with the deformation of the target object, which will generate failure or fault in the step of image matching.

SIFT is an algorithm in computer vision to detect and describe local features in images. The SIFT features are local and based on the appearance of the object at particular interest points, and are invariant to image scale and rotation[10]. They are also robust to changes in illumination, noise, and minor changes in viewpoint. In addition to these properties of SIFT algorithm, it is highly distinctive, predictably easy to match electrical equipment and hardware fittings such as steel towers, insulators, vibration dampers and spacers[11]. In this study, the aerial images of transmission line malfunctions are extracted via matching, correction, detection and diagnosis. The SIFT algorithm is used in the step of matching, which means finding a one-to-one correspondence between sample images and template images.

2.1 The SIFT Algorithm

As described in [12], the SIFT algorithm has four major stages: (1) scale-space peak selection; (2) keypoint localization; (3) orientation assignment; (4) keypoint descriptor. The main idea of SIFT is detecting local peaks in difference-of-Gaussian (DoG) images, and acquiring their locations, scales and orientations. The scale space $L(x, y, \sigma)$ of an image $I(x, y)$ is defined as (1).

$$L(x, y, \sigma) = G(x, y, \sigma) * I(x, y) \tag{1}$$

where $G(x, y, \sigma)$ is Gaussian function with variable scale σ.

$$G(x, y, \sigma) = \frac{1}{2\pi\sigma^2} e^{-\frac{x^2+y^2}{2\sigma^2}} \tag{2}$$

In order to detect stable keypoints in $L(x, y, \sigma)$, the DoG function, $D(x, y, \sigma)$, computed from the difference of two nearby scales, is used[12].

$$\begin{aligned} D(x, y, \sigma) &= (G(x, y, k\sigma) - G(x, y, \sigma)) * I(x, y) \\ &= L(x, y, k\sigma) - L(x, y, \sigma) \end{aligned} \tag{3}$$

Scale extrema are obtained by comparing each sample point to its 26 neighbors in a 3×3 cube which is centered on the sample point. In fact, these extrema that are close together are quite unstable to small perturbations of the image. We can determine the best choices through experiments of sampling frequencies.

By assigning a consistent orientation to each keypoint based on local image properties, the keypoint descriptor can be represented relative to this orientation and therefore achieve invariance to image rotation. An orientation histogram is formed from the gradient orientations of sample points within a region around the keypoint. The orientation histogram has 36 bins covering the 360 degree range of orientations. As shown in Fig.1, the highest peak in the histogram is detected to assign its orientation to keypoint.

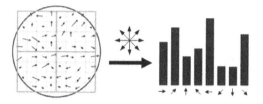

Fig. 1. Orientation histogram of sample points

The previous operations have assigned an image location, scale, and orientation to each keypoint. These parameters impose a repeatable local coordinate system in which to describe the local image region, and therefore provide invariance to these parameters. The next step is to compute a descriptor for the local image region that is highly distinctive yet is as invariant as possible to remaining variations. A keypoint descriptor is created by first computing the gradient magnitude and orientation at each image sample point in a region around the keypoint location. Then these samples are then accumulated into orientation histograms with 8 orientation bins. The descriptor is formed from a vector containing the values of all the orientation histogram entries, corresponding to the lengths of the arrows. The best results are achieved with a 4×4 array of histograms. Therefore, the experiments in this paper use a feature vector including 128 elements for each keypoint[13].

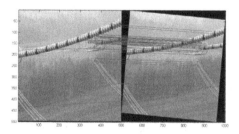

Fig. 2. SIFT features matching for transmission system

The Euclidean distance between two feature vectors is used to determine whether the two vectors correspond to the same keypoint in different images. The principle of correctly matching can be determined by the ratio among distance of the closest neighbor and the distance of the second closest[14]. A threshold is given in advance. If the ratio is lower than threshold, the matching is accepted. Smaller ratio threshold means less matching feature points but more reliability. Figure 2 shows that SIFT algorithm can get and match the features of the transmission line accurately, and is invariant to the scale and rotation changes to meet the requirements of transmission system inspection.

2.2 Image Registration Based on FMT

The model images and sample images are obtained in previous and next transmission system inspection respectively, which are different both in space and time. Usually the images of 3D objects are very complex[15]. In practice, the movement of objects such as zoom and rotation can be simplified to planar motion approximately, just like the case proposed above[16]. Following, FMT is used to realize calibration and registration on model and sample images after SIFT match[17].

Assume that after translations of $f_1(x, y)$ by x_0 and y_0 in directions x and y respectively, we get $f_2(x, y)$:

$$f_2(x, y) = f_1(x - x_0, y - y_0)$$ (4)

Assume that the Fourier transforms of $f_1(x, y)$ and $f_2(x, y)$ are $F_1(u, v)$ and $F_2(u, v)$ respectively, the relationship of $F_1(u, v)$ and $F_2(u, v)$:

$$F_2(u, v) = F_1(u, v) e^{-j(ux_0 + vy_0)}$$ (5)

So the cross-power spectrum of $f_1(x, y)$ and $f_2(x, y)$ can be written in (6):

$$\frac{F_1(u, v) \cdot F_2^*(u, v)}{\left| F_1(u, v) \cdot F_2^*(u, v) \right|} = e^{j(ux_0 + vy_0)}$$ (6)

where, $F_1^*(u, v)$ and $F_2^*(u, v)$ are complex conjugates of $F_1(u, v)$ and $F_2(u, v)$ respectively. In space (x, y), there will be a unit impulse function at (x_0, y_0) by inverse Fourier transform of (6), as shown in Fig. 3(a). The impulse localization is just the relative translations x_0 and y_0[18]. Because of the partial occlusion of model and sample as well as noise and error, the correlation value distribution of inverse Fourier transform of (6) is shown in Fig. 3(b). On the contrary, if the planar motion transform of model and sample is not satisfied, there will not be an obvious peak of inverse Fourier transform of (3), as shown in Fig. 3(c).

For rotation and scale transform, the same results can be achieved through polar coordinate and logarithm change. Fig. 4(a) and Fig. 4(b) are the model image and the sample image respectively, and Fig. 4(c) is the image after calibration, where the angle correction and scale transform are both executed.

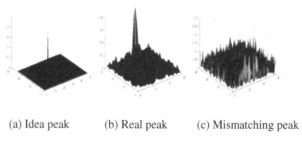

(a) Idea peak   (b) Real peak   (c) Mismatching peak

Fig. 3. Correlation value distribution of FMT image

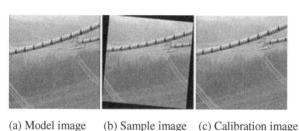

(a) Model image   (b) Sample image   (c) Calibration image

Fig. 4. FMT calibration for image of transmission system

2.3   Fault Image Recognition

After the image matching and accurate registration, in order to improve the computation speed, we get the image variation using the difference of two images. Allowing for the recognition reliability, the fault images can be recognized via setting proper threshold. Because of the accuracy error, model image and calibration image usually are not complete registration, as shown in Fig. 4. The calculated zoom coefficient is 1.0967, while the real zoom coefficient is 1.10. Fig. 5(a) shows the difference binary image of model image and calibration image.

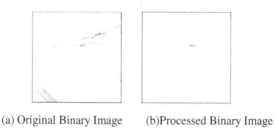

(a) Original Binary Image   (b)Processed Binary Image

Fig. 5. Difference of Model and Calibration binary images

Tiny line segments are detected through Hough transform in difference image[21]. And then, the necessary changes in difference image are left after deleting tiny line segments. Figure 5(b), which indicates a vibration damper loss, is the binary image of difference image after deleting tiny line segments.

3   Fault Diagnosis System

3.1   System Test

The fault diagnosis system of transmission system inspection contains 4 basic modules: image matching, image calibration, fault diagnosis and display[22-23]. System processing diagram is shown in Fig. 6.

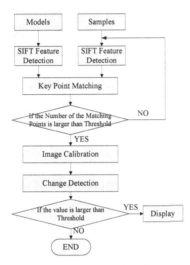

Fig. 6. System processing diagram

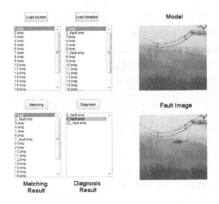

Fig. 7. GUI of fault diagnosis system

As shown in Fig. 7, model images and sample images are listed in listboxes 1 and 2 respectively, as well as matching result in listbox 3 and fault images in listbox 4. The fault image and corresponding model image are displayed on the right when we select the image name in listbox4.

3.2   Result Analysis

The whole sample set contains 500 aerial images obtained on UAV, and there are 50 fault samples and 450 normal samples respectively. The result shows that 55 samples are regarded as fault images by system, including 6 normal samples misdiagnosed. The great variation of those samples causes the misdiagnosis. One damaged cable fault is regarded as normal sample because of its linearity feature which is hardly detected.

Table 2. Test result of real UAV aerial images

Fault Samples	Fault Samples Recognized	Fault Samples Missed	Samples Misdiagnosed	Reliability Rate	Accuracy Rate
50	55	1	6	98%	98%

As shown in Table 2 reliability rate of the fault diagnosis is 98% while accuracy rate is 89%. Since the fundamental principle of the system is pick up the potential fault samples reliably from plenty of aerial images, the method mentioned in this paper is significant to the research on fault diagnosis of transmission system based on pattern recognition.

4   Conclusion

Contrast to artificial detection, transmission system inspection using UAVs has numerous advantages. However, it brings huge amounts of aerial images, which will also need intensive labor. A fault diagnosis method using SIFT and FMT is developed for solving this problem. The experiment of diagnosis system shows that the fault images can be recognized accurately and quickly, and the accuracy of the proposed method reaches 98% on real UAVs aerial images, which will assist pipeline riders to obtain the information of transmission lines condition in time.

References

1. Chen, X., Ma, Y., Xu, Z.: Research on Transmission Lines Cruising Technology with the Unmanned Aerial Vehicle. Southern Power System Technology 6, 59–61 (2008)
2. Ian, G., Dewi, J.: Corner detection and matching for visual tracking during power line inspection. Image and Vision Computing 21, 827–840 (2003)
3. Tong, W., Yuan, J., Li, B.: Application of Image Processing in Patrol Inspection of Overhead Transmission Line by Helicopter. Power System Technology 12, 204–207 (2010)
4. Li, B., Wang, Q., Wang, B., et al.: Applying Unmanned Autonomous Helicopter to Transmission Line Inspection. Shandong Power Technology 1, 1–4 (2010)
5. Ma, L., Chen, Y.: Aerial Surveillance System for Overhead Power LineInspection. Utah State University (2003)

6. Li, C., Yan, G., Xiao, Z.: Automatic Extraction of Power Lines from Aerial Images. Journal of Image and Graphics 6, 1041–1047 (2007)
7. Li, Z., Liu, Y., Hayward, R., et al.: Knowledge-based Power LineDetection for UAV Surveillance and Inspection Systems. In: The IEEE Conference on Image and Vision Computing, New Zealand (2008)
8. Huang, X., Zhang, Z.: A Method to Extract Insulator Image From Aerial Image of Helicopter Patrol. Power System Technology 1, 194–197 (2010)
9. Lowe, D.G.: Object Recognition from Local Scale-Invariant Features. In: International Conference on Computer Vision, Corfu, Greece (1999)
10. Li, L.: Study of the Key Technologies of the Feature-Point-Based Image Matching. Shandong University of Science and Technology, Shandong (2009)
11. Lowe, D.G.: Object Recognition from Local Scale-Invariant Features. In: The Proceedings of the Seventh IEEE International Conference, vol. 2, pp. 1150–1157 (1999)
12. Lowe, D.G.: Distinctive Image Features from Scale-Invariant Keypoints. International Journal of Computer Vision 2, 194–197 (2004)
13. Liu, X., Yang, J., Sun, J.: Image Registration Approach Based on SIFT. Infrared and Laser Engineering 37, 156–160 (2008)
14. Wang, L., Ma, S., Xue, H.: A New Improved Algorithm Based on SIFT Feature Matching. Journal of Inner Mongolia University 40, 615–619 (2009)
15. Kenneth, R.C.: Digital image processing. Prentice Hall, Englewood Cliffs (1996)
16. Liu, Y., Shi, W., Xu, Y.: Analysis and Recognition of Image Differences. Journal of Fudan University 39, 472–476 (2000)
17. Liu, W., Cui, J., Zhou, L.: Subpixel Registration Based on Interpolation and Extension Phase Correlation. Journal of Computer Aided Design & Computer Graphics 17, 1273–1277 (2005)
18. Gao, Y., Yang, J., Ma, X.: Interference Image Registration based on Fourier-Mellin Algorithm. Optics. and Precision Engineering 15, 1415–1420 (2007)
19. Hui, Z.: Research on SAR Image Change Detection. Xidian University, Xi'an (2008)
20. Shu, S.: Research on Remote Sensing Image Change. University of Science and Technology of China, Hefei (2008)
21. Aggarwal, N., Karl, W.C.: Line detection in images through regularized Hough transform. IEEE Transactions on Image Processing 15, 582–591 (2006)
22. Li, L., Zheng, S.: Application of MATLAB to Image Processing Technique. Micro Computer Information (2003)
23. Li, X.: Matlab Interface Design and Compilation Skills. Publishing House of Electronics Industry, Beijing (2006)

Research of Pneumatic Actuator Fault Diagnosis Method Based on GA Optimized BP Neural Network and Fuzzy Logic

Zhigang Feng[1], Xuejuan Zhang[1], and He Yang[2]

[1] Department of Automation, Shenyang Aerospace University, Shenyang, China
[2] Science and Technology Association, Shenyang Aerospace University, Shenyang, China
fzg1023@yeah.net, zxj0635@163.com

Abstract. In this paper, a pneumatic actuator fault diagnosis method based on GA optimized BP neural network and fuzzy logic is proposed. First of all, the Genetic algorithm is used to optimize the weights of BP neural network, overcoming the shortcoming of neural network including over learning and local optimum. Then the normal actuator model is trained by the GA optimized BP neural network using the health data of actuator. The residual is generated by comparing the output of the BP trained actuator model and the actual actuator, which is used to detect the fault. Finally, fuzzy logic reasoning is used to isolate the fault type of actuator. The simulation results based on DAMADICS valve model and Lublin Sugar Factory failure data indicate that the proposed method can detect and diagnosis fault of actuator fast and accurately.

Keywords: pneumatic actuator, fault diagnosis, GA, BP neural network, fuzzy logic.

1 Introduction

Pneumatic actuator relies on compressed air as its power, which can be simple to realize rapid linear circular movement, with its simple structure, convenient maintenance, no pollution, and can be used in all kinds of bad working environment such as an explosion protection requirements, dust or wet condition, is widely used in the field of industrial automation. It plays an important role in the process industry. As most actuators work in high temperature, high pressure and corrosive environment, and the movement is frequent, each component is prone to failure, so the research on actuator fault diagnosis technology is very important to the reliability of the whole system. Due to its own structure and materials problems, there are some characteristics such as the inherent nonlinear, time delay, parameter drift, static friction and dynamical friction and mechanical initial return error of pneumatic actuator [1]. Therefore, there are many difficulties and challenges in the research of pneumatic actuator fault diagnosis method.

All kinds of pneumatic actuator fault diagnosis methods have been proposed, mainly divided into three categories, namely, the methods based on the analytical

C. Guo, Z.-G. Hou, and Z. Zeng (Eds.): ISNN 2013, Part II, LNCS 7952, pp. 578–585, 2013.

model, neural network and signal processing [2]. These methods have themselves advantages and the combination of many different kinds of intelligence technologies will be a trend of actuator fault diagnosis, especially the combination of neural network, fuzzy logic and expert system [3]. In this paper, it takes the DAMADICS valve simulation model and Lublin Sugar Factory industrial control system as the application background, pneumatic actuator fault diagnosis method based on GA optimized BP neural network and fuzzy logic is proposed. Firstly, using GA to optimize weights and thresholds of BP neural network. Then training the health actuator by the optimized BP neural network, and compared with the output of actual actuator to get residuals of fault and completed fault detection. Finally, using the method of fuzzy logic reasoning to complete fault isolation.

2   BP Neural Network Optimizing Using Genetic Algorithm

At present, BP neural network is one of the most mature neural network models. With the abilities of self-learning, self-adaptability, robustness and generalization, the BP neural network has been widely used in pattern recognition, function approximation, image processing and other fields. But the BP neural network converges very slowly

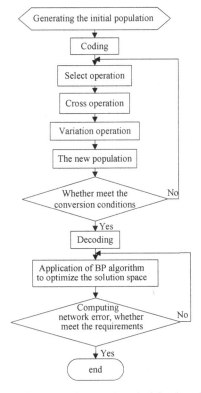

Fig. 1. The flow chart of BP neural network optimizing based on genetic algorithm

and is likely to fall into local minimum point. The genetic algorithm which can finish global parallel random search and converge to the global optimum, is a kind of bionic algorithm in macro significance. Thus, the combination of genetic algorithm and BP neural network is helpful to improve the global search ability and get stable network. The rate of convergence can be increased by using genetic algorithm to optimize weights and thresholds of BP neural network. Also it is able to overcome the disadvantage of easily getting into local minimum point, the number of the hidden layer neuron is solved according to optimizing the network structure. Then the optimized BP neural network based on genetic algorithm is established [4].

The step of the combination of genetic algorithm and BP neural network is as follows:

Step 1: The optimization parameters are encoded to generate initial population;

Step 2: The population is optimized by genetic algorithm;

Step 3: Computing the corresponding parameters and to see whether or not they meet the conversion condition between BP algorithm (in this paper, the conversion condition gm=100), if meet them step 4 will be carried out. Otherwise, step 2 is continued;

Step 4: The chromosome is decoded and the population is optimized by BP algorithm;

Step 5: Computing the target error of network. And the training will finish if the error satisfies the end condition. Otherwise, the population is optimized by BP algorithm unceasingly. The flow chart is shown in Fig. 1.

3   Fault Diagnosis of Pneumatic Actuator

3.1   The Structure and Common Fault Description of Pneumatic Actuators

Pneumatic actuator consists of three parts, pneumatic actuator, positioner and control valve [5], as shown in Fig. 2. Actuator keeping continuous and completing the given functional requirements in a certain period of working time, so that making the given

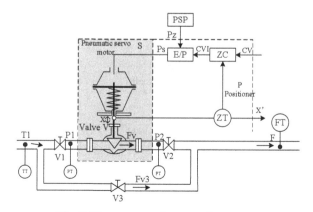

Fig. 2. The structure of pneumatic actuator

parameter value to maintain within the limits prescribed and this kind of performance is called the trouble-free performance, when the performance is damaged it will generate a fault. Literature [6] summarized the 19 kinds of typical fault, such as Table 1 shows:

Table 1. The 19 kinds of typical fault

Fault	Description	Fault	Description
f1	valve clogging	f11	servomotor spring fault
f2	valve or valve seat sedimentation	f12	electro-pneumatic transducer fault
f3	valve or valve seat erosion	f13	stem displacement sensor fault
f4	increase of valve friction,	f14	pressure sensor fault
f5	external leakage (leaky bushing, covers, terminals)	f15	positioner spring fault
f6	internal leakage (valve tightness)	f16	positioner supply pressure drop
f7	medium evaporation or critical flow	f17	unexpected pressure change across valve
f8	twisted servo-motor stem	f18	fully or partly opened bypass valves
f9	servomotor housing or terminal tightness	f19	flow rate sensor fault
f10	servomotor diaphragm perforation		

3.2   Fault Detection

Fig. 2 shows that the input and output of the pneumatic actuator have the following parameters:

Input: CV (Control Value): valve position command signal; P1 (Press inlet): pressures on valve: inlet; P2 (Press outlet): pressures on valve: outlet.

Output: X: valve plug displacement; F (Flow): actuator flow rate signal.

According to the principle of actuator action, it is known that there are some function relationships between the input and output of the actuator. Therefore, to establish the mathematical model as follows.

$$X = f(CV) \tag{1}$$

$$\hat{F} = f(X, P1, P2) \tag{2}$$

$$\hat{F} = f(CV, P1, P2) \tag{3}$$

$$\hat{F} = f(X) \tag{4}$$

$$\hat{F} = f(CV) \tag{5}$$

When fault occurs, fluid flow F or valve plug displacement X will be abnormal, the function relationships above will change. Actuator failures can be detected by detecting these changes. The intuitive method of abnormality detection is to detect the residuals of output of fault actuator and normal actuator. From the above five models draw the residuals as follows:

$$r_1 = X - \hat{X}(CV) \tag{6}$$

$$r_2 = F - \hat{F}\left(X, \sqrt{P1 - P2}\right) \tag{7}$$

$$r_3 = F - \hat{F}\left(CV, \sqrt{P1 - P2}\right) \tag{8}$$

$$r_4 = F - \hat{F}(X) \tag{9}$$

$$r_5 = F - \hat{F}(CV) \tag{10}$$

DAMADICS model is a classic valve fault simulation model [6], it can produce various types and intensities fault data, and also can import the actual industrial data into the simulation model. The model can generate the fault as in Tab. I . In this paper, actual industrial data which import into DAMADICS model is come from Lublin Sugar Factory, These data are produced in trouble-free operation. We conduct research for the valve of the Lublin Sugar Factory, and import P1, P2, X, F into model, and then add 19 kinds of fault to simulation model so as to get fault data. Each fault runs 100 times, each time runs 2000 seconds, and eventually we got the 100 samples.

Using GA-BP network to process the data sample, and choose appropriate neurons of hidden layer and output layer, and construct five neural networks. Using trouble-free sampling data of actuator as learning samples to train the network, and the GA was used to optimize the weights of neural network, so that the neural networks approach above five actuator models. Comparing the difference of the actual output and network output to produce residuals, and if one or a plurality of r1 ~ r5 abnormality, it suggests that there is a fault occurs.

For example, f7 and f13, its trend graph of r1 ~ r5 are shown in Fig.3 and Fig.4. From Fig. 3 and Fig. 4 we can see, when f1 or f7 occurs, r1 ~ r5 is abnormal, but the symbol of residuals is different. Some simulations were carried out for 19 kinds of fault, through analysis carefully. We found that once one or more of the residual signals abnormality, there must be a fault occurs.

3.3   Fault Isolation

When a fault occurs, the residuals will be abnormal, the absolute value of residuals will be larger. Thus, we put forward the fuzzy rules as follows:

If r_i is "big" then r_i=+1; (when r_i>0)

If r_i is "small" then r_i=-1; (when r_i<0)

If r_i is "appropriate" then r_i=0;

Through the analysis of the fault residuals, found that the combination of abnormal residuals is different, so the type of fault is different. For example, when the fault f7 and f13 respectively occurs, the trend performance of r1 ~ r5 is different (see Fig.3 and Fig.4). When f7 occurs, r1=-1, r2=r3=r4=r5=1. However, when f13 occurs, r1=0, r2=r3=r4=r5=-1. Thus, the fault isolation can be completed. According to the value of each group of residuals to separate the fault and get fault isolation table (see Table 2).

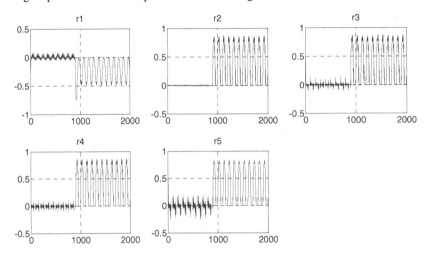

Fig. 3. r1 ~ r5 trend graph of fault f7

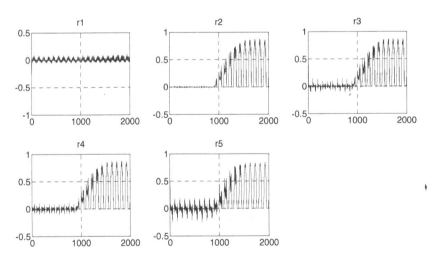

Fig. 4. r1 ~ r5 trend graph of fault f13

With r1, r2, r3, r4, r5 as input, FGN0, FGN1...FGN9 as output, constructing Fault Detection Diagnosis and Fuzzy Inference System (FDDFis) in Matlab FIS Editor. According to the discussion of residual fuzzification, using trapezoidal membership functions to fuzzy the input. According to the fault isolation table (see Table 2), can get the following diagnosis rules:

RULE1:IF r1=0 AND r2=0 AND r3=0 AND r4=0 AND r5=0 THEN FGN0=1 FGN1=0 FGN2=0 FGN3=0 FGN4=0 FGN5=0 FGN6=0 FGN7=0 FGN8=0 FGN9=0;RULE2:IF r1=-1 AND r2=1 AND r3=1 AND r4=1 AND r5=1 THEN FGN0=0 FGN1=1 FGN2=0 FGN3=0 FGN4=0 FGN5=0 FGN6=0 FGN7=0 FGN8=0 FGN9=0;RULE3:IF r1=0 AND r2=0 AND r3=0 AND r4=0 AND r5=1 THEN FGN0=0 FGN1=0 FGN2=1 FGN3=0 FGN4=0 FGN5=0 FGN6=0 FGN7=0 FGN8=0 FGN9=0;RULE4:IF r1=0 AND r2=1 AND r3=0 AND r4=0 AND r5=1 THEN FGN0=0 FGN1=0 FGN2=0 FGN3=1 FGN4=0 FGN5=0 FGN6=0 FGN7=0 FGN8=0 FGN9=0;

RULE5:IF r1=0 AND r2=0 AND r3=0 AND r4=0 AND r5=-1 THEN FGN0=0 FGN1=0 FGN2=0 FGN3=0 FGN4=1 FGN5=0 FGN6=0 FGN7=0 FGN8=0 FGN9=0;RULE6:IF r1=0 AND r2=-1 AND r3=-1 AND r4=-1 AND r5=-1 THEN FGN0=0 FGN1=0 FGN2=0 FGN3=0 FGN4=0 FGN5=1 FGN6=0 FGN7=0 FGN8=0 FGN9=0;RULE7:IF r1=0 AND r2=1 AND r3=1 AND r4=1 AND r5=0 THEN FGN0=0 FGN1=0 FGN2=0 FGN3=0 FGN4=0 FGN5=0 FGN6=1 FGN7=0 FGN8=0 FGN9=0;RULE8:IF r1=0 AND r2=1 AND r3=1 AND r4=1 AND r5=1 THEN FGN0=0 FGN1=0 FGN2=0 FGN3=0 FGN4=0 FGN5=0 FGN6=0 FGN7=1 FGN8=0 FGN9=0;

RULE9:IF r1=0 AND r2=0 AND r3=0 AND r4=1 AND r5=1 THEN FGN0=0 FGN1=0 FGN2=0 FGN3=0 FGN4=0 FGN5=0 FGN6=0 FGN7=0 FGN8=1 FGN9=0;RULE10:IF r1=1 AND r2=-1 AND r3=-1 AND r4=-1 AND r5=-1 THEN FGN0=0 FGN1=0 FGN2=0 FGN3=0 FGN4=0 FGN5=0 FGN6=0 FGN7=0 FGN8=0 FGN9=1;

Table 2. Fault isolation table

Fault Isolation Group	r1	r2	r3	r4	r5	Fault Type
FGN0	0	0	0	0	0	fault free
FGN1	-1	1	1	1	1	f1,f7,f10,f16
FGN2	0	0	0	0	1	f2,f3,f5,f6
FGN3	0	1	0	0	1	f4
FGN4	0	0	0	0	-1	f8,f9
FGN5	0	-1	-1	-1	-1	f11
FGN6	0	1	1	1	0	f12
FGN7	0	1	1	1	1	f13,f17f18,f19
FGN8	0	0	0	1	1	f14
FGN9	1	-1	-1	-1	-1	f15

Remark: FGN0(Fault Group Number 0), meaning fault isolation group 0.

Put these rules input the rule editor to generate FDDFis system. For f7, as an example, shows the working condition of the system. Making the fault residuals input to the fuzzy system FDDFis, got the fault diagnosis result. The output of FGN1 is 1, according to our fault separation matrix, f7 is belong to the type of FGN1. For other faults also made a similar simulation test, the result is consistent with the idea. It indicates that the algorithm is effective.

4   Conclusion

In this paper, a pneumatic actuator fault diagnosis method based on GA optimized BP neural network and fuzzy logic is presented. Firstly, it puts forward five actuator models and fault residuals models, then using BP network optimized by GA to structure the five models, and got the fault residuals, through detecting the residuals found that whether actuator failure or not, namely fault detection. Through the analysis of the fault residuals obtained the fault isolation matrix, according to the matrix abstracted out fuzzy rules of the fault isolation, then constructed Fault Detection Diagnosis and Fuzzy Inference System with the residuals for input and fault isolation group for output. Using GA to optimize the BP neural network improved the convergence speed and accuracy of network effectively. By a lot of simulation tests, the result shows the method can realize fault detection and fault isolation. But this method also has some deficiencies, to some faults such as f10, f16, the separation effect is ineffective, so the parameters of input membership function and neural network should be further improved.

Acknowledgment. The authors would like to thank the financial support of the financial support of the National Natural Science Foundation of China NSFC-61104023.

References

1. Shi, L., Fei, M.: Progess of Research on Fault Diagnostics and Tolerance Control for Pneumatic Actuator. Process Automation Instrumentation 25(8), 4–9 (2004)
2. Huang, X., Liu, J., Niu, Y.: Actuator Fault Detection and Diagnosis Methods in Control Systems. In: Control Engineering of China, vol. 10, pp. 165–169 (May 2003)
3. Sun, L.: The Method of Actuator Fault Diagnosis. China Water Transport 7(10), 172–175 (2007)
4. Ma, Q., Ma, S., Mao, Z.: Using Genetic Algorithm to Optimize BP Neural Network. Science Mosaic, 196–197 (December 2008)
5. Bartys, M., Patton, R., Syfert, M., de las Heras, S., Quevedo, J.: Introduction to the DAMADICS actuator FDI benchmark study. Control Engineering Practice 14(6), 609–619 (2006)
6. Bocaniala, C.D., da Costa, J.S.: Application of a novel fuzzy classifier to fault detection and isolation of the DAMADICS benchmark problem. Control Engineering Practice 14(6), 653–669 (2006)

Gaussian Function Assisted Neural Networks Decoding Algorithm for Turbo Product Codes[*]

Xingcheng Liu[1,2] and Jinlong Cai[1]

[1] School of Information Science and Technology, Sun Yat-sen University,
132 East Waihuan Rd, University Town, Guangzhou,
Guangdong 510006, China
[2] National Mobile Communications Research Laboratory, Southeast University,
2 Sipai Lou, Nanjing,
Jiangsu 210096, China
isslxc@mail.sysu.edu.cn, cjl26sysu@gmail.com

Abstract. We apply the radial basis functions (RBF) decoder adopting Gaussian function for the Turbo product codes (TPC). An extrinsic information extraction scheme based on RBF neural networks (NN) is suggested, and a novel RBF NNs decoding algorithm is proposed. The extrinsic information transfer (EXIT) charts have been used to analyze the convergence property of the TPCs. The EXIT chart analyses show that the proposed decoding algorithm could achieve convergence with about 5 iterations, and improve BER performance in low E_b/N_0 regions. Simulation results show that the proposed algorithm achieves promising BER performance while decreasing decoding computation compared with the maximum a posterior (MAP) algorithm.

Keywords: Artificial neural networks (ANN), iterative decoding, radial basis functions (RBF), Turbo product codes (TPC).

1 Introduction

In 1993, Berrou et al. proposed a class of powerful error-correction codes, Turbo codes, which make it possible to transmit data with code rate approaching channel capacity [1][2]. Turbo codes are capable of achieving exceptionally low bit error rate (BER) with the signal-to-noise ratio (SNR) per information bit (E_b/N_0), which is close to Shannon's theoretical limit over Gaussian channel [2][3]. In 1994, Pyndiah et al. proposed the class of block Turbo codes (BTC) [4][5], which attain low BER over Gaussian channels at a high code rate. Their studies were the extended to the other classes of codes, such codes as Reed-Solomon (RS) codes [6] and extended Hamming codes [6]. The applications of Turbo product codes (TPC) in wireless and optical communications were considered in [8] [9] [10].

---

[*] This work was supported by the National Natural Science Foundation of China (No.60970041 and No. 61173018), the Science and Technology Plan of Guangzhou City of China (No.2012J4300032), and the Open Research Fund of National Mobile Communications Research Laboratory, Southeast University (No. 2011D09).

C. Guo, Z.-G. Hou, and Z. Zeng (Eds.): ISNN 2013, Part II, LNCS 7952, pp. 586–596, 2013.
© Springer-Verlag Berlin Heidelberg 2013

However, all the above mentioned investigations were in the case of long codes associated with the decoding algorithm of maximum a posterior (MAP) or its modified versions, such as the algorithm proposed in [11]. Therefore, the complexity of the decoding algorithms is rather high and decoding delay is also long. In order to meet different requirements of Quality of Service (QoS), it is necessary to provide a decoding algorithm to get a trade-off between low BER and short decoding delay.

Employing artificial neural networks (ANN) for channel decoding is a smart technique to reduce decoding complexity for short and medium-sized TPC codes. El-Khamy et al. proposed an ANN for decoding block codes [12]. Due to their inherent parallel processing capability, ANN is recognized as a kind of promising techniques for error correction to meet the requirement of high data-rate transmission. Based on the error back propagation (EBP) algorithm, in [13] Annauth et al. proposed a scheme for decoding Turbo codes. However, the proposed approach in [13] yields poor error performance, though the decoding complexity is low. The ANN based decoder for Turbo codes is more reliable to get a better BER value in wireless channels [14]. Besides decoding, ANN has been applied in channel equalization [15][16].

The hidden layers of radial basis function (RBF) networks [12][17] are capable of transforming the input vectors from a lower-dimensional vector space to the corresponding vectors in a higher-dimensional space, where they can be classified, even though they are unable to be classified in the lower-dimensional one. The applications of RBF in communications are explored. For example, for channel equalization, Yee and Hanzo et al. [17] proposed an algorithm in which the information exchange between Soft-Input Soft-Output RBF (SISO-RBF) equalizer and SISO channel decoder was implemented to accomplish iterative processing. Deng et al. [18] proposed a complex-valued minimal RBF NN in channel equalization of QAM. Liu et al. [19] proposed a decoding algorithm for BTC based on RBF NN since the decoding algorithm for these codes has become a focus in the area [20].

In this paper, the decoding of TPCs using the RBF NNs is further considered. Since the RBF NNs have the characteristics of simple model and parallel computation, employing the RBF NNs for decoding Turbo product codes is capable of simplifying the decoding model and decreasing the decoding complexity compared with MAP algorithm. Furthermore, EXIT analyses and experiments show that the proposed algorithm can achieve promising BER performance.

2   Traditional Extrinsic Information Computation

TPCs are based on product codes. The product codes $P(n, k, d)$ is constructed by two component codes $C_1(n_1, k_1, d_1)$ and $C_2(n_2, k_2, d_2)$, forming $P = C_1 \otimes C_2$, where $n = n_1 \times n_2$, $k = k_1 \times k_2$ and $d = d_1 \times d_2$ [5]. The code rate R of $P(n, k, d)$ is $R = R_1 \times R_2$, where R_i is the code rate of $C_i(n_i, k_i, d_i)$ ($i=1, 2$). Provided that a SISO decoder for decoding the rows and columns of $P(n, k, d)$, we can iterate the sequential decoding of $P(n, k, d)$ and thus reduce the BER by more iterations.

At each iteration of the iterative decoding, the extrinsic information is forwarded to the followed iterations to improve the performance [2][5]. For TPCs, the traditional method of extrinsic information computation was proposed by Pyndiah et al. [4][5]. Hence, the notations used in [4] are adopted here for convenience.

Consider a transmitted codewords vector E encoded by using a linear block code C with parameters (n, k, d). Suppose the code vector is transmitted over Additive White Gaussian Noise (AWGN) channels using binary phase-shift keying (BPSK), such that $0 \rightarrow -1$ and $1 \rightarrow +1$. Then, for the transmitted codeword sequence $E=(e_1,...,e_i,...e_n)$, the received sequence $R=(r_1,...r_i,...,r_n)$ at the output of the Gaussian channels can be expressed by

$$R = E + G, \tag{1}$$

where $G=(g_1,...,g_l,...,g_n)$ and its components g_l are AWGN samples with mean zero and standard deviation σ. Denote the decoded codeword at the decoder output as $D=(d_1,..., d_j,...d_n)$. Assuming that different codewords are uniformly distributed and applying Bayes' rule, the reliability of d_j is measured by log-likelihood ratio (LLR) of decision d_j as

$$\Lambda(d_j) = \ln \left(\frac{\Pr\{e_j = +1 | R\}}{\Pr\{e_j = -1 | R\}} \right) = \ln \left(\frac{\sum\limits_{C_i \in S_j^{+1}} \Pr\{R | E = C_i\}}{\sum\limits_{C_i \in S_j^{-1}} \Pr\{R | E = C_i\}} \right), \tag{2}$$

where S_j^{+1} and S_j^{-1} are the sets containing the indices of codewords with $c_{ij} = +1$ and $c_{ij} = -1$, respectively, $C_i=(c_{i1},..., c_{il},...c_{in})$ is the ith ($i=1, 2,..., 2^k$) codeword of C, and $\Pr\{E=C_i | R\}$ is the probability of $E=C_i$ when the received sequence R is given. $\Pr\{R | E = C_i\}$ is the probability density function of R conditioned on $E = C_i$, expressed by Eq.(9) in [4]. Combining the Eq. (2) and the approximation used in Eqs. (10) to (13) in [4] and, an approximation expression for the LLR of decision d_j is obtained by

$$\Lambda(d_j) = \frac{2}{\sigma^2}(r_j + z_j) = \ln \frac{\sum\limits_{C_i \in S_j^{+1}} p_i\{E = C_i | R\}}{\sum\limits_{C_i \in S_j^{-1}} p_i\{E = C_i | R\}}, \tag{3}$$

where z_j is referred to as extrinsic information and its value can be computed by the formula

$$z_j = \sum_{l=1,l \neq j}^{n} r_l c_l^{+1(j)} p_l. \tag{4}$$

where $p_l=0$ if $c_l^{+1(j)} = c_l^{-1(j)}$, or $p_l=1$. Finally, based on $\Lambda(d_j)$, a decoded codeword can be decided by:

$$d_j = \begin{cases} 1, & \text{if } \Lambda(d_j) \geq 0 \\ 0, & \text{if } \Lambda(d_j) < 0 \end{cases}. \tag{5}$$

Let us now consider the iterative decoding algorithm designed based on the RBF NNs.

3   Iterative Decoding Algorithm Based on RBF NNs

Based on Eq. (4) of the previous section, a novel extrinsic information computation is firstly derived after considering the characteristics of RBF NNs. A modified RBF NN model is proposed for the decoding algorithm. Then, a novel iterative decoding algorithm for TPCs, based on RBF NNs is presented in detail. Finally, EXIT charts are used to analyze the proposed decoding algorithm.

In this correspondence, the RBF model is considered as a multi-layer feed-forward NN which with structure shown in the 1st figure in [19] (i.e., Fig. 1 in [19]). For (n, k) linear block codes, there are 2^k centers denoted as t_{mi} ($m = 1, 2, \dots, n$, $i = 1, 2, \dots, n$). The connection weights between the hidden layer and the output layer are denoted as ω_{ij} ($i = 1, 2, \dots, 2^k$, $j = 1, 2, \dots, n$). Hence, as shown in the 1st figure in [19] (i.e., Fig. 1 in [19]), the input layer in the RBF model has n neurons, indexed by m. The hidden layer has 2^k neurons, which are indexed by i, and the "Basis Function" of the ith hidden neuron is denoted by (R, t_i). Assume that the "Basis Function" consists of the Gaussian function, which is represented as

$$\varphi(R, t_i) = G(|R - t_i|) = \exp(|R - t_i|^2 / 2\sigma_i^2) = \exp\left(-\sum_{m=1}^{n}(r_m - t_{im})^2 / 2\sigma_i^2\right), \tag{6}$$

where $R = [r_1, \dots, r_m, \dots, r_n]$ is the input vector of the RBF network, $t_i = [t_{i1}, t_{i2}, \dots, t_{i,n}]$ is the RBF center receiving the values of the permissible codewords of the (n, k) linear block codes, and σ_i^2 is the variance of the AWGN noise. The output of the RBF NN is given by

$$R' = G \cdot W, \tag{7}$$

where, the $1 \times n$ matrix $R' = [r_1', \dots, r_m', \dots, r_n']$ is the output of the RBF NN, $G = [(R, t_1), \dots, (R, t_i), \dots (R, t_i)]$ is a 1×2^k matrix containing the "Basis Functions" of RBF NN in the form of Eq. (6), and $W = [\omega_{ij}]$ is the $2^k \times n$ dimensional weight matrix of the RBF NN.

Note that in our previous research [19], the RBF decoder shown in the 1st figure in [19] is used instead of the traditional decoder for Turbo codes, and a RBF NNs assisted decoding scheme for BTC is proposed in [19]. In that scheme, the weights of the RBF NNs were obtained by training or by computing (fixed). By contrast, for the decoding algorithm using fixed weights, the weights-matrix can be constructed by the centers of the RBF networks, with their values set to +1 or -1. In the following, the extrinsic information computation and the decoding algorithm will be considered.

3.1   Extrinsic Information Computation Based on RBF NNs

As shown in Eq. (6), the "Basis Function" of RBF NNs is assumed to be the Gaussian function, so

$$G(|R - C_i|) = \exp\left(-\frac{|R - C_i|^2}{2\sigma^2}\right). \tag{8}$$

Therefore, the reliability of d_j can be denoted in terms of the "Basis Function" of RBF NNs. Combining Eqs. (3) (8) then rearranging the result, the extrinsic information z_j, can be obtained as

$$z_j = \frac{\sigma^2}{2} \ln \left(\frac{\sum_{C_i \in S_j^{+1}} G(|R - C_i|)}{\sum_{C_i \in S_j^{-1}} G(|R - C_i|)} \right) - r_j .$$

(9)

Based on [5], the normalized LLR r_j' can be taken as the soft output of the decoder, which is the sum of r_j of the soft input of the decoder and z_j of the extrinsic information. r_j' is expressed by

$$r_j' = r_j + z_j = \frac{\sigma^2}{2} \ln \left(\frac{\sum_{C_i \in S_j^{+1}} G(|R - C_i|)}{\sum_{C_i \in S_j^{-1}} G(|R - C_i|)} \right).$$

(10)

So far, the soft output r_j' of decoder and the extrinsic information z_j have been expressed by the "Basis Function". However, Eq. (10) is not the output of traditional RBF NNs, and here can not be implemented by traditional RBF NNs. Therefore, it is required to modify the model of traditional RBF networks to construct new RBF NNs for implementing Eq. (10), which are now analyzed.

From Eq. (7), we can see that the weight matrix W is constructed by the centers of the RBF networks. In other words, the matrix W can be constructed with the permissible codeword C_i, with its components equaling to +1 or -1. In Eq. (10), $\sum_{C_i \in S_j^{+1}} G(|R - C_i|)$ is in fact the sum of all the RBF NNs' "Basis Functions" belonging to the set S_j^{+1}. Therefore, the numerator of Eq. (10) is equal to the sum of all the RBF NNs' "Basis Functions" with their weights equaling +1,

$$\sum_{C_i \in S_j^{+1}} G(|R - C_i|) = \sum_{W_{ij} = +1} G(|R - C_i|) .$$

(11)

Similarly, the denominator of Eq. (10) is equal to the sum of all the RBF NNs' "Basis Functions" with their weights equaling -1,

$$\sum_{C_i \in S_j^{-1}} G(|R - C_i|) = \sum_{W_{ij} = -1} G(|R - C_i|) .$$

(12)

Upon splitting each neuron at output layer of the 1st figure in [19] into two neurons, the soft outputs of the RBF decoder are

$$r_j' = \frac{\sigma^2}{2} \ln \left(\frac{y_{j,+1}}{y_{j,-1}} \right) = \frac{\sigma^2}{2} \ln \left(\frac{\sum_{W_{ij} = +1} G(|R - C_i|)}{\sum_{W_{ij} = -1} G(|R - C_i|)} \right).$$

(13)

$y_{j,+1}$ and $y_{j,-1}$ are the outputs of two neurons, where one deals with the weights equaling +1 and the other one with the weights equaling -1. From Eq. (10) the extrinsic information can be expressed as

$$z_j = \frac{\sigma^2}{2} \ln\left(\frac{y_{j,+1}}{y_{j,-1}}\right) - r_j .$$ (14)

In summary, a novel RBF NNs for SISO decoding and extrinsic information computation can be shown as Fig. 1. Compared with the previous model of the 1st figure in [19] (i.e., Fig. 1 in [19]), there are two different components in the novel model shown in Fig. 1. First, each of the neurons at the output layer is split into two sub neurons. Second, dividers are applied to compute the extrinsic information. The next part details the iterative decoding algorithm.

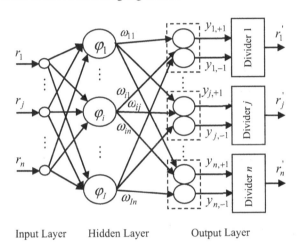

Fig. 1. The novel RBF NNs decoding mode

3.2   Iterative Decoding Algorithm Based on RBF NNs

The iterative RBF NNs decoding algorithm for TPCs is illustrated in Fig. 2. As shown in Fig. 2, the notations L^C and L^R are used to denote the LLR reliability values output by the RBF column decoder and the RBF row decoder, respectively. The subscripts e, s, a, and p are used to specify the extrinsic LLR, the combined channel and extrinsic LLR, the a priori LLR and the a posteriori LLR, respectively. L_e, L_s and L_p are defined as $L_e = [z_1, ..., z_j, ..., z_n]$, $L_s = [r_1, ..., r_j, ..., r_n]$ and $L_p = [r_1', ..., r_j', ..., r_n']$, respectively. Referring to Fig. 2, the RBF column decoder processes the channel outputs R and the a priori information of the column decoder L_a^C which is provided by the extrinsic information of row decoder L_e^R multiplied by the scaling factor α which is called scaling factor (or weighting factor) as explained in [4], generating the a posteriori LLR values of the column decoder L_p^C. Therefore, the input of column decoder L_s^C and row decoder L_s^R are given separately by

$$L_s^C = R + L_a^C = R + \alpha \cdot L_e^R ,$$ (15a)

$$L_s^R = R + L_a^R = R + \alpha \cdot L_e^C .$$ (15b)

From Eq. (8), the "Basis Function" of the RBF column decoder can be denoted as

$$G(|L_s^C - C_i|) = \exp\left(-\frac{|L_s^C - C_i|^2}{2\sigma^2} \right).$$ (16)

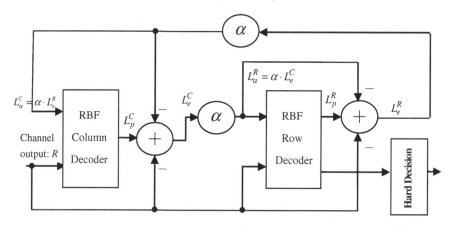

Fig. 2. Iterative decoding of BCH$(n, k)^2$ Turbo product codes with RBF NNs

Let the output of the RBF column decoder be $L_p^C = [r_1', ..., r_j', ..., r_n']$. From Eqs. (13) and (16), r_j' can be computed by

$$r_j' = \frac{\sigma^2}{2} \ln\left(\frac{\sum_{C_i \in S_j^{+1}} G(|L_s^C - C_i|)}{\sum_{C_i \in S_j^{-1}} G(|L_s^C - C_i|)} \right).$$ (17)

Before passing the a posteriori LLR L_p^C generated by the RBF column decoder to the RBF row decoder, as shown in Fig. 2, the contribution of the row decoder denoted by L_s^C in Eq. (15b) accruing from the previous column decoding must be removed, in order to yield the extrinsic information L_e^C of the RBF column decoder. In other words, the a priori information L_a^C of column decoder and the channel output R are removed at this stage in order to prevent the decoder from processing its own output information again, which would result in excessive influence on the decoder's current reliability estimation for the coded bits. So, after the RBF column decoding, with the aid of Eq. (15a) and Eq. (15b) the extrinsic information L_e^C of the RBF column decoder and L_e^R of the RBF row decoder can be given separately as

$$L_e^C = L_p^C - L_s^C = L_p^C - \alpha \cdot L_e^R - R , \qquad (18a)$$

$$L_e^R = L_p^R - L_s^R = L_p^R - \alpha \cdot L_e^C - R . \qquad (18b)$$

Here, L_a^C, which is yielded by multiplying L_e^R with the scaling factor α, is used as the a priori information in the next iteration of decoding as shown in Fig. 2. By now, the first iteration of decoding is completed.

It is important to note that only the extrinsic information, L_e^R in L_p^R, is fed back to the RBF column decoder in order to minimize the inter-dependence between the a priori information L_a^C used by the column decoder and the previous decoding information. Furthermore, after the final iteration, the a posteriori LLR L_p^R of an information symbol is computed and the hard decision is yielded according to Eq. (5).

By comparing of the RBF NN decoding scheme with the traditional Turbo decoding scheme, the major difference in the RBF NN decoding processes lies in the fact that the RBF column and row decoders employ the RBF NNs assisted decoding.

3.3   EXIT Charts of RBF NN Based on Iterative Decoding

The Extrinsic Information Transfer (EXIT) chart technique is very useful in performance analysis of iterative decoding [21], which can be employed for predicting the convergence behavior of Turbo codes. Compared with Monte Carlo simulations of the BER vs E_b/N_0 performance, the EXIT charts need much fewer frames to predict the three regions which are the low E_b/N_0 region, the "waterfall" region and the error-floor region [22] in the respective BER performance figures. The EXIT chart is beneficial to the analysis of the bit error rate prediction of Turbo codes [23]. In this part, the EXIT chart is applied to analyze the performance and convergence behaviors of the TPCs with proposed decoding algorithm.

As shown in the previous sections, the extrinsic information L_e^C of the RBF column decoder and L_e^R of the RBF row decoder are computed based on Eq. (18a) and Eq. (18b), respectively. The extrinsic information L_e, representing L_e^C or L_e^R, of the RBF column decoder or row decoder are considered as a prior input to the decoder. The method proposed in [22] could be applied to compute mutual information in decoding. The extrinsic information, I_e^R or I_e^C computed according to [22] is the mutual information of the RBF row decoder or the RBF column decoder.

In this paper, the mutual information of the TPC BCH code $(15, 7)^2$ is considered, and hence shown in Table 1, as well as in Fig. 3. Based on the results shown in Table 1, the three performance regions mentioned before can be roughly identified for the BCH $(n, k)^2$ product codes according to the change tendency of mutual information [22].

The EXIT charts in different iterations with different E_b/N_0 are constructed to find the number of iterations at which the convergence occurs, as shown in Fig. 3. The trajectory indicates how mutual information changes during decoding. If the mutual information in some iteration is close to the one in some other iteration, we predict that the iterative decoding is to converge. For BCH $(15, 7)^2$ product code, the iterative decoding trajectories are shown in Fig. 3. At $E_b/N_0=1$dB, the decoding is to converge after 5 iterations, as shown in the small figure located at the lower right corner in Fig. 3. In error-floor region, the number of iterations for convergence is less than 5. For BCH$(8, 4)^2$ product code, iterative decoding experiments were carried out and similar results could be obtained. The related details of this TPC are ignored here due to the limited space.

The EXIT analyses show that the proposed decoding algorithm could achieve fast decoding convergence, which needs about five iterations in "waterfall" region. When meeting low E_b/N_0, the proposed decoding algorithm could improve BER performance as the "waterfall" region begins from a low E_b/N_0.

According to [19], there are 2^k hidden neurons and n output in proposed decoding network. For each hidden neuron, there are n additions/subtractions and $n+2$ multiplications/divisions. For each output, there are 2^k additions and 2^k multiplications. The comparison of the number of operations between the MAP and the proposed decoding algorithm is shown in Table 2.

The performance gap between the MAP decoding algorithm and proposed decoding algorithm over the same channel is very narrow. Considering the complexity of the MAP algorithm employed in [7], the proposed algorithm has a better trade-off between BER performance and computation complexity.

Table 1. Mutual info of RBF NN row decoder, I_e^R, for BCH $(15, 7)^2$

E_b/N_0	-2 dB	-1 dB	0 dB	1.0 dB	1.5 dB	2.5 dB
Itr 1	0.444	0.508	0.619	0.689	0.739	0.822
Itr 2	0.533	0.630	0.755	0.828	0.858	0.890
Itr 3	0.617	0.723	0.831	0.879	0.887	0.900
Itr 5	0.871	0.890	0.904	0.907	0.908	0.909
Itr 8	0.896	0.903	0.908	0.909	0.910	0.910

Table 2. The number of operations of the MAP and the proposed algorithm

Algorithms	MAP	Proposed
Additions/Subtractions	4×2^n	$2n\times2^k$
Multiplications/Divisions	$6\times2^n+1$	$(2n+2)\times2^k$

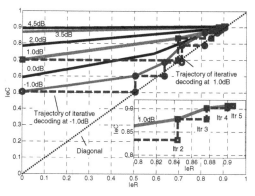

Fig. 3. Extrinsic information transfer characteristics of Turbo BCH code $(15, 7)^2$ for a set of E_b/N_0 values, and two decoding trajectories at -1.0dB and 1dB

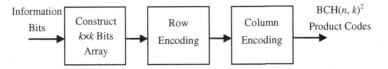

Fig. 4. BCH(n, $k)^2$ Turbo product codes encoding scheme

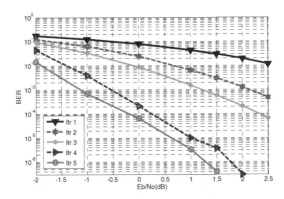

Fig. 5. BER performance of BCH $(15, 7)^2$ Turbo codes

4   Conclusion

A novel iterative decoding algorithm based on RBF NNs assisted by the Gaussian function has been proposed for decoding $(n, k)^2$ TPCs, and the EXIT charts have been used to analyze the convergence property of the TPCs. The EXIT chart analyses show that the proposed decoding algorithm could achieve convergence with about 5 iterations, and improve BER performance in low E_b/N_0 regions. Simulation results show that for the BCH(15, $7)^2$ TPC, the BER performance of 3.3×10^{-6} can be achieved at $E_b/N_0 = 1.0$dB after 5 iterations. The proposed decoding algorithm provides at least 50% reduction in computation compared with the MAP algorithm [7]. The ratio of computation reduction of proposed decoding algorithm is larger as the code length increases. For BCH(8, $4)^2$ TPC, the similar results could be obtained based on our experiments performed. Hence, our proposed decoding algorithm is of lower computation complexity and is capable of providing a better trade-off between BER performance and decoding complexity. This characteristic makes the proposed decoding algorithm a good candidate for communications, where real-time requirements are necessary.

References

1. Berrou, C., Clavieux, A., Thitimajshima, P.: Near Shannon limit error-correcting coding and decoding: Turbo-codes. In: IEEE Int. Conf. on Comm (ICC 1993), Geneva, Switzerland, vol. 2, pp. 1064–1070 (May 1993)
2. Berrou, C., Glavieux, A.: Near optimum error correcting coding and decoding: Turbo-codes. IEEE Trans. on Comm. 44(10), 1261–1271 (1996)

3. Sklar, B., Harris, F.: The ABCs of linear block codes. IEEE Signal Processing Magazine 21(4), 14–35 (2004)
4. Pyndiah, R., Glacieux, A., Picart, A., Jacq, S.: Near optimum decoding of product codes. In: IEEE Global Telecomm. Conf. (GLOBECOM 1994), San Francisco, CA, vol. 1, pp. 339–343 (November/December 1994)
5. Pyndiah, R.: Near-optimum decoding of product codes: block Turbo codes. IEEE Trans. on Commun. 46(8), 1003–1010 (1998)
6. Aitsab, O., Pyndiah, R.: Performance of Reed-Solomon block Turbo code. In: IEEE Global Telecomm. Conf. (GLOBECOM 1996), London, UK, vol. 1, pp. 121–125 (November 1996)
7. Xu, C., Liang, Y.C., Leon, W.S.: MAP Decoding Algorithm for Extended Turbo Product Codes over Flat Fading Channel. In: 40th Asilomar Conf. on Sig., Syst. & Comp. (ACSSC 2006), California, USA, pp. 2182–2184 (October/November 2006)
8. Zhang, X., Zhang, P., Li, D.: Overlapped Time Division Multiplexing System Using Turbo Product Code. In: IEEE 6th Int. onf. on Wireless Comm. Network. & Mobile Computing (WiCOM), Chengdu, China, pp. 1–5 (September 2010)
9. Sabbaghian, M., Kwak, Y., Smida, B., Tarokh, V.: Near Shannon limit and low peak to average power ratio Turbo block coded OFDM. IEEE Trans. on Comm. 59(8), 2042–2045 (2011)
10. Liao, H.X., Zhang, Q.J., Tan, L.Y., Yang, Z.H.: Performance Evaluation of Turbo Product Codes over the PPM-based Satellite Optical Channel. In: 2012 Int. Conf. on Comp. Sci. & Electron. Eng (ICCSEE), Hangzhou, China, vol. 3, pp. 72–75 (March 2012)
11. Yue, D.W., Nguyen, H.H.: Unified scaling factor approach for turbo decoding algorithms. IET Comm. 4(8), 905–914 (2010)
12. El-Khamy, S.E., Youssef, E.A., Abdou, H.M.: Soft decision decoding of block codes using artificial neural network. In: Proc. IEEE Symposium on Computers and Communications, Alexandria, Egypt, pp. 234–240 (June 1995)
13. Annauth, R., Rughooputh, H.C.S.: Neural network decoding of Turbo codes. In: Int. Joint Conf. on Neural Networks (IJCNN 1999), Washington D.C., USA, vol. 5, pp. 3336–3341 (1999)
14. Misra, A., Sarma, K.: ANN assisted Turbo coding for use with OFDM signals in wireless channels. In: 2011 Int. Conf. on Emerging Trends in Netw. & Comp. Comm. (ETNCC), Udaipur, India, pp. 65–69 (April 2011)
15. Zhao, H., Zhang, J.: Adaptively Combined FIR and Functional Link Artificial Neural Network Equalizer for Nonlinear Communication Channel. IEEE Trans. on NN 20(4), 665–674 (2009)
16. Yamashita, M., Osawa, H., Okamoto, Y., Nakamura, Y., Suzuki, Y., Miura, K., Muraoka, H.: Read/Write channel modeling and two-dimensional Neural Network Equalization for two-dimensional magnetic recording. IEEE Trans. on Magnetics 47(10), 3558–3561 (2011)
17. Yee, M.S., Yeap, B.L., Hanzo, L.: Radial basis function-assisted turbo equalization. IEEE Trans. on Comm. 51(4), 664–675 (2003)
18. Deng, J., Narasimhan, S., Saratchandran, P.: Communication channel equalization using complex-valued minimal radial basis function neural networks. IEEE Trans. on NN 13(3), 687–696 (2002)
19. Liu, X., Chen, Z., Wang, Z., Cull, P.: Decoding of Block Turbo Codes with RBF Networks. In: Int. Conf. on Sensing, Comp. & Automat. (ICSCA 2006), Chongqing, China (2006)
20. Le, N., Reza Soleymani, M., Shayan, Y.R.: Distance-Based decoding of Block Turbo Codes. IEEE Comm. Lett. 9(11), 1006–1008 (2005)
21. ten Brink, S.: Convergence of iterative decoding. IEEE Electron. Lett. 35(10), 806–808 (1999)
22. ten Brink, S.: Convergence behavior of iteratively decoded parallel concatenated codes. IEEE Trans. on Comm. 49(10), 1727–1737 (2001)
23. Tsai, T., Tseng, D., Han, Y.S., Pai, H.: Improved EXIT analysis for Turbo decoding. IEEE Comm. Lett. 15(9), 995–997 (2011)

A New BP Neural Network Based Method
for Load Harmonic Current Assessment

Ke Zhang, Gang Xiong, and Xiaojun Zhu

Guangdong Power Grid Corp. Foshan Power Supply Bureau,
528000 Foshan, China
{Ke Zhang,Gang Xiong,Xiaojun Zhu,huagonglily}@tom.com

Abstract. This paper proposed a new BP Neural Network (BPNN) based method for load harmonic current assessment where the nonlinearities of electricity loads have been modeled based on differential equations. With the trained BPNN, the load current due to fundamental voltage inputs can be well estimated and used to assess the harmonics components subsequently. The simulation results demonstrate that the proposed method can effectively estimate the total harmonic distortion of the load current when the supplied voltage is within the normal range of harmonic limits. The results also prove that the load harmonic current is nearly independent of load capacity and applied voltage, indicating its effectiveness to distinguish the responsibilities of harmonic pollution between the grid and load.

Keywords: Neural network, load harmonic current, harmonic control.

1 Introduction

Along with the widespread use of nonlinear load such as rectifier equipment, frequency conversion device, electric arc furnace, calcium carbide furnace and gas discharge based electric light, harmonic pollution in distribution systems is becoming more and more serious, which has caused increased attentions by power utilities and relevant industries[1], [2]. The harmonic current injected by the loads can be influenced by the network capacity and the harmonics of the supplied voltage. In order to facilitate the management of harmonic pollution by both the power utilities and the electricity users, , an assessment system should be developed to effectively quantify the level of the harmonics injected by the electricity loads.

In literature [3], [4], [5], [6], with the voltage and current measured within a practical system, the harmonic impedance of the system is obtained based on regression analysis, and then the harmonic voltage contribution from the load is determined. In literatures [7], [8], a difference time domain load model is established to estimate its linear component, and the load's nonlinear component can be estimated using the difference between the actual measured current and the estimated load current based on the linear model. In principle, the total harmonic distortion of load current under a pure sinusoidal voltage supply is an appropriate indicator of the load's harmonic contribution.

C. Guo, Z.-G. Hou, and Z. Zeng (Eds.): ISNN 2013, Part II, LNCS 7952, pp. 597–605, 2013.

In this paper, a BPNN is trained with the measured voltage and current data from a practical distribution system, so that the nonlinear response of the load can be well estimated by the BPNN, taking the fundamental voltage as the ideal sinusoidal voltage source applied to the load. Through proper training, the load current and its total harmonic distortion factor under various voltage supply conditions can then be effectively obtained by the BPNN.

2   The Difference Equations Based Load Model

An electricity load can generally consist of resistive, inductive and capacitive components as shown in Fig. 1.

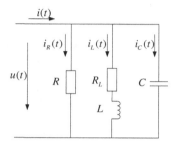

Fig. 1. A typical load model

The parameters R, R_L, L, C in Fig. 1 are constant, and the load can then be mathematically modeled in many ways including but not limited to the models based on state equations, transfer function, or difference equations etc[9]. In this paper, the difference equations based model is used since it can facilitate harmonics analysis using the practical measured voltage and current of electricity loads.

According to Fig. 1, the voltages and currents of the load should fulfill the following

$$\begin{cases} u(t) = L\dfrac{di_L(t)}{dt} + R_L i_L(t) \\[2mm] i_C(t) = C\dfrac{du(t)}{dt} \\[2mm] u = R i_R(t) \\[2mm] i(t) = i_R(t) + i_L(t) + i_C(t) \end{cases} \tag{1}$$

Takes $u(t)$ as the independent variable, $i(t)$, $i_R(t)$, $i_L(t)$, $i_C(t)$ as dependent variable, then $i(t)$ and $u(t)$ should fulfill the following

$$RLC\frac{d^2u(t)}{dt^2} + (RR_L C + L)\frac{du(t)}{dt}$$
$$+(R + R_L)u(t) = RL\frac{di(t)}{dt} + RR_L i(t) \tag{2}$$

Based on the implicit trapezoid integration method, the Equation (2) can then be reformulated into a difference equation with a time step of Δt,

$$i(k) = a_1 i(k-1) + b_0 u(k) + b_1 u(k-1) + b_2 u(k-2) \ , \tag{3}$$

where $a_1 = \dfrac{2L - \Delta t R_L}{2L + \Delta t R_L}$; $b_0 = \dfrac{\Delta t^2 (R + R_L) + 2(RLC + \Delta t(RR_L C + L))}{2\Delta t RL + \Delta t^2 RR_L}$;

$b_1 = \dfrac{\Delta t^2 (R + R_L) - 2(2RLC + \Delta t(RR_L C + L))}{2\Delta t RL + \Delta t^2 RR_L}$; $b_2 = \dfrac{2LC}{2\Delta t RL + \Delta t^2 R_L}$.

As seen from Equation (3), $i(k)$ is a linear function of $i(k-1)$, $u(k)$, $u(k-1)$, $u(k-2)$, where the current response of the load at time k are affected by the coefficients a_1, b_0, b_1, b_2. When R, R_L, L, C are related to $i(k)$, $i(k-1)$, $u(k)$, $u(k-1)$, $u(k-2)$, Equation (3) can be rewritten as

$$i(k) = f(i(k-1), u(k), u(k-1), u(k-2)) \ . \tag{4}$$

For the load model with higher complicities, Equation (4) can be used instead as

$$\begin{aligned} i(k) = f(i(k-1), i(k-2), ..., i(k-N_i), \\ u(k), u(k-1), u(k-2), ..., u(k-N_u)) \end{aligned} \tag{5}$$

where N_i and N_u are the delay steps of the current i and voltage u respectively.

3   BPNN for Load Harmonics Current Assessment

BP (Back Propagation) neural network is firstly proposed by the research group led by Rumelhart and McCelland in 1986, and it is the mostly frequently used NN algorithm due to its many advantages such as excellent generalization capability.

Through iterative learning procedure, BP neural network (BPNN) is able to approximate the complex mapping relationship between the inputs and outputs given sufficient amount of training samples, without explicit knowledge of the mapping beforehand[10]. To apply BPNN for load harmonic current assessment, the task is then for a BPNN to learn according to the load model in Equation (5), where $i(k-1)$, $i(k-2)$, ..., $i(k-N_i)$, $u(k)$, $u(k-1)$, $u(k-2)$, ..., $u(k-N_u)$ are the inputs, and $i(k)$ is the output of the BPNN.

A BPNN is basically a feed-ward neural network consisting of several layers of perceptrons or neurons which are connected through weighting factors between adjacent layers. The training of BPNN adopts a backward approach to propagate the errors between the BPNN outputs and the targets backward through layers to update the weights so that the errors can eventually be minimized through iterations. Notably to ensure the best performance, the cost function of BPNN training is usually a quadratic function based on the actual errors between the outputs and the targets. However, there can be many solutions of weights if the training objective is the minimization of errors alone. Generally speaking, to ensure BPNN to have better generalization capability over unseen data and to avoid the over-fitting problem, the values of weights should be minimized. Therefore in order to minimize the training errors and weights of BPNN simultaneously, the training objective function can be formulated as

$$F = \beta E_D + \alpha E_W \ , \tag{6}$$

where $E_D = \frac{1}{2}\sum_{k=1}^{p}(i_k - \hat{i}_k)^2$, $E_W = \frac{1}{2}\sum_{j=1}^{m}w_j^2$.

i_k and $\hat{i}_k$ are the target and actual output of BPNN for the kth training sample, p is the number of training samples, w_j stands for the jth weights of BPNN, m is the number of weights. α and β are so-called super parameters, which actually can control the training preference over the weighs and errors. If $\alpha \ll \beta$, the training will prioritize the minimization of training error; if $\alpha \gg \beta$, the training algorithm will prioritize the minimization of BPNN's weights otherwise.

David Mackay pointed out that both the training samples and BPNN's weights should fulfill a normal distribution, then the super parameters α, β can be determined using the Baye rule[11], [12] as follows,

$$\alpha = \frac{\gamma}{2E_W}, \beta = \frac{m - \gamma}{2E_D} \ , \tag{7}$$

where $\gamma = m - \alpha \cdot tr\left(\nabla^2 F\right)^{-1}$.

The above BP neural network training method is called the Baye rule based regularization algorithm, which is a iterative algorithm. Training using the Baye rule can start with supposing $\alpha=0$, $\beta=1$, then solve for the smallest w_{MP} to minimize the F in Equation (6). Subsequently, calculate the α, β given the w_{MP}, and then use resultant α, β to again solve Equation (6) to obtain BPNN's weights. The above procedure is repeated until the changes of α and β between iterations become small enough. In solving the w_{MP}, the L-M method[13] can be used, which can effectively guarantee satisfactory convergence speed at the vicinity of the optimal solution. Furthermore, use $(\beta J^T J + 2\alpha I)$ to replace the term $\nabla^2 F$ in Equation (7) can largely improve the computation speed for γ, where J is the Jacobian of $(i_k - \hat{i}_k)$ at w_{MP}, and I is the unitary matrix.

4   Load Harmonics Current Estimation under Ideal Voltage Supply

With the measured voltage and current data from practical distribution network, a BPNN can be trained using the Baye regularization method so that the knowledge of the load harmonics can be well captured by the BPNN weights. The training data inputted to BPNN includes the fundamental component of the measured voltage from practical network, as well as the corresponding load current. By doing so, the load harmonics current due to an ideal voltage supply can be well estimated.

Assuming the load parameters are constant or only dependable on suits voltage and current, $i(k)$ can be solely determined based on $i(k-1)$, $i(k-2)$, ..., $i(k-N_i)$, $u(k)$, $u(k-1)$, $u(k-2)$, ..., $u(k-N_u)$. However, the actual load is varying in real time which indicates the model (5) may not be applied directly. As such, using practical measurement data for training BPNN can result in unreliable outputs due to the load variations caused inconsistencies in the training data. In order to deal with the inconsistencies of training data, the proposed BPNN based load harmonics current assessment method can be trained and implemented in cycles. The detailed algorithm includes the following steps:

Step 1: Measure the voltage and current data from a practical distribution network for a fixed time duration;

Step 2: Based on the measured voltage data, estimate the fundamental frequency of the distribution system;

Step 3: Check if the voltage and current data are valid to be used for training BPNN. If not, go to Step 1;

Step 4: Use the measured data to train the BP neural network;

Step 5: Extract the fundamental voltage from the measurement data;

Step 6: Estimate the load current due to the fundamental wave voltage based on the trained BPNN;

Step 7: Carry out the total harmonics distortion (THD) analysis for the estimate load current, and then go to Step 1 for the next cycle.

More details about Step 2, Step 3, Step 5, and Step 6 will be provided in the following.

The fundamental frequency of the distribution system can be using the windowed discrete Fourier transform and double spectral line based interpolation algorithm[14].

Supposes the sampling rate is f_s, the sampled voltage is $u(n)$, the sample length is N, the sampling window function can be represented as $w(n)$ in time domain, then the discrete Fourier transform of the voltage can be expressed as

$$U(k) = \sum_{n=0}^{N-1} w(n)u(n)e^{-j2\pi nk\Delta f} \quad , \tag{8}$$

where $\Delta f = f_s / N$.

Because the fundamental frequency f_b should be close to system nominal frequency f_N, therefore the sampling rate can be selected as $f_s=pf_N$ (p is an integer), the sample length $N=qp$ (q is an integer). If $|f_b-f_N|<f_N/q$ ($q\geq2$), f_b must locate between the $(q-1)^{th}$ and $(q+1)^{th}$ spectral lines. When $|U(q-1)|>|U(q+1)|$, let $k_1=q-1$, $k_2=q$, or $k_1=q$, $k_2=q+1$. If the window function is the Blechman window function, f_b can be estimated as,

$$f_b = (k_1 + 0.5 + 1.96043163 \cdot \beta + 0.15277325 \cdot \beta^3 \tag{9}$$
$$+ 0.07425838 \cdot \beta^5 + 0.04998548 \cdot \beta^7) \cdot \Delta f$$

where $\beta = \dfrac{|U(k_2)| - |U(k_1)|}{|U(k_2)| + |U(k_1)|}$.

If the measured voltage varies periodically at the fundamental frequency, then the training samples for BPNN according to the Equation (5) should not have any inconsistencies, i.e. the measured data is qualified to be the training samples.

Perform linear interpolation[15] to $u(n)$ with the sampling frequency $f_s' = pf_b$, new voltage samples $u'(n)$ can be obtained. To determine if $u'(n)$ is periodical, we can split the samples into q groups with each group containing p samples. Then calculate the Pearson's correlation ρ_j between the sample groups $u_1'(m)$, $u_2'(m)$, ..., $u_q'(m)$, according to the following

$$\rho_j = \frac{\sum_{m=1}^{p} u_1'(m) u_j'(m)}{\sum_{m=1}^{p} u_1'^2(m)} \qquad (j = 2, 3, \cdots, q) \ , \tag{10}$$

when $0.98 < \rho_j < 1.02$, $u'(n)$ can be considered periodical.

Similarly, the same interpolation procedure can be repeated for $i(n)$ to obtain new current sample series $i'(n)$. Subsequently, the similar procedure to determine if the $i'(n)$ is periodical can be carried out.

For the measured voltage series that are verified periodical, its fundamental voltage can be extracted according to $A\sin(2\pi f_b t + \varphi)$, where

$$A = \frac{|U(k_1)| + |U(k_2)|}{N}(2.70205774 + 1.07115106 \cdot \alpha^2 \tag{11}$$
$$+ 0.23361915 \cdot \alpha^4 + 0.04017668 \cdot \alpha^6)$$

$$\varphi = \arg[U(k_1)] + \pi/2 - \pi \cdot (\alpha - 0.5) \ , \tag{12}$$

where $\alpha = f_b / \Delta f - k_1 - 0.5$.

The resulted $A\sin(2\pi f_b t + \varphi)$ can then be sampled at f_s' to obtain the fundamental voltage series $u_b'(n)$. To obtain the load current under fundamental voltage, e.g. $i_b'(0)$ at $t=0$, we need not only the fundamental voltage samples including $u_b'(0)$, $u_b'(-1)$, $u_b'(-2)$, ..., $u_b'(-N_u)$, but also $i_b'(-1)$, $i_b'(-2)$, ..., $i_b'(-N_i)$, which are however unknown at $t=0$. As a compromise, we can use $i'(-1)$, $i'(-2)$, ..., $i'(-N_i)$ to replace $i_b'(-1)$, $i_b'(-2)$, ..., $i_b'(-N_i)$ as the inputs to BPNN. Consequently, after the initial several cycles, $i_b'(n)$ can become precisely the current due to the fundamental voltage inputs.

Because $i_b'(n)$ is sampled current data, it can be used to calculate the total harmonic distortion by simply performing the discrete Fourier transform.

5   Simulation Analysis

For a practical distribution system with three phases, the load characteristic can still be described using difference equations for each phase respectively,

$$i_{abc}(k) = f(i_{abc}(k-1), i_{abc}(k-2), ..., i_{abc}(k-N_i), u_{abc}(k), \qquad (13)$$
$$u_{abc}(k-1), u_{abc}(k-2), ..., u_{abc}(k-N_u))$$

Because of three phases, the BPNN for three-phase load harmonics current assessment can have a large size than the single phase case. There can be more neurons inside the hidden layer, the input and output layers. In our case study, the delay orders for voltage and current are set as 3. To the knowledge of the load characteristics can be well acquired, a BPNN with two hidden layers is implemented, where each hidden layer consists of 10 neurons.

The method can be implemented to learn and estimate the load harmonics current in separate time intervals for online application. Because the fundamental frequency determined by the windowed discrete Fourier transform and the double special interpolation method can have slight difference, the averaged value of the two results is used as the fundamental frequency of the distribution network.

To test the proposed method, a three-phase full-controlled bridge rectifier is modeled and simulated according to Fig. 2.

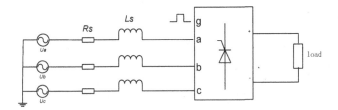

Fig. 2. Three-phase full-controlled bridge rectifier circuit

In Figure 2, the triggering pulses for the rectifier are generated using six group synchronization control method, and the load parameters are $R=100\Omega$, $L=3mH$, the three phase voltages of the distribution network are

$$\begin{cases} u_a = 311\sin \omega_1 t + A\sin \omega_2 t \\ u_b = 308\sin(\omega_1 t - 2\pi/3) + B\sin(\omega_2 t - 2\pi/3) \\ u_c = 309.5\sin(\omega_1 t + 2.01\pi/3) + C\sin(\omega_2 t + 2\pi/3) \end{cases} \qquad (14)$$

where $\omega_1 = 100\pi + \sin(10\pi t)$, $\omega_2 = 500\pi + 5\sin(10\pi t)$.

By changing the capacity of the power supply system and the harmonics in voltage, as well as the firing angle of the rectifier, the voltage and current singles at the connection point can be measured at the sampling frequency $f_s=10kHz$. The sample length is 4000 data points. If the obtained samples are qualified, apply them for BPNN training, and also determine the fundamental voltage of the distribution system. The fundamental voltage can be used to train a BPNN with characteristics of nonlinear load which can output the estimated load current. Further, the fundamental voltage as an ideal voltages supply can be used to train another BPNN. The resultant current is then the load current under the ideal voltage supply accordingly. Next the THDs of the obtained currents and the voltage measured at the connection of the load are calculated and listed in Table 1 below, where all THDs are expressed in %.

Table 1. Calculative results of current and voltage total harmonic distortion

the system internal resistance	the firing angle	the power harmonic amplitude	the measured current THD	the ideal load current THD	the estimated load current THD	the distribution network voltage THD
		A=B=C=0	29.05/29.40/29.62	30.30/31.22/30.46	30.39/31.32/30.55	1.86/1.59/1.69
	0°	A=B=C=5	30.82/28.76/29.77	30.30/31.22/30.46	30.41/31.33/30.56	2.35/2.01/2.33
R_s=0.05Ω		A=B=C=10	32.46/26.52/30.86	30.30/31.22/30.46	30.45/31.37/30.61	4.25/4.32/4.02
L_s=0.4mH		A=B=C=0	38.48/37.28/38.31	36.87/38.05/37.86	36.98/38.16/37.94	3.86/3.99/3.95
	30°	A=B=C=5	37.63/37.37/38.34	36.87/38.05/37.86	37.01/38.17/37.96	4.21/4.08/4.27
		A=B=C=10	38.51/37.68/38.22	36.87/38.05/37.86	37.45/38.50/38.32	5.60/5.26/5.65
		A=B=C=0	32.48/30.36/31.16	30.30/31.22/30.46	30.38/31.29/30.52	3.47/3.55/3.50
	0°	A=B=C=5	31.70/29.19/31.41	30.30/31.22/30.46	30.39/31.30/30.53	3.70/3.84/3.72
R_s=0.03Ω		A=B=C=10	32.55/27.37/32.47	30.30/31.22/30.46	30.58/31.68/30.72	5.04/5.29/5.14
L_s=0.2mH		A=B=C=0	38.59/36.22/38.98	36.92/38.09/37.91	37.01/38.20/38.02	3.07/2.45/2.59
	30°	A=B=C=5	37.66/37.40/38.37	36.92/38.09/37.91	36.99/38.21/38.03	2.64/2.54/2.68
		A=B=C=10	37.69/37.48/38.94	36.92/38.09/37.91	37.05/38.25/38.10	4.54/4.35/4.58

It can be seen from Table 1, under certain load situations, when the voltage distortion of the distribution network does not exceed 5%, the difference between the THDs based on the directly estimated load current and the estimated current due to fundamental voltage supply have very limited difference which is no more than 1%. When the voltage distortion goes above 5%, the differences between the two THDs become larger. This can be explained as the difference between the actual voltage and its fundamental component is too large to be ignored, so that different current THDs can be resulted when using directly estimated current and the current due to fundamental voltage supply in load current harmonics analysis respectively.

6   Conclusion

The simulation results demonstrate that the total harmonic distortion of the load current based on BP neural network which is proposed in this article is almost independent of power capacity and harmonic voltage within the range of utility grid harmonic voltage limits, which is beneficial to the division of harmonic responsibility and harmonic control.

References

1. He, J., Huang, Z.: Determination of emission limit of customer's harmonic current. J. Relay 35(1), 42–46 (2007)
2. Lang, W.: Generation and harm of harmonic in power supply system and the countermeasure. J. High Voltage Engineering 28(6), 30–31 (2002)

3. Zhang, W., Yang, H.: A method for assessing harmonic emission level based on binary linear regression. J. Proceeding of the CSEE 24(6), 50–54 (2004)
4. Che, Q., Yang, H.: Assessing harmonic emission level based on robust regression method. J. Proceeding of the CSEE 24(4), 39–44 (2004)
5. Huang, S., Xu, Y.: Assessing system harmonic impedence and harmonic emission level based on partial least-squares regression. J. Proceeding of the CSEE 27(1), 93–97 (2007)
6. Xie, S.: Harmonic impedance analysis and harmonic assessment based on rank regression. J. Electric Power Automation Equipment 30(11), 29–33 (2010)
7. Gong, H., Xiao, X.: Application of non-linearity degree method in harmonic sources detecting in distribution network. J. Power System Protection and Control 4(16), 30–34 (2010)
8. Moustafa, A.A.: Moustafa: Separation of customer and supply harmonics in electrical power distribution systems. In: 9th ICHQP Proceedings, pp. 1035–1040. IEEE Press, Orlando (2000)
9. Zhang, J., Yan, A., Zhang, C., et al.: Summary of load research in power system. J. Relay 35(6), 83–88 (2007)
10. Zong, X., Feng, H.: Neural network-based predictive control for time-delay systems. J. Control Theory and Application 24(12), 6–8 (2005)
11. David, J.C.: A practical Bayesian framework for back propagation networks. J. Neural Computation 4(3), 448–472 (1992)
12. Wei, D., Zhang, M.: Nonlinear model identification based on Bayesian neural network. J. Computer Engineering and Applications 41(11), 5–8 (2005)
13. Xiang, W., Zhang, H., Wang, H., et al.: Application of BP neural network with L-M algorithm in power transformer fault diagnosis. J. Power System Protection and Control 39(8), 100–103 (2011)
14. Pang, H., Li, D., Zu, Y., et al.: An improved algorithm for harmonic analysis of power system using FFT technique. J. Proceeding of the CSEE 23(6), 50–54 (2003)
15. Zhang, B., Jiang, G.: A new approach to frequency tracking and phase measurement. J. Journal of Tianjin University 39(2), 155–158 (2006)

Local Prediction of Network Traffic Measurements Data Based on Relevance Vector Machine

Qingfang Meng[1,2], Yuehui Chen[1,2], Qiang Zhang[3], and Xinghai Yang[1,2]

[1] School of Information Science and Engineering, University of Jinan, Jinan 250022, China
[2] Shandong Provincial Key laboratory of Network Based Intelligent Computing,
Jinan 250022, China
[3] Institute of Jinan Semiconductor Elements Experimentation, Jinan 250014, China
ise_mengqf@ujn.edu.cn

Abstract. In the reconstructed phase space, based on the nonlinear time series local prediction method and the relevance vector machine model, the local relevance vector machine prediction method was proposed in this paper, which was applied to predict the small scale traffic measurements data. The experiment results show that the local relevance vector machine prediction method could effectively predict the small scale traffic measurements data, the prediction error mainly concentrated on the vicinity of zero, and the prediction accuracy of the local relevance vector machine regression model was superior to that of the feedforward neural network optimized by PSO.

Keywords: small-time scale network traffic measurements data, nonlinear time series local prediction method, relevance vector machine.

1 Introduction

Network traffic analysis and modeling play a major role in charactering network performance, so it has been a focus of many researches. Models that accurately capture the salient characteristics of the traffic is useful for analysis and simulation, and they further our understanding of network dynamics. Complexity is a key issue in network geometry and information traffic. The complexity revealed from the traffic measurements has led to the suggestion that the network traffic cannot be analyzed in the framework of available traffic models[1]. Alternative reliable traffic models and tools for quality assessment and control should be developed[2].

In recent years, communication and network technologies are developing rapidly, which brings the traffic characteristics to change greatly. The research emphasis of the network traffic analysis and modeling has change from the large-time scale to the small-time scale. The researches have shown that the traffic characteristics of the small-time scale were different from those of the large-time scale[3]. So the large-time scale network traffic models can not suited to the small-time scale network traffic.

C. Guo, Z.-G. Hou, and Z. Zeng (Eds.): ISNN 2013, Part II, LNCS 7952, pp. 606–613, 2013.

There is growing evidence that the nonlinear deterministic component is exist in the network traffic series[4][5]. This paper apply the local prediction method to predict the small-time scale traffic data. The local prediction method can be successfully used for a deeper understanding of main features of the traffic data.

The nonlinear time series prediction methods can be grouped into two major classes: global prediction methods[6][7] and local prediction methods[13-17]. Recently, various golbal models, for example the neural networks [6] and the support vector machine SVM[7], have been used to predict nonlinear time series successfully. Reference [8] has applied the feed-forward neural network to predict the traffic measurements data. The flexible neural tree (FNT) was a special multi-layer feed-forward neural network which has drastically changed the neural network's optimizing and designing problem [9]. Reference [10] had applied the flexible neural tree (FNT) model to predict the traffic measurements data.

The relevance vector machine (RVM) was the sparse probability model based on a general Bayesian framework for obtaining sparse solutions to regression and classification tasks, which was proposed by Tipping [11], and attracted increasing attention in the field of statistical learning. The relevance vector machine model has been applied to the field of time series prediction and classification[12].

Based on the local prediction method and the relevance vector machine regression model, the local relevance vector machine prediction method was proposed in this paper and was applied to predict the small scale traffic measurements data.

2   Local Prediction of the Small-Time Scale Network Traffic Data

2.1   The Relevance Vector Machine (RVM) Model

Given a data set of input-target pairs $\{x_n, t_n\}_{n=1}^N$, considering scalar-valued target functions only, we follow the standard probabilistic formulation and assume that the targets are samples from the model with additive noise:

$$t_n = y(x_n; w) + \xi_n = \sum_{i=1}^N w_i \varphi_i(x) + w_0 + \xi_n \qquad (1)$$

where $w = (w_0, w_1, \cdots, w_N)$, ξ_n are independent samples from some noise process which is further assumed to be mean-zero Gaussian with variance σ^2, $\varphi_i(x)$ are the nonlinear basis function and $\varphi_i(x) \equiv K(x, x_i)$, $K(\bullet)$ is the kernel function. Thus $p(t_n | x) = N(t_n | y(x_n), \sigma^2)$, where the notation specifies a Gaussian distribution. Due to the assumption of independence of the t_n, the likelihood of the complete data set can be written as

$$p(t|w,\sigma^2) = (2\pi\sigma^2)^{-N/2} \exp\{-\frac{1}{2\sigma^2}\|t - \Phi w\|^2\} \tag{2}$$

where $t = (t_1, t_2, \cdots, t_N)^T$, and Φ is the $N \times (N+1)$ 'design' matrix with $\Phi = [\varphi(x_1), \varphi(x_2), \cdots, \varphi(x_N)]^T$, wherein $\varphi(x_n) = [1, K(x_n, x_1), K(x_n, x_2), \cdots, K(x_n, x_N)]^T$.

We encode a preference for smoother (less complex) functions by making the popular choice of a zero-mean Gaussian prior distribution over w :

$$p(w|\alpha) = \prod_{i=0}^{N} N(w_i|0, \alpha_i^{-1}) \tag{3}$$

with α a vector of $N+1$ hyperparameters. Importantly, there is an individual hyperparameter associated independently with every weight, moderating the strength of the prior thereon.

Having defined the prior distribution, Bayesian inference proceeds by computing, from Bayes' rule, the posterior distribution over the weights is thus given by:

$$p(w|t,\alpha,\sigma^2) = (2\pi)^{-(N+1)/2}|\Sigma|^{-1/2} \cdot \exp\{-\frac{1}{2}(w-\mu)^T \Sigma^{-1}(w-\mu)\} \tag{4}$$

where the posterior covariance and mean are respectively:

$$\Sigma = (\sigma^{-2}\Phi^T\Phi + A)^{-1} \tag{5}$$

$$\mu = \sigma^{-2}\Sigma\Phi^T t \tag{6}$$

with $A = \mathrm{diag}(\alpha_0, \alpha_1, \cdots, \alpha_N)$.

Based on the hyperparameter estimation, including alternative expectation-maximisation-based re-estimates, values of α and σ^2 can be obtained by the iterative re-estimation, that is:

$$\alpha_i^{new} = \frac{\gamma_i}{\mu_i^2} \tag{7}$$

Where μ_i is the i-th posterior mean weigh from (6) and we have defined the quantities γ_i by $\gamma_i \equiv 1 - \alpha_i N_{ii}$, with N$ii$ the i-th diagonal element of the posterior weight covariance from (5) computed with the current α and σ^2 values.

For the noise variance σ^2, differentiation leads to the re-estimate:

$$(\sigma^2)^{\text{new}} = \frac{\|t - \Phi\mu\|^2}{N - \sum_i \gamma_i} \tag{8}$$

Note that the N in the denominator refers to the number of data examples and not the number of basis functions.

At convergence of the hyperparameter estimation procedure, we make predictions based on the posterior distribution over the weights, conditioned on the maximizing values α_{MP} and σ_{MP}^2. Given a new input point x_*, predictions are made for the corresponding target t_*, and we can compute the predictive distribution:

$$p(t_*|t, \alpha_{MP}, \sigma_{MP}^2) = N(t_*|y_*, \sigma_*^2) \tag{9}$$

With $y_* = \mu^T \Phi(x_*)$ and $\sigma_*^2 = \sigma_{MP}^2 + \Phi(x_*)^T \sum \Phi(x_*)$. Then y_* can be the prediction value of t_*.

2.2   The Nonlinear Time Series Local Prediction Method

Various nonlinear time series prediction methods have been proposed [6-7][13-17], which can be grouped into two classes: global and local. The global methods[6][7] use the whole time series to model the true function on the whole attractor. This kind of method has the disadvantage that if new information is taken into account then all the parameters of the model may change, and that a long parameter estimation time may be required. The local prediction methods[13-17] overcome this drawback by building model only on the local attractor and utilizing only part of the past information. Farmer and Sidorowich have proved that the prediction performance of the local prdiction methods is superior than the global prdiction methods with the same embedding dimension[13].

The first step of the local prediction is to find the neighbor points of the current state $X(n)$ in the reconstructed phase space. According to the reconstructed trajectory, we first compute the distances between the current delay vectors $X(n)$ and its $n-1$ preceding delay vectors $X(i)$ (i=1,2,...,n-1) with an imposed metric,

$$d(i) = \|X(i) - X(n)\|_2 \tag{10}$$

and then find the k nearest neighbouring points $X(n_i)(i = 1,2,\cdots,k)$ of the current state $X(n)$. The local linear prediction model in an m-dimensional delay embedding space is an m-order autoregressive model, and the prediction model is a linear superposition of the m elements of delay vector, i.e.

$$\hat{x}_{n+T} = a_n X(n) + b_n \tag{11}$$

The coefficients an and bn are determined by using the least squared criterion.

$$\min \sum_{i=1}^{k} \left| x_{n_i+T} - a_n X(n_i) - b_n \right|^2 \tag{12}$$

2.3   The Small-Time Scale Network Traffic Data's Local Prediction

Most researchers only used the linear model in the local prediction method [13]. However, in recent years researchers have tried to use the nonlinear model to displace the linear model in the local prediction method[15][16].

In this paper we used the relevance vector machine model to displace the linear model and have proposed the local relevance vector machine prediction method, using the neighbor points to train this local relevance vector machine regression model. That was, we train the relevance vector machine, where the input was $X(n_i)(i=1,2,\cdots,k)$ and the output was $x_{n_i+T}(i=1,2,\cdots,k)$, and gets the corresponding relevance vector and weighting coefficient, then exploited this local relevance vector machine to compute the prediction value $\hat{x}_{n+T}$ of the future value.

In this paper, we applied the proposed local relevance vector machine regression model to predict the small-time scale network traffic data, and we applied the neighbor point selection method of local method[17], which was based on Bayesium information criterion, to choose the number of nearest neighbor points for the local relevance vector machine prediction method, which could be formulated as follows:

$$N(k) = \ln \varepsilon(k)^2 + (k+1)\frac{\ln N}{N} \tag{13}$$

where $\varepsilon(k)^2$ was the normalized variance of the prediction error. Then the proposed local relevance vector machine prediction method whose neighbor points have been optimized was applied to predict the small-time scale network traffic data.

3   Experiment Results and Analysis

In this paper we used the TCP traffic data which was issued by the Lawrence Berkeley Laboratory, DEC-Pkt1[18]. The traffic data aggregated with time bin 0.1s, that was the arrived package's amount within the 0.1s time interval.

Generally we may consider the traffic measurements as a sum of a regular process and a stochastic part, related to the high-frequency noise. In order to separate the regular component of the dynamical process from the stochastic noise component, we apply the wavelet soft threshold noise reduction method to the traffic measurements data. The filtered traffic measurements data were normalized to the interval [0, 1].

We applied the BIC-based neighbor point selection method[17] to choose the number of nearest neighbor points for the local relevance vector machine prediction method. The results were shown in Fig.1. One could see that when k=40, N(k) take the minimum, therefore, the number of nearest neighbor points should be 40.

We applied the proposed local relevance vector machine prediction method whose neighbor points have been optimized to predict the small-time scale network traffic measurements data, where the number of nearest neighbor points k was 40. The length of this traffic measurements data was 36000. The front 33000 data points were used as the training set, and the last 3000 data points were used as the test set. The normalized mean squared error was used to evaluate the prediction accuracy,

$$NMSE = \frac{(1/L)\sum_{i=1}^{L}\left|x_{N+i} - \hat{x}_{N+i}\right|^2}{(1/N)\sum_{i=1}^{N}[x_i - \overline{x}]^2} \tag{14}$$

Here L=3000, N=33000. The kernel function of the relevance vector machine model choose the Gauss kernel, the width value was 0.015 and the number of iterations was 100. The prediction results were shown in Fig.2. From Fig.2, it could be clearly seen that the local relevance vector machine prediction method whose neighbor points have been optimized could effectively predict the small-time scale traffic measurements data, the prediction accuracy was well, and the normalized mean squared error NMSE was very small and only equaled 0.0088.

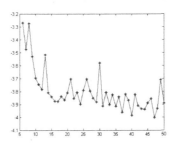

Fig. 1. *N(k)-k* for the traffic measurements data

For comparison purpose, the feed-forward neural network with the structure [10-12-1], trained by using PSO algorithm was also used to predict the same network traffic data. The NMSE for testing data sets was 0.0732. It could be seen that the prediction performance of the proposed local relevance vector machine model was better than the feed-forward neural network model.

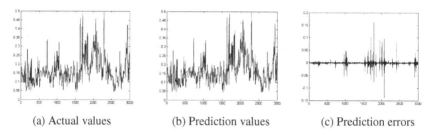

(a) Actual values   (b) Prediction values   (c) Prediction errors

Fig. 2. Prediction results of the local relevance vector machine prediction method for the traffic measurements data, $k=40$

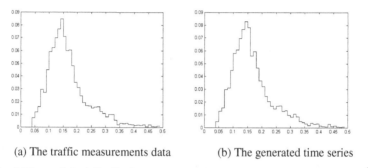

(a) The traffic measurements data   (b) The generated time series

Fig. 3. The statistical distributions of the traffic data and the generated time series

From Fig.3, it could be clearly seen that the time series generated by the local relevance vector machine prediction method had very similar probability distribution with the actual traffic time series. So the local relevance vector machine prediction method could reproduce quite well the statistical distribution of the real traffic data.

4   Conclusion

In summary we applied the local prediction method to the small-time scale traffic measurements data. The local relevance vector machine prediction method was proposed and used to predict the small-time scale traffic measurements data. The experiment results show that the local relevance vector machine prediction method whose neighbor points have been optimized could effectively predict the small-time scale traffic measurements data, the normalized mean squared error NMSE was very small, and the prediction accuracy of the proposed method was superior to that of the feedforward neural network optimized by PSO.

Acknowledgments. Project supported by the National Natural Science Foundation of China (Grant No. 61201428, 61070130), the Natural Science Foundation of Shandong Province, China (Grant No. ZR2010FQ020), the Shandong Distinguished Middle-aged and Young Scientist Encourage and Reward Foundation, China (Grant No. BS2009SW003), the China Postdoctoral Science Foundation (Grant No. 20100470081), the Shandong Province Higher Educational Science and Technology Program (Grant NO. J11LF02).

References

1. Orosz, G., Krauskopf, B., Wilson, R.E.: Bifurcations and multiple traffic jams in a car-following model with reaction-time delay. Physica D 211, 277–293 (2005)
2. Xie, Y.-B., Wang, W.-X., Wang, B.-H.: Modeling the coevolution of topology and traffic on weighted technological networks. Phys. Rev. E 75, 026111 (2007)
3. Zhang, Z.-L., Ribeiro, V.J., Mooon, S., Diot, C.: Small-time scaling behaviors of Internet backbone traffic: an empirical study. IEEE INFOCOM 3, 1826–1836 (2003)
4. Shang, P., Li, X., Kamae, S.: Chaotic analysis of traffic time series. Chaos Solitons Fractals 25, 121–128 (2005)
5. Shang, P., Li, X., Kamae, S.: Nonlinear analysis of traffic time series at different temporal scales. Phys. Lett. A 357, 314–318 (2006)
6. Ma, Q.L., Zheng, Q.L., Peng, H., Qin, J.W.: Chaotic time series prediction based on fuzzy boundary modular neural networks. Acta Phys. Sin. 58, 1410–1419 (2009)
7. Chen, Q., Ren, X.M.: Chaos modeling and real-time online prediction of permanent magnet synchronous motor based on multiple kernel least squares support vector machine. Acta Phys. Sin. 59, 2310–2319 (2010)
8. Akritas, P., Akishin, P.G., Antoniou, I., Bonushkina, A.Y., Drossinos, I., et al.: Nonlinear analysis of network traffic. Chaos Solitons and Fractals 14, 595–606 (2002)
9. Chen, Y.H., Yang, B., Abraham, A.: Flexible neural trees ensemble for stock index modeling. Neurocomputing 70, 697–703 (2007)
10. Chen, Y.H., Yang, B., Meng, Q.F.: Small-time scale network traffic prediction based on flexible neural tree. Applied Soft Computing 12, 274–279 (2012)
11. Tipping, M.E.: Sparse bayesian learning and the relevance vector machine. Journal of Machine Learning Research 3, 211–244 (2001)
12. Zio, E., Maio, F.D.: Fatigue crack growth estimation by relevance vector machine. Expert Systems with Applications 39, 10681–10692 (2012)
13. Farmer, J.D., Sidorowich, J.J.: Predicting chaotic time series. Phys. Rev. Lett. 59, 845–848 (1987)
14. Meng, Q., Peng, Y.: A new local linear prediction model for chaotic time series. Physics Letters A 370, 465–470 (2007)
15. Zhang, J.S., Dang, J.L., Li, H.C.: Local support vector machine prediction of spatiotemporal chaotic time series. Acta Phys. Sin. 56, 67–77 (2007)
16. Du, J., Cao, Y.J., Liu, Z.J., Xu, L.J., Jiang, Q.Y., Guo, C.X., Lu, J.G.: Local higher-order Volterra filter multi-step prediction model of chaotic time series. Acta Phys. Sin. 58, 5997–6005 (2009)
17. Meng, Q.F., Peng, Y.H., Qu, H.J., Han, M.: The neighbor point selection method for local prediction based on information criterion. Acta Phys. Sin. 57, 1423–1430 (2008)
18. Internet traffic archive, http://ita.ee.lbl.gov/

A Massive Sensor Sampling Data Gathering Optimization Strategy for Concurrent Multi-criteria Target Monitoring Application

Xin Song[1], Cuirong Wang[1], Zhi Xu[2], and Haiyang Zhang[1]

[1] Northeastern University at Qinhuangdao, Northeastern University, 066004,
Qinhuangdao, China
[2] Tianjin Electric Power Corporation, 300010, Tianjin, China

Abstract. The data gathering optimization of the large-scale, collaborative and concurrent multi-task in the sensing layer of internet of things is very important, especially in the environments where multiple geographically overlapping wireless sensor networks are deployed. In order to support large-scale, collaborative and concurrent multi-task monitoring, in this paper, we propose a massive sensor sampling data gathering optimization strategy in formed virtual sensor networks to meet various monitoring requirements from different kinds of application deployment and simplify the complexity of dealing with heterogeneous sensor nodes. Then, for the massive sensor sampling data gathering on the virtual sensor networks framework, the CH nodes set and update MinMax hierarchical thresholds to restrict the data transmission. Finally, the simulation results show that proposed strategy achieves more energy savings and increase the sensing layer lifetime of internet of things.

Keywords: Concurrent multi-criteria target monitoring, Massive sensor data gathering, Wireless sensor networks, Energy consumption optimization.

1 Introduction

The Wireless Sensor Network (WSN) is an important fundamental component in sensing layer of internet of things. For energy efficient optimization of the monitoring queries, the data collection task should be selectivity-aware. Physical phenomena are characterized by their spatial correlation; hence, when monitoring a physical phenomenon, some optimization schemes were proposed to minimize the use of non-selected parts of a WSN to achieve overall energy efficiency and extend the lifetime. Jabeen Farhana and Femandes Alvaro A. A. proposed an algorithmic strategy for in-network distributed spatial analysis in wireless sensor networks. The approach taken is algebraic, i.e., analyses are expressed as algebraic expressions that compose primitive operations. The main contributions are distributed algorithms for the operations in the proposed algebra and an empirical evaluation of their performance in terms of bit complexity, response time, and energy consumption [1]. In order to provide efficient monitoring service in a large scale sensor network, the

C. Guo, Z.-G. Hou, and Z. Zeng (Eds.): ISNN 2013, Part II, LNCS 7952, pp. 614–621, 2013.
© Springer-Verlag Berlin Heidelberg 2013

hierarchical organization of the sensors, grouping them and assigning those specific tasks into the group before transferring the information to higher levels, is one the mechanisms proposed to deal with the sensors limitations and is commonly referred to as clustering. In the last few years, a lot of clustering algorithms have been proposed in WSN. Most of these algorithms aimed at generating the minimum number of clusters that maximize the network lifetime and data throughput [2-5].The essential operation in such monitoring applications is data gathering, i.e., to collect the physical information from sensor nodes and transmit to the base station and control server for processing [6]. In Ref.[7], an energy-efficient hybrid data-gathering protocol based on the dynamic switching of reporting schemes was proposed. The novel aspect of their approach is that sensor nodes that seem to detect an event of interest in the near future, as well as those nodes detecting the event, become engaged in the time-driven data-reporting process. Although there are some research fruits, they are still immature and have wide unfathomed problem [8]. In this paper, our approach to cost reduction is through reducing the amount of data sent by each node where the accuracy can be pre-configured. The strategy implemented a multi-criteria target monitoring using MinMax operator in formed virtual sensor networks. Users can easily query and monitor the correlative data through this virtual sensor network. In this paper, we consider the problem of cooperation for heterogeneous WSN deployed on a physical monitoring domain but implementing different tracking tasks. We propose a massive sensor sampling data gathering optimization strategy for concurrent multi-criteria target monitoring using MinMax operator in formed virtual sensor networks.

The remainder of the paper is organized as follow: In Section 2, we describe and discuss implementation of the massive sensor sampling data gathering optimization strategy. The experimental setup and analytical results are discussed in Section 3. Finally, we conclude the paper in Section 4.

2　Implementation of Massive Sensor Sampling Data Gathering Optimization Strategy

Multi-Criteria target monitoring implies a making selection process of the most interested alternatives based on a series of monitoring target with multi-attribute. The Massive Sensor Sampling Data Gathering Optimization strategy includes three phases: distributed clustering process, the virtual CH tree formation and the implementation of the sensor data gathering optimization strategy using MinMax hierarchical thresholds. The distributed clustering process has introduced in the previous research works[9].

2.1　The Virtual CH Tree Formation

A virtual CH tree can be formed by providing logical connectivity among these collaborative communication CH nodes (the CH nodes can communicate directly each other). It is called virtual sensor networks (VSN). CH nodes can be grouped into different VSN based on the phenomenon they track (e.g., rockslides vs. animal crossing) or the task they perform. If the CH node detect the interested enent, it will send a virtual tree formation message to neighbor CHs and allows all the neighbors to join its child CH set. Each of those neighbor CHs then forwards the broadcast to the next set of neighbors. This process continues until the entire network is covered. The process achieved a virtual tree backbone. The second phase: whenever a node detects an interested event for the first time, it sends a discovery message towards the parent node, indicating that it apperceives the event and wants to collaborate with similar nodes. If cluster member node sent the message, the CH checks its array of VSN entries to determine whether it is already aware of the VSN. If not, the CH marks itself as being part of the VSN. The message is sequentially forwarded to its parent CH until the message is forwarded up to the root node. Note that, it is possible for a CH to be part of multiple VSN. The main message of CH nodes include the cluster members list (*Clu _ members*) , the number of VSN (*VSN _ number*), the type array of VSN (*VSN _ IDs*), the routing entries of VSN (*VSN _ REntries _ number*), the routing table of VSN (*VSN _ RTable*). Algorithm 1 shows the pseudocode description of the VSN formation at the CH node.

Algorithm 1. the VSN formation function

```
CH_Function(message)
//initialize the number of VSN and the routing entries of VSN
{
 VSN_number=0; VSN_REntries_number=0;
 if( message_S∈Clu_members)
 // if the messages are from the cluster member nodes
 {
  if(message_type!∈VSN_IDs)
  // if the message type does not belong to a known VSN
 {
  VSN_IDs(VSN_number)=Message_type; // flag the type array of VSN
  VSN_number++; //the increments of VSN number
  Forward_message(message,Parent_CH)
   // forward the message to the perant CH node
   }
  VSN_RTable(VSN_REntries_number)=(message_S,message_type);
  VSN_REntries_number++;
  // update the routing table of VSN
   }
  else              // if the messages are from the CH nodes
  if(message_type!∈VSN_RTable)
  // if the message type does not belong to a known VSN
  Forward_message(message,Parent_CH);
  // forward the message to the perant CH node
 VSN_RTable(VSN_REntries_number)=(Child_CH,message_type);
  VSN_REntries_number++;  // update the routing table of VSN
   }
```

2.2   The Implementation of the Sensor Data Gathering Optimization Strategy Using MinMax Hierarchical Thresholds

We formed the VSN as the target detection continuance in designing the multi-criteria monitoring strategy, every CH node installs a threshold that is n dimensional point (n is monitoring attribute number). Whenever the sensor node detects a relevant overstepping threshold event for the first time it sends the message towards the root node of the virtual CH tree. In order to further promote efficiency of threshold computation, we present method of threshold maintenance by using the MinMax operator in formed VSN backbone. We can monitor in real time all the most dangerous sites in terms of high temperature and low humidity. The definition of the MinMax operator described as equation (1). The input is a point set with the same monitoring attribute dimensionality and the output is a single point. Suppose data set $D = \{d_1, d_2, ..., d_n\}$, $MinMax(D)$ is defined as follow.

$$
\begin{aligned}
max_i &= MAX_{k=1}^{j} d_i[k](1 \leq i \leq n) \\
minmax &= MIN_{i=1}^{n} max_i \\
MinMax(D) &= \underbrace{\{minmax, minmax, ...minmax\}}_{j}
\end{aligned}
\tag{1}
$$

Where j denotes the monitoring attribute dimensionality [2]. An example of the MinMax operator was shown in Fig.1. Point set D consists of five 2-dimensional points. The five measurements are: (21,51), (28,45), (32,25), (41,20) and (53,18), according to the equation (1), we observe $max_i = d_i[y]$ when $i = 1,2$ and $max_i = d_i[x]$ when $i = 3,4,5$.

That   is, $minmax = MIN\{d_1[y], d_2[y], d_3[x], d_4[x], d_5[x]\} = d_3[x]$,   namely: $MinMax(D) = (d_3[x], d_3[x]) = (32,32)$.

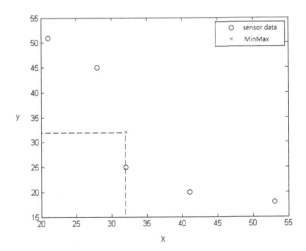

Fig. 1. The computation illustration of the MinMax operator

The implementation of the sensor data gathering optimization strategy using MinMax hierarchical thresholds can be illustrated by setting VSN with nine sensor node for monitoring the same interested enent. Fig. 2 shows the first phase, there is no historical information on each CH node, which gathering data toward the central monitoring node. Once the VSN was formed, the hierarchical threshold is initialized as the corresponding MinMax operator. The parent CH node computes $MinMax(D)$ among its own point and the reported points from its child node. Setting $n = 2$, the sensor data set D is (28,45), (33,27), (25,41), (56,35), (38,43), (23,31), (48,25), (29,35) and (51,35), then

$$minmax = MIN\{45, 33, 41, 56, 43, 31, 48, 35, 51\} = 31 \qquad (2)$$

Thus, $MinMax(D) = (31,31)$ is the initial hierarchical threshold of every CH node in VSN. Then, the parent CH node sends the multicast message with the initial threshold to all child nodes. In the next sampling round, as shown in Fig. 3, the CH node determined whether to forward a new sampling data to the parent node by judging the threshold domination.

Definition 1: Given sensor data set is $d_i = (d_i\_1, d_i\_2, \cdots, d_i\_attri\_n)$, the threshold is $MinMax(D)$, if in the sensor data attibute, $\forall d_i\_j, (j = 1, 2, \cdots, attri\_n)$,

$\exists d_i\_j > minmax$ (or $\exists d_i\_j < minmax$), then the sensor data d_i is dominated by the threshold $MinMax(D)$, the node will not forward the new sensor data to the parent node, or forwarding them.

Definition 2: Given sensor data set is $d_i = (d_i\_1, d_i\_2, \cdots, d_i\_attri\_n)$, the snesor data set is $d_s = (d_s\_1, d_s\_2, \cdots, d_s\_attri\_n)$ from the child node. If $\forall d_i\_j, (j = 1, 2, \cdots, attri\_n)$ and $\forall d_s\_k, (k = 1, 2, \cdots, attri\_n)$, $\exists d_i\_j > d_s\_k$ (or $\exists d_i\_j < d_s\_k$), then d_i_j is dominated by d_s_k, the node will forward d_s_k to the parent node, or forwarding d_i_j to the parent node.

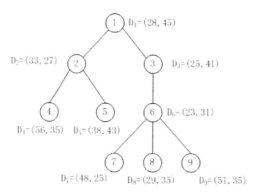

Fig. 2. The initialization state of virtual sensor network

According to the definiton 1, for the given VSN, suppose that the sampling data is dominated by the threshold when all sampling data attributes are greater than the threshold, the CH node will not forward the new sensor data to the parent node, or forwarding them. As shown in Fig. 4, the node 7 and node 8 forwarded the sensor data to the parent node 6, according to the definiton 2, the sensor data $D_8 = (28,36)$ is dominated by $D_6 = (25,32)$, so the node 6 will foward the sensor data $D_6 = (25,32)$ and $D_7 = (45,23)$ to the parent node 3, instead of fowarding $D_8 = (28,36)$ to node 8. Because the $D_3 = (26,40)$ is dominated by $D_8 = (28,36)$, the $D_6 = (25,32)$ and $D_7 = (45,23)$ were stored by the root node 1. For the other VSN branch, the node 2 transmited the new sampling data on the basis of the definiton 1 and definition 2.

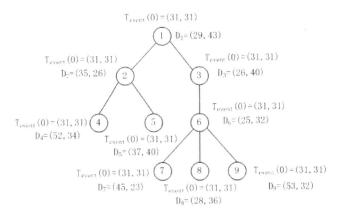

Fig. 3. The threshold *MinMax(D)* of virtual sensosr network

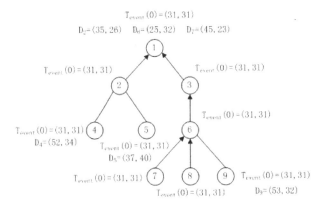

Fig. 4. The data gathering process with *MinMax(D)* threshold in VSN

3   Performance Evaluation

In this section, the energy consumption performance of proposed strategy was evluated. Consider two sensor-network based systems, one for warning of forest fire and the other for monitoring of chemical plume pollution, both to be deployed on a planar rectangle grid of 500m*500m, where 600 sensor nodes are randomly distributed with the transmission radius r = 100m. The initial energy of each node is 10J. The sink node is placed in the middle of the sensor field. The number of the conmunication hop is $Maxhops = 1$ in the cluster members. The number of formation and detection messages are 120 bit, the number of sensor data is 800 bit. the application of VSN formed was decided to compare it to the traditional dedicated sensor networks. Fig.5 shows the total energy consumption of the nodes obtained with two non-cooperation WSNs and two different VSN subsets with the $MinMax(D)$ threshold. Because the subset of nodes belonging to the VSN collaborates to carry out a given application data. VSN support will simplify application deployment and reduce energy consumption.

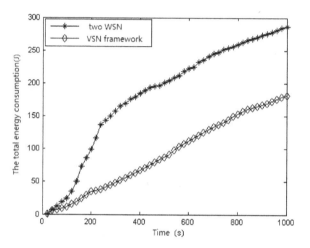

Fig. 5. The total energy consumption compare of two frameworks

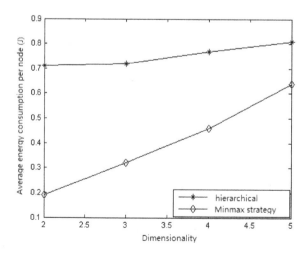

Fig. 6. The average energy consumption of variety dimensionality per node

Fig.6 shows the average energy consumption per node when the dimensionality of sensor sampling data increases using the common hierarchical framework and MinMax threshold strategy separately. MinMax threshold strategy is superior to the common scheme in any dimensionality, reducing the energy consumption. However, the difference shrinks with the increasing of the dimensionality because the dominance relationship between high-dimensional points is weakened.

4   Conclusion

In this paper, we have proposed and described a massive sensor sampling data gathering optimization strategy using MinMax threshold. The strategy is fully distributed, self-organizing and energy efficient, so that it is more suitable for a large scale sensor network in multi-criteria target monitoring. However, the scale of the sensor sampling data tends to increase with the emergence of multimedia sensor networks. We come up with future research issues related with the massive sensor data modeling and visualization using the novel computing technology for the sensing layer of internet of things.

Acknowledgment. The research work was supported by Young teachers' Scientific Research Project of the Fundamental Research Funds, Northeastern University and the Doctoral Fund of Northeastern University at Qinhuangdao under Grant No. XNB201301.

References

1. Farhana, J., Alvaro, A.A.F.: An Algorithmic Strategy for In-network Distributed Spatial Analysis in Wireless Sensor Networks. Journal of Parallel and Distributed Computing 72(12), 1628–1653 (2012)
2. Kemal, A., Fatih, S., Brian, M.: Clustering of Wireless Sensor and Actor Networks Based on Sensor Distribution and Connectivity. J. Parallel Distrib. Comput. 69, 573–587 (2009)
3. Xiang, M., Shi, W.R., et al.: Energy Efficient Clustering Algorithm for Maximizing Lifetime of Wireless Sensor Networks. Int. J. Electron. Commun. 64, 289–298 (2010)
4. Ali, C., Samuel, P.: A Distributed Energy-efficient Clustering Protocol for Wireless Sensor Networks. Computers and Electrical Engineering 36, 303–312 (2010)
5. Adamu, M.Z., Ang, L.-M., Kah, P.S.: Termite-hill: Performance optimized swarm intelligence based routing algorithm for wireless sensor networks. Journal of Network and Computer Applications 35(6), 1901–1917 (2012)
6. Xu, H.L., Huang, L.S., et al.: Energy-efficient Cooperative Data Aggregation for Wireless Sensor Networks. J. Parallel Distrib. Comput. 70, 953–961 (2010)
7. Lee, B.D., Lim, K.-H.: An Energy-Efficient Hybrid Data-Gathering Protocol Based on the Dynamic Switching of Reporting Schemes in Wireless Sensor Networks 6(3), 378–387 (2012)
8. Mini, R.A.F., Loureiro, A.A.F.: Energy-efficient Design of Wireless Sensor Networks Based on Finite Energy Budget. Computer Communications 35(4), 1736–1748 (2012)
9. Song, X., Wang, C.R., Wang, J.: A Multi-criteria Target Monitoring Strategy Using Minmax Operator in Formed Virtual Sensor Networks. In: Liu, D., Zhang, H., Polycarpou, M., Alippi, C., He, H. (eds.) ISNN 2011, Part III. LNCS, vol. 6677, pp. 407–415. Springer, Heidelberg (2011)

Improving the Performance of Neural Networks with Random Forest in Detecting Network Intrusions

Wenjuan Li[1] and Yuxin Meng[2]

[1] Computer Science Division, Zhaoqing Foreign Language College, Guangdong, China
wenjuan.anastatia@gmail.com
[2] Department of Computer Science, City University of Hong Kong, Hong Kong, China
yuxin.meng@my.cityu.edu.hk

Abstract. Neural Networks such as RBFN and BPNN have been widely studied in the area of network intrusion detection, with the purpose of detecting a variety of network anomalies (e.g., worms, malware). In real-world applications, however, the performance of these neural networks is dynamic regarding the use of different datasets. One of the reasons is that there are some redundant features for the dataset. To mitigate this issue, in this paper, we propose an approach of combining Neural Networks with Random Forest to improve the accuracy of detecting network intrusions. In particular, we design an intelligent anomaly detection system that uses the algorithm of Random Forest in the process of feature selection and selects an appropriate algorithm in an adaptive way. In the evaluation, we conducted two major experiments using the KDD1999 dataset and a real dataset respectively. The experimental results indicate that Random Forest can enhance the performance of Neural Networks by identifying important and closely related features and that our developed system can select a better algorithm intelligently.

Keywords: Neural Networks, Anomaly Intrusion Detection, Random Forest, Intelligent Applications.

1 Introduction

Network intrusion detection systems (NIDSs) are becoming an essential and important component in protecting network environments from various network attacks (e.g., Trojan, worms). Traditionally, these detection systems can be classified as misuse-based NIDS (or *signature-based NIDS*) [10, 18] and anomaly-based NIDS [4, 17]. The former attempts to detect an attack by comparing incoming packets with its pre-defined signatures[1] (or called *rules*), while the latter identifies an anomaly by pointing out great deviations between current events and pre-defined normal models[2]. The misuse-based NIDS can only detect known attacks whereas the anomaly-based NIDS has the capability of identifying potential novel attacks.

For an anomaly-based NIDS, a normal model should be established in advance by training with historical data. During the establishment, a machine learning algorithm is usually used to build this normal model. A lot of machine learning schemes have been

---

[1] In intrusion detection, a signature is a kind of descriptions for known threats.
[2] A normal model can be used to represent a normal connection or a normal event.

C. Guo, Z.-G. Hou, and Z. Zeng (Eds.): ISNN 2013, Part II, LNCS 7952, pp. 622–629, 2013.
© Springer-Verlag Berlin Heidelberg 2013

extensively studied in the field of network intrusion detection. For example, Kim and Park [8] proposed a method of applying Support Vector Machines (SVMs) to network-based Intrusion Detection System and their experimental results on KDD1999 datasets[3] indicated that their method was effective at detecting intrusions. Li *et al.* [11] proposed a novel algorithm of TCM-KNN (Transductive Confidence Machines for K-Nearest Neighbors) to help detect network attacks by using a small dataset. Their experiments were conducted on the same KDD1999 dataset and achieved a good result. Later, Meng and Kwok [16] applied a disagreement-based semi-supervised learning algorithm to detecting network intrusions in addition to false alarm reduction. Their experimental results on both KDD1999 and a real dataset indicated that the proposed scheme can achieve a high detection accuracy by only using a small labeled dataset compared to several traditional supervised learning algorithms (e.g., SVM).

Neural Networks such as Radial Basis Function Network (RBFN) and Back Propagation Neural Network (BPNN) are very popular for building supervised-based intrusion detection systems. But in real-world deployment, we find that its performance is fluctuant by testing with different datasets. In this paper, we propose a method of applying Random Forest algorithm to both RBFN and BPNN. In particular, we develop an intelligent anomaly detection system that uses the algorithm of Random Forest in the process of feature selection and selects a better algorithm intelligently. Accordingly, we conducted two major experiments to explore the performance of our developed system on the KDD1999 dataset and a private dataset respectively. The evaluation results demonstrate that the algorithm of Random Forest can improve the performance of neural networks (both RBFN and BPNN) in detecting network anomalies.

The rest of this paper is organized as follows: in Section 2, we introduce the background of Random Forest and review some related work about the applications of Neural Networks and Random Forest in intrusion detection; Section 3 describes our proposed method of combining Neural Networks with Random Forest and presents our developed intelligent system; we illustrate our experiments and the experimental results in Section 4; at last, Section 5 concludes our work and points out the future work.

2   Background and Related Work

This section introduces the background of Random Forest and reviews some related work regarding Neural Networks and Random Forest and their applications in the field of intrusion detection.

2.1   Random Forest

Random Forest (or *random forests*, shortly *RF*) is an ensemble classifier that is composed of many decision trees and outputs the class that is the mode of the classes output by individual trees [2]. Random Forest has low classification (and regression) errors comparable to SVM [9]. It has several merits described as follows: 1) it is one of the most accurate learning algorithms available and it can produce a highly accurate classifier for various datasets [3]; 2) it runs efficiently on large databases and can handle

---

[3] http://kdd.ics.uci.edu/databases/kddcup99/kddcup99.html

thousands of input variables without variable deletion; 3) it generates an internal unbiased estimate of the generalization error as the forest building progresses; 4) It gives estimates of what variables are important in the classification and provides methods for balancing errors in class population unbalanced datasets; and 5) it offers an experimental method for detecting variable interactions.

Specifically, Random Forest usually works as follows [2]: when the training set for the current tree is drawn by sampling with replacement, nearly one-third of the cases are left out of the sample. This oob (out-of-bag) data is used for obtaining a *running unbiased estimate* of the classification error as trees are added to the forest. It can also be used for obtaining estimates of variable importance. After each tree is built, all of the data is run down the tree, and proximities are computed for each pair of cases. If two cases occupy the same terminal node, their proximity is increased by one. At the end of the run, the proximities are normalized by dividing the number of trees. Proximities can be used to replace missing data, locate outliers, and produce low-dimensional views of the data. Each tree can be constructed as below:

- Let N denotes the number of training cases and M denotes the number of variables in the classifier.
- Choosing the number m of input variables which is used to determine a node of the tree (note that m should be much less than M).
- Choosing a training set for this tree by choosing n times with replacement from all N available training cases (i.e., take a bootstrap sample). Using the rest of the cases to estimate the error of the tree through predicting their classes.
- For each node of the tree, randomly choosing m variables and calculating the best split based on these variables in the training set.
- Each tree is fully grown and not pruned.

2.2   Related Work

Neural Networks have been widely studied in the area of intrusion detection, especially aiming to improve the detection efficiency of anomaly-based NIDSs. A lot of Neural-Network-based algorithms have been developed.

Han and Cho [6] first identified that neural networks were good at learning system-call sequences and proposed a novel intrusion-detection technique based on evolutionary neural networks (ENNs). The proposed method could consume less time to obtain superior neural networks than using other conventional approaches. The reason is that they discover the structures and weights of the neural networks simultaneously. Their experiments on the DARPA 1999 dataset showed that the time required for learning could be reduced without any loss of detection performance. Cha *et al.* [5] presented a method of combining neural networks and fuzzy membership function, and used the Soundex algorithm to conduct feature selection and to change variable length data into a fixed length learning pattern. They further compared their approach with N-gram technique and indicated that their method achieved a higher detection rate. Later, Yang and Karahoca [22] proposed an anomaly-based network intrusion detection based on Cellular Neural Networks (CNN) model, which featured with multi-dimensional array of neurons and local interconnections among cells. Their experiments on the KDD1999 dataset indicated that the CNN model was effective in intrusion detection and exhibited an excellent performance with a higher detection rate and a lower false positive rate in contrast to back propagation neural network. Several other applications of Neural Networks in detecting anomalies can be referred to [1, 7, 12–14, 19–21].

Random Forest (RF) has also been applied to intrusion detection. For instance, Kim *et al.* [9] proposed an approach of building lightweight Intrusion Detection System (IDS) based on Random Forest (RF). Their experiments on the KDD1999 dataset indicated the feasibility of their approach. That is, the approach could not only guarantee high detection rates but also figure out important features of audit data. Later, Malik *et al.* [15] proposed and implemented a hybrid classifier based on binary particle swarm optimization (BPSO) and random forests (RF) to classify PROBE attacks in a network environment. In the evaluation on the KDD1999 dataset, the results showed that their proposed classifier outperformed 8 well-known classifiers such as SMO (SVM), PART (C4.5), Bagging, etc.

3 Our Proposed Method

In real-world applications, we find that Random Forest is good at determining important features and this observation was also validated in [9]. Regarding the anomaly-based detection, there may exist a lot of features extracted from a network environment (i.e., each record in the KDD1999 dataset contains 41 features), in which some features are not important and essential. In this work, we therefore attempt to develop an intelligent anomaly detection system which employs Random Forest to conduct feature selection for Neural Networks.

The architecture of our developed intelligent anomaly detection system is described in Fig. 1. The term of *intelligent* in this architecture means that the system can compare the performance of algorithms and determine the best one in an adaptive way . In particular, we select two neural-network algorithms of RBFN and BPNN in this work as these two algorithms are very popular in real deployment.

- *RBFN:* is an artificial neural network (ANN) which employs radial basis functions as its activation functions. This algorithm can be used to model complex mappings.
- *BPNN:* can propagate input forward and perform backward passes through the network and calculate appropriate weights.

In total, there are four major components in the architecture: *Feature Selection, Decision Component, Detection Component* and *Extended Algorithm Pool.*

Feature selection. In this component, we use Random Forest to conduct feature selection. Before that, network events should be converted to pre-defined features (i.e., the features in the KDD1999 dataset). For different sets of features, Random Forest is expected to decide the most important and relevant features.

Decision component. In this component, the system can compare the performance of two specific algorithms of RBFN and BPNN in the aspect of detecting anomalies. As a case study, we measure the algorithm performance based on *detection accuracy*. That is, the algorithm with a better detection accuracy can be output from this component.

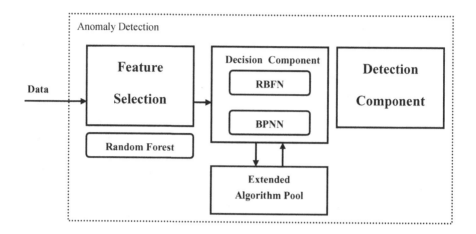

Fig. 1. The architecture of our developed intelligent anomaly detection system

Detection component. In this component, the system can use the output algorithm (e.g., RBFN or BPNN) to detect network intrusions. In addition, the system can require to re-select the algorithm after a fixed time.

Extended algorithm pool. This component is designed to integrate and store other potential neural-network algorithms. By adding new algorithms (i.e., not only neural-network algorithms) into this pool, the system can consider more algorithms and provide a larger comparison among different algorithms in the decision component.

4   Evaluation

In this section, we explore the performance of our developed intelligent system using the KDD1999 dataset and a real dataset respectively.

4.1   KDD1999 Dataset

The KDD Cup 1999 Dataset (KDD1999) derives from the DARPA packet traces and includes a wide variety of intrusions simulated in a military network environment. Although this dataset is decade-old, it is the only well-known and widely available dataset in the area of intrusion detection. The full KDD1999 dataset includes 4,898,431 records and each record contains 41 features. As a baseline experiment, we evaluate our system with a random selection of 45,069 records (each with 41 features) and 4 categories of attacks such as *Probe, DOS, U2R* and *R2L*. These 41 specific features (a total of 4 categories) in the KDD1999 dataset are shown in Table 1.

The results of detection accuracy are shown in Table 2. This table shows that RBFN and BPNN can achieve a detection accuracy of 98.54% and 98.65% respectively, while our system can achieve a detection accuracy of 99.12%. These results indicate that Random Forest can improve the performance of Neural Networks by identifying important

Table 1. Features and categories in the KDD1999 dataset

Category	Specific Featrues
TCP connection basic characteristic	duration, protocol_type, service, flag, src_bytes, dst_bytes, land, wrong_fragment, urgent
TCP connection content characteristic	hot, num_failed_logins, logged_in, num_compromised, root_shell, su_attempted, num_root, num_file_creations, num_shells, num_access_files, num_outbound_cmds, is_hot_login, is_guest_login
Time-based network traffic	count, srv_count, serror_rate, srv_seror_rate, rerror_rate, srv_rerror_rate, same_srv_rate, diff_srv_rate, srv_diff_host_rate
Host-based network traffic	dst_host_count, dst_host_srv_count, dst_host_same_srv_rate, dst_host_diff_srv_rate, dst_host_same_src_port_rate, dst_host_srv_diff_host_rate, dst_host_serror_rate, dst_host_srv_serror_rate, dst_host_srv_serror_rate, dst_host_rerror_rate, dst_host_srv_rerror_rate

Table 2. The results of detection accuracy

10-folder	RBFN	BPNN	Our System
Detection Accuracy (%)	98.54	98.65	99.12

and closely related features. In particular, in this experiment, Random Forest selects 18 features out of these 41 features. Through training with the most important and related features, the detection accuracy of our developed system is higher than that of those original neural network algorithms.

4.2 Real Dataset

In this experiment, we replace the KDD1999 dataset with a real trace collected in a network environment. Specifically, the real dataset was collected by a Honeypot[4] deployed in a public domain, in which anyone can access from anywhere on the Internet. The real dataset contains 58,232 records and each record has 22 features (see *TCP connection basic characteristic* and *TCP connection content characteristic* in Table 1). The dataset was labeled by means of expert knowledge (i.e., labeling as *attack* or *normal*).

The results of detection accuracy are shown in Table 3. This table shows that RBFN and BPNN can achieve a detection accuracy of 89.23% and 88.74% respectively, whereas our system can improve the detection accuracy to 93.23%. In the component of feature selection, 12 features were selected out of the 22 features. These results validate that our approach is encouraging in real settings.

To explore the intelligent nature of our developed system, we divided the real dataset into five parts (called *P1, P2, P3, P4* and *P5*). We present the output results of decision component in Table 4. This table shows that our developed system can intelligently

---

[4] http://www.honeybird.hk/

Table 3. The results of detection accuracy

10-folder	RBFN	BPNN	Our System
Detection Accuracy (%)	89.23	88.74	93.23

Table 4. The results of detection accuracy

Data	P1	P2	P3	P4	P5
Output	RBFN	RBFN	BPNN	BPNN	RBFN

select and output the algorithm which can achieve a better detection accuracy. For instance, RBFN is selected for *P1*, *P2*, *P5* while BPNN is selected for *P3* and *P4*.

5   Conclusions and Future Work

Neural Networks have been widely used to help establish a normal profile in the field of anomaly-based intrusion detection. But the performance of these neural networks is dynamic varied with different datasets. We identify that one of the main reasons is due to the redundancy of data features. To mitigate this issue, in this work, we attempt to conduct the process of feature selection by means of Random Forest. In particular, we develop an intelligent anomaly detection system that enables to decide and output a better algorithm in an adaptive way. In the evaluation, we conducted two major experiments on the KDD1999 dataset and a real dataset respectively, the experimental results demonstrate that our proposed method can overall enhance the performance (e.g., detection accuracy) of RBFN and BPNN in detecting network anomalies.

This work presents some initial results of our developed system. Future work could include conducting more analysis of feature selection (i.e., what features are most likely to be selected) and involving a larger dateset to validate our system. Future work could also include combining and comparing other algorithms in our system to provide a more comprehensive evaluation of our approach.

References

1. Barreto, G.A., Aguayo, L.: Time Series Clustering for Anomaly Detection Using Competitive Neural Networks. In: Príncipe, J.C., Miikkulainen, R. (eds.) WSOM 2009. LNCS, vol. 5629, pp. 28–36. Springer, Heidelberg (2009)
2. Breiman, L.: Random Forest. Machine Learning 45(1), 5–32 (2001)
3. Caruana, R., Karampatziakis, N., Yessenalina, A.: An Empirical Evaluation of Supervised Learning in High Dimensions. In: Proceedings of the 25th International Conference on Machine Learning (ICML), pp. 96–103 (2008)
4. Cha, B.R., Lee, D.S.: Network-Based Anomaly Intrusion Detection Improvement by Bayesian Network and Indirect Relation. In: Apolloni, B., Howlett, R.J., Jain, L. (eds.) KES 2007, Part II. LNCS (LNAI), vol. 4693, pp. 141–148. Springer, Heidelberg (2007)
5. Cha, B.R., Park, K.W., Seo, J.H.: Neural Network Techniques for Host Anomaly Intrusion Detection Using Fixed Pattern Transformation. In: Gervasi, O., Gavrilova, M.L., Kumar, V., Laganá, A., Lee, H.P., Mun, Y., Taniar, D., Tan, C.J.K. (eds.) ICCSA 2005. LNCS, vol. 3481, pp. 254–263. Springer, Heidelberg (2005)

6. Han, S.-J., Cho, S.-B.: Evolutionary Neural Networks for Anomaly Detection based on the Behavior of a Program. IEEE Transactions on Systems, Man, and Cybernetics, Part B: Cybernetics 36(3), 559–570 (2005)

7. Gao, M., Tian, J., Zhou, S.: Community Intrusion Detection System Based on Radial Basic Probabilistic Neural Network. In: Yu, W., He, H., Zhang, N. (eds.) ISNN 2009, Part II. LNCS, vol. 5552, pp. 745–752. Springer, Heidelberg (2009)

8. Kim, D.S., Park, J.S.: Network-Based Intrusion Detection with Support Vector Machines. In: Kahng, H.-K. (ed.) ICOIN 2003. LNCS, vol. 2662, pp. 747–756. Springer, Heidelberg (2003)

9. Kim, D.S., Lee, S.M., Park, J.S.: Building Lightweight Intrusion Detection System Based on Random Forest. In: Wang, J., Yi, Z., Żurada, J.M., Lu, B.-L., Yin, H. (eds.) ISNN 2006. LNCS, vol. 3973, pp. 224–230. Springer, Heidelberg (2006)

10. Kruegel, C., Toth, T.: Using Decision Trees to Improve Signature-Based Intrusion Detection. In: Vigna, G., Kruegel, C., Jonsson, E. (eds.) RAID 2003. LNCS, vol. 2820, pp. 173–191. Springer, Heidelberg (2003)

11. Li, Y., Fang, B.-X., Guo, L., Chen, Y.: TCM-KNN Algorithm for Supervised Network Intrusion Detection. In: Yang, C.C., et al. (eds.) PAISI 2007. LNCS, vol. 4430, pp. 141–151. Springer, Heidelberg (2007)

12. Linda, O., Vollmer, T., Manic, M.: Neural Network based Intrusion Detection System for Critical Infrastructures. In: Proceedings of the 2009 International Joint Conference on Neural Networks (IJCNN), pp. 1827–1834 (2009)

13. Liu, G., Yi, Z.: Intrusion Detection Using PCASOM Neural Networks. In: Wang, J., Yi, Z., Żurada, J.M., Lu, B.-L., Yin, H. (eds.) ISNN 2006. LNCS, vol. 3973, pp. 240–245. Springer, Heidelberg (2006)

14. Liu.W., Duan, H.-X., Ren, P., Wu, J.-P.: IABA: An Improved PNN Algorithm for Anomaly Detection in Network Security Management. In: Proceedings of the 2010 International Conference on Natural Computation (ICNC), pp. 335–339 (2010)

15. Malik, A.J., Shahzad, W., Khan, F.S.: Binary PSO and Random Forests Algorithm for PROBE Attacks Detection in a Network. In: Proceedings of the 2011 IEEE Congress on Evolutionary Computation (CEC), pp. 662–668 (2011)

16. Meng, Y., Kwok, L.-F.: Intrusion Detection using Disagreement-based Semi-Supervised Learning: Detection Enhancement and False Alarm Reduction. In: Xiang, Y., Lopez, J., Kuo, C.-C.J., Zhou, W. (eds.) CSS 2012. LNCS, vol. 7672, pp. 483–497. Springer, Heidelberg (2012)

17. Paxson, V.: Bro: A System for Detecting Network Intruders in Real-Time. Computer Networks 31(23-24), 2435–2463 (1999)

18. Roesch, M.: Snort: Lightweight Intrusion Detection for Networks. In: Proceedings of the 1999 Usenix Lisa Conference, pp. 229–238 (1999)

19. Sun, N.-Q., Li, Y.: Intrusion Detection Based on Back-Propagation Neural Network and Feature Selection Mechanism. In: Lee, Y.-h., Kim, T.-h., Fang, W.-c., Ślęzak, D. (eds.) FGIT 2009. LNCS, vol. 5899, pp. 151–159. Springer, Heidelberg (2009)

20. Tian, D., Liu, Y., Li, B.: A Distributed Hebb Neural Network for Network Anomaly Detection. In: Stojmenovic, I., Thulasiram, R.K., Yang, L.T., Jia, W., Guo, M., de Mello, R.F. (eds.) ISPA 2007. LNCS, vol. 4742, pp. 314–325. Springer, Heidelberg (2007)

21. Wang, Y., Gu, D., Li, W., Li, H., Li, J.: Network Intrusion Detection with Workflow Feature Definition Using BP Neural Network. In: Yu, W., He, H., Zhang, N. (eds.) ISNN 2009, Part I. LNCS, vol. 5551, pp. 60–67. Springer, Heidelberg (2009)

22. Yang, Z., Karahoca, A.: An Anomaly Intrusion Detection Approach Using Cellular Neural Networks. In: Levi, A., Savaş, E., Yenigün, H., Balcısoy, S., Saygın, Y. (eds.) ISCIS 2006. LNCS, vol. 4263, pp. 908–917. Springer, Heidelberg (2006)

Displacement Prediction Model of Landslide Based on Functional Networks

Jiejie Chen[1,2], Zhigang Zeng[1,2], and Huiming Tang[3]

[1] Department of Control Science and Engineering,
Huazhong University of Science and Technology, Wuhan 430074, China
[2] Key Laboratory of Image Processing and Intelligent Control of Education Ministry
of China, Wuhan 430074, China
[3] Faculty of Engineering, China University of Geosciences, Wuhan 430074, China
chenjiejie118@gmail.com

Abstract. In this paper, a novel computational intelligence scheme is proposed to forecast landslide based on functional networks. Two types functional networks, general functional networks with two variables basis function (GFN) and separable functional networks (SFN) are applied to predict a real-world example. In addition, the experiments reveal that the landslide prediction using functional networks is reasonable and effective, and GFN are consistently better than SFN in terms of the same measurements.

Keywords: Functional networks, Two variables basis function, Landslide prediction, Separable.

1 Introduction

Landslide [1,2] are serious nature disasters which cause major socioeconomic disruptions, extensive property damage, and casualties. In order to avoid or reduce the harm in advance, the prediction of landslide is essential for carrying out quicker and safer mitigation programs. Hence, a number of methods have been tried in the problem of displacement of landslide forecasting.

Recently, functional networks [3,4] have been introduced as an extension of neural networks. Functional networks provide simple and valid techniques to model nonlinear system. In this paper, considering that the broad application prospects of functional networks [5,6,7], we apply them in landslide forecasting. Moreover, applying two types function networks such as GFN and SFN into research of landslide in the Three Gorges reservoir of China , the validity and practical value of these new framework can be demonstrated.

2 Functional Networks

Functional networks may be considered as more problem-driven than as data-driven, so the initial architecture is designed based on a problem in hand. As it is

C. Guo, Z.-G. Hou, and Z. Zeng (Eds.): ISNN 2013, Part II, LNCS 7952, pp. 630–637, 2013.

shown in Fig. 1, which represents a typical architecture of a functional network illustrating its principal components. A functional networks consist of several elements including:

a) One layer of input storing neurons of the data set $\{x_1, x_2, x_3, x_4\}$.

b) None, one or several layers to store intermediate information, there one layer $\{x_5, x_6\}$.

c) One or several layers of processing neurons or computing units that evaluate a set of input values and delivers a set of output values f_i, there are two layers, the first layer of neurons contains neurons f_1, f_2, f_3 and the second layer of neurons contains neurons f_4.

d) One layer of output storing neurons reduce to units O.

e) A set of direct links between them.

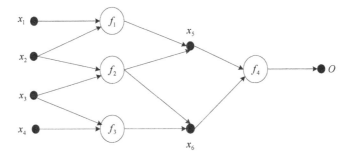

Fig. 1. Functional networks

Generally speaking, the neuron functions in functional networks are unknown functions from a given family, such as, polynomial, exponential,Fourier and so on. The SFN introduced by Castillo [3] is an interesting family with many applications. The SFN uses a functional expression that combines the separated efforts of input variables. It contains two inputs and one output, where x_1, x_2 are the two input variables and O is the output of the functional network, f_i and y_i are unknown neuron functions. The relationship between O, $\{x_1, x_2\}$ can be defined mathematically as follows:

$$O = F(x_1, x_2) = \sum_{i=1}^{n} f_i(x_1) y_i(x_2). \tag{1}$$

For illustrative purposes, considering the simplest architecture from this family, which neglects double interactions by separating the contributions of each of the inputs in the form, where $n = 2$ and $f_2(x_1) = y_1(x_2) = 1$,

$$O = F(x_1, x_2) = f_1(x_1) + y_2(x_2) := f(x_1) + y(x_2). \tag{2}$$

3   Analytical Method

Training is an very important part in the application of a functional network. Similar to the works Castillo et al [3] and Bruen and Yang [6], we consider the general Function networks in Fig. 2. It is easily know that the network structure is multiple input and multiple output. This network is composed of many layers, where each layer has many functional neurons. The number of every layer functional neurons may not be same, which only dependent on the initial topology. The set $\{x_1, \ldots, x_k\}$ as a variable is the first input to the network, with every layer functional neurons take previous neurons as a new input with different function equation to calculate, and the last output layer reduces to the unit $O_1, \ldots, O_n$.

$$
\begin{cases}
y_1 = f_1(x_1, \ldots, x_k), \\
y_2 = f_2(x_1, \ldots, x_k), \\
\quad \vdots \\
y_{l-1} = f_{l-1}(x_1, \ldots, x_k), \\
y_l = f_l(x_1, \ldots, x_k), \\
z_1 = g_1(y_1, \ldots, y_l), \\
z_2 = g_2(y_1, \ldots, y_l), \\
\quad \vdots \\
z_{m-1} = g_{m-1}(y_1, \ldots, y_l), \\
z_m = g_m(y_1, \ldots, y_l).
\end{cases}
\tag{3}
$$

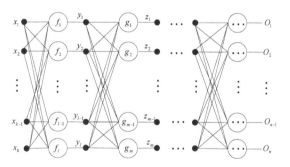

Fig. 2. Generalized functional networks

A neuron functional $f(x_1, x_2, \cdots, x_k)$ can be denoted as:

$$
\hat{f}(x_1, x_2, \cdots, x_k) = \sum_{j=1}^{m} \alpha_j \theta_j(x_1, x_2, \cdots, x_k),
\tag{4}
$$

where $\{\theta_1, \cdots, \theta_m\}$ is a linear independent combination of a standard functional form for each neuron function, the coefficients α_j are the parameters of the functional networks. In this paper, a new linear independent combination of a standard functional form with multiple basis functions are proposed $\theta_j = \{x_1^{m_1} \times x_2^{m_2} \times \cdots \times x_k^{m_k}, m_s = 0, 1, 2, \cdots, n; s = 1, 2, \cdots, k\}$.

The objective function used here is the sum of square errors, which should be minimised between the theoretic output and the output calculated by the network. The error can be measured as in

$$E = \sum_{i=1}^{t} \hat{e}^2 = \sum_{i=1}^{t} [O_i - \hat{f}(x_{i1}, x_{i2}, \cdots, x_{ik})]^2, \tag{5}$$

or

$$E = \sum_{i=1}^{t} [O_i - \sum_{j=1}^{m} \alpha_j \theta_j (x_{i1}, x_{i2}, \cdots, x_{ik})]^2. \tag{6}$$

For a unique representation of the functional networks, it should be add an initial functional condition. In this case, it uses either of the one initial conditions:

$$f(x_{01}, x_{02}, \ldots, x_{0k}) = \lambda, \tag{7}$$

where t is the training sample size, $x_{01}, \cdots, x_{0k}$ and λ are given constants to ensure the unique representation of functional networks. Therefore, we want to find the optimum coefficients α_j, and minimize the sum of square errors of the approximation $\hat{E}$. Using the Lagrange multipliers, it build the auxiliary function:

$$\hat{E} = \sum_{i=1}^{t} [O_i - \sum_{j=1}^{m} \alpha_j \theta_j (x_{i1}, x_{i2}, \cdots, x_{ik})]^2$$
$$+ \sigma (\sum_{j=1}^{m} \alpha_j \theta_j (x_{01}, x_{02}, \cdots, x_{0k}) - \lambda). \tag{8}$$

Minimisation of the above function $\hat{E}$ is equivalent to solving the set of derivative equations of $\hat{E}$ with respect to parameters α_j and multiplier σ. Then, for $j = 1, \cdots, m$:

$$\frac{\partial \hat{E}}{\partial \alpha_j} = -2 \sum_{i=1}^{t} [O_i - \sum_{j=1}^{m} \alpha_j \theta_j (x_{i1}, x_{i2}, \cdots, x_{ik})]$$
$$\times \theta_j (x_{i1}, x_{i2}, \cdots, x_{ik}) + \sigma \theta_j (x_{01}, x_{02}, \cdots, x_{0k}) = 0,$$
$$\frac{\partial \hat{E}}{\partial \sigma} = \sum_{j=1}^{m} \alpha_j \theta_j (x_{01}, x_{02}, \cdots, x_{0k}) - \lambda = 0. \tag{9}$$

The minimum is obtained by solving the system of linear equations, then the coefficients α_j are derived.

4   Application to Landslide Forecasting

4.1   Data Collection

The Baishuihe landslide occurred in the town of Zigui, on the south bank of the Yangtze River, and its 56 km away from west of the Three Gorges Dam of China. The landslide covers an area of (1260×10^4). The general grade is about 30. It is a high-speed loess landslide, in which the formation of the landslides is mainly related to the easy sliding loess material, complex geologic structure, precipitation and human engineering activities. The bedrock geology of the study area consists mainly of sandstone and mud stone, which is an easy slip stratum. The slope is of the category of bedding slopes. However, the land is limit, there are massive farmlands and people near the slope. When the slope failure of the warning area occurs, the landslide often caused serious damages to the inhabitant in reservoir area, the river side highway and even affected the Three Gorges Dam.

In this paper, two types of functional networks was used to predict the displacement of Baishuihe landslide. First, we propose a general functional networks with two variables basis function. And approximate $f(x_1, x_2)$:

$$\hat{f}(x_1, x_2) = \sum_{j=1}^{10} \alpha_j \theta_j(x_1, x_2), \tag{10}$$

where the coefficients α_j are the parameters of the functional networks. Choose a new linear independent combination of a standard functional form for each neuron function with binary basis function $\theta_j(x_1, x_2) = \{1, x_1, x_2, x_1^2, x_2^2, x_1 x_2, x_1^2 x_2, x_1 x_2^2, x_1^3, x_2^3\}$. j is the orders of two variables function binary basis function, and 10 is total the number of elements in the combination of sets of linearly independent θ_j.

Second, we choose a simple SFN with two inputs and one output, by (1) and (2) to predict the displacement of Baishuihe landslide.

$$\hat{f}(x_1) = \sum_{j=1}^{4} \alpha_j \phi_j(x_1), \tag{11}$$

$$\hat{y}(x_2) = \sum_{j=1}^{3} \beta_j \varphi_j(x_2), \tag{12}$$

where $\phi_j(x_1) = \{1, x_1, x_1^2, x_1^3\}$ and $\varphi_j(x_2) = \{1, x_2, x_2^2\}$. j is the orders of polynomial family and the total number of ϕ_j and φ_j sets of linearly is 4 and 3.

4.2   Results and Discussion

Eleven GPS deformation monitoring points layout in the landslide surface. And we choose the displacement data of ZG118 monitoring point as a case study. From Fig. 3, we can see the measurement data between January 2008 to June 2009, and then select the data between January 2008 to December 2008 as training

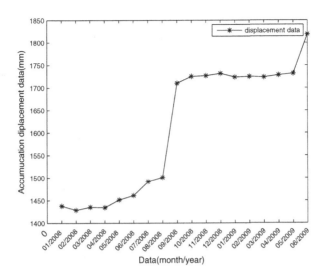

Fig. 3. Baishuihe landslide accumulation displacement data

data in order to construct the forecasting model and the rest of 6 groups of data from January 2009 to June 2009 as testing data. It is very obvious that the deformation of the landslide has obvious ascend trend with small fluctuations, which are related to the seasonal change.

In order to functional network to work well, the input data between January 2008 to December 2008 were normalized in a specific range, in between 0 and 1 based on (13) to make sure that the utilized input variables were independent of measurement units.

$$x_i^{(new)} = \frac{(x_i^{(old)} - min(x_i))}{(max(x_i) - min(x_i))},$$

(13)

where $x_i^{(old)}$ and $x_i^{(new)}$ means input and output data, $min(x_i)$ and $max(x_i)$ are respectively minimization and maximization of the input data.

Then, carrying out three functional networks compute themselves models. First, we obtained the following models with GFN by (10):

$$\begin{aligned}
f(x_1, x_2) = {}& -0.050422917999985 + 1.595658859300810x_1 \\
& + 2.210746424005230x_2 + 0.012027056091536x_1x_2 \\
& + 0.000034253302232x_1^2 - 2.220803919057385x_2^2 \\
& - 3.248695211763328x_1^2x_2 + 0.011012083105003x_1x_2^2 \\
& + 1.895126430244329x_1^3 + 0.810383805161921x_2^3.
\end{aligned}$$

(14)

Second, we got a SFN by (11) and (12).

$$f(x_1) = 1.080820780744473 \times 10^3$$
$$+0.005743071593316 \times 10^3 x_1$$
$$-0.012645877551293 \times 10^3 x_1^2$$
$$+0.006889501059795 \times 10^3 x_1^3,$$
$$y(x_2) = -1.081040041133778 \times 10^3$$
$$+0.002349151558379 \times 10^3 x_2$$
$$-0.002102411096194 \times 10^3 x_2^2. \tag{15}$$

And then, we unnormalized the data which were computed by three models value use (16) to get Tables 1.

$$x_i^{(old)} = x_i^{(new)}(max(x_i) - min(x_i)) + min(x_i). \tag{16}$$

Table 1. Comparision between predicted values and measured values

Time	Value	GFN(mm)	GFN relative error(%)	SFN(mm)	SFN relative error(%)
2009.01	1723.7	1736.2	0.73	1442.0	16.3
2009.02	1724.9	1732.2	0.72	1449.1	16.0
2009.03	1723.7	1740.4	0.97	1443.5	16.3
2009.04	1728.3	1739.7	0.66	1446.9	16.3
2009.05	1731.9	1737.8	0.34	1437.1	17.0
2009.06	1819.0	1728.2	5.0	1435.5	21.0

The simulation results illustrate two types functional network have well ability for prediction, and successfully predict the obvious deformation from January 2009 to June 2009. As shown in Table 1, GFN had the predicting values and actual measurement values are very close for every calculation, and the relative error falls into 5 percent in Table 1, the predicting precision is high enough which can satisfy the request of deformation prediction of landslide in medium-term and demonstrate the superiority of the functional network approach to solve real-time landslide forecast. And GFN had a lower relative error than SFN. Because, GFN have more input of variables and more complicated compute, which benefit functional networks to give more accurate prediction.

5   Conclusion

In this paper, the real case of Baishuihe landslide of the Three Gorges reservoir of China was utilized to investigate the capabilities of two kinds of functional networks intelligence predictive models to forecast landslide deformation.The functional network can work out the complex nonlinear relation by learning models and using the present data. And it is possible that a more detailed study with more complex forms of the model will improve even further on these results.

Acknowledgements. The work is supported by the Natural Science Foundation of China under Grant 61125303, the 973 Program of China under Grant 2011CB710606, the Fund for Distinguished Young Scholars of Hubei Province under Grant 2010CDA081.

References

1. Cubito, A., Ferrara, V., Pappalardo, G.: Landslide Hazard in the Nebrodi Mountains (Northeastern Sicily). Geomorphology 66(1-4), 359–372 (2005)
2. Chen, H., Zeng, Z.: Deformation Prediction of Landslide Based on Improved Back-propagation Neural Network. Cognitive Computation 5(1), 56–62 (2013)
3. Castillo, E.: Functional Networks. Neural Process Letters 7(3), 151–159 (1998)
4. Castillo, E., Cobo, A., Gomez-Nesterkin, R., Hadi, A.S.: A General Framework for Functional Networks. Networks 35(1), 70–82 (2000)
5. Castillo, E., Gutiérrez, J.M.: Nonlinear Time Series Modeling and Prediction Using Functional Networks. Extracting Information Masked by Chaos. Physics Letters A 244(1-3), 71–84 (1998)
6. Bruen, M., Yang, J.: Functional Networks in Real-time Flood Forecasting–A Novel Application. Advances in Water Resources 28(9), 899–909 (2005)
7. El-Sebakhy, E.A.: Software Reliability Identification Using Functional Networks: A comparative Study. Expert Systems with Applications 36(2), 4013–4020 (2009)
8. Castillo, E., Gutierrez, J.M., Cobo, A., Castillo, C.: A Minimax Method for Learning Functional Networks. Neural Processing Letters 11(1), 39–49 (2000)

Target Recognition Methods
Based on Multi-neural Network Classifiers Fusion

Huiying Dong and Shengfu Chen

College of Information Sciences and Technology, Shenyang Ligong University
Shenyang 110159, China
{huiyingdong,chshfmail}@163.com

Abstract. In the paper, three kinds of classifiers are fused they are BP network classifier, self-organizing feature map network classifier and RBF network classifier and the moment invariant features as well as roundness features as the inputs of the fused neural network. Given targets are recognized by the majority voting method and self-adapts weighted fusion algorithm of the fused classifier, and also by the three network classifiers respectively. The recognition results of single neural network and fusion algorithm are analyzed and compared. The results indicate that the recognition rate of multi-neural networks fusion algorithm is higher than any single neural network, and also show that the fusion algorithm has the significance for improving the accuracy of recognition.

Keywords: multi-neural network classifier, target recognition algorithm, multi-classifier fusion, BP network, SOM network, RBF network.

1 Introduction

It is information fusion that the multiple sensors or multi-source information are processed comprehensively and the results are getting more accurate and more reliable. So the best experimental method will be selected for solving a target recognition problem. However, a large number of experiments have shown that although the overall performance of a particular method is best, but other methods may correctly recognize the false recognition samples of this method. That is to say, for samples to be recognized, the different recognition methods may result in complementary information. For the complementary information of classifier fusion, it will improve the recognition performance by fusing multi-classifiers organically. In this paper, BP network classifier, self-organizing feature map network classifier and RBF network classifier are fused. Given targets are recognized by majority voting algorithm andself-adaptive weighted fusion algorithm, they can improve the effect of target recognition effectively.

2 Multi-neural Network Classifiers Fusion System

Different neural network classifiers have different performance in target recognition. If use a single classifier to recognize targets, we can't achieve good effect. With the

C. Guo, Z.-G. Hou, and Z. Zeng (Eds.): ISNN 2013, Part II, LNCS 7952, pp. 638–647, 2013.

development of fusion technology, target recognition rate is improving. The composition of multi-neural network target recognition system is shown in Figure 1. It consists of the following components: feature extraction, neural network classifiers, decision fusion, output of targets classification.

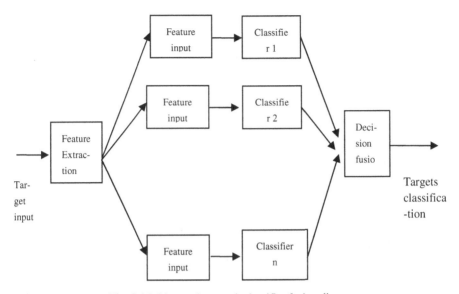

Fig. 1. Multi-neural network classifier fusion diagram

3   Neural Network Classification Algorithm

Currently, there are more than 40 kinds of neural network model. Three typical neural networks, which are BP neural network, SOM neural network and RBF neural network, are used in the paper. In the experiment, 8 moment invariant features which contain a roundness feature and 7 moment features are used as the input of each neural network; At the same time, referencing a experience rule that training sample number is 5 to 10 times the total number of the network connection weights[1] and the actual situation of networks and samples, each type sample is selected 5 samples. So, each network input is 50 samples; Finally, the trained networks above are used to recognize 10 categories targets.

3.1   BP Neural Network

BP neural network is a one-way communication multilayer feedforward network. So far, the multilayer feedforward network by using BP algorithm is the most widely used in neural network. In multilayer feedforward network applications, single hidden layer network is the most widely. Generally, single-hidden layer feedforward network is known as three-layer feedforward network or three-layer sensor, which is shown in Figure 2.

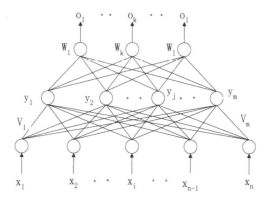

Fig. 2. Three layers BP network

The BP neural network structure which is used in the experiment is as follows:

　　1) Input layer and output layer node: Input layer have 8 neurons(which correspond 8 invariant features) and output layer have 10 neurons (which correspond 10 categories targets).

　　2）Initial weight value: According to the experience, hidden layer initial weight value is the random value between 0 to 1, output layer initial weight value is randomly assigned +1 or -1, and both number is equal.

　　3）Hidden layer number and hidden node: A single hidden layer BP network is used in the experiment and the trained network achieve the desired accuracy.

According to past designers accumulating many experiences, we can get determine hidden layer number empirical formula[2]:

$$m = \sqrt{n+l} + \alpha \tag{1}$$

In the formula, m is hidden node number, n is input node number, l is output node number, α is a constant between 1 to 10.

　　According to the specific circumstances of experimental data and combining with the empirical formula, hidden node number is determined as 30. BP network with the activating function $f(x) = 1/(1 + \exp(Bx))(B > 0)$ （Log-Sigmoid type）, in the experiment, $B = -1$.

3.2　Self-Organizing Feature Map Neural Network

SOM neural network is self-organizing feature map neural network. The network formed by input layer and competitive layer which is shown in Figure 3.

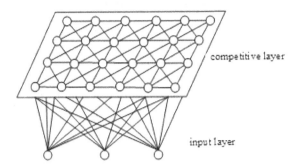

Fig. 3. Two-dimensional SOM topology structure

The basic thought of SOM neural network is[3]: competitive layer neurons in the network compete the response opportunities of the input pattern. Finally only one neuron become the winner of the competition and its connection weight value adjust to a more favorable its competition direction. Thus, winning neuron express the classification of input patterns.

The SOM neural network structure which is used in the experiment is as follows:

1) Input layer and output layer: Input layer have 8 neurons and output layer have 10 neurons.

2) Initial weight value : In the experiment, m input samples are randomly selected from the training setting as the initial weights[4], that is

$$W_j(0) = X^{rand} \quad , \quad j = 1,2,...,m \tag{2}$$

In the formula, X^{rand} is random sample vector.

3) Neighborhood : Neighborhood radius is used to express the size of the better neighborhood $N_{j^*}(t)$, and radius $r(t)$ is usually taken as[5]:

$$r(t) = C(1 - t/t_m) \tag{3}$$

In the formula, C is constant, t_m is pre-selected maximum training number. In the experiment, setting $t_m = 2000$, .

4) Learning rate : In the experiment, network learning rate $\delta(t)$ is used as the function as follows[5]:

$$\delta(t) = ce^{-bt} \tag{4}$$

In the formula, c is a constant between 0 to 1, b is a constant that greater than one.

3.3   Based on Radial Basis Function Neural Network

RBF neural network is based on radial basis function neural network. The network is a triple-layer feedforward neural network. Input layer is formed by the source node; The second layer is hidden layer and its unit number depend on the needs of the description problem; The third layer is output layer that response the input pattern. Supposing the network has m input neurons, l hidden neurons and n output neurons. The topology structure is shown in Figure 4.

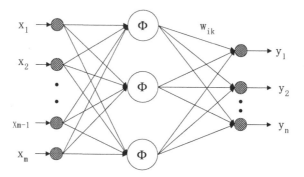

Fig. 4. RBF topology structure

In the experiment, input layer have 8 neurons and output have 10. According to empirical formula (1) and testing results, hidden layer are set 13 neurons. $\Phi(x) = e^{-x^2}$ is taken as radial basis function. In the formula, function center is origin, width is 1. Weight value is set a value between 0.1 to 1 and $\eta = 0.5$ is taken as learning rate. Then, weight the 13 radial basis function. Using above parameters trained the network then use it to recognize targets.

4   Multiple Classifiers Fusion Algorithm

Multiple classifiers fusion results not only depend on the selection of fusion algorithm, but also the classifier performance and classification data, and so on. Although various fusion algorithm have been proposed, most of the studies are limited to verify the effectiveness of these algorithms from the application, not demonstrated the conclusion from theoretical point. Therefore, it has great theoretical and practical significance to generalize the meaning of fusion, then reveal the nature of fusion from mathematics. There are some commonly multi-classifier fusion operator that including majority voting method, average method (simple average and weighted average), the multiplication rule, maximum and minimum rules, Bayesian method[6], and so on. Majority voting method and dynamic adaptive weighting method are mainly used in this article.

4.1   Majority Voting Fusion Algorithm

Majority voting method [7]' use single classifier for output category of given test samples, the test samples are divided into a class that multiple classifier with the same decision-making, does not require any training, simple and effective. The method is similar to the election of voting process. The category that most classifier agree with is taken as fusion classification result. Supposing there are L neural network classifier and target categories have M. The l th ($l = 1, ..., L$) neural network classifier mark the i th target P_i recognition result as follows:

$$V_{lm}(P_i) = \begin{cases} 1 & \text{when the lth neural network determine } p_i \text{ as the mth target} \\ 0 & \text{when the lth neural network doesn't determine } p_i \text{ as the mth target} \end{cases}$$

(5)

Here $m = 1, ..., M$. so, L classifier for the total votes number of the i th target determined as m class is as follows:

$$V_m(P_i) = \sum_{l=1}^{L} V_{lm}(P_i)$$

Therefore, target recognition result by neural network classifiers majority voting method is as follows:

$$D(P_i) = n \quad , \quad if \quad V_n(i) = \max_{m=1...M} V_m(P_i) \ , n = 1, ..., M \quad (6)$$

Thus, P_i is the n th class. Then request $V_1(P_i), ..., V_M(P_i)$ variance. If the variances less than given threshold, refuse to recognize.

4.2   Self-Adaptive Weighted Fusion Algorithm

Classical weighted algorithm propose a computational model, then, according to the training sample estimates the model parameter. Such as, the weight of each classifier in linear model, fuzzy measure in fuzzy integral, etc. Then optimize the model to best in statistical sense. Although many scholars from different angles have proved that the fusion method will increase system performance under certain conditions[8], the method has clear defect that the classifier is only given a weight value(the value reflect the importance of classifier in fusion) base on the statistical performance and doesn't consider each sample specific circumstances. That is to say, the classifier output are fused together by fixed weight value. Obviously, this is unreasonable. If we can estimate weight value in the line with each sample specific circumstance, multiple classifiers fusion method of adaptive changing weight could be realized. The method used in this article is based on classical weighted average model. Setting up, there are L classifiers in the system and each classifier output are denoted as $f_l (l = 1, 2, ...L)$. So, the fused result is :

$$g = \sum_{l=1}^{L} w_l f_l \tag{7}$$

In the formula, w_l is the weight value of the l th classifier. However, the difference is that the different value of w_l is base on different sample. Weight value calculating uses the following method. From the single classifier network recognition method of the first few chapters, it can be seen that the network has an actual output or the center of clustering after use the samples with selected target feature to train the network. The article use the Euclidean distance between the target eigenvector, which to be recognized, and sample eigenvector of network output to recognize. If the Euclidean distance between the target that to be recognized and the sample target of a category is least, the target belongs to this category. If the distance greater than certain threshold, refuse to recognize. Similarly, this distance judging method is also used in adaptive fusion algorithm. Using a monotonic decrease function, take the variable that convert d_l to a value between 0 to 1 as each classifier weight value. In the article, use the following function:

$$w_l = 1 - 1/(1 + \exp(-\alpha(d_l - \beta))) \tag{8}$$

In the formula, α and β is the parameter to be determined. When given a test sample, first calculate the Euclidean distance d_l between the test sample in each neural network classifier and training sample. Taking this as a basis, generate each classifier weight value, thus realize the fusion method of adaptive changing weight value. It can be seen that this method can adjust independently the weight value. From the following experiment, it also can be seen that this method can achieve better classification effect.

5   Multi-neural Network Classifiers Fusion Experimental Results

5.1   The Recognition by Majority Voting Fusion Algorithm

According to the principle of majority voting that described above, fuse the recognition results of BP neural network, SOM neural network and RBF neural network: The recognized target is m, when the three networks determine the target as the same m target; The target is also recognized as target m, while any two network determine the target as m target, but the other is target n; When a network refused to recognize, but the other two network have been recognized the target as m, the target is recognized as target m; When a network refused to recognize, and the other two recognized the target as different targets, then refused to recognize; The target is recognized as m when there are two network refused to identify, but the third recognize the target as target m; When the three networks refused to identify, then refused to identify. Recognition results are shown in Table 1.

Table 1. The target recognition results of majority voting fusion algorithm

%	T 1	T 2	T 3	T 4	T 5	T 6	T 7	T 8	T 9	T 10
Recognition rate	87.45	90.31	92.64	88.26	89.77	91.79	93.65	88.96	92.30	91.78
Error rate	5.98	2.57	3.19	6.79	3.68	2.14	1.36	5.61	3.24	4.25
Rejection rate	6.57	7.12	4.17	4.95	6.55	6.07	4.99	5.43	4.46	3.97

5.2 Recognition by Adaptive Weighted Fusion Algorithm

By using the fusion algorithm in Section 4.2, and adopting the equation (8) to give the weight value separately to BP neural network, SOM neural network and RBF neural network, while calculating the target feature vectors of each neural network and the Euclidean distance d_l of the classifier training samples. Then according to equation (7) to obtain recognition results after fusing. The weight function of the experiment is shown in Figure 5, $\alpha=2$、 $\beta=0.25$, the horizontal coordinates denotes the Euclidean distance, while the vertical coordinates stands for weight value. Rejection threshold $J=0.1$. Recognition results are shown in Table 2.

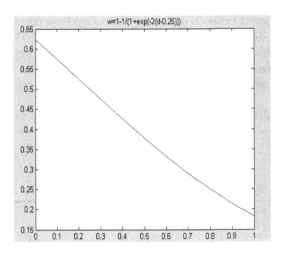

Fig. 5. Adaptive fusion weight function

Table 2. Target recognition results by adaptive weighted fusion algorithm

%	T 1	T 2	T 3	T 4	T 5	T 6	T 7	T 8	T 9	T 10
Recog-ni-tion rate	91.79	94.26	92.18	93.42	92.69	93.70	94.03	90.43	93.58	95.26
Error rate	2.44	2.57	1.96	2.57	1.38	2.52	1.21	3.64	2.24	1.65
Rejec-tion rate	5.77	3.17	5.86	4.01	5.93	3.78	4.76	5.93	4.18	3.09

5.3   The Comparison of Each Recognition Methods

The target recognition results of 10 categories by BP neural network, SOM neural network, RBF neural networks, majority voting fusion algorithm and adaptive fusion algorithm respectively, are averaged which are shown in Table 3. It can be seen from Table 3, when using a single neural network classifier to recognize, the recognition rate of BP algorithm is slightly higher, while the recognition rate of RBF neural network is lower, the recognition rate of SOM neural network is between the above two methods. Because the weight value of RBF network is partly modified, the training time of RBF neural network is shorter. The recognition rate of majority voting fusion algorithm is slightly lower than BP neural network and SOM neural network. It is because the method treats each classifier "vote" equally without considering these classification differences in performance. But the recognition rate of adaptive fusion algorithm is significantly higher than single neural network recognition rate.

Table 3. Comparison of target recognition results by each methods

%	BP	SOM	RBF	Majority voting fusion algorithm	Adaptive fusion algorithm
Recogni-tion rate	91.94	91.21	89.64	90.69	93.13
Error rate	2.38	3.25	4.04	3.88	2.22
Rejection rate	5.68	5.54	6.32	5.43	4.65

6   Conclusion

Using different features or different classifiers could lead different classification results in target recognition and there is very strong complementary between these results. Therefore fusing multi-classifier recognition results can improve target classification recognition effect effectively. The recognition results of BP neural network, SOM neural network and RBF neural network are fused by majority voting and adaptive fusion algorithm in this paper. The results indicate that the recognition rate of multi-neural networks fusion algorithm is higher than single neural network. Adaptive fusion algorithm overcomes the disadvantage and classical weighted algorithm is fixed weight value. Both of them have the advantages for self-adjusting the weight value as well as for improving the recognition rate.

References

1. Jia, C.-C., Yu, X., Zhang, J.-T.: Two-dimensional invariant target recognition based on BP network. Journal of Dalian University of Technology 37(2) (1997)
2. Qi, X.-X., Duan, Z.-M.: Auto Type Passive Noise Recognition Based on BP Neural Network. Instrumentation and Measurement Technology 23(2) (2004)
3. Huang, J.: Self-organizing Neural Network Used In Air Target Recognition. Xinyang Normal College (Natural Science) 12(1), 81–83 (1999)
4. Fei, S., Ke, J.: MATLAB 6.5 Analysis and Design Of Supportive Neural Network. Electronics Industry Press, Beijing (2003)
5. Han, L.-Q.: Leather Textures Classification Based on SOM Neural Network. China Leather 26(6), 11–13 (1997)
6. Liu, R.-J., Yuan, B.-Z., Tang, X.-F.: A New Multi-classifier Fusion Algorithm Based on Clustering. Computer Research and Development 38(10), 1236–1241 (2001)
7. Zhao, Y.-H., Chen, G.-H., Shi, X.-Z.: A New Weighting Algorithm In Multi-classifier Fusion. Journal of Shanghai Jiaotong University 36(6), 765–768 (2002)
8. Wang, H.-Y., Pan, Q., Zhang, H.-C.: An Improved Weighted Fusion Algorithm. Computer Engineering and Applications (2003)

Author Index